WATER QUALITY

WATER QUALITY
Prevention, Identification, and Management of Diffuse Pollution

Vladimir Novotny
Marquette University
Milwaukee, Wisconsin

Harvey Olem
Center for Watershed Protection
Olem Associates, Inc.
Herndon, Virginia

VNR VAN NOSTRAND REINHOLD
_____ New York

Copyright © 1994 by Van Nostrand Reinhold

Library of Congress Catalog Card Number 93-14571
ISBN 0-442-00559-8

All rights reserved. No part of this work covered by the copyright hereon may be reproduced or used in any form or by any means—graphic, electronic, or mechanical, including photocopying, recording, taping, or information storage and retrieval systems—without the written permission of the publisher.

I(T)P Van Nostrand Reinhold is an International Thomson Publishing company.
ITP logo is a trademark under license.

Printed in the United States of America

Van Nostrand Reinhold
115 Fifth Avenue
New York, NY 10003

International Thomson Publishing GmbH
Königswinterer Str. 418
53227 Bonn
Germany

International Thomson Publishing
Berkshire House, 168-173
High Holborn, London WC1V 7AA
England

International Thomson Publishing Asia
221 Henderson Bldg. #05-10
Singapore 0315

Thomas Nelson Australia
102 Dodds Street
South Melbourne 3205
Victoria, Australia

International Thomson Publishing Japan
Kyowa Building, 3F
2-2-1 Hirakawacho
Chiyoda-Ku, Tokyo 102
Japan

Nelson Canada
1120 Birchmount Road
Scarborough, Ontario
M1K 5G4, Canada

BBR 16 15 14 13 12 11 10 9 8 7 6 5 4 3 2

Library of Congress Cataloging-in-Publication Data
Novotny, Vladimir, 1938-
  Water quality: prevention, identification, and management of
diffuse pollution/Vladimir Novotny and Harvey Olem.
    p. cm.
  Includes bibliographical references and index.
  1. Water—Pollution. 2. Water quality management. I. Olem,
Harvey. II. Title. III. Title: Diffuse pollution.
TD420.N69   1993
628.1'68—dc20   93-14571

```
TD       Novotny, Vladimir
420
.N69     Water quality
1994

628.168 N859w
```

Contents

Preface, ix

1 Introduction   1
Historical Perspectives and Trends in Environmental Degradation, 1
Definitions, 11
Environmental Systems and Their Changes, Interactions, and Impact on Water Quality (Pollution), 22
Pollution by Toxic Chemicals and Metals, 41
Important Examples of Systems Affected by Diffuse Pollution, 46

2 Laws, Regulation, and Policies Affecting Water-Pollution Abatement   69
Social Causes of Pollution, 70
Environmental Policies of Diffuse Pollution Abatement, 74
Water Uses and Water Rights, 83
Laws Affecting Pollution Abatement, 85

3 Hydrologic Considerations   101
Precipitation–Runoff Relationship, 102
Rainfall Excess Determination: Surface Runoff, 129
Overland Routing of the Precipitation Excess, 144
Ground-Water Systems, 171

4 Atmospheric Deposition   185
Interdependence between Air and Water Quality, 186
Toxic Compounds in the Atmosphere, 217

5 Erosion and Sedimentation   237
Definition and Description of the Erosion Process, 238
Sediment Problem, 250
Estimating Sediment Yield, 251
Sediment Delivery and Enrichment Processes during Overland Flow, 270
Sediment Transport in Streams, 292

vi Contents

6 Pollutant Interaction with Soils and Sediments   311
Loading Functions, 317
The Fate of Contaminants in Soils: Three-Phase Approach, 323
Specific Pollutant Interactions, 337

7 Ground-Water and Base-Flow Concentration   387
Ground Water (Base Flow) and Diffuse Pollution, 387
Ground-Water Movement, 392
Origin of Natural Ground-Water (Base-Flow) Quality, 406
Impact of Diffuse Pollution on Ground Water and Base Flow, 415
Ground-Water Quality Models, 433

8 Urban and Highway Diffuse Pollution   439
Land-Use Effects on Urban Nonpoint Pollution Loads, 439
Urban Land Use and Magnitude of Diffuse Pollution, 443
Individual Sources of Pollution, 451
Runoff Pollution Generation Process, 461
Sources of Specific Pollutants in Urban Runoff and CSOs, 476
Statistical Quality Characteristics of Urban Runoff, 484
Combined Sewer Overflows, 495

9 Modeling and Monitoring Diffuse Pollution   507
General Concepts, 507
Simple Statistical Routines and Screening Models, 519
Deterministic Hydrologic Simulation Models, 528
Stochastic and Neural Network Models, 546
Real-Time Control Models for Urban Sewerage Systems, 552
Calibration and Verification of Deterministic Models, 554
Monitoring and Data Acquisition, 556

10 Control of Urban Diffuse Pollution   573
Pollution-Control Measures, 573
Urban Drainage and Runoff Pollution Control, 576
Source-Control Measures, 577
Collection System Control and Reduction of Delivery of Pollutants, 602
Detention–Retention Facilities, 616
Treatment, 637
Efficiency of Best Management Practices for Control of Priority Pollutants, 642
Real-Time Control of CSOs and Pollution Load Trade-Offs, 649

11 Agricultural Issues   673
Extent of American Agriculture, 675
Agriculture and Its Effect on the Environment, 681

Best Management Practices: Their Implementation and
  Effectiveness, 689
Roles, Responsibilities, and Programs of Federal, State, and
  Local Agencies and Groups, 711
Balancing Agricultural Production with Environmental
  Protection, 711

12 Receiving Water Impacts 735

Assessment of Water Quality Problems, 735
The Acquatic Ecosystem, 740
Dissolved Oxygen Problem, 764
Nutrient Problem of Lakes: Eutrophication, 783
Modeling Fate of Conventional Pollutants and Nutrients, 796
pH and Acidity, 802
Background (Natural) Water Quality, 806

13 Toxic Pollution and Its Impact on Receiving Waters 817

Toxicity Concepts, 817
Toxicity and Its Measurement, 821
Water Quality and Sediment Criteria and Standards for
  Toxicity, 838
Modeling Fate of Toxic Compounds in Receiving Water
  Bodies, 852

14 Wetlands 861

Definitions and Types of Wetlands, 863
Wetland Function, 869
Constructed Wetlands, 889
Aquatic Plant Systems, 917

15 Management and Restoration of Streams, Lakes, and Watersheds 927

Restoration Versus Reclamation, Rehabilitation, and
  Management, 928
Restoration Techniques for Rivers and Streams, 932
Restoration Techniques for Lakes and Reservoirs, 949
Integrated Aquatic Restoration: Case Studies, 972

16 Integrated Planning and Control of Diffuse Pollution—Watershed Management 981

Pollution Load Trade-Offs and the Permit Process, 983
Water Uses and Use Attainability, 985
Financing Diffuse-Pollution Abatement Programs:
  Who Should Pay? 993
Institutional Issues, 1002

viii   Contents

Epilogue 1011

Appendixes 1017

Index 1035

Preface

Man has lost the capacity to foresee and to forestall. He will end by destroying the earth.

A quote by Albert Schweitzer in Rachel Carson's *Silent Spring*

The nation behaves well if it treats the natural resources as assets which it must turn over to the next generation increased and not impaired in value.

Theodore Roosevelt

In the 1960s the environmental awareness of the U.S. population was greatly awakened. Many attributed this to a book by Rachel Carson entitled *Silent Spring* (Houghton Mifflin, Boston, 1961) that described, among other hazards, the effect of emissions of chemicals and other pollutants on the environment and the potential destruction of the ecology and mankind. But for whatever reasons, the environmental era of the late 1960s was a wide, grassroots movement of activists, many of them ill-informed but extremely enthusiastic. The movement caused several environmental protection legislative acts, of which the 1972 Water Pollution Control Act Amendments was the most far-reaching law (see Chapter 2 for pertinent water quality laws dealing with diffuse pollution). Earth Day of 1970 was the culminating event of this period.

Some well-known traditional environmental scientists and engineers called the environmental period of the 1960s and early 1970s "The Age of Unreason" or "an ecology binge in which relatively few activists aroused the concerns of millions but hopelessly ill-equipped Americans over the future of our environment and indeed our own existence" (Schroepfer, *Journal of the Water Pollution Control Federation*, September 1978). Some of this "unreason" was incorporated into laws that called for "zero pollution discharge,"* but in actuality advocated shifting the pollution

---
*It is interesting to note that the calls for zero discharge, this time on a more rational and scientific basis, have been renewed in mid-1990s.

disposal from surface water resources onto soils, and eventually into the ground water. Indeed, this environmental period of the 1960s and early 1970s was replaced in the 1980s by consumerism and a general lack of enthusiasm of the population about the state of the environment. In the same period, however, the global problems, unseen but predicted to some degree by Carson and others, became widely publicized. The first large-scale global environmental problem of diffuse nature was acid rainfall and its detrimental effects on North American and Scandinavian lakes (see Chapter 4 for a detailed discussion of acid rainfall). Even though natural rainfall is acidic due to the dissociation of dissolved carbon dioxide from the atmosphere (pH of rainfall in equilibrium with the atmospheric CO_2 is about 5.6), rainfall with much lower pH—down to less than 3 in some areas—is now falling over larger geographical areas of North America and Europe. This elevated acidity is due to sulfuric and nitric acids formed from fossil-fuel burning (sulphur-containing coal) and emissions from all kinds of motorized vehicles. There is now even evidence that, in addition to acidifying lakes, acid rainfall is also damaging soil fertility to such an extent that in some parts of the world it poses a threat to future agricultural production and its sustainability (see Chapter 6). Toxic compounds present in water are more toxic in acidic water than in neutral and slightly basic water bodies (see Chapter 8). Other problems of global proportions include the "greenhouse effect"—that is, the increased atmospheric content of carbon dioxide and other gases—and the ozone-hole problem caused by the discharge of fluorocarbons into the atmosphere. Fluorocarbons, which originate mostly from leaking cooling systems, are also a water quality problem. Consequently, solving the ozone-hole problem by banning the production and use of damaging fluorocarbons may also have a water quality benefit.

Also since the 1970s, beaches of the North, Mediterranean, and Adriatic seas became clogged by algae to the point that swimming was not possible except in swimming pools that the hotels and casinos had to build.

On a smaller but still widespread scale, soil loss from farms and construction sites is alarming as well as being detrimental to surface waters. In order to replace the lost plant nutrients and to protect the monocultural crops from insects and weeds, farmers use more and more chemicals, a practice that results in severe ground and surface water contamination. In many places, ground-water resources have been contaminated by nitrates and pesticides to a point where they cannot be used as a drinking water supply, with the result that potable water must be trucked in or provided in bottles.

As the countries of eastern Europe opened their doors to the West, environmental problems on a monumental scale were discovered. It was known that these countries were a major cause of some global problems, such as acid rain and atmospheric PCB emissions, but the extent of the damage to their soil, air, and water resources had not been known and was kept secret. The most serious pollution problems are diffuse and widespread in some countries (central and eastern Europe and some developing countries), where they have reached locally catastrophic proportions.

These large-scale environmental scares, plus a still very remote but real possibility that man actually can bring about his own destruction by contamination of the environment and the atmosphere, have generated a new environmental awareness in the 1990s. How correct was Rachel Carson decades ago.

This book does not want to promote gloom and doom, nor does it want to cause unreasonable limitations on resources. The objective of this book is rather to identify the environmental problems caused by pollution, some of them of a global nature, and to suggest possible feasible and economical solutions. The authors wanted to approach the task of preparing the book in a positive, problem-solving fashion, with emphasis on the sustainable use of water and soil resources, their protection, and rehabilitation. As man can destroy the earth, he can also save it and live in harmony with the environment.

This approach requires some rethinking of the philosophies with which environmental engineers and scientists used to approach such problems. For example, a change is needed from the traditional "sanitary engineering," mostly structural approaches that advocated the removal and transport of pollutants and excess water from the affected areas in the fastest way possible to less structural and more ecologically oriented approaches that rely on water conservation and retention. This rethinking will involve a shift from traditional curb-and-gutter storm sewer drainage of urban and urbanizing areas to drainage maximizing the use of natural (grassed) waterways, retention, and infiltration; a change from draining wetlands for monocultural agriculture and urban development to preserving and retaining them; from lining urban and rural streams with concrete to preserving the streams in their natural state or restoring them so they can support an aquatic habitat; from deforestation to forest preservation and reforestation; from intensive agriculture relying on the heavy use of chemicals to sustainable agriculture.

A new branch of environmental engineering called *ecological engineering* is emerging. An ecological engineer knows how to balance pollution discharges with the waste-assimilating capacity of the environment. Then the environment can receive the residual waste loads without harm to the ecology, aquatic biota, and beneficial downstream uses of water resources. If this capacity of water bodies to receive pollutants is not sufficient, engineering techniques can be employed to both reduce the destructive quantity of the pollutants and to increase the ability of the environment to accept potential pollutants without harm. This ability of the environment to accept limited amounts of pollutants without harm is a great economic asset that must be included in all considerations of environmental protection and restoration, but it must not be exceeded. Most environmental restoration and protection projects do not require a zero-pollution discharge approach.

This book essentially is a follow-up of the *Handbook of Nonpoint Pollution: Sources and Management* by Novotny and Chesters (Van Nostrand Reinhold, 1981). The authors followed also the ideas and solutions presented in others of their previous books (Krenkel and Novotny: *Water Quality Management*, Academic Press, New York, 1980; Novotny et al.: *Karl Imhoff's Handbook of Urban Drainage and Wastewater Disposal*, John Wiley & Sons, New York, 1989; Olem: *Liming Acidic Surface Waters*, Lewis Publishers, 1991). However, the emphasis on the sustainable use of water resources and the ecological approaches to solving the problems of diffuse pollution are somewhat new in the field of environmental engineering and pollution control.

The situation today is different from what it was at the end of the 1970s. Prior to the mid-1970s, diffuse pollution was an unknown phenomenon to the general population and its representatives. Environmental engineering and science was almost exclusively oriented toward wastewater conveyance, treatment and disposal, and water supply. Urban engineering was promoting such approaches as curb-and-gutter storm sewers, sewer separation, and lining streams for drainage. Today, however, there are very active and quite large groups of professionals interested in solutions for urban or agricultural diffuse pollution that would be harmonious with the ecological principles and would lead to the preservation and enhancement rather than the destruction of ecosystems. Many excellent examples of developments that have incorporated the nature and protected the habitat have recently emerged. Specialized groups have been formed by most major environmental professional associations, including the International Association on Water Quality (formerly the International Association for Water Pollution Research and Control), the Water Environment Federation (formerly the Water Pollution Control Federation), the American Water Resources Association, the American Society of Civil Engineers, the American Society of Agricultural Engineers, the Soil Science Society of America, the North American Lake Management Society, and the American Society of

Agronomy. Their journals and proceedings are now a major source of information on topics related to the diffuse-pollution problem and its solutions, and the authors wish to acknowledge the positive role these associations and their publications are playing in the recognition of the problems in this area, as well as to their solution. The authors are also indebted to several foundations and other sponsoring agencies (The National Science Foundation, the Water Environment Research Foundation, the U.S. Environmental Protection Agency, among others) that provided funding for research and enabled the authors to gather the knowledge necessary for preparing this book. The environmental education efforts on diffuse pollution by some of these professional organizations (AWRA, Air and Waste Management Association) have reached elementary and secondary schools throughout the country.

The authors would also like to recognize the important contributions of the reviewers of the manuscript of the book, Professor Peter A. Krenkel of the University of Nevada and Thomas Davenport of the U.S. Environmental Protection Agency. Credit should also be given to Paul and Eric Novotny for their computer art work.

The book is primarily for graduate students and practitioners in the environmental areas. Unlike the previous books by the authors, this one should be considered both a textbook and a handbook. It is intended to teach the recognition of problems and the finding of solutions as well as presenting facts and methodologies. Metric (SI) units are used throughout the book. Conversions into the customary U.S. units are included in Appendix A.

1

Introduction

Originally all pollution was of nonpoint (diffuse) nature. It became "point" pollution when years ago people in urban and industrial areas collected urban runoff and wastewater and brought it, at a great expense, to one point for disposal.

> Paraphrase of a statement made by a well-known urban environmental economist (Gaffney, 1988) that introduces the topics to be presented in this chapter.

HISTORICAL PERSPECTIVES AND TRENDS IN ENVIRONMENTAL DEGRADATION

From the Romans to Earth Day

> It is an irony of history that semi-desert conditions now prevail in much of the region once known as the Fertile Crescent. . . . Moreover, the earlier peoples had on the whole a higher standard of living than most of the present inhabitants. The degradations of the region came about almost entirely because of human discord and neglect. The ancient people had ingeniously developed the lands of the Fertile Crescent by intelligent use of meager water resources. . . . Then invaders laid waste to the region and a long decline set in. A succession of indolent and mutually intolerant people allowed the cisterns and reservoirs to fall into ruin, the irrigation channels and terraces to crumble, the trees to be cut down, the low vegetation to be destroyed by sheep and goats and the land to be scoured by erosion. (*Copyright © 1965 by Scientific American, Inc. All rights reserved.*)

These statements are a portion of the introduction by Maurice A. Garbell (1965) in his discussion of "The Jordan Valley Plan." The history of the

Middle East shows that if land stewardship is absent, the well-being of the people who misuse the land and water resources declines. At some point these adverse effects and deterioration become irreversible.

However, in the eighteenth century when the first Europeans arrived in the Piedmont area of the American Southeast they found rivers and lakes "crystal clean," without visible pollution, water transparent and abundant with fish (Clark, Haverkamp, and Chapman, 1985). At about the same time and earlier throughout the Middle Ages salmon migrated during the spawning season all the way into the headwater streams of central Europe to the delight of fishermen in Prague, in the present Czech Republic (located in the very center of Europe, several hundred kilometers from the North Sea into which the rivers flow).

One would be greatly mistaken if these statements about the cleanliness of the rural, mostly uninhabited environment were taken as a general rule about the environment of the ancient world and the Middle Ages. The pristine clean state of the rural environment centuries ago was contrasted by the filth and uncleanness of urban centers. The streets of medieval cities (and the same may be true for large urban centers of ancient Rome and other great historical centers) were covered by garbage, manure, and human excreta. In medieval Paris and other cities piles of garbage and manure in the streets were one meter or more high. The smell was strong and nauseating. Terrible epidemics plagued medieval cities, and even the rural population was not spared. There is no doubt that the medieval urban governments developed some kind of street-sweeping–cleanup or disposal services, and cleanliness or filth varied from city to city. Many urban dwellers themselves also tried to keep the streets in front of their houses clean. In addition the water supplies used by urban dwellers were much smaller than they are today, resulting in less pollution generation. However, rainfall and urban surface runoff were the primary and sometimes the only means of disposal of accumulated street surface pollution. Evidently, problems with urban runoff are not new. Consequently, storm sewers were built, primarily for storm water disposal.

When in the middle of the 1800s it was realized that the filth of the cities and contamination of the water supplies were the major reasons for such waterborne epidemics as cholera and typhoid fever, the first major period of environmental awareness was born. It was born because life in growing industrialized urban centers with medieval drainage became unbearable to the population and its governments. The first urban sewer system in the United States was planned in Chicago in 1855, although sewers had been built in Europe decades and in ancient Rome thousands of years before. The mixture of urban runoff and wastewater was brought by the sewers to the nearest watercourse, and the dilution by the flow of the receiving water body was considered satisfactory for controlling pollution. It is interesting to note that until the 1950s many European receiving water standards were based on dilution (for example, according to the British water quality standards, one part of untreated sewage discharge required 500 parts of receiving stream flow). Many rivers soon became heavily overloaded and gave off an obnoxious stench, which was caused by the anoxic decomposition of sewage and garbage.

The period between 1880 and 1920 marks the beginning of major concerns about water quality, especially drinking water. In 1910, in Essen, Germany, one of the most industrialized areas of the world, the first water quality management agency was established to provide safe urban runoff and wastewater disposal. A few communities added treatment plants at the end point of their sewer systems to purify the discharged sewage. Almost all sewer systems built in this period carried a mixture of sewage and urban runoff. These systems are called *combined sewers* in contrast to newer and more expensive *separate sewer systems*, which employ dual sewers, one for sanitary sewage and the other for urban runoff (see Chap. 8 for discussion of urban drainage and its water quality impact). Even though the sewer systems were called combined, they were designed to carry primarily sewage and industrial wastewater—so-called *dry-weather flow*. A typical design capacity of combined sewers was six to eight times the dry-weather flow. However, this excess capacity was greatly insufficient for storage and conveyance of rainfall-generated runoff. Similarly, treatment plants were designed mostly for the dry weather flow (a typical design capacity was about four times the dry-weather flow). When, as a result of a rainfall event, the capacity of the sewers or of the treatment was exceeded, an untreated mixture of sewage and rain water was allowed to discharge into the nearest watercourse.

In rural areas, family farming using organic (manure) fertilizers flourished until the middle of the 1900s. This type of farming, in spite of its appearance and sometimes odor, causes less harm to surface and ground-water resources, although localized pollution problems from barnyard wastes were common. Farmers did not use chemical insecticides and fertilizers until the late 1950s.

After the epidemics of the Middle Ages were largely eliminated public interest in the environment subsided until the late 1960s. Meanwhile, however, pollution of the environment in the first half of the 1900s increased rapidly. The pollution of many urban rivers was again becoming unbearable. For example, every summer from the nineteenth century to the middle of the twentieth century the stench of the Thames River in London became so unbearable that the British Parliament recessed during the affected periods.

Man-made chemicals were introduced in the middle of the 1900s and many of them have found their way into the environment where they caused great and almost irreparable harm. Such were the cases of DDT (dichloro-diphenyl-trichloro-ethane), which was originally heralded as a savior from malaria and every possible obnoxious insect, including lice, and PCBs (poly-chlorinated bi-phenyls), which is a group of very useful industrial chemicals. Both types of chemicals were later found to be greatly damaging, persistent, and bioaccumulating environmental contaminants. Many other chemicals were developed during and since World War II, which now have contaminated soils, water, and the air. The spread of man-made toxic chemicals and the potential dangers led to the second environmental activist period, the impetus for which was the book *Silent Spring* by Rachel Carson (1962). In 1970, the Earth Day celebration emphasized public concern about the state of the environment and initiated calls for action. This period was also marked by the rapid expansion of personal and commercial vehicular traffic, spurred by the building of freeways. Automobile and truck traffic is a major source of toxic chemicals as well as the activities associated with the expansion of the freeway system and suburban developments (urban sprawl).

In the United States and elsewhere, however, some progress in the abatement of municipal wastewater collected by sewers was made, and between 1920 and 1970 treatment plants were built at a rapid pace. By 1977 in the United States, 95% of the (156 million) people residing in sewered communities received some form of treatment of their wastewater while 70% received secondary biological treatment predominantly of the dry weather—sewage and wastewater—flows (Schroepfer, 1978). By 1970, the River Thames in London was alive again and fish have been caught there since. In 1972, the United States enacted the Water Pollution Control Act Amendments, which was the most far-reaching environmental legislative act to solve environmental problems.

However, even in the United States, and more so in Europe, many rivers and lakes could still not support viable fishery, being so polluted that fish were absent and their bottoms were covered with mud contaminated with toxic substances of unnatural man-made origin. For example, Lake Erie of the Great Lakes system was dying. Even where fish were present, carcinogenic compounds discharged into the receiving waters in the post–World War II period had stressed the aquatic population, fish had become unfit for human consumption, and water recreation had been reduced or had ceased.

In addition to pollution, activities that lead to habitat destruction should also be considered and remedied. Typically in the past, in the jargon of water resource developers, "channel improvement" meant lining a stream with concrete and cutting down stream-bank vegetation, and "beneficial use of water" meant diverting flows from streams and lakes to the point that no flow was left during some periods. These activities caused severe damage, if not elimination, of aquatic habitat.

Wetland Drainage

Until recently in the United States as well as throughout the world, wetlands were considered as a source of disease (malaria) and as an obstacle to man's use of land resources for growth, agriculture, and economic development. Early "wetland management programs" in the United States, both governmental and private efforts, concentrated on drainage, filling, and conversion to agriculture and urban uses. Furthermore, most extensive insecticide applications (DDT spraying) on wetlands were to control malaria and mosquitoes, which are vectors of the disease. The "success" of these earlier programs has been documented by the U.S. Fish and Wild Life Service (a government agency under the U.S. Department of the Interior), which estimated that more than one-half of the approximately 1 million wetlands in the conterminous 48 states have been lost between the arrival of the first settlers and the present time.

However, it is known today that wetlands are ecological and hydrological assets and should be protected and restored rather than drained and destroyed. They provide habitat for waterfowl, animals, and wetland vegetation. Wetlands have the capability of storing water and of purifying polluted waters. Control of mosquitoes can be achieved by planting mosquito-eating fish (*Gambusia affinis*) or by fish management.

Fortunately, Congress and the federal government have realized the very high ecological, hydrological, and pollution-control benefits of wetlands and have enacted both federal and state wetland protection and rehabilitation acts. For example, in Florida developers must now reestablish 2 hectares (ha) of artificial or restored wetland for each hectare of natural wetland lost due to development. Draining or filling wetlands is now prohibited in many states, and efforts to reestablish formerly drained wetlands are pending. Valuable wetlands cannot be destroyed even if a replacement is offered.

Present Status of Water Quality Abatement

Since the passage of the Water Pollution Control Act Amendments in 1972, hundreds of billions of dollars have been spent on the cleanup of pollution, primarily that caused by sewage and industrial wastewater discharges. But at the same time as the money was beginning to be spent on this type of cleanup, it was realized that these efforts may be insufficient. In addition to pollution from sewage and industrial wastes, pollution from land and from the activities by man occurring on the land would cause the cleanup goals not to be met in spite of the vast expenditures.

The land-use activities that create pollution cause other damage in addition to pollution. For example, soil losses that cause the pollution of receiving waters by sediment and associated pollutants, diminish the agricultural productivity of soils. An authoritative publication by the Conservation Foundation (Clark, Haverkamp, and Chapman, 1985) estimated that damage due to erosion and soil loss of cropland amount to 2.2 billion per year (in 1980 dollars).

Great Lakes Studies

In 1972 the Pollution from Land Use Reference Group (PLUARG) of the International Joint Commission (IJC) was established for the purpose of determining the levels and causes of pollution from land-use activities. A large group of scientists in Canada and the United States studied the pollution of surface runoff and found that indeed the land runoff, in addition to pollution from the atmosphere and from the traditional wastewater sources, was a significant and often major source of pollutant loads to the Great Lakes. PLUARG was a major international cooperative effort undertaken from 1972 to 1978. The resulting studies provided the most exhaustive review conducted up to that time, and to date it remains the most definite data base and reference source for many aspects of diffuse pollution in the Great Lakes and elsewhere (Nonpoint Source Control Task Force, 1983; Novotny and Chesters, 1981).

These studies have found that the most serious pollution of the Great Lakes arises from land areas of intensive agriculture and urban use. The most significant and damaging pollutants to the lakes from these sources were phosphorus, sediments, and pesticides, in addition to a number of toxic industrial compounds. Phosphorus is of concern in the Great Lakes because it is the principal factor causing and controlling accelerate eutrophication (Chapter 12), symptoms of which are the excessive development of algae, increase of turbidity, and general water quality deterioration. Land-use activities contributed from a third to a half of the total phosphorus loads (IJC, 1980). Also, pollution by toxic and hazardous substances from land drainage is an equal if not greater threat to the Great Lakes ecosystem. About 400 organic toxic compounds were identified in the Great Lakes ecosystem, including persistent pesticide compounds; specifically, aldrin-dieldrin and chlordane continue to appear in the Great Lakes biota in spite of the fact that their use in the Great Lakes basin has been banned or severely restricted. In addition to land drainage and wastewater sources, significant loads of many pollutants are transported to the lakes via the atmosphere (IJC, 1980).

In 1978, the governments of the United States and Canada revised and signed a Great Lakes Water Quality Agreement. In this agreement the governments reaffirmed their determination to restore and enhance water quality in the Great Lakes system.

US EPA Activities

Based on PLUARG and several other studies, and recognizing the fact that abatement of traditional municipal and industrial wastewater sources will not alone achieve the water quality goals of the Clean Water Act, Congress requested the U.S. Environmental Protection Agency to prepare a comprehensive report on the sources of pollution other than traditional wastewater sources—the nonpoint sources.[1] The report was submitted to the Congress in 1984 (U.S. EPA, 1984). The report pointed out that virtually every state identified some sort of nonpoint pollution problem. The principal sources of nonpoint pollution were identified as *agricultural activities*—including those resulting from tillage practices and animal waste management—which were the most pervasive polluting activities reported from every part of the United States, followed by *urban runoff* and pollution from *mining*. Nonpoint pollution from *silvicultural activities* was considered substantial but localized. The large amounts of sediment associated with *construction activities* were recognized as causes of localized water quality problems in those parts of the United States experiencing significant development pressures (Southeast, mid-South, and Northwest). The EPA report to Congress emphasized voluntary approaches to the abatement of nonpoint sources.

In 1976 the Chesapeake Bay Program of the EPA (see the section "Important Examples of Systems Affected by Diffuse Pollution" in this chapter) was established. The National Urban Runoff Project (NURP) also sponsored by the EPA, was a large research project carried out in 28 localities throughout the United States from 1978 to 1983 (U.S. EPA, 1983). This program had the following objectives:

- To investigate and establish quality characteristics of urban runoff, and similarities or differences at different urban locations;
- To identify the extent to which urban runoff is a significant contributor to water quality problems across the nation;

---

[1] See the following section for the definition of nonpoint sources of pollution.

- To establish performance characteristics and the overall effectiveness and utility of management practices for the control of pollutant loads from urban runoff.

From its findings, it concluded that

- Urban runoff contains high concentrations of toxic metals; the so-called "priority pollutants" (toxic, mostly organic chemicals) were also detected in significant quantities;
- Urban runoff is contaminated by coliform and pathogenic (disease causing) bacteria;
- Urban runoff carries high quantities of sediment.

Other pollutants, such as those causing depletion of the dissolved oxygen or algal growths in the receiving waters, were present but were less significant. The NURP program also tested various abatement measures, some of which they found to be effective and therefore recommended, such as the use of retention ponds and grassed waterways (swales), while some were found to be ineffective for water quality control, such as street sweeping (U.S. EPA, 1983).

Rural Programs

The Clean Water Act of 1972 authorized the EPA to carry out demonstration projects of pollution-control technologies in the Great Lakes basin. Several experimental watersheds were established, from which the Black Creek basin (Alen County, Indiana) received the highest funding. The program demonstrated and addressed in a research fashion several pioneering pollution-control technologies, such as conservation tillage, including no-till, and animal-waste management as well as soil conservation.

The Black Creek project paralleled the PLUARG studies conducted in the neighboring experimental watersheds of the Great Lakes basin. The program spent more than $800,000 as a cost-sharing for conservation treatment on the 4000-ha watershed, but could not show significant water quality improvement (Humenik, Smollen, and Dressing, 1987). The experience clearly indicated that the traditional "first come-first serve" cost-sharing incentive programs are not efficient for water quality improvement. The major reason was the fact that many soil-conservation and erosion-control practices do not yield significant water quality improvement (see Chapter 5). Consequently, the project managers and other researchers participating in the PLUARG projects developed a concept of targeting the areas that needed management that would yield the greatest water quality improvement benefit. The Black Creek project also demonstrated institutional aspects of the nonpoint pollution programs, that is, that the Soil Conservation Service of the U.S. Department of Agriculture was capable of administering a rural nonpoint pollution-abatement program (Humenik, Smollen, and Dressing, 1987).

The Rural Clean Water Program (RCWP), which began in 1980, has funded 21 watershed projects whose objectives were to improve water quality, to help agricultural homeowners and operators to use pollution-control practices, and to develop programs, policies, and procedures for the control of agricultural nonpoint source pollution (*Federal Register*, 1979). It is a long-term (10 to 15 years) demonstration project, so the critical targeting of lands producing pollution is required, water quality objectives are clearly defined, and water quality is monitored. Hence, the program has a much greater water quality emphasis than previous programs, which focused primarily on soil conservation. Approved management practices include water management systems, animal-waste management systems, fertilizer and pesticide management, in addition to erosion control and soil conservation.

At the end of the 1980s, approximately seven to ten years after the start of the program, most projects exceeded their goals of implementing pollution management practices in 75% of the critical watershed areas. Projects that have achieved a high level of farmer participation have been successful because they offered cost sharing for practices farmers want, such as animal-waste storage, conservation tillage, and irrigation system improvement. It should be pointed out that the willingness of farmers to install pollution abatement under voluntary cost-sharing scenarios required that these practices include other benefits, such as the reduction of the cost of fertilizers and chemicals, which would compensate the farmer. Water quality improvements have been documented in several RCWP projects; however, it commonly took more than 5 years to document water quality improvement in these pilot watersheds.

Present and Future Trends and Needs

In 1990 there was worldwide recognition of the twentieth anniversary of the Earth Day. This recognition renewed environmental awareness in the public. The Earth Day movement of the 1960s and 1970s had an impact in the United States, and produced a landmark legislation, the 1972 Water Pollution Act Amendments. It also initiated extensive environmental remediative actions. The present environmental movement is concerned with global as well as local pollution problems and environmental contamination. Also the attitude of lawmakers is changing worldwide. The emergence of Green political parties in Europe which, although still small, keep the balance of power in some European countries. Public pressures have also led to the further enactment of environmental protection laws.

In the United States it was realized that despite requirements incorporated in the 1972 and 1977 versions of the Clean Water Act, progress in reducing diffuse pollution has been extremely slow. It was demonstrated that in the 15-year period since the 1972 passage of the Water Pollution Control Act Amendments very little had been accomplished in reducing the diffuse pollution problem (Thompson, 1989). As a matter of fact, some problems, especially those related to the use of agricultural and industrial chemicals, have intensified. Therefore Congress in the 1987 Water Quality Act shifted from 15 years of nonpoint source pollution planning, studying, and problem identification to a new National Nonpoint Pollution Control Program (U.S. EPA, 1989). It stated that:

> It is the national policy that programs for the control of nonpoint sources of pollution be developed and implemented in an expeditious manner so as to enable the goals of this act to be met through the control of both point and nonpoint sources of pollution.

Far-reaching regulations are being implemented to control diffuse pollution, including urban and industrial runoff and agricultural diffuse pollution. Federal and state laws are also being implemented that would stop the loss of wetlands that are so vital in the ecological balance and for the control of diffuse pollution.

In spite of some monumental efforts in the United States and elsewhere to remedy pollution, environmental degradation is still continuing. The diffuse-pollution problem is only one component of the overall problem with the earth's ecological system. The *State of the Environment* (OECD, 1991) pointed out three major problems now facing the earth's ecosystem:

- Stratospheric ozone depletion;
- The greenhouse effect;
- The global spread of pollution, primarily by atmospheric currents.

These global issues are linked in a number of ways. The causes and transport of diffuse pollution by surface water bodies may not be separated from these global issues, but are a part of the overall system.

As forests shrink at an alarming rate thus contributing to the greenhouse effect by emissions from slush burning, soil is simultaneously eroded from the deforested lands, causing siltation of bodies of water and other water quality problems. As the environment is being filled with man-made chemical compounds, such as fluorocarbons and other halogenated hydrocarbons, many of them carcinogenic, these emissions are also contributing to the depletion of the protective ozone layer. Nitrous oxides emitted from vehicular traffic and other sources contribute to both the greenhouse effect and the pollution of surface and ground water.

There is no doubt that in spite of the expenditures on pollution, the worldwide state of the environment in the 1990s is worse than it was during the Earth Day of 1970. In the *State of the World*, the Worldwatch Institute (Brown et al. 1990) points out that only a monumental effort can reverse the deterioration of the planet. As the East–West ideological conflict vanishes, time and energy of the population and its political leaders can be concentrated on environmental threats to security. The present and future years will be a period of reordering priorities, providing resources to reforest the earth, and conversion to sustainable development with less harmful pollution problems, as well as avoiding pollution catastrophes during the second half of the twentieth century.

In the past and even today, economic development has caused environmental degradation. While on one hand food supplies and economic output have grown tremendously since the end of World War II, at the same time the world has lost nearly one-fifth of the topsoil from its cropland and one-fifth of its tropical rain forest plus tens of thousands of its plant and animal species (Brown et al., 1990). Rapid increases in pollutant levels in ground water, particularly nitrate and pesticide pollutants, are directly tied to the increase in crop yields that has been stimulated by agricultural chemicals. However, at least in some parts of the world, the reverse is beginning to emerge, that is, environmental trends are beginning to shape economic trends.

This combination of public and government awareness of environmental dangers of diffuse pollution and their willingness to do something about it and pay for it, gives environmental engineers and scientists an opportunity to design abatement programs that would bring about true progress in the reversal of past degradation trends.

DEFINITIONS

Water Quality and Pollution

Any treatise dealing with water quality and pollution needs definitions. In the minds of the public water quality is often synonymous with pollution and, similarly, water quality management, including that related to diffuse sources, is equated with pollution control.

Water quality reflects the composition of water as affected by natural causes and man's cultural activities, expressed in terms of measurable quantities and related to intended water use. Water quality is perceived differently by different people, for example, a public health official is concerned with the bacterial and viral safety of water used for drinking and bathing; fishers are concerned that the quality of water be sufficient to provide the best habitat for fish; and aquatic scientists are concerned with the health of aquatic habitats, including fish, plankton, and other plants and organisms. For each intended use and water quality benefit, different parameters express best water quality.

The term *pollution* is derived from a Latin word (*pollu'ere*), which means "to soil" or "to defile." The terms *pollution, contamination, nuisance,* and *water (air, land) degradation* are often used synonymously to describe faulty conditions of surface and ground water. Various definitions have been offered to define pollution and other related terms (Krenkel and Novotny, 1980; Vesilind, 1975; Henry and Heinke, 1989). However, these definitions are not identical and, in a legal sense, not even similar. Probably the definition of pollution most accepted by scientists is "unreasonable interference with the beneficial uses of the resource." However, the perception of beneficial use is again different to different people, which could be a problem. For example, from the economic standpoint, the greatest "beneficial use" of water and air resources is to provide an inexpensive way to dispose of wastes, in which case fishing and swimming may be perceived by these "economical users" as interfering with their "beneficial use." Indeed, during discussions of the implications of the Water Pollution Control Act Amendments of 1972 some people with good intentions tried to put dollar values on the cost of reducing pollution versus the market value of fish in the receiving water body.

Fortunately, such interpretations are not acceptable now, but they show the possible problems with simple definitions and with perceptions. Today's interpretations put a very high value on the protection of the environment, and supersede any economic savings that might be achieved by allowing injurious discharges of pollutants.

The statutory definition of pollution is included in the Clean Water Act, Sec. 502-19 (U.S. Congress 1987):

The term "pollution" means man-made or man-induced alteration of chemical, physical, biological, and radiological integrity of water.

The term *integrity* means "being unimpaired;" therefore, *alteration of integrity* means impairment or injury. An alternate working definition (the one that will be followed and expanded on in this book) is:

Pollution is a change in the physical, chemical, radiological, or biological quality of the resource (air, land or water) caused by man or due to man's activities that is injurious to existing, intended, or potential uses of the resource.

According to this definition, pollution is differentiated from changes in the quality of the environment due to natural causes, such as volcanic eruptions, deposition of fly ash from natural forest fires, natural erosion, weathering of rocks, and natural elutriation of minerals, even though these events may have the same actual or potential adverse impact on water use or of a resource as does pollution. These changes could be considered *nuisance* or *undesirable quality modifications* or even *Acts of God* by the legal profession, but in a technical engineering sense they mostly do not require abatement or abatement is not technically and economically feasible.

Consider as an example the Rio Puerco in New Mexico (Fig. 1.1). This river drains a highly erodible, sparsely populated arid watershed in the northwestern quadrant of the state. Consequently, extremely high concentrations of suspended solids exceeding 200,000 mg/l are measured during erosive flows. There is no doubt that man has contributed to these high loads. Before ranchers moved in more than 100 years ago, the

FIGURE 1.1. Rio Puerco in New Mexico. (Photo: V. Novotny.)

watershed was more stable. However, overgrazing on the fragile arid lands of the Rio Puerco for the last 70 to 100 years has been a big contributor to the high "natural loads."

Similarly, the Yellow River in China, the Nile River in Egypt, and many other rivers in arid areas of the world are known to carry high amounts of sediments and other water quality constituents. Yet the authorities do not plan any remedial actions at this time, mainly because of the extremely high cost of the watershed reclamation and a lack of concern among the local population. On the other hand, sediment concentrations two orders of magnitude lower due to farming and urbanization in the Great Lakes region and elsewhere are not being tolerated, and plans are on the way to remedy the situation. As a matter of fact, the building of the Aswan High Dam on the Nile River drastically reduced the natural sediment loads of the river, thus depriving the farmers in the Nile Delta region of the natural fertilizer brought onto their fields in flood waters, which resulted in the subsequent damage done to them and to the economy of Egypt. Similarly, the present decrease in sediment loads in the Mississippi River caused by settling of sediment in upstream navigation pools is causing increased erosion and loss of land along coastal Louisiana. The present rate of wetland loss in Louisiana, caused primarily by the reduction of sediment input from the Mississippi River, represents about 80% of coastal wetland losses in the United States (Rooney, 1989).

Other examples of undesirable natural water contamination, but not pollution by man, include the high carbon dioxide content of some ground water which is injurious to building materials and elutriation of humic organics from decaying aquatic vegetation which impairs the suitability of water for water supply. Hence, water quality composition contains both constituents that may be pollution and constituents from natural sources (Fig. 1.2). The use of the term *background pollution* for natural water quality composition is not correct because this pollution is not caused by man.

Pollution may result from causes other than discharges of wastewater or soil losses, the two most commonly mentioned causes. Cutting down a forest, channel modifications (lining), draining wetlands, and similar activities commonly result in undesirable changes in water quality. Hence, such activities create pollution.

Water quality and pollution are determined and measured by comparing physical, chemical, biological, microbiological, and radiological quantities and parameters to a set of standards and criteria. The difference between the *standards* and *criteria* should be explained.

A *criterion* is basically a scientific quantity upon which a judgment can

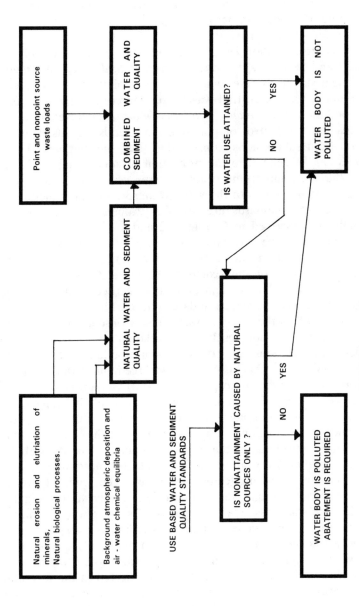

FIGURE 1.2. Concept of water quality.

be based. It is usually developed from scientific experiments. A water quality criterion can be based on morbidity or chronic toxicity of various substances to man or aquatic life, or it can be related to technical methods of removing the substances from water. A *standard* applies to a definite rule, principle, or measure established by an authority.

The water quality criteria and standards currently used by water-pollution-control authorities throughout the world, as well as by engineers and scientists, are either *stream standards* or *effluent standards*. The effluent standards determine how much pollution can be discharged from municipal and industrial wastewater sources and by some types of diffuse pollution. *Performance standards*, which are equivalent to effluent standards for the control of pollution from lands, are used to control pollution from subdivisions, construction sites, and mining. The *stream standards* can be related to the protection of aquatic habitats, which is one of the most important beneficial uses of water, and/or to other existing or intended uses of the water resource. Actual numerical values of water quality standards are given in Appendix 1, and their fundamentals are described in Chapters 12 and 13. The designated use of a water body must be attainable (see Chap. 16 for a discussion of use attainability). Standards and criteria may be numerical, chemical-based, or narrative or based on the toxicity of the entire effluent or water body.

In water quality planning and evaluation, exceeding the water quality parameters over one or more standards (criteria) implies an injury to the water use for which the standard was issued. Consequently, a wastewater discharge that does not result in a violation of a standard may be considered noninjurious, as it does not cause pollution. The quantity of potential pollutants that can be discharged into the environment (receiving water body, atmosphere, or land) is then called the *waste-assimilative capacity*. Determining the waste-assimilative capacity of the receiving water or air body is one of the most important steps in any environmental protection study. Not taking the waste-assimilative capacity into consideration during the planning of pollution abatement or water quality restoration would lead to uneconomical wasteful approaches or even ineffective solutions. The concept of waste-assimilative capacity and its determination is shown on Figure 1.3. Typically, the waste-assimilative capacity of surface water bodies might be higher for decomposable organic matter, but it is very low to nil for some toxic chemicals that bioaccumulate in tissues of aquatic organisms and become injurious to animals and man using them as food. Consequently, this detrimental capacity is reflected in the magnitude of the standards. Determination of the waste-assimilative capacity is a part of the water quality that is based on pollution control, and is now

Definitions 17

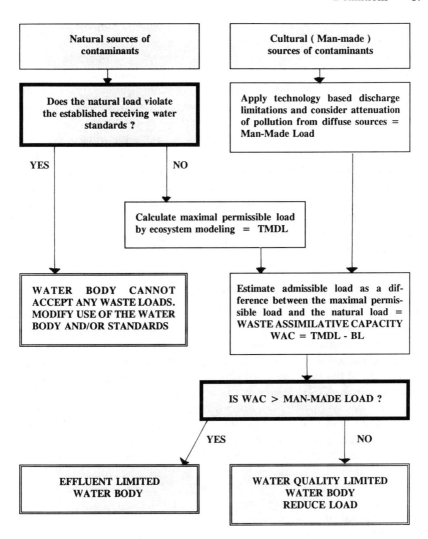

FIGURE 1.3. Waste-assimilative capacity concept.

becoming the most significant part of pollution-control efforts. In water-quality-based controls, effluent standards are tied to the the waste-assimilative capacity of the receiving water body. The process of integrated water-quality-based (point and nonpoint) pollution abatement is presented in Chapter 16. Existing waste-assimilative capacity of a water body can be increased by water body restoration and other measures (see Chap. 15).

According to these definitions a waste constituent becomes a pollutant if it is discharged into the environment in quantities that are injurious to or impair the beneficial uses of environmental resources. Many waste constituents in small quantities are not injurious, and some of them may even be beneficial in low quantities, becoming injurious (toxic) only in quantities that exceed the waste-assimilative capacity. For example, some metals that in higher concentrations are known to be toxic, such as zinc, are in smaller trace quantities, necessary nutrients for aquatic life.

Best management practices (BMP) are methods, measures, or practices selected by an agency to meet its nonpoint (diffuse) source control needs. BMPs include, but are not limited to, structural and nonstructural controls and operations and maintenance procedures. BMPs can be applied before, during, and after pollution-producing activities to reduce or eliminate the introduction of pollutants from diffuse sources into receiving waters.

Definition of Pollution Sources

From the previous discussion it follows that pollution is caused by man and results in an undesirable or harmful change in the quality of the resource, whether water, soil, or air. The sources or causes of pollution can be classified as either *point* or *nonpoint sources of pollution*. During publishing of the previous books by the first author on this topic (Krenkel and Novotny, 1980; Novotny and Chesters, 1981) the term *diffuse pollution* was synonymous to nonpoint pollution. After the U.S. Congress passed the Clean Water Act in 1977 and the Water Quality Act of 1987, it became necessary to redefine these characterizations.

Point sources of pollution were originally defined as pollutants that enter the transport routes at discrete, identifiable locations and that can usually be measured. Major point sources under this definition included sewered municipal and industrial wastewater and effluents from solid waste disposal sites. Nonpoint sources were simply "everything else," and included diffuse, difficult to identify, and intermittent sources of pollutants, usually associated with land or land use. These definitions led to some legal ramifications for abatement efforts. According to the U.S. Constitution, the government can mandate the control of point sources that enter so-called navigable waters,[2] while use of land is considered sacred making enforcement of nonpoint pollution control impossible. Hence the new definitions broadened the category of *point sources*.

---

[2] Under legal interpretations in the United States, any body of water on which a canoe can float, even potentially, is considered navigable.

Today's statutory definition of point sources is as follows (Water Quality Act, Sec. 502–14, U.S. Congress, 1987):

> The term "point source" means any discernable, confined and discrete conveyance, including but not limited to any pipe, ditch, channel, tunnel, conduit, well, discrete fissure, container, rolling stock, concentrated animal feeding operation, or vessel or other floating craft from which pollutants are or may be discharged. This term does not include agricultural stormwater and return flows from irrigated agriculture.

This definition does not leave much space for "everything else" being nonpoint sources. This ambiguity or lack or definition led the National Resources Defence Council to call nonpoint pollution "poison runoff" (Thompson, 1989).

Rather than looking for some kind of exact legal definition of point and nonpoint sources we attempt to categorize the sources of pollution into point and nonpoint categories according to the latest statutory regulations (see the next chapter for details). The statutory point source category in the United States today includes the following sources:

- Municipal and industrial wastewater effluents;
- Runoff and leachate from solid waste disposal sites;
- Runoff and infiltrated water from concentrated animal feeding operations;
- Runoff from industrial sites not connected to storm sewers;
- Storm sewer outfalls in urban centers with a population of more than 100,000;
- Combined sewer overflows;
- Leachate from solid waste disposal sites;
- Runoff and drainage water from active mines, both surface and underground, and from oil fields;
- Other sources, such as discharges from vessels, damaged storage tanks, and storage piles of chemicals;
- Runoff from construction sites that are larger than 2 hectares (5 acres).

Bypasses of untreated sewage (when the capacity of a treatment plant would be exceeded) are not allowed by law, and hence cannot be considered to be a legal source of pollution.

Two common characteristics of these statutory point sources are that they do indeed enter the receiving water bodies at some identifiable single- or multiple-point location and that they carry pollutants. Three more common characteristics of these point sources are that in the United States they are regulated, their control is mandated, and a permit is required for the waste discharges. The statutory nonpoint sources ("everything else") then include:

- Return flow from irrigated agriculture (specifically excluded from point source definition by Congress);
- Other agricultural and silvicultural runoff and infiltration from sources other than confined concentrated animal operations;
- Unconfined pastures of animals and runoff from range land;
- Urban runoff from sewered communities with a population of less than 100,000 not causing a significant water quality problem;
- Urban runoff from unsewered areas;
- Runoff from small and/or scattered (less than 2 hectare) construction sites;
- Septic tank surfacing in areas of failing septic tank systems and leaching of septic tank effluents;
- Wet and dry atmospheric deposition over a water surface (including acid rainfall),
- Flow from abandoned mines (surface and underground), including inactive roads, tailings, and spoil piles;
- Activities on land that generate wastes and contaminants, such as:
    deforestation and logging
    wetland drainage and conversion
    channeling of streams, building of levees, dams, causeways and flow-diversion facilities on navigable waters
    construction and development of land
    interurban transportation
    military training, maneuvers, and exercises
    mass outdoor recreation.

Some of these "nonpoint"sources are either locally or federally regulated. For example, in many states developers are required to implement erosion-control practices, wetland protection laws regulate drainage of wetlands, before individual septic tank systems can be installed local or state public health and/or land-use authorities must issue permits. In the United States no potentially polluting activities on navigable waters (such as dredging, channel construction, dams) can proceed until the U.S. Army Corps of Engineers has issued the necessary permits. This agency is also responsible for issuing permits for any alterations to wetlands.

From the previous discussion and definition of pollution, one might also conclude that not all lands and land-use activities are polluting. For example, diffuse-waste emission from low-density suburban developments are very close to the background emissions of most important constituents. Waste emitted from some disturbed lands located far from a watercourse is redeposited between the source and the recipient, hence the land may not cause impairment of water quality (pollution). For this reason some soil-conservation practices might not have a significant water quality improvement benefit. The lands that are most polluting within a watershed (and require pollution elimination) are called *hazardous* or *critical lands*. Determination and location of hazardous (critical) lands is one of the most important tasks in planning a diffuse-pollution abatement.

As can be seen, the division of sources into "point" and "nonpoint" is not as straightforward as it might have been during the 1970s when the *Handbook of Nonpoint Pollution: Sources and Management* (Novotny and Chesters, 1981) was written. Today these definitions have more statutory and legal than technical meanings.

The fact remains, however, that the "traditional" point sources of wastewater, that is, municipal, industrial, and agricultural (farm), are different from *diffuse* sources, which according to the statutory definition, may be both point and nonpoint.

The traditional point sources strictly include wastewater effluents from municipal and industrial areas. The flow and pollution loads from these sources vary; however, in most cases they are continuous, uninterrupted discharges, variability is not greatly related to meteorological factors, and the variability is not great. The primary parameters of interest for control and regulation are the degradable organics (measured as BOD_5 or COD), pH, suspended solids, nitrogen, and phosphorus and toxic compounds (both organic and inorganic). The prevalent method of control is treatment.

Diffuse sources can be characterized as follows:

- Diffuse discharges enter the receiving surface waters in a diffuse manner at intermittent intervals that are related mostly to the occurrence of meteorological events.
- Waste generation (pollution) arises over an extensive area of land and is in transit overland before it reaches surface waters or infiltrates into shallow aquifers.
- Diffuse sources are difficult or impossible to monitor at the point of origin.
- Unlike for the traditional point sources, where treatment is the most effective method of pollution control, abatement of diffuse land is focused on land and runoff management practices,
- Compliance monitoring is carried out on land rather than in water.
- Waste emissions and discharges cannot be measured in terms of effluent limitations.

22 Introduction

- The extent of diffuse waste emissions (pollution) is related to certain uncontrollable climatic events, as well as geographic and geologic conditions, and may differ greatly from place to place and from year to year.
- The most important waste constituents from diffuse sources subject to the management and control are suspended solids, nutrients, and toxic compounds.

ENVIRONMENTAL SYSTEMS AND THEIR CHANGES, INTERACTIONS, AND IMPACT ON WATER QUALITY (POLLUTION)

Ecological Systems

There are five biological systems—croplands and grasslands, forests, urban, air systems, and aquatic systems—that support the world economy and provide the means for sustaining biological life on the earth. In addition, all biological life takes place in the atmosphere (including that of soil) or in water. These five systems are interconnected in a balance that is often disrupted by overuse of the resources and/or by their loss and degradation. Figure 1.4 shows the interconnections and pathways of

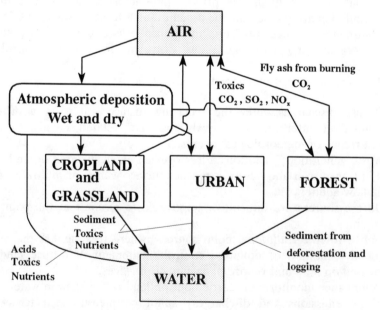

FIGURE 1.4. Pathways of pollutants between basic ecological systems.

contaminants between the five basic biological (ecological) systems. Atmospheric pollution impacts on water quality, not counting the harm done to human health, deforestation (loss of land resources) has a tremendous effect on water resources as well as on the atmosphere (greenhouse effects by slash burning and emissions of CO_2); urbanization (another loss of resources) is a source of many types of pollution, including contamination of air, water, and soil as well as soil loss. Soil loss from agriculture and construction diminishes the use of water resources by clogging the receiving water bodies with excessive sediments and depriving the aquatic biota of their natural habitat. Sediment is also a carrier of many pollutants, including those that cause excessive algal growths (nutrients) and those that cause toxic contamination.

These five systems are not stagnant, but continuously evolve and change. Until the twentieth century most of the changes that adversely impacted the environment were of natural origin, such as the erosion of land, volcanic eruptions, and natural flooding. There have been incidences throughout history where man has adversely and detrimentally impacted entire regions. For example, the rich Mesopotamian civilization flourished until its vast irrigation systems were destroyed by sediment from erosion when the forests on the surrounding hills were cut down and the fertility of the soil was destroyed by salts deposited by the irrigation water. Syria, which is now mostly arid or desert, thousands years ago was forested and enjoyed more humid hydrological conditions. The first deforestation of southern Europe took place during the Roman period, followed by reforestation after the fall of the Roman Empire in the fifth century. Then in the Middle Ages deforestation was repeated by Venetians who cut down the forests of Dalmatia on the Adriatic Sea (present Croatia), which resulted in tremendous soil loss and changed the entire hydrology of the region. However, these instances were mostly localized. As long as the equilibrium between the five ecological systems is not disrupted, the adverse ecological consequences are minimal.

Several major factors caused the environmental (ecological) balance to be severely disrupted in the second half of the twentieth century, resulting in accelerated increases of environmental pollution. They are

- population increase (sometimes termed explosion),
- deforestation,
- conversion of land to intensive agriculture,
- urbanization and industrialization,
- increased living standard, resulting in an increased per capita use of natural resources.

TABLE 1.1 World Population Growth

Year	Population (billions)	Population Increase by Decade (millions)
1650	0.5	—
1800	1.0	—
1930	2.0	—
1950	2.515	—
1960	3.019	504
1970	3.698	679
1980	4.450	752
1990	5.292	842
2000	6.251	959

Source: From United Nations, Department of International Economic and Social Affairs, New York (1988).

Population Increase

Table 1.1 shows the approximate world population. It can be seen that it took about one million years for the world population to reach the first one billion. About the same population increase may occur in the 10 years between 1990 and 2000. The population increase is not evenly distributed throughout the world. While in most developed, industrialized countries population growth is declining or the population is remaining stagnant, most of the growth is occurring in less developed countries. Because the less developed countries lack adequate sanitation, the increase in population results in the increase of diffuse pollution. The population pressures are also detrimental to forests and other resources that are diminishing at an accelerated rate, which results in more pollution.

In the context of sustainable resources development and containment of environmental damage, each additional person represents additional demands on productive resources, and hence additional wastes, plus wastes caused by the maintenance of the life processes.

Increased population also forces accelerated land-use changes. Since rural areas cannot economically sustain all families, the excess rural population is directed toward urban areas in search of employment and even, as in developing countries, for survival.

The truth is, however, that in many developing countries, and to some extent even in developed countries, urban areas cannot absorb the influx of population plus their own population increase, which leads to homelessness and shantytowns that, without sanitation, create high levels of diffuse pollution. The population increase of some world urban centers has been phenomenal and has greatly outpaced world population growth. For example, according to United Nations statistics, the population of Sao Paulo in Brazil, which in 1950 was about 2.4 million, is now over 20 million; Mexico City's population increased by 19 million between 1950 and 1990; the population of Paris increased by 4 million (from 7.4 million in 1950 to about 11 million in 1990); and the population of London remained almost constant.

Watershed in Transition

Land Use and Diffuse Pollution

It has to be realized that in most cases it is not the land or land use *per se* that causes pollution. Land use is a simple term to describe the prevailing activity occurring in the area. As such it bears little relationship to the pollution generated from the area. Land-use zoning and separation of activities on land is more unique to the United States than to other countries. In many European and world urban centers, use of urban land is mixed, allowing commercial and sometimes industrial activities to exist side-by-side with residential developments.

Many land uses, such as for economic and living purposes and for development, are justified. In most cases, it is the misuse or excessive use of land that causes excessive pollution.

In the United States, land uses have been generally divided into the following categories (Novotny and Chesters, 1981):

General Urban Lands

Residential (low, medium, and high density)
Commercial
Industrial
Other developed (large parking areas, sports complexes)
Open (parks, golf courses)
Transportation (airport, rail, vehicular)

Rural (Nonurban) Lands

Cropland (irrigated and nonirrigated)
Improved pasture and rangeland
Woodland and silviculture
Concentrated animal feedlots
Idle land (includes open water surfaces and native lands)
Land used for waste disposal.

This "static" categorization of land uses has been questioned as insufficient for estimating nonpoint loading (Novotny, 1985; Marsalek,

1978). Wide variations of loads within each land-use category have been observed (IJC, 1980; Marsalek, 1978; Novotny and Chesters, 1981; Beaulack and Reckhow, 1982; Novotny, 1985).

Since it is the use of land by man and his/her associated activities that generate pollution, diffuse pollution has been related to the land-use type and to the transition from one land use to another.

Land Use and Land Transformation

There are four basic types of native undisturbed lands (excluding mountains): arid land (including deserts), prairie, wetland, and woodland. These four types have various forms, depending on the geographical location and the elevation.

The transition process, which today is mostly cultural (anthropogenic), is depicted in Figure 1.5. It begins with one of the four natural lands and progresses through several intermediate stages to a fully urbanized, sewered watershed, which represents the end of the transition process. Most of this transition is irreversible. Just as the natural morphological processes that carve the basins by erosion and the elutriation of minerals are responsible for the background water quality composition, cultural processes of land changes are responsible for pollution. These cultural land-use modifications and land development processes are also illustrated on Figure 1.5.

Deforestation

Deforestation is an environmentally devastating and polluting land-use activity, which today is recognized as having very serious global consequences. Ancient examples are the Middle East, where deforestation changed the woodlands of Syria and other Middle East countries into desert, and forest cutting by the Romans and Venetians, which changed most of Dalmatia into a barren, soil-poor country. These examples are now being followed on a much larger scale by the disappearance of the tropical rain forest.

Deforestation deprives the soil of its protective, vegetative cover, and hence increases soil loss through erosion by several orders of magnitude. As a consequence, the quality of receiving water bodies is also dramatically reduced by sediments, organic compounds, nutrients (nitrogen and phosphorus), and possibly other pollutants (Fig. 1.6).

Slush burning practices used today to clear rain forests for farming and rangeland also add significantly to the carbon dioxide levels in the atmosphere, thus increasing the *greenhouse* phenomenon. Brazil, for example, contributes some 336 million tons of carbon to the atmosphere each year through deforestation. This contribution also increases

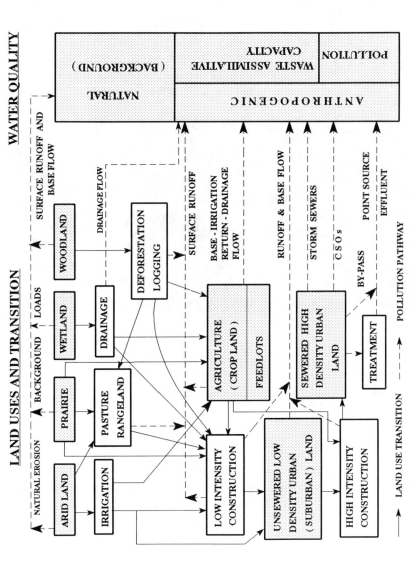

FIGURE 1.5. Land-use transformation and the generation of pollution.

28   Introduction

FIGURE 1.6.   Deforestation.

atmospheric pollution, which by dry and wet deposition then affects the quality of water and terrestrial resources. It should also be pointed out, however, that the United States and Soviet Union are the largest sources of atmospheric emissions of carbon (1224 and 1000 million tons per year, respectively, in 1987), mostly from burning fossil fuel and vehicular traffic. These emissions far exceed those from the destruction of rain forests in developing countries (Brown et al., 1990). Developed countries are also guilty of large-scale deforestation. For example, the disappearance of forests in Florida due to urban development is of the same order of magnitude as the deforestation of the Amazon rain forest in Brazil.

Wetland Alteration

About 6% of the total land surface is wetland (Mitsch and Gosselink, 1986); however, like rain forest, wetlands are disappearing rapidly due to drainage and conversion to agricultural and urban uses. Wetlands represent a natural sink for various potential pollutants, including sediment, nutrients, and organic compounds.

Wetland alteration or destruction can take several forms; however, drainage and filling are the most common. The fertile soil of drained prairie wetlands provided excellent crop yields, but also deprived the watershed of natural hydrologic buffers and water quality preservation. In addition to drainage for agricultural uses, other activities that adversely alter or destroy wetlands include their conversion to urban uses, transportation, peat mining and mineral extraction, flood control, navigation, and industrial activities.

Construction

Newly developing urban lands should receive special attention. For all land uses, this stage of land is characterized by high production of suspended solids caused by erosion of unprotected exposed soil and soil piles. Extremely high pollutant loads (see the subsequent sections) are produced from construction sites if no erosion-control practices are implemented. Therefore, in establishing pollutant loading related to land uses one must first determine whether the area is fully developed or if it is a developing area and/or if significant construction activities are taking place therein. An area is considered fully developed and established one year after completion of the development.

Conversion to Intensive Agriculture

The first conversion of American prairies to an agricultural use was the use of the land for *pasture* or *range land*. It should be noted that the native prairies (as well as prairie wetlands and woodlands) were actually sinks of potential pollutants. Timmons and Holt (1977) studied organic pollution and nutrient (nitrogen and phosphorus) losses from native prairies in west central Minnesota and found that annual losses of nutrients in runoff were less than the nutrient inputs in precipitation.

Today about 40% or 3.6 millions km^2, of the land area of the United States is used for grazing livestock. These are mostly unconfined systems. Although the trend is to confine cattle operations, unconfined production is expected to continue to predominate the beef and sheep industries.

The differences between *confined* and *unconfined animal operations* are the same as the division of pollution sources into point and nonpoint. Unconfined operations, that is, operations where cattle graze more-or-less freely over an extended pasture area are strictly nonpoint, diffuse sources. Confined operations are statutory point sources; however, the largest waste loads occur during storms and are carried by surface runoff, which is a characteristic of diffuse sources.

As grazing animals traverse pasture and rangeland, the top soil becomes more compacted. Trampling by cattle may reduce infiltration and increase surface runoff (Robbins, 1985); however, such increases were not significant for pastures with light and moderate intensities of grazing (Moore et al., 1979). In his studies Robbins (1985) also claimed that distinguishing between the water quality of streams affected by light to medium grazing and unaffected streams is difficult or impossible. He concluded that when management is directed to optimize forage production, pollutant yields from unconfined animal operations are not greater than would have occurred under native conditions.

However, it is not the grazing intensity on the land that determines the degree of pollution from rangeland, it is the unrestricted access of cattle to the watercourses that has the major impact. This is particularly true in high-quality cold water fishery areas of the Rocky Mountains and in slow-moving warm streams of the deep southwest. For example, a 100-hectare ranch with 200 steers that are excluded from the stream may have much less impact than a 100-hectare hobby farm with 50 head of cows/calves and steers that have a direct stream access.

Until the 1950s the expansion of *agricultural land for crop production* has more or less kept pace with the population increase. Most farming in the United States and the world was done on small family farms without an excessive use of chemicals. However, around 1950 farming began to qualitatively change. Dramatic changes happened in Eastern Europe, which followed the Soviet example of a forceful change to large collective farming, and also in the United States, where small family farming began to change into larger monocultural enterprises. Concurrently, farming began to rely more and more on the use of chemical fertilizers to increase plant yields and on pesticides for insect and weed control. Monocultural planting requires more chemical use than rotational planting.

Figure 1.7 shows a typical trend in U.S. agriculture. Until 1950 most increases in agricultural production were due to the conversion of native lands to agriculture. However, since 1950 the cultivated land area in Iowa has remained about the same, while in some other states it has decreased due to urban pressures. The increased farm output was solely due to the increased use of chemicals.

The primary agricultural activity that causes the elevated emissions of potential pollutants is the practice of disturbing the soil by tillage, which increases sediment losses, in comparison to the original native lands, by several orders of magnitude. Increased use of chemicals results in more losses into the environment. As noted by Alberts, Schuman, and Burnwell (1978), over 90% of nutrient (nitrogen and phosphorus) losses are associated with soil loss. Although nutrient losses represent only a

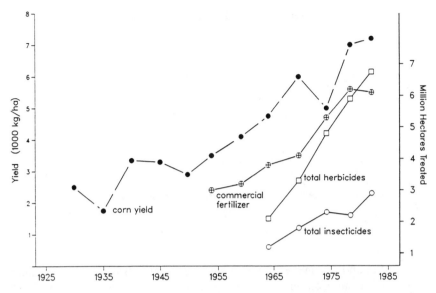

FIGURE 1.7. Trends in the use of chemicals in the agriculture (corn yield) of Iowa.

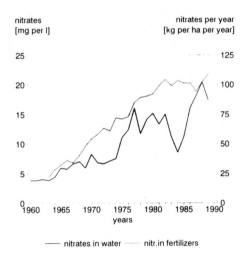

FIGURE 1.8. Relation of nitrate concentration (such as NO_3^-) in the Vltava River in the Czech Republic and fertilizer use in the watershed. (From Czechoslovak Academy of Science, 1992.)

small portion of the applied fertilizer, their contribution to the receiving waters almost always exceeds the standards accepted for preventing accelerated eutrophication of surface waters. Furthermore, the loss of nitrates into the ground water in areas of intensive agriculture are

TABLE 1.2 Typical Concentrations of Pollutants in Rural Runoff in the Midwestern U.S.

	Suspended Solids (mg/l)	BOD$_5$ (mg/l)	COD (mg/l)	Total Nitrogen (mg/l)	Total Phosphorus (mg/l)	Total Coliforms (MPN/100 ml)
Background levels[a]	5–1000	0.5–3	NA	0.05–0.5	0.01–0.2	10–10^2
Cropland[b]	(780)	NA	(80)	(9)	(1.2)	NA
Grazed pasture[c]	NA	(13)	NA	4.5	(7)	10^5
Feedlots[d]	(30)	1000–11,000	31,000–41,000	920–2100	290–380	NA

Note: ( ) = mean; NA = data not available or insufficient.
[a] Lager and Smith (1974).
[b] Wisconsin Priority Watersheds, Wisconsin Department of Natural Resources.
[c] Robbins (1985).
[d] Loehr (1972).

substantial and can make ground water unsuitable for drinking. Alarming levels of nitrate contamination of ground- and surface-water resources have been reached in Eastern and Central Europe. Figure 1.8 shows trends of nitrate concentrations in the Vltava River upstream of Prague, Czechoslovakia. The agricultural use of pesticides (atrazine in Central Wisconsin, the Po River valley in Italy, and elsewhere) is also responsible for contamination of ground and surface receiving waters. Table 1.2 presents typical concentration ranges of pollutants in agricultural runoff.

According to the U.S. EPA (1984), at the beginning of the 1980s cropland, pastureland, and rangeland contributed over 6.8 million tons of nitrogen and 2.6 million tons of phosphorus to conterminous U.S. surface waters per year. The Corn Belt (Illinois, Indiana, Iowa, Missouri, and Ohio) used 39% of the nation's phosphorus fertilizer and 32% of its nitrogen fertilizer.

Delivery losses of pesticides applied to agricultural lands to control insects and weeds amount on average to about 5% of the applied pesticide. Most of the loss ends as pollution of ground and surface waters. However, if rainfall occurs shortly after a pesticide is applied, losses can be substantial and can result in fish kills. Herbicides (weed control) are the most used pesticides in U.S. agriculture. In 1980, farmers used about 200,000 tonnes of herbicides and 140,000 tonnes of insecticides. Since 1980, these uses have doubled. Figure 1.9 shows total trends in pesticide use in the United States.

Agricultural use of arid lands requires irrigation. Civilizations flourished and vanished because of pollution due to irrigation. The problem is the salt content of irrigation water and the elutriation of

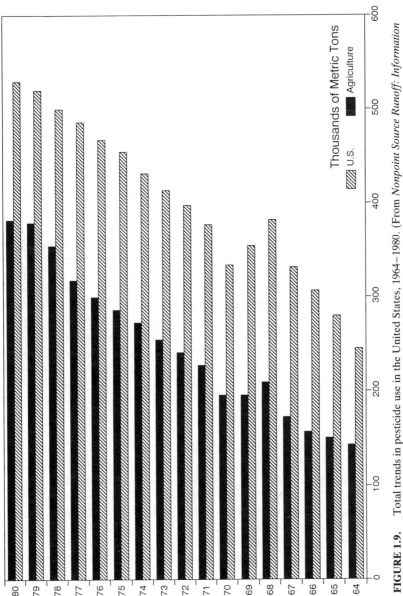

FIGURE 1.9. Total trends in pesticide use in the United States, 1964–1980. (From *Nonpoint Source Runoff: Information Transfer System*, EPA, Office of Water, July 1983, p. 2.7.)

minerals from soils during irrigation. Water lost by evapotranspiration into the atmosphere does not contain salt, hence in order to control salt buildup in soils, excess irrigation water over evapotranspiration losses must be applied. This water excess, called *irrigation return flow*, is considered pollution that gets deposited in surface or ground waters. In some irrigated parts of the arid southwest (such as the lower Colorado River Basin), salinity and pollution of streams receiving irrigation return flows have reached such high levels that the water is not suitable for further use. By an agreement between the United States and Mexico, excess salt from the Colorado River must be removed at great expense to the U.S. taxpayers.

Feedlots and Barnyards

Feedlots and barnyards, statutory agricultural point sources, are the land uses that exhibit the highest amounts of diffuse waste loads, mostly carried by surface runoff. With the advent of improved feeding techniques, farmers prefer not to put cattle to pasture, but to hold them in relatively small areas. The high organic content of the surface crust protects against erosion, and consequently sediment losses from feedlots are lower than those from uncompacted bare land surfaces. Nevertheless, as shown in Table 1.2, barnyard and feedlot runoff has extremely high concentrations of BOD_5, COD, organic nitrogen, and phosphorus. Even with these extremely high concentrations, land management techniques typical for diffuse sources and some "point source" structural abatement (lagoons and manure storage basins) are a more effective means of control than a point-source-type treatment would be (see Chapter 11).

Urbanization

Probably the greatest adverse change in water quality is due to urbanization. The population pressures mentioned previously, migration of population, and economic development result in urbanization.

Urbanization, that is, the transformation of land use from natural or agricultural lands, occurs in several steps. Urbanization changes atmospheric composition, the hydrology of the watershed (Fig. 1.5), receiving streams, and other water bodies, and soil. Native ecological systems are replaced by urban ecology. Waste emissions increase dramatically, and the sources of these contaminants are diverse, such as industries, household heating, transportation, sewage conveyance and disposal, garbage collection and disposal (landfills, incinerators), litter deposition, fallen leaves on impervious surfaces, and street salting. Chapter 8 discusses these sources in detail.

Watershed during Urban Development (Urbanization)

The soil loss from construction sites can reach magnitudes of over 100 tonnes/ha/year. In urbanizing watersheds, a few percent of the watershed under construction can contribute a major portion of the sediment carried by the streams. Often the streams themselves are affected and irreversibly changed by urbanization, regulation, straightening and lining, which destroys their natural habitat so that they can no longer sustain fish and other biotic populations. More surface runoff and hence flooding results from increased imperviousness that also makes surface runoff more flushy and higher in volume. On the other hand, ground-water recharge is reduced. Other profound and adverse changes in hydrology and water quality are caused by draining wetlands. Pollution is mostly nonpoint, unless a sewer system is in place and the drainage is located in a large urban area and/or industrial zone, which would legally reclassify the runoff pollution as a point source.

Unsewered Urban Development

Disposal of sewage into soils (septic tanks or small Imhoff tanks followed by infiltration) eliminates or reduces only pollutants that can be filtered out, decomposed, and/or adsorbed onto soil particles. Mobile pollutants such as nitrates can cause severe contamination of ground water and, subsequently, of the base flow of the streams and other water bodies. When the adsorption capacity of the soil-disposal system is exhausted, contamination of surface waters by organics and pathogenic microorganisms by surfacing sewage may occur and be severe. Failures of older septic tank systems are common. However, in the absence of such failures, the pollution potential of established low-density residential and commerical areas without storm or combined sewer systems is generally low and not much above the background water quality contributions from a natural prairie watershed.

Pollution from Fully Developed, Sewered Urban Watersheds and Transportation Corridors

As the imperviousness of the urban watershed increases, the watersheds become more hydrologically active, which will impact both pollutant loadings and flooding potential. The surface runoff events that carry the heaviest pollution loads become more frequent. Pollutants that accumulate on the surface from traffic, litter, street dust, and other sources are then washed off by the surface runoff into the drainage (sewer) system.

Novotny and Chesters (1981, 1989) have pointed out that the type of drainage can greatly affect the pollution load within the same land-use category. Residential areas with natural (swale) drainage produce

TABLE 1.3 Comparison of the Strength of Point and Nonpoint Urban Sources

Type of Wastewater	BOD$_5$ (mg/l)	Suspended Solids (mg/l)	Total Nitrogen (mg/l)	Total Phosphorus (mg/l)	Lead (mg/l)	Total Coliforms (MPN/100 ml)
Urban storm-water[a]	10–250 (30)	3–11,000 (650)	3–10	0.2–1.7 (0.6)	0.03–3.1 (0.3)	10^3–10^8
Construction site runoff[b]	NA	10,000–40,000	NA	NA	NA	NA
Combined sewer overflows[a]	60–200	100–1100	3–24	1–11	(0.4)	10^5–10^7
Light industrial area[c]	8–12	45–375	0.2–1.1	NA	0.02–1.1	10
Roof runoff[c]	3–8	12–216	0.5–4	NA	0.005–0.03	10^2
Typical untreated sewage[d]	(160)	(235)	(35)	(10)	NA	10^7–10^9
Typical POWT effluent[d]	(20)	(20)	(30)	(10)	NA	10^4–10^6

Note: ( ) = mean; NA = not available; POWT = Publicly owned treatment works with secondary (biological) treatment.
[a]Novotny and Chesters (1981) and Lager and Smith (1974).
[b]Unpublished research by Wisconsin Water Resources Center.
[c]Ellis (1986).
[d]Novotny, et al. (1989).

pollution loads that are approximately one order of magnitude lower than pollution loads from similar land with storm (separate) sewers.

Sewer systems can be either *combined* or *separate*. The pollution potential of urban runoff carried by separate storm sewers is similar to treated sewage, while that of combined sewer overflows is between treated and untreated municipal wastewater (Field and Turkeltaub, 1981). Rohmann, Lyke, and Hoban (1988) claimed that urban nonpoint source released 760 times more lead than the load from point sources. Table 1.3 shows the approximate concentration ranges for some components of diffuse urban sources compared to the strength of municipal wastewater.

In combined sewer systems, wet-weather pollution loads are divided into two components. The first component is the load that is conveyed by the combined sewer interceptors (due to their excess capacity over the dry-weather flows) toward the treatment plant. At the plant this load represents a point source and should not be considered in nonpoint pollution studies. The second component is conveyed by combined sewer overflows (CSO) that occur during wet weather when the interceptor capacity is exceeded. The load conveyed by overflows, referred to as *wet-weather* load, is of diffuse (nonpoint) nature and is considered in this study. Legally, however, the U.S. Environmental Protection Agency and state agencies consider the combined sewer overflows as a point source, for which a National Pollution Discharge Elimination System (NPDES) permit is required.

A part of the wet-weather pollution carried by combined sewers is diverted to the treatment plant. Even though the concentrations of pollutants in CSOs are greater than those in flows from separate storm sewers, the volume of discharged CSOs is smaller. Hence, European experience and measurements indicate that overall, the total pollution loads—including dry-weather point loads—from urban zones with separate and combined sewers are about the same (Novotny et al., 1989b).

In addition to the traditional pollutants reported in Table 1.3, urban runoff contains a variety of toxic (so-called priority) pollutants, such as oils, polyaromatic hydrocarbons (PAH), PCBs, and lead. With the ban on the use of leaded gasoline in the United States, the lead level in urban runoff has dropped significantly. Urban diffuse sources have been identified as a major cause of pollution in surface water bodies by the US EPA (Myers et al., 1985).

Using a hypothetical example, Pitt and Field (1977) showed that in an urban area of 100,000 people, the COD contribution from urban runoff is approximately 50% of that from raw sewage. These results, together with

TABLE 1.4 Comparison of Areal Loadings of Pollutants from a Hypothetical American City of 100,000 People (tonnes/year)

Pollutant	Storm Water	Raw Sewage	Treated Sewage
Total solids	17,000	5,200	520
COD	2,400	4,800	480
BOD_5	1,200	4,400	440
Total Phosphorus	50	200	10
Total Kjeldahl Nitrogen	50	800	80
Lead	31		
Zinc	6		

Source: After Pitt and Field (1977).

comparisons for some other constituents, are shown in Table 1.4. It is also shown that if sanitary sewage is receiving adequate treatment (90 to 95% removal of solids and organic compounds), almost all—approximately 97%—of the total solids and 70% of the BOD reaching the receiving waters comes from urban runoff.

Example 1.1

Compare volume and pollution load (BOD_5) generated by sewage and by a 50-mm (2-in) 24-hr storm, after it comes from an urban catchment that is 70% impervious. The population density in the catchment is 100 persons/ha, with an average per capita sewage flow of 300 l/(cap-day).

Referring to Table 1.3 the average BOD_5 of urban runoff is 30 mg/l, and that for raw municipal sewage is 160 mg/l. Treated sewage has a BOD_5 of about 20 mg/l. Then the flow and volumes are:

Sewage

$$\text{Flow} = 300 \text{ l/(cap-day)} \times 100 \text{ persons/ha}$$
$$= 30,000 \text{ l/(ha-day)} = 30 \text{ m}^3/\text{(ha-day)}$$
$$\text{Raw sewage } BOD_5 = 30 \text{ m}^3/\text{(ha-day)} \times 160 \text{ g of } BOD_5/\text{m}^3$$
$$= 4800 \text{ g/(ha-day)}$$
$$\text{Treated sewage } BOD_5 = 30 \times 20 = 600 \text{ g/(ha-day)}$$

Stormwater

$$\text{Flow} = 0.7 \times 50 \text{ mm/day} \times 10,000 \text{ m}^2/\text{ha} \times 0.001 \text{ mm/m} \times 1000 \text{ l/m}^3$$
$$= 350,000 \text{ l/(ha-day)}$$
$$= 350 \text{ m}^3/\text{(ha-day)}$$
$$\text{Stormwater } BOD_5 \text{ load} = 350 \text{ m}^3/\text{(ha-day)} \times 30 \text{ g/m}^3$$
$$= 10,500 \text{ g/(ha-day)}$$

The stormwater BOD load from this storm would be twice as large as the 24-hr BOD load by raw wastewater.

Pollutant Loads from Combined Sewer Overflows

Combined sewers have commonly been designed with a capacity large enough to accommodate flows that are 4 to 8 times the peak dry-weather flow. In Europe, combined sewer designs are now being implemented to handle a critical rainfall of 15 l/sec-ha (5.4 mm/hr) plus the dry-weather flow. If the rainfall exceeds the critical rate, overflow from the sewer system is anticipated. To mitigate pollution by the combined sewer overflow (CSO), over 10,000 storage tanks have been built in Germany alone and another 10,000 are planned (Geiger, 1990).

Using this criterion for a typical city located in Switzerland, out of about 1000 hours of rainfall in a typical average year, overflows would occur for about 80 hours (Krejci, 1988). In Milwaukee, Wisconsin, in an average year there were over 40 overflows from combined sewers in the older sections of the city. A large tunnel-interceptor to be completed in 1994 will store the overflows for subsequent treatment.

The pollution strength (Table 1.3) of the overflows is about the same or only marginally less than the strength of raw sewage. However, due to the larger volumes when compared to typical dry-weather sewage flows, overflow from a larger storm represents a pollution shock to the receiving water body that may greatly exceed, during the time of overflow, the load by sewage.

As stated by Imhoff and Imhoff (1990) and documented in Table 1.5, which is taken from Waller and Hart (1986), suspended-solids loadings from separate and combined sewers are about the same, while nutrient loadings from an urban watershed with combined sewers and a treatment

TABLE 1.5 A Comparison of Pollutant Loadings (kg/ha-yr) from Separate and Combined Sewer Systems for Ontario Urban Areas

	Separate Systems			Combined Systems		
Pollutant	Dry-weather STP Effluent	Wet-weather STP Effluent	Surface Runoff	Dry-weather STP Effluent	Wet-weather STP Effluent	Combined Sewer Overflows
Suspended solids	194	24	553	383	66	490
Total Nitrogen	133	16	11	253	44	25
Total Phosphorus	8	0.9	1.1	15	2.5	4.5

Source: After Waller and Hart (1986). (Copyright © 1986 by Springer Verlag; reprinted with permission.)

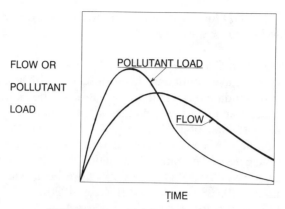

FIGURE 1.10. The first flush concept.

plant, but without storage for the overflows, are about twice that if the same watershed has separate sewers. Sewage solids that accumulate on the bottom or as a slime on the walls of the combined sewers during a dry period preceding a storm may contribute a higher pollutant load to the first portion of the overflow, the so-called "first flush." If the drainage system has a significant supply of readily washable solids located especially in the sewer system, the peak concentrations and pollutant loads will precede the peak flow and volume (Fig. 1.10). However, Ellis (1986), in summarizing primarily the European experience, stated that the occurrence of first flush is not a consistent feature of either the separate or combined sewer systems. Nevertheless, controlling pollution in the first portion of the runoff hydrograph makes sense (see Chap. 10).

Mining Nonpoint Sources

Unlike construction activities, mining cannot be viewed as a homogenous source of nonpoint pollution. The most common minerals extracted by mining are coal and metallic ores. Mining nonpoint sources include discharges from inactive mining operations as well as runoff from roads, old tailings, and spoil piles. Active mines are considered as point sources for which a discharge permit is required.

Although mining is not as widespread as agriculture, water quality impairment resulting from mining is usually more harmful; sediment discharges and concentrations from mines can be extremely high. Furthermore, entire streams may be biologically dead as a result of acid mine drainage (U.S. EPA, 1984). Erosion and sedimentation problems are associated with almost every abandoned surface coal mine. Other pollutants associated with mining operations can have an even more serious water quality impact than those associated with sediments.

POLLUTION BY TOXIC CHEMICALS AND METALS

Distribution and Types of Toxic Chemicals

Diffuse pollution may be responsible for the major part of the contamination of the environment by toxic pollutants. Toxic chemicals and components contaminating the environment are either *inorganic* or *organic*. Inorganic contaminants are mostly in a category of trace metals, which may be natural or anthropogenic (man-made) or both. Other inorganic nonmetallic toxic compounds detected in the aquatic environments are unionized ammonia (NH_3), cyanides, asbestos, hydrogen sulfide (H_2S), and low or high pH. Most of the organic toxic contaminants are anthropogenic, a category that includes organic chemicals such as pesticides, PCBs, organochlorine chemicals, solvents, and polyaromatic hydrocarbons (PAH).

As is pointed out in detail in Chapter 13, almost any compound may become toxic to aquatic biota, water fowl, or man when a certain tolerance threshold level is exceeded. From the many thousands of potentially toxic compounds and elements, the category of the so-called *priority pollutants* includes the toxic compounds that can be found in the environment in concentrations and quantities that can be toxic. These pollutants represent a risk to ecology and human health. The priority pollutants category includes both inorganic and organic (the majority) toxic compounds. Many but not all priority pollutants are carcinogenic.

Marsalek (1986) studied the distribution of toxic compounds in Ontario. He found that in or near urban areas trace metallic elements were the most prevalent toxic contaminants in runoff and stream sediments. The most frequently detected elements were zinc (98% of all water samples), copper (93%), nickel (87%), and lead (78%). Similar results were also found by the National Urban Runoff Project in the United States (Table 1.6). In both studies the frequencies of detection of trace elements in sediments were higher than those in water columns.

Among organic chemicals, pesticides were most frequently detected in the Canadian and U.S. studies. Marsalek (1986) found that in Ontario two organochlorine compounds, α-BHC, and γ-BHC (lindane) were found most frequently (98% and 86% of all water samples, respectively). The detection frequencies for chlorinated benzenes varied from 3% to 64% of all water samples. The detection frequencies of PAH compounds

TABLE 1.6 Most Frequently Detected Toxic (Priority Pollutants) in NURP Urban Runoff Samples

Detection rate[a]	Inorganic	Organic
Detected in 75% or more of NURP samples	Lead (94%) Zinc (94%) Copper (91%)	None
Detected in 50%–74% of the NURP samples	Chromium (58%) Arsenic (52%)	None
Detected in 20%–49% of NURP samples	Cadmium (48%) Nickel (43%) Cyanides (23%)	Bis(2-ethylhexyl) phthalate (22%) α-Hexachlorocyclohexane (20%)
Detected in 10%–19% of the NURP samples	Antimony (13%) Beryllium (12%) Selenium (11%) γ-Hexachlorcyclohexane (Lindane) (19%) Pyrene (15%) Phenol (14%) Phenanthrene (12%) Dichloromethane (methylene-chloride) (11%) 4-Nitrophenol (10%) Chruysene (10%) Fluoranthene (16%)	α-Endosulfan (19%) Pentachlorphenol (19%) Chlordane (17%)

Source: U.S. EPA (1983).

[a] Percentages indicate frequency of detection, not concentrations.

varied from 2% to 19% in water samples and from 1% to 37% in sediment samples. The high frequency of detection of some of these toxic organic and inorganic chemicals today confirmed Rachel Carson's worries in *Silent Spring*. The origins of these toxic compounds are very diversified. With the exception of trace elements that have a natural occurrence, organic toxic chemicals are man-made and entered the environment mostly after World War II.

The example of PCBs shows how a chemical component, originally considered to be very useful for a number of applications and that represents only a minor traditional point source problem, could contaminate the environment and become a serious diffuse-pollution problem. A similar fate has occurred to asbestos, which until the 1960s, was a material widely used for insulation, fire protective materials, tires, sewer pipes, brake linings, and many other applications. Both compounds

Pollution by Toxic Chemicals and Metals

have subsequently been recognized as both a carcinogenic threat to humans and a persistent widespread pollutant.

Sources and Properties of PCBs

PCBs are in a class of chlorinated organic compounds. These substances have been manufactured in the United States since 1929 by Monsanto Chemical Company. Other PCB-producing nations included Germany, France, Japan, Italy, the USSR, and Czechoslovakia.

PCBs are prepared individually by the reaction of biphenyl with anhydrous chlorine, and are available as liquids, resins, or solids. The most important physical properties of PCBs are their low vapor pressure (high boiling point), low water solubility, and high dielectric constant. They can be dissolved in many organic solvents, including some humic acids. The PCBs manufactured in the United States were sold under the trade name Aroclor. Table 1.7 shows some physical properties of PCBs as compared to another environmentally serious contaminant—DDT. Since 1929 over 600,000 tonnes of PCBs have been produced in the United States, with annual production reaching its maximum of 38,000 tonnes in 1970. Figure 1.11 shows the approximate distribution of PCBs that are said to have contaminated the environment.

From an environmental viewpoint, the commercial use of PCBs can be divided into three categories:

- *Controllable closed systems*: PCBs used as dielectrics in transformers and large capacitors have a life equal to that of the equipment, and with proper design environmental contamination should not occur.
- *Uncontrollable closed systems*: PCBs are used in heat transfer and hydraulic systems that permit leakage.

TABLE 1.7 Physical Properties of Some PCBs and DDT

	AROCLOR				
	1242	1248	1254	1260	DDT
Molecular weight					
Range	154–338	222–358	290–3982	324–460	352
Average	262	288	324	370	352
Chlorine (%)	42	28	54	60	50
Solubility in H_2O (µg/l)	200	100	50	25	0.7
Vapor pressure at 20°C mmHg	10^{-4}	3×10^{-5}	3.6×10^{-6}	1.5×10^{-7}	

Source: After Nisbet and Sarofim (1972).

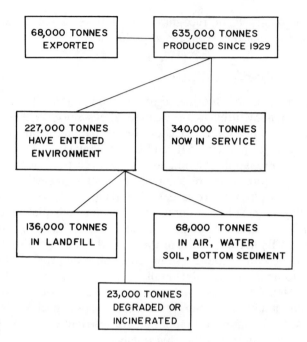

FIGURE 1.11. Distribution of PCBs. (From the Committee on the Assessment of PCBs in the Environment, 1979.)

- *Dissipative use*: PCBs in paint, lubricants, and plasticizers are in direct contact with the environment.

The major source of PCBs that have contaminated the environment are (Fig. 1.12):

- Leaks from sealed transformers and heat exchangers;
- Leaks of PCB-containing fluids from hydraulic systems that are only partially sealed;
- Spills and losses in the manufacture of either PCBs or PCB-containing fluids;
- Vaporization or leaching from PCB-containing formulations;
- Disposal of waste PCBs or PCB-containing materials and the migration of PCBs from landfills and freshwater sediments.

The chemical property of extreme stability that made PCBs an ideal industrial and commercial compound also made them persistent and cumulative toxic components in the environment. Because PCBs are soluble in lipid tissue, these components have been found to accumulate

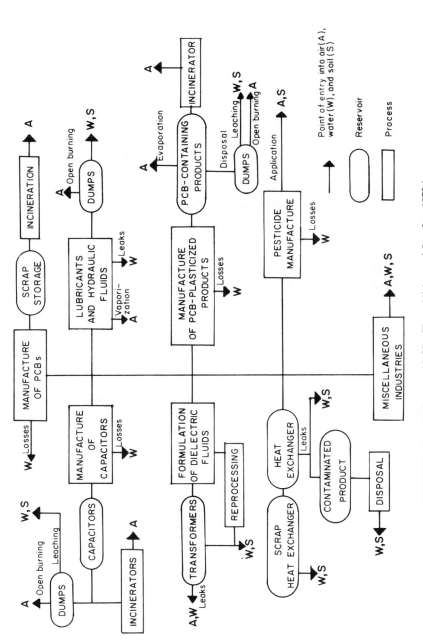

FIGURE 1.12. Pathways of PCBs (From Nisbet and Sarofim, 1972.)

in the fat of living organisms including humans. Salmonoid fish are very susceptible to PCB accumulation because these species have high body fat. Because of the toxic (carcinogenic) nature of PCBs, the U.S. Food and Drug Administration established a tolerance level of 2 mg/kg of PCB in fish tissue. Fish testing has shown that most large Lake Michigan fish species (trout and salmon) contain PCB levels exceeding this limit.

PCBs, which are in a way similar to DDT in the 1950s and 1960s, have entered into global transport routes and can be detected in extremely remote areas. Their removal rate from the environment is very slow, and the primary sink in aquatic ecosystems is burial with organic sediments.

IMPORTANT EXAMPLES OF SYSTEMS AFFECTED BY DIFFUSE POLLUTION

Chesapeake Bay

The Chesapeake Bay is the largest estuary on the east coast of the United States. It has been one of the world's most productive water bodies, providing habitat to fish and shellfish. However, water quality in this relatively shallow body of water has been declining. Submerged aquatic vegetation has been disappearing; fishers have been landing fewer of certain freshwater spawning fish; and oyster harvests have been declining. These problems were traced to excess levels of nutrients and toxic pollutants in the bay system. A 1983 EPA study concluded that these contaminants were also causing, among other phenomena, depressed oxygen concentrations in the water column, algal blooms, increased turbidity, and high concentrations of heavy metals in sediments. The study also found that diffuse (nonpoint) sources of pollution were among the chief causes of the bay's decline.

As a result of these developments, the governors of Pennsylvania, Maryland, and Virginia, the Mayor of the District of Columbia, and the Administrator of the U.S. Environmental Protection Agency in December 1983 pledged to address the problem of nonpoint pollution as well as other sources of pollution in order to restore and protect the Chesapeake Bay. This commitment, known as the Chesapeake Bay Agreement of 1983, established the Chesapeake Executive Council to coordinate all bay cleanup efforts undertaken by the signatories of the agreement. Implementing programs to reduce nonpoint source pollution is one of the most significant elements of the cooperative effort.

The Problem

Excessive nutrients appeared to account for much of the decline in living resources as well as many of the trends in water quality deterioration

TABLE 1.8   Nutrient Loads Reaching the Chesapeake Bay from Diffuse Sources

Land-use Type	Total Nitrogen from Diffuse Sources (%)	Total Phosphorus from Diffuse Sources (%)
Cropland	45–70	60–85
Pasture	4–13	3–8
Forest	9–30	4–8
Urban/suburban	2–12	4–12
Subtotal for Agriculture (Cropland + Pasture)	49–83	63–93

Source: U.S. Environmental Protection Agency (1988).

specified in the preceding paragraph. A watershed model was used to identify the sources. The model calculations indicated that in a year of average rainfall, diffuse (nonpoint) sources contribute 67% of the nitrogen and 39% of the phosphorus entering the bay and that the traditional point sources (wastewater effluents) account for the rest. Runoff from cropland was identified as the largest diffuse source of nutrients in the bay (Table 1.8). Urban runoff loads were minor and caused only localized problems. However, due to urban development pressures, the urban runoff contribution is expected to increase.

The watershed model, along with other information, has provided the basis for understanding the relative contributions of point and nonpoint by major river basins, and linked the nutrient loadings with specific areas where nutrient and dissolved oxygen concentrations potentially limited the aquatic resources.

The inorganic toxic loads to the bay system are also of concern. These toxics affect the aquatic life and habitats, especially near urban centers. The first EPA study (1983) found that in some areas toxic concentrations in sediments had reached magnitudes that reduce the hatching and survival of aquatic organisms, cause gross effects such as lesions or fin erosion in fish, and eventually could destroy an entire population of some sensitive species. Toxic discharges of metals from traditional point sources and from urban runoff appear to be most significant in urbanized/industrialized areas such as Baltimore, Maryland, Norfolk, Virginia, and Washington, D.C. As far as pollution by organic chemicals is concerned, it was concluded that herbicides were not the primary culprit in the decline of the bay ecological system.

Abatement Program

The Chesapeake Bay Agreement has established a framework for cooperation. Concurrently, the U.S. Environmental Protection Agency, which is a partner in the program, has established a liaison office in Annapolis, Maryland. The federal support for the program provided implementation funds to each state and to Washington to carry out the program. The bay states and Washington, D.C., have developed a variety of approaches to address the diffuse pollution problem; however, their approaches reflect the diverse problems and priorities in each of the jurisdictions. Furthermore, each jurisdiction began with a different base of laws and regulations for the control of the sources of pollution and with varying amounts of resources.

In the agricultural sector, the Chesapeake Bay states have been relying primarily on voluntary cost-sharing programs to carry out their program objectives. These programs are helping farmers reduce soil and associated nutrient losses into the bay. The programs actually build upon the soil-erosion control programs begun in the 1930s by establishing soil- and water-conservation districts. All states targets the diffuse sources at several levels:

- First, the states have targeted general geographic areas where each will emphasize implementation of agricultural diffuse-pollution controls.
- Second, once a general area has been identified, all the states have procedures to target the critical areas and management needs within that area.
- Third, state and local staff identify cost-effective, site-specific management practices for individual landowners and users.

Urban programs to control diffuse sources within the bay watershed lean more toward regulation. The urban sediment control programs in the area have existed since the 1970s. Later sediment control regulations for developing areas were added.

Progress in the 1980s has been slow. For example, the percentage of highly erodible lands on which best management practices were implemented in the 1985/1986 period ranged from 2% to 12% between the states. Installations of animal-waste management (mostly winter storage) in the same period ranged from 0.5% in Pennsylvania to 10% in Maryland. By 1990 Maryland had shown the highest reduction of nutrient inputs (nearly 20% for both nitrogen and phosphorus), while reduction in the other two states was significantly lower. These reductions in nutrient loadings were insufficient to yield significant water quality improvement at the time this book was being written.

Wisconsin Priority Watershed Program

In 1978 the Wisconsin legislature created and funded the Wisconsin Nonpoint Source Water Pollution Abatement Program. The basic purpose of the program was to systematically control nonpoint source pollution so that the surface-water and ground-water quality goals specified by the Clean Water Act were met. The program was designed to deal with the varying nature of nonpoint pollution problems throughout the state. These problems include pollution from croplands, construction sites, stream-bank erosion, and nutrient loads from barnyard runoff, cropland erosion, manure spreading on croplands, and runoff from city lawns and streets (Konrad, 1985). In 1993, funding for this program was increased by the Wisconsin legislature.

The Wisconsin program concentrates available funds into selected hydrological units (watersheds) that exhibit large problems due to diffuse-pollution inputs. Such units are then called *priority watersheds*. This program allows all categories of urban and rural diffuse sources within the watershed area to be addressed and solved in a comprehensive and coordinated effort. Specific areas within the priority watershed that contribute significant amounts of pollutants to lakes and streams are collectively called *priority management areas*. This approach enables effective targeting of most significant pollution sources. The Priority Watershed Program concentrates available educational, financial, and technical resources in those critical areas where maximum water quality benefits will result from investing financial and human resources.

The selection of a Priority Watershed Project is followed by an 8- to 9-year planning and implementation process. An implementation plan, based on a detailed inventory and assessment of critical source areas in the watershed and the program objectives, is prepared, usually within the first year of the project. The plan guides the Priority Watershed Project and spells out procedures and responsibilities. Central to each Priority Watershed Project are the water quality objectives identified for the lakes and streams to be protected. The determination of critical pollutants, significant sources, the desired level of diffuse-pollution control (pollutant reduction), and the measurement of results are all based on the specific water quality objectives for each individual watershed.

Some of the water quality objectives identified for the Priority Watershed Projects are:

1. Protection of the near-shore waters of Lake Michigan;
2. Rehabilitation of warm-water fishery;
3. Rehabilitation of cold-water fishery, such as the upgrading of a trout stream through habitat improvement;
4. Protection of a desired warm-water fishery;
5. Protection of a desired cold-water fishery;
6. Inland lake rehabilitation;
7. Protection of an inland lake.

The State provides financial support in three major categories: (1) cost-share (50% to 70%) for landowners and municipalities to install management practices; (2) aids for local governments to fund additional technical assistance, education and information, and financial and project management; and (3) administrative and planning funds for state administration and the preparation of priority watershed plans.

The Milwaukee River Priority Watershed Program

The Milwaukee River Priority Watershed Program is an example of a watershed-wide approach to solving excessive nonpoint pollution problems (Gayan and D'Antuono, 1989). The Milwaukee River is the largest of the three tributaries of the Milwaukee, Wisconsin, Harbor, which is a freshwater estuary of Lake Michigan. The drainage area of the Milwaukee River watershed is about $1800\,km^2$, with a resident population of about 500,000. The average flow of the river at the entrance into the harbor is $5.6\,m^3/sec$, and the low flow characteristics for pollution abatement studies and planning (so-called Q_{7-10}, that is, the 7 days duration for 10 years expectancy of low flow) is $0.72\,m^3/s$.

The upper two-thirds of the watershed are mostly composed of agricultural and mixed land uses, while the lower third is made of suburban and urban (Milwaukee County) land uses. All sewage flow from the city of Milwaukee and its sewered suburban sections is collected and diverted to two regional treatment plants located outside the watershed boundary. As a result, biologically treated sewage (point pollution) from only about 31,000 inhabitants is currently discharged into the river. Phosphorus is removed from the point-source effluents to meet the standard of $1\,mg\,P/l$ for sources located in the Great Lakes Basin. Hence, most of the pollution of the river is of nonpoint origin. In the lower urban portion and in the harbor the river receives combined sewer overflows (CSO) from about $40\,km^2$ of the older section of the city of Milwaukee. Most of the city and all of sewered suburbs (more than 90% of the area served by the Milwaukee Metropolitan Sewerage District) are drained by a separate sewer (storm and sanitary) system. A large tunnel–interceptor, providing about 1.5 million m^3 of storage space for interception of CSO and infiltration flow excess, has been built in the dolomite formation 100 meters below the city surface. The total cost of the point source abatement program (deep tunnel, treatment plant, and sewer system rehabilitation) is over $2.5 billion.

With the upstream point sources greatly reduced or eliminated, the existing water quality problems must be attributed to diffuse sources. These problems include excessive algal growth in the upper reaches of the

Important Examples of Systems Affected by Diffuse Pollution 51

FIGURE 1.13. The Milwaukee River is affected by nutrient discharges. (Photo: V. Novotny.)

river (Fig. 1.13) and dissolved oxygen (DO) depletions in the impounded lower reaches of the river.

The mayor of Milwaukee, Wisconsin, found that after spending more than $2 billion (mostly subsidized by the federal government) on sewer rehabilitation and wastewater treatment, including expensive means of storage and treatment of overflows and by-passes of untreated wastewater, the receiving water bodies might not be clean and suitable for most of the uses specified by the Clean Water Act due to upstream mostly unquantified diffuse discharges of pollutants. He found this situation politically damaging and irresponsible to the taxpayers.

As a result of serious nonpoint pollution problems, caused by upstream nonpoint sources, that could diminish the water quality benefits of the very expensive point source pollution abatement program, the state legislature declared the Milwaukee River a priority watershed. The Priority Watershed Program focuses on the control of pollution loads caused by soil loss and from concentrated animal operations. The integrated water resource management plan has nine components and goals:

- Reduction of nutrient and sediment contributions to surface and ground water throughout the watershed by 30%. This goal is being accomplished by soil-conservation practices such as strip cropping and no-till planting in agricultural zones, erosion control of construction sites, and by sedimentation ponds in urban areas.
- Reduction of toxic contaminants in surface and ground water. Most of toxic contamination originates from abandoned landfills and urban runoff.
- Enhancement and protection of wetlands. Existing wetlands are now protected and previously drained wetlands adjacent to watercourses are being restored.
- Buffer strips along the watercourses. Five percent of existing agricultural lands are being converted to perennial grassland to provide protective buffer strips along watercourses.
- Conversion of highly erodible land to woodland. In addition to the funding provided by the state of Wisconsin from the Priority Watershed Program, additional funding and incentives to farmers to install soil-conservation practices and for barnyard waste management are provided by the Food Security Act of 1985. This act also provides compensation to farmers for taking highly erodible lands and lands adjacent to watercourses out of production.

The combined Milwaukee River water quality enhancement plan consists of:

- *The Priority Watershed Program* aimed at nonpoint sources of pollution in the upper, mostly rural and suburban (unsewered) parts of the basin;
- *Milwaukee Metropolitan Sewerage District Water Pollution Abatement Program*, an ambitious $2.5 billion action plan focusing on abatement of point source pollution, including CSO;
- *The Milwaukee Harbor Plan*, involving in-stream measures such as low-flow augmentation and aeration during low DO periods.

These programs could be considered an example of an integrated approach to a regional water quality problem.

Lagoon of Venice

The Lagoon of Venice, Italy, is a tidal embayment located in the northern Adriatic Sea. This relatively shallow water body has an area of $500\,km^2$. The total surface area of the basin is about $1700\,km^2$. The land distribution today is about two-thirds agricultural and one-third urban (metropolitan Venice and surrounding communities), with a resident population of 1,000,000. During the tourist season the population greatly increases due to influx of tourists.

The historical center of Venice (resident population of about 80,000 people plus tourists) is located on an archipelago of about 120 small islands inside the lagoon and is connected with the mainland by a causeway and railroad bridge.

Evolution of the Water Quality Problem

Until the last century the drainage basin of the Lagoon of Venice was composed primarily of lowland marshes and wetlands transected by canals and small tributary streams. Marshes and wetlands served as sinks for nutrients that promoted the growth of lush vegetation, while at the same time protecting the lagoon and other waterways from today's symptoms of eutrophication and hypereutrophication exhibited by extensive and obnoxious algal blooms throughout the lagoon.

The Lagoon of Venice is a dynamic ecological system that throughout the centuries has been subjected to cultural (anthropogenic) changes and evolution that have included large hydraulic works such as the dredging of navigational canals and the relocation of two major tributaries outside of the lagoon to reduce siltation, as well as daily maintenance of the canals. These works have been carried out by Venetians for centuries.

In some ways the evolution of the water quality problems of the lagoon is a consequence of the watershed transition delineated in Figure 1.5. This transition has been in progress for a period of 100 years, starting from a mostly rural natural lowland–wetland watershed and moving to an urbanized and agricultural drained basin of one million inhabitants.

The historical center city of Venice located inside the lagoon, has an interesting but difficult diffuse-pollution problem. The historic city, which currently has about 80,000 inhabitants, but which during the height of the Venetian Republic 500 years ago, had a population exceeding 250,000, has no sewers. All sewage from the present population and from thousands of tourists visiting this historical treasure, plus urban runoff from the city's almost 100% impervious lands, are discharged by individual houses and street outlets directly into the famous canals (Fig. 1.14). As a result the canals inside the city are and have been for centuries severely polluted, with sludge deposits accumulating on the bottom of the canals. Odors, due to anaerobic decomposition, have plagued the city since the Middle Ages. The introduction of flushing toilets and the influx of tourists in recent years have intensified these problems. Tides provide the only flushing action by which the pollutants can be carried away from the historic city into the lagoon and through the gaps between the beach islands into the Adriatic Sea. In 1990 only about

FIGURE 1.14. A canal in Venice. (Photo: V. Novotny.)

one-half of the urban sewage on the mainland received some treatment; the rest was discharged without treatment into mainland waterways and subsequently into the lagoon.

The drainage work that was begun in the basin of the Lagoon of Venice approximately in 1880 and still may continue, has transformed the wetlands of the basin into agricultural and urban dry lands. A complex network of drainage canals with pumping stations has dropped significantly the ground-water levels throughout the basin and large quantities of drainage and irrigation return flows rich with nutrients and residues of other agricultural and industrial chemicals are now directed toward the lagoon (Fig. 1.15).

Important Examples of Systems Affected by Diffuse Pollution 55

FIGURE 1.15. Terra Ferma (mainland) canal draining into the Lagoon of Venice, which is affected by the elevated nutrient concentrations. (Photo: V. Novotny.)

When the wetlands were drained the lagoon was deprived of its natural buffering system for nutrients, which before the drainage work retained them. Increased use of fertilizers by the agricultural sector, especially after the switch from organic (manure) fertilizers to chemicals, is another factor contributing to the greatly increased transport of nutrients from the basin to the lagoon. On top of these loads, the largest Italian producer of chemical nitrogen fertilizers (plus other chemicals) is located on the shore of the lagoon.

Originally, the lagoon itself was surrounded by brackish tidal marshes, which still remain in some parts of the lagoon. These marshes constituted an additional natural buffer, shielding the lagoon from the influx of nutrients by runoff and shallow ground-water flow. Impounding the marshes and forming fish ponds (valli da pesca) and draining to reclaim land limited this buffering capacity.

The watershed itself is mainly composed of flat lowlands, hence, erosion and soil losses are minimal (mostly from construction erosion in sewered urban watersheds). Most of the nutrient load from rural watersheds is carried by drainage flows.

Pollution Loads

In the agricultural areas annual average fertilizer applications range from 25 to 280 kg/ha of nitrogen and 15 to 140 kg/ha of phosphorus, depending on the type of crops. These fertilizer loads and their losses resulted in mean concentrations of nitrogen and phosphorus in the mainland drainage canals of between 1 and 20 mg/l of nitrogen and 0.1 and 4.5 mg/l of phosphorus (Bendoricchio, 1988). As a result the mainland canals produce high quantities of algal biomass, as documented on Figure 1.15.

Urban diffuse loads (excluding sewage and industrial wastes) are also significant; however, the greatest loads originate from the mainland industrial area where chemical fertilizers are produced. Unit loadings of pollutants from urban sources and their comparison to sewage loads are given in Table 1.9.

The annual nitrogen and phosphorus load to the lagoon from all sources has been estimated as from 10,000 to 15,000 tonnes of nitrogen/year and 1000 to 2000 tonnes of phosphorus/year from which agricultural nonpoint sources were responsible for about 50% of the load (Bendoricchio, 1988). The lagoon is greatly affected by these loads. It should be realized that, when converted to water-surface loading, these loads would be devastating to an average lake; however, the tidal water exchange with the Adriatic Sea results in a relatively short detention time for pollutants in the lagoon, which used to be adequate to cleanse most of these waters. Since the 1980s the coastal waters of the Adriatic have also become enriched by nutrients, mainly from the Po River, resulting in the rapid deterioration of the water quality of both the lagoon and coastal waters. Massive algal blooms are now an annual occurrence. The tidal exchange between the lagoon and the sea may be further affected by the planned construction of tidal barriers to alleviate the problem of tidal flooding.

TABLE 1.9 Unit Loads of Pollutants from Diffuse Sources in kg/ha/yr from the Mestre-Porto Marghera Urban Catchment of Metropolitan Venice and from Agriculture

Source	Suspended Solids	Nitrogen	Phosphorus
Untreated dry weather wastewater flow[a]	995	939	62
Wet weather diffuse urban loads	1241	223	26
Average agricultural loads		44–66	4–9

Source: After Novotny, Miller, and Zheng (1989).

[a] Includes primarily industrial loads from Porto Marghera. The dry weather flow and a small portion of the wet weather flow receives biological treatment with nitrification and denitrification.

Abatement Program

The Pollution Abatement Program approved for the lagoon by the Italian government envisions broad measures to protect the lagoon ecosystem and to reverse past trends. For the agricultural sector, which is responsible for about 50% of the total nutrient load to the lagoon, the action plan proposes the following:

1. Optimization of fertilizer application rates to minimize losses and matching applications to the nutrient requirement by crops.
2. Optimization of the temporal distribution of the application of fertilizers.
3. Use of soil drainage to increase denitrification of the soil.
4. Limitation and optimization of irrigation to reduce nutrient losses in the irrigation (surface and subsurface) return flow.
5. Substitution of organic slow-release fertilizers.
6. Increasing the organic content of the soil.
7. Soil-conservation practices on erodible lands to reduce loads of phosphorus, soil adsorbed, and organic nitrogen.
8. Suitable choice of the set-aside lands and conversion to less polluting crops according to the guidelines of the European Community.

The Pollution Abatement Program is also using wetlands situated along the boundary of the lagoon for the abatement of diffuse pollution loads. Point source abatement is focused on both dry weather and wet weather (diffuse) point loads. Finally, sewers will be installed in the historic city.

Lake Balaton

Evolution of the Water Quality Problem

Lake Balaton is located in the southwestern part of Hungary. It is the largest freshwater body in central Europe. On summer weekends up to one million tourists come to the shores to enjoy recreation. The lake was reasonably clean until 1965, since when increased agricultural use of chemical fertilizers in the drainage watershed, tourist use of the lake, which also increased pollution loads and overloaded existing sewage disposal systems, and industrial effluents increased the nutrient loads of the lake by an order of magnitude. As a result, algal biomass levels have increased dramatically and the lake has become eutrophic (Somlyódy and van Straten, 1986).

The progression of the intensification of land and lake use along with the phytoplankton biomass is shown on Figure 1.16. There were also

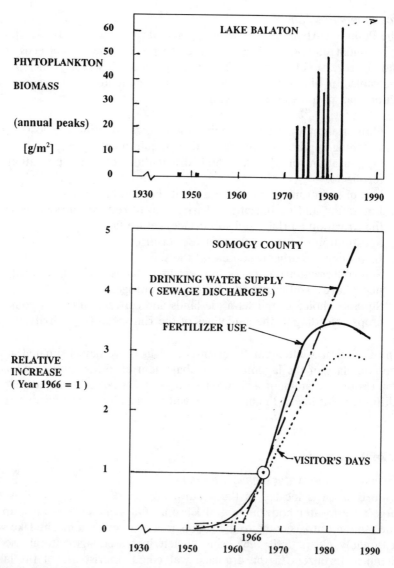

FIGURE 1.16. Nitrates, recreation use, and eutrophication of Lake Balaton, Hungary. (Data from Somlyody and van Straten, 1986.)

several major fish kills that received wide public attention. The first occurred in 1965 and was most likely the result of fish poisoning by pesticides. The second, which happened in 1975, was indicative of a collapsing ecosystem (Hock and Somlyódy, 1990). This occurrence was followed by mass proliferation of cyanobacteria in the summer of 1982, and the waters of the lake became mostly unsuitable for contact recreation.

The Action Plan
Following these catastrophic water quality events the authorities in early 1980s recognized the problem, and in January 1982, the Hungarian government decided to launch a series of extensive programs to analyze the problem and then suggest remedial measures. The remedial plan was adopted in 1983.

The goal of the plan is restoration of water quality to the levels that prevailed in the 1960s. This goal is to be achieved in stages, with target A corresponding to the water quality level of late 1970s, target B corresponding to water quality of late 1960s, and finally target C is restoration to early 1960s levels. The plan schedule is 1990 for target A, 1995–2000 for target B, and 2005–2010 for target C.

With the cooperation and sponsorship of the Food and Agriculture Organization (FAO) of the United Nations, Hungarian scientists embarked on an ambitions research project to identify the sources of nonpoint pollution and its management (FAO, 1986). A pilot experimental watershed was established on a small tributary of Lake Balaton. By establishing a data base on pollutant loads from various diffuse and traditional point sources, lake and watershed modeling, and extrapolation, scientists from the Water Resources Management Institute (VITUKI) and other institutes were able to present the authorities with targets, criteria, and possible abatement scenarios.

The first stage of the plan (target A) involves abatement of point sources, such as upgrading of biological treatment plants along the shore of the lake and installing tertiary phosphorus removals, and nonpoint source controls, including construction of sedimentation basins on several tributaries to reduce loads of particulate pollutants, dredging of polluted sediments, control of livestock wastes, improved farming methods, and a complete halt to building of summer homes along the shore.

The Kis–Balaton reservoir network was put in operation by 1990. The reservoir system is designed to filter out the nutrients and pollutants flowing into the western end of the lake from the Zala River. The pollutants are retained in the tributary reservoirs and do not reach the lake. Concurrently, phosphorus precipitation was introduced in 10 regional treatment plants. By 1990, the phosphorus input in the lake was cut in half, resulting in improved water quality in the eastern part of the lake. The rest of the lake, because of internal nutrients storage, will need a longer recovery time for a noticeable improvement.

The Emscher River

The Emscher River is a small river located mostly within the municipal boundaries of Essen in the Ruhr area of Germany. It is a historical river because the first watershed-wide water quality management agency in the world was established there in 1906. Karl Imhoff, a famous pioneer of modern environmental engineering, was then put in charge of the agency.

The drainage area of the Emscher River is 865 km^2, of which 20% is currently impervious. The present resident population within the Emscher River watershed boundaries is 2.5 million. Most of the pollution entering the river is of urban origin, with only minimal agricultural contributions.

The Emscher River Association (Emschergenossenschaft) is one of several river association that have been subsequently established in this highly industrialized area of Germany. In order to provide sanitation and a safe water supply by relatively few small rivers for about 8 million inhabitants and about one-third of the German heavy industry, Karl Imhoff and his coworkers devised a plan of primary water uses for each individual river. In this plan, the largest river, the Ruhr, became the primary source of water supply and recreation for the population, while the Emscher River assumed the sad task of conveying mostly untreated or only partially treated sewage and industrial wastewater.

In the late 1800s and early 1900s the Emscher River received mostly untreated sewage from the Essen industrial area. The river was mostly septic with thick sludge deposits on the bottom. In the early 1900s the goal of the new association was to keep the river fresh and avoid further deposits of sewage sludge. However, land subsidence of up to 20 meters due to deep coal mining made drainage by conventional enclosed sewers extremely difficult and costly. Hence, in the first period of the sanitation work in the Emscher River watershed, wastewater effluents received primary treatment (mostly in so-called Emscher or Imhoff primary settling tanks), after which the effluent was discharged into open, concrete-lined channels, and hence into Emscher River. Such open channels allowed for easier correction of the slopes that were disrupted by the subsidence.

To reduce public hazards and increase the aesthetics of these open sewers the channels were fully lined to provide higher velocities that would keep the flow fresh (aerobic) and control odors, the river banks were lined with trees, and the channels were fenced on both sides to prevent access (Fig. 1.17). A treatment plant to treat the entire river flow was built before the confluence with the Rhine River. The river thus became an example of the ultimate conversion of a natural small stream into an open sewer (Anon., 1986).

FIGURE 1.17. The Emscher River in Essen, Germany. (Photo: V. Novotny.)

Today, combined sewers replaced most of the open channels; however, until 1990 the river itself was still primarily unchanged. Hence, all of the pollution entering the river has been from combined urban and industrial sewer connections, and of a diffuse nature.

A Plan to "Renaturalize" the Emscher River and Drastically Reduce Pollutant Loads

A new plan for returning the river to a more naturally looking urban stream was prepared by the Emscher Association (Geiger, 1990). In 1990 the annual nutrient inputs in the river were estimated as 13,500 tonnes of nitrogen and 3600 tonnes of phosphorus, respectively. The treatment plant at the mouth of the river reduces these nutrient loadings from the Emscher and Rhine rivers (and the North Sea, which has a severe eutrophication problem caused by nutrient loads, primarily from the Rhine and Elbe Rivers) by about 15%. To reduce the pollutant loads to the river the association proposed an ambitious plan that considered installation of 200 storm-water-detention basins with a total volume of 1.3 mil m^3 combined with limited reuse and infiltration of storm water. The individual pretreatment of wastewater discharges should meet the effluent standards of 15 mg/l for nitrogen and 0.5 mg/l for phosphorus, respectively.

To further reduce flow and pollutant loads to the river, partial infiltration, separation, and reuse of storm water was considered. It was also suggested that storm-water discharges should be diverted to about 500 meters from tributary creeks, which would also enhance infiltration. Through modeling it was found that such measures could reduce peak storm-water flows and volumes by up to 85%. With these measures the nitrogen load from the watershed could be reduced by about 60%, and that of phosphorus by 92%. Along with the flow and pollutant reduction measures the river channel would be partially converted to a more naturally looking stream.

The Experimental Watersheds of the Shirako and Shakiji Rivers

Until 1964 (the year when the Olympic games were held in Japan) most of the watersheds of the Shirako and Shakiji Rivers located in the metropolitan Tokyo area were not sewered. Since 1965 sewerage projects

FIGURE 1.18. Shiroko River in Tokyo. (Photo provided by S. Fujita.)

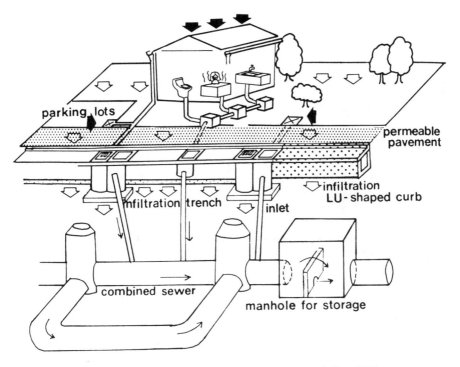

FIGURE 1.19. The ESS concept. (Replotted from Fujita, 1984.)

in Japan have accelerated, with the result that the two watersheds became fully urbanized by the mid-1980s. Consequently, the river channels, limited by urbanization, became insufficient to handle flows (Fig. 1.18). Flooding has thus become a major problem.

In 1982 the Tokyo metropolitan government selected the two rivers as experimental watersheds, in which various infiltration, storage, and other flood-control measures were to be installed and tested. The objective of the measures implemented within the experimental watersheds was to minimize flooding; the pollution control at the beginning of the project was secondary. Unlike U.S. and European practices, which usually do not infringe on private properties and all stormwater control is carried out outside of private homes, the Japanese plan included on-site household water infiltration and heavily relied on it (Fujita, 1984; Fujita and Koyama, 1990). Figure 1.19 shows the concept of the experimental sewer system (ESS). The extent of the ESS system as of 1990 is given in Table 1.10.

The ESS system relies heavily on site infiltration and to a lesser degree

TABLE 1.10 Characteristics and Extent of the ESS System

Parameter/Abatement	Shakiji River	Shirako River	Total
Watershed area (ha)	305	597	702
Resident population	39,600	72,100	111,700
Infiltration inlets	4,778	15,218	19,996
infiltration trenches (km)	43	84	127
Infiltration curbs (km)	22	49	71
Permeable pavements (1000 m^2)	200	408	608
Construction cost bil. Jap. ¥ (mil. U.S. $)	12.5 (83)	24.3 (162)	36.8 (245)

Source: Fujita and Koyama (1990).

TABLE 1.11 Comparison of Pollutant Loads from Tokyo Urban Watersheds with Combined Sewers with and without the ESS System for an Average Year

Constituent	Without ESS	With ESS	Percent Reduction
Total number of rainfalls	71	71	
Overflow frequency	36	7	81
Overflow loads (kg/ha)			
Suspended solids	223	19	91
BOD	103	4.7	95
Total annual load^a (kg/ha)			
Suspended solids	1057	589	45
BOD	788	528	33

Source: Fujita and Koyama (1990).

^{a} Includes treatment plant effluent with dry-weather flow.

on storage. Again this is in contrast to past sewerage practices that were designed to convey storm water as fast as possible from the site to the watershed outlet. The goal of the ESS system is to retain and infiltrate and minimize conveyance as much as possible. Although the sewer system is combined, wastewater and storm water enter the system separately.

Implementation of a system such as the ESS, which relies heavily on site measures and storm-water disposal, requires the cooperation of homeowners. An extensive educational effort was part of the program.

Water Quality Benefits of the ESS System

As stated the water quality improvement was not the primary objective of the program. However, the ESS system is an example of the dual benefits of urban storm-water management practices. The pollution loads and the frequency of overflows from the ESS systems, both of which were measured and simulated by a model, were compared to a similar watershed served by combined sewers without on-site storm-water management (Fujita and Koyama, 1990). The results of the comparison are given in Table 1.11.

References
Alberts, E. E., G. E. Schuman, and R. E. Burnwell (1978). Seasonal runoff losses of nitrogen and phosphorus from Missouri Valley loess watersheds, *J. Environ. Qual.* **7**:203–208.
Anonymous (1986). Karl Imhoff, ein Wegbereiter der Stadtentwässerung und der Gewässerreinhaltunng (Karl Imhoff, pioneer of urban drainage and wastewater treatment), *Herausgeber VDI*, Gesselschaft Bautechnik im Verein Deutscher Ingenieure, Düsseldorf, Germany.
Bendoricchio, G. (1988). The NPSP abatement program for the Lagoon of Venice, in *Proceedings, Nonpoint Pollution 1988—Policy, Economy, Management and Appropriate Technology*, V. Novotny, ed., American Water Resources Association, Bethesda, MD, pp. 241–260.
Beaulac, M. N., and K. H. Reckhow (1982). An examination of land use—nutrient export relationship, *Water Resour. Bull.* **18**(6):1013–1024.
Brown, L. R., et al. (1990). *State of the World, A Worldwatch Institute Report on Progress toward a Sustainable Society*, Norton, New York.
Carson, R. (1962). *Silent Spring*, Houghton Mifflin, Boston.
Clark, E. E., II, J. A. Haverkamp, and W. Chapman (1985). *Eroding Soils, The Off-Farm Impact*, The Conservation Foundation, Washington, DC.
Committee on the Assessment of PCBs in the Environment (1979). *Polychlorinated Biphenyls*, National Academy of Sciences, Washington, DC.
Czechoslovak Academy of Science (1992). *National Report of the Czech and Slovak Republic.* United Nations Conference on Environment and Development.
Ellis, J. B. (1986). Pollutional aspects of urban runoff, in *Urban Runoff Pollution*, H. C. Torno, J. Marsalek, and M. Desbordes, eds., Springer Verlag, Berlin, New York, pp. 1–36.
Garbell, M. A. (1965). The Jordan Walley plan, *Sci. Amer.* **221**(3):23–31.
Fed. Regist. (1979). **44**:76202–76209.
Field, R., and R. Turkeltaub (1981). Urban runoff receiving water impact: Program overview, *J. Environ. Eng. Div. ASCE* **107**(EE1):83–100.
Food and Agricultural Organization (1986). *Prevention of Non-point Source Pollution—Hungary*, United Nations, FAO, Rome.
Fujita, S. (1984). Experimental Sewer System for reduction of urban storm runoff, in *Proceedings, Third International Conference on Urban Storm Drainage*, International Association on Water Pollution Control and Research (IAWPRC)-International Association on Hydrological Research (IAHR), Chalmers University of Technology, Göteborg, Sweden, pp. 1211–1220.

Fujita, S., and T. Koyama (1990). Pollution abatement and the Experimental Sewer System, in *Proceedings, Fifth International Conference on Urban Storm Drainage*, International Association on Water Pollution Control and Research (IAWPRC)-International Association on Hydrological Research (IAHR), Suita-Osaka, University of Osaka, Japan, pp. 799–904.

Gaffney, M. (1988). Nonpoint pollution: An economical solution to an unconventional problem, in *Political, Institutional and Fiscal Alternatives for Nonpoint Pollution Abatement Programs*, V. Novotny, ed., Marquette University Press, Milwaukee.

Gayan, S. L., and J. D'Antuono (1988). The Milwaukee River East–West Branch Watershed Resource Management, in *Proceedings, Nonpoint Pollution 1988—Policy, Economy, Management and Appropriate Technology*, V. Novotny, ed., American Water Resources Research Association, Bethesda, MD.

Geiger, W. F. (1990). New drain—New dimensions in urban drainage, in *Proceedings, Fifth International Conference on Urban Storm Drainage*, International Association on Water Pollution Control and Research (IAWPRC)-International Association on Hydrological Research (IAHR), Suita-Osaka, University of Osaka, Japan, pp. 33–48.

Henry, J. G., and G. W. Heinke (1989). *Environmental Science and Engineering*, Prentice Hall, Englewood Cliffs, NJ.

Hock, B., and L. Somlyódy (1990). *Freshwater Resources and Water Quality*, VITUKI, Budapest.

Humenik, F. J., M. D. Smollen, and S. A. Dressing (1987). Pollution from nonpoint sources, *Environ. Sci. Technol.* **21**(8):737–742.

Imhoff, K., and K. R. Imhoff (1990). *Taschenbuch der Stadtentwässerung*, Oldenbourg Verlag, Munich.

International Joint Commission (1980). *Pollution in the Great Lakes Basin from Land Use Activities*, International Joint Commission, Windsor, Ontario.

Konrad, J. (1985). The Wisconsin nonpoint source program, in *Proceedings, National Conference on Perspectives on Nonpoint Source Pollution*, EPA 440/5-85-001, U.S. Environmental Protection Agency, Washington, DC.

Krejci, V. (1988). New strategies in urban and stormwater pollution control in Switzerland, in *Proceedings, Nonpoint Pollution: 1988—Policy, Economy, Management, and Appropriate Technology*, V. Novotny, ed. American Water Resources Association, Bethesda, MD, pp. 203–212.

Krenkel, P., and V. Novotny (1980). *Water Quality Management*, Academic Press, New York.

Lager, J. A., and W. G. Smith (1974). *Urban Stormwater Management and Technology: An Assessment*, EPA-670/2-74-040, U.S. Environmental Protection Agency, Cincinnati, OH.

Loehr, R. C. (1972). Agricultural runoff—characterization and control, *J. Sanitary Eng. Div. ASCE* **98**:909–923.

Marsalek, J. (1978). *Pollution due to Urban Runoff: Unit Loads and Abatement Measures*, International Joint Commission, Windsor, Ontario.

Marsalek, J. (1986). Toxic contaminants in urban runoff: A case study, in *Urban*

Runoff Pollution, H. C. Torno, J. Marsalek, and M. Desbordes, eds., Springer Verlag, Berlin, New York, pp. 38–55.

Mitsch, W. J., and J. G. Gosselink (1986). *Wetlands*, Van Nostrand Reinhold, New York.

Moore, E., et al., eds. (1979). *Livestock Grazing Management and Water Quality Protection, EPA-910/9-79-67*, National Technical Information Service, Springfield, VA.

Myers, C., J. Meek, S. Tuller, and A. Weinberg (1985). Nonpoint sources of water pollution, *J. Soil Water Conserv.* **40**(1):14–18.

Nisbet, C. T., and A. F. Sarofim (1972). Rates and routes of transport of PCBs in the environment, *Environ. Health Perspect.* **1**:21–38.

Nonpoint Source Control Task Force (1983). *Nonpoint Pollution Abatement in the Great Lakes Basin. An Overview of Post-PLUARG Developments*, International Joint Commission, Windsor, Ontario.

Novotny, V. (1985). Urbanization and nonpoint pollution, in *Proceedings, Fifth World Congress on Water Resources*, Paper No. 17, Aspect 6, International Water Resources Association. Brussels.

Novotny, V., and G. Chesters (1981). *Handbook of Nonpoint Pollution: Sources and Management*, Van Nostrand Reinhold, New York.

Novotny, V., and G. Chesters (1989). Delivery of sediment and pollutants from nonpoint sources: A water quality perspective, *J. Soil Water Conserv.* **44**(6):568–576.

Novotny, V., B. Miller, and S. Zheng (1989). *Urban Diffuse Pollution of Lagoon of Venice*, Technical Report to Vanezia Nuova, Department of Civil Engineering, Marquette University, Milwaukee.

Novotny, V., K. R. Imhoff, M. Olthof, and P. A. Krenkel (1989). *Karl Imhoff's Handbook of Urban Drainage and Wastewater Disposal*. Wiley, New York.

Organization for Economic Co-operation and Development (1991). *The State of the Environment*, OECD, Paris.

Pitt, R., and R. Field (1977). Water quality effects from urban runoff, *J. Am. Waterworks Assoc.* **69**:432–436.

Robbins, J. W. D. (1985). Best management practices for animal production, *Proceedings, Nonpoint Pollution Abatement Symposium*, V. Novotny, ed., Marquette University, Milwaukee, pp. P-III-C 1–11.

Rooney, P. (1989). Louisiana's wetland calamity, *EPA J.* **15**(5):37–39.

Schroepfer, G. J. (1978). A philosophy of water pollution control—past and present, *J. WPCF* **50**(9):2104–2109.

Somlyódy, L., and G. van Straten, eds. (1986). *Modeling and Managing Shallow Lake Eutrophication, with Application to Lake Balaton*, Springer Verlag, Berlin.

Thompson, P. (1989). *Poison Runoff. A Guide to State and Local Control of Nonpoint Source Water Pollution*, Natural Resources Defense Council, New York.

Timmons, D. R., and R. F. Holt (1977). Nutrient losses in surface runoff from a native prairie, *J. Environ. Qual.* **6**:369–373.

U.S. Congress (1987). *Water Quality Act of 1987*, P.L. 100-4, Washington, DC.

U.S. Environmental Protection Agency (1983). *Chesapeake Bay: A Framework for Action*, U.S. EPA, Region III, Philadelphia.
U.S. Environmental Protection Agency (1984). *Report to Congress: Nonpoint Source Pollution in the U.S.*, Water Planning Division, U.S. EPA, Washington, DC.
U.S. Environmental Protection Agency (1988). *Chesapeake Bay Nonpoint Source Program*, U.S. EPA Chesapeake Bay Liaison Office, Annapolis, MD.
U.S. Environmental Protection Agency (1989). *Nonpoint Sources: Agenda for the Future*, Office of Water, U.S. EPA, Washington, DC.
Vesilind, P. A. (1975). *Environmental Pollution and Control*, Ann Arbor Science, Ann Arbor, MI.
Waller, D. H., and W. C. Hart (1986). Solids, nutrients and chlorides in urban runoff, in *Urban Runoff Pollution*, H. C. Torno, J. Marsalek, and M. Desbordes, eds., Springer Verlag, Berlin, New York, pp. 59–85.

2
Laws, Regulation, and Policies Affecting Water-Pollution Abatement

Every person learns early in life that "self-preservation" is the first law of nature.

Henry P. Caufield, Jr. (1991)

Often, engineers and scientists dealing with pollution abatement may not realize that they are working and acting within a social and legal system that is relatively complex and that imposes legal constraints on what can and cannot be done to resolve or limit the problem of pollution. These legal problems and social issues have been partially addressed in the previous chapter. The laws and legal rules that affect or may affect discharges of pollution and their abatement and control are numerous. In order to comprehend the complexity and ramifications of the legal rules one should become familiar with U.S. and international legal systems and with policies that rule pollution-abatement programs. Pollution abatement is both a technological and political–economical–legal problem. However, it has become clear that managing water resources and maintaining and achieving acceptable water quality is far less a technological issue than a political, institutional, and economical one. Technical and scientific solutions of the problem and the economical feasibility of various alternatives alone will not automatically lead to implementation of pollution-abatement programs. Although technical solutions to the problem of diffuse pollution have been researched and are available, implementing diffuse-pollution programs is not easy, even if financial resources are available (Novotny, 1988a, 1988b). This book addresses both technological and nontechnological means of environmental protection and abatement.

The planners, engineers, and scientists involved in pollution abatement must also deal with the legal profession, since environmental law is now a well-established branch of our legal system. Special courses on environmental law are taught in law schools and many law firms now specialize in handling environmental cases. Several international treaties include transboundary environmental issues, including pollution.

SOCIAL CAUSES OF POLLUTION

Living and production to sustain living both produce waste. According to the definition of pollution in the previous chapter generation of pollution results from the use of resources in a fashion that is detrimental to the environment or to the beneficial uses of environmental resources. However, man has three distinct roles in this process. On one side, man is a *producer* (developer) and *consumer*, hence the polluter. On the other side, man and ecology are adversely impacted by pollution, hence man is a *sufferer* of pollution. The conventional explanation for pollution by economists is that it is the least expensive way for consumers and producers to get rid of waste products (Braden, 1988). Excess pollution arises when the waste-disposal capacity of the environment is provided free of charge (Solow, 1971) and/or when the consumers–producers do not incorporate into their economic considerations the cost of the damage caused by pollution. (Recall from Chapter 1 that the definition of pollution implies damage and subsequent cost to the society.)

The basic driving force for production and consumption is sustainment of living processes first and production profit second. In the societies of the Middle Ages and ancient times the producers, consumers, and sufferers of pollution were small groups of people living in a relatively small confined area. Most likely they were unable to recognize the link between pollution and disease, but certainly they could smell and had some sense of aesthetics, therefore, they dumped their waste and that of their domestic animals at some distance from their living areas. The relationship between the living and economical process and pollution unaffected by regulations and pollution-control laws can be illustrated by the following realistic example.

In the historic walled city of Fez in Morocco (one of the world's historical treasures, with about 200,000 inhabitants) time has stopped and people live there in the same way as they did 500 years ago (Fig. 2.1). People living in the city are brought up and educated in the Moslem religion. The oldest Islamic university was founded in the city in the

FIGURE 2.1. Historic Fez in Morocco. Famous outdoor tanneries. (Photo: V. Novotny.)

eighth century, long before the first European university was established in Bologna, Italy (founded in the thirteenth century). The streets of Fez are reasonably clean because homeowners and merchants clean the narrow streets and alleys around their premises. Hence, there is no major urban runoff pollution problem, even though instead of automobiles (which cannot enter the historical city) donkeys and mules are used for transporting goods and people. The simple reason is that the sufferer of pollution is also its cause and he/she may realize the linkage between living and economic activities and the damage done by consequent pollution (for example, no one would come to an artisan shop if the area around it was unclean and smelly). In this way the cost of the damage by pollution is incorporated in the citizens' economic reasoning.

This kind of more or less voluntary participation in cleanup efforts is successful if everybody participates. Let us consider a situation in which several citizens of the city do not participate and let their garbage, refuse, and dirt accumulate. By doing so they cause harm to their neighbors by bad smells, disease, and rats, thus keeping away customers for their products. In the absence of regulations and laws the sufferers have no legal recourse. This is the problem of economic externalities, which will be explained throughout this book. However, those who want to keep their neighborhood clean can do the following: (1) express their displeasure and try to persuade the polluter not to pollute; (2) put moral and economic pressure on the polluters (for example, boycott their merchandise); (3) go to court; or (4) set up an enforcement scheme by asking their legislative body to pass a law preventing this kind of pollution-generating activity.

In order to set up enforcement they first have to state that pollution is a problem and has to be limited. Hence, they have to formulate a pollution-control policy. In the policy statement they have to say what is acceptable and what is not. Hence, they have to define pollution and pollution-causing activities. They may even say how much pollution is tolerable. Subsequently, they must formulate the penalties for violations and a mechanism for collecting and enforcing the penalties. These are typically the simplest components of a policy to control pollution. Such systems of self-control are feasible only if the sufferers of pollution are a majority in a group of producers and consumers—the polluters and their own pollution directly affects them.

In the same city, however, there is also a small stream that transects it. The stream receives all pollution from households, commerce, and the famous outdoor tanneries, as well as urban runoff. In a section between the city walls about one kilometer long the water quality of the stream changes from marginally good to an awful open sewer, leaving the stream heavily polluted and devoid of oxygen for many kilometers downstream from the city. However, a great majority of the populace of Fez rarely leaves the confines of the walls of the city. Hence, they do not realize the damage done to the stream, nor do they experience any of the economic damage caused by the heavy pollution of the stream. The damage by pollution has been done to downstream farmers who cannot use the stream water for irrigation or drinking, as the water is unfit for any use. The stream is smelly and unsightly, but there is no linkage between the damage done to downstream users and economic production and consumption in Fez where the pollution originates. Consequently, there is no abatement. Again the externality character of pollution prevents abatement if no enforceable regulations or law to control pollution are in place.

One does not have to go to north Africa to see the same effect. In the United States the water quality of Chesapeake Bay has been deteriorating rapidly, causing great economic harm to commercial fishing and the recreation industry. The causes of pollution are farming, point, and nonpoint urban sources throughout the watershed. Yet again there is no economic linkage nor mechanism by which the sufferers of pollution (fishers and recreationalists) can recover the damage from polluters. A farmer whose motives are economical (profit making) will use man-made chemicals to increase yields and pesticides to control weeds and insects. Consequently, nutrient and pesticide losses degrade both downstream water quality and ground-water resources. Yet those who are impacted by this pollution cannot recover the cost of damage from the farmer, and the farmer himself may not be economically impacted by the pollution. As pointed out in the preceding paragraph, this situation is creating a cost to someone else—including the damage to the resource, cost of resources forgone, and remedial costs—and the costs are transferred by physical means and not by market transaction. This is called *an externality*. The externality problem, especially in diffuse-pollution generation, is pervasive and general, and must be resolved before any meaningful plan of abatement is put in place (Novotny, 1988a, 1988b).

A prominent political economist (Solow, 1971) defined externality as follows: "One person's use of a natural resource can inflict damage on other people who have no way of securing compensation, and who may not even know that they are being damaged." Overcoming the externality problem and incorporating the cost of damage caused by pollution in the economic thinking of producers and consumers are the major objectives of pollution-abatement policies. Again quoting Solow: "We would like to insure that each resource is allocated to that use in which its net social value is highest," which is called by political economists *Pareto optimality*.

Ignoring externality may have serious consequences as it did in the countries of Eastern Europe where these economic principles were disregarded. (To a lesser degree many developing and less developed countries ignore them as well.) The implications of external effects must be traced further than just considering them a cause of pollution. They have secondary effects on production and resource allocation. If, for example, the cost of acid rainfall damage to lakes and soils (forests) is not included in the economic thinking of the producers of electricity, electric power is then "too cheap" to consumers, because they are not charged for the damage. Other commodities that are produced with the help of cheap electricity will also be cheaper, and they will be overproduced (Solow, 1971). This will then result in greater consumption and more damage. In this way, society will subsidize those using a lot of electricity. Similarly, it is known that automobiles cause significant pollution, especially of a toxic and acid nature (see Chapter 8 for details). If the cost of damage by the emitted pollutants was not included in the cost of driving an automobile, then society would subsidize automobile users, resulting in more automobile use, more urban sprawl, and consequently more urban erosion from building highways and urbanization. Very often the cost is hidden. Using again East Europe as an example, "free" health care and sending children from heavily polluted zones to sanatoria and health care resorts in Czechoslovakia was a substitution for "costly" pollution abatement during the period of the socialist totalitarian regime. Overall, the cost of the damage to the health of the populace and the destruction of the ecology was much greater than the investment in pollution abatement and the curtailment of subsidized but grossly polluting production processes.

ENVIRONMENTAL POLICIES OF DIFFUSE POLLUTION ABATEMENT

In the environmental policy arena there is a difference between the management tool and policy. A *tool* is a single element—effluent standard, enforcement technique, zoning restriction—devised to achieve a specific result or, in a few cases, several results. A *policy* is a set of one or more tools chosen to achieve an overall environmental objective. A management tool can be thought of an action that is taken in the hope of achieving a particular result. Taking the action involves *adverse effects* (cost of implementation, cost imposed on various participating parties), but the expected results will produce *beneficial effects* (improved environmental quality, lower cleanup cost, increased benefits to users) (Boland, 1991).

From the reasons stated in the preceding section, policies for use of resources, waste disposal, and protection of the environment represent a compromise between the producers in the economic production process and consumers on one side and those who suffer from the adverse effects of production and development and waste disposal on the other. This compromise is reached on several institutional levels. All three branches of government (legislative, judicial, and executive) are involved, as are many pressure and lobbying groups. Implementation of successful and efficient diffuse-pollution control programs requires that competing groups (farmers, urban dwellers, urban polluters, developers, industries) cooperate. Pollution abatement often involves conflicts with powerful interests, notably the farm lobby, chemical manufacturers and their lobbying associations, and developers.

Various policy options can be used for enactment and implementation of pollution control and water quality protection and restoration. As shown below, these options range from those that are simple but ineffective to those that are potentially effective (Carson, 1980; Novotny and Joeres, 1983; Novotny 1988a, 1988b):

- No action
- Moral persuasion and public pressure
- Court litigation:
    damage payments
    court established standards
- Regulation (laws and enforcement required):
    ban on harmful substances and chemicals
    mandatory control processes and performance standards and permits
    stream and effluent standards
- Economic incentives (laws and enforcement required):
    taxes and charges
    subsidies, tax write-offs, and payments for taking land out of production
    marketable discharge permits
- Government contributions for research, education, rehabilitation, and preservation

However, the effectiveness of a solution depends not only on technological or legal methodology but also on political conditions and the institutional framework in which the abatement alternatives are implemented.

Most efficient policy alternatives require regulation and enforcement. One may ask why any regulation and enforcement is needed for control and abatement of (diffuse) pollution. The economic system prevalent in the United States and many other countries is based on capitalistic–democratic free enterprises—the market system where market forces determine how much production is needed to satisfy the need of society. The reader should distinguish between the political system of the government (democracy, dictatorship, feudal authocracy) and the economic system (capitalist–free market, socialist, feudal, communist). History and recent social changes in eastern Europe have proved that the capitalist–democratic system is more efficient than any other economic–political systems. Then why not let the same market forces determine the level of pollution and the level of abatement. For one thing, political economists have found, demonstrated, and documented that market forces do not control pollution and do not stimulate pollution abatement (Kneese and Bower, 1968; Bator, 1958; Solow, 1971; Braden, 1988; Baumol and Oates, 1988). Bator describes the market's inability to control pollution as the failure of a more or less idealized system of price–market institutions to sustain "desirable" activities. Hence, to charge for the use of the environment for waste disposal and to collect damages may require regulation. The free market economy in industrialized countries without environmental regulation and enforcement will lead to a deteriorating environment and to environmental catastrophes as exemplified several decades ago in Japan (Minamata mercury poisoning) and currently in several less developed but rapidly industrializing countries throughout the world. On the other side of the spectrum are the authoritative political regimes, such as those in place until 1989 in Eastern Europe and the Soviet Union, where all activities were subject to some kind of regulation, yet the state of the environment was even more deplorable. The problem in these latter systems was the lack of market forces. Also, the value of environmental damage was not included in the economic thinking of policymakers, leading to the overuse of the environment and to the well-publicized environmental catastrophes primarily caused by diffuse pollution of immense proportions (for example, most soils and ground- and surface-water bodies in former East Germany are heavily contaminated by chemicals, acid rain is widespread throughout central and eastern Europe, irrigation and chemical uses in the Soviet Uzbekistan led to the disappearance of the Aral Sea, and excessive contamination of the entire region by dangerous chemicals, including DDT, which is now an ecological catastrophe).

The process of formulating an environmental policy compromise is neither gradual nor smooth. Throughout history there were a few landmark periods during which most important environmental regulations were enacted, followed by periods of partial regression. In the early 1970s, the first strong environmental legislation was enacted in the United States by a well-organized classic "iron triangle" of legislators, public agency officials, and environmentalists. These three groups are largely responsible for formulating environmental policies today. In Europe the presence of environmental political parties (the Greens) has had a profound impact on formulating environmental policies. These parties, although small in membership, often represent swing votes in the parliaments and cannot be ignored by the larger governing political parties.

Viessmann and Welty (1985) wrote that policymaking in the management of water resources and pollution control is an outcome of political forces operating in different political arenas. Conflict is an inherent element of these political processes, and it serves to ensure that a multiplicity of values is represented. Compromise is a partner to all policymaking, and the art of reaching compromise can be greatly enhanced by appropriate technological input and the use of state-of-the art analytical techniques. The formulation of environmental policies has been described by Caufield (1988, 1991). Environmental policy formulation and implementation is a process that is structured and occurs over time; it is societal in that it is a collective process in which many individual decisions play a role; and it is "authoritative" in the sense that outputs of the process are generally accepted and have the force of the law.

Caufield (1991) also described the process of bargaining and compromising among policymakers (legislators) that must take place if an environmental statute is to be enacted. In this process at least one interest group must be served by the proposed legislation, though more often the legislation serves a much wider segment of the population, if not the public as a whole. Most environmental legislation requires the active support of environmental interest groups. Legislators who support the views of the groups and seek to pass environmental legislation engage in vote trading.

Policies are built on tradition, scientific knowledge, and common sense. Some of this knowledge has been incorporated into certain generally accepted rules called *doctrines* or *imperatives*. These imperatives can be technological–physical (such as conservation of mass and/or energy), economic (a project will fail if the cost associated with the project is less than the benefit gains), or political. An example of political rule is the notion of private ownership of land—*Cujus est solem ejus ad collum at ad inferno*, or "He who owns the soil owns it from the heavens to the inferno." Using this doctrine until recently, the courts interpreted all polluting activities on private lands as not being subject to restrictions, which had a profound effect on diffuse-pollution abatement. This decision precluded any effective enforcement of pollution abatement on private lands, which meant that most of the pollution efforts of the past twenty years had to rely on the voluntary participation of landowners.

Other doctrines have been incorporated into legal documents. For example, if pollution discharge is causing harm to a navigable body of water, it can be regulated and/or restricted by the federal government. This right is derived from the Commerce Clause of the U.S. Constitution. The sociopolitical imperatives of diffuse pollution control are introduced later in this chapter. The bulk of this book is devoted to physical–technological rules and imperatives. The last chapter is devoted to institutional imperatives and solutions.

Rogers and Rosenthal (1988) defined several *policy imperatives* for control of pollution. These imperatives are

Equity. No group of individuals in society should bear a disproportionate cost in meeting environmental quality requirements. The levels of environmental quality chosen should be such that no additional benefit can be derived without making one group or individual worse off (Pareto optimality).

Irreversible impact. No actions may be permitted that would irreversibly harm the environment and natural resources. This concern for irreversible environmental deterioration and the consumption of nonrenewable resources has to be guarded against by society at large (intergenerational impact). This imperative is also known and is now being implemented as a requirement for *sustainable economic development.*

Regulations and statutes. Due to the failure of the general market to control the quality of the environment and protect the resources, there must be legislation and regulation. The regulations must be clear and easy to carry out.

Acceptance. There must be concurrence on the part of the people and groups being regulated that they will, by and large, obey the regulations.

The equity imperative is especially crucial to diffuse-pollution abatement. For example, consider two identical farmers, one of whom is located near a watercourse, while the other is far from it. A physical rule that will be described in subsequent chapters states that pollution is attenuated as it travels overland from the source to the receiving body of water. Hence, relating the required pollution abatement to the damage of the receiving body of water would impose a cost on the farmer located near the body of water that the other will not have to bear. A more equitable solution, at least subjectively, would be to share the cost; however, this would be impossible without tax incentives and subsidies, which can only be imposed by regulation.

All pollution management tools will likely redistribute benefits and cost nonuniformly across the population. Benefits might be enjoyed by recreational fishers and boaters, while the cost is borne by the residents in polluting communities installing point source abatement, by farmers implementing soil conservation, and by industries treating their wastewater. The situation where some must bear the cost so that others can benefit is described as a *redistribution of income.* The equity criterion implies that all those affected should bear the cost and share the benefits more or less equally. However, there has been a long-standing policy imperative used in water resource development projects that is also applied to foreign aid to developing countries, stating that those who are poor should receive proportionally more benefit than those who are more well off (Maas et al., 1970)

In most cases, however, the most equitable resolution of the conflict between the polluters and sufferers of pollution is—by regulation or voluntary participation—to make polluters pay for abatement. In economic terms this means that the price of goods and services should fully reflect both the cost of production and the cost of the resource used, including the use of the environment for waste disposal. In theory and practice, the polluter should pay the full cost of damages caused by his/her activity, which, in turn, will create an incentive for the reduction of polluting activities at least to the level where the cost of pollution reduction equals the cost of damage caused by pollution. This leads to another policy imperative:

Polluter Pays Principle (PPP). Polluters bear the primary responsibility for pollution and its abatement.

The member governments of the Organization for Economic Co-operation and Development (OECD, 1991) agreed to pursue environmental policies that would appropriately use economic instruments, alongside cost-effective regulatory instruments, which would be in accordance with the polluter pays principle. These economic instruments and types of payments for pollution and its abatement are explained in Chapter 16.

Pollution of the environment and adverse water quality are a result of many human activities. Generally, the sources of pollution are classified into point and nonpoint (see Chap. 1). In the past, water quality considerations were fragmented, and almost all financial resources were devoted to point source abatement. Sewage treatment, storm-water management, nonpoint source control, and point source programs were carried out separately from each other, despite the fact that these sources are interrelated and produce combined water quality deterioration. Also water quality has been generally evaluated as if air pollution, solid and hazardous waste disposal, and land- and water-use management and decisions were unrelated. The reverse is true. All these aspects of the pollution problem are interrelated and their combined effect must be considered. This leads to the next policy imperative:

Integrated Approach. The pollution problem must be resolved in an integrated manner, whereby the causes of pollution, all sources, and the combined environmental impact are considered, and the resulting combined solutions are therefore most equitable and efficient.

Single management tools (such as a treatment plant) may have a single medium result (improvement of water quality). Some solutions may produce results in more than one medium. Land-use restrictions, for example, can protect the quality of the air, as well as that of surface runoff and ground water. Phosphorus loads to the receiving water bodies originate from both point and nonpoint sources and to the biota of the receiving water body whose productivity is stimulated by the phosphorus input; it does not matter where the phosphorus load is coming from.

The OECD member governments will also pursue policies that will encourage the progressive adoption of anticipatory, rather than exclusively reactive, instruments of environmental policy. This means that more emphasis should be placed on *pollution prevention* in contrast to pollution abatement, which was prevalent throughout the 1970s and 1980s. Emphasis will be placed on not creating pollutants in the first place. Pollutants that must be created should be removed as close to the source as possible. Care must be taken to avoid shifting pollutants between media (air, water, soil, or underground formations) and to find the most appropriate (smallest risk) media into which unavoidable pollutants should be released. Therefore the last imperative is

Pollution Prevention. Environmental policies will promote economic development that will anticipate potential pollution problems and react to them by political and economic and technological means before they occur.

Criteria for Pollution Abatement Based on Equity and Irreversible Impact

As was pointed out in Chapter 1, two sets of criteria and standards are in force in the environmental area. The first set (the legal difference between standards and criteria was explained in Chap. 1) is designed to protect human health and the well being of fish and aquatic life. We can call them *ecological–toxicological* or *receiving water* standards and criteria, which will be introduced and discussed in Chapters 12 and 13 and in the Appendix.

The second group of standards is not directly related to water quality, though they generally apply to all sources within various source categories. Most of these standards are based on the equity imperative— that is, all polluters should reduce some part of their waste load regardless of whether or not harm is being done by the emissions to the receiving body of water. Other standards and criteria may be based on the avoidance of the irreversible harm to future generations by the

TABLE 2.1 Examples of Technology-based Standards

Land-use Type	Typical Standard	Policy Imperative
Urban runoff	Capture and treatment of first 1.2 to 2.5 cm of runoff[a]	Equity
	Mandatory street sweeping	Aesthetics
Pasture	Cattle density limitations, fencing along the streams	Irreversible impact, Public health
Agriculture	Soil erosion control	Irreversible impact,[b] equity
Construction sites	Soil erosion control	Equity, air pollution, irreversible impact
Suburban lands	Septic tank regulations	Public health, ground-water protection
	Pesticide and fertilizer sale regulations	Ground-water protection,[b] public health
Combined sewer overflows	Restriction on the number of overflows or mandatory capture and treatment of a certain portion of the wet-weather flow	Equity[b]
Surface mining	Land reclamation and restoration	Equity, irreversible impact
BOD and suspended solids standard for municipal point sources	Maximum effluent BOD_5 and suspended solids limits of 30 mg/l as monthly averages	Equity

[a] Typically it controls about 90% of the water and pollution load.
[b] In addition to water quality control benefits.

overuse of the resource, or the standards may be based on aesthetics, air pollution control, ground-water protection, or damages to infrastructures. Most of these standards require the mandatory application of certain technologies, hence, we call them *technology-based* standards. Examples of technology-based standards are given in Table 2.1. Such standards are also known as *effluent standards* and are based on the *best available technology economically achievable (BATEA)* used in point source abatement or on *performance standards* used for diffuse-pollution abatement.

Once issued, technology-based standards are easy to implement and monitor for compliance. Water quality standards, on the other hand, are difficult to enforce because there is rarely a direct and simple relationship between a pollution discharge and the water quality (pollution) of the receiving body of water. Water-quality-based approaches require models that may sometimes be inaccurate and unreliable. For example, nitrogen and phosphorus discharged from point and nonpoint sources cause

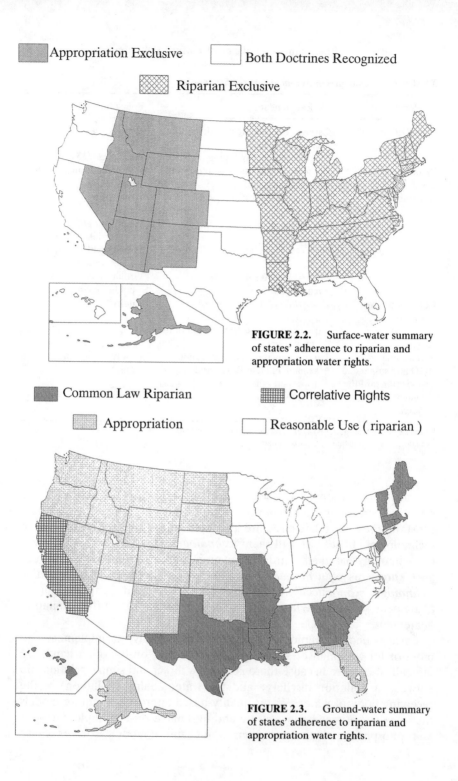

FIGURE 2.2. Surface-water summary of states' adherence to riparian and appropriation water rights.

FIGURE 2.3. Ground-water summary of states' adherence to riparian and appropriation water rights.

accelerated eutrophication, which is manifested by algal blooms and other symptoms (see Chapter 12), but the relation between the nutrient levels in the receiving bodies of water and algal biomass is very complex and to some degree speculative. Even more uncertainty is involved in modeling the fate and effects of toxic chemicals (see Chapter 13).

WATER USES AND WATER RIGHTS

The primary objective of all pollution-control efforts is to protect and enhance the use of bodies of water for present and future generations. There is a general consensus that water is public goods and everybody should have an access to and the right to use them. However, in the United States the uses and ownership of water rights (who can use water and in what quantity and quality) differ among the states, because water laws are under the administration of the individual states and consist primarily of two doctrines known as *riparian* and *appropriative rights*, as shown on Figures 2.2 and 2.3. In general, two kinds of laws have developed because of the two major problems occurring with water, use and drainage. The majority of water-use laws have developed in the West, where critical water shortages occur, while the majority of drainage laws have been promulgated primarily by the "wet" eastern states (Krenkel and Novotny, 1980). These two completely different water rights doctrines can make some eastern water laws inoperational in the West and vice versa. For example, maintaining minimum flow for waste assimilation is impossible in the West, while flows are guaranteed in the East. On the other hand, interbasin transfer of water is legally very difficult to impossible in the East while permitted in the West.

Riparian Water Laws

Nearly all of the states east of the Mississippi River follow the riparian doctrine. Its key features are:

> The owner of land adjacent to a stream is entitled to receive the full natural flow of the stream undiminished in quantity and unimpaired in quality. The riparian owner has a legal privilege to use the water at any time, subject only to the limitation that the use is reasonable. The right is a natural right that cannot be transferred, sold, or granted to another person as property. The legal body owning this right is then called a *riparian owner* and the property adjacent to the water body is then *riparian land*.

During a time of water shortage, all riparian owners have equal rights to the reasonable use of water and the supply is shared, although domestic use may have preference over other uses. As pointed out, the riparian rights do not allow transfer of water from one basin to another (a notable exception is a small transfer of flow from the Great Lakes basin to the Illinois River via the Chicago River and connecting canals that was approved by the governors of all the Great Lakes states and by two Canadian provinces and after a succession of litigations was finally approved by the courts). The riparian landowners and users are protected from withdrawals or uses of water that unreasonably diminish its quality or quantity.

The private riparian ownership of water rights may conflict with the rule of water as public goods and with the right of the federal government stemming from the Commerce Clause of the U.S. Constitution. Imagine a situation (that could be very common in the East) where one owner or a group of private owners acquires all riparian lands surrounding a lake or any other body of water. This would preclude public access and public use of that water. The courts, however, have allowed the states to become riparian owners, even by expropriation, of a small piece of land (for example, for a public boat landing) and have upheld the federal government's rights over all navigable waters. Hence, through a public agency becoming a riparian owner of the body of water the public interests are protected and the use of the body is in the public domain.

Use of so-called *riparian buffer zones* and *riparian wetlands* for the abatement of diffuse pollution (see Chapters 10, 11, and 14) necessitates either the acquisition of water rights of a strip of riparian lands along the affected body of water or, by regulation or persuasion, the imposition on the riparian landowners of the abatement in a strip of land adjoining the body of water.

Appropriation Water Laws

The system of water laws adopted by most western states is known as the law of appropriation, which is best stated as: "First in time is first in right." The basic tenets of the system are:

(1) A water right can be acquired only by the acquiring party diverting the water from the water course and applying it to a beneficial use, and (2) in accordance with the date of acquisition, an earlier acquired water right shall have priority over later acquired water rights. Water in excess of that needed to satisfy existing uses is viewed as unappropriated water, available for appropriation and application of the water to a beneficial use.

The process of appropriation can continue until all the water from the stream is subject to rights of use through withdrawals from the stream. In times of shortages, the earliest claimants take full share, and others may do without water. If the right is not used, it is lost. In addition, the right is not identified by ownership of riparian land.

There are several problems related to the application of the appropriation doctrine in western states. First, there is no natural flow notion (Beuscher, 1967). The appropriators can take as much water as they are entitled to, even though it exhausts the flow. This leads to situations where the entire river flow may be withdrawn by irrigators and other users, which means that the downstream flow, if any, is then composed of irrigation return flows with greatly elevated salinity and other pollutant content. Some western states, however, permit the states to file for, and ultimately acquire, the right to the unappropriated flow, and thus to preserve such flow, if desired.

Second, because water rights can only be acquired by diversion, protecting in-stream quality values and uses is difficult to impossible. If water is brought from another basin to augment the flow to improve water quality and increase the waste-assimilative capacity of the stream, the increased flow could be appropriated, diverted, and used by others, thus negating the diversion's purpose of improving water quality.

Third, some management practices that are very popular in the East, for example, those relying on infiltration, in the West may be seen as using water that is already appropriated. Hence, implementation of these practices may be prevented.

Fourth, many western rivers are subject to appropriation by Indian Nations (tribes) that are considered to be first claimants whose right to the water precedes any subsequent claims by white settlers. In such situations, any water quality management plans must include consultation, consent, and the cooperation of water managers from the Indian Nations whose water rights would be affected.

LAWS AFFECTING POLLUTION ABATEMENT

The term environmental law refers primarily to that body of law that seeks to prevent adverse environmental consequences by regulating individual, corporate, and governmental behavior. It is also a form of social control whose objective is to regulate the production and consumption of goods in order to preserve ecological balance, natural beauty, and protect public health and endangered species, so as to maintain a stable and satisfactory level of living and quality of life for present and future generations.

The U.S. legal system and the laws protecting the environment and governing environmental abatement efforts are based on legislative (statutory) enactments by legislative or other authoritative bodies or on judicial decisions in environmental litigations.

Statutory Laws

The first statutory environmental law, passed by Congress in 1899, was The Refuse Act, which stated that it is unlawful to place any material, except sewage and runoff, into a navigable waterway or tributary thereof without a permit. It is interesting to note that in the absence of any meaningful pollution-control legislation until 1972, this archaic statute was used in the late 1960s and early 1970s to control water pollution from industrial sources, although the original intent of the act was to protect navigation and not water quality.

The first Water Pollution Control Act was passed by Congress in 1948 (PL 80-845) and was amended several times between 1948 and 1972. The act and its subsequent amendments authorized the creation of environmental research centers, established a Division of Water Pollution Control within the Public Health Service, authorized grants to build public treatment plants, gave the federal government authority to abate interstate pollution, among other provisions. In general, these laws were ineffective in controlling pollution or enforcing abatement. Responsibility for pollution control, which was originally with the Division of Water Pollution Control, was transferred several times between the departments of the federal government until by the executive order of the president in 1970 the Environmental Protection Agency was created as an independent governmental agency.

The Water Pollution Control Act Amendments of 1972 (PL 92-500)

This grandiose legislative piece known as the Clean Water Act (CWA) had a monumental impact on environmental efforts in the field of water quality control. It was enacted by Congress over a presidential veto (as were several other subsequent amendments).

The declaration of Section 101(a) states:

> The objective of this Act is to restore and maintain the chemical, physical, and biological integrity of the Nation's waters.

The objective clearly specifies that ecological objectives and concerns should receive the highest priority. The objectives of the act also stated as a goal that the nation's (navigable) waters be suitable for contact recreation and provide for the protection and propagation of fish and aquatic wildlife.

Although PL 92-500 was intended to be a comprehensive water quality program, in practice, great emphasis was placed on controlling point sources, while diffuse-pollution control received far less attention and minimal financial resources. The act accomplished three basic tasks: (1) the regulation of discharges of point sources; (2) the regulation of oil spills and other hazardous substances; and (3) financial assistance for wastewater treatment plant construction.

The most significant and revolutionary contribution of the act is the establishment of an enforcement scheme that is built around the National Pollution Discharge Elimination System (NPDES). This system serves as the basic mechanism for enforcing the implementation of pollution abatement of point sources. As shown in Chapter 1 the definition of point sources has been gradually expanded so that many diffuse sources are now legally classified as point sources (animal feedlots, industrial and municipal storm sewers, combined sewer overflows, runoff and acid discharges from active mines) and require an NPDES permit. For point source discharges the permit, among other things, establishes specific effluent limitations and specifies compliance schedules that must be met by the discharger. It also requires compliance with other relevant state and local pollution control laws, if more stringent. The NPDES permit system is the most important tool for implementing the polluter pays principle.

Section 208 of the CWA. Section 208 of the original CWA of 1972 had a far-reaching impact because it enacted a land-use planning process. For the first time, it was realized that the control of point sources would not have solved all the pollution problems in the United States. Many excellent planning reports were produced by designated planning agencies. However, instead of developing a nonpoint source regulatory program that was deemed prohibitively expensive (Billings, 1976), Congress gave an incentive to the development of state- and areawide water quality management plans that would include all sources of pollution and water quality degradation. Incentives for treatment and penalties for noncompliance with the plan were included for point sources (which at that time excluded urban storm water and other diffuse sources currently defined as point sources), while no enforcement tools were available for nonpoint sources. Also after the plans were completed no mechanism for program implementation and maintenance were in place. Consequently, in many cases the plans were not pursued after their initial release and the effort has essentially never fully achieved its potential.

The Clean Water Act Amendments of 1977
(PL 95-217)

A major revision in the law allowed the EPA to add or remove toxic materials without first requiring a formal hearing. New deadlines were included for meeting the point source abatement requirements stipulated by the act.

The Water Quality Act (Clean Water Act) of 1987
(PL 100-24), Sections 319, 402, and 404

These three sections of the act are the most important tools for controlling diffuse pollution. Their highlights and other important issues of the 1987 CWA related to diffuse source pollution management are discussed by Berg (1988).

Section 319 requires the states to prepare State Nonpoint Source Assessment Reports and encouraged states to develop and implement management programs in order to be eligible for federal funds. The deadline for preparation of the reports was February 1989, and most of the states have complied. The management plans developed in the 319 program are now part of the states' water quality management agenda. Federal matching grants are provided to those states that qualify for assistance in implementing their nonpoint source management programs.

Section 402 establishes the permit program for discharges of pollutants from point sources. More specifically, Section 402(p) requires a NPDES permit for separate storm sewers.

Section 404 regulates the discharge of dredged and fill materials into waters of the United States and also establishes a permit program to ensure that such discharges comply with environmental requirements.

The permits for point source discharges under Section 402 are issued by states, while the permits under Section 404 are issued by the U.S. Army Chief of Engineers. Other provisions of the act such as the National Estuary Program (Section 3020), Clean Lakes Program (Section 314), and the Great Lakes Basin and Chesapeake Bay Programs also deal with diffuse source pollution management. The act also reauthorizes funding for the areawide water quality management plans under Section 208.

The Water Quality Act's jurisdiction extends to all waters of the United States. The phrase "all waters" includes waters that are currently used, were used in the past, or may be susceptible to use in interstate or foreign commerce (this jurisdiction is guaranteed by the Commerce Clause of the U.S. Constitution). Generically, such bodies of water include all navigable waters that have been legally understood as waters on which a canoe can be floated. These waters include:

- All waters that are subject to the ebb and flow of the tide;
- The territorial seas;
- Interstate waters and wetlands;
- All other waters (such as intrastate lakes, rivers, streams, and wetlands), if their use, degradation, or destruction could affect interstate or foreign commerce;
- Tributaries to waters or wetlands previously identified;
- Wetlands adjacent to waters previously identified.

Present Regulations for the Control of Urban Diffuse Sources Derived from the Clean Water Act

Control of Combined Sewer Overflows

The U.S. Environmental Protection Agency published its control strategy for combined sewer overflows (CSOs) in 1989 (*Federal Register*, August 10, 1989). The strategy relies on the NPDES permit system. The permit system is aimed at bringing all CSO discharges into compliance with the technology-based requirements of the Clean Water Act and applicable state standards, and to minimize water quality, aquatic biota, and human health impacts from wet-weather overflows.

All permits for CSO discharges should require the following technology-based limitations as a minimum: (1) proper operation and regular maintenance programs for sewer systems and combined sewer overflow points; (2) maximum use of the collection system for storage; (3) review and modification of pretreatment programs to assure that CSO impacts are minimized; (4) maximization of flow to the treatment plant for treatment; (5) prohibition of dry-weather overflows; and (6) control of solid and floatable materials in CSO discharges.

Additional CSO control measures are based on the potential impact on receiving water bodies that would bring CSO discharges in compliance with state standards. Additional control measures include improved operation, best management practices, supplemental pretreatment program modifications, sewer ordinances, local limits programs, identification and elimination of illegal discharges into sewer systems, specific pollutant limitations, compliance schemes, direct treatment of overflows, sewer rehabilitation, in-line and off-line storage, reduction of tide water intrusion, construction of CSO controls within the sewer system or at the discharge point, sewer separation, and new or modified treatment facilities. The compliance monitoring program should be described and included in the permit.

The strategy does not cover treatment plant bypasses, which are considered to be an "intentional diversion of waste streams from any portion of the treatment facility" that begins at the headwork of the facility. Bypasses are not allowed unless (1) they are unavoidable to prevent loss of life; and (2) there is no other feasible alternative to the bypass.

Storm Water (separate sewers) Permit Regulations
Overview. The storm-water control rules established by the EPA seek to establish NPDES permit application requirements for

- Stormwater discharges associated with industrial activity;
- Discharges from large separate storm sewer systems (systems serving a population of more than 250,000);
- Discharges from medium municipal separate storm sewer systems (systems serving a population of more than 100,000 but less than 250,000).

Discharges from other municipal separate storm sewers, including separate municipal systems serving a population of less than 100,000 and discharges associated with industrial activity connected to these systems are not subject to the permit system.

The rules are to be implemented in a phase-in approach. Storm-water discharges from industrial areas that are subject to the permit system have to comply with Sections 301 and 402 of the Clean Water Act, which require application of the best available treatment technology (BAT). Permits for discharges from municipal storm sewers include controls that reduce the discharge of pollutants to the maximum extent practicable (MET), as well as a requirement to effectively prohibit discharging nonstorm water (cross-connections) into the storm sewers.

The permit system requires industrial facilities that discharge storm water associated with industrial activity to submit sampling data, a description of storm-water management practices, and certification that the discharge does not contain processed water, domestic sewage, or hazardous wastes. Group applications, industry by industry, are permitted. Indirect discharges to municipal systems serving a population of 100,000 or more generally do not have to submit applications, but do have to notify the municipality of their discharge.

Permits are issued on a systemwide basis for municipal separate storm sewers. Municipalities are first required to describe their existing storm-water management program, identify all known outfalls, and conduct field screening for illicit connections. The municipalities are then required to verify illicit connections, conduct representative sampling, and describe priorities for storm-water management during the 5-year permit term.

The data collected during these phased tasks will allow the permit to be developed for site-specific conditions.

The EPA rules define storm water as follows:

"Storm water" means storm water runoff, snow melt runoff, surface runoff, street wash waters related to street cleaning or maintenance, infiltration (other than infiltration contaminated by seepage from sanitary sewers or by other discharges) and drainage.

Relation to ground-water quality. In Section 319 the Clean Water Act (1987) strengthens the regulatory link between diffuse (nonpoint) pollution and ground-water quality. Under the 1987 amendments, the CWA now specifically requires states to select best management practices, taking into account the impact of the practice on ground-water quality. The Senate Report explained (Thompson, Adler, and Landman, 1989):

States are required to consider impact of management on groundwater quality. Because of the intimate hydrologic relationship that often exists between surface and groundwater, it is possible that measures taken to reduce runoff of surface water containing contaminants may increase transport of these contaminants to groundwater. The State should be aware of this possibility, when defining best management practices, especially in aquifer recharge areas.

The 1972 Coastal Zone Management Act (PL 92-583) programs. This act was established in response to the high rate of development in coastal areas and out of concern about the environmental effects of this growth. These coastal areas encompass land and tidal zones, including those located in the Great Lakes basin. The National Oceanic and Atmospheric Administration (NOAA) is implementing programs mandated by the act.

The act specified that:

The habitat areas of the coastal zone, and the fish, shellfish, other living marine resources, and wildlife therein, are ecologically fragile and consequently extremely vulnerable to destruction by man's alterations (Section 302d);

Important ecological, cultural, historic, and aesthetic values in the coastal zone which are essential to the well-being of all citizens are being irretrievably damaged or lost (Section 302e).

Land uses in the coastal zone, and the uses of adjacent lands which drain into the coastal zone, may significantly affect the quality of coastal waters and habitats, and efforts to control coastal water pollution from land use activities must be improved (Section 302k).

Section 6217 (1990 amendments) of the act then delineates the program for controlling diffuse (nonpoint) pollution. Participation by states in the Coastal Nonpoint Pollution Control Program (CNPCP) is on a voluntary basis; however, the act provides participating states with funding for the program and enforceable policies with penalties for noncompliance. At a minimum, each state program is to provide for the implementation of management measures to protect coastal waters. These programs should contain the following components:

1. Identification of land uses that may individually or cumulatively cause or contribute to degradation of coastal waters.
2. Identification of critical coastal areas that should be subject to management measures.
3. Identification of management measures that are most appropriate for the threatened coastal areas.
4. Delineation of technical assistance.
5. Public participation.

The program is carried out in coordination with the Section 319 programs of the Clean Water Act. Since 1992, about $12 million per year has been authorized for the program.

The Clean Water Act of 1987 also recognized estuaries and other coastal water as a critical national resource whose health and productivity are increasingly threatened by coastal growth and development. The act then formally established the National Estuary Program (NEP) to demonstrate innovative approaches applicable to coastal areas nationwide. The NEP program covers 12 estuaries, 6 in the North Atlantic region, 1 in the mid-Atlantic zone, 2 on the Gulf Coast, and 3 on the West Coast. Most of the research projects are conducted by universities under NOAA's Sea Grant program. NOAA's Coastal Zone Management and the EPA's National Estuary Programs are coordinated under an agreement between the two agencies.

Wetland Protection

Wetlands have great value for the hydrological–ecological terrestrial system. Their benefits are numerous and include maintaining and storing flows, wildlife habitat, and attenuation of pollutants (see Chapter 14). Wetlands are protected under both Section 402 and 404 permits. The NPDES permit system regulates discharges from point sources, including CSO and storm-water discharges. Section 404 permits regulate drainage and filling of wetlands. These permits, which treat wetlands as receiving bodies of water, do not allow wide use of natural wetlands for treatment of discharges from diffuse sources, unless they comply with the stream and effluent standards. Specifically, no discharges into wetlands can be permitted that would violate other applicable laws, such as state water quality standards, toxic effluent standards, or the Endangered Species Act. These regulations do not apply to man-made wetlands that can be considered as land treatment.

The U.S. Army Corps of Engineers' evaluation of the Section 404 permit application is a two-part process that includes both the evaluation of the project's eligibility for a permit and environmental assessment. The key policies of the permit evaluation are

- Dredged or fill material should not be discharged into waters of the United States unless it can be demonstrated that such discharges will not have an adverse impact on the aquatic ecosystem.
- From a national perspective, the degradation or destruction of special aquatic sites, such as filling wetlands, is considered to be among the most severe environmental impacts addressed by Section 404.

Under the Food Security Act of 1985 (see the following section) the "Swampbuster" provision (Subtitle III) requires that USDA program benefits be withdrawn from farmers who convert any naturally occurring wetlands to cropland after 1985. Many states have enacted or are considering enacting more stringent regulations protecting wetlands (Florida, Wisconsin, and others).

The Food Security Act (The Farm Bill) of 1985 (PL 99-198). The major deficiency of the Clean Water Act is that it does not provide for enforcement of the abatement of agricultural nonpoint sources, which represent the most significant cause of water quality degradation of many receiving bodies of water. Almost exclusively, programs relied on the voluntary participation of farmers to implement pollution abatement. This lack of enforcement procedures is most likely related to the "sacred" right of unrestricted use of land for family farming. This argument may not be valid today, since more and more small family farms are acquired by large industrial operations. The most significant influences on state programs to control agricultural pollution have to come from federal assistance and land management programs, rather than from the EPA's water quality protection programs mandated by the Clean Water Act.

However, Congress has realized that farming in many cases greatly subsidized by the federal and state governments, does cause pollution. As a consequence, the legislators passed the Food Security Act, which creates two programs that are aimed at reducing pollution from agricultural operations. The Conservation Reserve Program (CRP) gives the Soil Conservation Service the authority to make annual rental payments for 10 years to farmers who retire highly erodible land and land bordering bodies of water from farming and plant it with such permanent cover crops as grasses, legumes, or trees. The CRP intends to remove more than 30 million hectares of the most erodible cropland from agricultural production. The CRP stream buffers can idle cropland for up to 30 meters (100 ft) from the water's edge. Under "Conservation Compliance" provisions, farmers who plant annually tilled crops on highly erodible lands must implement locally developed and approved conservation plans in order to remain eligible for price support, crop insurance, and other USDA program benefits. Under the "sodbuster" provision (Subchapter II), to retain USDA benefits farmers must follow an approved conservation system when plowing fields that were not in use for crop production between 1981 and 1985.

The CRP programs aimed at excluding highly erodible land have indirect water quality benefit, though not all lands included in the program pose a threat to water quality. On the other hand, buffer strips have a direct water quality improvement benefit. Further modifications of the CRP program to increase water quality benefits were being considered by the lawmakers during the time this book was being written. Future improvements may include adding an additional 10 million hectares to the CRP to address environmental problems, including lands with salinity, selenium, siltation, soil drainage, and other problems. The benefits of the CRP program were estimated by Ogg and Ribaudo (1988).

The Conservatioin Compliance (CC) program requires all farmers with row crops located on highly erodible land to establish a Soil Conservation Service approved plan or lose eligibility for federal support, including supplements and disaster assistance. In the absence of enforcing regulations, this program is the most important tool to persuade farmers to participate.

To implement these programs, employees of the Soil Conservation Service work with local units of state governments of the Soil and Water Conservation Districts (SWCD), which are managed by locally elected, unsalaried citizens (mostly farmers).

Other Federal Laws Affecting Diffuse Pollution and Water Quality Management. Among the most complicating factors in diffuse-pollution abatement and water quality management are the plethora of laws affecting the decision-making process and specifying various sometimes conflicting environmental policies. These legislative pieces include:

Environmental Laws
1. The National Environmental Policy Act
2. The Clean Air Act Amendments

3. The Federal Environmental Pesticide Control Act
4. The Rare and Endangered Species Act
5. The Safe Drinking Water Act
6. The Federal Insecticide, Fungicide, and Rodenticide Act
7. Toxic Substances Control Act
8. The Wild and Scenic River Act

Floodplain Management Laws

9. Flood Control Act and Amendments
10. National Flood Insurance Programs
11. Flood Disaster Protection Act

U.S. Department of Agriculture Laws

12. Rural Development Act

Mining

13. Surface Mining Control and Reclamation Act
14. Federal Land Policy and Management Act

State Statutory Laws

Many states have enacted effective pollution-control statutes and programs. It is beyond the scope of this book to describe the programs in each individual state, though a summary of several state laws may demonstrate some effective and successful approaches to the resolution of the diffuse-pollution problem. These programs are carried in addition to or as a supplement to the federally mandated programs. According to Section 319 of the Clean Water act (1987 Amendments), all states are required to prepare nonpoint pollution abatement plans that list the state programs. These plans are available from the state pollution-control agencies. A few examples are listed herein.

Florida has strong wetland protection ordinances aimed at the reduction of wetland losses throughout the state. In addition, Florida's storm-water management program, which applies to all new developments, is designed to ensure that the volume, rate, timing, and pollutant load of runoff after development do not cause a violation of state water quality standards. The Florida rule is essentially a performance standard specifying what is expected to achieve compliance with state water quality standards.

Iowa has developed requirements for all confined feedlots that apply to all open feedlots exposed to rainfall and to total confinement facilities where precipitation is not a factor. At a minimum, all open facilities must remove settleable runoff solids.

Maryland began a comprehensive storm-water management program in 1982. This program is administered by the state's Sediment and Stormwater Administration. Depending on the county, mandated local ordinances must at a minimum require that postdevelopment peak discharge for 2-year- and/or 10-year-frequency storm events be maintained at the predevelopment levels. Maryland also participates in the regional Chesapeake Bay clean-up program.

Additional state programs are listed in Chapter 1 (Wisconsin Priority Watershed Programs), in Thompson et al. (1989), CH2M/Hill (1990), and several EPA and state monographs.

Judicial (Judge-Made) Laws

Dating back to the Roman Empire, courts and judges have made rulings that, among other things, affected many aspects of the environment. Most of the water quality litigations are decided by a judge according to the Equity Law, and primarily involve injunction or specific orders from the courts restraining certain types of actions or regulating other actions. These rulings were based on imperatives and doctrines as previously elucidated.

Many court decisions involving water quality are based on the doctrine of nuisance and trespassing. In many cases, judges used these doctrines when actual damage had occurred and was proved by the plaintiff. However, in a landmark case Judge John Grady stated that this may no longer be true because "It is the ability of the courts of equity to give a more speedy, effectual, and permanent remedy in cases of public nuisance. They can not only prevent the nuisance that is threatened, and before irreparable mischief ensues, but arrest or abate those in progress and, by perpetual injunction, protect the public from them in the future" (*Illinois vs. the City of Milwaukee*, Federal Court, Chicago, 1977). In his ruling, Judge Grady issued stringent performance and effluent standards for the clean up of point and diffuse source pollution from the Milwaukee metropolitan area. Some of these standards were almost technically unattainable; however, the ruling was appealed and subsequently overturned by higher courts. If this ruling had been allowed to stand, it would have made a legal precedent for similar cases by other federal and state courts.

In today's legal environment in the United States, litigation is common; however, the large number of statutory laws in place and in force do not allow for frequent judicial rulings outside the statutory legal framework. Many legal cases of violations of pollution-control laws and ordinances are settled out of court.

References

Bator, F. M. (1958). "The anatomy of marker failure," *Quart. J. Econ.*, Aug., pp. 351–379. Included in *Pollution Taxes, Effluent Charges and Other Alternatives for Pollution Control*, U.S. Congress Rep. No. 95-5, 1978, Washington, DC.

Baumol, W., and W. Oates (1988). *The Theory of Environmental Policy*, 2d ed., Cambridge University Press, Cambridge.

Berg, N. (1988). "Policies and intergovernmental relations of nonpoint pollution abatement programs," in *Political, Institutional and Fiscal Alternatives for Nonpoint Pollution Abatement Programs*, V. Novotny, ed., Proceedings, Symposium, Marquette University Press, Milwaukee, WI.

Beuscher, J. H. (1967). *Water Rights*, College Printing and Typing, Madison, WI.

Billings, L. G. (1976). "The evolution of 208 water quality planning," *Civ. Eng.* **46**(11):54–55.

Boland, J. J. (1991). Unpublished definitions. Panel on policies, institutions and economics, National Research Council, Washington, DC.

Braden, J. B. (1988). "Nonpoint pollution policies and politics: The role of economic incentives," in *Nonpoint Pollution 1988: Policy, Economy, Management and Appropriate Technology*, V. Novotny, ed., American Water Resources Association, Bethesda, MD, pp. 57–65.

Carson, W. D. (1980). "Effluent charges for water pollution control in California," State Water Research Control Board, State of California, Sacramento.

Caufield, H. P., Jr. (1988). "The federal environmental legislative process," in *Nonpoint Pollution 1988: Policy, Economy, Management and Appropriate Technology*, V. Novotny, ed., American Water Resources Association, Bethesda, MD, pp. 1–8.

Caufield, H. P., Jr. (1991). "Problem of nonpoint source agricultural water pollution: Toward a hypothetical federal legislative solution, *Water Resour. Bull.* **27**(3):447–452.

CH2M/Hill (1990). *Nonpoint Source Impact Assessment*, Report No. 90-5, Water Pollution Control Federation Research Foundation, Alexandria, VA.

Kneese, A., and B. Bower (1968). *Managing Water Quality: Economics, Technology, Institutions*, Johns Hopkins University Press, Baltimore, MD.

Krenkel, P., and V. Novotny (1980). *Water Quality Management*, Academic Press, New York.

Maas, A., et al. (1970). *Design of Water Resources Systems*. Harvard University Press, Cambridge, MA.

Novotny, V. (1988a) "Diffuse (nonpoint) pollution—a political, institutional, and fiscal problem," *J. WPCF* **60**(8):1404–1413.

Novotny V., ed. (1988b). *Political, Institutional and Fiscal Alternatives for Nonpoint Pollution Abatement Programs*, Proceedings, Symposium, Marquette University Press, Milwaukee, WI.

Novotny, V., and E. Joeres (1983). "Planning and financing water resources development with emphasis on water pollution control and sanitation, in

Proceedings, Fourth World Congress IWRA, Buenos Aires, Argentina, Tycooly Int. Publ., Dublin.
Ogg, C. W., and M. C. Ribaudo (1988). "Economic implications of the Food Security Act of 1985's discretionary environmental provisions," in *Nonpoint Pollution 1988: Policy, Economy, Management and Appropriate Technology*, V. Novotny, ed., American Water Resources Association, Bethesda, MD, pp. 49–56.
Organization for Economic Cooperation and Development (1991). *The State of the Environment*, OECD, Paris, France.
Rogers, P., and A. Rosenthal (1988). "The imperatives of nonpoint source pollution control," in *Political, Institutional and Fiscal Alternatives for Nonpoint Pollution Abatement Programs*, V. Novotny, ed., Marquette University Press, Milwaukee, WI.
Solow, R. M. (1971). "The economist's approach to pollution and its control," *Science* **173**:497–503.
Thompson, P., R. Adler, and J. Landman (1989). *Poison Runoff*, Natural Resources Defense Council, Washington, DC.
Viessman, W., Jr., and C. Welty (1985). *Water Management—Technology and Institutions*, Harper & Row, New York.

Additional Reading

Dahlman, C. (1979). "The problem of externality," *J. Law Econ.* **22**:141–62.
Fisher, A., and F. Peterson (1976). "The environment in economics," *J. Econ. Lit.* **14**:1–33.
Freeman, A., III (1979). *The Benefits of Environmental Improvement*, Johns Hopkins University Press, Baltimore, MD.
Just, R., D. Hueth, and A. Schmitz (1982). *Applied Welfare Economics and Public Policy*, Prentice-Hall, Englewood Cliffs, NJ.
Kneese, A., and C. Schultze (1975). *Pollution, Prices, and Public Policy*, Brookings Institution, Washington, DC.
Kneese, A., R. Ayres, and R. d'Arge (1970). *Economics and the Environment: A Materials Balance Approach*, Johns Hopkins University Press, Baltimore, MD.
Krutilla, J., and A. Fisher (1975). *The Economics of Natural Environments*, Johns Hopkins University Press, Baltimore, MD.
Majone, G. (1986) "International institutions and the biosphere," in *Sustainable Development of the Biosphere*, W. C. Clark, and R. E. Munn, eds., Cambridge Univ. Press, Cambridge, England.
Mäler, K. G. (1974). *Environmental Economics: A Theoretical Inquiry*, Johns Hopkins University Press, Baltimore, MD.
Miller, W. L., and J. H. Gill (1976). "Equity considerations in controlling nonpoint pollution from agricultural sources," *Water Resour. Bull.* **12**:253–261.
Organization for Economic Cooperation and Development (1974). *Problems in Transfrontier Pollution*, OECD, Paris.
Organization for Economic Cooperation and Development (1976). *Economics of Transfrontier Pollution*, OECD, Paris.

Organization for Economic Cooperation and Development (1979). *OECD and the Environment*, OECD, Paris.
Pfeiffer, G. H., and N. K. Whittlesey (1978). "Controlling nonpoint externalities with input restrictions in an irrigated river basin," *Water Resour. Bull.* **14**:1387–1403.
Schultze, C. (1977). *Public Use of Private Interest*, Brookings Institution, Washington, DC.
Walter, I. (1975). *International Economics of Pollution*, MacMillan, London.
Williams, A. (1966). "The optimal provision of public goods in a system of local government," *J. Polit. Econ.* **74**:18–33.

3
Hydrologic Considerations

He who sees things grow from the beginning will have the best view of them.

Aristotle

Pollution from diffuse sources is driven by meteorological events, that is, precipitation. It is a known fact that there is a correlation between the pollutant loadings from a watershed and rainfall volume, infiltration and storage characteristics of the watershed, permeability of soils, and other hydrological parameters. This distinguishes diffuse pollution from traditional point source pollution, which bears little relation to watershed hydrology.

Diffuse pollution therefore has its beginning in the atmospheric transport of pollutants, and its occurrence and magnitude are closely related to the hydrologic cycle. Consequently, the pollutant load from nonpoint sources has a strong random (unpredictable) component. In addition, hydrologic modifications of watersheds can increase or decrease diffuse-pollution loads.

Most of the models used for simulating diffuse-pollution loadings are basically models of watershed hydrology or are closely related to it. Rainfall energy and the splashing effect of rain droplets liberate soil particles, which then become available for the transport by the overland flow. If, for example, the overland flow is diminished due to higher infiltration, so is the transport of particulate pollutants. If agricultural chemicals or organic fertilizers are placed on the land and surface overland flow is generated by a storm, a significant portion of these contaminants can be lost in surface waters. Mobile pollutants (that is, pollutants that are dissolved and move with water as ions or salts) can be leached into the ground-water zone and cause ground-water contamination and pollution. The highest pollutant loadings, and in many cases the highest concentrations of contaminants from diffuse sources, occur during high-flow and flood conditions. On the other hand, point source impact and the impact of pollutants carried into surface waters by subsurface flow are most severe during low-flow (drought) conditions.

To control and understand the generation and transport of diffuse pollution, one has to study the hydrologic process causing and contributing to diffuse pollution and to consider the various paths the contaminants travel from the source areas to the receiving water bodies. A large number of basic textbooks and handbooks dealing with hydrological processes and hydrologic cycle are available (Bedient and Huber, 1988; Bras, 1990; Chow, 1964; Chow, Maidment, and Mays, 1988; Linsley, Kohler, and Paulhus, 1982; Hall, 1984; Viessman, Lewis, and Knapp, 1989; Ponce, 1989; McCuen, 1989). The reader is referred to these texts for further reference.

PRECIPITATION–RUNOFF RELATIONSHIP

A classic representation of the rainfall transformation into runoff and the components of the hydrological cycle is shown on Figure 3.1. The first stage of the runoff formation is condensation of atmospheric moisture into rain droplets or snowflakes. During this process, water is in contact with atmospheric pollutants. The pollution content of rainwater can therefore reach high levels. In addition, rain water dissolves atmospheric carbon dioxide, sulfur and nitrogen oxides, and as a weak acid, it then reacts with soil, limestone, and dolomite geological formations. Many ancient statues and historical structures, for example, in Greece and Rome, were made of marble (a form of limestone) and have been deteriorating rapidly in the past few decades as a result of the greatly increased acidity of precipitation (see Chap. 4 for further discussion of the interaction of precipitation with atmospheric pollution).

Runoff formation begins after rain (snow) particles reach the surface. During the winter months runoff formation may be delayed by snowpack formation and subsequent melting. During the initial phase of runoff formation, rain energy liberates the soil particles and picks up the particulate and contaminants deposited on the surface and dissolves salts and other chemicals.

Runoff generated by precipitation has three components:

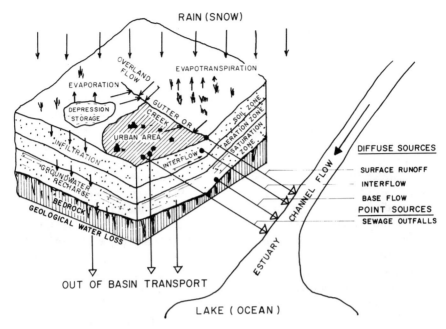

FIGURE 3.1. Schematic representation of watershed hydrology.

1. *Surface runoff* is a residual of precipitation after all loses have been satisfied. Numerical subtraction of the losses yields so-called *excess* or *net rain*. The losses include interception by surface vegetation, depression storage and ponding, infiltration into soils, evaporation from soils and open water surfaces, and transpiration by vegetation. *Evapotranspiration* is a term describing both evaporation from the soil and transpiration by plants. Since surface runoff is a residual of precipitation after all losses have been subtracted, a linear relationship between the volume of precipitation and runoff does not exist. The highest loads of particulate pollutants are carried by surface runoff.
2. *Interflow* is that portion of the water infiltrating into the soil zone that moves in a horizontal direction due to the lower permeability of subsoils. The amount of interflow is again a residual of infiltration after ground-water recharge, soil moisture storage, and evapotranspiration have been subtracted.
3. *Ground-water runoff* (base flow) is defined as that part of the runoff contribution that originates from springs and wells. In sewered urban areas one can also include infiltration inflow into sewers, which can be substantial. Most stream flow during prolonged drought periods can

be characterized as ground-water runoff. In some arid and semiarid regions, the natural base flow may be zero during certain times of the year, and the measured flow in streams may originate from sewage outfalls.

Streams that have measurable flow during the entire hydrological season (a hydrological year begins on October 1, and ends on September 30 of the following year) are called *perennial streams.* *Ephemeral streams* are streams without a measurable flow during certain times of the hydrological year. Urbanization and the accompanying discharges of wastewater effluents can change an ephemeral stream into a perennial one, for example, the Las Vegas, Nevada, Wash draining into the Colorado River is a man-made perennial stream–marsh system. Overuse and mining of ground-water resources will change a perennial stream into an ephemeral watercourse, as has happened in many places, for example, in Tucson, Arizona. From these examples one can see that man has a profound impact on the hydrology of a body of water. In fact, very few streams in populated areas of the world have truly natural flows. Runoff quantity and quality can be dramatically altered by changing the use of land by man, that is, urbanization, deforestation, storage reservoirs, and other land and stream modifications discussed in Chapter 1 (Fig. 1.5).

The quality of surface runoff can be related to the erosion intensity by precipitation and to the quantity of contaminants accumulated on the surface or in the top soils. Interflow and ground-water quality can be related to the amount of contaminants present in the soil, which also reflects the basic chemical composition of the soil, subsoil, and bedrock. Very often pollution of the interflow and ground-water (base) flow results from the excessive contamination of soils; for example, from overloaded septic tank seepage, overfertilization, and excessive chemical use on farms and urban lawns.

Mathematically, the runoff relation to precipitation can be expressed as

Surface runoff

$$R_s = P - \Delta S_i - \Delta S_d - f\Delta t \tag{3.1}$$

Interflow

$$R_i = (f - ET)\Delta t - \Delta S_s - q_g \tag{3.2}$$

Ground-water (base) flow

$$R_g = q_g - \Delta S_g - q_d \tag{3.3}$$

where
R_s = volume of the surface runoff in cm during a time interval Δt
P = precipitation volume (cm)
ΔS_i = change in available interception storage (cm)
ΔS_d = change in available depression surface storage (cm)
f = infiltration rate (cm/hr)
ET = evapotranspiration rate from the soil zone (cm/hr)
ΔS_s = soil moisture storage change (cm)
q_g = ground-water recharge (cm)
R_i = interflow (cm)
R_g = ground-water flow contribution (cm)
ΔS_g = ground-water storage change (cm)

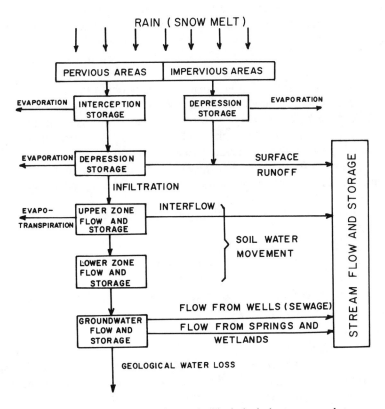

FIGURE 3.2. Block diagram of watershed hydrological processes and storage.

q_d = geological water loss (cm)
Δt = time interval (hr)

A block diagram of the rainfall–runoff transformation process is shown in Figure 3.2.

Components of the Rainfall–Runoff Transformation Process

Interception

A part of the precipitation volume is intercepted by vegetation where it adheres to the surface until a sufficiently heavy film is formed, at which point gravity begins to prevail over adhesion. Interception storage is that part of precipitation that wets or adheres to the surface of aboveground objects and vegetation and is returned to the atmosphere by evaporation. The amount that is intercepted depends on the type and intensity of the vegetation, intensity and volume of the rainfall, roughness of the surface, and the season of the year or growth stage of the vegetation.

The few models for interception storage reported in the literature are crude and inaccurate. Interception can be measured by comparing precipitation in gages or simple open buckets beneath the vegetation with that recorded nearby under the open sky. Generally, about 0.5 to 1.2 mm of rain can be held on foliage before an appreciable drip can take place (Viessman, Lewis, and Knapp, 1989). The total interception by an individual plant is directly related to the amount of foliage (its surface area per unit area of ground surface) and its character and orientation. About 1.2 to 1.8 mm of precipitation can be intercepted by grass and dense shrubbery.

A general form for interception was proposed as (Gray, 1973; Bras, 1990)

$$I = a + bP^n \tag{3.4}$$

where P is precipitation in centimeters. Values of the coefficients for some typical vegetal covers are given in Table 3.1.

Depression Storage

Water reaching the surface must first fill the surface depressions, forming small puddles, ponding, or adding to the general wetness of the area. Water stored in the depression storage either evaporates or percolates into the soil zone. Only when the precipitation rate exceeds infiltration

TABLE 3.1   Coefficients for the Interception Formula (Eq. (3.4))

	Coefficients		
Vegetal Cover	a	b	n
Orchards	0.1	0.18	1.0
Maple, beach, oak in forest	0.1–0.12	0.18	1.0
Beans, potato, cabbage, and other small crops	$0.05h^a$	0.45 h	1.0
Forage, alfalfa, etc.	0.025 h	0.30 h	1.0
Small grains, rye, wheat, barley	0.012 h	0.15 h	1.0
Corn	0.012 h	0.15 h	1.0

Source: From Gray (1973).

[a] h refers to the height of the plant in meters. Interception is in centimeters for P in centimeters.

and all surface storage (depression and interception) is exhausted will surface runoff result.

The character of depression storage as well as its magnitude depends largely on the surface characteristics that can be generally related to land use. The primary factors determining depression storage are surface character, roughness, and slope. An accurate estimation of depression storage is not possible, and little information is available that could serve as a guide for choosing the values of depression storage that would be based on physical measurements in the field.

In hydrological models interception storage is usually lumped together with depression storage into one surface storage (abstraction) parameter, which is determined by calibrating the model. After his first experience with the well-known *Stanford Watershed Model*, Linsley (1967) pointed out that the surface storage parameter is the key element in calibrating of the model for smaller watersheds (less than 50 km^2). Some information derived from water balance and/or modeling on the magnitude of the surface storage parameter has been published. For example, Tholin and Keifer (1960) estimated surface storage for Chicago's urban areas as being 6.25 mm ($\frac{1}{4}$ in.) on pervious areas with grass and 1.56 mm ($\frac{1}{16}$ in.) on impervious areas. Figure 3.3 relates surface storage to the slope for various agricultural land uses.

As stated before, the surface storage volume must be exhausted before surface runoff can begin. Therefore, it represents an *initial abstraction* from the gross rainfall input. On the other hand, surface storage is not uniform, even on small watersheds. Hence, for modeling, a range, which is then subtracted from the rainfall as shown on Figure 3.4, may be more appropriate.

Depression storage can be increased by engineering and agronomic

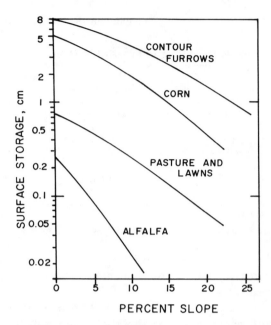

FIGURE 3.3. Relationship of the surface storage parameter to the slope of the land. (After Hiemstra, 1968.)

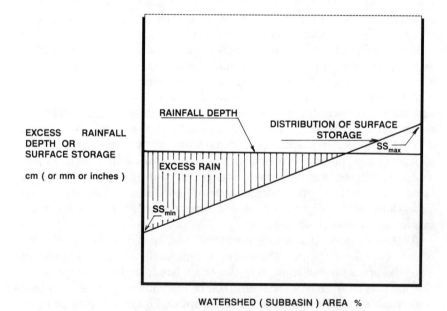

FIGURE 3.4. Initial subtraction of rainfall in a watershed hydrological concept, considering a statistical distribution of surface storage throughout the watershed.

practices, such as plowing, raking of the surface, mulching, and seeding and planting vegetation. Depression storage on agricultural fields is at maximum during planting, but decreases afterwards.

Soil Permeability and Infiltration

Permeability

Infiltration and permeability are not synonymous. Permeability is defined by Darcy's law and denotes the rate of water movement through the soil column under saturated conditions (all voids are filled by water and flow is primarily due to gravitational forces). Infiltration, on the other hand, is the rate at which water percolates from surface storage into the soil zone, and it is governed by the forces of gravity and capillary suction. The permeability of soils depends on such characteristics as texture, compactness, and organic and chemical composition.

As to their permeability and surface runoff potential the soils in the United States have been classified by the Soil Conservation Service (SCS) into four hydrologic groups:

1. Group A are soils with low total surface runoff potential due to high infiltration rates, even when thoroughly wetted. These soils consist chiefly of deep, well to excessively drained sands and gravel.
2. Group B are soils of low to moderate surface runoff potential that have moderate infiltration rates and have a moderately fine to moderately coarse texture.
3. Group C soils have high to moderate surface runoff potential and slow infiltration rates, and consist chiefly of soils with a layer that impedes downward movement of water, or soils with a moderately fine to fine texture.
4. Group D soils have high surface runoff potential and very slow infiltration rates, and consist chiefly of clay soils with a high swelling potential, soils with a permanently high water table, soils with a clay pan or clay layer at or near the surface, and shallow soils over nearly impervious material.

Soil classification and approximate permeabilities can be obtained from the SCS soil maps that are available for most counties of the conterminous United States.

Particle size distribution (texture), arrangement of soil particles, organic matter content, clay mineral content, exchangeable sodium content, and total concentrations of salts are the most important factors affecting permeability (Horn, 1971; Chow, Maidment, and Mays, 1988;

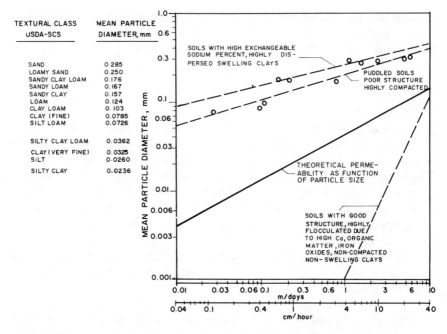

FIGURE 3.5. Relationship of permeability to soil texture. The circled points represent permeability of septic seepage fields after few years of operation. (After Horn, 1971, by permission of ASCE.)

TABLE 3.2 Permeability Classes According to the Soil Conservation Service, U.S. Department of Agriculture

Permeability Class	in./hr	cm/hr	m/day
A Very rapid	>10	>25	>6.2
+B Rapid	3.00–10.00	12.5–25.0	3.1–6.2
B Moderately rapid	2.5–5.0	6.3–12.5	1.5–3.1
+C Moderate	0.8–2.5	2.0–6.3	0.5–1.5
C Moderately slow	0.2–0.8	0.5–2.0	0.12–0.5
+D Slow	0.05–0.2	0.12–0.5	0.03–0.12
D Very slow	<0.05	<0.12	<0.03

Viessman, Lewis, and Knapp, 1989). In addition, permeability rates can be affected by soil compaction, cultivation, vegetation, and land cover.

Most guides developed as aids to estimating permeability rates are based on the relationship of permeability to the soil texture (Fig. 3.5).

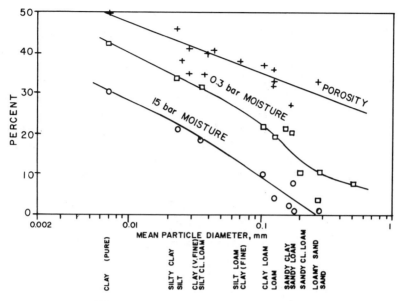

FIGURE 3.6. Moisture characteristics of soils.

Table 3.2 shows the permeability ranges for the hydrologic soil groups recognized by the U.S. Soil Conservation Service.

Soil Water Storage

Storage of soil moisture can be divided into two moisture classes: that held between saturation and 0.3-bar tension, and that held between 0.3-bar tension and 15-bar tension, respectively. The former moisture content is also called *field capacity*; moisture between the field capacity characteristics and full saturation can be drained by gravity. The 15-bar-tension moisture is called *wilting point* and represents the minimum soil moisture content that can be used by plants. Moisture content below 15-bar tension is not available to most crops and plants, and can be reduced only by evaporation (not by transpiration). Gravitational water (G) is then determined by subtracting 0.3-bar-moisture volume percentage from the total porosity (in percent). The plant available soil water capacity (AWC) is the difference between moisture content at 0.3-bar and 15-bar tensions, respectively. As long as the soil water content is between the field capacity and wilting point, transpiration is not affected. Prolonged saturation of soils may have an adverse effect on transpiration. Figure 3.6 shows the moisture characteristics related to the soil texture.

Infiltration

Infiltration is a function of the permeability of soils and subsoils, soil moisture content, vegetation cover, temperature, and possibly other parameters. During infiltration, water enters from surface storage into soils via the combined effects of gravity and capillary forces. The capillary forces are inversely proportional to the diameter of pores. As the process continues, the pore space becomes filled and the capillary tension decreases. Under saturated conditions, flow is mostly due to gravity.

The distribution of soil moisture within the soil profile during the downward movement of water is shown in Figure 3.7. Apparently, the downward movement of water is related to the advancement of the

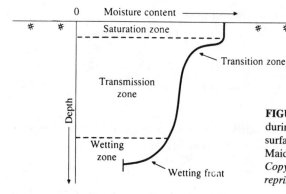

FIGURE 3.7. Soil moisture zones during infiltration from a ponded surface. (Concepts from Chow, Maidment, and Mays, 1988. Copyright © 1988 by McGraw-Hill; reprinted with permission.)

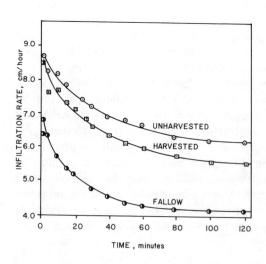

FIGURE 3.8. Characteristic relationship of infiltration rate vs. time of an agricultural field. (Data from Larsen, Axley, and Miller, 1972.)

wetting front, which is exhibited by a sharp and almost discontinuous difference between the advancing moisture driven by infiltration and the moisture below. As water percolates to greater depths, the resistance increases due to the increased length of the channels, decreased pore size from the swelling of clay particles, or the pressure on an impermeable barrier such as rock or clay. Consequently, the infiltration rate decreases as the time from the commencement of the storm increases, as shown on Figure 3.8. Depending on the depth of the soil column and the water supply from rainfall the wetting front may penetrate from a few centimeters to more than one meter into the soil (Hillel, 1980).

The *infiltration rate* f is expressed in cm/hr (in./hr). Potential infiltration is the maximum infiltration rate that presumes an excess supply of water at the surface, generally exhibited by ponding. Almost all infiltration formulas presented in the literature are for potential infiltration with ponding. Cumulative infiltration, F, is then obtained by integrating the infiltration rate over time, or $F(t) = \int_0^t f \, dt$.

Horton's Formula

Horton (1939a) more or less intuitively suggested an infiltration formula for exponentially decaying infiltration:

$$f = f_c + (f_0 - f_c)e^{-Kt} \tag{3.5}$$

where
f = rate of infiltration (cm/hr)
f_c = the infiltration rate assumed to be similar to the saturation permeability (cm/hr)
f_0 = the initial rate of infiltration (cm/hr)
K = a constant derived from soil and surface characteristics (hr^{-1})
t = time in hours from the beginning of infiltration

Although this equation at first appears to be a completely empirical model, it was pointed out that it does reflect the laws and basic equations of soil physics (Chow, Maidment, and Mays, 1988). The model assumes that the constant K is independent of the moisture content of the soil.

Holtan's Formula

Holtan (1961) proposed a formula that would relate the infiltration rate to the exhaustion of the available soil moisture storage. The formula was presented as follows:

$$f = a(S - F)^n + f_c = aF_p^n + f_c \tag{3.6}$$

TABLE 3.3 Estimates of the Vegetation Factor a in Holtan's Infiltration Equation for f in Centimeters per Hour and F_p in Centimeters

	Area Rating[a]	
	Poor Conditions	Good Conditions
Fallow[b]	0.07	0.2
Row crops	0.07	0.14
Small grains	0.14	0.20
Hay (legumes)	0.14	0.28
Hay (sod)	0.28	0.40
Pastures (bunchgrass)	0.14	0.28
Temporary pastures (sod)	0.28	0.40
Permanent pastures (sod)	0.55	0.68
Woods and forest	0.55	0.68

Source: From U.S. Department of Agriculture, Agricultural Research Service (1975).

[a] Adjustments needed for "weeds" and "grazing."
[b] For fallow land only, "poor conditions" means "after row crop," and "good conditions" means "after sod."

where
S = the volume of soil water storage above the control horizon (cm)
F = cumulative infiltration (cm)
F_p = a measure of the soil moisture remaining in the soil column at any time (cm)
a, n = coefficients

The coefficients a and n were empirically determined. While the value of n was nearly constant and equaled $n \approx 1.4$, the value of the multiplier a was related to the crop cover as shown in Table 3.3.

Equation (3.6) was modified later and included in the USDAHL watershed model in the form (Holtan and Lopez, 1973):

$$f = \text{GI } aF_p^{1.4} + f_c \quad (3.6a)$$

where a is the vegetation parameter defined in Table 3.3 and GI is the index in a fraction of maturity. Information on the estimated magnitudes of the growth index for various crops and growing seasons is available from the U.S. Department of Agriculture publication by Holtan and Lopez (1973).

The depth of the control horizon is supposed to coincide with the topsoil zone between the soil surface and the depth of cultivation, or subsoils for uncultivated soils. The latter corresponds to the thickness of the topsoil horizon A (see Chap. 6 for definition of soil horizons). Although Holtan's model is somewhat more complex than Horton's equation, it appears less physically based, since it relates the infiltration rate to the total moisture content in an arbitrarily chosen control layer and to the advancement of the wetting front in the unsaturated soil zone.

Example 3.1: Infiltration Rate by Holtan's Model

Estimate the infiltration rate curve for ponded soil with a saturation permeability of 2 cm/hr and soil moisture characteristics as follows:

porosity = 45%
0.3 bar moisture (field capacity) = 30%
15 bar moisture (wilting point) = 21%

The depth of the control horizon is assumed to be 50 cm and the antecedent moisture is equal to the field capacity (soil is drained). Then the inital available soil moisture storage capacity becomes

$$F_p(0) = (\text{porosity} - 0.13 \text{ bar moisture}) * 50 \text{ cm} = (0.45 - 0.3)50 = 7.5 \text{ cm}$$

Solution If the land surface is fallow, then the approximate magnitude of the vegetation parameter will be between 0.07 and 0.14. Select $a = 0.1$ and $GI = 1.0$. Then the initial infiltration rate at $t = 0$ from Equation (3.6a) becomes

$$f(0) = 0.1(7.5)^{1.4} + 2 = 3.68 \text{ cm/hr}$$

When $t \to \infty$ the infiltration rate would approach 2 cm/hr. However, due to the exponent n equaling 1.4 an exact solution of Equation (3.6a) is not possible. A simple numerical solution will be shown here. Since the infiltration rate is a function of the available water storage, a simultaneous solution of the storage equation must accompany the infiltration equation. Hence

$$\frac{dF_p}{dt} = f_r - f$$

where f_r = storage recovery rate.
If the storage recovery rate $f_r = f_c$ (which occurs when the permeability

of subsoils is more than or equal to the permeability of the top layer, and the moisture content is above the field capacity), then

$$\frac{dF_p}{dt} = aF_p^n$$

If the soil moisture content is below the field capacity value (0.3-bar moisture), then the recovery rate equals the evapotranspiration rate.

The preceding differential equation can be solved by simple numerical techniques such as Runge–Kuta, Euler, or Heund's methods. Heund's method would yield an equation

$$F_{pt+\Delta t} = F_{pt} \frac{\Delta t}{2}(a(F_{pt})^n + a(F_{pt} - a(F_{pt})^n \Delta t)^n)$$

and the solution for $\Delta t = 1$ hr is given in Table 3.4 and plotted on Figure 3.9.

Philip's Equation

Philip's infiltration model (Philip, 1957, 1969, 1983) is based on soil physics. The model and its derivation is very complex; however, its simplified final version has been widely accepted and incorporated into several common watershed hydrological models. The simplified Philip's equation is in the form

$$f(t) = \tfrac{1}{2}St^{-1/2} + K \tag{3.7}$$

TABLE 3.4 Solution of Holtan's Infiltration Equation

t (hr)	F_p (cm)	f (cm/hr)
0	7.5	3.68
1.0	5.94	3.11
2.0	4.89	2.92
3.0	3.84	2.66
4.0	3.26	2.52
5.0	2.79	2.42
6.0	2.41	2.34
8.0	1.89	2.24
10.0	1.49	2.17
15.0	0.93	2.09
20.0	0.62	2.05

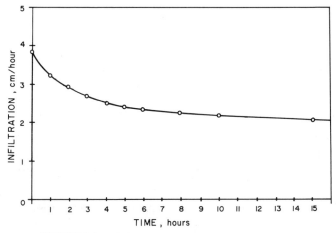

FIGURE 3.9. Infiltration rate curve for Example 3.1.

where
S = sorptivity of the soil
K = conductivity of the wetting front

Sorptivity is computed from soil moisture distribution and is generally difficult to define. As $t \to \infty$, f approaches K, which is loosely related to the saturation permeability parameter (K = 0.3 to 0.5 times saturation permeability). Cumulative infiltration is obtained by integrating Equation (3.7), which yields

$$F(t) = St^{1/2} + Kt \qquad (3.8)$$

Since the sorptivity parameter, S, is a function of the soil suction potential, which is in turn related to the dryness of the soil, S will vary with the soil moisture content. Methods of measurement of sorptivity and its relation to the moisture content were presented by Chong and Green (1983); the schematics are shown on Figure 3.10. It was also demonstrated that the sorptivity and conductivity parameters are not uniform, even in small watersheds (Brutsaert, 1976; Sharma, Barron, and Boer, 1983). Bras (1990) included a good discussion and semiempirical formulas for determining the parameters S and K in the Philip's equation. The spatial distribution of infiltration rates, even in small watersheds, are of a statistical nature, probably ranging from close to zero to rates that greatly exceed the mean rate. The statistical distribution, such as is shown

118   Hydrologic Considerations

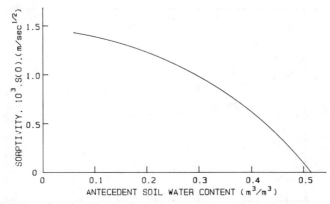

FIGURE 3.10.   Relationship of soil sorptivity parameter, S, to soil moisture. (Based on data from Chong and Green, 1983.)

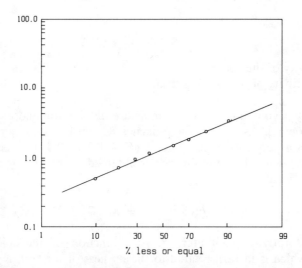

FIGURE 3.11.   Typical statistical distribution of infiltration rates. The relation for a watershed is typically log-normal.

in Figure 3.11 (log-normal probability distribution), should be considered in the hydrologic mathematical models.

Green–Ampt Equation

The Green and Ampt (1911) infiltration model assumes that water is moving from a ponded surface downward into the soil as a piston of saturated moisture. The concept follows the physics of soil water movement through the unsaturated zone shown on Figure 3.7; however, the model assumes saturated soil moisture at the wetting front. For a small ponding depth the Green and Amt equation becomes (Bras, 1990)

$$f = K_s + \frac{K_s S \Psi}{F} \qquad (3.9)$$

where
K_s = hydraulic conductivity (which is less than the saturated permeability defined earlier)
Ψ = suction sorptivity
F = total infiltrated water that equals (porosity − initial moisture) ∗ depth of the wetting front

The parameters K_s and Ψ usually have to be determined experimentally or by calibrating of the model. Rawls and Brakensiek (1983) provided procedures, graphs, and tables for determining these parameters from soil data.

Infiltration rates can be partially controlled by engineering and agricultural practices, such as tillage, raking of the surface, enrichment of soils and root systems of vegetation by organic residues, and chemical treatment of soils. Compaction by heavy machinery and by cattle will reduce the permeability and, hence, infiltration.

Infiltration into Frozen Soils

The physics of water movement into unsaturated soils at temperatures below the freezing point is not simple. The presence of ice in soils affects the soil water movement in two ways. First, permeability is reduced due to the reduction in pore size caused by ice crystals, which also cause a reduction in capillary suction. However, studies in Sweden (Lundin, 1989) pointed out several factors that must be considered. (1) Repeated freezing and thawing of clayey soils actually increases infiltration by creating more pore space due to the expansion of frozen water. (2) Water freezing in soils is not the same process as it is on the surface. Below freezing point temperatures, the interactions between the soil matrix and water result in free unfrozen water surrounding the soil particles. Romanov, Pavlova, and Kolyushnyy (1974) and others (Lundin, 1989) pointed out that unfrozen water content of about 10% to 16% may exist in soils at temperatures up to −20°C. The salt content of soil water further depresses the freezing point. The result is a gradual freezing of soil water, starting at temperatures that are below 0°C. Thus, the degree of freezing point depression depends on the soil characteristics and salinity of soil water. Consequently, in many cases infiltration is reduced by the creation of an ice crust on the surface caused by ponding rather than by a reduction in infiltration rates. If ponding and frozen surface crust do not occur, the speed of infiltration into fine-textured dry soils during subzero temperatures is not drastically reduced when compared to that at above zero temperatures.

Kane (1980) and Romanov, Pavlova, and Kolyushnyy (1974) showed that the amount of infiltration of snowmelt for shallow snowpack (<1-m snow depth) depends on the melt rate, soil type, and soil moisture status at the time of freeze-up. Dry soils exhibit essentially the same or only slightly reduced infiltration rates (after accounting for increased viscosity when compared to those into unfrozen soils). Kane and Stein (1983) then found that the infiltration rate into a dry silt loam frozen soil at 0°C was about 50% of that at 15°C. The 50% decrease reflects the increased viscosity. However, the infiltration rate was reduced by two orders of magnitude when the soil was saturated with water. The authors in another paper (Kane and Stein, 1983b) then concluded that the infiltration rate into frozen soils is inversely proportional to the total moisture content and that the infiltration rate is controlled by the ice content in the upper few centimeters of the soil system.

Burt and Williams (1976) measured hydraulic conductivites of frozen soils. They also noted that frozen soils may contain considerable amounts of unfrozen water. The hydraulic conductivity of frozen soils is about two or more orders of magnitude smaller than that for unfrozen soils, but still enough to produce appreciable infiltration. The measured hydraulic conductivities for several soils ranged from 10^{-8} cm/sec for unlensed silt soil to 10^{-10} cm/sec for unlensed sand. The presence of the lenses reduces the permeability of frozen soils.

Romanov, Pavlova, and Kolyushnyy (1974) noted that an impermeable frozen water layer forms in frozen clayey and loamy frozen soils, when the moisture content reaches 50% to 60% while in sands, the impermeable layer is formed when the moisture reaches 87% to 97% of the total pore volume.

Evaporation and Transpiration

Unlike the previously mentioned losses (surface storage and infiltration), which are directly subtracted from rainfall to produce net (excess) rain and surface runoff, the direct effect of evapotranspiration on the magnitude of surface runoff is not great. A rainfall event usually implies high humidity, which depresses the evapotranspiration rates. On the other hand, evapotranspiration determines the antecedent moisture conditions of soils and surface storage which, in turn, determine the magnitude of these losses. Hence the knowledge of evaporation and transpiration is important in estimating the rates of recoveries of surface storage and soil moisture storage capacity.

By definition, evapotranspiration represents water loss into the atmosphere by evaporation from both open water surface and soils, while transpiration refers to water drawn from the soil zone by the root systems of plants and vegetation and released to the atmosphere as a part of the life cycle of plants.

Evaporation

Potential evaporation is either measured or computed. Pan evaporation (measured) data are available from the U.S. Weather Service of the National Oceanic and Atmospheric Administration (NOAA). It must be realized that pan evaporation data may differ from the actual evaporation from larger water bodies and soil surfaces. Typically, pan evaporation measurements yield evaporation rates that are 20% to 40% higher than actual evaporation from land or water surfaces (Linsley, Kohler, and Paulhus, 1982).

The models used for estimating potential evaporation are either in the category of energy balance or aerodynamics. The energy balance estimates heat lost from the system by evaporation (about 590 calories are needed to convert 1 gram of liquid water into vapor). This high heat loss is balanced by other heat inputs and changes of temperature within the system. Using the energy balance equation requires measurements of several heat and energy inputs and changes of temperature. Some of these measurements may not be readily available.

The aerodynamic models essentially describe mass transfer across the air–water interface. The rate of vapor transfer by evaporation or condensation is proportional to the difference in vapor content (pressure) in the ambient air above the air–water boundary interface and the saturation vapor pressure in the air immediately at the water's surface. The rate of transfer depends on the degree of turbulence and mixing in the thin boundary layer just above the water's surface. The mixing rate can be determined empirically by relating it to the wind velocity above the surface and other parameters. Over 100 empirical and semiempirical formulas have been published in the literature (Helfrich et al., 1982; Bras, 1990).

A majority of the evaporation formulas are of the type (Bras, 1990):

$$E_v = \frac{Q_e}{L\rho} = \frac{(a + bU)(e_s - e_a)}{L\rho} \tag{3.10}$$

TABLE 3.5 Relation of Saturation Vapor Pressure to Temperature

Temperature (°C)	Saturation Vapor Pressure (e_s, millibars[a])
−10	2.86
0	6.11
5	8.72
10	12.27
15	17.00
20	23.37
25	31.70
30	42.43

Source: From Linsley, Kohler, and Paulhus (1982). (*Copyright © 1982 by McGraw-Hill; reprinted by permission.*)

[a] One millibar = $100 \, \text{N/m}^2$.

TABLE 3.6 Coefficients for Evaporation Rate Formula (Eq. (3.10))

Author or Measurement	a	b
Lake Hefner (Marciano and Harbeck, 1954) (water surface)	0	8.1
Harbeck (1962) for several lakes (water surface)[a]	0	$10.44 \times A^{-0.05}$
Penman (quoted in Priestley, 1959) (land surface)	0	5.87
Zaikov (from Braslavskii and Vikulina, 1963) (water surfaces)	6.45	4.64

[a] A = surface area of the lake in hectares.

where
E_v = evaporation rate (cm/day)
Q_e = heat loss by evaporation or gain by condensation (cal/cm²-day)
L = $597.3 - 0.57\,T$ = latent heat of vaporization (cal/gram of water)
ρ = specific density of water ($= 1 \, \text{g/cm}^3$)
U = wind speed in m/sec measured 2 m above the surface
T = temperature (°C)
e_s = saturation vapor pressure in milibars for the water surface temperature (Table 3.5)
e_a = ambient vapor pressure in milibars measured 2 m above the surface
a,b = coefficients given in Table 3.6

Note that 1 calorie is equivalent to 4.186 joules.

The saturation vapor pressure, e_s, which is a measure of the maximal content of vapor in the air, is related to the temperature as given in Table 3.5.

Example 3.2: Calculation of Evaporation

Compare evaporation rates estimated using the formulas of Zaikov (obtained from large reservoirs in the former USSR) and Harbeck for U.S. lakes. The following daily average values are given:

wind velocity measured at 2 meters above the surface $U = 2.5$ m/sec
relative humidity $r = 50\%$
ambient air temperature $T_a = 25°C$
water temperature $T_s = 20°C$
surface area of the lake $A = 500$ ha

Relative humidity is a ratio of the actual humidity of the air to the saturation value. At the ambient air temperature of 25°C the saturation humidity of the air is (Table 3.5) $e_{sa} = 31.7$ mbar and the ambient humidity is $e_a = r \times e_{sa} = 0.5 \times 31.7 = 15.85$ mbar. The saturation vapor pressure at the water's suface with a temperature of 20°C is 23.37 mbar.

Solution The latent heat of vaporization at the water's surface with a temperature of 20°C is

$$L = 597.3 - 0.57 * 20 = 585.9 \text{ cal/g}$$

Zaikov

$$E_v = \frac{(6.45 + 4.64 \times 2.5)(23.37 - 15.85)}{585.9 \times 1} = 0.23 \text{ cm/day}$$

Harbeck

$$E_v = \frac{(0 + 10.44 \times 500^{-0.05} \times 2.5)(23.37 - 15.85)}{585.9 \times 1} = 0.25 \text{ cm/day}$$

Transpiration and Evapotranspiration

Several methods have been developed for estimating the evapotranspiration requirements of crops and forested areas. As defined previously, evapotranspiration is a composite of transpiration by plants and crops and evaporation from the soil.

The actual rate of transpiration is a function of the type and growth stage of the crops, the soil moisture (below or above wilting point), and climatic conditions. Similar factors affect both transpiration and soil evaporation; therefore, evaporation data are sometimes used as a surrogate for the potential evapotranspiration rate. Evapotranspiration can be related to the potential evaporation by a parameter or a function such as that in the *Evaporation Index Method* described by McDaniel (1960):

$$ET = E_p \times KU \tag{3.11}$$

where

ET = crop evapotranspiration requirement on a monthly or shorter period basis

E_p = climatic index, which is identical to the evaporation potential from a shallow hypothetical lake situated at the locality under consideration

KU = crop-use coefficient, which reflects the growth and stage of crops. Average values for crop-use coefficients are presented in Table 3.7

A similar approach has been suggested by the USDA–Agricultural Research Service (Holtan, 1961), which proposed the following equation for the potential evapotranspiration:

$$ET = KU \times (E_p/k) \times \beta \tag{3.12}$$

where

ET = evapotranspiration potential (cm/hour or in./hr)

KU = growth index or crop-use coefficient defined previously and given in Table 3.7

k = ratio of ET to pan evaporation, usually 1.0 to 1.2 for short grasses, 1.2 to 1.6 for crops up to shoulder height, and 1.6 to 2.0 for forests. If evaporation is calculated, $k = 1.0$

E_p = calculated or measured (pan) evaporation potential (cm/hr or in./hr)

β = a moisture stress coefficient that expresses the reduction of evapotranspiration to plain evaporation at or below wilting point soil moisture

β = 1.0 if $\theta \geq \theta_{15\text{bar}}$ and

$\beta = \left(\dfrac{\theta}{\theta_{15\text{bar}}}\right)^{AWC/G}$ if $\theta < \theta_{15\text{bar}}$

TABLE 3.7 Crop-Use Coefficients for Use in Evaporation Index Method

Perennial Crops (Northern Hemisphere)

Average KU values by months

Crop	Jan.	Feb.	Mar.	Apr.	May	June	July	Aug.	Sept.	Oct.	Nov.	Dec.
Alfalfa	0.83	0.90	0.96	1.02	1.08	1.14	1.20	1.25	1.22	1.18	1.12	0.86
Grass pasture	1.16	1.23	1.19	1.09	0.95	0.83	0.79	0.80	0.91	1.91	0.83	0.69
Grapes	—	—	0.15	0.50	0.80	0.70	0.45	—	—	—	—	—
Citrus orchards	0.58	0.53	0.65	0.74	0.73	0.70	0.81	0.96	1.08	1.03	0.82	0.65
Deciduous orchards	—	—	—	0.60	0.80	0.90	0.90	0.80	1.50	0.20	0.20	—
Sugarcane	0.65	0.50	0.80	1.17	1.21	1.22	1.23	1.24	1.26	1.27	1.28	0.80

Annual Crops

KU values at listed percent of growing season

Crop	0	10	20	30	40	50	60	70	80	90	100
Field corn	0.45	0.51	0.58	0.66	0.75	0.85	0.96	1.08	1.20	1.08	0.70
Grain sorghum	0.30	0.40	0.65	0.90	1.10	1.20	1.10	0.95	0.80	0.65	0.50
Winter wheat[a]	1.08	1.19	1.29	1.35	1.40	1.38	1.36	1.23	1.10	0.75	0.40
Cotton	0.40	0.45	0.56	0.76	1.00	1.14	1.19	1.11	1.83	0.58	0.40
Sugar beets	0.30	0.35	0.41	0.56	0.73	0.90	1.08	1.26	1.44	1.30	1.10
Cantaloupes	0.30	0.30	0.32	0.35	0.46	0.70	1.05	1.22	1.13	0.82	0.44
Potatoes (Irish)	0.30	0.40	0.62	0.87	1.06	1.24	1.40	1.50	1.50	1.40	1.26
Papago peas	0.30	0.40	0.66	0.89	1.04	1.16	1.26	1.25	0.63	0.28	0.16
Beans	0.30	0.35	0.58	1.05	1.07	0.94	0.80	0.66	0.53	0.43	0.36
Rice[b]	1.00	1.06	1.13	1.24	1.38	1.55	1.58	1.57	1.47	1.27	1.00

Source: From *Handbook of Applied Hydraulics*, by Davis and Sorensen (1969). Copyright © 1969 by McGraw-Hill. Used with the permission of the McGraw-Hill Book Company.

[a] Data given only for springtime season of 70 days prior to harvest (after last frost).

[b] Evapotranspiration only.

θ = soil moisture
AWC = $\theta_{0.3bar} - \theta_{15bar}$ = plant available soil moisture capacity
G = $S - \theta_{0.3bar}$ = gravitational soil water moisture

Other methods for estimating evapotranspiration include the Blaney–Cridle, and the modified Penman and Lowry–Johnson methods. The reader is referred to standard hydrology texts (Bras, 1990; Chow, 1964; Davis and Sorensen, 1969; Gray, 1973; Chow, Maidment, and Mays, 1988) for further reference.

Snowpack Formation and Snowmelt

Knowledge of snow hydrology is important in diffuse-pollution studies. Snow, which in the northern latitudes of North America and Europe may stay on the ground for several months, is a surface trap of pollutants deposited from the atmosphere. In addition, by the process of repeated crystallization and the melting of snow in the snowpack, dissolved pollutants are rejected by the crystals and become available for fast pickup by melt water, resulting in a strong "first flush" effect. Snow precipitation has much less energy than rain droplets. Consequently, the erosion of soils during snowfall or snowmelt is minimal.

The main elements determining the amount of snowmelt are meteorological factors, such as temperature, solar radiation, and wind velocity. An accurate determination of snowpack accumulation and subsequent snowmelt is not simple. For example, in urban areas in addition to meteorological inputs, the heat balance of the snowpack and, hence, the quantity of snowmelt, are also affected by heat losses from buildings and traffic, by deicing operations (street salting), coloration of the snowpack, and by snow removal and dumping.

In the most simple concept, the volume of snowmelt from a snowpack is related to the ambient temperature deviation from a reference equilibrium temperature such as in the following degree-day formula

$$\Delta P_s = -SM = CD \times (T_a - T_r) \quad \text{for} \quad T_a > T_r$$
$$\Delta P_s = P \quad \text{for} \quad T_a \leq T_r \tag{3.13}$$

where
ΔP_s = change of water equivalent in the snowpack (cm/day)
SM = snowmelt (cm/day)
P = water equivalent in precipitation (cm/day)
T_a = ambient temperature (°C)

T_r = reference equilibrium temperature (°C)
CD = a proportionality coefficient (also called a degree-day coefficient) (cm/°C-day)

In most technical applications, the equilibrium reference temperature, T_r, is assumed to be 0°C. The proportionality (degree-day) coefficient, CD, is an empirical quantity loosely correlated to the meteorological conditions of the season. Empirical values of the degree-day coefficient reported in the literature ranged from 0.1 to 1 cm/°C-day, with values between 0.2 and 0.5 cm/°C-day the most common (Bengtsson, 1982; Westerström, 1984; Ellena and Novotny, 1985). The simple degree-day coefficient is also assumed to vary with the season, as shown on Figure 3.12. Due to its simplified nature, the degree-day model has been subjected to critique and modifications. For example, Bengtsson (1984) added a solar radiation component, which, as documented by Ellena and Novotny (1985), has markedly improved the performance of the model.

Advanced models for snowmelt estimation use some kind of the energy balance method. The most comprehensive study of the process of melting and components of the snowpack energy (heat) balance was undertaken by the U.S. Army Corps of Engineers (1956). Anderson and Crawford (1964) incorporated this concept into the Stanford Watershed Model, which is a predecessor to the HSP-F watershed model distributed by the U.S. Environmental Protection Agency (see Chapter 8 for a description of the model).

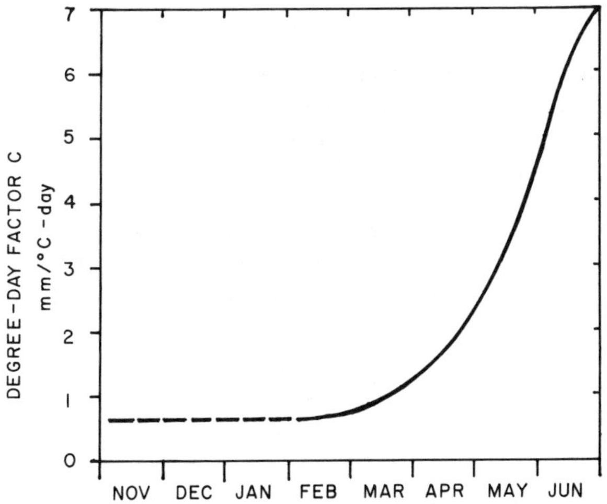

FIGURE 3.12. Degree-day factor for snowmelt computations. (From Gray, 1973.)

The energy balance concept is based on the fundamentals of the physics of melting. The heat balance equation for the snowpack can be written as follows (U.S. Army Corps of Engineers, 1956; Ellena and Novotny, 1985; Novotny, 1988):

$$\Delta Q = Q_{SAB} + Q_{ATM} + Q_E + Q_P + Q_C + Q_G \qquad (3.14)$$

where
ΔQ = net change of the heat content of the snowpack
Q_{SAB} = absorbed (net) direct solar radiation
Q_{ATM} = net atmospheric (long wave) radiation
Q_E = latent heat transfer due to the condensation–evaporation–sublimation process, which in the absence of direct measurements, can be estimated by Equation (3.10)
Q_C = sensible heat transfer between air and snow
Q_P = heat content of precipitation
Q_G = heat gain or loss due to other sources or sinks (in urban areas it also includes heat gains from vehicles, as well as from exothermic dissolution of salt for deicing)

The units of the energy (heat) balance equations are calories cm^{-2} hr^{-1} or joules m^{-2}. At the melting point, the heat content of the snowpack, Q_m, is

$$Q_m = T_m c_p \rho_s D \qquad (3.15)$$

where
T_m = the melting point temperature (°C)
c_p = specific heat calories/g or W/kg
ρ_s = specific density (g/cm^3 or kg/m^3)
D = snow depth in centimeters or meters

If, due to increased heat input, the energy (heat) content of the snowpack exceeds the energy needed to melt the snow (Q_m), 1 g of water will melt for each 79.7 calories of the heat excess, or

$$SM = \frac{Q - Q_m}{79.7 Q_t} \qquad (3.16)$$

where
SM = resulting melt in centimeters
Q_t = thermal quality of the snowpack ($\approx$ 1.0)

The melting point temperature is a function of salinity or the molar concentration of salts in the water (snow) solution. The melting point is lowered by approximately 0.7°C for each 10 g of salt added to 1 liter of water equivalent of snow. For salt-free snow $T_m = 0°C$, therefore, $Q_m = 0$. Furthermore, salt dissolution is an exothermic process that is heat generated by the dissolution of salt.

The magnitude and models of each component of the snowpack energy balance have been described in detail by Novotny (1985), Ellena and Novotny (1985), U.S. Army Corps of Engineers (1956), and others. As most of the energy components vary throughout the day, the snowmelt rates are also variable, so it may be inappropriate to use average daily values.

Of note is the approximate magnitude of snowmelt rates. Assuming a clear, sunny day in late February, the average incoming solar radiation at latitude 46° is about 20 cal/cm²-hr. During clear, dry conditions with the ambient air at 10°C, the net radiation heat loss and condensation heat gain are about equal. The albedo (reflectivity of direct solar radiation) of urban snow varies between 20% for dirty saturated snow to more than 85% for freshly fallen snow. From Equation (3.16) it follows that the maximum runoff rate from melting snow is only about 0.15 cm/hr. This is much lower than runoff rates resulting from typical design storms, and lower than most potential rates of infiltration into dry or moderately wet frozen soils. Hence, surface runoff generation from snowpack over a permeable soil is generally possible only if a frozen layer of solid ice is formed on the surface.

RAINFALL EXCESS DETERMINATION: SURFACE RUNOFF

Design Storm

In most designs of hydrologic systems the designer is faced with the task of selecting the so-called *design storm*. This type of input information is especially important for flood-mitigation projects. It will be pointed out throughout this book, however, that the abatement of diffuse pollution should be focused on precipitation events that are frequent, typically medium magnitude storms with rainfall depths ranging from 1.2 to 3.5 cm (0.5 to 1.5 in.), which would occur several times each year rather than rare large storms.

The isopluvial depths (isohyets) of a typical medium-size one-hour storm with a recurrence time of once per year are shown on Figure 3.13. Such maps for design storms in the United States are prepared by the

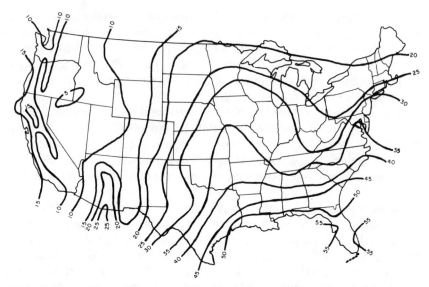

FIGURE 3.13. Isopluvial map of the United States for once-per-year, 1-hr-long rainfall in millimeters, published by the U.S. Weather Service (to convert from mm/hr to l/sec-ha, multiply by 2.78, 1 in. = 25.4 mm).

National Weather Service. Figure 3.14a is a normalized duration-intensity curve. It has been found (first by Karl Imhoff in the 1920s, see Novotny et al., 1989) that the relationship between the average intensity within a storm and the storm duration are similar for most of European and U.S. locations, with the exception of Pacific coastal areas and Hawaii. If the duration-intensity curves are normalized by the one-hour, once-per-year storm depths, a unified duration-intensity curve is obtained. Figure 3.14b is then used to modify the design storm for different recurrence intervals. The use of Figures 3.13 and 3.14 is illustrated on the following example.

Example 3.3: Selection of the Design Storm

Determine the intensity of a twice-a-year design that has a duration of 30 minutes. The watershed is located in the Chicago, Illinois, metropolitan area.

Solution Read the intensity of the standard 1-year, 1-hr-duration storm from the isopluvial map in Figure 3.13. Therein $r_1^1 = 32.5$ mm/hr. To convert this 1-year, 1-hr storm to the desired $\frac{1}{2}$-year, 30-min design storm, first read the magnitude of the duration multiplier from the upper portion of Figure 3.14. For Chicago $\phi_1 = 1.4$. Then from the lower portion of

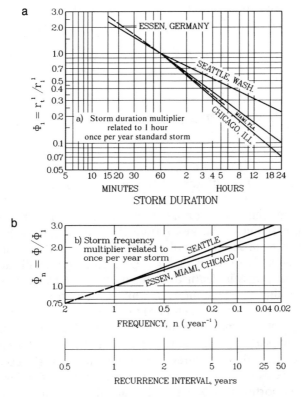

FIGURE 3.14. Normalized storm-duration–frequency–intensity curves for several geographical locations. (From Novotny et al., 1989. Copyright © 1989; reprinted by permission of John Wiley & Sons, Inc.)

Figure 3.14 read the frequency multiplier, ϕ_n, which is $\phi^n = 0.75$. Hence the average intensity of a design storm of 30-minutes duration that occurs approximately twice a year, is

$$I = \phi_1 \phi_n r_1^1 = 1.4 \times 0.75 \times 32.5 = 34.12 \, \text{mm/hr}$$

Rainfall Excess from Pervious Areas

Rainfall Excess by Subtracting Losses from Precipitation

From the foregoing discussion and referring to Equations (3.1) to (3.3), it is now clear that surface runoff, which is often the most polluted component of the total runoff, will be generated from a surface only when precipitation or snowmelt is greater than the losses, that is,

$$R_s = P - S_i - S_d - f\Delta t$$

if

$$P > S_i + S_d + f\Delta t$$

and

$$R_s = 0 \quad \text{for} \quad P \leq S_i + S_d + f\Delta t \tag{3.17}$$

The term *rainfall excess* or *net rain* is used to denote the simple numerical subtraction of the losses from the precipitation volume. This differentiates it from surface runoff, which refers to that part of the flow in the receiving body of water that was generated by rainfall excess. The unit for rainfall excess is depth of water on the surface from the excess rain generated during a time interval, while the unit for surface runoff is volume/time. A time lag between maximum rain excess and the peak of the surface runoff is typical for all but very small drainage areas. This time lag (called *peak time*) is due to overland and channel flow routing.

There are several procedures for estimating rainfall excess. The definition of excess rainfall excludes the use of the formulas that are based on a proportionality between the rainfall and runoff, such as the well-known rational formula presented in the next section. Two methods presented here are: (1) numerical subtraction of losses from precipitation, and (2) the Soil Conservation Service Runoff Curve Method. Any infiltration formula can be used for estimating infiltration losses; however, Holtan's infiltration equation is used in Example 3.4.

Example 3.4: Numerical Rainfall Excess Determination by Subtracting Losses

Determine the rainfall excess from a storm with the hyetograph given in the top portion of Figure 3.15. A dry period preceding the storm lasted 5 days. The evaporation rate during the dry period averaged 0.3 cm/day. The area is covered by small grains and has an average slope of 2%. For this slope and surface, the combined depression and interception storage can be read from Figure 3.3 as 0.62 cm. The soil and infiltration characteristics are similar to Example 3.1. Gravitational moisture content (field capacity) is $G = 50$ (porosity $-$ 0.3-bar moisture) $= 7.5$ cm.

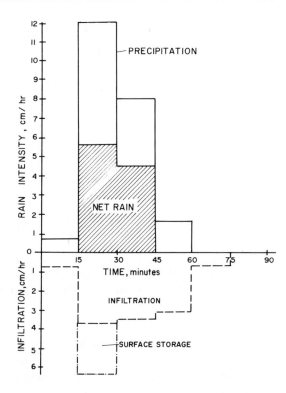

FIGURE 3.15. Rainfall hyetograph, subtractions, and excess rainfall for Example 3.4.

Solution The initial soil moisture and available storage capacity can be computed, assuming that the soil was saturated during the preceding rain and that the crop use factor, KU, is close to one. Since the saturation permeability is 2 cm/hr, the soil moisture reached 0.3-bar moisture shortly after the preceding rain. Hence, the initial available storage capacity becomes

$$F_p(0) = G + ET \times 5 \text{ days} = 7.5 + 0.3 \times 5 = 9.5 \text{ cm}$$

The parameter a for Holtan's infiltration equation can be read from Table 3.3. For small grains the parameter is between 0.07 and 0.14. Select $a = 0.1$. Then for $F_p(0) = 9$ cm, the initial infiltration rate is

$$f(0) = 0.1(9.0)^{1.4} + 2 = 4.17 \text{ cm/hr}$$

The excess rain is computed in 15-min time intervals by the simultaneous solution of the following equation (based on Eq. (3.1))

$$R_s = P - S_{id} - f\Delta t$$
$$\text{if } P > Sd + f\Delta t \text{ then } S_{id} = 0 \quad \Delta F_p = (f_r - f)\Delta t \qquad (3.18)$$

and

$$R_s = 0 \quad \text{if } P \leq S_{id} + f\Delta t$$

with

$$-\Delta S_{id} = P - f\Delta t$$
$$f\Delta t = P + \Delta S_{id} \quad \text{if } (aF_p^{1.4} + f_c) > (P + \Delta S_{id})/\Delta t$$
$$\Delta F_p = (f_r - f)\Delta t$$

and $f_r = 0$, if soil moisture is below 0.3-bar tension (of $F_p > G$).

During the rainy period the recovery of surface and soil water storage by evapotranspiration is minimal. The computation and results are given in Table 3.8 and Figure 3.15.

Rainfall Excess by the Soil Conservation Method

The Soil Conservation Service has developed a method for estimating rainfall excess that does not require computing infiltration and surface storage separately. Both runoff characteristics are included as just one watershed characteristic. The method has evolved from analysis of numerous storms under a variety of soil and cover conditions.

In the SCS method the excess rain volume, Q, depends on the volume of precipitation, P, and the volume of total storage, S, which includes both the initial abstraction and total infiltration, I_a. The relationship between rainfall excess and total rainfall (on a 24-hour basis) is thus (SCS, 1968; McCuen, 1982)

$$Q = \frac{(P - I_a)^2}{(P - I_a) + S} \qquad (3.19)$$

The initial subtraction operation is a function of land use, treatment, and condition; interception; infiltration; depression storage; and antecedent soil moisture. An empirical statistical relationship relates the initial subtraction to the total storage as

$$I_a = 0.2S \qquad (3.20)$$

TABLE 3.8 Rainfall Excess Calculation

Time Interval (min) (1)	Precipitation Intensity during the Time Interval (cm/hr) (2)	Precipitation Volume, P (cm) (3)	Available Depression and Interception Storage, S_{id} (cm) (4)	ΔS_{id}[a] (cm) (5)	Infiltration Volume[b] (cm) (6)	Soil Moisture Storage Capacity, F_p[c] (cm) (7)	Excess Rain Volume, R_s (cm) (8)	Net Rain Intensity, R_s/t (cm/hr) (9)
0–15	3.0	0.75	0.62	0.0	0.75	9.0	0.0	0.0
			0.62			8.25		
15–30	12.0	3.00	0.0	−0.62	0.96	7.40	1.42	5.68
30–45	8.0	2.00	0.0	0.0	0.90	7.00	1.10	4.40
45–60	1.6	0.40	0.47	+0.47	0.87	6.63	0.00	0.00
60–75	0.0	0.0	0.62	+0.15	0.15	6.98	0.00	0.00

[a] $\Delta S_{id} = f\Delta t - P$, min $S_{id} = 0$, max $S_{id} = 0.62$ cm.
[b] Infiltration volume $= f\Delta t$ or $= P + S_{id}$, whichever is less.
[c] $\Delta F_p = (f_r - f)\Delta t$, $f_r = f_c$ if $F_p < G$, $f_r = 0$ if $F_p > G$; rate of exhaustion of the soil moisture storage capacity reduced at $F_p = 7.5$ cm.

Substituting Equation (3.20) into (3.19) yields

$$Q = \frac{(P - 0.2S)^2}{(P + 0.8S)} \qquad (3.21)$$

The cumulative infiltration is also

$$F = (P - I_a) - Q$$

and, after substituting from Equation (3.19),

$$F = \frac{(P + I_a) \times S}{(P + I_a) + S} \qquad (3.22)$$

Note that Equation (3.21) has been reduced to only one unknown, the storage parameter S. This parameter (in millimeters) can be obtained from

$$S = \frac{25{,}400}{CN} - 254 \qquad (3.23)$$

where CN is the runoff curve number that can be obtained from Table 3.9. All parameters in Equations (3.19)–(3.23) are in millimeters; P and Q represent daily precipitation and runoff volumes.

The relation between excess rain (runoff) volume, precipitation, and storage for different runoff curve numbers is plotted on Figure 3.16. The watershed soil moisture conditions for determining the runoff curve numbers have been classified by the SCS as follows:

AMC I: A condition of watershed soils where the soils are dry but not to the wilting point, and when satisfactory plowing or cultivation takes place.

AMC II: The average case for annual floods, that is, an average of the conditions that have preceded the occurrence of the annual flood on numerous watersheds.

AMC III: If heavy rainfall or light rainfall and low temperatures have occurred during the 5 days prior to the given storm and the soil is nearly saturated.

Table 3.9 gives the curve numbers for the average antecedent soil moisture AMC II. The corresponding curve numbers for AMC I and

TABLE 3.9 Runoff Curve Numbers for Hydrologic Soil Cover Complexes (Antecedent Soil Moisture Conditions AMC = II)

Land-Use Description and Cover			Hydrologic Soil Groups			
			A	B	C	D
Residential[a]						
Average Lot Size	Average Imperviousness[b] (%)					
0.05 ha (1/8 acre)	65		77	85	90	92
0.10 ha (1/4 acre)	38		61	75	83	87
0.15 ha (1/3 acre)	30		57	72	81	86
0.20 ha (1/2 acre)	25		54	70	80	85
0.4 ha (1 acre)	20		51	68	79	84
Paved parking lots, driveways, etc.[c]			98	98	98	98
Streets and roads						
Paved, with curbs and storm sewers			98	98	98	98
Gravel			76	85	89	91
Dirt			72	82	87	89
Commercial and business	85 (average)		89	92	94	95
Industrial districts	72		81	88	91	93
Open spaces, lawns, golf courses, cemeteries, etc.						
Good condition, grass cover on 75% or more of the area			39	61	74	80
Fair conditions, grass cover on 50% to 75% of the area			49	69	79	84
		Hydrologic Conditions				
Fallow	Straight row	—	77	86	91	94
Row crops	Straight row	Poor	72	81	88	91
	Straight row	Good	67	78	85	89
	Contoured	Poor	70	79	84	88
	Contoured	Good	65	75	82	86
	Contoured and terraced	Poor	66	74	80	82
	Contoured and terraced	Good	62	71	78	81
Small grain	Straight row	Poor	65	76	84	88
	Straight row	Good	65	75	83	87
	Contoured	Poor	63	74	82	85
	Contoured and terraced	Poor	61	72	79	87
	Contoured and terraced	Good	59	70	78	81
Close-seeded legumes[d] or rotational meadow	Straight row	Poor	66	77	85	89
	Straight row	Good	58	72	81	85
	Contoured	Poor	64	75	83	85
	Contoured	Good	55	69	78	83
	Contoured and terraced	Poor	63	73	80	83
	Contoured and terraced	Good	51	67	76	80

TABLE 3.9 *Continued*

Land-Use Description and Cover			Hydrologic Soil Groups			
			A	B	C	D
Pasture or range		Poor	68	79	86	89
		Fair	49	69	79	84
		Good	39	61	74	80
	Contoured	Poor	47	67	81	88
	Contoured	Fair	25	59	75	83
	Contoured	Good	6	35	70	79
Meadow		Good	30	58	71	78
Woods or forest land		Poor	45	66	77	83
		Fair	36	60	73	79
		Good	25	55	70	77
Farmsteads		—	59	74	82	86

Source: After Soil Conservation Service (1968).

[a] Curve numbers are computed assuming the runoff from the house and driveway is directed toward the street with a minimum of roof water directed to lawns where additional infiltration could occur.
[b] The remaining pervious areas (lawns) are considered to be in good pasture condition for these curve numbers.
[c] In some warmer parts of the country a curve number of 95 may be used.
[d] Close-drilled or broadcast.

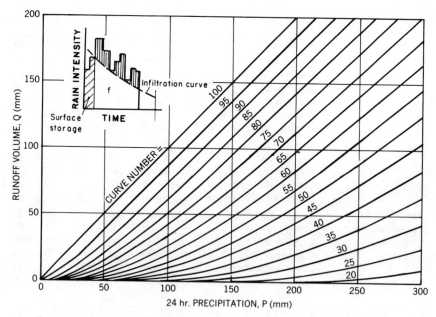

FIGURE 3.16. SCS rainfall–runoff transformation concept and runoff curves.

TABLE 3.10 Runoff Curve Number Correction for AMC I and AMC III

CN for AMC Condition II	Corresponding CN for AMC	
	Condition I	Condition III
100	100	100
95	87	99
90	78	98
85	70	97
80	63	94
75	57	91
70	51	87
65	45	83
60	40	79
55	35	75
50	31	70
45	27	65
40	23	60
35	19	55
30	15	50
25	12	45
20	9	39
15	7	33
10	4	26
5	2	17
0	0	0

Source: After Soil Conservation Service (1968).

AMC III can be read from Table 3.10 if the CN for AMC II are known.

The correction for the curve numbers can also be calculated from empirical equations given by Chow, Maidment, and Mays (1988) as

$$CN(I) = \frac{4.2\,CN(II)}{10 - 0.058\,CN(II)} \quad (3.24a)$$

and

$$CN(III) = \frac{23\,CN(II)}{10 + 0.13\,CN(II)} \quad (3.24b)$$

Table 3.11 provides seasonal rainfall limits for the three antecedent soil moisture conditions.

TABLE 3.11 Determination of Antecedent Soil Moisture Conditions

AMC	Total 5-day Antecedent Rainfall (mm)	
	Dormant Season	Growing Season
I	Less than 12.5	Less than 35
II	12.5–28	35–53
III	Over 28	Over 53

Source: After Soil Conservation Service (1968).

Example 3.5: Estimation of Daily Excess Rain by the SCS Method

Estimate the rainfall excess for the storm using the SCS method. The following information is given:

Total precipitation $P = 61.5$ mm
Hydrologic soil group C
Antecedent 5-day rainfall (dormant season) 35 mm
Surface cover—grass

Solution Using the information given in Table 3.11 the antecedent soil moisture conditions are AMC III. From Table 3.9, which gives the information on runoff curve numbers for AMC II the runoff curve number for grass (meadow) and soil hydrologic group C is CN = 71. The correction for AMC III is read from Table 3.10. By interpolation, the corresponding curve number for AMC III is 83.

The storage parameter is then (Eq. (3.23))

$$S = \frac{2540}{83} - 254 = 52 \text{ mm}$$

The depth of excess rainfall becomes (Eq. (3.21))

$$Q = \frac{(61.5 - 0.2 \times 52)^2}{(61.5 + 0.8 \times 52)} = 25.3 \text{ mm}$$

Excess rain can also be read from Figure 3.16.

TABLE 3.12 Calculation of Excess Rain from a Hyetograph

Time Interval (min) (1)	Rainfall Intensity (mm/hr) (2)	Cumulative Precipitation (volume, mm)		$f(P_{t-\Delta t})$ (mm) (5)	$f(P_t)$ (mm) (6)	Excess Rain, ΔQ (mm) (7)	Cumulative Q (mm)
		$t - \Delta t$ (3)	t (4)				
0–15	30.0	0.0	7.5	0.0	0.0	0.0	0.0
15–30	120.0	7.5	37.5	0.0	9.31	9.31	9.31
30–45	80.0	37.5	57.5	9.31	22.39	13.08	22.39
45–60	16.0	57.5	61.5	22.39	25.33	2.94	25.33

Note: Column (4) = column (3) + column (2) × Δt [hr].

Example 3.6: Excess Rain by the SCS Method from a Hyetograph

The SCS method also enables a detailed but approximate estimation of excess rainfall from a hyetograph, such as the one shown on Figure 3.15. In this example, the excess rainfall in a 15-minute interval will be determined following the relationship expressed by Equation (3.21). From the equation

$$\Delta Q = f(P_t) - f(P_{t-\Delta t})$$

where f is the function expressed by Equation (3.21). The calculation is shown in Table 3.12. In the calculation, $S = 52$ mm (from the previous example) and $Q = 0$ if $P < 0.2S = 0.2 \times 52 = 10.4$ mm.

Rainfall Excess from Impervious Areas

Because of the surface impermeability and small-depression storage, the impervious areas (asphalt and concrete pavements, rooftops, etc.) appear to be 100% active (i.e., they generate surface runoff even during small rains). However, not all of the rainfall excess will appear as surface runoff. Thus the impervious area from which excess rain overflows onto adjacent pervious surfaces and, subsequently, into soils is considered to be not directly connected. Such cases include roof drains and driveways overflowing onto lawns, roadways and other impervious surfaces with poor or no apparent drainage, and parking lots separated by pervious areas. The rainfall excess generated on impervious areas that overflows onto adjacent pervious areas is added to the hydrological balance of the pervious area, that is, the depth of precipitation should be increased by the corresponding amount of overflow from adjacent impervious surfaces.

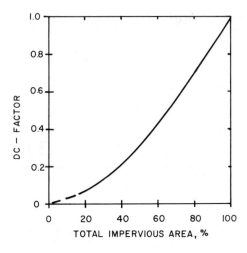

FIGURE 3.17. Relation of the fraction of the impervious surface directly connected to the channel drainage system (DC parameter) to the total imperviousness of the area. (Adapted from Anon., 1972.)

It is apparent that a fraction of the impervious areas directly connected to the drainage system increases with the degree of urbanization reflected in the total imperviousness of the area. Rainfall excess on impervious surfaces in predominantly rural areas will mostly overflow onto adjacent pervious surfaces or pervious road ditches. In densely built-up urban zones, there may not be pervious surfaces available, so all runoff will be connected to a sewer or channel drainage system. An approximate relation of the fraction of the directly connected impervious surfaces (DC) to the total imperviousness of the area is shown on Figure 3.17.

In rural zones, during medium- and low-intensity storms, when most of the surface runoff originates from impervious surfaces, the volume of the surface runoff might be sensitive to the magnitude of the parameter DC. It is therefore recommended that the value of the DC factor be estimated and verified by comparison of the computed surface runoff volumes to measured field observations. Commonly during modeling, the DC factor is a calibration parameter. Areas with storm or combined sewers will have a much higher DC factor than unsewered areas.

Hydrologically Active Areas

Based on the net rain estimation, it is evident that not all areas within the watershed will generate surface runoff and, thus the diffuse pollution associated with it. The areas that will produce surface runoff are called *hydrologically active*, while the rest of the basin contributes only to interflow and base flow (they may also be contaminated by dissolved

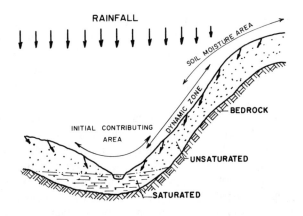

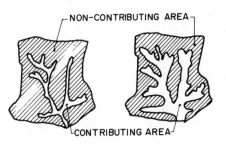

FIGURE 3.18. A hydrologically active area concept for natural watersheds. (After Engman, 1974.)

2-YEARS RECURRENCE INTERVAL   10-YEARS RECURRENCE INTERVAL

pollutants and ions). The areas showing the highest hydrological activity are obviously connected to impervious surfaces (even in this instance the depression storage must be subtracted), followed by clayey soils with low permeability, frozen soils with a high moisture content, soils with a high ground-water table, and highly compacted soils.

Areas with high surface storage, such as woods and flat cropland and, generally, soils with high permeability rates, have the lowest hydrologic activity, and often generate surface runoff only during extreme storms. Remember that the hydrological activity of an area is a stochastic hydrologic phenomenon and that the surface runoff is not generated uniformly over the entire watershed. The extent of the hydrologically active portion of the watershed also changes with the rainfall characteristics, antecedent soil conditions, and surface characteristics. The contributing hydrologically active area will be smaller for storms with a small recurrence interval and will increase with the magnitude of the storms and their recurrence interval. Figure 3.18 shows an example of contributing areas for 2-year- and 10-year- recurrence storms in a watershed.

Identification of the areas that have a tendency to be hydrologically active is a necessary step in the abatement of diffuse pollution. These areas contribute to sediment and all surface runoff pollution. Hydrologically active areas can be determined by field surveys, plane and satellite photogrammetry and imagery, and by hydrological modeling. Topsoil distribution of a radioactive isotope cesium 137, which originated in the atmospheric nuclear weapon tests of the 1950s, will indicate erosive, hence hydrologically active areas (Ritchie, Sparberry, and McHenry, 1974). Chapter 5 contains a more detailed discussion on the methods of identifying erosive lands.

Reducing hydrological activity, for example, by increasing surface storage or permeability, or by draining high ground-water levels, is one of the most effective measures to abate diffuse pollution potential. Similarly, "disconnecting" the connected impervious areas is one of the most effective measures of controlling both flooding and pollution by urban runoff (Livingston and Roesner, 1991). This control can be accomplished by letting the roof drains overflow onto the adjacent pervious areas, incorporating infiltration into the drainage system, and other means that are discussed in Chapter 10.

OVERLAND ROUTING OF THE PRECIPITATION EXCESS

Excess or net rain can be imagined as the depth of water on the surface contributing directly to surface runoff. *Overland flow routing* is a process by which excess rain is transformed into surface runoff flow. In contrast, *channel flow routing* is a process by which a flow or flood wave is modified as it moves downstream through the channel system. It is defined as a procedure whereby the time and magnitude of a flood wave is determined at a point in a stream from the known or assumed data at one or more points upstream (Chow, 1964).

The watershed size and the length of the overland flow, along with the roughness and slope characteristics, volume and intensity of precipitation, and percent of imperviousness, seem to be the most important factors affecting the shape and magnitude of the surface runoff hydrograph. For larger drainage areas, the hydrograph curve is also affected by channel routing. The channel portion of the routing process may not be significant for small watersheds of up to 5 km². Channel routing is a process that depends on the hydraulic characteristics of the channel.

Although the hydrograph shape seems to be of lesser importance in diffuse-pollution control than the determination of rainfall excess, the hydrograph evaluation is necessary if the impact of diffuse pollution on receiving waters is studied. Furthermore, it is demonstrated in Chapter 5 that the delivery of pollutants from the source area to a receiving body of water is affected by the characteristics of the hydrograph, namely, by the slope of the receding portion of the hydrograph.

The literature on storm-water and flow routing is quite extensive, and the most recent references include those by Bras (1990), McCuen (1982), Chow, Maidment, and Mays (1988), Gray (1973), Hall (1984), and Viessman, Lewis, and Knapp (1984).

Early overland flow-routing models are known as the *rational formula* and the *unit hydrograph*. Both concepts are oversimplified, though found to be theoretically sound.

Rational Formula
The origin of the rational formula, known in Great Britain as the Lloyd-Davis Formula, can be dated back to the second part of the nineteenth century, since when it has been used for the design of storm and combined sewer and drainage systems. It relates the peak flow of runoff in a sewer or drainage basin outlet to the rain intensity as

$$Q_p = CIA \qquad (3.25)$$

where
Q_p = peak runoff discharge (l/min or cfs)
C = a runoff coefficient depending on the characteristics of the drainage area
I = the average rainfall intensity during a specified time interval called the time of concentration (mm/min or in./hr)
A = area (m² or acres).

The time of concentration is the time required for surface runoff to travel from the remotest part of the drainage area to the point of consideration (sewer inlet or sewer or watershed outlet). It consists of the overland flow time (inlet time) and sewer or channel flow time. The flow time in sewers may be estimated from the hydraulic properties of the conduit. The overland flow time is related to the watershed slope, roughness, length of the overland flow, and storm characteristics. The

TABLE 3.13 Runoff Coefficients for the Rational Formula

Description of the Area	
Urban Areas	Runoff Coefficient
Busines	
Downtown	0.7–0.95
Neighborhood	0.5–0.7
Residential	
Single family	0.3–0.5
Multiunits—detached	0.4–0.6
Multiunits—attached	0.6–0.75
Residential—suburban	0.25–0.4
Apartments	0.5–0.7
Industrial	
Light	0.5–0.8
Heavy	0.6–0.9
Pavements	
Asphalt and concrete	0.7–0.95
Bricks	0.7–0.95
Roofs	0.75–0.95
Lawns—sandy soils	
Flat, slope 2% or less	0.05–0.10
Average, slope 2%–7%	0.10–0.15
Steep, greater than 7%	0.15–0.20
Lawns—tight soils	
Flat, slope 2% or less	0.15–0.17
Average, 2%–7%	0.18–0.22
Steep, greater than 7%	0.25–0.33
Rural areas	Value of C^a
Topography	
Flat land with slopes less than 1%	0.3
Rolling land with average slopes 1%–3%	0.2
Hilly land with average slopes of 3%–6%	0.1
Soil	
Tight, impervious clay	0.1
Medium, combination of clay and loam	0.2
Open, sandy loam	0.4
Cover	
Cultivated land	0.1
Woodland	0.2

Source: Data for urban areas from American Society of Civil Engineers (1982) and for rural areas from Gray (1972).

[a] The magnitude of the runoff coefficient, C, is obtained by adding values of C's for each of the three factors (topography, soil, and cover) and subtracting the sum from unity. For example, for flat cultivated watershed with medium soils $C = 1 - (0.3 + 0.2 + 0.1) = 0.4$.

nature of the time of concentration and its relation to the peak time of the hydrograph are discussed in the next section. In sewered watersheds, however, the overland flow time to the nearest inlet may be very short, ranging between 5 and 30 minutes. In well-developed areas with closely spaced storm inlets, an inlet time of 5 minutes is common, whereas in flat residential districts with widely spaced street inlets, inlet times of 20 to 30 minutes are customary (Novotny et al., 1989).

The runoff coefficient, C, of the rational formula given in Table 3.13 is a ratio of the peak runoff flow to the rainfall intensity of a constant, longer duration rainfall. This is not hydrologically correct since runoff is a residual of precipitation after losses are subtracted, which was documented in the previous section. Thus the coefficients given in Table 3.13 are only approximations. The rational formula estimates directly the peak runoff rate and indirectly rainfall excess, since under certain circumstances the coefficient C also expresses the relation between rainfall excess and total rainfall. Theoretically, C would equal one if rainfall excess during the time of concentration was used instead of the total rainfall.

A good discussion on the nature of the rational formula and its coefficients is given in Hall (1984), who states that the runoff coefficient should be considered as an "impermeability" factor, which is more logical than to consider it as a proportionality factor. An obvious simplifying assumption is to use the ratio of paved (impervious), connected surfaces to the total area as an estimate of the runoff coefficient.

If the drainage area is not homogenous, the area must be first divided into homogenous segments (roofs, streets, lawns) with partial areas A_1, A_2, A_3, etc. The average runoff coefficient, C, can then be computed using the coefficients for the partial surface areas, C_1, C_2, C_3, etc., as follows

$$C = \frac{A_1 C_1 + A_2 C_2 + A_3 C_3 + \cdots}{A_1 + A_2 + A_3 + \cdots} \qquad (3.26)$$

An accurate estimation of the runoff coefficient is the most important task of the entire calculation, as can be seen when considering the large differences between the values in Table 3.13. The values of the coefficients, C, in the table correspond to rather flat catchments; larger runoff coefficients should be selected for steeper catchments.

Example 3.7: Rational Formula

Determine peak runoff and approximate runoff hydrographs from a uniform rainfall with an average intensity of 20 mm/hr, lasting two hours.

148   Hydrologic Considerations

Solution The residential watershed with an area of 50 hectares is composed of the following:

Roofs: 25% ($C = 0.9$)
Asphalt pavements: 25% ($C = 0.8$)
Lawns on flat tight soils: 50% ($C = 0.15$)

The watershed is drained by a 0.6-km-long storm sewer with a design flow velocity of 1.5 m/sec.

1. Estimated average runoff coefficient

$$C = \frac{0.25 \times 0.9 + 0.25 \times 0.8 + 0.5 \times 0.15}{0.25 + 0.25 + 0.5} = 0.505$$

2. Estimated time of concentration

    Overland flow time: assume 30 min

    Sewer time = length/velocity = 750(meters)/1.5(m/sec)
    = 500 sec = 8.33 min

Total time of concentration is 38.3 minutes which is less than 2 hrs. The peak flow is

$Q_p = 0.505 \times 20\text{(mm/hr)} \times (1/60\,\text{hr/min}) \times 50\text{(ha)} \times 10^4\text{(m}^2\text{/ha)}$
  $= 84\,166(\text{l/min})/[60(\text{sec/min})/1000(\text{l/m}^3)]$
  $= 1.402\,\text{m}^3/\text{sec}$

The approximate surface runoff hydrograph is plotted in Figure 3.19. Note that the time length of the ascending and recessed limbs of the hydrograph are approximately equal to the time of concentration.

The Unit Hydrograph

The concept of the unit hydrograph was conceived by hydrologists more than 60 years ago (Sherman, 1932), mostly from observations without theoretical mathematical development, which followed later.

Chow (Chow, Maidment, and Mays, 1988) defined two basic differential equations governing overland flow. The first equation is an equation of continuity, such as

$$\frac{dS}{dt} = I - Q \qquad (3.27)$$

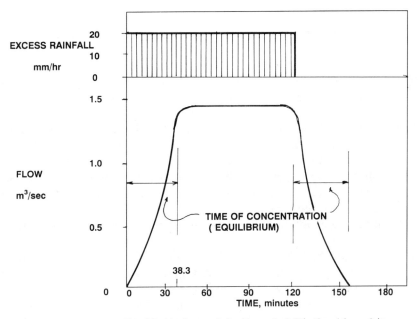

FIGURE 3.19. Simplified hydrograph for Example 3.7 (rational formula).

where
S = water storage within the watershed system
I = input (excess rainfall)
Q = output (runoff flow)
t = time

The second equation relates the outflow rate to the amount of storage and to the inflow. At this point this equation will be represented by a general equation, such as

$$Q = f(S, I) \tag{3.28}$$

Chow and coworkers have then shown that this general storage–input–output relationship can be expressed using the so-called convolution integral

$$Q(t) = \int_0^t h(\tau)I(t - \tau)\,d\tau \tag{3.29}$$

where
$h(\tau)$ = the ordinate of the instantaneous unit hydrograph (IUH)
τ = lag time

The instantaneous unit hydrograph in Equation (3.29) is the watershed response to a short-duration unit rainfall. Theoretically, to conform with the mathematics, the duration of the rainfall pulse defining the unit hydrograph should be infinitesimally small to instantaneous. The input–output relationship expressed by Equations (3.24) to (3.26) is typical for linear systems, that is, for systems where the shape of the unit hydrograph does not depend on the magnitude of the input or output of the system. The unit hydrograph concept defined herein corresponds to the linear system representation.

Under the assumption of linearity the following two theorems hold:

1. If the input of Equation (3.28) is multiplied by a constant, such as $c \times I$, then the output is also multiplied by the same constant, or $c \times Q$ (*principle of proportionality*).
2. If two solutions $f_1(Q)$ and $f_2(Q)$ of Equation (3.29) are added, the resulting solution $f_1(Q) + f_2(Q)$ is also a solution of the equation (*principle of superposition*).

Based on these two theorems any excess rainfall input can be broken down into the rainfall pulses of the same duration as the unit rainfall input defining the unit hydrograph. The volume of the rainfall pulse is then the multiplier determining the magnitude of the individual response hydrograph. All individual response hydrographs are then summed up as shown graphically on Figure 3.20. The same principle is incorporated into several hydrologic models as an overland flow-routing routine.

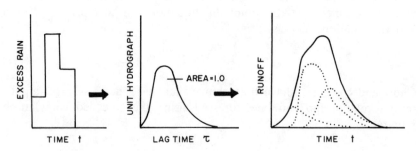

FIGURE 3.20. Graphical convolution of excess rainfall into runoff using unit hydrograph concept.

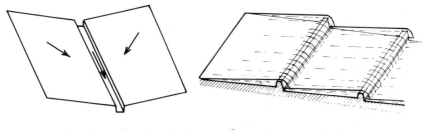

V-shaped watershed   Cascade watershed

FIGURE 3.21. Representation of watersheds in simple lumped-parameter rainfall–runoff models.

The principle of linearity implied in the unit hydrograph concept was subsequently questioned. Horton (1939b) and Izzard (1946) showed that the ordinates of the unit hydrograph, its peak time, as well as the time of concentration in the rational formula, depend on the intensity of the rainfall excess.

Synthetic Unit Hydrographs

The functions describing the instantaneous unit hydrograph can be based on two simplified watershed representations (Fig. 3.21). The first representation is a two-plane V-shaped watershed, as shown on the left side of Figure 3.21. The hydrograph solution for this watershed can be obtained by kinematic wave approximation (Henderson and Wooding, 1964; Wooding, 1965; Overton and Meadows, 1976). Equations (3.27) and (3.28) are represented in this concept by

$$\frac{\partial H}{\partial t} + \frac{\partial q}{\partial x} = i - f$$

and

$$q = \alpha H^\beta \tag{3.30}$$

where
H = water depth
q = the discharge rate/unit width
$i - f$ = rainfall excess
x = distance measured downstream from the top of the catchment
α, β = empirical coefficients

The kinematic wave model can be applied to small drainage areas of uniform slope and surface characteristics, such as parking lots and highways. The numerical solutions and various types of hydrographs are discussed in Overton and Meadows (1976).

The second watershed representation breaks the watershed into a series of overflowing pool-cascades and uses the continuity equation (Eq. (3.27)) for routing the flow. For a single reservoir the storage–outflow relationship (Eq. (3.28)) is expressed by

$$S = K \times Q \qquad (3.31)$$

where K is called the single-reservoir watershed constant. Substituting Equation (3.31) into the continuity equation (Eq. (3.27)) will yield to the following one-dimensional differential equation

$$K\frac{dQ}{dt} + Q = I$$

the solution of which for a unit rainfall input pulse is

$$h(t) = \frac{1}{K}e^{-t/K} \qquad (3.32)$$

The cascade watershed instantaneous unit hydrograph function was developed by Nash (1957) as

$$h(t) = \frac{1}{K_N} \frac{e^{-t/K_N}}{\Gamma(n)} \left(\frac{t}{K_N}\right)^{n-1} \qquad (3.33)$$

where
K_N = multiple basin storage constant
n = a watershed characteristic representing approximately the number of reservoirs
$\Gamma(n)$ = the gamma function of n

Note that if $n = 1$, then the formula will yield the single reservoir model expressed by Equation (3.32). In the watershed modeling process, the principal question is how many equal linear "pools" are needed for an adequate model? Based on the authors' experience, $n = 1$ gives satisfactory results for a small watershed of up to $10\,\text{km}^2$. For larger watersheds, the ranges of n may be higher; however, as pointed out by Overton and Meadows (1976), the model quickly approaches translation (that is, the results are not much different from $n \to \infty$) for $n > 5$. An analysis of a very large number of storm hydrographs by Holtan and Overton (quoted in Overton and Meadows, 1976) indicated that $n = 2$ produced optimum results in fitting computed runoff hydrographs to measured data.

The constants K and K_N and the reservoir characteristics that can be corelated to the travel time of water from the most remote point on the watershed to the watershed outlet. This time parameter is called the *time to equilibrium* or *time of concentration*, t_e. According to Rao, Delleur, and Sarma (1972)

$$t_e = K = n \times K_N \tag{3.34}$$

The time of concentration as previously defined is the theoretically correct time parameter that should also be used in the rational formula discussed in the preceding section. As shown on Figure 3.22, the unit hydrograph concept can be used to derive theoretical justification for the rational formula, which in this case is the watershed response to uniform rainfall that is equal or greater in length than the time of concentration.

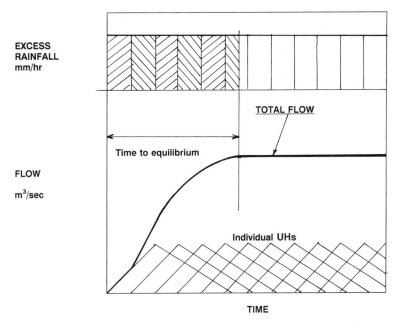

FIGURE 3.22. Rising (S-curve) hydrograph and time to equilibrium estimated by convolution of a long-duration uniform excess rainfall.

The rising limb of the hydrograph (so-called S-curve) is obtained by a summation of individual unit hydrographs multiplied by the average rainfall within the storm. It can be seen that equilibrium is reached when the time approaches the time of concentration.

Henderson and Wooding (1964) used the kinematic wave approximation for a V-shaped watershed and developed the following equation for t_e:

$$t_e = 6.9 \frac{(L \times n_M)^{0.6}}{i^{0.4} S^{0.3}} \tag{3.35}$$

where
t_e = time to equilibrium (min)
L = length of the overland flow (m)

TABLE 3.14 Manning's Roughness Coefficient, n_M for Overland Flow

Ground Cover	Manning's n_M
Urban zones	
Smooth asphalt	0.012
Street pavement	0.013
Asphalt or concrete paving	0.014
Packed clay	0.03
Light turf	0.20
Dense turf	0.35
Dense shrubbery or forest litter	0.40
Short grass	0.03–0.035
High grass—submerged flow	0.025–0.05
Heavy weeds—scattered grass	0.05–0.07
Nonurban zones	
Fallow field	
Smooth—rain packed	0.01–0.03
Medium—freshly disked	0.1–0.3
Rough—freshly disked	0.4–0.7
Cropped field	
Grass and pasture	0.05–0.03
Clover	0.08–0.25
Small grains	0.1–0.4
Row crops	0.07–0.2
Woods	
Light underbrush	0.4
Dense underbrush	0.8

Source: Compiled from several engineering texts and handbooks.

i = rainfall intensity (mm/hr)
S = slope (m/m)
n_M = Manning's surface roughness factor (Table 3.14)

An almost identical formula was developed by Morgali and Linsley (1965). Rao, Delleur, and Sarma (1972) statistically analyzed the hydrograph curves for several urbanizing watersheds. The authors investigated the effects of many variables on the shape of the runoff hydrograph, and only those that were found to be statistically significant were included in the final formula. Based on their work, t_e and the reservoir number n can be estimated from

$$t_e = 304 \frac{(AW)^{0.458}(TR)^{0.104}}{(1 + U)^{1.662}(i)^{0.269}} \qquad (3.36)$$

and

$$n = 2.23 \frac{(AW)^{0.069}}{(1 + U)i^{0.155}} \qquad (3.37)$$

where
t_e = time to equilibrium (hr)
AW = watershed area (km^2)
i = rainfall intensity (mm/hr)
U = fraction of the impervious area of the total watershed area
TR = rain duration (hr)

Note that minimum $n = 1$.

Gray (1961) developed the following relationship for the watershed constant, K_N, which is applicable to small rural watersheds (0.6 to 80 km^2) in Wisconsin, Illinois, central Iowa, and Missouri:

$$K_N = \frac{t_c}{n} = 7.33 \left(\frac{L}{\sqrt{S_c}}\right)^{0.562} \qquad (3.38)$$

where
t_c = time of concentration (min)
S_c = average watershed slope (%)
L = length of watershed (km), which includes overland-flow length and the length of the longest watershed channel

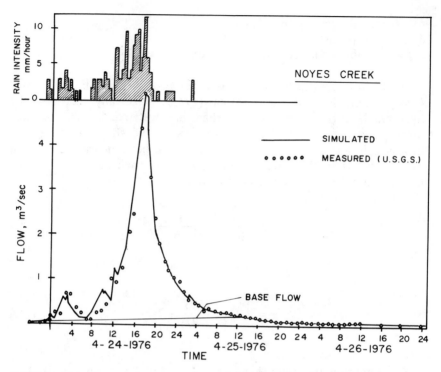

FIGURE 3.23. Comparison of measured and computed flows by a rainfall–runoff model for a small urban (mixed land use) watershed in southeastern Wisconsin.

The formulas for the time of concentration are not overly sensitive to the shape of the watershed.

Equations (3.35) and (3.36), which were obtained by a solution of the kinematic wave equation for the watershed overland flow or by statistically analyzing the hydrographs, indicate that the watershed constants depend on the magnitude of the input (rainfall intensity), which violates the assumption of linearity. It means that each partial rainfall input will be convoluted with different unit hydrographs. This should impose no problem in computer modeling applications and the results of convolution should be fairly accurate for reasonably small watersheds, as shown on Figure 3.23.

Example 3.8: Overland-Flow Routing by a Synthetic Hydrograph

For the rainfall excess computed in Example 3.3 and reported in Table 3.8, estimate the magnitude and shape of the surface runoff hydrograph if the drainage area is $1\,km^2$, the slope is 2%, the surface area is covered by grass, and 3% of the watershed is impervious.

Overland Routing of the Precipitation Excess   157

Solution Since the watershed size is less than $10 \, km^2$, the watershed reservoir number characteristic is selected as $n = 1$. This can be checked from Equation (3.37) using the smaller rainfall intensity reported in Table 3.8, column 9. Hence, $i = 44 \, mm/hr$. Then

$$n = \frac{2.23(1.0)^{0.069}}{(1 + 0.03)44^{0.155}} = 1.2$$

Therefore the single reservoir routing formula (Equation (3.32)) can be used to represent the unit hydrograph. The Manning roughness factor for grass is close to 0.35. For $i = 57 \, mm/hr$

$$K = t_c = 6.9 \frac{(0.5 \times 1000)^{0.8} 0.35^{0.6}}{57^{0.4} 0.02^{0.3}} = 98.2 \, min = 1.64 \, hr$$

For $i = 44 \, mm/hr$

$$K = t_c = 6.9 \frac{(0.5 \times 1000)^{0.6} 0.35^{0.6}}{44^{0.4} 0.02^{0.3}} = 99.3 \, min = 1.65 \, hr$$

Since both storage constants are similar, linear response may be assumed. The hydrograph computation using the following equation and convolution are shown in Table 3.15 and are plotted in Figure 3.24.

TABLE 3.15 Hydrograph Calculation for Example 3.8

Time (min) t or τ	Unit Hydrograph at Time τ	Ordinates of the Hydrograph (m³/sec)		
		$P = 57 \, mm/hr$	$P = 44 \, mm/r$ at Time t	Total
0	0	0	0	0
15	0.56	0	0	0
30	0.48	2.22	0	2.22
45	0.41	1.90	1.71	3.61
60	0.35	1.62	1.47	3.09
75	0.31	1.38	1.26	2.64
90	0.26	1.22	1.07	2.29
105	0.23	1.03	0.95	1.98
120	0.19	0.91	0.79	1.70
150	0.15	0.67	0.64	1.31
180	0.11	0.49	0.46	0.95
210	0.08	0.38	0.34	0.72

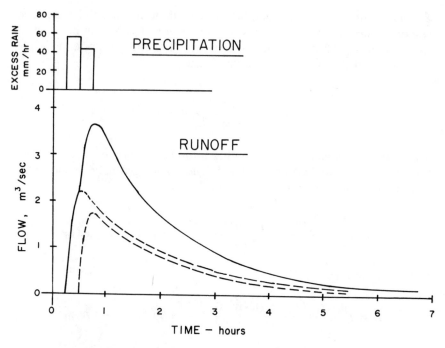

FIGURE 3.24. Rainfall–runoff convolution for Example 3.8.

$$\text{Flow}(\text{m}^3/\text{sec}) = \text{Area}(\text{m}^2) \times \frac{\text{m}}{1000\,\text{mm}} \times \frac{\text{hr}}{3600\,\text{sec}}$$

$$\times \sum_{\tau=0}^{t} [X_{t-\tau} \times h(\tau) \times \Delta t]$$

where
X = rain intensity (P in mm/hr)
h = hydrograph ordinate
Δt = time increment

Overland Routing by the SCS Method

The U.S. Soil Conservation Service hydrologic method (Soil Conservation Service, 1968) provides a methodology for overland routing of excess rainfall. Excess rain is determined, for example, in hourly intervals, as shown in Example 3.6.

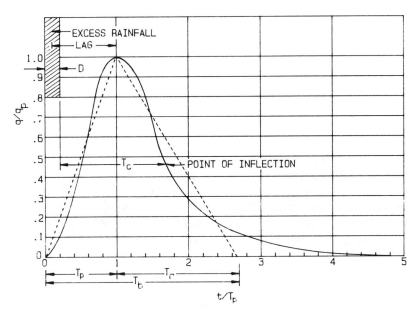

FIGURE 3.25. The Soil Conservation Service dimensionless unit hydrograph.

The Soil Conservation Service (SCS) proposed the dimensionless triangular hydrograph shown on Figure 3.25. This shape is based on the SCS analysis of numerous measured hydrographs under varying watershed and rainfall conditions. The peak time is the only parameter determining the shape of the hydrograph.

The peak time, t_p, according to Figure 3.25, is as follows:

$$t_p = D/2 + t_l \qquad (3.39)$$

where
t_l = the lag time from the centroid of the rainfall pulse
D = duration of the unit rainfall pulse

The hydrograph is approximated by a triangle with its peak at t_p and its base $t_b = \frac{8}{3} t_p$, which the SCS suggests is the best approximation of a unit hydrograph for typical rural watersheds. The recession time of the hydrograph is then $t_r = t_b - t_p = \frac{5}{3} t_p$.

The area under the unit hydrograph then equals the unit volume of the rainfall excess, or

160   Hydrologic Considerations

$$1 = \tfrac{1}{2} q_p (t_p + t_r) \tag{3.40}$$

or

$$q_p = \frac{1}{t_p}\left(\frac{2}{1 + t_r/t_p}\right) = \frac{K}{t_p} \tag{3.41}$$

where q_p is the peak runoff rate of the unit hydrograph.

For unit rainfalls given in millimeters and flow in cubic meters/second the watershed constant K has to include the watershed area and a conversion between the units. For watershed area, A, in hectares, rainfall excess in millimeters, t_p in hours, and $t_r/t_p = \tfrac{5}{3}$, as obtained from the dimensionless unit hydrograph on Figure 3.25, Equation (3.41) for peak runoff of the unit hydrograph becomes

$$q_p = \frac{A}{(480 \times t_p)}$$

The lag time, t_l is estimated from

$$t_l[\text{hr}] = \frac{1}{7053} \frac{L^{0.8}(S + 25.4)^{0.7}}{(\text{Sl})^{0.5}} \tag{3.42}$$

where
L = the length of overland flow to divide in meters
S = watershed storage defined by Equation (3.23)
Sl = percent slope of the watershed

In order to include the effect of imperviousness in urban or suburban watersheds and transportation corridors, the lag time is adjusted (i.e., multiplied) by the lag factor LF defined as

$$\text{LF} = 1 - \text{PRCT}(-0.006789 + 0.000335\,\text{CN} - 0.0000004298\,\text{CN}^2 \\ - 0.00000002185\,\text{CN}^3) \tag{3.43}$$

in which CN is the runoff curve number (Table 3.9) and PRCT is the percent imperviousness of the area or percent of the drainage system that has been channelized and lined.

The average lag time is 0.6 × time of concentration or $t_c = 1.666 \times t_l$. Then the duration of the rainfall pulse for convolution can be related to the time of concentration and peak time as follows

$$D = 0.133 t_c$$

and

$$D = 0.2 t_p$$

Small variations of D are permitted, but they should not exceed $0.25 t_p$ or $0.17 t_c$ (Viessman, Lewis, and Knapp, 1989).

Example 3.9: Determination of the SCS Unit Hydrograph

Determine the triangular SCS unit hydrograph for a 250-ha watershed that has been developed into a residential subdivision. The flow length is 2000 m, the slope is 3%, the watershed is 25% impervious, and the runoff curve number is CN = 80. Determine the uncorrected time of concentration.

Solution From Equation (3.23) $S = 25400/80 - 254 = 63.5$; then

$$t'_l = \frac{(2000)^{0.8}(63.5 + 25.4)^{0.7}}{7053 \times 3^{0.5}} = 0.83 \, \text{hr}$$

Correct t_l for imperviousness

$$LF = 1 - 25 \times (-0.006789 + 0.000335 \times 80 - 0.0000004298 \times 80^2 \\ - 0.00000002185 \times 80^3) = 0.848$$

$$t_l = LF \times t'_l = 0.848 \times 0.83 = 0.70 \, \text{hr}$$

The time of concentration $t_c = 1.666 \times t_l = 1.17 \, \text{hr}$.
 The duration of the rainfall pulse for convolution should be about

$$D = 0.1333 t_c = 0.1333 \times 1.17 = 0.15596 \, \text{hr} = 9.35 \, \text{min}$$

Select $D = 10 \, \text{min} = 0.16667 \, \text{hr}$. The peak time is then

$$t_p = \frac{D}{2} + t_l = \frac{0.1667}{2} + 0.7 = 0.87 \, \text{hr} = 52.2 \, \text{min}$$

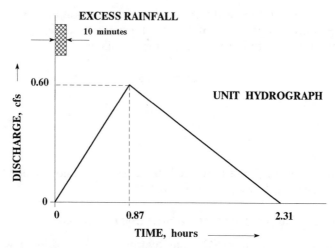

FIGURE 3.26. SCS unit hydrograph for Example 3.9.

also

$$t_r = \tfrac{5}{3}t_p = 1.44 \,\text{hr}$$
$$t_b = \tfrac{8}{3}t_p = 2.31 \,\text{hr} = 138.6 \,\text{min}$$

The peak of the unit hydrograph is then

$$q_p = 0.0020833 \times \frac{A}{t_p} = 0.0020833 \times \frac{250}{0.87} = 0.60 \,\text{m}^3 \,\text{sec}^{-1} \,\text{mm}^{-1}$$

The plot of the unit hydrograph is in Figure 3.26.

Example 3.10: Convolution with the SCS Unit Hydrograph

The rainfall hyetograph given below is to be convoluted into surface runoff by the SCS method. The watershed characteristics and the UH are given in Example 3.9. The excess rainfall is estimated in Table 3.16 by Equation (3.21).

From Example 3.9

$$D = 10 \,\text{min}$$
$$t_p = 0.87 \,\text{hr} = 52.2 \,\text{min}$$
$$t_b = 2.31 \,\text{hr} = 138.6 \,\text{min}$$
$$q_p = 0.6 \,\text{m}^3 \,\text{sec}^{-1} \,\text{mm}^{-1}$$

TABLE 3.16 Excess Rain Calculation

Time Interval, t (min) (1)	Rainfall Intensity (Hyetograph), i (mm/hr) (2)	Cumulative Rainfall, TR (min) (3)	Cumulative Excess Rain, Q (mm) (4)	Excess Rain 10 min Pulse, QRP (mm) (5)
0–10	36	6	0	0
10–20	36	12	0	0
20–30	54	21	0.95	0.95
30–40	60	31	4.09	3.14
40–50	30	36	6.25	2.16
50–60	0	36	6.25	0

Notes: Column (3) TR = i/D; column (4) Q from Equation (3.21); column (5) $QRP = Q_t - Q_{t-D}$.

The excess rainfall in column (5) is then convoluted with the SCS triangular UH with parameters calculated in Example 3.9. The ordinates of the UH are calculated in Table 3.17.

The hydrograph ordinates in columns (II), (III), and (IV) are obtained by multiplying the ordinates of the UH by the excess rain pulse shifted in time to place the begging of the hydrograph at the beginning of the excess rain pulse. Columm (V) is the summation of columns (I) to (IV).

Statistically Estimated Unit Hydrograph

The most desirable unit hydrographs (UH) are those estimated from monitoring data. Engineering manuals and hydrology textbooks recommend that the hydrograph be determined from hydrographs of several storm events. As pointed out previously, due to the nonlinearity of the process, determination of one uniform UH is not possible. Preferably the UH is determined from a hydrograph resulting from a short-duration, uniform, intense storm. If such a hydrograph is available, then the UH is determined from the measured hydrograph by dividing the ordinates of the hydrograph by the total hydrograph volume, expressed in millimeters (inches), of the runoff volume distributed over the watershed area. Before the UH is estimated, flow components other than those directly driven by the rainfall—including infiltration into sewers, ground-water base flow and interflow—must be subtracted from the measured hydrograph, which is neither easy nor straightforward (see, for example, Linsley, Kohler, and Paulhus (1982) for the procedures of a single storm–single runoff peak UH determination).

Chow, Maidment, and Mays (1988) and Bras (1990) describe several procedures of deconvolution for more complex hydrographs resulting

TABLE 3.17 Calculation of the Coordinates of the Unit Hydrograph and Convolution of Excess Rainfall into Surface Runoff

Lag Time τ (min) (1)	Unit Hydrograph			Runoff Time (min) (I)	Convolution			
					Real Storm Time (min)			
					20–30	30–40	40–50	Total Runoff (m³/s) (V)
					Excess Rain (mm)			
					0.95	3.14	2.16	
	τ/τ_π (2)	q/q_p (3)	q (m³/sec-mm) (4)		Partial Hydrograph (m³/s)			
					(II)	(III)	(IV)	
0	0	0	0	0	0	0	0	0
10	0.192	0.192	0.115	20	0	0	0	0
20	0.383	0.383	0.230	30	0.109	0	0	0.109
30	0.575	0.575	0.345	40	0.218	0.361	0	0.579
40	0.766	0.766	0.460	50	0.327	0.722	0.248	1.297
50	0.958	0.958	0.574	60	0.437	1.083	0.496	2.017
52.2	1.000	1.000	0.600	70	0.545	1.444	0.745	2.734
60	1.149	0.910	0.546	80	0.519	1.802	0.994	3.314
70	1.341	0.795	0.477	90	0.453	1.714	1.240	3.407
80	1.533	0.679	0.408	100	0.387	1.497	1.179	3.063
90	1.724	0.564	0.338	110	0.321	1.281	1.030	2.632
100	1.915	0.465	0.279	120	0.265	1.061	0.881	2.207
110	2.107	0.333	0.200	130	0.190	0.876	0.730	1.796
120	2.298	0.217	0.130	140	0.123	0.628	0.603	1.353
130	2.490	0.102	0.061	150	0.058	0.408	0.432	0.898
138.6	2.66	0.000	0.000	160	0.0	0.191	0.281	0.472
				170	0.0	0.0	0.132	0.132
				180	0.0	0.0	0.0	0.0

from more complex rainfall. Generally, the deconvolution is obtained from a set of equations expressed in matrix form

$$[R][h] = [Q] \qquad (3.44)$$

where **[R]** is the vector of rainfall data, **[Q]** is the corresponding flow vector, and **[h]** is the unit hydrograph expressed here as a polynomial. Several numerical procedures for deconvolution were presented by Bras (1990).

However, Chow et al. (1988) warned that given **[R]** and **[Q]**, there is usually no direct solution for **[h]**. The solution must be found by minimizing the least-square errors between the measured and computed **[Q]**; however, the solution is not easy because the many repeated and blank entries in **[R]** create computational difficulties.

The limitations and difficulties of the UH determination from measured data were summarized by Novotny and Zheng (1989) as follows:

1. Direct estimation of UH is difficult, especially when the hydrograph contains flow components other than surface runoff. The separation of hydrograph components into surface runoff, interflow, and base and sewage flow is generally inaccurate and arbitrary.
2. The UH is commonly estimated from a few characteristic storms. However, since only net rain contributes to the surface runoff component of the hydrograph, all precipitation losses (infiltration, evapotranspiration, surface storage) must be subtracted or the net rain will be determined from the surface runoff volume only. The losses are generally highly variable, almost random processes; therefore, deterministic (not considering the random component) net rain estimates usually carry large errors.
3. There are also physical inconsistencies of such functions, and some portions of the estimated hydrographs have negative and undulating ordinates.

Somewhat better results can be obtained by using dynamic hydrologic models where the output from the watershed hydrologic model is matched with a time series of measured data. The model must have a synthetic UH function incorporated in the structure of the program, and the parameters of the function are obtained by calibration and verification processes (see Chap. 8 for a description of the models and of their calibration and verification). The estimates of the UH function relies heavily on the adequacy of the function itself, on the accuracy of the

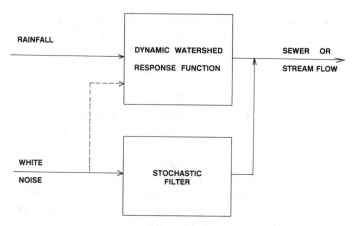

FIGURE 3.27. Stochastic rainfall–runoff transformation process that includes deterministic stochastic transfer function and input random noise.

separation of nonsurface flow contributions, and on the accuracy of the data. Furthermore, calibration and verification, which are often accomplished by eye fitting the measured and computed data, do not give an unbiased UH function estimate.

Novotny and Zheng (1989) proposed an ARMA–transfer-function model to represent the relationship between the rainfall and flow time series. Following standard ARMA modeling concepts (Box and Jenkins, 1976; Bras and Rodriguez-Iturbe, 1985), a standard rainfall–runoff transformation model consists of two parts, as shown on Figure 3.27: (1) a stochastic transfer of the input rainfall into the output hydrograph; and (2) a noise term that is a filtered (transformed) uncorrelated random series generically called *white noise*.

The ARMA modeling uses a backshift operator B introduced by Box and Jenkins for convenience in expressing the time series, such as that for any series $BZ_t = Z_{t-\Delta t}$, $B^2 Z_t = Z_{t-2\Delta t}$, and so forth. Then the input–output rainfall–runoff transformation model can be written as

$$Q = h(B)R_t + N_t \tag{3.45}$$

where
$h(B)$ = a polynomial representing the UH function
R_t = the rainfall or excess rainfall series
N_t = the noise component

In this simple representation it is not necessary to focus a priori on the surface runoff component only if it is assumed that N_t is another stochastic model representing flows other than those due to rainfall. In sewered watersheds, N_t could be considered as dry-weather sewage and wastewater contribution in combined sewers, cross-connections, and illegal dry-weather flow connection in separate sewers, or base flow in unsewered watersheds.

Also R_t does not have to be limited to net rain, since net rain is correlated to the total rainfall by some unspecified transfer function. Furthermore, the amount of infiltration is also correlated to rainfall. Thus, the transfer function $h(B)$ can contain both overland and subsurface routing components that, theoretically, could be separated and identified. Hence, $h(B)$ may or may not be similar to the theoretical synthetic UH formula described previously.

Since N_t is a filtered random series, it contains autocorrelative and random components. These must be filtered out from the measured series before the UH function $h(B)$ can be estimated. Novotny and Zheng (1989) developed a simple multiple-regression procedure that accomplishes filtering. In this procedure the measured hydrograph flows are correlated to rainfall and to the past flow series as follows

$$Q_t = I_1 Q_{t-1} + I_2 Q_{t-2} + I_3 Q_{t-3} + \cdots + I_m Q_{t-m} + \omega_0 R_t + \omega_1 R_{t-1} \\ + \omega_3 R_{t-3} + \cdots + \omega_n R_{t-n} + \Theta_0 \qquad (3.46)$$

The constant Θ_0 is included to account for the fact that the noise term (dry weather or base flow) may not have a zero mean. The polynomial of I coefficients is called the inverse Green's function. The size of polynomials I and ω is determined by observing the decrease in the mean residual error and by the requirement that the error series represents an uncorrelated series–white noise. The error series is created by generating a series of one-step-ahead forecasts using the model and subtracting the measured values from the forecasts as described by Box and Jenkins (1976).

The identification of coefficients $I(B)$ and $\omega(B)$ and the constant Θ_0 is done by standard linear multiple-regression routines. Statistical autocorrelation routines are used for the analysis of the series of residuals (errors).

The polynomial of coefficients ω contains the UH function $h(B)$; however, it is "corrupted" by the autocorrelative and moving average terms of the noise. The filtering process involves the division of polynomials $\omega(B)$ and $I(B)$ to yield the UH $h(B)$ polynomial, or

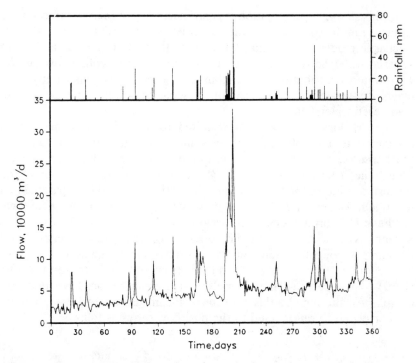

FIGURE 3.28. Rainfall and influent flows time series for an urban watershed with combined sewers (Fusina, Region Veneto in Italy).

$$h(B) = \frac{\omega(B)}{I(B)} \tag{3.47}$$

Example 3.11: Estimation of the UH Function by Stochastic Modeling

The method of Novotny and Zheng is used to estimate the UH function for a combined sewer system. The daily rainfall and corresponding flow series are shown in Figure 3.28.

The multiple regression of the time series of rainfalls and flows was performed (Equation (3.46)) with varying magnitudes of the maximum lags, m and n. In the analysis, m and n were kept equal. Using the F-test the significance of the reduction of the mean square residuals was tested along with the testing of the residual series as to whether they are *white noise*. Table 3.18 shows the results of the estimation. It can be seen that a transfer-function model of the order $m = n = 7$ provides an adequate model. Figure 3.29 shows the autocorrelation function of the residuals.

TABLE 3.18 Coefficient Estimation for Fusina System

Parameters (1)	ARTF Order							
	(1, 1) (2)	(3, 3) (3)	(5, 5) (4)	(7, 7) (5)	(9, 9) (6)	(11, 11) (7)	(13, 13) (8)	(15, 15) (9)
I_0	1.00	1.00	1.00	1.00	1.00	1.00	1.00	1.00
I_1	0.77	0.55	0.52	0.52	0.52	0.51	0.51	0.53
I_2	—	0.16	0.15	0.18	0.18	0.17	0.18	0.15
I_3	—	0.14	0.11	0.15	0.15	0.15	0.13	0.13
I_4	—	—	0.09	0.12	0.10	0.11	0.11	0.12
I_5	—	—	0.08	0.09	0.08	0.07	0.07	0.05
I_6	—	—	—	−0.11	−0.12	−0.13	−0.12	−0.11
I_7	—	—	—	−0.08	−0.07	−0.08	−0.06	−0.06
I_8	—	—	—	—	0.07	0.06	0.03	0.02
I_9	—	—	—	—	−0.03	−0.02	−0.06	−0.03
I_{10}	—	—	—	—	—	0.10	0.07	0.06
I_{11}	—	—	—	—	—	−0.04	−0.07	−0.07
ω_0	0.23	0.23	0.23	0.23	0.23	0.23	0.23	0.23
ω_1	−0.10	−0.06	−0.05	−0.05	−0.05	−0.05	−0.05	−0.05
ω_2	—	−0.04	−0.04	−0.05	−0.05	−0.04	−0.04	−0.04
ω_3	—	−0.05	−0.05	−0.06	−0.05	−0.05	−0.05	−0.05
ω_4	—	—	−0.03	−0.04	−0.04	−0.04	−0.03	−0.04
ω_5	—	—	−0.01	−0.03	−0.03	−0.02	−0.02	−0.02
ω_6	—	—	—	0.04	0.04	0.05	0.05	0.05
ω_7	—	—	—	0.01	0.01	0.01	0.01	0.01
ω_8	—	—	—	—	−0.02	−0.02	−0.01	−0.01
ω_9	—	—	—	—	−0.01	−0.01	0.00	0.00
ω_{10}	—	—	—	—	—	−0.03	−0.02	−0.02
ω_{11}	—	—	—	—	—	0.00	0.01	0.01
Θ_0	0.98	0.62	0.54	0.62	0.56	0.51	0.44	0.43
SSR	499	499	443	424	418	411	396	389
MSR	1.38	1.25	1.24	1.19	1.18	1.16	1.13	1.11
F-test	—	19	3.2	7.8	2.5	2.9	2.1	3.0

Note: The adequate model is ARTF (7, 7).

If the autocorrelation coefficients of the residuals for all lags greater than zero are close to zero, then the series is a white noise. Figure 3.30 then shows the filtered UH function $h(B)$, which has the character of a multiple reservoir function for surface runoff. It is interesting to note that the UH function shows a secondary peak with a much longer response time. Since both surface runoff and infiltration (interflow + ground-water inputs) are correlated with rainfall with different response times, the UH response function is composed of two parts, one with a shorter response

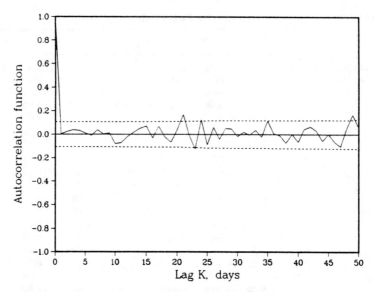

FIGURE 3.29. Autocorrelation function of residuals for Fusina time series, indicating that the residuals are a random white noise.

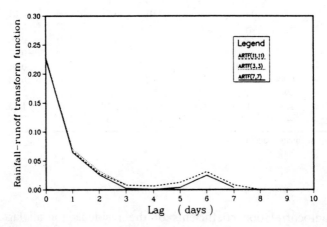

FIGURE 3.30. Filtered unit hydrograph obtained from the Fusina rainfall and runoff time series.

time for surface runoff, and the second, longer response component for infiltration. Hence, the rainfall–runoff transfer function obtained by this method can be used for quantitative separation and evaluation of surface runoff, ground-water inputs, and dry-weather sewage-flow components.

The polynomial $I(B)$ contains most of the information on the dry-weather component that is not related to the rainfall (Zheng, 1989).

III-5 Interflow

Interflow is that part of the subsurface flow that moves at shallow depths and reaches the surface channels in a relatively short period of time. It is therefore commonly considered part of the direct surface runoff (Viessman, Lewis, and Knapp, 1989). Although the quantity of interflow may represent only a small portion of total runoff, for some mobile pollutants its pollution effect may be of the same order of magnitude as the surface runoff. Interflow pollution originates mainly from salts and pesticides deposited in soils. On the other hand, interflow may be free of suspended pollutants.

The occurrence of interflow may be observed in areas where the permeability of the subsoils is less than that of the upper soil zone, causing horizontal movement of water in the upper soil zone. Lateral water movement in soils is especially significant during the snowmelt process, when subsoils are still frozen.

Theoretically, the amount of interflow could be computed from Darcy's law if the depth of the saturated upper zone storage, the saturation permeability, and the slope of the piezometric soil water surface were known. The Stanford Watershed Model (Crawford and Linsley, 1966) approximated the outflow from the interflow storage as

$$\text{INTF} = \alpha \text{SRGX} \qquad (3.48)$$

where
SRGX = current volume of water in interflow storage
α = an empirical coefficient

The volume of the interflow storage is continuously calculated by the model using mass continuity equations in which the infiltration from the surface is the source of water, and the ground-water recharge and evapotranspiration are the losses or outflow.

GROUND-WATER SYSTEMS

Ground-water movement and occurrence are an integral part of the hydrologic cycle. Almost all ground water originates from infiltrated precipitation after subtraction of the surface losses—surface runoff, evapotranspiration, and interflow. During prolonged dry periods, most of

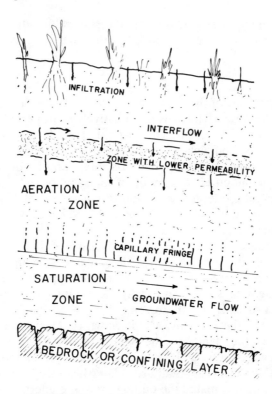

FIGURE 3.31. Soil and ground-water flow zones.

the natural flow in surface waters, as well as infiltration into deep sewers, originates from ground-water systems and is referred to as *base flow* or *ground-water runoff*.

Subsurface Distribution of Water

Water below the ground's surface occurs in four zones, which are shown in Figure 3.31 (Todd, 1959):

1. *Soil water zone*, which begins at the surface and extends downward to the end of the root zone. This zone is commonly unsaturated, except during periods of heavy infiltration. Its depth varies, but is generally from a fraction of a meter to a few meters thick.
2. *Intermediate zone* extends from the bottom of the soil zone to the top of the capillary fringe zone.
3. *Capillary zone* extends from the ground-water table, and its height is

determined by the capillary rise of water in pores of the soil. Due to capillary suction forces, water in this zone occurs at less than piezometric atmospheric pressures.
4. *Saturated zone.* The pores are completely filled with water, and water exists at pressures that are greater than or equal to the atmospheric pressure. The *ground-water table* is defined as the plane of a fully saturated zone where water occurs at atmospheric pressure.

The soil and intermediate zones are called *vadose zones* (in Latin *vadosus* means shallow) or zones of aeration. The zone of aeration contains voids and cracks that are partially occupied by water and air. In the saturated and capillary zones, all interstices are filled by water.

The direction of water movement in the vadose zone is primarily vertical, except when the topsoil zone becomes saturated and the interflow moves in a lateral direction. The direction of water movement in the saturated zone is lateral (nearly horizontal). The top of the saturated zone—the ground-water table—can be found in wells and borings that penetrate into the saturation zone.

The vadose zone is the place where most of the pollutant–soil interactions take place. It is a transition zone between surface contamination and ground-water pollution. Many pollutants are effectively adsorbed by soil particles and/or are decomposed in soils and will not penetrate the vadose zone, but other pollutants will penetrate the vadose zone and may potentially contaminate ground-water resources.

The vadose zone can be absent under the high ground-water conditions exhibited especially in wetland areas, or may be several hundred meters thick in arid and semiarid regions. Part of the water in the intermediate zone is held by hygroscopic and capillary forces. The excess moves downward by gravity. Water that can be drained from soil by gravity is known as *specific yield* and is given as the volume of water that can be drained by gravity to the gross volume of the soils. The specific yield depends on the type of soil.

Saturated Zone, Aquifers, and Aquitards

A geological formation saturated by water that yields appreciable quantities of water that can be economically used and developed is called an *aquifer* (Todd, 1959). An aquifer can be either confined or unconfined (Fig. 3.32). An unconfined aquifer is one in which the upper boundary of the saturation zone is the same as the water table. Confined aquifers, also known as artesian or pressure aquifers, are overlain by an impermeable geological stratum that keeps water under pressure. Water enters an

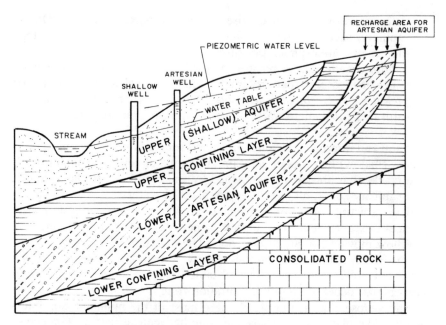

FIGURE 3.32. Representation of aquifers.

aquifer through a *recharge area*, which is an area where the waterbearing stratum is exposed to the atmosphere or is overlain by a permeable zone of aeration.

Aquitards are geological formations that are not permeable enough for the economic development as a ground-water source. An *aquiclude* is a formation that stores water, but is incapable of transmitting (e.g., clays). A solid rock formation that neither transmits nor stores water is an *aquifuge*.

Aquifers and aquitards can exist in layers with an unconfined aquifer on the top, and underlain by one or more confined zones. The top unconfined aquifer, often called a *shallow aquifer*, is most susceptible to diffuse pollution and contamination. Ground water that can be recovered by springs and wells represents the ground-water runoff or base flow.

Relationship between Surface and Ground-Water Systems

Ground- and surface-water systems are interrelated through two processes: *recharge* and *discharge*. There are two major sources of natural recharge to an aquifer. The first is the residual of precipitation that infiltrates through the unsaturated (vadose) zone, the second is freshwater inflow from surface-water bodies, such as streams, rivers, lakes, and wetlands. In addition, aquifers may be recharged by septic tanks, irrigation, artificial recharge, and sewer leakage. Natural discharge from aquifers occurs through springs, spring-fed lakes, wetlands, and oceans. In addition, large plants known as phreatophytes, whose roots extend to the water table, extract water from the aquifer by the transpiration process. Man extracts water from aquifers by pumping from wells, by intercepting ground water in galleries and drainage pipes, and by other more sophisticated systems. Under natural conditions, natural recharge and discharge are commonly in balance. Perennial streams in more humid regions are connected with ground-water aquifers. In some places aquifers discharge into streams, providing base flow. In some other areas streams and lakes discharge into aquifers. Ephemeral streams in arid regions are not connected with a discharging aquifer, that is, the ground-water table is well below the bottom elevation of the body of water and can only recharge the aquifer.

Depending on where recharge and discharge take place, ground-water flow systems can be divided into three types; local, intermediate, and regional (Toth, 1963). A local flow system has its recharge area in a topographic high and discharges into an adjacent topographic low. The

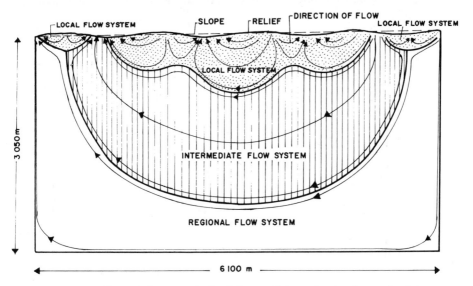

FIGURE 3.33. Theoretical patterns for local, intermediate, and regional ground-water flow systems with recharge and discharge. (Adapted from Toth, 1968. *Copyright © 1968 by the American Geophysical Union; reprinted with permission.*)

176   Hydrologic Considerations

intermediate flow system is viewed as having recharge and discharge areas that are not in adjacent topographic highs and lows. The discharge area for the intermediate system may be located several subbasins downstream. The regional system has its recharge in the regional topographic high, while its discharge occupies the topographic low for the basin (Fig. 3.33).

The flow in ground-water systems is generally slow, and the response to surface and subsurface pollution loadings is often gradual. The residence of water in local ground-water systems may range from days to months, intermediate flow systems may have residence times ranging from months to years, and the residence time of water in regional systems may reach centuries or more. A major part of ground water in regional aquifers of the midwestern United States originates from glaciers that melted during the postglacial period thousands of years ago. In most cases, surface contamination by pollutants affects only local ground-water systems and shallow aquifers.

The three systems have well-defined boundaries that theoretically identify changes in flow patterns, water quality, and the relative rate of ground-water movement. The water movement in a local system is commonly faster, which produces water quality that is different from the regional system. Due to the longer residence time and slower flow rates, regional systems produce ground water with higher concentrations of

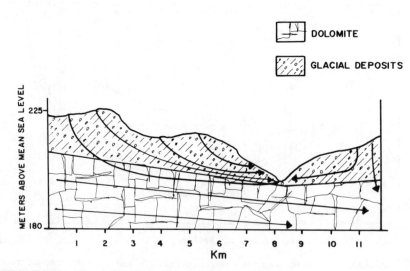

FIGURE 3.34.   Generalized flow patterns for local flow systems in a shallow glacial aquifer underlain by a dolomite aquifer. (After Eisen and Anderson, 1978.)

dissolved solids and ions than occur in ground water moving through a local system.

The depths at which the boundaries between the flow systems develop are dependent upon the following geomorphological factors: local topographic relief and its steepness, distribution and depth of bedrock and impermeable geological strata, aquifer thickness, and general topographic and geomorphological characteristics of the basin.

The computer simulation by Toth (1963) indicated that the development, intensity, and depth of local flow systems could be directly related to increased local topographic relief, and inversely to increased regional slope. Strong regional flow systems developed where local relief is negligible and where the ratio of total aquifer thickness to local relief was high. A weak local flow system will have only a portion of the total aquifer contributing discharge to that subbasin. Where there is pronounced local relief, mainly local systems develop. A strong local system has most or all of the aquifer discharging into that subbasin. Figure 3.34 shows typical flow patterns for a glacial aquifer overlaying dolomite and the flow patterns in the two aquifers.

Ground-Water Hydrological Balance

Figure 3.35 shows the relation between the recharge and discharge of a surface–ground-water system. Note that the boundaries of the systems are determined by the extent of the ground-water system. Hence several surface basins may be included for intermediate and regional systems.

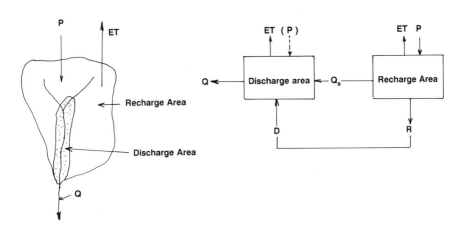

FIGURE 3.35. Recharge–discharge relationships for ground-water systems. (After Freeze and Cherry, 1979.)

The separation of the surface of the system into the recharge and discharge areas enables us to consider the surface and subsurface components of the total discharge from the system (Freeze and Cherry, 1979) and to formulate hydrologic budget for the joint system expressed first in this chapter by Equations (3.1)–(3.3). The overall hydrological balance of the joint system then becomes (all units in centimeters over the watershed area)

$$P = Q + \text{ET} + \Delta S_s + \Delta S_g \qquad (3.49)$$

where
P = precipitation
Q = runoff (mostly surface, since the area is recharging ground water)
ET = evapotranspiration
ΔS_s = change in surface storage
ΔS_g = change in ground-water storage

If the balance is averaged over a period of several years, then $\Delta S_s = 0$. Hence in the recharge area

$$P = Q_s + R + \text{ET} \qquad (3.50)$$

where
R = recharge rate (infiltration)
Q_s = surface runoff only because base flow is not discharging

At the watershed outlet in the discharge area

$$Q = Q_s + D - \text{ET} + P \qquad (3.51)$$

where D is the discharge rate (exfiltration).

By comparing Equations (3.50) and (3.51) with the overall balance (Eq. (3.49)), it follows that $\Delta S_g = R - D$. Under steady-state balanced conditions over a long period of time $\Delta S_g = 0$, and hence, $R = D$. In a dry year or when discharge is increased by man (for example, by excessive withdrawals for irrigation and water supply), $D > R$ and $\Delta S_g < 0$, and the ground-water table is decreasing. If the decrease is caused by excessive use of the ground-water resource by man, then the aquifer is being mined. An amount of water that can be withdrawn from the aquifer while maintaining the aquifer storage constant is called *safe yield*.

By setting $Q_g = D - ET$ as the base flow provided by the aquifer into the river system, and neglecting P in the discharge area, Equation (3.51) becomes

$$Q = Q_g + Q_s$$

In the analysis of water quality pollution and contamination from diffuse sources, one has to determine whether the source of contamination is surface or ground-water discharge. It is not proper—as was common in nonpoint pollution studies of the 1970s and early 1980s—to focus only on the surface runoff component.

Ground-Water Hydrological Models

The state of the art of ground-water hydrological modeling is well advanced, and numerous models have been developed to represent the flow and quality conditions of aquifers. Some fundamental models are introduced and discussed in Chapter 7. Comprehensive reviews were prepared by Kisel and Duckstein (1976), Anderson (1979), and Bedient, Borden, and Leib (1985), among others. Also most of the basic hydrological texts mentioned in this chapter contain discussions on the fundamentals of ground-water modeling and descriptions of the most common ground-water flow and quality models. These models can be either distributed-parameter flow models or lumped-parameter aquifer models. The lumped systems consider aquifers as homogenous, and the lumped values can be obtained by averaging the aquifer characteristics at a few discrete points or by estimating the parameters from the response of the aquifer to a known input, for example, a dye injection.

Many models are three-dimensional, box-grid representations that use finite-difference approximations of the flow equations (Wang and Anderson, 1982). Due to the linear nature of the basic flow equations, many ground-water problems can be solved by electric analog simulations. For diffuse pollution studies, ground-water flow models provide basic information on the residence times of contaminants in ground-water systems, subsurface flow patterns and dispersion of contaminants, base flow and drainage infiltration pollutant loads, and other valuable data needed in comprehensive diffuse-pollution studies.

References

American Society of Civil Engineers (1982). *Design and Construction of Sanitary and Storm Sewers*, ASCE-WPCF Manual No. 9, ASCE, New York.

Anderson, E. A., and N. H. Crawford (1964). *The Synthesis of the Continuous Snowmelt Runoff Hydrographs on a Digital Computer*, Tech. Rep. No. 26, Dept. of Civil Eng., Stanford University, Palo Alto, CA.

Anderson, M. P. (1979). Using models to simulate the movement of contaminants through groundwater flow systems, *CRC Crit. Rev. Environ. Control* 9(2):97–156.

Anonymous (1972). *Hydrocomp Simulation Programming Operation Manual*, Hydrocomp International, Palo Alto, CA.

Bedient, P. B., and W. C. Huber (1988). *Hydrology and Floodplain Analysis*, Addison-Wesley, Reading, MA.

Bedient, P. B., R. C. Borden, and D. L. Leib (1985). Basic concepts for ground water transport modeling, in *Ground Water Quality*, C. H. Ward, W. Giger, and P. L. McCarty, eds., Wiley, New York.

Bengtsson, L. (1982). Snowmelt-generated runoff in urban areas, in *Urban Stormwater Hydraulics and Hydrology*, B. C. Yen, ed., Water Resources Publ., Littleton, CO, pp. 442–451.

Bengtsson, L. (1984). Modeling snowmelt induced runoff with short time resolution, *Proceedings, Third International Conference on Urban Storm Drainage*, P. Balmér, P. A. Malmqvist, and A. Sjöberg, eds., Chalmers University of Technology, Göteborg, Sweden, pp. 305–314.

Box, G. E. P., and G. M. Jenkins (1976). *Time Series Analysis: Forecasting and Control*, Holden-Day, Oakland, CA.

Bras, R. L. (1990). *HYDROLOGY: An Introduction to Hydrologic Science*, Addison-Wesley, Reading, MA.

Bras, R. L., and I. Rodriguez-Iturbe (1985). *Random Functions and Hydrology*, Addison-Wesley, Reading, MA.

Braslavskii, A. P., and Z. A. Vikulina (1954). *Evaporation Norms for Water Reservoirs*, Gidrometeorologicheskoe izdatelstvo, Leningrad. (English translation available from Israel Program for Scientific Translations, Jerusalem, 1963.)

Brutsaert, W. (1976). The concise formulation of diffusive sorption of water in a dry soil, *Water Resour. Res.* **12**:1118–1124.

Burt, T. P., and P. J. Williams (1976). Hydraulic conductivity in frozen soils, *Earth Surf. Processes* **1**:349–360.

Chong, S. K., and R. E. Green (1983). Sorptivity measurement and its application, in *Advances in Infiltration*, ASAE Publ. No. 11-83, American Society of Agricultural Engineers, St. Joseph, MI, pp. 82–91.

Chow, V. T., ed. (1964). *Handbook of Applied Hydrology*, McGraw-Hill, New York.

Chow, V. T., D. R. Maidment, and L. W. Mays (1988). *Applied Hydrology*, McGraw-Hill, New York.

Crawford, N. H., and R. K. Linsley (1966). *Digital Simulation in Hydrology, Stanford Watershed Model IV*, Tech. Rep. No. 39, Dept. of Civil Eng., Stanford University, Palo Alto, CA.

Davis, C. V., and K. E. Sorensen (1969). *Handbook of Applied Hydraulics*, McGraw-Hill, New York.

Eisen, C. E., and M. P. Anderson (1978). *Field Data Quantifying Groundwater Surface Water Interaction*, Final Report Menomonee River Pilot Watershed Study, International Joint Commission, Windsor, Ontario.

Ellena, M., and V. Novotny (1985). Computer simulation of the urban snowmelt process, *Proceedings, 1985 International Symposium on Urban Hydrology, Hydraulic Infrastructures and Water Quality*, University of Kentucky, Lexington, KY.

Engman, E. T. (1974). Partial area hydrology and its application to water resources, *Water Resour. Bull.* **10**(3):512–521.

Freeze, R. A., and J. A. Cherry (1979). *Groundwater*, Prentice-Hall, Englewood Cliffs, NJ.

Gray, D. M. (1961). Synthetic unit hydrograph for small watersheds, *J. Hydraul. Div. ASCE* **87**:33–54.

Gray, D. M., ed. (1973). *Handbook on the Principles of Hydrology*, Water Information Center, Pt. Washington, NY.

Green, W. H., and G. A. Ampt (1911). Studies in soil physics. The flow of air and water through soils, *J. Agr. Sci.* **4**:1–24.

Hall, M. J. (1984). *Urban Hydrology*, Elsevier, Amsterdam.

Harbeck, G. E., Jr. (1962). *A Practical Field Technique for Measuring Reservoir Evaporation Utilizing Mass Transfer Theory*, U.S. Geological Survey Prof. Paper 272-E, Washington, DC.

Helfrich, K. R., E. E. Adams, A. L. Godbey, and D. R. Harleman (1982). *Evaluation of Models for Predicting Evaporation Water Loss in Cooling Impoundments*, Interim Rep. No. CS-2325, Electric Power Research Institute, Palo Alto, CA.

Henderson, F. M., and R. A. Wooding (1964). Overland flow and groundwater flow from a steady rainfall of finite duration, *J. Geophys. Res.* **69**:114–121.

Hiemstra, L. (1968). *Frequencies of Runoff for Small Basins*, Ph.D. thesis, Colorado State University, Fort Collins, CO.

Hillel, D. (1980). *Applications of Soil Physics*, Academic Press, Orlando, FL.

Holtan, H. N. (1961). *A Concept for Infiltration Estimates in Watershed Engineering*, USDA Agricultural Research Service, Washington, DC, pp. 41–51.

Holtan, H. N., and N. C. Lopez (1973). *USDAHL-73 Revised Model of Watershed Hydrology*, Rep. No. 1, USDA Agricultural Research Service, Plant Physiology Institute, Washington, DC.

Horn, M. E. (1971). Estimating soil permeability rates, *J. Irrig. Drainage Div. ASCE* **97**:263–274.

Horton, R. E. (1939a). Approach toward a physical interpretation of infiltration capacity, *Proc. Soil Sci. Soc. Am.* **5**:399–417.

Horton, R. E. (1939b). The interpretation and application of runoff plot experiments with reference to soil erosion problem, *Proc. Soil Sci. Soc. Am.* **3**:340–349.

Huggins, L. F., T. H. Podmore, and C. F. Hood (1976). *Hydrologic Simulation Using Distributed Parameters*, Purdue University, Water Resources Research Center Rep. No. 82, West Lafayette, IN.

Izzard, C. F. (1946). Hydraulics of runoff from developed surfaces, *Proc. Highway Res. Board* **26**:129–150.

Johanson, R. C., J. C. Imhoff, J. L. Kittle, and A. S. Donigian (1984).

Hydrologic Simulation Program—Fortran (HSPF): User's Manual, EPA-600/3-84-006, U.S. Environmental Protection Agency, Athens, GA.

Kane, P. L. (1980). Snowmelt infiltration into seasonally frozen soils, *Cold Region Sci. Tech.* **3**:153–161.

Kane, P. L., and J. Stein (1983). Physics of snowmelt infitration into seasonally frozen soils, in *Advances in Infiltration*, ASAE Publ. No. 11-83, American Society of Agricultural Engineers, St. Joseph, MI, pp. 178–187.

Kisel, C. C., and L. Duckstein (1976). Groundwater models, in *Systems Approach to Water Management*, A. K. Biswas, ed., McGraw-Hill, New York.

Larsen, V., J. H. Axley, and G. L. Miller (1972). *Agricultural Wastewater Accommodation and Utilization of Various Forages*, WRRC A-006-Md, 14-01-0001-790, University of Maryland, College Park, MD.

Linsley, R.K. (1967). The relation between rainfall and runoff, *J. Hydrol.* **5**(4):297–311.

Linsley, R. K., M. A. Kohler, and J. L. H. Paulhus (1982). *Hydrology for Engineers*, 3d ed., McGraw-Hill, New York.

Livingston, E. H., and L. A. Roesner (1991). Stormwater management practices for water quality enhancement, in *Manual of Practice—Urban Stormwater Management*, American Society of Civil Engineers, New York.

Lundin, L. C. (1989). *Water and Heat Flow in Frozen Soils. Basic Theory and Operational Modeling*, Acta Universitatis Upsaliensis, Upsala, Sweden.

McCuen, R. H. (1982). *A Guide to Hydrologic Analysis Using SCS Method*, Prentice-Hall, Englewood Cliffs, NJ.

McCuen, R. H. (1989). *Hydrologic Analysis and Design*, Prentice-Hall, Englewood Cliffs, NJ.

McCuen, R. H., and W. M. Snyder (1986). *Hydrologic Modeling. Statistical Methods and Applications*, Prentice-Hall, Englewood Cliffs, NJ.

McDaniel, L. L. (1960). *Consumptive Use of Water by Major Crops in Texas*, Texas Water Devel. Board Bull. 6019, Austin, TX.

Marciano, T. T., and G. E. Harbeck (1954). *Mass Transfer Studies in Water Loss Investigation, Lake Hefner Studies*. U.S. Geological Survey Professional Paper No. 269. Washington, DC.

Morgali, J. R., and R. K. Linsley (1965). Computer analysis of overland flow, *J. Hydraul. Div. ASCE* **91**:81–100.

Nash, J. E. (1957). The form of the Instantaneous Unit Hydrograph, *Bull. Int. Assoc. Sci. Hydrol.* **111**:114–121.

Novotny, V. (1986). *Effects of Pollutants from Snow and Ice on Quality of Water from Urban Drainage Basins*, Tech. Report., Dept. of Civil Eng, Marquette University, Milwaukee, WI (available from National Technical Information Service, Springfield, VA).

Novotny, V. (1988). Modeling urban runoff pollution during winter and off-winter periods, in *Advances in Environmental Modeling*, A. Marani, ed., Elsevier, Amsterdam, pp. 43–57.

Novotny, V., and S. Zheng (1989). Rainfall-runoff transfer function by ARMA modeling, *J. Hydraul. Eng. ASCE* **115**(10):1386–1400.

Novotny, V., K. R. Imhoff, M. Olthof, and P. A. Krenkel (1989). *Karl Imhoff's Handbook of Urban Drainage and Wastewater Disposal*, Wiley, New York.
Overton, D. E., and M. E. Meadows (1976). *Stormwater Modeling*, Academic Press, New York.
Philip, J. R. (1957). The theory of infiltration: 1. The infiltration equation and its solution, *Soil Sci.* **83**(5):345–357.
Philip, J. R. (1969). Theory of infiltration, in *Advances in Hydroscience*, vol. 5, V. T. Chow, ed., Academic Press, New York, pp. 215–296.
Philip, J. R. (1983). Infiltration in one, two, and three dimensions, in *Advances in Infiltration*, ASAE Publ. No. 11-83, American Society of Agricultural Engineers, St. Joseph, MI, pp. 1–23.
Ponce, V. M. (1989). *Engineering Hydrology*, Prentice Hall, Englewood Cliffs, NJ.
Priesley, C. H. B. (1959). *Turbulent Transfer of the Lower Atmosphere*, University of Chicago Press, Chicago, IL.
Rao, R. A., J. W. Delleur, and B. S. P. Sarma (1972). Conceptual hydrologic model for urbanizing watersheds, *J. Hydraul. Div. ASCE* **98**:1205–1220.
Rawls, M. J., and D. L. Brakensiek (1983). A procedure to predict Green and Ampt infiltration parameters, in *Advances in Infiltration*, ASAE Publ. No. 11-83, American Society of Agricultural Engineers, St. Joseph, MI, pp. 102–112.
Ritchie, J. C., J. A. Sparberry, and J. R. McHenry (1974). Estimating soil erosion from the redistribution of fallout of Cs-137, *Soil Sci. Soc. Am. Proc.* **38**:283–286.
Romanov, V. V., K. K. Pavlova, and I. L. Kolyushnyy (1974). Meltwater losses through infiltration into podzolic soils and chernozems, *Soviet Hydrology: Selected Papers, No. 1*, pp. 32–42.
Sharma, M. L., R. J. W. Barron, and K. S. DeBoer (1983). Spatial structure and variability of infiltration parameters, in *Advances in Infiltration*, ASAE Publ. No. 11-83, American Society of Agricultural Engineers, St. Joseph, MI, pp. 113–121.
Sherman, L. K. (1932). Streamflow from rainfall by unit-graph method, *Eng. News Rec.*, p. 501.
Soil Conservation Service (1968). Hydrology, supplement A to sec. 4, *Engineering Handbook*, USDA-SCS, Washington, DC.
Tholin, A. L., and C. J. Keifer (1960). The Hydrology of Urban Runoff, Trans. ASCE, paper 3061.
Todd, D.K. (1959). *Groundwater Hydrology*, Wiley, New York.
Toth, J. (1968). A theoretical analysis of groundwater flow in small drainage basins, *J. Geophys. Res.* **68**:4795–4812.
U.S. Army Corps of Engineers (1956). *Snow Hydrology*, North Pacific Division, Portland, OR.
Viessman, W., Jr., G. L. Lewis, and J. W. Knapp (1989). *Introduction to Hydrology*, 3d ed., Harper & Row, Cambridge, Philadelphia.
Wang, H. F., and M. P. Anderson (1982). *Introduction to Groundwater*

Modeling, Finite Difference and Finite Element Methods, W. H. Freeman, San Francisco.

Westerström, G. (1984). Snowmelt runoff from Porsön residential area, Luleå, Sweden, *Proceedings, Third International Conference on Urban Storm Drainage*, P. Balmér, P. A. Malmqvist, and A. Sjöberg, eds., Chalmers University of Technology, Göteborg, Sweden, pp. 315–324.

Wooding, R. A. (1965). A hydraulic model for the catchment-stream problem, Parts I and II, *J. Hydrol.* **3**:254–282.

Zheng, S. (1989). Stochastic Modeling of Urban Drainage Systems, Ph.D. thesis, Department of Civil Engineering, Marquette University, Milwaukee, WI.

4
Atmospheric Deposition

The best never let a little rain stand in their way.

Gene Kelly

The atmosphere is the portion of the environment where some of the most severe diffuse pollution problems originate, and, in fact, the magnitude of diffuse pollution often can be correlated with contamination of the atmosphere. Emissions of sulfur dioxide and nitrogen oxides by coal-burning processes and vehicular traffic cause the phenomenon of *acid rain*, which is having a severe adverse effect on many bodies of water throughout the world. Particulate aerosols in the atmosphere contain appreciable quantities of sulfur, toxic metals, pesticides, and other toxic organic compounds, fungi, pollen, soil, fly ash, nutrients, tar; and a variety of other chemical compounds, such as oxides, nitrites, nitrates, chlorides, fluorides, fluorocarbons, ozone, and silicates. Several extensive treatises have been devoted to the air-pollution problem (for example, Stern, 1976).

The adverse effects of the deposition of man-induced emissions to the atmosphere have been recognized for over a century. In the 1872 book *Air and Rain: The Beginnings of a Chemical Climatology* the British chemist Robert Angus Smith wrote that there were "three kinds of air" in and around the industrial town of Manchester, England. He described them as "... *that with carbonate of ammonia in the fields at a distance, ... that with sulphate of ammonia in the suburbs, ... and that with sulfuric acid, or acid sulphate, in the town.*" He pointed out that acid air in the town bleached the colors of fabrics and attacked metal surfaces, that the acid rain damaged vegetation and materials, and that substances such as arsenic, copper, and other metals were precipitated with the rain upon industrial regions.

This chapter covers the major pollution sources from the atmosphere that can be transported long distances and fall to earth in rain, snow, mist, fog, and in dry form as gases and particulates. One of the best known pollutants in atmospheric deposition is acidity. Acidity occurs when nitrogen and sulfur oxides are emitted from fossil-fuel combustion. Nitrogen in deposition can also affect the productivity of surface waters, particularly in the coastal zone.

Fossil-fuel burning also emits other substances, such as mercury, that fall to earth and can affect aquatic ecosystems. Other major atmospheric pollutants that affect water quality are trace metals, such as lead and agricultural chemicals, such as pesticides and herbicides.

INTERDEPENDENCE BETWEEN AIR AND WATER QUALITY

Many pollution studies are media specific, dealing with either water or air, but not both. However, there are many instances where the interdependence between media is important. Figure 4.1 illustrates the interdependence between water and air quality. For example, wind, temperature, and mixing patterns of both the atmosphere and a body of water can influence the concentrations of pollutants. Also, large lakes can influence precipitation patterns, and hence the rate of wet deposition of pollutants.

Atmospheric Sources of Water Pollutants

Airborne chemicals come from many natural and human sources. Before the 1800s, natural processes such as photosynthesis, decomposing organic matter, fire, volcanic activity, and wind erosion contributed gases, particles, and other byproducts to the atmosphere. Since that time, fossil-fuel burning, release of other industrial airborne chemicals, automobile exhausts, and intensive agriculture and forestry have emitted more substances to our atmosphere. In 1985, human activity accounted for an estimated 44 million tons of sulfur and nitrogen oxides released into the atmosphere (Irving, 1991). These same sources also release millions of tons of toxic metals and organic substances into the air each year. Air-pollution sources are global—emitted pollutants travel long distances

Interdependence between Air and Water Quality   187

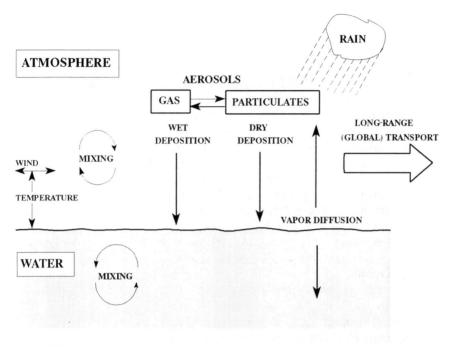

FIGURE 4.1.   Water/atmospheric-interacting physical and chemical processes.

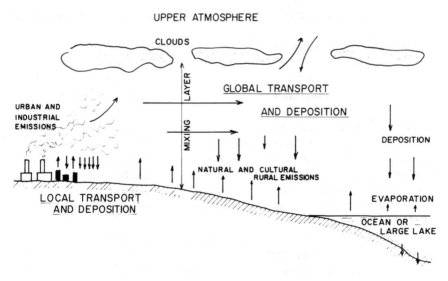

FIGURE 4.2.   Local and distant sources of atmospheric deposition.

and deposition occurs over a large regional or global scale—or local (Fig. 4.2).

Major global sources of air pollution include:

1. Urban and industrial emissions resulting from human activities, such as industrial processes and domestic burning.
2. Agricultural and forest emissions resulting from such human activities as:
    a. Soil erosion by wind during dry weather.
    b. Slash burning, which in many parts of the world is still recommended to prevent and/or reduce the spread of disease.
    c. Fertilizer components reaching the atmosphere through wind erosion and/or volatilization (such as the volatilization of ammonium from soils with higher pH).
    d. Pesticides entering the atmosphere from drift during application, by wind erosion, and by volatilization.
    e. Decomposing farm wastes and animal operations releasing ammonium, hydrogen sulfide, methane (cows), and mercaptans to the atmosphere.
3. Naturally occurring emissions on a global scale, including:
    a. Dust blown from arid and desert areas.
    b. Forest, brush, and grass fires.
    c. Volcanic eruptions, which are a source of sulfuric compounds and ash.
    d. Volatile hydrocarbons emitted from forests and other silvicultural activities.
    e. Sea spray, which is a significant source of salt and other particulates.
    f. Evaporation from large bodies of water, which can contribute significant quantities of volatile compounds and trace gases.

Local sources of atmospheric pollution and deposition include most of the sources just mentioned plus vehicular traffic. The magnitude of pollution, and hence atmospheric deposition, is magnified by several orders of magnitude in the vicinity of some sources, but fall off rapidly to background global levels as distance from the source is increased. For example, most of lead deposited from automobile exhaust is found within 100 m of the roadway.

Emissions of sulfur dioxide and nitrogen oxides are more thoroughly researched than other pollutants known to affect water quality, such as metals and pesticides. The annual contribution of sources of sulfur dioxide and nitrogen oxides in the United States and Canada in 1985 are

TABLE 4.1 U.S. and Canadian Emissions of SO_2, NO_x and Volatile Organic Carbon from Anthropogenic Sources Based on the 1985 NAPAP Emissions Inventory

Source	Emissions (Tg/yr[a])		
	SO_2	NO_x	VOC
U.S. Sources			
Electric utilities	14.6	6.0	neg[b]
Industrial combustion	2.4	2.9	neg
Commercial/residential/other combustion	0.6	0.7	1.7
Industrial/manufacturing processes	2.7	0.8	3.4
Transportation	0.8	8.0	8.0
Other	neg	0.1	6.9
Total[c]	21.6	18.6	20.0
Canadian Sources			
Electric utilities	0.7	0.2	neg
Industrial combustion	0.3	0.2	neg
Commercial/residential/other combustion	neg	0.1	0.1
Industrial/manufacturing processes	2.5	0.1	0.1
Transportation	0.1	1.2	1.0
Other	0.0	neg	0.7
Total[c]	3.7	1.9	2.2
Total United States and Canada	24.7	20.5	22.3

Source: From Irving, 1991.
[a] Tg = Teragrams = 10^6 metric tons.
[b] Neg = negligible < 0.1.
[c] Values may not sum to totals due to independent rounding.

shown in Table 4.1 (Irving, 1991). Total annual U.S. emissions in 1985 were estimated to be 24.7 million tons for sulfur dioxide and 20.5 million tons for nitrogen oxides. Canadian emissions in 1985 were estimated to be 3.7 and 1.9 million tons of sulfur dioxide and nitrogen oxides, respectively. The largest source of sulfur dioxide emissions in the United States is from the electric utility industry (70% of total). In Canada, the largest source is from industrial/manufacturing processes (67%). Nitrogen oxides, on the other hand, are primarily contributed by mobile sources. Transportation accounts for 43% and 63%, respectively, of total emissions in the United States and Canada. Natural sources are estimated to contribute only between 1% and 5% of total U.S. emissions of sulfur dioxide and nitrogen oxides.

The largest regional sources of SO_2 emissions in the United States are from the Northeast (10.72 million tons in the Northeast versus 4.8

TABLE 4.2 Regional Anthropogenic Emissions of SO_2, NO_x, and VOC

Federal Region	Emissions (Tg/yr)		
	SO_2	NO_x	VOC
1. New England	0.55	0.52	0.86
2. New York/New Jersey	0.77	0.89	1.44
3. Middle Atlantic	2.93	1.97	1.98
4. Southeast	4.76	3.41	3.76
5. Great Lakes	6.46	3.84	3.75
6. South Central	2.21	3.81	3.34
7. Central	1.53	1.26	0.98
8. Mountain	0.65	1.00	0.80
9. West	0.89	1.46	2.27
10. Northwest	0.23	0.48	0.86
Subtotals:			
Northeast (1–3 and 5)	10.7	7.2	8.0
Southeast (4)	4.8	3.4	3.8
West (6–10)	5.5	8.0	8.2
Total	21.0	18.6	20.0

Source: From Council on Environmental Quality (1990).

and 5.5 million tons, respectively, for the Southeast and West). On the other hand, the West proved to be the largest source of NO_x emissions (8.0 million tons versus 7.2 and 3.4 million tons, respectively, for the Northeast and Southeast). The high NO_x emissions for the West represent higher sources of emissions compared to stationary sources (Table 4.2).

Nriagu and Pacyna (1988) calculated the worldwide emissions of trace metals to the atmosphere using emissions factors and statistics on global production or consumption of industrial goods (Table 4.3). Table 4.4 compares atmospheric fallout to aquatic sources of anthropogenic inputs of trace metals into aquatic systems. The atmosphere is an important source of all metals listed when compared to domestic and industrial effluents. Atmospheric inputs are about 10% of the total liquid discharges.

Goolsby (1991) measured herbicide concentrations in rainwater from a 23-state area, principally in the Midwest and Northeast. He found traces of herbicides in all 23 states and in all but two of the 81 collection sites. Pesticide concentrations in wet deposition were measured by Glotfelty et al. (1990) in Maryland. The researchers calculated the projected amounts of pesticides entering the Chesapeake Bay from precipitation.

TABLE 4.3 Worldwide Sources of Trace Elements to the Atmosphere (10^3 kg/yr)

Source Category	Global Production/Consumption (10^9 kg/yr^{-1})	As	Cd	Cr	Cu	Hg	In	Mn	Mo
Coal combustion	[15.5 × 10^9 MJ]								
—Electric utilities		232–1550	77–387	1240–7750	930–3100	155–542		1080–6980	232–2320
—Industry and domestic	990	198–1980	99–495	1680–11,880	1390–4950	495–2970		1485–11,880	396–2480
Oil combustion	[5.8 × 10^9 MJ]								
—Electric utilities		5.8–29	23–174	87–580	348–2320			58–580	58–406
—Industry and domestic	358	7.2–72	18–72	358–1790	179–1070			358–1790	107–537
Pyrometallurgical nonferrous metal production									
—Mining		40.0–80	0.6–3		160–800			415–830	
—Pb production	3.9	780–1560	39–195		234–312	7.8–16			
—Cu–Ni production	8.5	8500–12,750	1700–3400		14,450–30,600	37–207	8.5–34.0	850–4250	
—Zn–Cd production	4.6	230–690	920–4600		230–690		2.3–4.6		
Secondary nonferrous metal production			2.3–3.6		55–165			1065–28,400	

TABLE 4.3 (*Continued*)

Source Category	Global Production/Consumption (10^9 kg/yr^{-1})	As	Cd	Cr	Cu	Hg	In	Mn	Mo
Steel and iron manufacturing	710	355–2480	28–284	2840–28,400	142–2840				
Refuse incineration									
—Municipal	140	154–392	56–1400	98–980	980–1960	140–2100		252–1260	
—Sewage sludge	3	15–60	3–36	150–450	30–180	15–60		5000–10,000	
Phosphate fertilizers	137		68–274		137–685				
Cement production	890	178–890	8.9–534	890–1780					
Wood combustion	600	60–300	60–180		600–1200	60–300			
Mobile sources	647 (gasoline)								
Miscellaneous Total, emissions		1250–2800 12,000–25,630	3100–21,040	7340–53,610	19,860–50,870	910–6200	11–39	10,560–65,970	793–5740
Median value		18,820	7570	30,480	35,370	3560	25	38,270	3270

Source Category	Ni	Pb	Sb	Se	Sn	Tl	V	Zn
Coal combustion								
—Electric utilities	1395–9300	775–4650	155–775	108–775	155–755	155–620	310–4650	1085–7750
—Industry and domestic	1980–14,850	990–9900	198–1480	792–1980	99–990	495–990	990–9900	1485–11,880
Oil combustion								
—Electric utilities	3840–14,500	232–1740		35–290	348–2320		6960–52,200	174–1280
—Industry and domestic	7160–28,640	716–2150		107–537	286–3580		21,480–71,600	358–2506

Source								
Pyrometallurgical nonferrous metal production								
—Mining	800	1700–3400	18–176	18–176				
—Pb production	331	11,700–31,200	195–390	195–390			310–620	195–468
—Cu–Ni production	7650	11,050–22,100	425–1700	427–1280	425–1700	43–85	4250–8500	
—Zn–Cd production		5520–11,500	46–92	92–230			46,000–82,800	
Secondary nonferrous metal production		90–1440	3.8–19	3.8–19			270–1440	
Steel and iron manufacturing	36–7100	1065–14,200	3.6–7.1	0.8–2.2		71–1420	7100–31,950	
Refuse incineration								
—Municipal	98–420	1400–2800	420–840	28–70	140–1400	300–2000	2800–8400	
—Sewage sludge	30–180	240–300	15–60	3–30	15–60		150–450	
Phosphate fertilizers	137–685	55–274		0.4–1.2			1370–6850	
Cement production	89–890	18–14,240				2670–5340	1780–17,800	
Wood combustion	600–1800	1200–3000					1200–6000	
Mobile sources		248,030						
Miscellaneous	24,150–87,150	3900–5100	1480–5540	1810–5780	1470–10,810	3320–6950	30,150–141,860	1724–4783
Total, emissions	55,650	288,700–376,000	3510	3790	6140	5140	86,000	70,250–193,500
Median Value		332,350						131,880

Source: From Nriagu and Pacyna (1988).

TABLE 4.4 Anthropogenic Inputs of Trace Metals into the Aquatic Ecosystems (10^6 kg/yr)

Source Category	Annual Global Discharge[a] ($10^9 m^3$)	As	Cd	Cr	Cu	Hg	Mn	Mo	Ni	Pb	Sb	Se	V	Zn
Domestic wastewater[b]														
—Central	90	1.8–8.1	0.18–1.8	8.1–36	4.5–18	0–0.18	18–81	0–2.7	9.0–54	0.9–7.2	0–2.7	0–4.5	0–2.7	9.0–45
—Noncentral	90	1.2–7.2	0.3–1.2	6.0–42	4.2–30	0–0.42	30–90	0–1.8	12–48	0.6–4.8	0–1.8	0–3.0	0–1.8	6.0–36
Steam electric	6	2.4–14	0.01–0.24	3.0–8.4	3.6–23	0–3.6	4.8–18	0.1–1.2	3.0–18	0.24–1.2	0–0.36	6.0–30	0–0.6	6.0–30
Base metal mining and dressing	0.5	0–0.75	0–0.3	0–0.7	0.1–9	0–0.15	0.8–12	0–0.6	0.01–0.5	0.25–2.5	0.04–0.35	0.25–1.0	—	0.02–6
Smelting and refining														
—Iron and steel	7						14–36			1.4–2.8				5.6–24
—Nonferrous metals	2	1.0–13	0.01–3.6	3–20	2.4–17	0–0.04	2.0–15	0.01–0.4	2.0–24	1.0–6.0	0.08–7.2	3.0–20	0–1.2	2.0–20
Manufacturing processes														
—Metals	25	0.25–1.5	0.5–1.8	15–58	10–38	0–0.75	2.5–20	0.5–5.0	0.2–7.5	2.5–22	2.8–15	0–5.0	0–0.75	25–138
—Chemicals	5	0.6–7.0	0.1–2.5	2.5–24	1.0–18	0.02–1.5	2.0–15	0–3.0	1.0–6.0	4–3.0	0.1–0.4	0.02–2.5	0–0.35	0.2–5.0
—Pulp and paper	3	0.36–4.2	—	0.01–1.5	0.03–0.39	—	0.03–1.5	—	0–0.12	0.01–0.9	0.01–0.27	0.01–0.9	—	0.09–1.5
—Petroleum products	0.3	0–0.06	—	0–0.21	0–0.06	0–0.02	—	—	0–0.06	0–0.12	0–0.03	0–0.09	—	0–0.24
Atmospheric fallout[c]		3.6–7.7	0.9–3.6	2.2–16	6.0–15	0.22–1.8	3.2–20	0.2–1.7	4.6–16	87–113	0.44–1.7	0.54–1.1	1.4–9.1	21–58
Dumping of sewage sludge[d] [6×10^9 kg]		0.4–6.7	0.08–1.3	5.8–32	2.9–22	0.01–0.31	32–1.06	0.98–4.8	1.3–20	2.9–16	0.18–2.9	0.26–3.8	0.72–4.3	2.6–31
Total input, water		12–70	2.1–17	45–239	35–90	0.3–8.8	109–414	1.8–21	33–194	97–180	3.9–33	10–72	2.1–21	77–375
Median value		41	9.4	142	112	4.6	262	11	113	138	18	41	12	226

Source: From Nriagu and Pacyna (1988).

[a] The discharges given represent contaminated process waters, and do not include cooling waters.

[b] The wastewater production figure corresponds to about 60 m^3/capita/yr multiplied by the 2.4×10^9 residents in urban and rural areas of the world. The other discharge figures likewise have been derived from the reported water demand per unit tonne of metal smelted or goods manufactured.

[c] We have assumed that 70% of each metal emitted to the atmosphere is deposited on land, and that the remaining 30% is deposited in the aquatic environments.

[d] Worldwide sewage sludge production is estimated to be 30 million tonnes, assuming average sludge production rate of 30 g/capita/day in urban and rural communities. It is believed that 20% of the municipal sludge is directly discharged or dumped into aquatic ecosystems, about 10% is incinerated, and the rest is deposited on land.

Long-Term Trends in Atmospheric Emissions

Total emissions of sulfur dioxide and nitrogen oxides generally increased in the United States during the period from 1900 to 1970 (Fig. 4.3). During this time, annual emissions of sulfur dioxide increased by a factor of 3 and nitrogen oxides increased almost tenfold.

Tall stacks constructed by electric utilities and smelters in the last 25 years have caused atmospheric pollutants to be transported long distances from their emission sources by emitting them into the upper air currents. Prior to 1970, there were only two stacks in existence taller than 150 meters (500 ft); in the United States today there are more than 175 such stacks. These stacks were built to reduce ambient air-pollutant concentrations in the vicinity of the stacks, but their net effect has been to spread air pollutants over long distances.

Emission-control plans adopted since passage of the U.S. Clean Air Act in 1970 have been effective in reducing the growth of national emissions of sulfur dioxide and nitrogen oxides (Fig. 4.3). Sulfur dioxide emissions decreased an estimated 30% between 1970 and 1988, while nitrogen oxide emissions were about 10% lower in 1988 than their peak in 1978.

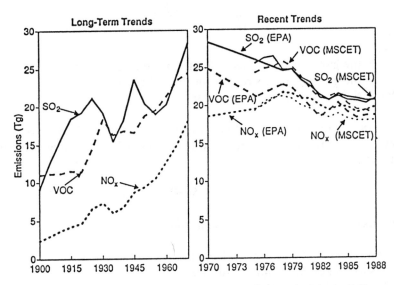

FIGURE 4.3. Historical trends in SO_2, NO_x, and volatile organic carbon emissions. Sources for 1940–1970 trends: Gschwandtner, Wagner, and Husar 1988; for 1970–1987 trends: U.S. EPA, 1990; for 1975–1988 trends: Kohout et al., 1990. (From Irving, 1991.)

It is expected that the Clean Air Act Amendments passed by the U.S. Congress in 1990 will result in regulations that lower sulfur dioxide and nitrogen oxides emissions even further. Also, the act offers a comprehensive plan for significantly reducing emissions of hazardous air pollutants, also referred to as air toxics. The law lists 189 pollutants and requires major sources to reduce emissions according to a prescribed schedule. Unfortunately, we do not know as much about air toxics and their effects on bodies of water compared to our knowledge of acidification.

Global Transport of Pollutants

Measurements at the most remote points on the earth, such as Antarctica, indicate that many pollutants, such as organic chemicals (e.g., PCBs), enter the global cycle and are deposited in appreciable quantities anywhere on the earth. Most of the discussion on global transport is taken from Junge (1977).

The concentration of pollutants in the atmosphere is determined by the mass balance between global sources and sinks of the pollutant. Mathematically, one can write that

$$\frac{dM}{dt} = Q - S(M) \qquad (4.1)$$

where
M = the global mass of the pollutant in the atmosphere
Q = the global source strength for the pollutant
$S(M)$ = the global sink of the pollutant

Under a steady-state assumption, which can be applied only to time intervals of more than one year, and fairly steady inputs, the left side of Equation (4.1) becomes zero and

$$Q_{ss} = S(M) \qquad (4.2)$$

The sinks of atmospheric pollutants include:

1. Deposition (wet and dry) on land and sea surfaces.
2. Adsorption on land and sea surfaces.
3. Decomposition by atmospheric chemical and photochemical processes.
4. Emissions into the stratosphere.

The global removal (sink) rate is a function of the mass of pollutants present in the atmosphere (or its concentration), and if deposition prevails as in the cases of some relatively inert components (DDT, PCBs), it can be approximated by

$$S(M) = v_d \times C_M \tag{4.3}$$

where
v_d = depositional velocity (m/day)
C_M = average global (background) concentration of the pollutant, typically measured at some remote point unimpacted by cultural emissions

The average residence time of a pollutant in the atmosphere under steady-state conditions is given by

$$T = \frac{M}{S(M)} = \frac{C_M V_A}{v_d C_M A_G} = \frac{H}{v_d} \tag{4.4}$$

where
A_G = global surface area
V_A = volume of the atmosphere within the mixing layer
H = average depth of the surface air boundary layer (typically of the order of about 1000 m)

The most effective natural removal process is the attachment of pollutants to atmospheric aerosols and their subsequent removal by dry and wet fallout on land and sea surfaces. For many pollutants, the sea is the final sink, since pollutants deposited on land can be reentrained or can reenter the atmosphere by volatilization, wind erosion, and other processes described previously.

If the input of a pollutant into the global transport system is instantaneous, as occurs during an explosion, volcanic eruption, or one-time widespread pesticide application, and if the sink function is linearly proportional to the mass of the constituent in the atmosphere ($S(M) = S_0 \times M$), Equation (4.1) can be solved to yield

$$M(t) = \frac{Q_0}{V_A} e^{-S_0 t} = \frac{Q_0}{V_A} e^{-t/T} \tag{4.5}$$

where Q_0 = the mass of the instantaneous input.

Example 4.1: Global Pollution Transport

Background (steady-state) concentrations of an "inert" pesticide in the atmosphere are measured at about $C_M = 0.1 \text{ ng/m}^3 = 10^{-10} \text{ g/m}^3$. The worldwide production of the pesticide is about 1000 tonnes/yr = 10^9 g/yr from which about 40% is lost to the atmosphere during and after application. Estimate the average residence time and deposition velocity of the pesticide. Assume an average depth of the mixed air boundary layer of $H = 3000$ m.

Solution Atmospheric input of the pesticide:

$$Q = 0.4 \times 10^9 \text{ (g/yr)} = 4 \times 10^8 \text{ g/yr}$$

To determine the amount of the pesticide in the atmosphere it is necessary to know:

Volume of the atmospheric mixing layer (earth radius $r = 6.3 \times 10^6$ m):

$$V_A = 4\pi r^2 H = 4\pi (6.3 \times 10^6)^2 \times 3000 = 1.5 \times 10^{18} \text{ m}^3$$

Mass of the pesticide in the atmosphere:

$$M = C_m V_A = 10^{-10} \text{ (g/m}^3) \times 1.5 \times 10^{18} \text{ (m}^3) = 1.5 \times 10^8 \text{ g}$$

Average residence time:

$$T = \frac{M}{Q} = \frac{1.5 \times 10^8}{4 \times 10^8} = 0.375 \text{ yr} = 137 \text{ days}$$

Deposition velocity:

$$v_d = \frac{H}{T} = \frac{3000}{0.375} = 8000 \text{ m/yr}$$

Entry of Atmospheric Pollutants into Surface Waters

Atmospheric pollution consists of gases and aerosols or atmospheric particulates. The particulate matter ranges in size from 6×10^{-4} to 10^3 µm. The term aerosol should be differentiated from dust. Dust contains particles that are mostly insoluble, while aerosols contain also water-soluble materials (about 50 according to Paterson and Junge

(1971)). Besides direct emissions from terrestrial sources, aerosols can be formed in the atmosphere by precipitation, absorption, and chemical reactions.

Removal of particles (aerosols and dust) from the atmosphere is due to:

1. Dry deposition by sedimentation.
2. Removal by rainfall and snowfall.
3. Dry deposition by impact on vegetation and rough surfaces.

Removal of gases occurs primarily by:

1. Removal during periods of precipitation.
2. Absorption at the earth's surface.
3. Adsorption of aerosolic particles and subsequent deposition.

Dry Deposition

Dry deposition is an important process for the removal of gases and airborne particles. While data on wet deposition are relatively abundant, dry deposition data are sparse. Sisterson et al. (1990) evaluated a number of data sources on the dry deposition of sulfur and nitrogen species. Wet deposition was found to account for most of the total deposition of sulfur and nitrogen species for regionally representative sites. Individual sites that were more heavily impacted by sulfur emissions tended to have a relatively larger dry deposition in relation to the total contribution than sites not impacted by local emissions. This means that dry deposition is probably much greater than wet deposition in and near urban areas. When urban areas are excluded, dry deposition of sulfur species is estimated to be 30% to 60% of the total (wet plus dry) deposition. Under the same conditions, the percentage for nitrogen species ranges from 30% to 70%.

The rate of dry fallout from the atmosphere is primarily determined by the force of gravity, but other effects, such as surface impaction, electrostatic attraction, adsorption, and chemical interactions, may explain why the deposition rate of small particles (order of magnitude of 1 μm or less) onto the ground is often greater than can be expected from the pull of gravity.

The rate of deposition of aerosol particles can be related to their average above-ground concentration:

$$D_d = v_d C(x, y, z) \qquad (4.6)$$

where
D_d = amount of aerosols removed per unit area per unit time (e.g., g/m²-day or tonnes/km²-month)
$C(x, y, z)$ = average concentration of aerosols at x, y, and z locations from the source or coordinate origin (g/m³)
v_d = deposition velocity of particles (m/day or m/month)

Depositional velocity differs from the physical settling velocity expressed by Stokes' law. Figure 4.4 shows the depositional velocity of particles as estimated by measurements. MacMahon, Dension, and Fleming (1976) reported that the depositional velocities of gases range from 0.5 to 2.5 cm/sec, depending on the ground surface.

Wet Deposition and Composition of Precipitation

Due to the fact that precipitation scavenging is one of the most effective processes for cleansing the atmosphere, rain and snow contain many pollutants in quantities that may be harmful to terrestrial and aquatic ecosystems. Pollutants included in wet precipitation are acidity, toxic metals, organic chemicals, phosphates, and nitrogen compounds. In some areas pollution from rain has been devastating to surface water biota and often leads to acidification of lakes, fish kills, and severe reduction in the productivity of lakes. Acid rainfall also leaches cations from soils (such as aluminum) and from urban infrastructures (damage to concrete and elutriation–corrosion of metals).

Contamination can be incorporated into precipitation within or below clouds. In-cloud scavenging is called *rainout* or *snowout*, the below-cloud process of enrichment is called *washout*.

Washout Function. In many cases the amount of pollutants deposited by wet fallout can only be estimated from known atmospheric concentrations, because the rainwater concentrations are either not known or are unreliable. The process of scavenging of pollutants by raindrops during washout or rainout (snowout) is basically an exponential function (Slade, 1968):

$$C_w = C_{w,0} \exp(-\lambda t) \tag{4.7}$$

where
C_w = the atmospheric concentration of the contaminant after the rain
$C_{w,0}$ = the atmospheric concentration of the contaminant before the rain
t = duration of the rain
λ = the washout coefficient

Interdependence between Air and Water Quality   201

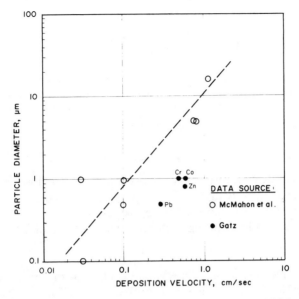

FIGURE 4.4. Deposition velocity for particles <10 μm. *Sources*: ○ McMahon, Dension, and Fleming, 1976; ● Gatz, 1975.

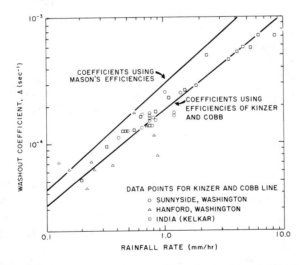

FIGURE 4.5. Rain washout coefficient λ related to rainfall intensity. (From Slade, 1968.)

The wet fallout per unit area, D_w, is then

$$D_w = (C_{w,0} - C_w)H = C_{w,0}(1 - \exp(-\lambda t))H \qquad (4.8)$$

where $H =$ the depth of atmosphere thorough which the pollutant plume is mixed. The magnitude of the washout coefficient, λ, is of the order $10^{-4} \sec^{-1}$ and is a function of rain intensity (Fig. 4.5). The magnitudes of λ are similar both for gases and particulates.

Example 4.2: Concentration of Pollutants in Precipitation

The atmospheric concentration of phosphate before rain was estimated or measured as $C_{w,0} = 10\,\mu g/m^3$. The depth of the mixed atmospheric layer extended about 1000 meters above the ground surface. Estimate the amount of wet fallout during a storm with a volume of 20 mm lasting 2 hours.

Solution From Figure 4.5 the washout coefficient for the storm intensity $i = 10\,\text{mm}/2\,\text{hr} = 5\,\text{mm/hr}$ is $\lambda = 7 \times 10\,\sec^{-1} = 2.52\,\text{hr}^{-1}$. Then the mass of deposited phosphate becomes (Eq. (4.3)):

$$\begin{aligned}D_w &= C_{w,0}(1 - e^{-\lambda t})H = 10(1 - e^{-2.52 \times 2})1000 \\ &= 9935\,\mu g/m^2 = 9.935\,mg/m^2\end{aligned}$$

The phosphate concentration in rainwater (rain volume $V_r = 10\,\text{mm} = 0.01\,m^3/m^2$) is

$$C_r = \frac{D_w}{V_r} = \frac{9.93\,mg/m^2}{0.01\,m^3/m^2 \times 1000\,l/m^3} = 0.99\,mg/l$$

Acidity

All rainfall is by nature somewhat acidic. The principal factor in rainfall's acidity is carbon dioxide. Pure water in equilibrium with the atmosphere would have a pH of 5.6. The accumulation of acidic chemicals in the atmosphere is also contributed to by decomposing organic matter, volcanic eruptions, and movements of the sea.

In some parts of the world the acidity of rainfall is considerably lower than pH 5.6, primarily because of the emission of sulfur dioxide (SO_2) and nitrogen oxides (NO_x). The areas of maximum deposition of chemical species related to acidity (H^+, SO_4^{--}, NO_3^-) in North America are located in the northeastern United States and southeastern Canada (Figs.

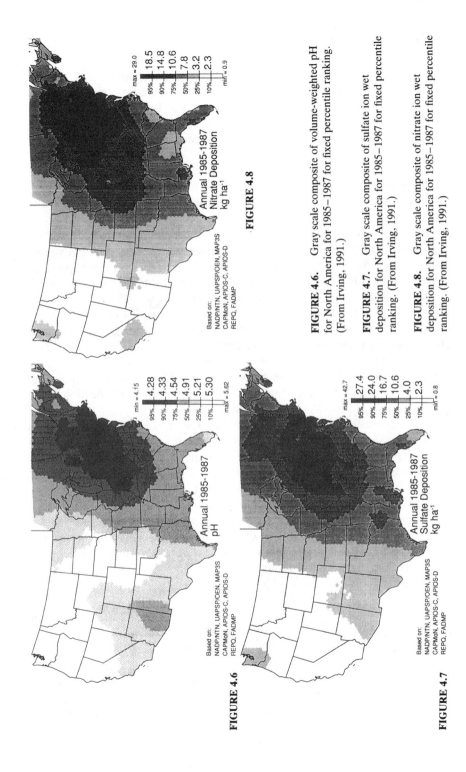

FIGURE 4.6. Gray scale composite of volume-weighted pH for North America for 1985–1987 for fixed percentile ranking. (From Irving, 1991.)

FIGURE 4.7. Gray scale composite of sulfate ion wet deposition for North America for 1985–1987 for fixed percentile ranking. (From Irving, 1991.)

FIGURE 4.8. Gray scale composite of nitrate ion wet deposition for North America for 1985–1987 for fixed percentile ranking. (From Irving, 1991.)

4.6–4.8). The pattern for annual average pH of wet deposition for the United States shows that large areas in and downwind of emission zones have depositions in the pH range of 4.1 to 4.5. Individual precipitation events can produce rainfall as acidic as pH 3 or below. These areas have considerably higher levels of atmospheric deposition (as measured by acidifying species) than remote parts of the world. Table 4.5 summarizes wet deposition pH values from remote regions of the world. When areas with possible anthropogenic influence are taken into account, it is apparent that the average pH of precipitation in remote regions are closer to 5.0 than 5.6, the pH initially thought to represent "clean" rain.

In areas where precipitation pH is below about 4.6, we also find surface waters that are acidic (considered here to be those with pH values below about 5). Some surface waters are more sensitive to precipitation acidity than others because of differences in the neutralizing capacity of watersheds. The National Surface Water Survey (NSWS) was initiated in 1986 to establish the current chemical status of surface waters in the United States in relation to known deposition levels and the biogeochemical environment (Linthurst, Landers, and Eilers, 1986). The key elements of the survey were three major synoptic surveys of lakes and streams in eight geographic regions of the United States believed to contain the majority of surface waters susceptible to acidification (Fig. 4.9).

The survey found acidic lakes and streams in some of those regions, although some of them were naturally acidic or acidic for reasons other than deposition (Baker et al., 1990). The majority of the surface waters were circumneutral but weakly buffered, with about half the lakes and streams having acid-neutralizing capacities (ANC) $\leq 200\,\mu eq/l$ (Tables 4.6 and 4.7). A relatively small percentage (4.2%) of the 1181 lakes investigated were acidic. Most acidic lakes were found in the Northeast, Florida, and the upper Midwest. Of the total stream length in the survey, 2.7% (5506 km) was acidic. Most of the acidic stream length was in the mid-Appalachian and mid-Atlantic coastal plain regions. The researchers concluded that atmospheric deposition was the dominant source of SO_4 in most surface waters sampled.

Baker et al. (1990) conducted a literature review of acidic lakes in the high deposition areas of Canada, Finland, Norway, Sweden, and the Kanai Peninsula of Alaska, plus several acidic lakes in low deposition areas. As in the National Survey, there was a positive relationship between SO_4^{--} deposition and lake water SO_4^{--} concentrations in Canada and Scandinavia. Nitrate in deposition was found to play a more important role in lakes in southern Norway and parts of Europe than was observed in lakes in the United States and Canada.

TABLE 4.5 Summary of Wet Deposition pH Values from Remote Regions of the World

Location	N^a	Period	pH Min	pH Max	pH Avg	Other Chemical Measurements	Sampling Protocol[a]	Possible Anthropogenic Influence
American Tropics								
Rain Forest								
Manaus, Brazil	53	1966–1968	3.6	5.4	4.6^b	None reported	UNK	Urban (definite)
Adolfo Ducke Forest, Brazil	2	UNK	—	—	4.38^c	None reported	UNK	Urban (possible)
Amazon River from coast of Brazil to Colombia, Peru, Bolivia	31	1976–1977	4.71	5.67	5.03^c	Full inorganic	F/B-wet	Unlikely
La Selva, Costa Rica	UNK	1973	4.40	4.90	4.66^c	None reported	UNK	Unlikely (volcano influenced)
San Carlos, de Rio Negro, Venezuela	70	1979–1980	4.0	6.7	4.69^d	None reported	Bottles	Unlikely
Cloud Forest								
Alto de Pipe, Venezuela	19	UNK	4.21	5.90	5.03^c	None reported	UNK	Urban (definite)
San Eusebio, Venezuela	UNK	UNK	3.82	6.21	4.55^c	None reported	UNK	Unlikely
Savannah								
Calabozo, Venezuela	151	1981–1983	4.8	6.9	5.8^d	NH_4^+	F/B-bulk	Unlikely (wildfires)
Camburito, Venezuela	18	1983–1984	4.0	5.2	4.4^d	None reported	F/B-wet	Unlikely (wildfires)
Joaquin del Tigre, Venezuela	17	1984–1985	4.2	5.8	5.1^d	None reported	F/B-wet	Urban (possible)
La Paragua, Venezuela	14	1985	4.0	5.6	4.8^d	None reported	F/B-wet	Unlikely (wildfires)
Lake Valencia, Venezuela	92	1977–1978	3.2	7.7	5.9^e	Full inorganic	F/B-bulk	Light industry (possible)

TABLE 4.5 (*Continued*)

Location	N^a	Period	pH Min	pH Max	pH Avg	Other Chemical Measurements	Sampling Protocol[a]	Possible Anthropogenic Influence
Greenland								
East coast	10	1981–1983	4.40	6.00	5.13[e]	Full inorganic	Bulk (long exposure)	Local source (possible)
Israel								
Negev Desert	30	1978–1983	—	—	7.9[f]	All major ions but NO_3^-, NH_4^+, pH	F/B-wet	Urban (possible)
Portugal								
Coimbra	195	1978–1980	3.5	7.7	4.75[g]	Conductance	F/B-bulk	Urban (definite)
Australia								
Hunter Region, New South Wales	UNK	1984–1986	4.0	7.0	5.0[d]	Conductance	Bulk/wet only	Unlikely
New Zealand								
Maimai, South Island	10	1985	6.2	5.2	5.6[e]	Full inorganic	F/B-bulk	Unlikely

Source: From Sisterson et al. (1990).

Note: Probable local anthropogenic influence category is deduced from information provided in the original studies.
[a] N is the number of samples, and UNK is unknown; F/B is funnel/bottle sampling.
[b] Median of simple monthly average pH values.
[c] Simple average pH value.
[d] Volume-weighted pH value.
[e] Median value.
[f] pH calculated from average bicarbonate concentration.
[g] pH calculated from average H^+ concentration.

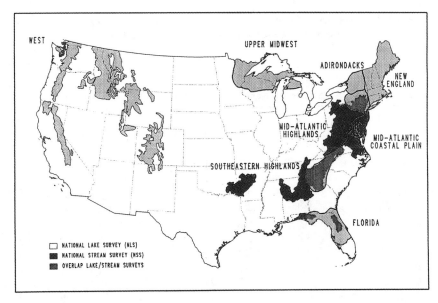
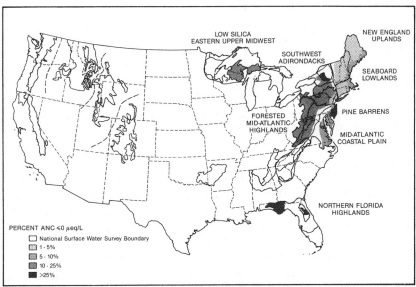

FIGURE 4.9. National Acidic Precipitation Assessment Program—Survey of surface bodies of water; (a) Areas included in the survey; (b) surface waters identified as affected. (From Baker et al., 1990.)

TABLE 4.6 Population Estimates of the Percentage of the Total Number of NSWS Lakes and Streams with ANC, pH, and Al in Reference Ranges[a]

Region	Lake or Stream[b]	Total Number	ANC (μeq/l) ≤0	≤50	≤200 %	pH ≤5.5	≤6.0	Al$_{MIBK}$ (μg/l)[c] >50
Northeast	L	7,096	6	22	6	19	13	6
	S-u	3,235	6	23	48	7	13	5
	S-d	3,235	1	5	30	<1	3	<1
Mid-Appalachians	S-u	21,527	5	20	49	9	15	6
	S-d	21,527	2	9	43	2	6	2
Mid-Atlantic coastal plain	S-u	11,284	12	30	56	24	49	37
	S-d	11,284	7	20	41	13	22	20
Interior Southeast	L	258	1	1	34	1	<1	<1
	S-u	18,598	1	7	52	2	9	<1
	S-d	18,598	1	6	47	<1	3	<1
Florida	L	2,098	23	40	55	21	33	6
	S-u	1,274	25	66	77	39	69	3
	S-d	1,274	11	59	82	23	59	3
Upper Midwest	L	8,501	3	16	41	4	10	3
West	L	10,393	<1	16	66	<1	1	<1
All NSWS	L	28,346	4	19	56	5	9	3
	S-u	55,917	2	11	44	4	9	5

Source: From Baker et al. (1990).
[a] Based on fall index chemistry (lakes) and spring baseflow (streams).
[b] L = lakes; S-u = streams—upstream ends of reaches; S-d = streams—downstream ends of reaches.
[c] MIBK: methyl-isobutyl-ketone (method that measures total monomeric Al).

Acidic lakes and streams are not common in areas where acidic deposition is low. Where they are found, they are typically associated with areas where there are large amounts of dissolved organic carbon (DOC). Other reasons for highly acidic surface waters in areas of low deposition include acid mine drainage, geothermal springs, high chloride clearwater coastal lakes, and lakes with extremely low conductivity.

As has been pointed out the acidity of precipitation is attributed to the presence of sulfates and sulfides (SO_4^{--} and SO_3^{--}) and nitrates (NO_3^-) in the atmosphere. Sulfur is one of the elements that is always found in the atmosphere, and it occurs as SO_4^{--} and SO_3^{--} in aerosols and SO_2 and H_2S gases. Hydrogen sulfide in air is normally oxidized to SO_2, which is then oxidized to SO_3. The oxidation reaction proceeds quickly if such metallic catalysts as iron and manganese oxides are present (Stern, 1976).

TABLE 4.7 Population Estimates of the Percentage of NSWS Lake Surface Area (km²) and Stream Length (km) with ANC, pH, and Al in Reference Ranges

Region	Lake or Stream	Total Resource[b]	ANC (μeq/l) ≤0	≤50	≤200 %	pH ≤5.5	≤6.0	Al$_{MIBK}$ (μg/l)[c] >50
Northeast	L	4,279	2	16	68	4	6	3
	S	15,144	4	11	36	6	9	6
Mid-Appalachians	S	54,425	3	11	47	7	13	7
Mid-Atlantic coastal plain	S	40,296	6	24	52	24	47	48
Interior Southeast	L	243	<1	<1	55	<1	<1	<1
	S	86,938	<1	5	48	<1	7	1
Florida	L	662	20	49	64	24	33	9
	S	3,848	12	61	76	44	74	40
Upper Midwest	L	5,015	<1	4	15	<1	2	<1
West	L	1,819	<1	8	35	<1	<1	<1
All NSWS	L	12,016	2	11	40	3	5	2
	S	200,652	3	12	48	8	18	13

Source: From Baker et al. (1990).
[a] Based on fall index chemistry (lakes) and spring baseflow (streams).
[b] Total resource for lakes is expressed as surface area (km²); total resource for streams is expressed as length (km).
[c] Total monomeric Al (as measured by the MIBK (methyl-isobutyl-ketone) method; see Section 2.5.3.4 of the full Report).

These metallic compounds are commonly emitted by the burning processes in fly ash. Formation of sulfuric acid is greatly enhanced by the moisture emitted from cooling towers (Fig. 4.10).

The SO_4^{--} and NO_3^- anions in the air are balanced by cations, primarily NH_4^+, Ca^{2+}, Mg^{2+}, and N^+. The major sources of these compounds are sea spray, soil dust, and ammonia volatilization from soils. Since Na^+ from sea spray is already balanced by Cl^- in the absence of other buffering agents in the air, there may not be other cations available to balance additional SO_4^{--} and NO_3^- ions, so they can only react with H^+ to produce acid rain (Fig. 4.11).

Example 4.3: Acidity of Precipitation

The ambient concentration of sulfur trioxide and sulfates (SO_3 and SO_4^{--} is about 30% of the ambient SO_2 concentration. Estimate the approxi-

FIGURE 4.10. Formation of sulfuric acid from the SO_2 emissions is greatly enhanced when stack effluent combines with the vapor drift from cooling towers. (Photo: V. Novotny.)

mate pH of rainwater resulting from 5 mm of rain lasting 5 hours if the ambient SO_2 concentrations is $20 \mu g/m^3$. The mixed atmosphere depth is $H = 1000$ meters.

Solution From Figure 4.5 the washout coefficient $\lambda = 2 \times 10^{-4} \, \text{sec}^{-1} = 0.72 \, \text{hr}^{-1}$. Rainfall volume $V_r = 5 \, \text{mm} = 5 \, \text{l/m}^2$. Mass of sulfates washed out:

$$D_w = (30/100)C_{w,0}(1 - e^{-\lambda t})H = 0.3 \times 20(1 - e^{-0.72 \times 5}) = 5836 \, \mu g/m^2$$

Sulfate concentration in rainwater

$$C_r = \frac{D_w}{V_r} = \frac{5.836 \, \text{mg/m}^2}{5 \, \text{l/m}^2} = 1.17 \, \text{mg/l}$$

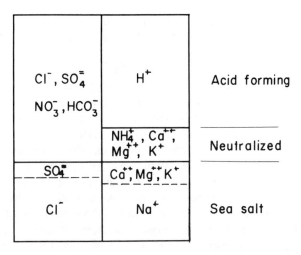

FIGURE 4.11. Atmospheric cation–anion balance resulting in an increase of H^+ ions in rainwater.

Change C_r to equivalent SO_4^{--} (equivalent weight of SO_3 = [32 + 3 × 16]/2 = 40 g/eq = 40,000 mg/eq).

$$\{SO_4^{--}\} = \frac{C_r}{EW(SO_3)} = \frac{1.17 \text{ mg/l}}{40,000 \text{ mg/eq}} = 2.92 \times 10^{-5} \text{ eq/l}$$

Each equivalent of SO_4^{--} must be balanced by one equivalent of $[H^+]$ or any other available atmospheric cation (such as NH_4^+). The $[H^+]$ concentration of rainwater not contaminated by SO_x (pH = 5.6) is $[H^+] = 10^{-pH} = 10^{-5.6} = 2.5 \times 10^{-6}$ mole/l, which is more than one order of magnitude less than the equivalent weight of $[H^+]$ required to balance SO_4^{--} ion. Since almost all of the sulfuric acid in water is dissociated, the resulting pH is roughly

$$\text{pH} = \log\frac{1}{[H^+]} = \log\frac{1}{\{SO_4^{--}\}} = \log\frac{1}{2.92 \times 10^{-5}} = 4.5$$

Effects of Precipitation Acidity on Drinking Water

Concerns have been raised about the possibility that acidic deposition might contribute to levels of chemical contaminants in untreated or partially treated drinking water in excess of established or contemplated drinking water standards. Background information is presented on concepts important to the understanding of chemical contaminants in drinking water that may be related to acidic deposition.

Instances of elevated concentrations of chemical contaminants exceeding drinking water standards have been reported for precipitation collected in cisterns, shallow wells and springs, and surface waters. The elevated concentrations are usually the result of pipe corrosion and other materials in the distribution system that come in contact with the water.

In the context of this chapter corrosion is defined as the deterioration of a pipe or fixture by electrochemical reaction with its environment (Patterson and O'Brien, 1979). Water's corrosive tendency depends on its physical, chemical, and biological characteristics and the nature of the material with which it comes in contact. An example of physical action is the erosion or wearing away of a pipe elbow because of excess flow velocity in the pipe. A simple example of chemical action is the oxidation or rusting of an iron pipe. An example of biological action is the release of corrosive by-products by iron-oxidizing and sulfate-reducing bacteria (Singley, Beaudet, and Markey, 1984).

There are several piping materials used in drinking water distribution and home plumbing systems. Common piping materials are composed of copper with lead-based solder, galvanized iron, plastic, and asbestos-cement. Besides the material itself, contaminants in it, such as cadmium and lead, can leach into the water. Even certain plastic piping may contain contaminants that can leach into the water due to corrosion. For example, some plastic pipes have been reported to contain lead stearate as a stabilizing agent in its manufacture (Jacks, 1984). Little is known, however, about the degree of leaching of lead from plastic piping. Further, the brands of plastic that contain lead stearates are trade secrets, thus, the leaching potential of a particular brand cannot be ascertained without experimentation.

The contaminants of interest that may leach from piping due to corrosion are arsenic, asbestos, cadmium, copper, lead, selenium, and zinc. Aluminum, mercury, and nitrate are not significantly leached from piping due to the corrosivity or aggressive properties of water. The results of three U.S. surveys and one in Canada are summarized below for lead, the contaminant of most concern in corrosive drinking water supplies. The impact of rainfall acidity on elutriation of metallic ions (zinc) from tin roofs and downspouts is discussed in Chapter 8.

New England Water Works Association Survey. Taylor et al. (1984) sampled raw and treated surface water and ground water for lead. One of the 328 samples (0.30%) of untreated source waters had lead levels above the current U.S. MCL of $50\,\mu g/l$. This was a ground-water sample. Treated waters were sampled at the treatment facility before distribution through piping and plumbing systems. Of the 134 treated surface water samples two (1.5%) contained lead levels above the current MCL, while one of the 22 treated ground-water samples (4.5%) contained lead levels above the MCL.

National Inorganics and Radionuclides Survey. The EPA collected tap water samples from nearly 1000 ground-water drinking water supplies in the United States on which they conducted analyses for selected inorganics and radionuclides (Logtin, 1988). Of the 983 sites sampled for lead, the concentrations in tap water ranged from <5 mg/l to 240 µg/l. A total of 55 samples (5.6%) were above the minimum reported detection limit of 5 µg/l, which is also the proposed MCL. Only four of the 983 samples (0.41%) were above the current MCL of 50 µg/l. All of the systems exceeding the current MCL were those serving less than 2500 households.

Rural Water Survey. The rural water survey described previously sampled tap water at 2654 households for some constituents, including lead concentrations (Francis et al., 1984). A total of 16.6% of the supplies contained lead levels above the MCL. The median concentration was 8 µg/l. The highest recorded lead concentration was 970 µg/l, nearly twenty times higher than the current MCL. The authors attributed the high lead levels to leaching of lead from distribution and plumbing systems, but also noted potential lead contamination from sample bottles.

Canadian Survey. Meranger, Subramanian, and Chalifoux (1979) reported that mean lead levels in treated water samples in 70 Canadian cities were equal to or less than 1 µg/l, and none of the samples in nine out of ten provinces exceeded 3 µg/l. The maximum lead concentration found in these source waters was 7 µg/l, one-seventh the current U.S. MCL.

Based on U.S. surveys of lead in ground-water supplies, the EPA (1988) estimated that approximately 900 ground-water suppliers (about 1%) may have water leaving the treatment plant at levels greater than 5 µg/l. Based on this information and other surveys, the EPA (1988) estimated that about 99% of the 219 million people in the United States using both surface and underground public water supplies are exposed to distributed water levels between 0 and 5 µg/l and about 1% (2 million people) are served by water with lead levels greater than 5 µg/l.

The occurrence of lead as a by-product of corrosion can be an important source of lead in drinking water at the user's tap. Three overall factors are particularly important in determining the degree of corrosion in the water distribution and home plumbing systems. First, the type and age of piping are important factors in the levels of lead leached from the plumbing system. For instance, the use of copper pipes joined with solder containing lead can result in elevated lead levels due to galvanic corrosion. This corrosion is especially reactive in the case of newly installed solder. Second, the corrosivity of the water toward lead is a major factor influencing the levels of lead at the consumer's tap. The corrosivity of water, in turn, is dependent upon many other factors. Third, regardless of the age of the piping, water that has been in contact with the pipe for a period of time will contain higher lead levels than flushed water from the same pipe.

For this reason, recent surveys of drinking water at the consumer's tap have focused on samples collected in the morning prior to any use of tap water in the house that day and/or samples collected after flushing, usually for 30 to 60 seconds. The first-draw samples provide a worst case estimate of contaminant concentrations, while the flushed samples may be more representative of average concentrations of water consumed during the day. The following three studies reported results of standing water analyses.

Patterson (1981) collected 782 random daytime grab samples flushed for 30 seconds from drinking water taps in 58 cities in 47 states. The percentage of samples collected from each state generally reflected the state population. It was found that 60% of all samples contained lead levels less than or equal to $10\,\mu g/l$, 84% less than $20\,\mu g/l$, and 97% less than $50\,\mu g/l$. Unfortunately, information on lead levels below $10\,\mu g/l$ were not available from this study, since the reported analytical detection limit was $10\,\mu g/l$.

Taylor and Symons (1984) sampled 36 treated surface waters and ground waters in seven northeastern states for lead at various points in the system. Samples were collected after standing overnight in plumbing, from the service line, and from the water main. None of the 36 samples from the water main had lead levels above the current MCL. One of the 36 samples from the service line (3%) had lead levels above the current MCL. Treated waters sampled at the tap first thing in the morning exceeded the lead MCL in 6 of the 36 samples (17%). The higher frequency of exceedences for the samples that involved additional contact with plumbing systems indicated corrosion of piping materials.

The American Water Works Service Company (1988) conducted a study of lead levels in 1484 drinking water supplies throughout the American water system. Subsequent evaluation indicated a general association between higher lead levels and several factors, such as pH and plumbing age. It was also observed that the lead levels at specific sites or within specific systems could not be predicted based on these factors.

Assessment

Although there is some likelihood that metals such as lead can be mobilized by acidity in drinking water systems, Grant et al. (1990) concluded that measurable increases in exposure will occur only under conditions related to specific, isolated exposure scenarios, such as individual drinking water systems that use surface ponds, shallow wells, or cisterns. Even assuming extreme-case conditions, such as acidic depositions resulting in a drinking water pH of 4.5, these researchers concluded that adverse health effects are unlikely except possibly for lead.

In the case of lead, there is a special reason for concern, which is that many individuals are already exposed to lead from multiple sources at levels that cause their blood lead to approach or exceed the lowest observed adverse effect level. As a result, any small increment attributed to acidic deposition might be sufficient to place some people in the range of exposure associated with increased risk of adverse health effects. On the other hand, increments in lead exposure projected to cause health effects are predicated on the assumption that acidic deposition can shift pH levels to the extreme levels necessary for lead mobilization. These extreme pH shifts are not confirmed by current research.

Nitrogen in Atmospheric Deposition

Nitrogen in deposition not only contributes to acidity, but it can also adversely affect the productivity of surface waters, particularly in the coastal zone. This section describes the current understanding of the atmospheric sources of nitrogen and its effects on near-shore waters. There remains considerable uncertainty about the importance of this source of nitrogen.

For many years, scientists believed that the portion of nitrogen entering near-shore waters was trivial compared to nitrogen from other sources, particularly land runoff and sewage treatment plant effluents. In recent years, however, numerous reports have suggested that atmospheric nitrogen contributes significantly to surface water quality.

Fisher et al. (1988) used an empirical approach to describe a cause–effect association between the amount of atmospheric nitrogen and its effects on the Chesapeake Bay. It was estimated that atmospheric nitrate and ammonia contribute 35 and 19 million kilograms, respectively, of biologically available nitrogen to the bay annually. Atmospheric nitrate and ammonia were estimated to account for 25% and 14%, respectively, of the total nitrogen loading to the bay.

Many scientists initially discounted the validity of the findings because the calculations necessarily required considerable conjecture and numerous simplifying assumptions. However, an objective analysis supported through the National Acid Precipitation Assessment Program reinforced the hypothesis that atmospheric nitrogen is likely a major contributor to surface water quality (Turner et al., 1990). Also, a report by the EPA suggested that at least 13% of the total nitrogen loading to the bay is from atmospheric deposition.

The results from the studies of the Chesapeake Bay may not be indicative of other near-shore waters. The fate of the ecosystem and the effects of nitrogen inputs vary substantially across ecosystems and sites. Information also suggests that a watershed has a high capacity to retain or denitrify the nitrogen entering it, and that this capacity is highly variable among watersheds.

Turner et al. (1990) assessed the effects of atmospheric nitrogen on surface waters according to ecosystem components. A summary is presented below:

1. Forest ecosystems
   - Biological fixation and nitrogen deposition are the two major processes providing nitrogen inputs to forests.
   - Little nitrogen is normally exported from forests because they tend to be nitrogen poor. The majority of atmospherically deposited nitrogen is assimilated by forested watersheds.

2. Agricultural ecosystems
   - Insufficient information is available to determine the effects of a change in agricultural management practices on the export of nitrogen.

3. Wetlands
   - Removal and/or storage processes by wetlands in the Chesapeake Bay could prevent the transport of atmospherically deposited nitrogen from reaching the bay.
   - Approximately 20% of the Chesapeake Bay drainage area passes through wetlands.

4. Freshwater ecosystems
   - A significant amount of surface-water nitrogen is removed within the stream–river ecosystem, mostly through denitrification.

5. Estuaries and near-coastal waters
   - The rates of nitrogen processes in these waters increase as nitrogen deposition loadings increase.
   - Nitrogen deposition onto watersheds is an important contributor to the total nitrogen loading in estuaries; however, the assumption that nitrogen export from the watershed is proportionate to nitrogen inputs is not necessarily valid.
   - There is some uncertainty whether nitrogen is largely retained or largely lost by the mechanisms of denitrification and advective transport.

Although most of the information presented in this section is based on calculations in the Chesapeake Bay watershed, recent evidence suggests that atmospheric nitrogen plays an important role in the water quality of Long Island Sound (New York), the Neuse River (North Carolina), Narragansett Bay (Rhode Island), Ochlockonee Bay (Florida), the Upper Potomac River Basin (Maryland and Pennsylvania), the New York Bight, and the Great Lakes (Turner et al., 1990).

TOXIC COMPOUNDS IN THE ATMOSPHERE

Lead in Atmospheric Deposition

The debate over SO_2 emissions and acidification has slowed since passage of the 1990 Clean Air Act, but the issue of toxic air pollution has intensified. For example, the effects of toxic metals emissions, such as lead, on aquatic ecosystems have received considerable interest recently. Less is known about air toxics and their effects on bodies of water than is now known about acidic deposition and surface-water acidification.

Not only can lead leach from acidified water distribution systems, but there is considerable lead emitted into the atmosphere. Lead (Pb) is a member of the Group IV elements (C, Si, Ge, Sn, and Pb) of the periodic table. Lead is truly metallic compared to carbon and silicon. Lead has stable (+2) and (+4) oxidation states. With the exception of nitrate and acetate, most (+2) lead salts, such as halides, hydroxides, sulfates, and phosphates, are only slightly soluble in water. Lead is one of the oldest metals known to man and, since medieval times lead has been used in building materials, piping, paint solders, ammunition, and castings. In modern times, lead has also been used in batteries, chemicals, and pigments.

Emissions of lead into the atmosphere have increased sharply during

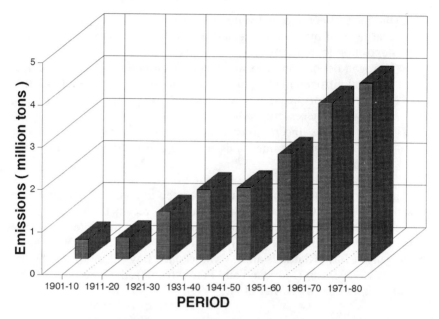

FIGURE 4.12. Global anthropogenic emissions of lead. (After Nriagu and Pacyna, 1988.)

the twentieth century (Fig. 4.12). Since the peak emissions of the 1970s (4265×10^3 tonnes), lead emissions have declined. Anthropogenic emissions greatly exceed those from natural sources (Table 4.8). Although mining discharges contribute the majority of lead to the land, atmospheric deposition is usually the most important source of lead in surface waters.

The emission of lead into the atmosphere is largely the result of

TABLE 4.8 Comparison of Annual Anthropogenic and Natural Emissions of Lead

Source	Global Production (10^6 metric tonnes)	Global Emissions (10^3 metric tonnes)
Anthropogenic emissions	Not available	4265
Wind-blown dust	500	16
Forest fires	36	0.5
Volcanic particles	10	6.4
Vegetation	75	1.6
Seasalt spray	1000	0.2

Source: From Nriagu (1979).

vehicular emissions, though smaller amounts are emitted from stationary fossil-fuel combustion sources. The emission rates of lead from automobiles using leaded gasoline vary from 30% to 90% of the lead content, depending on vehicle's speed, operation, and exhaust system. Average emission rates for lead are 60% for urban driving and 80% for highway driving (Provenzano, 1978). Lead emissions have decreased dramatically in the United States and other developed countries in the past few years with the ban of lead as an automotive fuel additive. Lead particles generally coagulate to a diameter of 0.1 to 1.0 µm, and thus are capable of being transported long distances.

According to the EPA (1985), atmospheric lead concentrations range from $0.000076\,\mu g/m^3$ in remote areas to $10\,\mu g/m^3$ near point sources. Average annual lead concentrations in air in most areas were reported to be below $1.0\,\mu g/m^3$. The EPA calculated the average intake of lead from respiration to be approximately 1 µg/day. This is very low compared to the maximum drinking water intake, which would be 100 µg/day, assuming there are 50 µg/l of lead present and a daily water intake of 2 liters.

Wet deposition of lead may roughly equal dry deposition. Lindberg et al. (1982) found nearly equal wet and dry annual fluxes of lead in the Walker Branch watershed, Tennessee. Peirson et al. (1973), however, showed that dry deposition of lead accounts for less than 10% of the total deposition at a rural site in England. This area, however, had much higher precipitation amounts. Lindberg (1982) showed that lead is higher in summer months due to increased air stagnation and a lower precipitation volume. Galloway, Eisenreich, and Scott (1980) reported the results of 32 studies of lead in wet deposition, wherein lead levels ranged from 0.6 to 64 µg/l, with a median of 12 µg/l. Direct atmospheric deposition of lead to a lake has been generally found to exceed watershed sources because lead levels in rocks and soils are generally low and lead is retained in soils by various mechanisms.

A significant amount of the total lead present in surface waters exists as suspended matter. The level of lead drawn from rivers and streams is generally slightly lower than the concentrations present in lakes and ground waters. The EPA (1988) reported the results of a survey by Fishman and Hem, who found that 86% of raw surface waters sampled contained less than 10 µg/l lead and that less than 1% had over 50 µg/l. Clarkson, Baker, and Sharpe (1984) reported that most natural ground waters have concentrations ranging from 1 to 10 µg/l.

In urban areas lead becomes incorporated into street dust and accumulates within 1 meter of the curb (see Chapter 8 for a more detailed discussion of atmospheric sources of toxic contaminants in urban areas).

Health Effects

The health effects of lead are generally correlated with blood lead levels. High lead exposure is associated with the interference with the heme synthesis necessary for formation of red blood cells, anemia, kidney damage, impaired reproductive function, interference with vitamin D metabolism, impaired cognitive performance, delayed neurological and physical development, and elevations in blood pressure (U.S. EPA, 1988). Human absorption of lead is apparently independent of the chemical form of lead. Effects on children generally occur at lower blood lead levels compared to adults.

Because there are several sources of lead exposure other than drinking water, it is important to examine the relative contribution of water lead to health effects. The EPA (1988) reviewed a number of such studies and suggested that 50 µg/l of lead (the current MCL) in morning first-draw drinking water contributes less than the lowest observed effect level in children, but that this contribution could be important to the total body burden of lead from all sources. Other major sources of lead exposure in humans, including air, soil, food, cigarettes, paint, and other specific sources, can increase the range of mean blood lead levels tenfold.

Mercury in Atmospheric Deposition

Mercury (Hg) is perhaps the toxic pollutant most widely discussed with regard to emissions and effects on bodies of water. One reason is that mercury has a long residence time in the atmosphere and can be transported to remote and pristine ecosystems.

Mercury is a member of Group IIB of the periodic table. It is a silver-white, heavy, mobile liquid at 25°C with a freezing point of −38.9°C. It is the most volatile metal known, with a vapor pressure of approximately 2×10^{-3} torr at 25°C. The principal valence states of mercury are +1 (mercurous) and +2 (mercuric).

Mercury exists in two basic forms, inorganic salt and organic mercury compounds (methylated mercury). The major anthropogenic impact of mercury occurs when inorganic mercury in soils and sediments is methylated to a more toxic methyl-mercury (see Chapter 13).

Sources of Mercury

Mercury is found naturally in soil in the range of 30 to 300 mg/kg (Shacklette and Boerngen, 1984). In addition, there are known areas with substantially elevated mercury in geologic materials. For example, in the United States mercury in soils is elevated in certain parts of California and Nevada.

The major uses of mercury in industry are in electrical equipment (batteries, lamps, switches, and rectifiers) and in the chloralkali industry as a flowing cathode for electrolytic deposition of salt brine into chlorine, sodium hydroxide, and hydrogen. Mercury is used in agriculture as a fungicide. The medical profession uses mercury in dental and pharmaceutical products.

In addition to these sources, mercury may enter the environment from mining, smelting, and fossil-fuel combustion. Mercury is present in coal in the range of 10 to 46,000 mg/kg, though generally it is in the range of 200 to 400 mg/kg (U.S. EPA, 1985). Because of its high volatilization, most mercury in smelting and combustion processes is emitted to the atmosphere. On the other hand, mineral extraction can result in mercury contamination of soil and ground water from improper mining and reclamation practices. Soil and ground water can also become contaminated from improper disposal practices in industries that use mercury and burn fossil fuels. Natural processes such as volcanic activity, geothermal activity, and volatilization from mineral deposits also result in mercury entering the atmosphere.

Most atmospheric mercury has a long residence time in the atmosphere—on the order of months—and consequently is dispersed globally (Lindqvist et al., 1984). Still, a small but significant fraction of what is emitted at a particular source is deposited in its vicinity. Several studies of mercury deposition around chloralkali plants clearly show that deposition is augmented well above the background within 5 or 10 km from the plants. In fact, Lindqvist et al. (1984) calculated that 10% to 20% of the emissions are deposited within 10 km of the plants.

Mercury Exposure to Humans

The most significant pathways of mercury into humans include inhaled air, dust, food, water, and medical care. These exposure pathways are represented in Figure 4.13 and described further in the following list.

Table 4.9 shows estimates of natural and baseline exposures to mercury for children and adults. These estimates are based on consumption factors in inhaled air, dust, food, and drinking water. Estimates for natural exposures represent background exposure to natural global sources of mercury with no input from anthropogenic sources or localized elevated natural sources. The baseline exposures represent background anthropogenic global sources. No estimates are provided for medical exposure because of the specific nature of this route of exposure (for example, medications used, dental fillings present, and other dental treatments).

222   Atmospheric Deposition

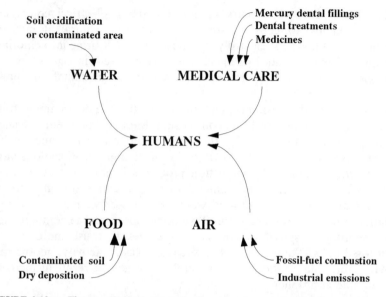

FIGURE 4.13.   The quantitatively most significant exposure pathways of mercury in human uptake.

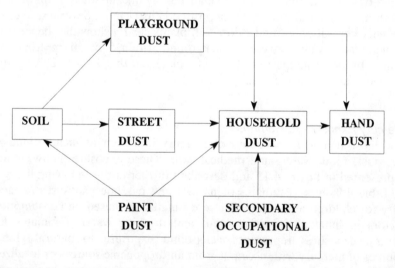

FIGURE 4.14.   Intermediate pathways and source of exposure to mercury through dust.

TABLE 4.9 Exposure of Children and Adults of Natural and Baseline Concentrations of Mercury (μg/day)

	Children			Adults		
Source	Factor	Natural	Baseline	Factor	Natural	Baseline
Inhaled air	17 m^3/day	0.00	0.00	20 m^3/day	0.00	0.00
Dust	0.2 g/day	0.20	0.80	0.2 g/day	0.02	0.08
Food	1120 g/day	1.12	1.63	2200 g/day	2.20	3.20
Water	0.5 l/day	0.00	0.05	2.4 l/day	0.00	0.10
Total		1.32	2.48		2.22	3.38

Source: After Grant et al. (1990).

Inhaled air. Mercury is introduced into the air from both natural and anthropogenic sources, as described in the previous section. Methylmercury is absorbed from the lungs after inhalation and through the skin. Because mercury and its compounds tend to bind to soil and sediments in the environment, direct inhalation is a minor source of exposure compared to the other sources. Atmospheric concentrations may be as low as 0.01 ng/m^3 for remote sources and 0.07 ng/m^3 for rural areas. Levels may increase to as high as 4 ng/m^3 near industrial emissions, but this level is still relatively low compared to other exposure pathways. Therefore, mercury uptake through inhaled air is considered an unmeasurable source at natural and background exposures compared to the others included in Table 4.9.

Dust. Mercury can enter dust from a variety of sources (Fig. 4.14). The dust is ingested and is absorbed from the gastrointestinal tract. The origin of most dust in the human environment is ultimately the soil. Mercury is deposited mainly as metallic metal and divalent salts to the soil. The divalent salts dominate because they are water soluble and are washed out of the atmosphere with wet precipitation. For children, almost all exposure to dust is by hand-to-mouth activity. Because hand-to-mouth contact is considerably higher in children, they have higher total exposure to mercury from this source than do adults. The exposure levels for dust are an order of magnitude higher for children compared to adults. Therefore, exposure to elevated levels of mercury in dust is an important pathway for children.

Food. Food is a primary source of exposure to mercury. Mercury is ingested with food and absorbed from the gastrointestinal tract. Humans absorb about 90% to 100% of the methyl-mercury that they ingest. Meat, fish, and poultry are the only important sources of mercury in food, and the major dietary source appears to be the consumption of fish (Hoffman, 1989). Fruits, vegetables, and dairy products contain relatively little mercury (Gartrell et al., 1986), because methyl mercury is taken up by insects in the terrestrial environment and bottom feeders in the aquatic food chain and are passed up the trophic levels to carnivores and then to humans. Bioaccumulation also occurs along the way. Based on a number of studies, the dietary intake of mercury is estimated to be 3.2 µg/day for the adult male (excluding beverages), of which about 80% comes from meat, fish, and poultry categories.

Drinking water. In the soil, divalent mercury is absorbed by organic matter in ion exchange sites, because it is easily mobilized during soil acidification. Several biochemical processes subsequently transform the mercury into organically bound forms that are not very mobile. Almost all mercury detected to date in drinking water is in the form of inorganic mercury, and the levels are normally below 0.5 µg/l. The estimated total mercury exposure to adults from drinking water is 0.1 µg/day.

Medical and dental exposure. A number of studies document that dental mercury amalgam fillings can cause substantial leaching of mercury, primarily through vapor emissions that are efficiently absorbed in the lungs or soft tissue (Sverdrup, 1990). There is a great deal of controversy concerning the relation between mercury poisoning and dental fillings. The evaluation and interpretation of all the results reported in the literature is beyond the scope of this discussion. The values reported in the literature indicate that the average mercury uptake to the body due to fillings is in the range of 1 to 50 µg/day. Other medical and dental sources of mercury include pharmaceutical products, such as various anointments and disinfectants for surficial wounds, eyedrops, and vaccines. Several of these products have been recently removed from the market by the manufacturer.

Guidelines and Standards for Maximum Human Intake

Many governments have set occupational exposure limits, drinking water concentration limits, and daily intake limits for mercury. Table 4.10 shows selected limits taken from the U.S. Environmental Protection Agency, World Health Organization, Swedish Food Administration and Environmental Protection Board, and the Japanese Food Administration. The actual limits were adapted as needed to standardize the numbers for comparative purposes. For example, the WHO limits for intake were

TABLE 4.10 Exposure Limits for Mercury Taken from Selected Government Guidelines and Standards

Mercury Form	Organization	Occupational 8 hr/5 day Air Limit ($\mu g/m^3$)	Environmental 24 hr/7 day Air Limit ($\mu g/m^3$)	Drinking Water Limit ($\mu g/l$)	Maximum Daily Uptake Limit ($\mu g/day$)	Maximum Daily Intake Limit ($\mu g/day$)
Methylmercury	WHO	25	1.3	1	22	25
	EPA	10	0.5	2	22	25
	Sweden	10	0.1	—	30	30
	Japan	—	—	—	35	35
Mercury vapor	WHO	50	—	—	50–80	60–100
	EPA	50	12	—	30	35
	Sweden	50	2.5	—	50–80	60–100
	Japan	50	—	—	—	—

Source: After Sverdrup (1990).

given per unit body weight, and for comparison have been presented here based on an average adult weight.

The toxic effects of mercury are related to the chemical form of the mercury compound. Based on toxicological characteristics, there are three important forms of mercury: elemental, inorganic, and organic compounds. Methylmercury is a highly toxic organic compound and is the chemical species of most concern in terms of potentially increased human health risk due to mercury exposure. Humans absorb about 90% to 100% of the methylmercury that they ingest. Methylmercury is resistant to environmental degradation and is capable of passing through important biological barriers such as the blood/brain membranes and the placenta. Inorganic mercury, on the other hand, is poorly absorbed through the gastrointestinal tract, does not penetrate cell membranes rapidly, and is less toxic than methylmercury.

Summary of lowest observed adverse effect levels. The lowest observed adverse effect levels for adults and for women during pregnancy are summarized in Table 4.11, which is based on the mercury poisonings mentioned previously and other clinical studies. By applying a safety factor of 10, tolerable daily intake becomes 30 μg Hg/day for adults and 8 μg Hg/day for women during pregnancy. As can be seen from a comparison of these figures with the baseline exposures described previously, the tolerable intake level for pregnant women does not allow for a large intake of mercury-contaminated food.

Environmental Effects

Effects on plants. Mercury can be introduced to plants from soils contaminated by atmospheric deposition or the application of wastewater treatment sludge. Fungicides based on mercury were used in the past for treating stored seed grain against molding, and the mercury may remain in the soil in the regions where it was used. No reports could be

TABLE 4.11 Lowest Observed Adverse Effect Levels for Mercury for Adults and for Women during Pregnancy

	Critical Effects	
Mercury Levels (as methylmercury)	Women during Pregnancy	Adults
Exposure (μg/kg/day)	0.8–1.7	4.3
Red blood cell mercury (μg Hg/l)	40–80	400
Hair (μg Hg/g)	10–20	50–100

Source: After Grant et al. (1990).

found of even these high levels of mercury in soils causing effects on plants.

Ecological effects due to soil and ground-water contamination. Typical levels of mercury in the soil are 0.1 to 1.2 mg/kg in the upper soil horizon (O-horizon), 0.0005 to 0.002 mg/kg in the A and B horizons, and as low as 0.003 to 0.008 mg/kg in the mineral soil (Sverdrup, 1990). The high mercury levels in the upper soil horizons indicate that mercury is not highly mobile in soils. This observation also suggests that it is not likely that mercury contamination results in high levels in ground waters.

Low ground-water mercury concentrations have been confirmed through surveys conducted by the U.S. Environmental Protection Agency and others (U.S. EPA, 1985). Compliance monitoring results from thousands of ground-water supplies in the United States show only 12 supplies with levels at any time above the drinking water limit for mercury of 2 µg/l. Of 95 ground-water samples collected by in the mid-1980s in acidified regions of the northeastern United States, the maximum mercury concentration was 1.1 µg/l (Taylor and Symons, 1984).

Pesticides and Other Organic Chemicals in Atmospheric Deposition

Generally, runoff from agricultural areas produces locally high concentrations of pesticides and herbicides, while atmospheric transport and deposition results in low-level but widespread contamination. It is generally accepted that certain organochlorine chemicals, such as DDT, PCBs, and toxaphene, can contaminate surface waters some distance from the source of application (Hites and Eisenreich, 1987; Mackay, 1990). These chemicals evaporate from their point of production or use, are transported in the atmosphere, and then enter surface waters by

- direct absorption
- wet and dry deposition
- washoff from the surrounding watershed.

The atmosphere becomes contaminated with pesticides and other organic chemicals by drift, wind erosion, and postapplication volatilization (Glotfelty et al., 1990). Drift is that portion of the spray that is moved away from the target area by wind. Aerial spraying (Fig. 4.15) contributes a large portion of the atmospheric drift input. Chesters and Simsiman (1974) estimated drift losses for pesticides applied aerially as ranging from 25% to 75% of the quantity applied. The extent of drift

FIGURE 4.15. View of pesticide spray operation by plane. (Photo: USDA.)

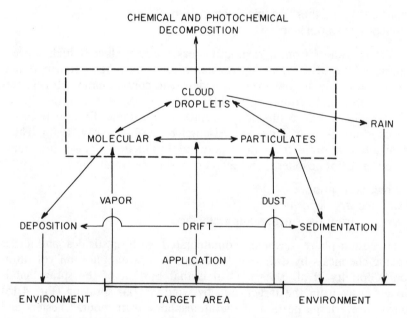

FIGURE 4.16. Pesticide input and removal from the atmosphere. (From Glotfelty, 1978.)

and the subsequent dispersion and transport of pesticides in the air are governed primarily by prevailing atmospheric conditions, the chemical constitution of the pesticide, and the method of application. High temperature and windy conditions (unstable atmosphere) accentuate drifting. The quantity and extent of drift losses can be reduced by considering a combination of interrelated factors such as spray formulation, type of equipment used, meteorological factors, and spray method used.

Volatilization has been recognized as a pathway of pesticide loss from soil, plant, and water surfaces. Once in soil, subsequent losses can differ depending whether the pesticide is surface applied or incorporated into the soil. Chapter 6 describes the interactions of pesticides and organic chemicals with soils. Largely due to their volatility, the chemicals also actively cycle between the atmosphere, soils, and water bodies. The cycle of pesticides between the soil and atmosphere is shown on Figure 4.16.

Although any organic chemical can cause a local diffuse-pòllution problem, only persistent compounds such as organochlorine pesticides, DDT and its derivatives, aldrin, dieldrin, and PCBs have global significance. The discovery of DDT and PCBs in Antarctic seals and snow (Peterle, 1969; Sladen, Menzie, and Reichel, 1966; Cramer, 1973; and Rosenbrough et al., 1976) suggests that both components have become a part of global atmospheric pollution and travel with the air to the most remote points of the earth surface. Paasivirta (1990) compared ratios of organochlorine concentrations in fish and predatory birds from the Baltic versus the Arctic in an attempt to rank the organochlorine contaminants from most locally to most globally dispersed in the atmosphere (Fig. 4.17). Toxaphene was found to be the organochlorine compound most globally dispersed, while HexaCN was most locally dispersed. No natural loading exists for these substances.

Goolsby (1991) measured herbicide concentrations in rainwater in a 23-state area, principally in the Midwest and Northeast. He found traces

FIGURE 4.17. General ranking of organochlorine contaminants from mostly locally to mostly globally dispersed, based on tissue ratios from fish and predatory bird species from Baltic and Arctic environments. (After Paasivirta, 1990.)

TABLE 4.12 Projected Amounts of Pesticides Entering Chesapeake Bay via Summer (May–August) Rainfall

	Wet Deposition (metric tonnes)[a]		
Pesticide	1981	1982	1984
Atrazine	0.88	1.2	0.64
Simazine	0.13	0.11	0.14
Alachlor	3.7	2.4	9.8
Metolachlor[b]	1.7	1.1	4.6
Toxaphene	0.54	1.1	na[c]

Source: From Glotsfelty et al. (1990).

[a] Pesticide data from Table 6, page 215, of Glotfelty et al. (1990). Area of Chesapeake Bay and tributaries is $11.9 \times 10^9 \, m^2$, from W. B. Cronin, "Volumetric, areal, and tidal characteristics of the Chesapeake Bay estuary and its tributaries," Special Report 20, Johns Hopkins University, Baltimore, Md. (Ref No. 71-2), 1971.
[b] Only a partial set of measurements is available for metolachlor; the projected amounts of metolachlor entering Chesapeake Bay are based upon the relative concentrations of metolachlor and alachlor in the samples for which data are available.
[c] na = not analyzed.

of herbicides in all 23 states and in all but two of the 81 collection sites. Pesticide concentrations in wet deposition were measured by Glotfelty et al. (1990) in Maryland. These latter researchers calculated the projected amounts of pesticides entering the Chesapeake Bay from precipitation. Table 4.12 gives the projected amounts of atrazine, simazine, alachlor, and toxaphene that entered the bay with rainfall during the summers of 1981, 1982, and 1984. Quantities of pesticides entering the bay ranged from a low of 110 kg of simazine in 1982 to a high of 9.8 tonnes of alachlor in 1984.

Strachen and Eisenreich (1990) attempted to calculate mass balances for selected chemicals entering the Great Lakes. The atmosphere was found to play an important, and often dominant role in the loading of many toxic chemicals into the Great Lakes. For example, 97% to 98% of the DDT loading into the upper Great Lakes was from the atmosphere. In the lower Great Lakes, with considerable toxic inputs from the land, the atmosphere played a smaller, but still important role, contributing 22% to 31% of the total DDT load.

Rapaport and Eisenreich (1986) reported that toxaphene deposition in peat bogs in northern Minnesota and northern Ontario reached a maximum in the 1970s that corresponded to the maximum production and use of the pesticide in the United States. Derek et al. (1990) demonstrated that toxaphene is also present in remote lakes in northern Ontario and Manitoba due to long-range atmospheric transport and deposition.

Residence Time of Organic Chemicals in the Atmosphere and their Deposition

In the absence of a strong local source, the transport of persistent chemicals is of global nature. Despite its simplistic nature, the global model (Eqs. (4.1) to (4.4)) can be applied with some accuracy to estimate global (background) loadings, deposition rates, and residence times. For a strong local source a Gaussian dispersion model should be used (see Novotny and Chesters, 1981, or Zanneti, 1992). Atmospheric layers up to the middle latitudes become mixed relatively quickly; air parcels move around the world in 3 to 4 weeks; and a quasi-steady state is reached in months.

As shown previously, the residence time of a chemical component in the atmosphere depends on its form and sinks. There is a great deal of controversy as to the average global residence time and, hence, deposition rates of the most important, persistent chemicals. Because the vapor pressure of these chemicals indicates that they most likely exist in the vapor phase, the residence times were compared to those for gases, which are in years. For example, Woodwell et al. (1971) estimated that the residence time of DDT is 4 years; however, because the ambient concentrations of DDT were very low, Glotfelty et al. (1990) and Orgil et al. (1976) speculated that at least a part of the vaporized chemical becomes adsorbed by atmospheric aerosols and is deposited at a much faster rate and with much shorter residence times.

Junge (1977) estimated that the overall average residence time for atmospheric aerosols is about 7 days, and if ϕ denotes the fraction of an organic chemical present in the air adsorbed on aerosols, the residence time is approximated as $7/\phi$. For DDT the value of ϕ was estimated in the range 0.28 to 0.48, and ϕ for PCBs seems to be <0.01. This would indicate that while the average residence time for DDT and similar organic components could be in months, the residence time of PCBs may be >1 year.

CONCLUSIONS

There are several major pollution sources from the atmosphere that can be transported long distances and fall to earth in rain, snow, mist, fog, and in dry form as gases and particulates. One of the best-known pollutants in atmospheric deposition is acidity. Acidity occurs when nitrogen and sulfur oxides are emitted from fossil-fuel combustion. Nitrogen emissions have also been shown to directly affect bodies of water, particularly in coastal areas, through increases in fertilization and resultant eutrophication. The widespread concern for acidification has abated in the United States since passage of the Clean Air Act of 1990. The act has major provisions to control acidifying emissions and the general public believes the problem is being adequately addressed.

Fossil-fuel burning also emits other substances, such as mercury, that fall to earth and can affect aquatic ecosystems. Other major toxic atmospheric pollutants that affect water quality are trace metals, such as lead, and agricultural chemicals such as pesticides and herbicides. There has been less public debate over the issue of air toxics, but these pollutants are not as well understood as those involved in acidification and they may not be adequately addressed in current regulations.

References

American Water Works Service Company (1988). *Lead at the Tap—Sources and Control. A Survey of the American Water System.* American Water Works Service Co., Voorhees, NJ.

Baker, L. A., P. R. Kaufmann, A. T. Herlihy, and J. M. Eilers (1990). Current status of surface water acid-base chemistry, in *Acidic Deposition: State of Science and Technology*, Vol. II, *Aquatic Processes and Effects*, P. M. Irving, ed., National Acid Precipitation Assessment Program, Washington, DC, pp. 9-1–9-367.

Chesters, G., and G. Simsiman (1974). *Impact of Agricultural Use of Pesticide on the Water Quality of the Great Lakes*, Task A-5, Water Resources Center, University of Wisconsin, Madison, WI.

Clarkson, T. W., J. P. Baker, and W. E. Sharpe (1984). Indirect effects on health, in *The Acidic Deposition Phenomenon and Its Effects*, Committee on Monitoring and Assessment of Trends in Acid Deposition, National Research Council, National Academy Press, Washington, DC, pp. 6-1–6-66.

Council on Environmental Quality (1990). *Environmental Quality*, 20th annual report of the Council on Environmental Quality, with the President's message to Congress, Washington, DC.

Cramer, J. (1973). Model of the circulation of DDT on earth, *Atmos. Environ.* 7:241–256.

Derek, C. G., N. P. Grift, C. A. Ford, A. N. Reiger, M. R. Hendel, and W. L. Lockhart (1990). Evidence for long-range transport of toxaphene to remote arctic and sub-arctic waters from monitoring of fish tissues, in *Long Range Transport of Pesticides*, D. A. Kurtz, ed., Lewis Pub., Chelsea, MI, pp. 329–346.

Fisher, D., J. Caraso, T. Mathew, and M. Oppenheimer (1988). *Polluted Coastal Waters: The Role of Acid Rain*, Environmental Defense Fund, New York.

Francis, J. D., B. L. Brower, W. F. Graham, O. W. Larson, III, J. L. McCaull, and H. M. Vigorita (1984). *National Statistical Assessment of Rural Water Conditions: Executive Summary*. U.S. Environmental Protection Agency, Office of Drinking Water, Washington, DC.

Galloway, J. N., S. J. Eisenreich, and B. C. Scott, eds. (1980). *Toxic Substances in Atmospheric Deposition: A Review and Assessment*, Report NC-141, National Atmospheric Deposition Program, Charlottesville, VA.

Gartrell, M. J., J. C. Craum, D. S. Podrebarac, and E. L. Gunderson (1986). Pesticides, selected elements, and other chemicals in infant and toddler total diet samples, *J. Assoc. Off. Anal. Chem.* **69**:123.

Gatz, D. F. (1975). Pollutant aerosol deposition into southern Lake Michigan, *Water Air Soil Pollut.* **5**:239–251.

Glotfelty, D. E. (1978). The atmosphere as a sink of applied pesticide, *J. Air Pollut. Assoc.* **28**:917–921.

Glotfelty, D. E., G. H. Williams, H. P. Freeman, and M. M. Leech (1990). Regional atmospheric transport and deposition of pesticides in Maryland, in *Long Range Transport of Pesticides*, D. A. Kurtz, ed., Lewis Publ., Chelsea, MI, pp. 199–221.

Goolsby, D. A. (1991). *Regional Studies of Agricultural Chemicals in Water Resources of the Midcontinental United States*, progress report, U.S. Geological Survey, Lakewood, CO.

Grant, L. D., R. A. Goyer, H. Olem, W. J. Nicholson, and R. W. Elias (1990). Indirect health effects associated with acidic deposition, in *Acidic Deposition: State of Science and Technology*. Vol. III. *Terrestrial, Materials, Health, and Visibility Effects*, P. M. Irving, ed., National Acid Precipitation Assessment Program, Washington, DC.

Gschwandtner, G., J. K. Wagner, and R. B. Husar (1988). *Comparison of Historic SO_2 and NO_x Emission Data Sets*, EPA-600/7-88-009a, U.S. Environmental Protection Agency, Research Triangle Park, NC.

Hites, R. A., and S. J. Eisenreich, eds. (1987). *Sources and Fates of Aquatic Pollutants*, ACS Advances in Chemistry Series, No. 216, American Chemical Society, Washington, DC.

Hoffman, S. (1989). *Acid Precipitation and Human Health*, Rep. EN-6492, Electric Power Research Institute, Palo Alto, CA.

Irving, P. M., ed. (1991). *Acidic Deposition: State of Science and Technology*, Summary Report, U.S. National Acid Precipitation Assessment Program, NAPAP, Washington, DC.

Jacks, G. (1984). Trace elements in water supplies, in *Trace Elements in Health and Disease*, Skandia International Symposia Proceedings, Almquist and Wiskell International, Stockholm, pp. 40–48.

Junge, C. E. (1977). Basic considerations about trace constituents in the atmosphere as related to the fate of global pollutants, in *Fate of Pollutants in the Air and Water Environment—Part I*, I. H. Suffert, ed., Wiley Interscience, New York.

Kohout, E. H., D. J. Miller, and L. Nieves (1990). *Month and State Current Emission Trends for NO_x, SO_x, and VOC: Methodology and Results*, Argonne National Laboratory, Argonne, IL.

Lindberg, S. E. (1982). Factors affecting trace metal, sulfate and hydrogen ion concentrations in rain, *Atmos. Environ.* **16**:1701–1709.

Lindqvist, O., A. Jernelov, K. Johansson, and H. Rodhe (1984). *Mercury in the Swedish Environment: Global and Local Sources*, Rep. 1816, Swedish Environmental Protection Board, Solna.

Linthurst, R. A., D. H. Landers, and J. M. Eilers (1986). *Characteristics of Lakes in the Eastern United States.* Vol. I. *Population Descriptions and Physico-Chemical Relationships*, EPA/600/4-86-007a, U.S. Environmental Protection Agency, Washington, DC.

Logtin, G. (1988). *National Inorganics and Radionuclides Survey*, U.S. Environmental Protection Agency, Cincinnati, OH.

Mackay, D. (1990). Atmospheric contributions to contamination of Lake Ontario, in *Long Range Transport of Pesticides*, D. A. Kurtz, ed., Lewis Publ., Chelsea, MI.

McMahon, T. A., P. J. Dension, and R. Fleming (1976). A long distance air pollution transportation model incorporating washout and dry deposition component, *Atmos. Environ.* **10**:751–761.

Meranger, J. C., K. S. Subramanian, and C. Chalifoux (1979). A national survey for cadmium, chromium, copper, lead, zinc, calcium, and magnesium in Canadian drinking water supplies, *Environ. Sci. Technol.* **13**:707–711.

Novotny, V., and G. Chesters (1981). *Handbook of Nonpoint Pollution: Sources and Management*, Van Nostrand Reinhold, New York.

Nriagu, J. O. (1979). Global inventory of natural and anthropogenic emissions of trace metals to the atmosphere, *Nature* **279**:409–411.

Nriagu, J. O., and J. M. Pacyna (1988). Quantitative assessment of worldwide contamination of air, water, and soils by trace metals, *Nature* **333**:134–139.

Orgil, M. M., A. Schnel, and M. R. Petersen (1976). Some initial measurements of airborne DDT over Pacific Northwest forest, *Atmos. Environ.* **10**:827–834.

Paasivirta, J. (1990). Predicted and observed fate of selected persistent chemicals in the environment, in *Dioxin '90, Short Papers*, Vol. 7, Bayreuth, FRG, pp. 367–374.

Patterson, J. W. (1981). *Corrosion in Water Distribution Systems*, Report to U.S. Environmental Protection Agency, Office of Drinking Water, Washington, DC.

Patterson, J. W., and C. E. Junge (1971). Sources of particulate matter in the atmosphere, in *Man's Impact on the Climate*, W. H. Mathew, et al., eds., MIT Press, Cambridge, MA, pp. 310–320.

Patterson, J. W., and J. E. O'Brien (1979). Control of lead corrosion, *J. Am. Water Works Assoc.* **71**:264–271.

Peirson, D. H., P. A. Cawse, L. Salmon, and R. S. Cambray (1973). Trace elements in the atmospheric environment, *Nature* **241**:252–256.

Peterle, T. J. (1969). DDT in Antarctic snow, *Nature* **224**:629–630.

Provenzano, G. (1978). Motor vehicle lead emissions in the USA: An analysis of important determinants, geographic patterns, and future trends, *J. Air Pollut. Assoc.* **28**:1193–1199.

Rapaport, R. A., and S. J. Eisenreich (1986). Atmospheric deposition of toxaphene to eastern North America derived from peat accumulation, *Atmos. Environ.* **20**:931–941.

Rosenbrough, R. W., et al. (1976). Transfer of chlorinated biphenyls to Antarctica, *Nature* **264**:738–739.

Shacklette, H. T., and J. G. Boerngen (1984). *Element Concentrations in Soils and Other Surficial Materials of the Conterminous United States*, U.S. Geological Survey, Prof. Paper 1270, Denver, CO.

Singley, J. E., B. A. Beaudet, and P. H. Markey (1984). *Corrosion Manual for Internal Corrosion of Water Distribution Systems*, Rept. No. ORNL/TM-8919, Oak Ridge National Laboratory, Oak Ridge, TN.

Sisterson, D. L., et al. (1990). Deposition monitoring: Methods and results, in *Acidic Deposition: State of Science and Technology*, Vol. I: *Emissions, Atmospheric Processes, and Deposition*, P. M. Irving, ed., National Acid Precipitation Assessment Program, Washington, DC, pp. 6-1–6-338.

Slade, D. H., ed. (1968). *Meteorology and Atomic Energy—1968*, U.S. Atomic Energy Commission, Washington, DC.

Sladen, W. J. L., C. M. Menzie, and W. L. Reichel (1966). DDT residues in Adelic penguins and a crabeater seal from Antarctica, *Nature* **210**:670–671.

Stern, A. C., ed. (1976). *Air Pollution*, Vol. 1–5. Academic Press, New York.

Strachan, W. M. J., and S. J. Eisenreich (1990). Mass balance accounting of chemicals in the Great Lakes, in *Long Range Transport of Pesticides*, D. A. Kurtz, ed., Lewis Pub., Chelsea, MI, pp. 291–301.

Sverdrup, H. U. (1990). *Acidification and Health Effects*, Report, Dept. of Chemical Engineering, Lund Institute of Technology, Lund, Sweden.

Taylor, F. B., and G. E. Symons (1984). Effects of acid rain on water supplies in the Northeast, *J. Am. Water Works Assoc.* **76**:35.

Taylor, F. B., J. A. Taylor, G. E. Symons, J. J. Collins, and M. R. Schock (1984). *Acid Precipitation and Drinking Water Quality in the Eastern United States*, EPA-600/2-84-054, U.S. Environmental Protection Agency.

Turner, R. S., et al. (1990). Watershed and lake processes affecting surface water acid-base chemistry, in *Acidic Deposition: State of Science and Technology*. Vol. II. *Aquatic Processes and Effects*, P. M. Irving, ed., National Acidic Precipitation Assessment Program, Washington, DC.

U.S. Environmental Protection Agency (1985). National Primary Drinking Water Regulations. Synthetic organic chemicals, inorganic chemicals, and microorganisms; proposed rule, *Fed. Regist.* **50**:46936.

U.S. Environmental Protection Agency (1988). Drinking water regulations; maximum contaminant level goals and National Primary Drinking Water Regulations for lead and copper; proposed rule, *Fed. Regist.* **53**(160):31516–31578.

U.S. Environmental Protection Agency (1990). *National Air Pollutant Emission*

Estimates, 1940–1988, EPA-450/4-90-001, U.S. Environmental Protection Agency, Research Triangle Park, NC.

Woodwell, G. M., P. P. Craig, and H. A. Johnson (1971). DDT in the biosphere: Where does it go? *Science* **174**:1101–1107.

Zannetti, P. (1992). *Air Pollution Modeling*, Van Nostrand Reinhold, New York.

5
Erosion and Sedimentation

My earliest lessons on environmental protection were about the prevention of soil erosion on our family farm.

Al Gore

"If we assume that the average worldwide rate of erosion is on the order of, say 2.7 cm per thousand years the average land elevation of the continents (840 m) could be reduced to sea level in about 31 million years" (Leopold, Wolman, and Miller, 1964). This approximate estimate by these well-known geomorphologists (scientists who study the process of the evolution of the earth surface) indicates the scope and magnitude of natural erosion that has been carving the earth for billions of years. If indeed erosion was the only process shaping the earth surface, the whole world would have already been reduced to a featureless marshy flat plain billions of years ago. However, the erosion in the denudation process is countered by tectonic uplifts, hydrostatic buoyancy uplift of land masses, subsidence, and other processes. Furthermore, erosion is not uniform and depends on slope, climate, properties of the parent geological material, and weathering and other factors (Leopold, Wolman, and Miller, 1964). Walling and Webb (1983) estimated the average annual rates of sediment yield (sediment from erosion reaching streams) throughout the world. They reported that the rates of natural erosion and sediment yields range from less than 1 tonne $km^{-2} y^{-1}$ for some rivers in Poland and New South Wales in Australia to more than 10,000 tonnes $km^{-2} yr^{-1}$ for some tributaries of the Yellow River in China and some rivers in Kenya, New Zealand, Taiwan, Java in Indonesia, and New Guinea. The discharge of a sediment-laden stream into one of the Great Lakes is shown on Figure 5.1.

Geomorphologists have been interested in the process of erosion and deposition for a long time. Only recently, however, have water quality and pollution dimension been added to the science of geomorphology

FIGURE 5.1. Sediment discharge from a stream into one of the Great Lakes. (Photo: University of Wisconsin, Madison, Department of Agricultural Journalism.)

(Wolman, 1977; Walling and Webb, 1983; Walling, 1983; Novotny and Chesters, 1989).

DEFINITION AND DESCRIPTION OF THE EROSION PROCESS

Denudation is a geomorphologic process consisting of weathering or breakdown of the parent rock material, entrainment of the weathered debris, and transportation and deposition of that debris. The word *erosion* is often used synonymously with denudation, although erosion classically applies to entrainment and transportation, but not to weathering (Leopold, Wolman, and Miller, 1964). Portions of this chapter were previously published in Novotny and Chesters (1981).

Erosion and weathering are a result of stresses applied on the surface of the earth. These stresses can be gravitational, molecular, or chemical. Water and air movement over the surface are sources of the gravitational stresses; however, other stresses, such as ruptures, also cause the movement of geological materials. Chemical weathering, for example, due to natural or anthropogenic acidity of the rainfall, is also of great interest, since it is a primary cause of the background chemical quality of ground and surface waters. Molecular and chemical stresses are caused primarily by weathering, while gravitational stresses are generally responsible for entrainment, transport, and deposition. The properties of geological materials considered under gravitational stresses, primarily due to water and air action, range from rigid resistant hard rocks to flowing fluids such as mudflows.

Erosion processes can be divided into sheet and rill erosion (Figs. 5.2 and 5.3), gully erosion (Fig. 5.4), stream and floodplain scour (shown on Figs. 5.5 and 1.1), and shoreline erosion (Fig. 5.6). Typical roadside erosion (Fig. 5.7) could be classified as gully erosion. Erosion from farms, construction, and surface mines is commonly of the sheet and rill type, with gully formation in the drainageways.

Hydrologically, these processes can be classified into sheet (upland) erosion and stream or channel erosion and transport. Thus, the total amount of on-site sheet and channel erosion is gross (potential) erosion.

Erosion, usually expressed in tonnes per km² or hectare per time (year), should be distinguished from *sediment yield*, which is the flow of sediment measured in the receiving body of water in the same time period. As will be described (the section titled "Sediment Delivery and Enrichment Processes during Overland Flow"), significant differences between erosion rates summed up over the watershed area and sediment yield should be expected. A *delivery ratio* (DR) is then a dimensionless ratio of the sediment yield divided by the total potential erosion in the contributing watershed, or

$$\text{DR} = \frac{Y}{A} \qquad (5.1)$$

where
DR = delivery ratio (dimensionless)
Y = sediment yield
A = gross erosion in the watershed

The ease with which materials give way during erosion is called *erodibility*. Erodibility, as will be shown, depends on both the composition of the geological material and the state of its consolidation. Consolidated rocks obviously have low erodibility. For unconsolidated geological materials (soils, river deposits, sand dunes, etc.) erodibility depends on particle size and the texture of the material, water content, composition

240   Erosion and Sedimentation

FIGURE 5.2. Agricultural rill and interrill (sheet) erosion. (Photo: University of Wisconsin, Madison, Department of Agricultural Journalism.)

FIGURE 5.3. Sheet erosion of construction sites. (Photo: USDA—SCS.)

Definition and Description of the Erosion Process 241

FIGURE 5.4. Gully erosion. (Photo: University of Wisconsin, Madison, Department of Agricultural Journalism.)

FIGURE 5.5. Streambank erosion. (Photo: USDA—Soil Conservation Service.)

FIGURE 5.6. Shoreline erosion. (Photo: V. Novotny.)

FIGURE 5.7. Roadside erosion. (Photo: V. Novotny.)

of the material, and the presence or absence of protective surface cover such as vegetation. Leopold, Wolman, and Miller (1964) stated that even a small amount of clay in the unconsolidated mixture markedly improves the cohesion of the soil. Cohesion is also increased by organic materials and by chemical bonding.

Vegetative cover is also extremely important, since it provides additional resistance to shear stresses caused by falling and running water and wind. Water is the primary agent responsible for erosion. As pointed out by Leopold, Wolman, and Miller, only in regions with virtually no precipitation can wind be expected to be the dominant erosive force. Hence there is direct proportionality between the annual precipitation and erosion. However, the intensity of the vegetative cover that reduces erosion is also proportional to the amount of precipitation. This leads to the relation between the annual precipitation and sediment yield published by Langbein and Schumm (1958) shown on Figure 5.8. This figure is only a conceptual schematic since other factors such as the slope of the relief and man-induced changes in vegetation and surface character have not been included. As a matter of fact, Walling and Webb (1983) documented that global data do not follow this concept, and obtained a large scatter of sediment yields on the Lanbein–Schumm chart. However, they have pointed out that the Lanbein–Schumm concept is physically sound and meaningful for an undisturbed watershed in continental climates, such as the United States.

Erosion, sediment transport, and deposition are to a large degree natural processes that have been occurring throughout the geological ages. In many areas the top surface layers are very young, that is, years to thousands of years. These areas are commonly mildly sloping lowlands, alluvial fans, and alluvial deposits located where sediment is being deposited by surface flow or wind. The process of the deposition of sediment and build up of sedimentary deposits is called *aggradation*. Areas on exposed high slopes are the principal sources of sediment. *Degradation* is a process by which sediment is eroded away from the area. The processes of aggradation and degradation are shown in Figure 5.9.

Due to the facts that erosion is a natural process and significant quantities of sediments are being moved as a result of natural denudation, it would be unrealistic to expect or require complete control or elimination of sediment loads to receiving waters. Such "wall-to-wall" control measures would be technically and economically impossible. However, it is feasible to control or manage excessive sediment loadings that have resulted from man's land use activities that would be detrimental to the quality of the receiving bodies of water and to the aquatic and terrestrial habitat.

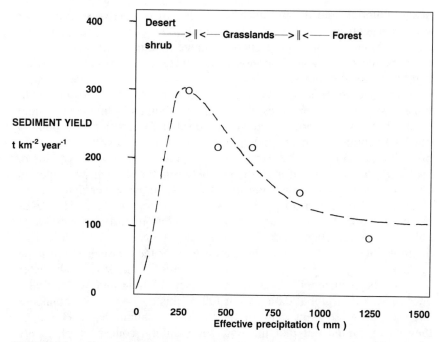

FIGURE 5.8. The relationship between sediment yield and annual precipitation proposed by Langbein and Schumm (1958). (After Walling and Webb, 1983.)

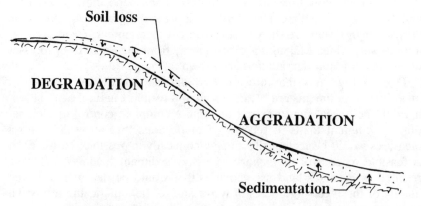

FIGURE 5.9. Watershed aggradation and degradation concept.

Classification of Eroded Sediment

Eroded soil materials are classified according to their grain size in clay, silt, sand, and gravel. Clay particles range from 10^{-4} to 10^{-3} mm, silt ranges from 10^{-3} to 10^{-2} mm, sand from 10^{-2} to 10^{0} mm, and gravel particles are greater than 10^{0} mm. Table 5.1 provides information on the distribution of sediment particle sizes.

Soil materials are classified as cohesive and noncohesive sediments. Cohesive sediments include clays and other fine particles, including organic materials. These fractions can form bonds between particles that resist scour, cause flocculation during overland and channel flows, and adsorb pollutants. Noncohesive sediments, mostly particles of sand and gravel size, are transported as individual particles, the movement of which depends solely on such particle properties as mass, shape, size, and relative position of particles with respect to surrounding particles. Their sorption potential is small compared to cohesive sediments.

Once in motion (overland or channel flow), sediments transported by the flow follow two modes (Vanoni, 1975; Shen, 1982):

1. as suspended load (*washload*), where the sediment is moved away from the bed and supported mainly by the turbulence of the flow;
2. as *bedload*, where the sediment is moved near the bed and supported most of the time by the bed.

Washload is carried by streams in suspension, while bedload movement consists of shifting bottom sediments due to the intensity of the turbulence and mean stream velocity. It has been found by Gottschalk

TABLE 5.1 Scale of Sediment Particle Sizes

Class	Particle Size (mm)	
	AGU Scale	USDA Scale
Cobbles and boulders	>64	>10
Gravel	2–64	2–10
Very coarse sand	1–2	1–2
Coarse sand	0.5–1.0	0.5–1.0
Medium sand	0.25–0.5	0.25–0.5
Fine sand	0.125–0.25	0.1–0.25
Very fine sand	0.062–0.125	0.05–0.1
Silt	0.004–0.062	0.002–0.05
Clay	<0.004	<0.002

Note: The scale is according to the American Geophysical Union (AGU) and the U.S. Department of Agriculture (USDA).

(1964) and Leopold, Wolman, and Miller (1964) that most of the sediment carried by streams in humid areas originates from sheet erosion and is in the form of fine sediments or washload. Transport of sediments in channels also depends on whether the sediments are cohesive or noncohesive. Cohesive fine sediments, with the exception of impounded sections with very low velocity, are mostly incorporated into washload. There is an exchange of noncohesive sediments between the bottom (bedload) and suspended (washload) forms; however, it is suspected that such continuous exchanges between bedload and washload sediments do not exist for cohesive fine sediments (see the section titled "Sediment Transport in Streams," describing channel transport of cohesive and noncohesive sediments).

Enrichment ratio refers to the difference in particle size distribution and sorbed pollutant content of washload particles and the soils from which the sediment originated. Erosion and sedimentation are the selective processes by which clay and fine particles are first liberated from soils while they are deposited last during redeposition in overland flow.

Erosion control implies the reduction of soil loss by land management and the reduction of the delivery of sediment from the source area to the receiving body of water that can be accomplished by land management, buffer strips, channel modification, sediment traps, and other structural and nonstructural measures and practices. Chapters 10 and 11 contain descriptions and design procedures for sediment management aimed at water quality control. A publication of Goldman, Jackson, and Bursztynsky (1986) and a number of manuals available from the Soil Conservation Service of the USDA, state pollution control agencies, and other sources provide ample information on erosion control. Construction erosion control is now mandated in many states.

Effect of Cultural (Man's) Land-Use Activities on Erosion and Sediment Yields

It has been already pointed out in this book and several other publications (for example, The Committee on International Soil and Water Research and Development of the National Academy of Sciences, 1991) that deforestation in areas of high precipitation and high slope can be devastating both to water quality and flooding. Catastrophic depositions of sediments and increased flooding in Bangladesh are a result of deforestation in the high precipitation regions of the Indian Himalayan mountains.

Changing climate conditions are often related to the amount of disturbance of the watershed. Man-made (cultural) impacts in various portions of the world may have lasted for the thousands of years (Egypt, Middle East, China, India) or for a hundred years or less (North America). There are historically documented occurrences where man has dramatically increased erosion such as the deforestation of the Middle East, which converted Syria, among other countries, into highly erosive arid regions. Another example is the Adriatic coast of Dalmatia, which was deforested by the Venetians during the Middle Ages. However, the intensity of watershed disturbances has increased between then and now.

Based on measurements by the agricultural research centers in the United States following the devastating "dust bowl" erosion of farmland during the 1930s, which then culminated in the development of the Universal Soil Loss Equation by Wischmeier and Smith (1965), it was found that general land disturbance by agriculture or construction can increase erosion rates by two or more orders of magnitude. Patrick (1975) quoting works by Leopold (1968) and Borman and Likens (1969) pointed out that cutting down a forest—reducing the forest from 80% to 20%—increased sediment yield in a riverine system eight times. Walling and

TABLE 5.2 Increases in Sediment Yield Caused by Land-Use Changes

	Land-Use Change	Increase in Sediment Yield	Source
Rajasthan, India	Overgrazing	× 4–18	Sharma and Chatterji (1982)
Utah, U.S.	Overgrazing of rangeland	× 10–100	Noble (1965)
Oklahoma, U.S.	Overgrazing and cultivation	× 50–100	Rhoades, Welch, and Coleman (1975)
	Cultivation	× 5–32	
Texas, U.S.	Deforestation and cultivation	× 340	Chung, Roth, and Hunt (1982)
Northern California, U.S.	Conversion of steep forest to grassland	× 5–25	Anderson (1975)
Mississippi, U.S.	Deforestation and cultivation	× 10–100	Ursic and Dendy (1965)
South Brazil	Deforestation and cultivation	× 4500	Bordas and Canali (1980)
Oregon, U.S.	Deforestation (clear-cutting)	× 39	Frederiksen (1970)
Ontario, Canada[a]	Urbanization	× 26	Ostry (1982)
	Agriculture	× 14	

Source: After Walling and Webb (1983).

[a] Compared to undisturbed forested land in the same area.

Webb (1983) compiled data from selected watershed studies to document the impact of a particular land-use activity on the increase of sediment yields (Table 5.2). Walling (1979), and Ostry (1982) conducted studies on two parallel watersheds, one undisturbed and one disturbed. These studies confirmed that the suspended sediment loads of many rivers may have increased by an order of magnitude or more as a result of cultural land-use changes within the watershed.

The land-disturbing activities were listed in Chapter 1. In addition to deforestation, construction-site erosion and intensive agriculture on highly erodible lands are the activities that will result in the highest rates of erosion.

According to the definition of pollution introduced in Chapter 1, sediment from natural undisturbed watersheds, volcanic eruptions, and the natural weathering of soils and minerals are not considered pollution, although sediment yields may sometimes reach large proportions. One example of this is the sediment load of the Columbia River in the Northwest United States following the eruption of Mount St. Helen in 1980. Pollution in the form of elevated sediment yields results from man-made modification of watersheds.

The following land-use activities by man are producing large sediment loads (Clark, Haverkamp, and Chapman, 1985):

1. Agricultural erosion is a major sediment source due to the large areas involved and the land-disturbing effects of cultivation. Most sediment loading from agricultural zones takes place during the spring, especially when spring rains fall on still frozen soils. Reported erosion rates range from 100 to 4000 tonnes $km^{-2} yr^{-1}$ (Brant et al. 1972; Clark, Haverkamp, and Chapman, 1985).
2. Urban erosion. Sediment originates mainly from exposed soils in areas under construction and from street dust and litter accumulation on impervious surfaces. Sediment yields from developing urban areas can be extremely high, sometimes reaching values up to 50,000 tonnes $km^{-2} yr^{-1}$ (Brant et al. 1972; Novotny, 1980; Novotny and Chesters, 1981).
3. Highway erosion is associated with the stripping of large areas of their vegetative protection during road construction. Uncompacted or unsettled fills may be subject to intensive rill erosion and, eventually, to landslides. The magnitudes of sediment yields from these areas are similar to those from urban areas under construction.
4. Silvicultural erosion is caused by clear-cutting practices, road building, and timber transportation. Due to the significant quantities of vegetation residues remaining on the surface, erosion rates from silviculture

are deemed to be one to two orders of magnitude less than that from cropland (Brown, 1985).
5. Erosion of strip mines is similar in nature and magnitude to erosion from large construction sites. The potential for erosion is created as soon as vegetation is removed to construct roads and facilitate prospecting and mining. The mining process itself also creates an even greater erosion potential.
6. Stream bank, channel, and shoreline erosion results from concentrated water flows and wave action in channels and floodplains. Channel erosion usually is a source of large-grain-sediments, and in most cases represents a smaller portion of the total sediment loading unless the stream is undergoing structural channelization with excavation. Sediment loadings originating from bank and channel erosion of some streams in arid and semiarid regions (Fig. 1.1) may be quite high (Nordin, 1962).

Steward et al. (1975) listed the following sources and conditions that cause excessive sediment movement:

Farming on long slopes without terraces or runoff diversions
Row cropping up and down moderate or steep slopes
Bare soil surface following seeding of crops
Bare soil between harvest and the establishment of a new crop canopy
Intensive cultivation close to a stream
Intensive runoff from upslope pasture or rangeland that traverses areas of row crops
Poor crop stands
Gully formation
Poorly managed idle or wooded land
Unstable stream banks
Surface mining
Feedlots close to a stream
Long exposure of bare soil resulting from a land use

As was pointed out in the preceding section, sediment produced at the source is not qualitatively or quantitatively the same as sediment measured in the receiving body of water. Many factors, such as distance of the source from the receiving body of water, vegetative buffers, slope and roughness characteristics of the land, and ponding and the presence of depositional areas during overland flow, can affect the delivery of the sediment from the source to the receiving body of water.

SEDIMENT PROBLEM

The publication by Clark, Haverkamp, and Chapman (1985) is an authoritative treatise on the impact of erosion and sedimentation on the environment in general and water quality in particular. Soil erosion is the major cause of diffuse pollution and sediment is also the most visible pollutant. The effects of excessive sediment loading on receiving waters include the deterioration or destruction of aquatic habitat, deterioration of aesthetic value, loss of storage capacity in reservoirs, and accumulation of bottom deposits that inhibit normal biological life. Sediment can destroy spawning areas, food sources as well as directly harming fish, and other aquatic wildlife. Nutrients carried by the sediment can stimulate algal growths and, consequently, accelerate the process of eutrophication. Chapter 12 contains a more detailed discussion of the impact of sediment discharge on water quality.

Sediment per se can be considered a major pollutant of bodies of water. Furthermore, sediment—especially its fine fractions—is a primary carrier of such other pollutants as organic components, metals, ammonium ions, phosphates, and many organic toxic compounds. For example, persistent organochlorine compounds, such as aldrin and dieldrin pesticides, DDT, and PCBs, have low solubility in water but are readily adsorbed by suspended sediment. Phosphates from fertilizer applications and pollution discharges are also adsorbed by soils and suspended sediment. This strong affinity of fine sediments—primarily clay and organic particulates—to adsorb and make biologically unavailable the pollutants is considered by some as a partial water quality benefit of sediment discharges.

Large amounts of particulate materials (sediment) originate from urban areas. Urban sediment sources are known to contribute significant amounts of particulate pollutants containing toxic metals, organochlorine pesticides, asbestos, and other compounds originating from traffic, elevated air-pollution levels, urban litter accumulation, and erosion of pervious urban surfaces, which is accelerated by construction.

Many kinds of environmental damage by sediment are still weakly documented. However, in the United States the damage that has been documented amounts to billions of dollars. Annual in-stream damage by sediments is between $3.2 and $13 billion (Clark, Haverkamp, and Chapman, 1985). Recreational uses of bodies of water are most impacted by sediment.

In an extreme, reduction in the discharge of sediment may have a negative effect. For example, farmers in the delta region of the Nile River in Egypt relied on the deposited sediment brought from the Sudan and

Ethiopia for a source of fertilizing nutrients and silt. By building the Aswan High Dam in Upper Egypt, which acts as a sediment trap, significant damage has been done to delta farmers who now, at great expense and greater environmental consequences, must substitute chemical fertilizers. Similarly, building navigation dams on the Mississippi River has dramatically reduced sediment loads into the Mississippi Delta in Louisiana, which is suspected of causing shoreline and wetland losses by subsidence that otherwise would have been compensated for by new sediment brought by the river into the Delta region. The sediment, and especially its clay and organic factions, can also adsorb and immobilize potentially toxic compounds and make them biologically unavailable. These toxic compounds may then be buried with the sediments (see Chapter 13).

ESTIMATING SEDIMENT YIELD

Sediment yield from a watershed can be estimated by one of the following methods (Gottschalk, 1964):

1. The stream-flow sampling method, which yields a sediment/flow relationship (Fig. 5.10) commonly called the rating curve. Typically, this relationship is plotted on logarithmic paper as the sediment concentration or sediment flow (concentration × flow) versus discharge. Once the long-term sediment/flow relation is established, it is combined with a long-term flow frequency curve to obtain average annual yields.

   A word of caution is needed for the users of standard sediment flow: water discharge curves. These correlations could easily be spurious and misleading since for large discharges and small concentrations the relationship

   $$\log Q \text{ vs. } \log(Q \times C) = \log Q + \log C$$

   where Q is flow and C is concentration, could easily become a logarithmic regression of Q vs. Q, which will yield very high apparent coefficients of correlation. Yet the relationship between C and Q may be very poor or not exist, so the estimate of the annual sediment discharge would essentially be the average sediment concentration times the average flow.

2. The reservoir sedimentation survey method, which allows sediment input to be determined if accumulated sediment and reservoir trap efficiencies are known and/or measured.

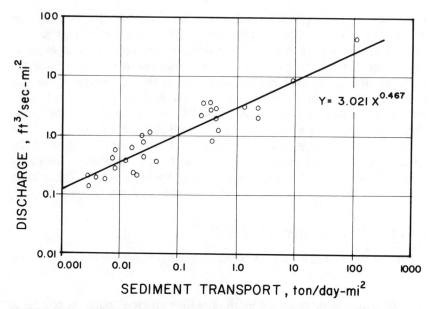

FIGURE 5.10. The relationship between suspended sediment transport and discharge for the Menomonee River in Wauwatosa, Wisconsin. (Compiled from the U.S. Geological Survey, Wisconsin Department of Natural Resources, and Southeastern Wisconsin Regional Planning Commission data. To convert from ft^3/sec/mi^2 to m^3/s-km^2, multiply by 0.011; to convert from U.S. ton/day/mi^2 to SI tonnes/day-km^2, multiply by 0.35.)

3. The sediment delivery method, in which sediment yield to some downstream section or deposition point is based on an estimate of the total upstream (upland) erosion factored down to account for the loss (or gain) of sediment during overland and channel transport. This method requires expressing the ratio between the sediment yield and gross upstream erosion, usually based on an empirical or semi-empirical formula. Both determination of the delivery ratio (DR) and upland gross erosion estimates are still unreliable if calibration and verification survey data are not available.
4. Bedload function methods use mathematical equations developed for calculating the rate and quantity of sediment materials in the bed portion of alluvial (sediment carrying and depositing) streams. Application of these equations requires information on sediment particle size, channel gradient and cross sections, and flow-duration curves. The equation can be used when sediment transport is not limited by the upstream supply, but depends solely on the transport capacity of

the flow. These models are mostly applicable to noncohesive, larger grain size sediments.
5. Methods using empirical equations relating sediment yield (directly measured by methods 1 or 2) to watershed hydrologic and/or morphologic characteristics. Most of these empirical equation have severely limited applications, even in the region of origin (Foster, Meyer, and Onstad, 1977).
6. Simulation watershed sediment load models, which usually are attached to a watershed hydrologic model. The watershed models are capable of simulating individual storm events or seasonal water and sediment yields. The hydrologic portion is necessary for determining so-called hydrologically active areas, that is, areas from which most intensive surface runoff and erosion occur. Chapter 9 includes an expanded discussion of mathematical modeling.

Estimating Upland Erosion

Variables influencing upland (sheet) erosion are climate, soil properties, vegetation, topography, and human activities. Rainfall, snowfall, and temperature are the primary climatic factors. Soil particles are detached and transported by the impact of rainfall energy, resulting in eroded materials being carried by surface runoff. Freezing temperatures and snow cover affect permeability and reduce the energy of precipitation. Conversely, spring rains occurring when subsoils are still frozen may cause high sediment yields due to reduced soil permeability.

The major soil properties related to erosion are soil texture and composition. Soil texture determines the permeability and erodibility of soils. Permeability and infiltration determine whether or not the soil surface is hydrologically active (that is, yielding or not yielding surface runoff). Erosion occurs only when surface runoff is generated or when wind picks up loose particles.

The chemical composition of soils (minerals, clay, and organic matter) provide bonding of soil particles, and thus affects erosion rates. Loose soils (silt and fine sand) with low chemical and clay content have the highest erodibility.

Vegetation influences sediment yields by dissipating rainfall energy, binding the soil, and increasing porosity by its root system, reducing soil moisture by evapotranspiration, thereby increasing infiltration. Organic residues also provide better texture and reduce erodibility.

The topographic factors of greatest importance are slope and the path length traversed by sediment generating flow. Human activities are related mostly to agricultural land use and construction.

Since many of the just-mentioned factors are seasonal and vary throughout the year, it follows that sediment yield and its chemical composition from a watershed also vary with time. Walling and Kane (1983) demonstrated that certain mineralogical characteristics of liberated sediment, such as Si, Al, Fe, Ti, and K, remain relatively constant over a range of hydrological conditions, while others, particularly those associated with organic fractions, may exhibit considerable variations.

Universal Soil Loss Equation

The universal soil loss equation (USLE) is the most common and best known estimator of soil loss caused by upland erosion. The equation and its development utilized more than 40 years of experimental field observations gathered by the Agricultural Research Service of the USDA. The USLE, formulated by Wischmeier and Smith (1965), predicts primarily sheet and rill erosion. The equation is

$$A = (R)(K)(LS)(C)(P) \tag{5.2}$$

where
A = calculated soil loss in tonnes/ha for a given storm or period
R = rainfall energy factor
K = soil erodibility factor
LS = slope-length factor
C = cropping management (vegetative cover) factor
P = erosion-control practice factor

The equation expresses soil loss/unit area due to erosion by rain. It does not include wind erosion and it does not give direct sediment yield estimates.

The *rainfall energy factor* (R_r) is equal to the sum of the rainfall erosion indices for all storms during the period of prediction. For a single storm it is defined as follows:

$$R_r = \Sigma[(2.29 + 1.15 \ln X_i)D_i]I \tag{5.3}$$

where
i = rainfall hyetograph time interval
D_i = rainfall during time interval i (cm)
I = maximum 30-min rainfall intensity of the storm (cm/hr)
X_i = rainfall intensity (cm/hr)

Average annual rainfall energy factors were determined for the eastern portions of the United States, and were later developed for the remainder

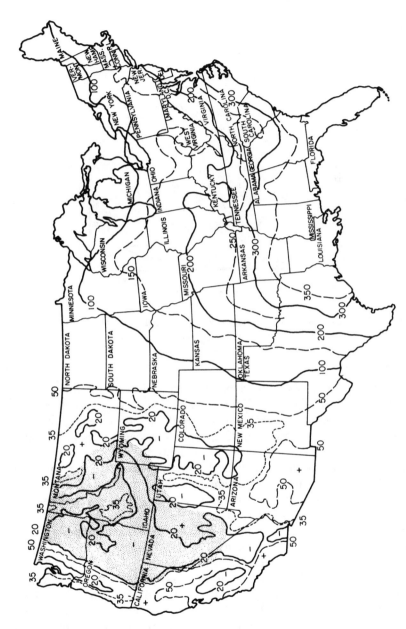

FIGURE 5.11. Values of the annual rainfall energy factor (R) in tons/acre. To convert to tonnes/ha, multiply by 2.24. (From Stewart et al., 1975.)

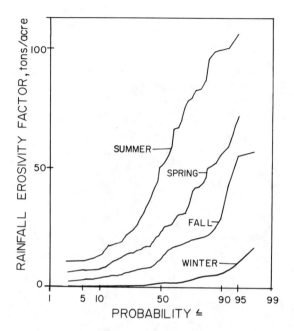

FIGURE 5.12. Seasonal cumulative frequency of the rainfall energy factor (R) for southeastern Wisconsin in tons/acre. To convert to tonnes/ha, multiply by 2.24.

of the country. The distribution of average R_r factors for the 48 conterminous states are shown in Figure 5.11. These curves can help as first estimates of the gross erosion potential, but as shown on Figure 5.12, significant yearly and seasonal differences in the magnitudes of the rainfall energy factor R_r may be typical. In the midwestern part of the United States, summer rains have the highest rainfall erosive energy, while the effect of winter precipitation on sediment yields is minimal.

The distribution of erosive rain differs significantly for different regions of the country. In the western plains and Great Lakes regions, from 40% to 50% of the erosive rainfall normally occurs during a 2-month period following spring planting when soils have the least protection. In most Corn Belt areas and the eastern parts of Kansas, Oklahoma, and Texas, the value is about 35%, while, for the lower Mississippi Valley and southeastern United States, the value is 20% to 23%. In the dry-land grain-growing region of the Pacific Northwest, about 80% to 90% of the annual erosion occurs in the winter months when the soil has little crop cover, since grain is seeded late in the fall (Steward et al., 1975).

Both erosion by rainfall energy (interrill erosion) and detachment of soil particles by overland runoff (rill erosion) contribute to soil loss (Foster, Lombardi, and Moldenhauer, 1982). Thus, the rainfall factor (R) should also include the effects of runoff. A modification of Equation (5.3) was proposed by Foster, Meyer, and Onstad (1977):

$$R = aR_r + bcQq^{1/3} \qquad (5.4)$$

where
a and b = weighting parameters ($a + b = 1$)
c = an equality coefficient
Q = runoff volume (cm)
q = maximum runoff rate (cm/hr)

The weighting factor compares the relative amounts of erosion caused by rainfall and runoff under unit conditions. It was suggested that the detachment of particles by runoff and rain energy is about evenly divided ($a = b = 0.5$). The equality coefficient in metric (SI) units is about 15. Substituting values for a, b, and c into the USLE, the overall rainfall factor (R) becomes

$$R = 0.5R_r + 7.5Qq^{1/3} \qquad (5.5)$$

However, the proportion between the rainfall and runoff erosivity may vary greatly between regions. Wischmeier (1976) pointed out that, for example, in the Palouse region of the Pacific Northwest, 90% of erosion is caused by runoff from thaw and snowmelt. A discussion of various functional forms of the rainfall erosivity factors—both rill and interrill effects—is included in Foster, Lombardi, and Moldenhauer (1982).

Williams and Berndt (1977) proposed a modified soil loss equation (MUSLE) that replaced the rainfall energy factor by a runoff energy parameter that is proportional to $(Q \times q)^{0.56}$. Equations (5.4), (5.5), and MUSLE of Williams and Bernt provide soil loss estimates in tonnes per storm (per area). Williams and Berndt pointed out that although q and Q are correlated, the runoff flow rate is more related to detachment, while the peak flow rate defines sediment transport. If flow is retarded by vegetation or other means, the peak flow rate is reduced, thus reducing sediment transport.

The soil erodibility factor (K) is a measure of potential erodibility of soil and has units of tonnes/unit of rainfall erosion index for a 22-m-long overland flow length on a 9% slope in clean-tilled continuous fallow

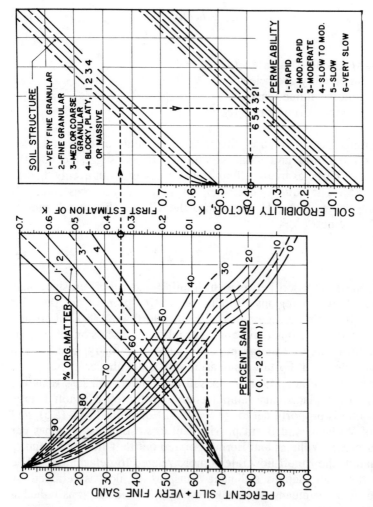

FIGURE 5.13. Soil erodibility nomograph for determining (K) factor for U.S. mainland soils. (Reprinted with permission from W. H. Wischmeier, C. B. Johnson, and B. V. Cross, *J. Soil Water Conserv.* **26**:189–193. Copyright © 1971 by the Journal of Soil and Water Conservation.)

TABLE 5.3 Magnitude of Soil Erodibility Factor, K

	K for Organic Matter Content (%)		
Technical Class	0.5	2	4
Sand	0.05	0.03	0.02
Fine sand	0.16	0.14	0.10
Very fine sand	0.42	0.36	0.28
Loamy sand	0.12	0.10	0.16
Loamy fine sand	0.24	0.20	0.16
Loamy very fine sand	0.44	0.38	0.30
Sandy loam	0.27	0.24	0.19
Fine sandy loam	0.35	0.30	0.24
Very fine sandy loam	0.47	0.41	0.35
Loam	0.38	0.34	0.29
Silt loam	0.48	0.42	0.33
Silt	0.60	0.52	0.42
Sandy clay loam	0.27	0.25	0.21
Clay loam	0.28	0.25	0.21
Silty clay loam	0.37	0.32	0.26
Sandy clay	0.14	0.13	0.12
Silty clay	0.25	0.23	0.19
Clay		0.13–0.2	

Source: After Steward et al. (1975).

Note: The values shown are the estimated averages of broad ranges of specific soil values. When a texture is near the borderline of two texture classes, use the average of the two K values.

ground (Wischmeier and Smith, 1965; Wischmeier, 1976). It is a function of soil texture and composition. The soil erodibility nomograph shown on Figure 5.13 is used to find the appropriate values of the soil erodibility factor using five parameters: percent silt and fine sand, that is, 0.05 to 0.1 mm fractions; percent sand >0.1 mm; percent organic matter; textural class; and permeability. The general magnitudes of the soil erodibility factors are given in Table 5.3.

The slope-length factor (LS) is a function of overland runoff length and slope. It is a dimensionless parameter that adjusts the soil loss estimates for the effects of length and the steepness of the field slope. The general magnitudes of the LS factor are given in Figure 5.14. For slopes >4%, the LS factor can be estimated as follows:

$$LS = L^{1/2}(0.0138 + 0.00974S + 0.00138S^2) \qquad (5.6)$$

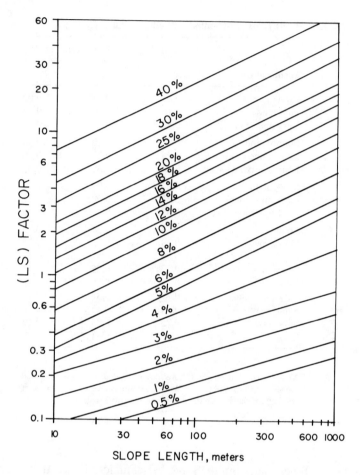

FIGURE 5.14. Slope-length factor (LS) for different slopes. (From Stewart et al., 1975.)

where
L = length in meters from the point of origin of the overland flow to the point where the slope decreases to the extent that deposition begins or to the point at which runoff enters a defined channel
S = the average slope (%) over the runoff length

Values of the LS factor estimated for length >100 meters or slopes >18% are extrapolated beyond the experimental data from which the magnitudes of the factor was determined.

If the average slope is used in calculating the LS factor, predicted erosion differs from actual erosion when the slope is not uniform. The equation for LS factors shows that when the actual slope is convex, the average slope will underestimate predicted erosion. Conversely, for a concave slope, the equation will overestimate actual erosion. To minimize these errors, large areas should be broken up into areas of fairly uniform slope. If sediment moves from an area with steep slope to an area of less steep slope, the smaller LS factor will control the amount eroded and the excess sediment is likely to be deposited.

The cropping management factor (C), also called the vegetation cover factor, estimates the effect of ground cover conditions, soil conditions, and general management practices on erosion rates. It is a dimensionless quantity with a value of one corresponding to continuous fallow ground, which has been defined as land that has been tilled up and down the slope and maintained free of vegetation and surface crusting. The effect of vegetation on erosion rates results from canopy protection, reduction of rainfall energy, and protection of soil by plant residues, roots, and mulches.

TABLE 5.4 Values of C for Cropland, Pasture, and Woodland

Land Cover or Land Use	C
Continuous fallow tilled up and down slope	1.0
Shortly after seeding prior to harvesting	0.3–0.8
For crops during main part of growing season	
Corn	0.1–0.3
Wheat	0.05–0.15
Cotton	0.4
Soybean	0.2–0.3
Meadow	0.01–0.02
For permanent pasture, idle land, unmanaged woodland	
Ground cover 85%–100%	
As grass	0.003
As weeds	0.01
Ground cover 80%	
As grass	0.01
As weeds	0.04
Ground cover 60%	
As grass	0.04
As weeds	0.09
For managed woodland	
Tree canopy of 75%–100%	0.001
40%–75%	0.002–0.004
20%–40%	0.003–0.01

Sources: Based on data from Stewart et al. (1975); Wischmeier and Smith (1965), and Wischmeier (1972).

TABLE 5.5 C-Values and Slope-Length Limits (LS) for Construction Sites

Mulch				
Type	Application (tonnes/ha)	Slope (%)	C	LS
No mulch or seeding		All	1.0	
Straw or hay tied down by anchoring and tracking equipment used on slope	2.25	<5	0.2	60
	2.25	6–10	0.2	30
	3.4	<5	0.12	90
	3.4	6–10	0.12	45
	4.5	<5	0.06	100
	4.5	6–10	0.06	60
	4.5	11–15	0.07	45
	4.5	16–20	0.11	30
	4.5	21–25	0.14	23
Crushed stone	300	<15	0.05	60
	300	16–20	0.05	45
	300	21–33	0.05	30
	540	<20	0.02	90
	540	21–35	0.02	60
Wood chips	15	<15	0.08	23
	15	16–20	0.08	15
	27	<15	0.05	45
	27	16–20	0.05	23
	56	<15	0.02	60
	56	16–20	0.02	45
	56	21–33	0.02	30
Asphalt emulsion 12 m^3/ha			0.03	

		During first 6 weeks of growth	After the 6th week of growth
Temporary seeding with grain or fast-growing grass with			
No mulch		0.70	0.10
Straw	2.25	0.20	0.07
Straw	3.4	0.12	0.05
Stone	300	0.05	0.05
Stone	540	0.02	0.02
Wood chips	15	0.08	0.05
Wood chips	27	0.05	0.02
Wood chips	56	0.02	0.02
Sod		0.01	0.01

Source: After Ports (1975).

Table 5.4 shows the general magnitudes of C for agricultural land, permanent pasture, and idle rural lang. Grassed urban areas have C factors similar to those for permanent pasture. The C factor for construction sites can be reduced if the surface is protected by seeding or the application of hay, asphalt, wood chips, or other protective covers. The effects of these protective practices on C are given in Table 5.5.

The erosion control practice factor (P) accounts for the erosion-control effectiveness of such land treatment as contouring, compacting, established sedimentation basins, and other control structures. Terracing does not affect P because the soil loss reduction by terracing is reflected in the value of LS. Generally, C reflects protection of the soil surface against the impact of rain droplets and subsequent loss of soil particles. On the other hand, P involves treatments that retain liberated particles near the source and prevent further transport.

Values of P for various farm and urban practices are given in Tables 5.6 and 5.7, respectively. It should be pointed out that these coefficients are highly empirical and may be used only as a first approximation. More accurate models are available for several practices, included in the P factor, such as models for the removal efficiency of sedimentation ponds and buffer strips. These concepts are discussed in Chapters 9 and 10, respectively.

Reliability of the USLE. The universal soil loss equation was subjected to a lot of testing and criticizing; however, it withstood the test of time and today it is the only widespread and tested model. It has been used in many applications. The author of the equation (Wischmeier, 1976) reported the results of testing on the reliability of the equation. He pointed out that when the USLE is used for estimating the annual soil loss on many experimental testing plots, the average prediction error (coefficient of variation = deviation/mean estimate) was about 12%. Larger errors

TABLE 5.6 Values of P for Agricultural Lands

		Strip Cropping and Terracing	
Slope (percent) Crops	Contouring	Alternate Meadows	Closegrown
0–2.0	0.6	0.3	0.45
2.1–7.0	0.5	0.25	0.40
7.1–12.0	0.6	0.30	0.45
12.1–18.0	0.8	0.40	0.60
18.1–24.0	0.9	0.45	0.70
>24		–1.0–	

Source: After Wischmeier and Smith (1965).

TABLE 5.7 Values of *P* for Construction Sites

Erosion Control Practice	P
Surface Condition with No Cover	
Compact, smooth, scraped with bulldozer or scraper up and down hill	1.30
Same as above, except raked with bulldozer and root-raked up and down hill	1.20
Compact, smooth, scraped with bulldozer or scraper across the slope	1.20
Same as above, except raked with bulldozer and root raked across the slope	0.90
Loose as a disked plow layer	1.00
Rough irregular surface, equipment tracks in all directions	0.90
Loose with rough surface >0.3-m depth	0.80
Loose with smooth surface <0.3-m depth	0.90
Structures	
Small sediment basins	
0.09 ha basin/ha	0.50
0.13 ha basin/ha	0.30
Downstream sediment basin	
With chemical flocculants	0.10
Without chemical flocculants	0.20
Erosion-control structures	
Normal rate usage	0.50
High rate usage	0.40
Strip building	0.75

Source: After Ports (1973).

should be expected if the equation is used for predicting the soil loss of individual storms.

The accuracy of the model is increased if it is combined with a hydrological excess-rainfall model. Note that the rainfall erosivity factor, R, has a value greater than zero for every rainfall, hence, erosion and soil loss are anticipated by the soil loss equation for any precipitation. A hydrological excess-rainfall model in combination with the USLE would eliminate erosion by rainfalls with no excess rain.

The USLE was tested and specifically designed for the following applications (Wischmeier, 1976):

1. Predicting average annual soil movement from a given field slope under specified land-use and management conditions and from construction, rangeland, woodland, and recreational areas.
2. Guiding the selection of conservation practices for specific sites.
3. Estimating the reduction of soil loss attainable from various changes that a farmer might make in his cropping system or cultural practices.

4. Determining how much more intensively a given field could be safely cropped if contoured, terraced, or stripcropped.
5. Determining the maximum slope-length on which given cropping and management can be tolerated in a field.
6. Providing local soil loss data to agricultural technicians, conservation agencies, and others to use when discussing erosion-control plans with farmers and contractors.

The USLE will not provide direct estimates of the sediment yield and cannot be used for calculations of soil losses from spring snowmelt.

Example 5.1: Estimation of Annual Soil Loss

An erosive 100-ha farm field in southeast Wisconsin is situated on silt loam soil with a slope classification B (3% to 6% slope). The farmer is growing corn and plowing up and down the slope. Estimate the average annual soil loss per hectare without soil conservation and with contour plowing. The field has a square shape with a drainage ditch located on the side of the field. The overland slope is toward the drainage ditch.

Solution From Figure 5.10 the average annual rainfall erosivity for southeast Wisconsin is $R_r = 125$ U.S. tons/acre $\times$ 2.24 = 280 tonnes/ha. From Table 5.3 the average soil erodibility factor for silt loam soil is $K = 0.42$. The slope-length factor (LS) can be read from Figure 5.14 (overland flow length $L = 1000$ m, and average slope $S = 45\%$) as LS = 1.

The plowing practice is to till up and down the slope of the continuous fallow field. Consequently the slope has a C factor of $C = 1$ after plowing and 0.1 to 0.3 for corn during the main growing season (Table 5.4), respectively. Average C for no soil conservation planting is assumed to be $C = (1 + 0.3)/2 = 0.65$. Since no erosion control is implemented, then $P = 1$. The average annual soil loss without soil conservation is then

$$A = (R)(K)(LS)(C)(P) = 280 \times 0.42 \times 1 \times 0.65 \times 1$$
$$= 76.4 \text{ tonnes/ha}$$

Implementing contour plowing will reduce the P factor to 0.5 (Table 5.7). Hence the soil loss will then be

$$A = 280 \times 0.42 \times 1 \times 0.65 \times 0.5 = 38.2 \text{ tonnes/ha}$$

Example 5.2: Soil Loss from a Construction Area for a Design Storm

A 50-ha land area is to be developed into a single family residential area. The soil map indicates that the soil is loam with the following composition:

Clay 20%
Silt 35%
Fine sand 20%
(Silt + fine sand) 55%
Coarse sand and gravel 25%

The organic content of the soil is 1.5%.

The lot has a square shape with a drainage ditch in the center. It has been proposed to replace the ditch with a storm sewer. The average slope of the lot toward the ditch is 2.4%.

Determine soil loss (potential erosion) for a storm for which the hyetograph is given in Figure 5.15. Soil loss should be determined from the pervious areas for the two periods, namely, during construction when all vegetation is stripped from the soil surface (100% pervious) and subsequent to construction when 25% of the area is impermeable (streets, roofs, driveways, etc.).

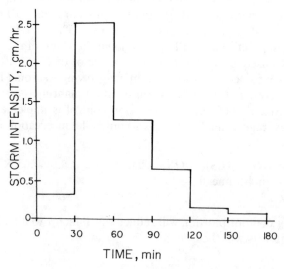

FIGURE 5.15.   Storm hyetograph for Example 5.2.

Solution The rainfall energy factor R_r is determined from the hyetograph shown on Figure 5.15. From this information it can be determined that the maximum 30-min rainfall intensity is 2.5 cm/hr.

Utilizing Equation (5.4)

$$\begin{aligned}R_r = [&(2.29 + 1.15 \ln 0.3)0.15 \\ + &(2.29 + 1.15 \ln 2.5)1.25 \\ + &(2.29 + 1.15 \ln 1.25)0.6175 \\ + &(2.29 + 1.15 \ln 0.7)0.35 \\ + &(2.29 + 1.15 \ln 0.2)0.1 \\ + &(2.29 + 1.15 \ln 0.1)0.05]2.5 = 16.4\end{aligned}$$

The soil erodibility factor is determined from Figure 5.13, assuming that the soil texture is fine grained and the permeability is moderate, giving a K value of 0.33.

To determine the LS factor for a 50-ha area with a ditch or storm sewer in the middle, the length of the overland flow $L = 0.5\sqrt{50 \times 100 \times 100} = 353.5$ m. With the use of Figure 5.14 or Equation (5.6), the LS factor for $L = 353$ and $S = 2.4\%$ becomes

$$\text{LS} = (353.5)^{1/2}[0.138 + (0.00974 \times 2.4) + (0.00138 \times 2.4^2)] = 0.47$$

Factors R_r, K, and LS are the same for both alternatives. The remaining factors, C and P must be evaluated for each alternative (P only if erosion-control practices are implemented during construction). For the period during construction (alternative 1), C is estimated assuming no vegetative protective cover and bulldozed soil. In this case, C is approximately the same as for bare fallow ground, that is, $C = 1$. In the absence of erosion-control practices, $P = 1$. Thus soil loss for this storm is

$$A = 16.4 \times 0.33 \times 0.47 \times 1 \times 1 = 2.54 \text{ tonnes/ha}$$

Thus, for 50 ha, total soil loss from the storm is

$$50 \times 2.54 = 127.3 \text{ tonnes.}$$

For the period after construction (alternative 2) and assuming that the pervious areas are covered by lawns, C is reduced to 0.01 and the soil loss/ha is

$$A = 16.4 \times 0.33 \times 0.47 \times 0.01 \times 1 = 0.025 \text{ tonnes/ha.}$$

Given that 75% of the area is subject to soil loss, the total sediment generation from pervious areas is

$$0.75 \times 50 \times 0.025 = 0.93 \text{ tonnes}$$

In order to complete the analysis, sediment generated from connected impervious areas (street dust and dirt—see Chapters 8 and 9) would have to be added to the amount just given. This estimate is also subjected to the condition that the storm will generate appreciable surface runoff.

Modifications of the USLE

The USDA-Agricultural Research Service has improved and simplified the annual form of the USLE by expanding the erosivity map for the western United States, improving information on soil erodibility factor, and by simplifying the slope-length factor (Renard et al., 1991).

For the CREAMS-GLEAMS model (Knisel, 1980), Foster et al. (1981) developed an erosion detachment–redeposition model that is derived from the steady-state mass balance equation of the sediment detachment and transfer by overland flow. The sediment movement downslope obeys continuity expressed by

$$\frac{dq_s}{dt} = D_L + D_F \tag{5.7}$$

where

q_s = sediment load (mass/unit width-time)
D_L = lateral inflow of sediment (mass/unit area-time)
D_F = detachment or deposition by flow

Detachment on interrill areas and transport and deposition by flow in rills are then modeled by the following equations:

$$D_L = \beta_1 EI(q_p/Q)(s_o + 0.014)KCP \tag{5.8a}$$

and

$$D_F = \beta_2 \eta Q q_p^{1/3}(/22.1)^{n-1}(q_p/Q)KCP \tag{5.8b}$$

where
D_L = interrill detachment rate (g/m² of land surface-sec)
D_F = capacity rate for rill detachment (g/m² of land surface-sec)

EI = rainfall erosivity factor (defined by Eq. (5.3))
x = distance downslope (meters)
η = slope length exponent for rill erosion
s_o = slope
Q = overland flow (volume/unit area-unit time = m/sec)
q_p = peak runoff rate (m/sec)
K, C, P = factors defined previously

β_1 and β_2 are numerical conversion factors depending on the selected units.

The CREAMS-GLEAMS erosion modeling concept, based on the steady-state mass balance of sediment transport, detachment, and redeposition in rills and interrills areas was further improved and modified and subsequently incorporated in the Water Erosion Prediction (WEPP) hillside erosion model (Nearing et al., 1989).

The *Negev sediment generation model* (Negev, 1967) simulates the generation and transport of soil by raindrop impact and overland flow. The production of fine soil particles by raindrop splash is determined for each computational interval and unit area as

$$A(t) = (1 - \text{COV}) \times K_N \times P(t)^{RER} \qquad (5.9)$$

where
A = fine soil particles produced during time interval i
COV = fraction of vegetative cover as a function of the relative value during the growing season
K_N = the coefficient of soil properties
P = precipitation during the time interval Δt
RER = an exponent

The model represents the production of fine soil particles, that is, silt and clay fractions, of the total watershed.

The fine particles produced by raindrop impact are available for transport by overland flow if it occurs during the time interval. If overland flow does not occur, for example, during the initial or final stages of the storm, the fine particles accumulate at the soil surface and represent a reservoir of liberated fine material available for transport by subsequent overland flow. The mechanism is modeled by

$$SER(t) = KSER \times SRER(t - 1) \times ROSB(t)^{SR} \qquad (5.10)$$

where
- SER = fine particles transported during time interval Δt
- $KSER$ = coefficient of transport
- $SRER(t - 1)$ = reservoir of deposited fine particles existing at the beginning of the time interval Δt
- $ROSB$ = the overland flow occurring during time interval Δt
- SR = an exponent

The Negev model was incorporated into the Stanford Watershed Model, and subsequently into the EPA's Hydrocomp Simulation Program (HSPF). This model requires knowledge of several empirical factors and coefficients. Unlike the USLE, the model has not been substantiated by experimental field data and measurements, and the coefficients must be estimated by calibrating the model against an extensive set of field data. Possible conversions between the USLE and parameters of the Negev model have been suggested by Fleming and Leytham (1976).

SEDIMENT DELIVERY AND ENRICHMENT PROCESSES DURING OVERLAND FLOW

As stated previously in this chapter sediment yield measured at a watershed outlet or a point on the water's course is not equal to the upland erosion. The state of the art for estimating delivery ratios has not progressed much beyond the long established empirical relationships, and available lumped models are still inaccurate.

As pointed out by Wischmeier (1976) the universal soil loss equation was developed from measurements on small plots with an average length of 22 meters. Extrapolation of erosion estimates by the USLE beyond this scale was not substantiated by large-scale field data and was not recommended by the authors of the equation. However, in the nonpoint pollution studies of the 1970s nonpoint pollution was equated with soil loss and the delivery ratio parameter obtained by calibration of the watershed models containing the USLE with measured field data, was used to overcome the differences between the upland erosion estimates and measured sediment yields. In this sense the delivery ratio factor became a scaling parameter by which the erosion estimate was adjusted to the measured sediment yields. Hadley and Shown (1976) have documented that only a fraction (about 30%) of the eroded sediment in several tributary basins (0.5 to 5.2 km^2) of the Ryan Gulch Basin in northwest Colorado found its way to the main basin and, subsequently, only 30% was transported to the mouth of the 125-km^2 basin. A nationwide study of 105 agricultural watersheds in the United States un-

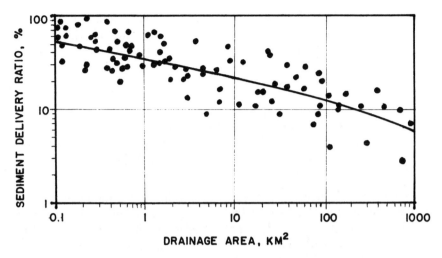

FIGURE 5.16. Relation of sediment delivery ratio to the watershed area. (From Roehl, 1962.)

dertaken by Wade and Heany (1978) reported sediment delivery ratios ranging from 1% to 38% of the gross erosion.

Golubev (1982) provided estimates on the components of the overall delivery process as delineated on Figure 5.9. In an investigation in the Oka River Basin in the central European part of the Russian Republic it was found that only 10% of the gross erosion is transported to the larger rivers and that 60% is deposited on the lower parts of the slopes, 20% in ephemeral channels, and 10% in minor streams of the network.

Other geomophologists and sedimentologists provided answers on the magnitude of the delivery ratio related to watershed characteristics. The most widely published was a relationship established by Roehl (1962) that relates the delivery ratio to the watershed size (Fig. 5.16). A statistical equation also related the delivery ratio to several geomorphological parameters, such as basin area, relief-to-length ratio, and bifurcation ratio. These relationships were obtained by comparing quantities of sediment deposited in a number of southern reservoirs with upstream erosion potential.

However, Beer, Farnham, and Heinemann (1966) found that for reservoirs draining loessial watersheds located in Iowa and Missouri the drainage area and delivery ratio were poorly correlated. In another geomorphological approach, Hadley and Shown (1976) related the delivery ratio to the degree of gullying, drainage density and soil texture (McElroy et al., 1976), and watershed size and runoff (Mutchler and

Bowie, 1976), while for construction sites the delivery ratio was related to the area under construction (Holberger and Truett, 1976). Williams (1975) related the delivery ratio to the travel time of the sediment in the channels and particle size of the sediment.

Wolman (1977) stated that

> The relationship between quantities eroded from the land surface and the quantities delivered to some distance downstream is exceedingly tenuous. The sediment delivery ratio provides a cover for real physical storage processes as well as for errors in estimates of the amount eroded and for temporal discontinuities of the process.

Walling (1983) divided the various approaches to the sediment delivery problem into several types, including a black box approach, by which sediment yield is correlated to upland erosion without considering the processes involved, delivery estimates based on channel characteristics, and estimates based on watershed characteristics. He argued that spatial and temporal lumping (or averaging) of the parameter limits most delivery estimates. As stated by Wolman (1977), the concept of the delivery ratio, as developed by geomorphologists, is indeed a cover for a number of processes that contribute to temporal or permanent deposition of sediments in an eroding watershed. These processes are highly variable and intermittent, and can be described only in the statistical sense. Novotny and Chesters (1989) followed with a comprehensive review of the sediment and pollutant delivery process and of the factors affecting the dynamic magnitude of the delivery parameter as related to water quality.

Factors Affecting Sediment Delivery

Any hydrologic process, such as delivery of sediments or pollutants from diffuse sources, is stochastic (Fig. 5.17). A stochastic process has deterministic and random components. The stochasticity and randomness of the process obviously vary. Flow in a uniform, open channel or runoff formation from a small parking lot are mostly deterministic processes, while a time series of annual rainfalls or stream flows is mostly random with some degree of autocorrelation. A delivery process falls somewhere in between (Fig. 5.18). For example, the correlation of the delivery ratio with runoff coefficients indicates a marked deterministic effect of infiltration and such losses as hydrologic storage and evaporation on the magnitude of the delivery ratio. Yet the scatter of the data around the line of the best fit might indicate a random component. Also, the statistical cumulative probability distribution of the delivery ratio is close to lognormal, which is the most common distribution for several hydrologic processes (Fig. 5.19).

Sediment Delivery and Enrichment Processes during Overland Flow 273

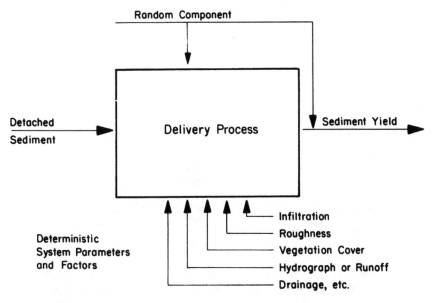

FIGURE 5.17. Stochastic representation of the delivery ratio process.

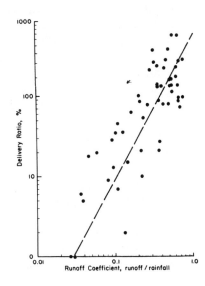

FIGURE 5.18. Statistical relationship of delivery ratio to runoff coefficient. (Data from Piest, Kramer, and Heinemann, 1975.)

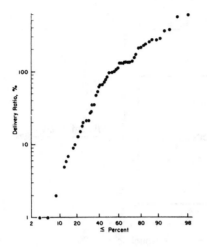

FIGURE 5.19. Statistical cumulative probabilistic distribution of delivery ratio for a loessial watershed in Iowa. (Data from Piest, Kramer, and Heinemann, 1975.)

The hydraulic (deterministic) factors affecting delivery can be divided into two categories: those that affect sediment and pollutant movement during overland flow, and those that affect sediment and pollutant movement during channel flow, including movement and deposition in floodplain.

Overland flow effects. There are six major overland flow factors and processes that affect delivery (Novotny, 1980; Novotny, Simsima, and Chesters, 1986; Novotny and Chesters, 1989).

- *Rainfall impact* detaches soil particles and keeps them in suspension for the duration of the storm. When the energy from rainfall ends or is reduced, the excess particles in suspension are deposited.
- *Overland flow energy* detaches soil particles from small rills and together with some interrill contribution, the particles remain in suspension as long as overland flow persists. Sediment content of overland flow is at or near saturation, and the sediment carrying capacity of overland flow is directly proportional to the amount of flow. If the flow is reduced during the receding portion of the hydrograph, excess sediment is deposited.
- *Vegetation* slows flow and filters out particles during shallow-flow conditions. If the vegetation is not submerged, the flow is mostly laminar, hence its sediment carrying capacity is very low.
- *Infiltration* filters out the particles from the overland flow.
- *Small depressions and ponding* allow particles to be deposited because of reduced flow velocity.
- *Change of slope of overland flow* because of drainage area concavity

often flattens the slope near the drainage channel and steepens the slope uphill. The sediment and pollutant transport capacity of overland flow changes according to slope.

The principal relationship between the detachment, deposition, and sediment carrying capacity of overland flow derived from the steady-state sediment transport mass balance concept (Eq. (5.7)) was defined by Foster and Meyer (1972a) as follows:

$$\frac{D_F}{D_c} + \frac{G_F}{T_c} = 1 \qquad (5.11)$$

where
D_F = detachment or deposition rate at a location (mass/unit area/time)
C = coefficient relating rate of deposition or detachment to the difference between transport capacity and sediment load
G_F = sediment load of flow at any location on the slope (mass/unit width/time)
T_c = flow transport capacity at a location on a given land profile (wt/unit width/time)
D_c = the detachment capacity of fow (mass/unit area/time)

Equation (5.11) represents the basis for understanding the process of attenuation and detachment of sediment and particulate pollutants from nonpoint sources. If D_c and T_c are similarly related to a function involving the shear stress of the overland flow and other parameters (that is, if $D_c \approx \alpha T_c$), the detachment or deposition rate defined in Equation (5.11) becomes

$$D_F = \alpha(T_c - G_F) \qquad (5.12)$$

where α = a coefficient.

Equation (5.12) essentially states that when the sediment load reaches the transport capacity of flow, there is no detachment, and if G_F by some way becomes greater than T_c, D_F becomes negative (deposition). Consequently if $G_F < T_c$, the flow is erosive.

As shown by Shen (1982) and illustrated in Figure 5.20, the concept can be generalized. Shen called the runoff rate at which no deposition or erosion takes place (at which $T_c = G_F$) "a critical runoff rate." As is shown later, these concepts are applicable primarily to noncohesive sediments. For cohesive sediments and soils the critical flow rates for detachment and deposition may not coincide.

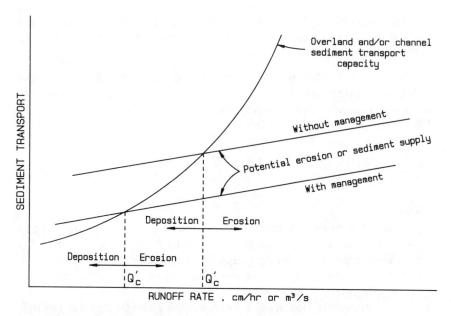

FIGURE 5.20. Shen's (1982) concept of the sediment carrying capacity of flow. *(Reprinted with permission from Water Resources Publications, POB 260026, Littleton, CO 80126-0026.)*

Figure 5.20 and Equation (5.11) point out an interesting fact on the impact of sediment load reduction in a watershed. When the erosion rate is reduced by land management, resulting in a reduced supply of sediment, the erodibility of runoff downstream increases, thus, to some degree, negating the erosion management effort. This conclusion would be only true to the fullest extent for noncohesive sediment, and its applicability to cohesive sediments and to pollutions carried by the sediment is uncertain.

The transport capacity of overland flow during rainfall can be estimated by the USLE (Eqs. (5.2) to (5.5)), and it is believed that the USLE truly represents the sediment transport capacity of flows in small eroded plots. However, with increased source area after the rain ends, a significant portion of the overland flow still remains on the surface and moves to the nearest established channel. This portion is contained in the recession portion of the hydrograph that does not have the same energy as the runoff during the period of the storm. Consequently, the excess sediment content must be deposited. A major portion of this deposition can occur even before the runoff leaves the field (Fig. 5.21).

The sediment carrying capacity of the runoff flow can be expressed best by Yalin's (1963) equation, which was reported in the following form:

FIGURE 5.21. During the delivery process of sediments most of the sediment is redeposited near the source. When the rainfall energy component of the overland flow terminates, the excess sediment settles out. (Photo: V. Novotny.)

$$P = 0.635s\left[1 - \frac{1}{\alpha s}\ln(1 + \alpha s)\right] \qquad (5.13)$$

where P is the dimensionless particle transport given by

$$P = \frac{p}{(\rho_s - \rho)Dv_*} \qquad (5.14)$$

The variables were defined as follows:

p = particle transport per unit width of flow (kg/m-sec)
ρ_s = particle density (kg/m^3), typically 2500 kg/m^3
ρ = density of water (kg/m^3) = 1000 kg/m^3
D = particle diameter (m)
v_* = $\sqrt{gHS}$ = shear velocity (m/sec)
g = gravity acceleration (= 9.81 m/sec^2)
H = depth of flow (m)

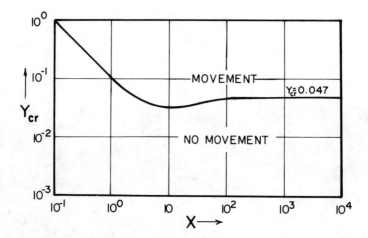

FIGURE 5.22. Shields's (1936) diagram for particle bedload reactive force.

S = slope of energy gradient (m/m)
$s = (Y/Y_{cr}) - 1$ = dimensionless excess tractive force (when $Y < Y_{cr}$, then $s = 0$)
$\alpha = 2.45\,(\rho/\rho_s)^{0.4}\sqrt{Y_{cr}}$
$Y = v_*^2/[(\rho_s - 1)gD]$ is the densimetric particle Froude number

The critical tractive force at which sediment movement begins, Y_{cr}, can be found from the Shields diagram (Fig. 5.22), which is based on the particle Reynolds number

$$X = \frac{Dv_*}{v}$$

where v is kinematic viscosity of the flow (m²/sec).

By rearranging Equation (5.13) and introducing the Manning equation for overland flow rate q

$$q = \frac{1}{n}H^{5/3}S^{1/2}$$

where n is the Manning's roughness coefficient (Table 3.14), and the saturated overland sediment load, p, or concentration, C_e, can be related to the flow rate as

$$p \sim D(qnS^{7/6})^\beta \tag{5.15}$$

or

$$C_e \sim D(nS^{7/6})^\beta q^{\beta-1} \tag{5.16}$$

where the exponent β could theoretically vary in range from 6/5 to 3/4.

Hence, proportionalities (5.15) and (5.16) indicate that the sediment concentration in the receding portion of the flow hydrograph decreases with $\beta - 1$ power as the flow decreases or with 7 $\beta/6$ power if the slope decreases.

Yalin's equation was suggested and used by Foster and Meyer (1972b) for estimating soil loss and transport of soil particles when detachment and transport is primarily due to runoff. Yalin himself pointed out that the equation is best used for the estimation of the saturated sediment flow

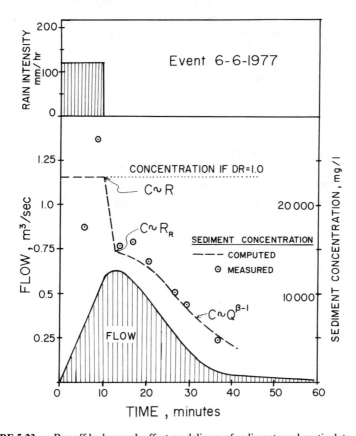

FIGURE 5.23. Runoff hydrograph effect on delivery of sediments and particulate pollutants during overland flow. (From Novotny, 1980, by permission of the American Water Resources Association.)

when most of the particle flow occurs near the bed, which means for the shallow overland and channel flows. It must be also realized that Yalin's equation was developed for fully turbulent flow conditions and for noncohesive sediments. Under shallow flow conditions or when overland flow is through nonsubmerged vegetation, all or a portion of the hydrograph tail can be under laminar conditions (Overton and Meadows, 1976). Under these circumstances, the equilibrium of the sediment concentration, C_e, may approach zero for all fractions. As a result, in a plot of C_e versus q, the exponent can have larger values than the theoretical estimate $(\beta - 1)$.

Figure 5.23 and Example 5.3 taken from Novotny (1980) illustrate this concept. In the figure, note that the sediment concentration that would result from both rain and runoff energy can persist only during rainfall. After the rain ends, the excess sediment content settles out, and the sediment concentrations reflect runoff energy. According to the principle of mass continuity it follows that only when the sediment concentration remains approximately constant during and after rainfall (for a uniform rainfall) would the delivery ratio be one.

Example 5.3: Delivery of Sediments from a Construction Site Basin

A 9.7-ha experimental watershed was located in southeastern Wisconsin. A storm with a duration of 10 minutes and storm depth of 2.03 cm occurred on July 6, 1977. The storm hyetograph and resulting flow hydrograph and sediment-graph are shown in Figure 5.23. Estimate the approximate delivery ratio for the storm and calculate the sediment graph.

Solution The parameters for the universal soil loss equation for the construction site basin are

K factor (silt loam) – $K = 0.48$
LS factor for $L = 100$ m and $S = 2.5\%$ – LS = 0.34
Cover factor (bare soil): $C = 1.0$
Erosion control practice factor: $P = 1.0$
Infiltration rate: assume 2.5 cm/hr

The storm characteristics are:

Volume of rain: $P_t = 2.03$ cm
Duration: $T = 10$ min
Maximum 30-min intensity: I_{30} 4.06 cm/hr

Calculated rainfall energy factor (Eq. (5.3)): $R_r = 33.20$
(Rain–infiltration) during 10-min rainfall period $= 2.03 - 2.5/6 = 1.61$ cm
$= 1561$ m^3

Runoff characteristics

Excess rainfall = Volume of runoff by integrating the hydrograph

$$Q = 907 \text{ m}^3/(9.7 \text{ [ha]} \times 10{,}000 \text{ [m}^2\text{/ha]}) \times 100 \text{ (cm/m)} = 0.93 \text{ cm}$$

Peak flow $q_p = 0.68$ m^3/s $= 0.68$ [m^3/s] $\times 100$ [cm/m] $\times 3600$ [sec/hr]/
$(9.7 \text{ [ha]} \times 10{,}000 \text{ m}^2\text{/ha}) = 2.52$ cm/hr
Runoff erosivity: $R_Q = 0.5 \times 15 \times Qq_p^{0.33} = 9.46$

Combined rainfall erosivity factor

$$R = 0.5 \times R_r + R_Q = 0.5 \times 33.2 + 9.46 = 26.06$$

Total sediment generated (Eq. (5.2))

$$A_{\text{total}} = 26.06 \times 0.48 \times 0.34 \times 1 \times 1 = 4.25 \text{ tonnes/ha}$$

For the entire area, assuming that 90% of the watershed contributed the sediment,

$$A_{\text{total}} = 0.9 \times 4.24 \text{ tonnes/ha} \times 9.7 \text{ ha} = 37.05 \text{ tonnes.}$$

Sediment concentration during the rainfall

$$C_{e1} = \frac{\text{Total sediment loss}}{\text{Volume of rain–infiltration}}$$
$$= \frac{37.05 \text{ tonnes} \times 10^6 \text{ g/tonne}}{1561 \text{ m}^3}$$
$$= 23{,}734 \text{ mg/l } (= \text{g/m}^3)$$

This high sediment concentration can be maintained only as long as the rain energy persists. After the rain ends, the total sediment carrying capacity of the runoff drops. At the peak runoff rate, the sediment carrying capacity is approximately proportional to the runoff energy term. Hence, the concentration of sediment becomes

$$C_{e2} = \frac{\text{Sediment carrying capacity of runoff}}{\text{Volume of runoff}}$$

$$= \frac{(R_Q)(K)(KS)(C)(P) \times (0.9 \times \text{area})}{Q}$$

$$= \frac{\{9.46 \times 0.48 \times 0.34 \times 1 \times 1 \times 0.9 \times 9.7\}[\text{tonnes}] \times 10^6 \text{g/tonne}}{907 \, \text{m}^3}$$

$$= 14{,}860 \, \text{mg/l}$$

The runoff sediment carrying capacity is decreasing as the hydrograph decreases. Therefore

$$C(t) = \max C_{e2} \times \left(\frac{q(t)}{q_{\text{peak}}}\right)^{\beta-1}$$

The recession coefficient (β-1) in this case is close to 0.5.

The calculated sediment concentration is plotted as the dashed line on Figure 5.23. The sediment yield can be calculated by integrating the product of multiplying the sediment concentration with the flow. The estimated and measured sediment yield from this watershed during the storm was 16 tonnes. Since the sediment load estimated by the USLE was 37.5 tonnes, the delivery ratio for this storm becomes

$$\text{DR} = \frac{16}{37.05} \times 100 = 43.2\%$$

Effect of drainage on sediment and pollutant delivery. Several small experimental urban and rural watersheds were monitored in southeastern Wisconsin from 1975 to 1980 (Novotny et al., 1979). A hydrological model, calibrated and verified by field measurements from very small, uniform subbasins, was used to estimate upland loadings of sediment and phosphorus. Hydrology (water yield) and sediment and phosphorus yields were calibrated and subsequently simulated. The simulation showed that the delivery ratio for sediment ranged from 0.01 for perviously undeveloped portions of the basin, to 1.0 for completely sewered (separate storm sewers) urban basins. Table 5.8 shows the results of the study. It can be concluded that the delivery ratio depended largely on the degree of storm sewering.

Sediment flow–discharge relationship. Traditionally, suspended sediment concentration versus flow relationship—known as a rating curve—is obtained by plotting sediment concentrations versus flow on logarithmic

TABLE 5.8 Estimated Sediment Delivery Ratios from Pervious Areas for Various Land Uses in Subbasins of the Menomonee River, Wisconsin

Subbasin Type	Impervious Area (%)	Degree of Storm Sewering (%)	Sediment Delivery Ratio (%)
Agricultural	<5	0	1–30
Developing–construction	<5	20–50	20–50
Low-density residential, unsewered	<20	0	<10
Parks	<10	0	<3
Medium-density residential, partially sewered	30–50	<50	30–70
Medium-density residential, sewered	30–50	>50	70–100
Commercial, high-density residential, sewered	>50	80–100	100

Source: From Novotny et al. (1979).
Note: The delivery ratio for impervious connected areas = 100%.

paper (Fig. 5.9). Generally, according to Walling and Webb (1983), and based on a number of observations of this relationship by the authors, the suspended solids concentration increases with flow, and the logarithmic slope of this relation is between 1 to 2, or, mathematically,

$$\text{Concentration} = a \times \text{flow}^b$$

where b is the slope of the relationship.

However, the pattern of the sediment flow–discharge relationship measured at the flow-quality gauging stations, reflects patterns discussed in Chapter 3 and throughout this chapter. It was shown that the rising limb of the hydrograph is fed by flows during the rainfall event, and the receding limb is due to the remaining overland flows after the rainfall. Hence flows that have generated the hydrographic rise have a higher sediment carrying capacity than do the flows of the end of the hydrograph. This difference between the sediment supply during the rising and ending portions of the hydrograph creates a phenomenon of apparent "first flush" in which the sediment concentrations tend to peak before the peak of the hydrograph occurs (in larger watersheds). This was shown on Figure 5.23. Consequently, the sediment concentration flow relation has a typical "loop" feature, as shown on Figure 5.24. Such hysteric loops have been observed by many observers. Walling and Webb (1983) provided several interpretations of this phenomenon, and based on the work by

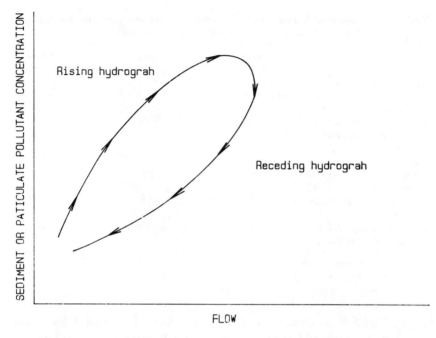

FIGURE 5.24. Histeric loops in the sediment transport–flow relationship.

Novotny (1980), stated that the hysteric loops of the sediment flow relationship during storm events could be interpreted in terms of reduced detachment or transport of soil particles after the cessation of rainfall. However, Walling and Webb also pointed out that other interpretations are possible, especially during complex multiple and/or prolonged rainfalls.

Long-term measurement of the delivery ratio. Long-term (twenty to thirty years) annual averages of erosion and sediment delivery may be estimated by measuring the cesium 137 isotope concentration in the soil profile (Ritchie, Spraberry, and McHenry, 1974; Ritchie and McHenry, 1975, 1978; Wilkin and Hebel, 1982).

During the period of intensive testing of nuclear weapons by atmospheric detonations, an artificial isotope cesium 137 was deposited more or less uniformly over the earth's surface. This testing period lasted for about twelve years and ended at the beginning of the 1960s. Cs-137 is known to be firmly adsorbed into the soil, and its activity (concentration) can be easily measured. There are no background concentrations of this isotope from a period prior to the nuclear weapons tests. Most of the Cs-

137 in untilled soil is located in the upper 10 cm. Tillage distributes the isotope uniformly throughout the tillage depth.

Ritchie and coworkers (quoted earlier) pioneered much of the work of using Cs-137 to quantitatively trace erosion, delivery, and sediment yields. The basic idea is that since the uniform distribution of Cs-137 during the 1950s the isotope has been redistributed with the sediment. Hence, in areas of higher erosion (degradation), from where the topsoil sediment has been removed, Cs-137 concentrations are lower than in depositional (aggradation) areas. Alternatively, following Figure 5.8, there will now be less Cs-137 in areas that were degraded than in aggradation areas. For example, Wilkin and Hebel (1982) found that in eroding areas, Cs-137 activity was about 0.7 to 0.9 CPMG (count per minute per gram), while in clearly depositional areas of midwestern streams the counts were generally 1 CPMG or greater (these activity counts correspond to a period of sediment redistribution of about 20 years, from about 1960 to 1980, but more profound differences would be found today). The activity of Cs-137 is determined by gamma-ray spectrometric analyses using a 1024-channel pulse-height analyzer and a solid-state detector.

A regression relationship of the type

$$E = a \times CA - b \tag{5.17}$$

where E is the soil loss in tonnes/ha and CA is the Cs-137 activity count of the top soil sample in CPMG, has been established by Wilkin and Hebel (1982), who also found that for midwestern watersheds $a = 214$ and $b = 201$. If E is positive, the watershed is eroding, while for depositional areas, E is negative.

For a known or measured change of Cs-137 Ritchie, Spraberry, and McHenry (1974) found that soil loss can be very closely correlated ($r^2 = 0.95$) to the Cs-137 activity change

$$A = 0.50 \times \Delta CA^{1.47} \tag{5.18}$$

where
A = total soil loss from 1960 in tonnes/ha
ΔCA = Cs-137 activity loss $n\%$ of the original input

Using Cs-137 measurements, Wilkin and Hebel found that much of upland erosion from agriculture was intercepted by upland depressions, grassed border fields, fencelines, hedgegrowths, and roadside ditches. The contribution of floodplains to the sediment problem was also investigated by this method. Cropped floodplains were eroding at excessive rates and delivering all eroded particles to downstream flow. Forested floodplains, on the other hand, even when grazed, appeared to be removing significant amounts of sediment from downstream transfer.

The Chernobyl nuclear power plant accident (Ukraine Republic) in 1986 caused a deposition of significant quantities of Cs-137 throughout northern and eastern Europe (Haldin, Rodhe, and Bjurman, 1990). Therefore the mass balance of Cs-137 in topsoils affected by the Chernobyl deposition must be recalculated, although the parameters for Equations (5.17) and (5.18) may not apply to the affected areas.

Enrichment of Sediments by Clay and Contaminants

The detachment of sediments and their deposition processes are selective. First, detachment of sediments and contaminants from the parent soil is selective for dissolved pollutants and for the fine soil fractions and pollutants adsorbed in or complexed by them. As will be shown in Chapter 6, most soil contaminants are adsorbed by clay and organic matter because of their high surface area, leading to strong adsorption bonds. When rainwater reaches the surface horizon of the soil, some contaminants are desorbed and go into solution; others remain adsorbed and move with the soil particles. The contaminant content of runoff sediment is then higher than in the parent soil. The difference is termed *enrichment ratio* (ER), and is defined as follows

$$\mathrm{ER} = \frac{C_r}{C_s} \tag{5.19}$$

where
C_r = contaminant concentration of the runoff per gram of sediment
C_s = contaminant content of the parent soil per gram

The enrichment concept can be applied to clay, organic matter, and all contaminants adsorbed by soil particles, which include phosphates, ammonium, metals, and pesticides. It is not appropriate to apply the enrichment ratio to contaminants that are mobile in soils, such as nitrates or soluble pesticides. Reviews of the enrichment problems have been published by Novotny and Chesters (1981, 1989) and by Walling (1983).

The enrichment ratio refers to the difference in particle size distribution and associated or adsorbed contaminant content of washload particles and the soils from which the sediment originated. Enrichment of sediments by clay is a two-step process: enrichment during particle pick-

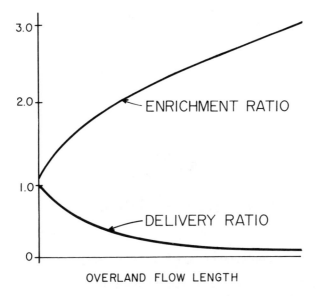

FIGURE 5.25. Relation of the sediment delivery to enrichment of sediment by clays.

up and enrichment during redeposition of the coarser particles during delivery in overland and channel flows. Thus, as the delivery ratio decreases with the increasing watershed area or time of overland flow the enrichment ratio of the washload increases as shown in Figure 5.25.

Pickup of Fines by Erosion

Two variables are necessary for determining the enrichment of sediment picked up by runoff (Free, Onstad, and Holtan, 1975). The variables are the specific surface (SS) of soils or sediment and the clay ratio (CR) defined as

$$SS = [200(\%\text{clay}) + 40(\%\text{silt}) + 0.5(\%\text{sand})]/100 \qquad (5.20)$$

and

$$CR = \frac{(\%\text{clay})}{(\%\text{silt}) + (\%\text{sand})} \qquad (5.21)$$

Free, Onstad, and Holtan proposed the following relationship, expressing the preceding variables in eroded material and in their soils of origin as

$$SS_e = 14.6 + 0.84 SS_m \qquad (5.22)$$

and

$$CR_e = 0.021 + 1.08 CR_m \qquad (5.23)$$

The initial enrichment ratio for sediment picked up by overland flow is then

$$ER = \frac{(\%\text{clay})_e}{(\%\text{clay})_m} \qquad (5.24)$$

where the subscripts m and e refer to the soil matrix and eroded material, respectively.

Example 5.4: Computation of the Enrichment Factor

Determine the approximate initial gradation of clay enrichment of the sediment eroded from a loam soil.

Solution The texture of soils can be approximately estimated from maps with the aid of Figure 5.26. In this case, the loam soil has an average composition of clay, 20%; silt, 40%; and sand, 40%. The specific surface according to Equation (5.20) is

$$SS_m = [(200 \times 20) + (40 \times 40) + (0.5 \times 40)]/100 = 56$$

and the clay ratio from Equation (5.21) is

$$CR_m = \frac{20}{40 + 40} = 0.25$$

From Equations (5.22) and (5.23) the specific surface and clay ratios of the eroded material are

$$SS_e = 14.6 + (0.84 \times 56) = 62$$
$$CR_e = 0.021 + (1.08 \times 0.25) = 0.29$$

The textural composition of the eroded material can be obtained by solving the three equations

$$[200(\%\text{clay})_e + 40(\%\text{silt})_e + 0.5(\%\text{sand})_e]/100 = 62$$

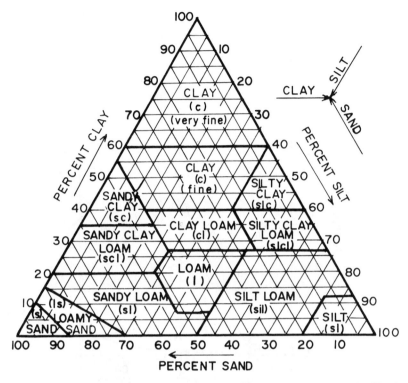

FIGURE 5.26. Guide for the USDA Soil Conservation Service soil texture classification.

$$\frac{(\%\text{clay})_e}{(\%\text{silt})_e + (\%\text{sand})_e} = 0.29$$

$$(\%\text{clay})_e + (\%\text{silt})_e + (\%\text{sand})_e = 100$$

which gives an initial textural composition for the eroded material of clay, 22%; silt, 43%; and sand, 35%. The clay enrichment ratio is then

$$\text{ER} = \frac{(\%\text{clay})_e}{(\%\text{clay})_m} = \frac{22}{20} = 1.1$$

Relationship of Enrichment to Delivery

During the delivery process, the texture of the sediment changes; eventually it is enriched by fine fractions. For a delivery ratio DR, a fraction of sediment equivalent to $(1 - \text{DR})$ is redeposited during transport. The redeposition is selective, that is, according to the bottom shear stress of

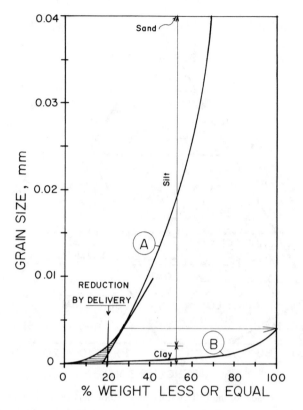

FIGURE 5.27. Graphical determination of washload composition and enrichment ratio from grain-size distribution curve.

flow, sand fractions are deposited first, followed by silt, and then, only when the flow slows down to a very low velocity and low Reynolds number, clays may deposit.

The deposition of clays is enhanced if the clay particles move in the form of aggregates. Dong, Chesters, and Simsiman (1983) have proposed a technique for determining an index for clay dispersion. Walling (1983) assumed that the probability of clay deposition is very low, hence the delivery ratio for clays and their enrichment are interrelated as follows

$$\text{DR} = \frac{\text{Clay content (soil)}}{\text{Clay content of sediment}} = \frac{1}{\text{ER}} \qquad (5.25)$$

An approximate graphical procedure based on the grain-size cumulative distribution curve can also be used to determine changes in sediment texture during overland transport, as shown on Figure 5.27.

Foster, Young, and Neibling (1985) defined the enrichment ratio as a ratio of specific surface area in the eroded sediment and parent soils, and related the enrichment ratio, ER, to the delivery ratio, DR, by the following empirical relationship:

$$ER = 1 + \alpha \exp(-\beta DR) \qquad (5.26)$$

where the coefficients α and β were dependent on soil texture.

Dong, Simsiman, and Chesters (1983) and Dong, Chesters, and Simsiman (1984) studied particle-size distribution and the composition of soils and sediments from the 350-km^2 Menomonee River basin in southeastern Wisconsin. Data from a 21-km^2, predominantly agricultural subwatershed located in the upper reaches of the river provided a direct determination of sediment enrichment by clays and contaminants. Particle-size distribution of soils in the subbasin was 27%, 49%, and 24% for clay, silt, and sand, respectively. Suspended sediment samples collected near the outlet of the subbasin (bedload movement is small compared to washload) during runoff events showed a particle-size distribution 91% clay and 9% silt.

Using the relationship between clay enrichment and the delivery ratio, DR = 27/91 = 0.29 and ER = 91/27 = 3.4. Delivery ratios for the same watershed established by modeling (Table 5.8) were in the range of 0.2 to 0.3.

The enrichment factor is a more dynamic phenomenon than sediment delivery. In addition to storm-to-storm variations this factor also exhibits seasonal and spatial variations for some constituents, while for some others, such as some metals eroded from soils and minerals, it could remain approximately constant (Walling and Kane, 1982). Most typical potential pollutants, including nitrogen, phosphorus, pesticides, and metals, exist in soils in adsorbed and dissolved forms (see Chapter 6 for details). The adsorbed materials move with sediment, and the fraction dissolved in soil water moves with runoff and/or ground-water recharge. The continuous process of adsorption and desorption–dissolution takes place during the overland and channel flow transport of the constituents. The concept of enrichment is not applicable to some dissolved or dissociated constituents, such as nitrates, which are repulsed by the electrostatic charge of the soil particles.

Enrichment of snowmelt. A different process of enrichment takes place during spring snowmelt. Snow pack is an effective filter of suspended particles; however, the snow crystal lattices during freezing and refreezing effectively reject many dissolved contaminants, which then remain available for pick up by melt water. For this reason, the first portion of snowmelt is highly enriched by dissolved and dissociated pollutants (nitrates, H^+, and others) with ER values much greater than one, while particulate pollutants remain in place, which results in ER factors of less than one (Johannessen and Henricksen, 1978; Colbeck, 1981).

Example 5.5 Determination of Texture Changes in Sediments During Delivery

A watershed with a drainage area of 21 km² had a measured sediment yield of 10,000 tonnes during a large spring storm. Estimated potential erosion from mostly loam agricultural soils was 50,000 tonnes. Assume that the texture of the soils and the eroded material are similar to those in Example 5.4. The eroded material therefore has the following initial texture: clay, 22%; silt 43%; sand 35%.

Solution   The delivery ratio is

$$\mathrm{DR} = \frac{Y}{A} = \frac{10{,}000}{50{,}000} = 0.2$$

An approximate cumulative grain-size distribution curve of the eroded soil is shown on Figure 5.27 as curve A. The delivery ratio of 20% indicates that most of the washload will be composited from clays. However, some coarser particles from areas closer to the receiving body of water can still remain in suspension. This can be represented by a tangent drawn to curve A, as shown in the figure. The triangular area on the left from the vertical line denoting the 20% delivery and the area between the tangent and the original curve on the right from the vertical delivery line are about the same. The shaded area represents the cumulative percentages of the remaining fractions in the washload, with the tangent as the baseline. To obtain the grain-size distribution curve of the washload, the vertical coordinates of the remaining fractions must be transformed to cover 100% scale (curve B).

From Figure 5.27, the composition of the washload will be: clay, 91%; silt, 9%; sand, not detectable.

The ER for a soil originally containing 20% clay is

$$\mathrm{ER} = \frac{91}{20} = 4.5$$

SEDIMENT TRANSPORT IN STREAMS

In channels (ephemeral and perennial streams or impoundments) flow is concentrated and has greater depth and velocity. While the overland sheet flow is mostly laminar and is able to support concentrated sediment flows only by the combination of rainfall and flow energy, flow in channels is turbulent and able to carry sediment flows on its own.

To a geomorphologist, the channel phase represents only temporary storage of transported sediment. Playfair's Law, one of the basic premises of stream morphology, states that over a long period of time a natural stream must transport essentially all sediment delivered to it (Boyce, 1975). One could argue with these geomorphological postulates because apparent deposition occurs in slow sections of rivers and impoundments, floodplains, and deltas. This results in distinct alluvial deposits, the best example being the mouth (delta) of the Mississippi River. Wilkin and Hebel (1982), using the Cs-137 indicator, found significant floodplain degradation on the Illinois River. Leopold, Wolman, and Miller (1964) concluded that materials eroded from a drainage basin are only temporarily stored in floodplains and that floodplain aggradation is essentially balanced by floodplain degradation. This means that a long-term (say 100 years or more) delivery ratio for streams is essentially one.

However, this is of little concern in pollution and water quality studies. For example, organic materials that are carried by and buried with the sediment in the alluvial deposits become harmless fossils in hundreds of years, long before they may be resuspended by catastrophic floods, yet, meanwhile, they may cause serious degradation of water quality by causing sediment oxygen demand (SOD) and other problems. Thus on the shorter time period (10 years or less) typical for water quality studies, deposition, resuspension of sediments, their contamination, and processes occurring in the deposited sediment layer are important and should be considered. The delivery ratio of many streams is much less than one when considering runoff magnitudes with a recurrence interval of less than 10 years.

Deposition of sediments and particulate pollutants occurs primarily in three places:

- In slow sections of streams, such as impoundments, estuaries, and deltas, and along the stream banks;
- Temporarily in a floodplain. For many pollutants that can be decomposed or degraded, a floodplain is a terminal sink.
- In ephemeral or recharging streams by filtration of particulates during recharge. This storage on the channel bed is only temporary, and the particles remain on the stream bottom available for resuspension by the next runoff event.

Contamination of bottom sediments by toxic and other polluting substances affects the quality of the water column and represents the entry of toxics into the food chain. It is a known fact that the pollutants that have been previously deposited into sediments contribute to the degradation of water quality and adversely affect aquatic ecosystems in many water bodies, including, among others, Green Bay and other bays of the Great Lakes system, Chesapeake Bay, Long Island Sound, and many other inland and near-shore aquatic systems. Milwaukee (Wisconsin), Grand Calumet Indiana Harbor (Indiana), and Waukeegan (Illinois) harbors, as well as many other urban estuaries, exhibit sediment contamination that prevents beneficial uses of these bodies of water. However, due to the binding of many pollutants with the sediments, the adsorbed or complexed pollutants become biologically unavailable and, hence, not toxic. Therefore knowledge of the relation between immobilized (biologically unavailable) and free dissolved (available) fractions of the pollutant in the sediment is important in pollution impact studies. See Chapter 6 for more discussion of this subject.

Bottom sediments become contaminated with toxic substances by the deposition of particulate fractions of pollutants and/or by the deposition of fine sediment particles, both organic and inorganic, that have an affinity to adsorb dissolved pollutants and pollutant ions in the water column.

After deposition, several processes may cause the deposited pollutants to reenter the water column, including the resuspension of sediment particles caused by turbulence and biological benthic activities, diffusion from the bottom layer, uptake and subsequent decomposition of rooted macrophytes, and several other processes. Reentry may be temporal, such as with a particle that had been temporarily resuspended, or semipermanent, such as by diffusion, in the water column.

In the sediment layer, many pollutants may undergo chemical and biological modifications, including chemical and biological degradation, conversion between inorganic and organic forms, and adsorption (fixation) –desorption reactions.

Sediment Transport Process

Sediment transport, erosion, and sedimentation have been a popular subject of research for more than 80 years (Vanoni, 1975). Sediment in streams is transported either as washload, or suspended sediment, or bedload, or sediment, and it moves near or at the bottom. In the classic representation of sediment movement, the suspended sediment contains mostly fine sediment particles, while normally bedload may be a composite of sand and gravel. Several sedimentologists showed a relation between the washload and bedload (Einstein, 1950; Yalin, 1963; Vanoni,

1975). In this representation there is a continuous exchange of particles between the bedload and the washload, and the vertical concentration distribution of the washload is related to the bedload transport. Most of these classic studies assumed that there is no interaction between the sediment particles other than the impact of collision. Such sediments are called *noncohesive sediments*, and include primarily sand and gravel fractions. Noncohesive sediments have a very low to zero affinity to adsorb pollutants, and transport of pollutants is generally not affected by their presence.

In contrast, *cohesive sediments*, which include primarily clay and organic fractions of the bedload and washload, have a high sorptive capacity for many chemicals and act as carriers for contaminant transport in riverine and estuarine systems. As shown in a publication by Novotny and Chesters (1981), clay and organic matter can effectively adsorb such pollutants as phosphates, ammonia, and a variety of organic chemicals, from which PCBs exhibit the highest affinity for adsorption. In addition, some pollutants, such as organic matter, nitrogen and phosphates, some metals, and some organic chemicals, exist both in solid and liquid phases (precipitated and dissociated), and are an integral part of the washload.

Organic-rich sediments (mud), such as those generated in the productive sections of streams by photosynthesis (Novotny and Bendoricchio, 1989), have the most severe implications on water quality conditions. The processes of decomposition that occur in these sediments impose a demand for oxygen on the overlying water. Under anaerobic conditions phosphates and ammonia are released back into the water and chemical adsorption equilibria are modified in the reducing environment of anaerobic sediments.

The classic literature of sedimentology has not addressed the problem and processes associated with the movement of cohesive sediment and their impact on water quality and pollutant transport. Only recently have sedimentologists become interested in the subject.

Transport of Cohesive Sediments

Figure 5.28 shows the interrelationship between the basic transport processes of cohesive sediments in riverine and estuarine systems. While in noncohesive sediment transport, bedload and washload are fairly distinct with a clear boundary interface. Cohesive sediment may exit in the four states shown in Figure 5.28: (1) a mobile suspension where particles move primarily in a horizontal direction; (2) a horizontally stationary high suspension; (3) a consolidating bed; and (4) a settled compacted bed in which movement of particles and water has ceased (Mehta et al., 1989).

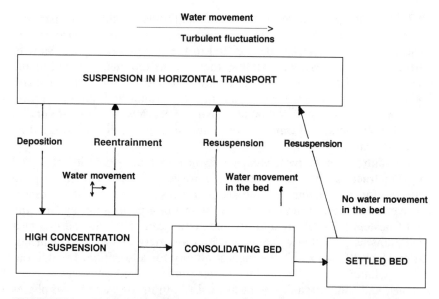

FIGURE 5.28. Physical states and processes governing cohesive sediment transport. (After Mehta et al., 1989, reprinted by permission of ASCE.)

Of interest in this concept is the movement of water. When particles are in suspension, water movement is primarily in a horizontal direction; however, at washload concentrations of solids, turbulent fluctuations may be less then those for clear water (Xingkui and Ning, 1989). In high concentration suspension, the horizontal movement of particles diminishes, and if the net particle flux is down, the replaced water moves up. This replaced water then mixes with water in the flow above, which should be included in the mass balance equations of pollutants. What it means is that if there is net settling of particles, there is a net water movement up, which can carry dissolved pollutants with it. There is similar water movement in the compacting layer where, as the layer reduces its thickness and increases its density, water is pushed upward from the sediment layer. In both cases, the sediment layer is a source of dissolved pollutants.

Higher concentrations of suspended solids modify the density of the sediment–water mixture. Similarly, to heat or salinity transport, suspended sediments in impoundments and estuaries can travel in density currents (Akiyama and Stefan, 1985).

Typically, cohesive sediments (mud) in riverine and estuarine systems are composites of clay, silt, and organic particulates. Cohesion of particles (floc formation) is more pronounced for clay particles (<2 μm) than for more coarse silt particles (2 to 60 μm). In fact, Mehta et al. (1989) speculated that the cohesion of silt particles into larger flocs and compacted matter is by clays. When large quantities of coarse materials (sand, larger organic detritus) are present in the sediment, its behavior and interactions become quite complex, and are currently not well understood (Mehta et al., 1989).

Aggregation of cohesive sediments is a complex problem that depends on several environmental factors. Some of the factors are well understood from water treatment technology, such as the effects of calcium and other cations, anions, total salt concentration, and pH. Other factors, such as the effect of turbulent mixing, settling characteristics of particles in the aquatic environment, and particle interactions are less well understood. The biological effects of bottom dwelling and feeding fish and invertebrates, or the mixing and resuspension of particles by the evolution of gases from the benthic layer are speculated about only qualitatively.

Settling velocity of cohesive sediments. In turbulent flow, suspended particles are supported and distributed in the flow by the mechanism of turbulent exchange, which acts against the gravity force of the particles. The fall velocity of the individual particles is given by the well-known Stokes' law. However, cohesive sediment particles will aggregate under favorable chemical conditions, thus increasing the settling velocity over that given by Stokes' law. Table 5.9 shows the effects of aggregation on the increase of settling velocity.

The preceding results of the effect of aggregation on settling velocity were obtained by experiments in settling tubes (Migniot 1968; Chase, 1979). Reported values of settling velocities for marine and estuarine sediments ranged from 10^{-4} to 10^0 mm/sec.

Effect of hindered settling. At a certain concentration of suspended particles, settling becomes hindered. This is analogous to the classic concept of the settling of flocculent particles in settling tanks and settling

TABLE 5.9 Settling of Individual Particles and Aggregate Settling

Original Particle Diameter (μm)	Stokes Settling Velocity (mm/sec)	Aggregate Settling Velocity (mm/sec)	Aggregate Diameter (μm)
20	$2.4 * 10^{-1}$	$2.7 * 10^{-1}$	88
2	$2.4 * 10^{-3}$	$1.7 * 10^{-1}$	56
0.2	$2.4 * 10^{-5}$	$1.1 * 10^{-1}$	34

Source: After Mehta et al. (1989). *(Reprinted by permission of ASCE.)*

columns. During hindered settling the probability of particle collision is so high that it slows down settling and the particles may settle as one mass (zone settling). This zone settling pattern may loosely correspond to the high concentration suspension layer depicted on Figure 5.28.

Tesarik and Vostrcil (1970) proposed an equation for the settling velocity of flocculated flocs as

$$W = \frac{\alpha}{c^\beta} \qquad (5.27)$$

where α and β are coefficients. The values of the coefficients for the bentonite clay were reported as $\alpha = 12.8$ and $\beta = 2.06$ for W in mm/sec and c in g/l.

The transition between hindered settling and discrete settling occurs approximately at a volumetric floc concentration of 0.005 (Camp, 1947), which is comparable to a concentration of 1000 to 5000 mg/l of silt and clay particles in turbid water.

Generally, the high concentration (hindered settling) zone should exhibit a distinct interface between the more clear turbid water above where settling is discrete. This is due to the fact that if a floc leaves the high concentration layer, the settling velocity ceases to be hindered and increases. Thus the likelihood of the particle to settle back into the high concentration zone also increases. Another reason for a distinct interface is the fact that turbulence fluctuations in the high concentration zone are reduced; therefore, at the interface there is less chance for a particle to move upward from the high concentration zone than there is for it to move downward from the discrete overlaying zone. If the turbulent energy of flow increases above a certain yet unspecified level, this increase may lead to a collapse of the high concentration zone and the flow may become uniformly mixed.

The surface stresses of the high suspension (hindered settling) layer are very small and in general are not measurable (Been and Sills, 1981). Hence, if under some flow conditions (presumably at low energy levels) the probability of particles to remain in the high concentration zone is greater than it is to escape from the zone, the particles will remain in the layer and will require a discrete resuspension energy to rise up from the layer.

What this means is that there appears to be a critical shear stress for formation of the high concentration layer that is an integral component of the deposition process of cohesive sediments. The existence of this layer may also have significant water quality implications. As water that was enriched by the high rate of desorption is pushed out of the high concentration layer by the settling of particles, it may become the primary source of contamination in the overlying water.

The energy level can be related to the bottom shear stress of flow. Data in the literature indicate (Mehta et al., 1989) that the shear stress for deposition (deposition of cohesive sediments will not take place if the shear stress is above this value) is about $\tau_c = 0.06$ to $0.08 \, \text{N/m}^2$. This correlates well with the experiments of Partheniades, who found that the critical shear stress for deposition is around $0.1 \, \text{N/m}^2$. For mixed sediments with broad size distribution, Mehta and Partheniades (quoted in Mehta et al., 1989) found that τ_c ranged from about 0.18 to $1.1 \, \text{N/m}^2$. A well-known formula for estimation of shear stress is

$$\tau = \gamma R S_e \tag{5.28}$$

where
γ = specific density of water (N/m^3)
R = hydraulic radius or depth (m)
S_e = the energy gradient of flow (dimensionless)

Below the high concentration (hindered settling) zone is a layer of consolidating sediments. In the theory of settling this layer could correspond to compression settling. The particles in this layer form a solid compacting mass that settles due to the gravity of particles and the water above. This consolidation by compressed settling begins when surface stresses become noticeable. Primary consolidation ends when the excess pore water pressure has completely dissipated. Secondary consolidation, which is the result of the plastic deformation of the bed under the constant overburden, begins during primary consolidation and may typically continue for many weeks or months after primary consolidation ends (Mehta et al., 1989).

Erosion and Resuspension of Cohesive Sediments

The erosion potential of consolidated sediments is again related to bottom shear stress. The critical shear stress for erosion, τ_s, has been related to the specific density of compacted sediments and to the pressure of the overlying water (depth of the water column) as

$$\tau_s = \xi \rho_D^\delta \tag{5.29}$$

If τ_s is in N/m^2, then the coefficient ξ is about 6×10^{-6} to 8×10^{-6}, and the exponent δ is about 2.3 to 2.4 (Mehta et al., 1989). ρ_D is the dry density of consolidated sediment in kg/m^3, which is a function of time.

The dry sediment density can be calculated from the bulk density as follows (Mehta et al., 1989):

$$\rho_D = \frac{(\rho_B - \rho_w)\rho_s}{(\rho_s - \rho_w)} \quad (5.30)$$

where
ρ_B = bulk (wet) density of the sediment
ρ_s = sediment density
ρ_w = water density

The critical shear stress for the erosion of consolidated sediments is one to two orders of magnitude greater than that for deposition. The scouring rate of sediment can be related to the bottom shear stress, as pointed out by Mehta et al. (1989) and Partheniades (1965).

Three modes of erosion have been identified: (1) reentrainment of stationary suspension (as discussed earlier), which occurs at shear stress rates between τ_c and τ_s; (2) aggregate or gradual erosion of a bed at shear stresses greater than τ_s and a certain critical value of bed shear stress τ_b; and (3) mass erosion of a bed when bottom shear stress energy is sufficient to lift the bulk of the sediments.

Experimental work (Lee, Kang, and Lick, 1981; Ziegler and Lick, 1988) leads to a formulation of a relationship between the erosion of fine-grained cohesive sediment and bottom shear stress as follows (Ziegler and Lick, 1988):

$$\begin{aligned}\varepsilon &= \alpha \left[\frac{\tau - \tau_s}{\tau_s}\right]^m \quad \text{for } \tau \geq \tau_s \\ &= 0 \quad \text{for } \tau < \tau_s\end{aligned} \quad (5.31)$$

where
ε = net amount of resuspended sediment in Kg/m^2
$\alpha \approx 0.08$
$m \approx 2$

There are several interesting conclusions from the past research on deposition, formation, consolidation, and erosion of cohesive sediments:

1. The "classic" sedimentology typical of noncohesive sediments is not fully applicable to cohesive sediments. For noncohesive sediments there is a certain limiting shear stress value τ_c above which deposition will not occur.

2. At low shear stresses, formation of a high concentration–hindered settling layer is likely. Even though this layer may not show any measurable shear stress resistance, energy (shear stress) in excess of τ_c is needed to break the high concentration layer into a dispersed flow form. The magnitude of this energy is not known and will most likely be between the critical shear stress for deposition and that for the erosion of a consolidated bottom. At certain small shear stress values above τ_c the erosion of the high concentration layer may be gradual; however, experience shows that it may be possible to anticipate that the surface of the high concentration layer will move upward in response to increased shear stress until the concentrations within the layer cease to be hindered and/or the top of the layer reaches the flow surface.
3. The high concentration layer provides a medium for water to come in contact with contaminated sediments even when resuspension does not occur. Because of the settling of particles there is a distinct upward velocity component within the layer that brings the dissolved pollutants to the overlying water. At some critical yet unknown shear stress the high concentration layer will break and become completely mixed with water.
4. The consolidated layer requires higher energy for breaking and resuspension. Below this energy level, the mechanism of pollutant transfer from (or into) the layer is by diffusion through the boundary layer.
5. At shear stresses greater than the critical erosive shear stress, τ_s, the bottom is resuspended and most likely completely mixed with the overlaying water (or hypolimnion of stratified flows). The rate of resuspension is related to the bottom shear stress. At certain high shear stress rates the compacted layer may be eroded at very high rates (Parker, 1982).

Kinetic Model for Sedimentation of Cohesive Sediments

The author previously presented a simple model for the sedimentation of cohesive sediments (Krenkel and Novotny, 1980). The model was based on classic sedimentology and may not be fully applicable today in view of new information published later in the 1980s. However, the basic concepts after modification are applicable. In view of the new information it is necessary to separate the flow of cohesive sediments and pollutant transport into three distinct zones (Fig. 5.28):

1. *Zone of dispersed flow or discrete settling zone in which classic sedimentology is partially applicable.* Turbulence is not greatly affected by

sediment concentrations, particle concentration might follow the classic Einstein distribution (Vanoni, 1975; also published in Krenkel and Novotny, 1980, and Novotny and Chesters, 1981), and dissolved pollutant transport is by convection and turbulent dispersion.
2. *Zone of hindered settling (high concentration) in which classic sedimentology is not applicable.* The turbulence level is affected by particle concentrations, the horizontal movement of particles is diminished, there is a net upward vertical component of water, and water is rich with desorbed pollutants or in a higher equilibrium with adsorbed fractions. This transitional zone may change thickness in response to the shear stress. This layer will exist when shear stress is less than a certain critical value that is smaller than the critical shear stress for erosion of a compacted bed.
3. *Compacting sediment zone.* There is no horizontal movement of sediment; however, there still may be some upward movement of water through the sediment layer during compaction. This convective transport adds to the diffusion of pollutants, which is controlled by diffusion through the boundary layer. When the shear stress exceeds the critical shear for erosion, sediment particles are released from the layer.

Discrete settling zone. In this zone, the classic laws of sediment transport are applicable. Define M as the scour rate of cohesive sediments or the reentrainment rate of particles from the high concentration layer, and W_s as a settling rate under aggregation. Then the concentration distribution can be described by Einstein's equation (see Novotny and Chesters, 1981). Also under steady-state conditions

$$\frac{dg}{dx} = M - W_s C_b \qquad (5.32)$$

where
C_b = near bottom concentration of the sediment
g = the sediment flux

However, Partheniades (1977) documented that this equation is not fully valid for the transport of cohesive sediments. It was also documented in the previous discussion that in cohesive sediment transport erosion and sedimentation do not occur simultaneously. Furthermore, according to the Partheniades theory and experimental results the erosion rate of cohesive sediments from the compacted layer should remain constant regardless of the sediment concentration in water. This may lead at high bottom shear stresses to very high sediment concentrations that are related to the supply of sediment rather than to a saturated sediment flux. Consequently, on the other hand, if the shear stress is below the critical shear stress for deposition, there is no erosion, and hence the minimal concentration of sediments is zero, provided that enough time is available for the particle to settle out.

High concentration zone. The classic sedimentology of discrete noncohesive sediment flow fails in the high concentration flow, that is, when volumetric concentrations of particulates are above $c = 0.5\%$. It should be pointed out that the high concentration (hindered settling) flow zone can extend all the way to the surface, as documented on flows occurring in some high turbidity streams of some southwestern regions, such as Rio Puerco in New Mexico.

Compaction zone. During deposition solids from the high concentration zone will reach the compaction zone and undergo compressed settling, or at some very low turbulence level the entire high concentration zone may begin to compress. During compressed settling water is released upward.

From the foregoing discussion it appears that no erosion of the compacted zone takes place as long as there is a high concentration zone above, since the collapse of the high concentration zone or its extending to the flow surface may occur at lower shear stress values than the minimum shear stress for erosion of compacted sediment layers.

Furthermore, as pointed out by Partheniades (1977), no simultaneous deposition and erosion of cohesive sediment takes place, and in the absence of erosion the only mechanism by which pollutants can be released from the compacted zone is by diffusion and by convection by water removed from the layer by compaction.

This concept of three-layered cohesive sediment transport and sedimentation, which has been proved by experiments of Mehta et al. (1989) and earlier by Partheniades deviates from some modeling concepts, such as those incorporated in the EPA model WASP (see Chapters 7 and 8), which presumes the existence of a certain unspecified constant layer of sediment that mixes with the overlying water. The variable high concentration layer that can exist only at bottom shear stresses below those for erosion ($\tau < \tau_s$) is the nearest approximation of such a mixing layer. At shear stresses above τ_s it is likely that there is no mixing layer that the entire bulk of the water may have a uniform concentration of sediment, and that, the flux of pollutants from the bottom sediment layer is related to the rate of erosion. This high concentration mixing layer may reappear when the sediment concentration increases to a level at which turbulence is reduced and/or the flow rate drops so that the bottom shear stress is reduced below that which allows deposition.

Example 5.6: Erosivity of a Channel Flow

A wide (width $>10 \times$ depth) rectangular channel is carved in cohesive sediments. The depth of flow in the channel is $H = 0.1$ meters and the velocity is $v = 0.3$ meters, the Manning roughness factor for the channel is $n = 0.02$. Determine whether the flow is erosive or depositional.

Solution In order to estimate the bottom shear stress one has to calculate the energy gradient of the flow. From Manning's equation and assuming that for a wide channel the hydraulic radius approximately equals the depth of flow, the energy gradient, S_e, is

$$S_e = \frac{(v \times n)^2}{H} = \frac{(0.3 \times 0.02)^2}{0.1} = 0.00036$$

The bottom shear stress (Eq. (5.28)) is (for $R = H$ and $\gamma = 9810\,\text{N/m}^3$)

$$\tau = \gamma H S_e = 9810 \times 0.1 \times 0.00036 = 0.35\,\text{N/m}^2$$

The shear stress of flow is above or near the critical shear stress for deposition ($\tau_c \approx 0.1\,\text{N/m}^2$); however, it appears to be below the critical shear stress for scouring. The critical bottom shear stress for scouring can be estimated by using Equation (5.29). For example, if the specific density of the bottom sediment is $\rho_s = 2500\,\text{kg/m}$, the specific density of water $\rho_w = 1000\,\text{kg/m}^3$, and the bulk density of the sediment $\rho_B = 1200\,\text{kg/m}^3$, then the dry density of the sediment is (Eq. (5.30)) $\rho_D = (1200 - 1000) \times 2500/(2500 - 1000) = 333$. Assuming that $\xi = 7 \times 10^{-6}$ and $\delta = 2.4$, then the critical bottom shear stress for scouring is

$$\tau_s = 7 \times 10^{-6} \times 333^{2.4} = 7.94\,\text{N/m}^2$$

One could conclude that the flow is either very mildly depositional or that no deposition or scour will occur. The delivery ratio would be near unity.

References

Akiyama, J., and H. Stefan (1985). Turbidity current with erosion and deposition, *J. Hydraul. Eng. ASCE*, **111**(12):1473–1496.

Anderson, H. W. (1975). Sedimentation and turbidity hazards in wildlands, in *Proceedings, Symposium of the Irrigation and Drainage Division*, ASCE, New York, pp. 347–376.

Been, K., and G. C. Sills (1981). Self-weight consolidation of soft soils, *Geotechnique* **31**(4):519–535.

References 305

Beer, C. E., C. W. Farnham, and H. G. Heinemann (1966). Evaluating sedimentation predictions in western Iowa, *Trans. ASAE* **8**(9):828–833.
Bordas, M. P., and G. E. Canali (1980). The influence of land use and topography on the hydrological and sedimentological behavior of basins in the basalic region of South Brazil, in *The Influence of Man on the Hydrological Regime with Special Reference to Representative and Experimental Basins*, Publ. No. 130, International Association of Hydrological Sciences, Washington, DC, pp. 55–60.
Borman, F. H., and G. E. Likens (1969). Biotic regulation of particulate and solution losses from a forested ecosystem, *Biosciences* **19**(7):600–610.
Boyce, R. C. (1975). Sediment routing with sediment delivery ratios, in *Present and Perspective Technology for Predicting Sediment Yields and Sources*, ARS-S40, U.S. Department of Agriculture, Washington, DC, pp. 61–65.
Brant, G. H., et al. (1972). *An Economic Analysis of Erosion and Sediment Control for Watersheds Undergoing Urbanization*, Dow Chemical Co., Midland, MI.
Brown, G. W. (1985). Controlling nonpoint source pollution from silvicultural operations: What we know and don't know, in *Perspectives on Nonpoint Source Pollution*, Proceedings, National Conference, Kansas City, EPA 440/S-85-001, U.S. Environmental Protection Agency, Washington, DC, pp. 332–334.
Camp, T. R. (1947). Sedimentation and the design of settling tanks, *Trans. ASCE* **3**:895–958.
Chase, R. R. P. (1979). Settling behavior of natural aquatic particulates, *Limnol. Oceanogr.* **24**:1417–1426.
Chung, M., F. A. Roth, and E. V. Hunt (1982). Sediment production under various forestsite conditions, in *Recent Development in the Explanation and Prediction of Erosion and Sediment Yield*, D. E. Walling, ed., Publ. No. 137, International Association of Hydrological Sciences, Washington, DC, pp. 13–23.
Clark, E. H., Jr., J. A. Haverkamp, and W. Chapman (1985). *Eroding Soils. —The Off-Farm Impacts*, The Conservation Foundation, Washington DC.
Colbeck, S. C. (1981). The simulation of the enrichment of atmospheric pollutants in snow cover runoff, *Water Resour. Res.* **17**(5):1383–1388.
Committee on International Soil and Water Research and Development (1991). *Toward Sustainability. Soil and Water Research Priorities for Developing Countries*, National Academy Press, Washington, DC.
Dong, A., G. Chesters, and G. V. Simsiman (1983) Soil dispersibility, *Soil Sci.* **136**(4):208–212.
Dong A., G. Chesters, and G. V. Simsiman (1984). Metal composition of soil, sediment, and urban dust and dirt samples from the Menomonee River Watershed, Wisconsin, U.S.A., *Water, Air, Soil Pollut.* **22**:257–275.
Dong, A., G. V. Simsiman, and G. Chesters (1983). Particle size distribution and phosphorus levels in soil, sediment and urban dust and dirt samples from the Menomonee River Watershed, Wisconsin, *Water Res.* **17**(5):569–577.
Einstein, H. A. (1950). *The Bed Load Function for Sediment Transportation in*

Open Channel Flow, Tech. Bull. No. 1026, U.S. Department of Agriculture, Washington, DC.

Fleming, G., and K. M. Leytham (1976). The hydrologic and sediment processes in natural watershed areas, in *Proceedings, Third Federal Interagency Sedimentation Conference*, Denver, CO, Water Resources Council Sedimentation Committee, Washington, DC, pp. 1-232–1-246.

Foster, G. R., and L. D. Meyer (1972a). A closed form erosion equation for upland areas, in *Sedimentation*, H. W. Shen, ed., Fort Collins, CO.

Foster, G. R., and L. D. Meyer (1972b). Transport of soil particles by shallow flow, *Trans. ASAE* **13**(1):99–102.

Foster, G. R., F. Lombardi, and W. C. Moldenhauer (1982). Evaluation of rainfall-runoff erosivity factors for individual storms, *Trans. ASAE* **25**(1):124–129.

Foster, G. R., L. D. Meyer, and C. A. Onstad (1977). An erosion equation devised from basic erosion principles, *Trans. ASAE* **20**:678–682.

Foster, G. R., et al. (1981). Estimating erosion and sediment yield on field-sized areas, *Trans. ASAE* **24**(5):1253–1262.

Foster, G. R., R. A. Young, and W. H. Neibling (1985). Sediment composition for nonpoint source pollution analyses, *Trans. ASAE* **28**(1):133–139.

Frederiksen, R. L. (1970). *Erosion and Sedimentation Following Road Construction and Timber Harvest on Unstable Soils in Three Small Western Oregon Watersheds*, Research Paper, PNW 104, U.S. Forest Service, Washington, DC.

Free, M. H., C. A. Onstad, and H. N. Holtan (1975). *ACTMO—An Agricultural Chemical Transport Model*, Report No. ARS-H-3, U.S. Department of Agriculture, Washington, DC.

Goldman, S. J., K. Jackson, and T. A. Bursztynsky (1986). *Erosion & Sediment Control Handbook*, McGraw-Hill, New York.

Golubev, G. N. (1982). *Soil Erosion and Agriculture in the World: An Assessment and Hydrological Implications*, Publ. 137, International Association of Hydrological Science, Washington, DC.

Gottschalk, L. C. (1964). Sedimentation. Part I. Reservoir Sedimentation, in *Handbook of Applied Hydrology*, V. T. Chow, ed., McGraw-Hill, New York.

Hadley, R. F., and L. M. Shown (1976). Relation of erosion to sediment yield, in *Proceedings, Third Federal Inter-agency Sedimentation Conference*, Denver, CO, U.S. Water Resources Council, Washington, DC, pp. 1-132–1-139.

Haldin S., A. Rodhe, and B. Bjurman (1990). Urban storm water transport and wash-off of Cesium-137 after the Chernobyl accident, *Water, Air Soil Pollut.* **49**:139–158.

Holberger, R. L., and J. B. Truett (1976). Sediment yield from construction sites, in *Proceedings, Third Federal Inter-agency Sedimentation Conference*, Denver, CO, U.S. Water Resources Council, Washington, DC, pp. 1-47–1-58.

Johannessen, N., and A. Henricksen (1978). Chemistry of snowmelt—Changes in concentration during melting, *Water Resour. Res.* **14**(4):615–619.

Knisel, W. G. (1980). *CREAMS: A Field Scale Model for Chemicals, Runoff, and Erosion from Agricultural Management Systems*, U.S. Department of Agriculture, Conservation Research Report No. 26, Washington, DC.

Krenkel, P. A., and V. Novotny (1980). *Water Quality Management*, Academic Press, New York.
Langbein, W. B., and S. A. Schumm (1958). Yield of sediment in relation to mean annual precipitation, *Trans. Am. Geophys. Union* **39**:1076–1084.
Lee, D. Y., S. W. Kang, and W. Lick (1981). The entrainment and deposition of fine-grained sediments, *J. Great Lakes Res.* **7**:224–233.
Leopold, L. B. (1968). Hydrology for urban planning—a guidebook on the hydrologic effects of urban land use, *Geol. Surv. Circular 584*, U.S. Geological Survey, Washington, DC, pp. 1–18.
Leopold, L. B., M. G. Wolman, and J. P. Miller (1964). *Fluvial Processes in Geomorphology*, W. H. Freeman, San Francisco.
McElroy, A. D., et al. (1976). Loading functions for assessment of water pollution from nonpoint sources, EPA 600/2-76-151, U.S. Environmental Protection Agency, Washington, DC.
Mehta, A. J., et al. (1989). Cohesive sediment transport. I: Process description, *J. Hydraul. Eng. Div. ASCE* **115**(8):1076–1093.
Migniot, C. (1968). A study of physical properties of different very fine sediments and their behavior under hydrodynamic action, *La Houille Blanche* **7**:591–620.
Mutchler, C. K., and A. J. Bowie (1976). Effect of land use on delivery ratios, in *Proceedings, Third Federal Inter-agency Sedimentation Conference*, Denver, CO, U.S. Water Resources Council, Washington, DC, pp. 1-11–1-21.
Nearing, M. A., G. R. Foster, L. J. Lane, and S. C. Finkner (1989). A process-based erosion model for USDA-Water Erosion Prediction Project technology, *Trans. ASAE* **32**(5):1587–1593.
Negev, M. (1967). A sediment model on a digital computer, *Tech. Rep. No. 62*, Dept. of Civil Eng., Stanford University, Palo Alto, CA.
Noble, E. L. (1965). Sediment reduction through watershed rehabilitation, in *Proceedings, Federal Inter-agency Sedimentation Conference 1963*, Misc. Publ. 920, U.S. Department of Agriculture, Washington, DC, pp. 114–123.
Nordin, C. F. (1962). Study of channel erosion and sediment transport, *J. Hydraulic Div. ASCE* **90**(Hy4):172–192.
Novotny, V. (1980). Delivery of suspended sediment and pollutants from nonpoint sources during overland flow, *Water Resour. Bull.* **16**(6):1057–1065.
Novotny, V., and G. Chesters (1981). *Handbook of Nonpoint Pollution: Sources and Management*. Van Nostrand Reinhold, New York.
Novotny V., and G. Chesters (1989). Delivery of sediment and pollutants from nonpoint sources: A water quality perspective, *J. Soil. Water Conserv.* **44**(6):568–576.
Novotny, V., and G. Bendoricchio (1989). Water quality: Linking diffuse pollution and deterioration, *Water Environ. Technol. WPCF* **1**(3):400–407.
Novotny V., G. V. Simsima, and G. Chesters (1986). Delivery of pollutants from nonpoint sources, in *Proceedings, Symposium on Drainage Basin Sediment Delivery*, Internat. Assoc. Hydrol. Sci., Publ. No. 159, Wallingford, England.
Novotny, V., et al. (1979). *Simulation of Pollutant Loadings and Runoff Quality*, EPA 905/4-79-029, U.S. Environmental Protection Agency, Chicago, IL.
Ostry, R. C. (1982). Relationship of water quality and pollutant loads to land uses

in adjoining watersheds, *Water Resour. Bull.* **18**(1):99-104.
Overton, D. E., and M. E. Meadows (1976). *Stormwater Modeling*, Academic Press, New York.
Parker, G. (1982). Condition for the ignition of catastrophically erosive turbidity currents, *Marine Geol.* **46**:307-327.
Partheniades, E. (1965). Erosion and deposition of cohesive soils, *J. Hydraul. Div. ASCE* **91**:105-139.
Partheniades, E. (1971). Unified view of wash load and bed material load, *J. Hydraul. Div. ASCE* **103**(HY9):1037-1057.
Patrick, R. (1975). Some thoughts concerning correct management of water quality, in *Urbanization and Water Quality Control*, W. Whipple, ed., American Water Resources Association, Bethesda, MD.
Piest, R. F, L. A. Kramer, and H. G. Heinemann (1975). Sediment movement from loessial watersheds, in *Proceedings, Present and Prospective Technology for Predicting Sediment Yield and Sources*, ARS S40, U.S. Department of Agriculture, Washington, DC, pp. 162-176.
Ports, M. A. (1973). Use of universal soil loss equation as a design standard, Water Resources Engineering Meeting, ASAE, Washington, DC.
Ports, M. A. (1975). Urban Sediment Control Design Criteria and Procedures, paper presented at the Winter meeting, ASAE, Chicago, IL.
Renard K., G. Foster, G. A. Weesies, and J. P. Porter (1991). RUSLE—Revised universal soil loss equation, *J. Soil. Water Conserv.* **46**(1):30-33.
Rhoades, E. D., N. H. Welch, and G. A. Coleman (1975). Sediment-yield characteristics from unit source watersheds, in *Present and Prospective Technology for Predicting Sediment Yield and Sources*, ARS Publ. No. ARS-60, U.S. Department of Agriculture, Washington, DC, pp. 125-129.
Ritchie, J. C., and J. R. McHenry (1975). Fallout of Cs-137: A tool in conservation research, *J. Soil. Water Conserv.* **40**:283-386.
Ritchie, J. C., and J. R. McHenry (1978). Fallout Cesium-137 in cultivated north central United States watersheds, *J. Environ. Qual.* **7**(1):40-44.
Ritchie, J. C., J. A. Spraberry, and J. R. McHenry (1974). Estimating soil erosion from the redistribution of fallout ^{137}Cs, *Soil. Sci. Soc. Am. Proc.* **38**:137-139.
Roehl, J. W. (1962). *Sediment Source Areas, Delivery Ratios and Influencing Morphological Factors*, Publ. No. 59, Internat. Assoc. Hydrol. Sci., pp. 202-213.
Sharma, K. D., and P. C. Chatterji (1982). Sedimentation in Nadis in the Indian arid zone, *Hydrol. Sci. J.* **27**:345-352.
Shen, H. W. (1982). Some basic concepts on sediment transport in urban storm drainage systems, in *Urban Stormwater Quality, Management and Planning*, B. C. Yen, ed., Water Resources Publications, Littleton, CO, pp. 495-501.
Shields, A. (1936). "Änwendung der Ahnlichkeits-mechanik und der Turbulenzforschung auf der Geschiebebewegung," Preuss. Versuchanstalt für Schiffbau, Berlin.
Steward, B. A., et al. (1975). Control of pollution from cropland, U.S. EPA Report No. 600/2-75-026 or U.S.D.A. Rep. No. ARS-H-5-1, Washington, DC.

Tesarik, I., and J. Vostrcil (1970). Hindered settling and thickening of chemically precipitated flocs, in *Proceedings, Fifth International Conference IAWPR*, Pergamon Press, Oxford.

Ursic, S. J., and F. E. Dendy (1965). Sediment yields from small watersheds under various land uses and forest covers, in *Proceedings, Federal Inter-agency Sedimentation Conference 1963*, USDA Misc. Publ. 970, Washington, DC, pp. 47–52.

Vanoni, V. (1975). *Sedimentation Engineering*, ASCE, New York.

Wade, J. C., and E. O. Heany (1978). Measurement of sediment control impact on agriculture, *Water Resources Res.* **14**:1–8.

Walling, D. E. (1979). Hydrological processes, in *Man and Environmental Processes*, K. J. Gregory and D. E. Walling, eds., Butterworth, London.

Walling, D. E. (1983). The sediment delivery problem, *J. Hydrol.* **65**:209–237.

Walling, D. E., and P. Kane (1983). Temporal variation of suspended sediment properties, *Recent Developments in the Explanation and Prediction of Erosion and Sediment Yield*, Proceedings, Exter Symposium, IAHS Publ. No. 137, Washington, DC, pp. 409–419.

Walling, D. E., and B. W. Webb (1983). Patterns of sediment yield, in *Background to Paleohydrology*, K. J. Gregory, ed., Wiley, New York.

Wilkin, D. C., and S. J. Hebel (1982). Erosion, redeposition and delivery of sediment to midwestern streams, *Water Resour. Res.* **18**(4):1278–1282.

Williams, J. R. (1975). Sediment routing for agricultural watersheds, *Water Resour. Bull.* **11**(5):965–974.

Williams, J. R., and H. D. Berndt (1977). Sediment yield prediction based on watershed hydrology, *Trans. ASAE* **20**:1100–1104.

Wischmeier, W. H. (1972). Estimating the soil loss equation's cover and management factor for undisturbed areas, Proceedings, Sediment-Yield Workshop, U.S. Department of Agriculture Sedimentation Laboratory, Oxford, MS.

Wischmeier, W. H. (1976). Use and misuse of the universal soil loss equation, *J. Soil Water Conserv.* **31**(1):5–9.

Wischmeier, W. H., and D. D. Smith (1965). *Predicting Rainfall-Erosion Losses from Cropland East of Rocky Mountains*, USDA Agricultural Handbook No. 282, Washington, DC.

Wischmeier, W. H., C. B. Johnson, and B. V. Cross (1971). "A soil erodibility nomograph for farmland and construction sites," *J. Soil Water Conserv.* **26**: 189–193.

Wolman, M. G. (1977). Changing needs and opportunities in the sediment field, *Water Resour. Res.* **13**:50–54.

Xingkui, W., and Q. Ning (1989). Turbulence characteristics of sediment laden flows, *J. Hydrol. Eng. ASCE* **115**(6):781–800.

Yalin, M. S. (1963). An expression for bed load transportation, *J. Hydraul. Div. ASCE* **89**:221–250.

Ziegler, C. K., and W. Lick (1988). The transport of fine-grained sediments in shallow waters, *Environ. Geol. Water Sci.* **11**(1):123–132.

6
Pollutant Interaction with Soils and Sediments

Future historians may be amazed by our distorted sense of proportions. How could intelligent beings seek to control a few unwanted species by a method that contaminated the entire environment and brought the threat of disease and death even to their own kind?

Rachel Carson, *Silent Spring*

All the human and animal manure, if returned to land instead being thrown into the sea, would suffice to nourish the world.

Victor Hugo, *Les Miserables*

Soils can retain, modify, decompose, or adsorb (immobilize) pollutants. Each year an enormous amount of organic material, atmospheric contaminants, and liquid and solid wastes are deposited and incorporated into soils. As a matter of fact, the focus of early "zero discharge" efforts after the passage of the 1972 Water Pollution Control Act Amendments was to shift the burden of residual waste discharges from water to soil.

Biodegradable organic materials deposited into soils are decomposed largely into such safe products as CO_2, methane, nitrogen, and phosphorus compounds. Due to the large number of microorganisms residing in typical fertile top soils, the decomposition processes under suitable environmental conditions represent the best natural recycling process. A properly balanced and managed soil system does not represent a great threat to water quality. In a balanced system, most of the nutrients and organic matter added to the soil, in amounts normally applied to increase crop production, will remain in the upper soil layer and/or will be taken up by the crops.

However, it was pointed out by a panel of distinguished soil scientists (Stigliani, 1991) that harmful pollutants due to man's activities had been gradually accumulating in soils and sediments for the past 2000 years. Early sources of such soil pollution were primarily metal mining operations. Since the beginning of the Industrial Revolution, the scale and pace of environmental contamination due to industrial, agricultural, commercial, and domestic activities has accelerated, and in the last 40 years has reached a pace that may be threatening to the health and existence of the human population, at least in some stressed areas of the world.

As stated previously, man always had a tendency to return waste, by-products of living and production processes, and the dead first to the soil and then to the water. The capacity of soil to assimilate pollutants was considered infinite, and discharging waste to soils was deemed beneficial (note the Victor Hugo quote at the beginning of the chapter). However, this capacity though large, is limited.

Today, large amounts of commercial chemical fertilizers are applied to agricultural and grassed urban lands in the United States and elsewhere. Typical application rates range from 20 to 200 kg N/ha and from 10 to 50 kg P/ha. To control pests and weeds farmers and urban dwellers have been using chemical pesticides in amounts that are environmentally damaging and defying the purpose of control. For example, to maintain a "perfect" suburban lawn and eradicate a few dandelions a typical suburban homeowner in the United States uses chemical pesticides and fertilizers in excessive, environmentally damaging quantities. During the 1960 to 1990 period, in order to sustain higher crop yields, agricultural cooperatives in the central and eastern European countries were applying chemicals at such excessive rates that most of the receiving surface- and ground-water bodies in the region had been severely damaged with a consequent impairment of their beneficial uses (for example, unsuitability of water resources for water supply, recreational uses, and severe ecological damages).

Due to the capacity of soils and sediments to store and immobilize toxic chemicals in so-called chemical sinks, the direct effects of pollution may not be initially directly manifested. However, there is now direct evidence that the capacity of soils to safely dispose of pollutants has been exhausted in some parts of the world. Excessive contamination of surface- and ground-water bodies by nitrates, pesticides, and other chemicals is a result of the failure of the soils to which these chemicals were applied to retain them, modify them, and transfer them to their target media (crops for fertilizing chemicals, weeds and insects for pesticides).

Because loading of the chemicals to the environment may occur long before the adverse consequences are noticed, scientists have used the term *chemical time bomb* to describe the present and future threats by the past, present, and future excessive uses of various chemicals and waste applications to soils.

> A chemical time bomb (CTB) is a concept that refers to a chain of events resulting in the delayed and sudden occurrence of harmful effects due to the mobilization of chemicals stored in soils and sediments in response to slow alteration of the environment (Stigliani, 1991; Stigliani et al., 1991).

Typically, the loading of soils by pollutants occurs at lower rates but is widespread (diffuse pollution). In some other instances, chemical loading was very high and localized, such as in the cases of hazardous waste disposal sites and landfills. In many instances, when the delayed environmental effects of the "overload" of soils and sediments with pollutants are manifested, the impact is widespread and may be devastating. Forest diebacks in Europe in the early 1980s, due at least in part to soil acidification by acid rainfall and atmospheric deposition, is a recent example of sudden, unanticipated, and delayed response. Similarly, the rapid explosive eutrophication of lakes (see the example of Lake Balaton in Chapter 1) and ground-water contamination by nitrates in Central Europe and elsewhere are a result of the excessive application of fertilizing chemical at rates that far exceed plant uptake.

When the waste-assimilative or buffering capacity of soils and sediments is gradually exceeded, the result is CTB. The buffering (waste-assimilative) capacity of soils is due to the microbial degradation of chemicals and pollutants, plant uptake and removal with crops, immobilization by adsorption and complexation on soil particles and colloids, chemical reactions, volatilization, and other processes by which pollutants are removed, degraded, and immobilized.

Contrary to the sudden occurrence of the time bomb, which is the effect of an exceedence of the waste-assimilative (buffering) capacity of soils and surface-water bodies, remediation and restoration of damaged ecosystems is gradual and may last years and even centuries. There are no lasting "quick fixes."

As the contaminated soils are eroded, aquatic sediments are also being contaminated at a rather fast pace. Aquatic sediment may also interact with the pollutants dissolved in water in the same manner as soil would

314   Pollutant Interaction with Soils and Sediments

with pollutants contained in soil moisture. As a matter of fact, the processes of pollutant interactions with soils and aquatic and wetland sediments are either the same or similar.

Soil and its buffering capacity is the only barrier between surface contamination and pollutant deposition and ground-water resources for which the buffering is either zero or very small. This difference between the buffering capacities of topsoils and ground water is due to a typically rich microbial population residing in the topsoil layer, higher content of organic matter, a greater proportion of weathered small-grain minerals, such as clays, different chemical environment, and other factors.

The processes and pollutant interactions in aquatic sediments and wetland substrates are similar to those in saturated (wet) soils. Therefore, both media are discussed in this chapter.

The Soil Profile

The soil profile is divided into three layers called horizons A, B, and C (Fig. 6.1). A typical profile may also contain a thin layer of decaying

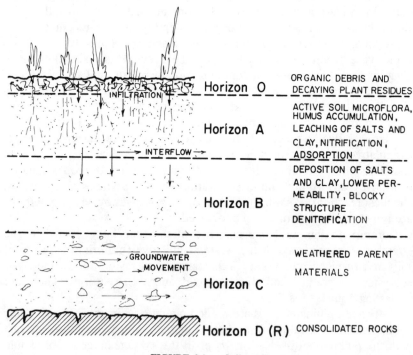

FIGURE 6.1.   Soil profile.

organic debris with little soil. The surface layer of organic debris is called horizon O.

The A horizon, usually several centimeters to a fraction of a meter thick, is the soil layer of the greatest concern, since roots, soil microorganisms, and organics can be found there in great densities. It is also a layer of considerable leaching. The B horizon, underlying the A horizon, is a subsoil where most of the leached salts, chemicals, and clay may deposit. It usually has little organic matter and few plant roots (only large plants—macrophytes—have root systems penetrating into subsoils). The C horizon extends from the bottom of the B horizon to the top of the parent bedrock from which the soil evolved by weathering.

The A horizon is of considerable importance in diffuse pollution studies since it is the soil layer where most of the adsorption and biochemical degradation of pollutants takes place. The microbiological processes by which pollutants and nutrients (nitrogen and phosphorus) are decomposed or transformed are mostly confined to the A horizon. Only soluble (mobile) pollutants can penetrate into deeper soil zones and eventually pollute ground water.

Soil nomenclature and soil maps provide a wealth of information on the types of soils, their texture, composition, and properties. The soil texture triangle presented in the preceding chapter (Fig. 5.26) and the U.S. Soil Conservation Service soil maps (Fig. 6.2) are for delineating the distribution of soils. In the classification of soil the name of the soil as reported in the soil map has two parts: (a) a local name that also includes textural classification, and (b) slope. For example, in the soil map of southeastern Wisconsin a code OuB signifies Ozaukee silt loam in slope category B (2% to 6% slope).

Soil (Sediment) Organic Matter

Organic matter is an integral part of soils. The organic matter content varies from <1 to >40% in some organic soils, feedlot soils, wetland, and aquatic sediments. Organic matter content is usually reported as organic matter (%) or organic carbon (%). Organic matter (%) $\approx 1.67 \times$ organic carbon (%).

The organic content of soils and sediments, often called humus, is a product of the biodegrading processes by microorganisms. It is rich in nutrients and remains for long periods as an important food supply for microorganisms. A significant part of the organic matter is not biodegradable.

Almost all of the organic matter in soils is contained in the A and O horizons of the soil profile. Organic-rich aquatic sediments may be much

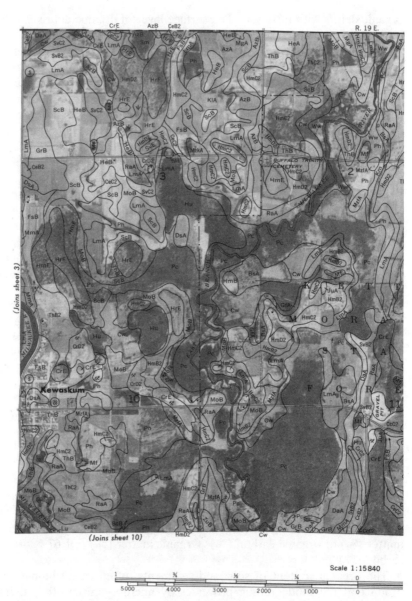

FIGURE 6.2. Soil map by the U.S. Soil Conservation Service. Soils are coded by their local name and slope.

deeper. As to their organic content, soils are divided into mineral soils that have low organic fraction and organic soils. A similar classification can also be applied to aquatic sediments. However, for sediments the organic content is affected by the hydraulics of the body of water.

High organic content is not synonymous with fertility in soils. Peat (wetland) soils that have an organic fraction are not suitable for growing most crops (with the exception of rice) that require aerated drained soils, even though wetlands can produce high quantities of organic matter. When peat soils are drained, the excess organic matter is converted under aerobic conditions to CO_2, and the soil's suitability for growing crops is improved. That was the major reason for draining wetlands during their conversion to agricultural use. Consequently, soil aeration is a factor affecting the soil organic content.

LOADING FUNCTIONS

The Concept of Enrichment

The transport of pollutants from topsoil is affected by the composition of the soil, slope, and other factors. Pollutant transport follows the three transport routes identified in Chapters 3 and 5, depending on whether the pollutants are associated with the sediment or are dissolved in soil water:

1. Soil erosion and overland transport will carry pollutants adsorbed or complexed by soil particulates. Their history of transport is similar (but not identical) to that of soil particles.
2. Pollutants contained in soil water can be transported either by surface runoff or by interflow. These pollutants are in the form of dissolved compounds or ions. For many compounds a dynamic equilibrium exists between the dissolved and adsorbed (complexed–particulate) fractions in soil water and in surface runoff).
3. Only dissolved or ionized pollutants can reach ground-water zones and reappear on the surface with the base-flow–ground-water discharge.

Many contaminants in soils (nutrients and organic chemicals) are subject to bacterial degradation, volatilization, and chemical breakdown. Those that will persist are classified as *conservative* materials, while those that change their total mass with time (excluding the effect of convection and dispersion) are *nonconservative*.

Figure 6.3 shows the major pathways of transport and major transformation processes of soil contaminants between the source and receiving bodies of water. Water is the primary transporting vector.

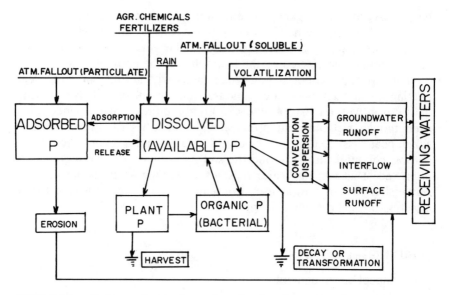

FIGURE 6.3. Pathways of contaminates through soil-environment–soil-contaminant interactions.

Since particulate pollutant transport is a part of sediment erosion and movement processes many studies and models use an arbitrary proportionality factor called the *potency factor* or *transmission coefficient* to equate sediment loading to that of other contaminants. The relationship can be expressed by the equation

$$Y_i = p_i Y_s \tag{6.1}$$

where
Y_i = the loading or concentration of contaminant i
Y_s = the loading or concentration of sediment
p_i = the potency factor for the contaminant

The potency factor can be related to the concentration of the contaminant in the parent topsoil and the enrichment factor for the contaminant, that is,

$$p_i = S_{is} \text{ER}_i \tag{6.2}$$

where
S_{is} = the concentration of the contaminant

ER_i = the enrichment ratio for the contaminant between the source and the point of interest or watershed outlet

The enrichment ratio is then the concentration of the contaminant in the eroded material on the sediment in runoff divided by its concentration in the parent topsoil expressed on an over-dried basis.

The use of Equations (6.1) and (6.2) presumes knowledge of the soil-contaminant concentration, S_{is}, and the enrichment ratio, ER_i. The U.S. EPA Screening Procedure (Mills et al., 1985) for estimating pollutant loads from nonpoint sources is based on the preceding concept and the documentation contains information on estimating the needed parameter values. This publication appears to be the most comprehensive treatise on the subjects. Also the publication by McElroy et al. (1976) on which the EPA screening procedure is based contains some information on the magnitudes of potency factors, enrichment ratios, and pollutant content of parent soils. Information may be available for soil content of organic matter, nutrients, and some other contaminants such as metals (Tables 6-1, 6-2, and 6-3); however, the nature of the enrichment factor is not well understood and its estimation is, at best, an approximation. For contaminants associated with fine fractions and particulate organic matter, the enrichment ratio may be inversely related to the delivery ratio as shown for clays in the preceding chapter. Table 6-3 shows the enrichment ratios

TABLE 6.1 Metal Concentrations (μg/g) in Surficial Materials in the United States

Element	Geometric Means Arithmetic Analysis		Conterminous United States	West of the 97th the Meridian	East of 97th the Meridian
	Average	Range			
As		0–1000			
Ba	534	15–5000	430	560	30
Cd		0–20			
Ce	86	<150–300	75	74	78
Cr	53	1–1500	37	38	86
Co	10	<3–70	7	8	7
Fe	25,000	100–100,000	18,000	20,500	15,000
Pb	20	<10–700	14	18	10
Mn	560	<1–7000	340	389	385
Ni	20	<5–700	14	16	13
Hg[a]		0.01–0.03			
Cu[a]		2–100			
Se[a]		0.1–2			

Source: After Shacklette et al. (1971).

[a] After Lindsay (1979).

TABLE 6.2 Concentrations of Toxic Metals (in µg/g) in Unpolluted Parent Soils and Sediments in Rhine River Basin

Metal	Soils	Subrecent Rhine Sediments	Lacustrine Sediments
Cd	0.82	0.3	0.40
Co	12	16	16
Cr	84	47	62
Cu	25.8	51	45
Hg	0.1	0.2	0.35
Mn	760	960	700
Ni	33.7	46	66
Pb	29.2	30	34
Zn	59.8	115	118

Source: From Salomons and Förstner (1984).

TABLE 6.3 Enrichment of Suspended Sediment by Clay and Associated Pollutants

Constituent[a]	Soil	Suspended Sediment	Enrichment Ratio
Clay	27	91	3.4
Total phosphorus	810	1700	2.1
Lead (Pb)	19	39	2.0
Cadmium (Cd)	0.3	0.31	1.0
Zinc (Zn)	69	280	4.0
Copper (Cu)	25	45	1.8
Aluminum (Al)	22,000	49,000	2.2
Iron (Fe)	21,000	46,000	2.2
Manganese (Mn)	700	730	1.0
Chromium (Cr)	29	56	1.9
Nickel (Ni)	17	45	2.6

Sources: Based on data from Dong, Simsiman, and Chesters (1983), and Dong, Chesters, and Simsiman (1984).

[a] Clay is in percent and phosphorus and metals are in µg/g of sediment.

of contaminants determined from the soil and sediment data of the Menomonee River experimental watershed.

Equations (6.1) and (6.2) are static, that is, they allow neither the prediction of future conditions nor the evaluation of the effects of future management practices. However, in some cases when a contaminant is conservative and immobile, approximate concentrations can be estimated from the soil-contaminant mass balance.

Since almost all particulate contaminants are usually associated with fine soil particles, and since erosion and sediment transport are selective for fine materials, a simple relation or guidelines for determining the potency factor apparently do not exist. Furthermore, the quantity is highly variable, depending on soil characteristics, storm and overland flow characteristics, channel flow hydraulics, the presence of sediment, and pollutant sinks, such as grassed buffer strips, riparian wetlands, forest litter, and the nature of the contaminant. Also, the sorption characteristics and the capacity of soil particles to retain pollutants are different, depending on whether the particles are in soil or are a part of the sediment washload and bedload in streams and other surface-water bodies (McCalister and Logan, 1978; Salomons and Förstner, 1984). Following are several processes that result in higher concentrations of contaminants in the eroded material and aquatic sediments than in the parent soil and, consequently, enrichment ratios of greater than one:

1. Selective removal of fine materials with higher pollutant concentrations than the remainder of the soil material, which is expressed as the clay enrichment ratio (see Chapter 5).
2. Diffusion of pollutants and salts from the topsoils into surface runoff.
3. Desorption of pollutants from soil particles due to low dissolved concentrations in runoff water.
4. Flotation of low-density materials such as organic components from soil into surface water.
5. Deposition of coarse fractions containing few sorbed pollutants during overland and channel flow and in man-made depositional facilities (retention ponds, buffer strips) that are more effective for removal of coarse heavier particles.

The transport and loading of dissolved contaminants is more complex because they can be transported by each of the three hydrological components, namely, surface runoff, interflow, and ground water. It is therefore evident that a simplistic approach to the determination of pollutant loadings from pervious surfaces (soils) may fail to provide adequate results, and prognoses using such approaches must be treated with caution and accepted only as rough estimates. Fortunately, several models have been developed in the past that in a comprehensive manner describe the transport and interactions of several typical contaminants in the soil and water environments. The U.S. EPA's computerized loading procedures (Mills et al., 1985) present methodologies and numerous supporting parameters for estimating loadings from diffuse sources and their impact on receiving bodies of water.

Example 6.1: Estimating Pollutant Concentration in Soils from Mass Balance

Application of 100 kg/ha of a conservative chemical that is immobile in soils was made to an area. Estimate the approximate concentration of the chemical in the soil particles if the pollutant is assumed to be uniformly distributed throughout the top 30 cm of the soil (approximate till depth). The specific density of the soil is 1.8 g/cm³. Hence

$$S_{is} = \frac{\text{Chemical application mass}}{\text{Soil mass}} = \frac{100\,(\text{kg/ha}) \times 10^9\,(\mu g/kg)}{30\,\text{cm} \times 10^8\,(\text{cm}^2/\text{ha}) \times 1.8\,(\text{g/cm}^3)}$$
$$= 18.5\,\mu g/g$$

Detachment and Enrichment of Runoff by Soil Organic Matter

Most of the organic matter in soils is in a particulate form and, as such, transport of organic matter from soils and sediments can be related to erosion by applying Equations (6.1) and (6.2). However, values of the enrichment factor are difficult to assess. Massey and Jackson (1952) found that the value of the enrichment ratio (ER) for organic matter increases with sediment concentration in runoff and the rate of erosion. The average enrichment ratio for three silt loam soils in Wisconsin measured by Massey and Jackson (1952) was 2.1. Young and Onstad (1976) proposed the following relationship for the enrichment ratio (ER_{or}) of organic matter based on the regression analyses of Indiana and Minnesota soils

$$ER_{or} = \frac{0.3}{(\%\text{ organic matter})} + 1.08 \qquad (6.3)$$

The organic content increases linearly with finer textures, so Equation (6.3) indicates lower ER values for clayey soils and higher values for sandy soils. Since the specific density of organic particles is much less than mineral soil particles, the ER may increase significantly during overland and channel flow.

The organic content of soils lost in agricultural erosion can be restored by the application of organic fertilizers-manure, man-made organic mixtures, sewage sludges, and by keeping plant residues in place. Heavy reliance on fertilizing chemicals only will degrade the soil quality. Increased organic content improves soil permeability and reduces erodibility. As will be shown in later sections of this chapter, a significant portion of the organic matter in fertile soils consists of soil microflora.

Detachment of Dissolved Chemicals: Mixing Layer

Many chemicals, including phosphorus, metals, ammoniacal nitrogen, and organic chemicals exist in soils and sediments in two or even three phases (see the next section for a detailed discussion). The phases are solid—adsorbed or precipitated chemicals, liquid chemicals dissolved or dissociated in pore water, and gaseous chemicals in the vapor phase in aerated soils. When runoff is generated on the soil surface by rainfall, solid-phase chemicals move with eroded sediment, while liquid-phase chemicals in the upper surface layer mix with runoff and are in this way leached from the soil.

Some hydrologic models of soil–water–chemical interaction assume that rainfall and runoff interact within a thin zone of the surface soil. The rainfall is assumed to mix completely and uniformly with the dissolved chemicals in the upper soil zone. The amount of a soluble chemical is then divided proportionately between the soil zone water storage, infiltration, and surface runoff in the water retained in each mode. The thickness of the mixing zone can be obtained by calibration or be fixed. In the ARM model (Donigian et al., 1977) the thickness of the upper soil mixing zone varied between 1.6 and 6 mm, with the smaller values giving better results under field conditions. In the CREAMS-GLEAMS model (Knisel, 1980) the thickness of the mixing depth was selected as 1 cm; however, it is assumed that only a fraction of the chemical in the mixing zone interacts with rainfall water.

Sharpley, Ahuja, and Menzel (1981) and Ahuja et al. (1982) measured and calculated the mixing depth for soils within the range of 2 to 6 mm with typical values between 2 and 3 mm. As expected, the thickness of the mixing depth was proportional to rainfall intensity, and decreased below 1 mm when protective vegetative covers were present.

For aquatic sediments the thickness of the upper mixing zone is affected by the action of benthic organisms. Typically for modeling purposes, the thickness of the interstitial sediment–water mixing zone is in centimeters.

THE FATE OF CONTAMINANTS IN SOILS: THREE-PHASE APPROACH

Chemicals or other pollutants such as nutrients may exist in soils, sediments, and sediment-laden waters in several phases. They can be precipitated and/or strongly adsorbed to particles (solid phase), be dissolved or dissociated in water or soil moisture (liquid phase), volatilize or be gasified. For example, ammonium may exist in soil adsorbed on soil particles, dissociated in water as an ammonia ion (NH_4^+) or as ammonium

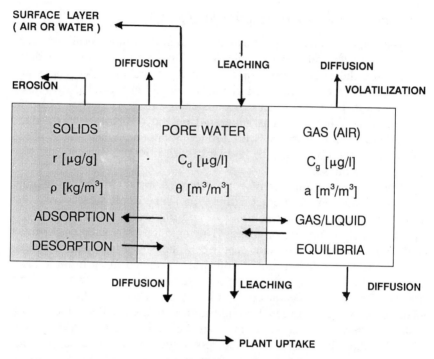

FIGURE 6.4. Compartmentalized mass balance concept of the distribution of a contaminant in soils and sediments. (After Burns et al., 1990.)

(NH_3). Generally, adsorbed or precipitated chemicals are considered immobile in soils and nonmoving sediments and biologically unavailable to plants and microorganisms. Dissolved or dissociated chemicals move with water, hence they are mobile and biologically available.

The fate of a chemical in soil and sediments is determined by the transformation processes. A chemical can volatilize to the atmosphere, be adsorbed, be taken by plants where it may or may not be degraded, and hence it might be returned to the soil–sediment in plant residue, be degraded by chemical, photochemical, and/or microbiological processes, be leached to lower depths by water movement and diffusion, be transported by erosion/runoff, and redeposited on land or reach surface water bodies. In many cases the fate of a particular chemical pollutant is determined by a number of processes (Chesters, 1986).

Figure 6.4 shows the processes that determine the fate of a single chemical in soil or a sediment environment. Typically, mass balance equations must be written for each compartment in order to estimate the overall fate of the chemical (Burns et al., 1990).

Sorption, Precipitation, and Partitioning Processes

The term *partitioning* implies division of the total pollutant mass between the particulate (soil particles) and dissolved (dissociated) fractions in soil or sediment pore water. Adsorption or complexation of the contaminant into the particulate form also immobilizes the contaminants and makes them biologically unavailable in most cases. Dissolved or dissociated (ionized) contaminants are mobile with soil and sediment pore water and can enter the tissues of animal and plant life. Division of the total pollutant mass into its particulate (immobile) and dissolved (mobile) fractions is thus very important in assessing the contaminant impact on the ecosystem and its toxicity and, conversely, it affects the waste-assimilative capacity of the soil–sediment ecosystem to safely accept and dispose of the pollutants and determines the amount of the contaminants that can leach to ground water. A report by Schnoor et al. (1987) describes the processes involving toxic metals and organics and also includes a long list of the numerical values of coefficients and parameters. A book by Salomons and Förstner (1984) is one of the most comprehensive treatises on the fate and transport of toxic metals in the ecosystems.

Generally, water-soluble compounds are weakly adsorbed on soil or sediment particles, hence they have a higher bioavailability and leach more easily into ground water. Due to the ease of leaching these compounds do not persist nor accumulate in soils, sediments, and the tissues of organisms. Water-insoluble compounds, on the other hand, are immobile in soils and sediments; however, they accumulate in soils and sediments and may bioaccumulate in organisms and biomagnify in the food chain (see Chapter 14).

The mobility of a chemical in soils and sediments is related to the so-called *octanol–water partitioning* of the chemical expressed by the *octanol/water partitioning coefficient*, K_{ow}. The coefficient is thus the accepted measure of the solubility of the chemical in water and, consequently, mobility in soils and sediments. The laboratory procedure for measuring K_{ow} is (Lyman, 1982; Schnoor et al., 1987):

1. A chemical is added to a mixture of pure octanol (a nonpolar solvent) and pure water (a polar solvent). The volume ratio of octanol and water is set at the estimated K_{ow}.
2. The mixture is agitated until equilibrium is reached.
3. The mixture is centrifuged to separate the two phases, and the phases are analyzed for the chemical.
4. K_{ow} is the ratio of the chemical concentration in the octanol phase to the chemical concentration in the water phase. From this experiment

K_{ow} is dimensionless, but it may be commonly reported as liters/ kilogram.

It has been found that for nonionic chemicals and some other chemicals the adsorption characteristics (partitioning) may be correlated to the octanol partitioning coefficient, K_{ow}. The values of K_{ow} and other important parameters for some environmentally important chemicals (priority pollutants) are given in Table A1 in the Appendix. The report by Schnoor et al. (1987) has an extensive compilation of partition coefficients for many pollutants. The parameter K_{ow} is one of the most important determinants of the fate of a chemical in the soil–sediment environment and its bioavailability. The partitioning coefficents for a large number of polar potentially toxic contaminants have been compiled by the Risk Reduction Engineering Laboratory in a computerized data base.

Sorption results from a variety of different types of attractive forces between the dissolved molecules or ions and the sorbing material (Weber McGinley, and Katz, 1991). These forces are of a chemical, electrostatic, and physical nature. The most significant properties for sorption and desorption of a soil–sediment system are those related to the surface at which accumulation occurs. Quartz particles (sand) do not provide adequate adsorption sites and generally are not considered as sorbents. Clay and particulate organics, on the other hand, provide the most suitable adsorption. In addition, the soil–sediment properties that exert the greatest influence on sorption–desorption processes are pH, temperature, and moisture content (Chesters, 1986).

In precipitation reactions, the proportion of dissolved (ionized) and precipitated fractions of a contaminant is strongly affected by the pH and the presence of complexing compounds—ligands. The precipitating–complexing reactions are important for toxic metals, while organic chemical reactions with soil–sediment particles are mostly adsorption–desorption processes. The presence or absence of complexing ligands is affected also by the oxidation-reduction status in the soil or sediment environment. For example, in anoxic soils and sediments, which occurs primarily when the medium is saturated with water, insoluble metal-sulfide complexes are formed. Such complexes do not form in aerated soils or aerobic sediments, and other precipitating and often less favorable reactions with clays and organic matter dominate. Therefore oxidation-reduction status affects the precipitation process of metals. On the other hand, adsorption of pesticides is not generally affected by the presence or absence of oxygen, and they primarily interact with the soil organic matter. Phosphates and ammonia interact with clays and organic particulates.

The preferred mathematical form for describing the proportion between the dissolved (ionized) and absorbed (complexed) particulate fractions is to relate the concentration of the contaminant adsorbed on the soil or sediment particles, r (in μg of contaminant per gram of soil, μg/g) to the equilibrium solution concentration, C_e (mg/l) at a fixed temperature (isothermic reaction). Several mathematical formulations of adsorption equilibria (isotherms) have been proposed from which the Langmuir and Freudlich isotherms are most widely accepted.

The Langmuir adsorption model is valid for monolayer adsorption, and is expressed in the form

$$r = \frac{Q^0 b C_e}{1 + b C_e} \tag{6.4}$$

where
Q^0 = the adsorption maximum at the fixed temperature (μg/g)
b = a constant related to the energy of net enthalphy of adsorption (l/mg or l/μg)
r = adsorbed concentration of the contaminant (μg/g)
C_e = dissolved (free) concentration of the contaminant water (μg/l)

Although the Langmuir isotherm is assumed to be valid for monolayer adsorption, it adequately represents soil adsorption processes for such pollutants as phosphorus and many organic chemicals.

The Freundlich isotherm is useful if the energy term, b, in the Langmuir isotherm varies as a function of surface coverage, r. The Freundlich equation has the general from

$$r = K C_e^{1/n} \tag{6.5}$$

where K and n are constants.

Parameters describing the isotherm equation for various pollutants are statistical quantities. In some cases, laboratory adsorption studies may be available that can provide the magnitude of the adsorption coefficients. In these experiments, soil or sediment samples are put in aqueous solutions with different concentrations of the compounds until an equilibrium between the sorbed and dissolved fractions is reached. Thereafter, the sorbed fraction is $r = (C_{\text{initial}} - C_e)$/dry weight of soil. To obtain the Langmuir isotherm parameters $1/r$ is plotted versus $1/C_e$ on an arithmetic plot. For the slope of the line of best fit, slope = $1/(Q^0 b)$ and the intercept = $1/Q^0$.

Freundlich isotherm parameters are then obtained by plotting r versus C_e on log-log graph paper. The logarithmic intercept (when X-ordinate equals unity) is K and the logarithmic slope ($sl = \log\{[r_1 - r_2]/[C_{e1} - C_{e2}]\}$, where 1 and 2 are arbitrarily chosen points on the line of best fit, equals $1/n$.

Linear isotherm. For low concentrations of contaminants, the Langmuir and Freundlich isotherms can be simplified to

$$r = \Pi c_e \tag{6.6}$$

where Π = partition coefficient, l/g.

By comparing the Langmuir isotherm with the linear isotherm for low concentrations of the chemical, it can be seen that $\Pi \approx Q^0 \times b$. The total concentration of the pollutant is then a sum of the dissolved (c_d) and particulate concentrations (c_p). Hence, if $c_p = m_{ss} \times r$ where m_{ss} is the concentration of solids in g/l, then

$$c_T = \theta c_d + c_p = \theta c_d + m_{ss} \times r = c_d(\theta + \Pi m_{ss}) \tag{6.7}$$

where θ = the water content of the soil or sediment as a fraction of the volume (for water $\theta = 1$).

The relation between the dissolved (pore water) concentration of the chemical and the total concentration in the soil, sediment-laden water, or sediment is then

$$c_d = \frac{1}{\theta + \Pi m_{ss}} c_T \tag{6.8}$$

For some chemicals (for example, metals) the term *precipitation* is commonly used. However, Salomons and Förstner (1984) quoting several other authors pointed out that there should be no difference between *sorption* and *coprecipitation*.

The isothermal concept for soil–sediment chemical interaction presumes either instantaneous adsorption or conditions of equilibria. The instantaneous adsorption assumption is adequate for modeling slowly varying processes in soil. For more dynamic conditions, the kinetic of adsorption can be expressed as

$$\text{Soluble chemical} \underset{1-K}{\overset{K}{\rightleftarrows}} \text{Adsorbed chemical} \tag{6.9}$$

Volatilization

Volatilization is defined as the loss of chemicals in vapor form from soil or water surfaces to the atmosphere. The process is limited by the chemical concentrations in water and air interfaces above the soil of water surface (Jury, 1986). Based on this definition volatilization does not occur from submerged aquatic sediments.

The potential of volatilization is related to the saturated vapor pressure of the chemical in the air above the interface; however, actual volatilization from soil is also affected by many other factors, such as atmospheric air movement, temperature, and soil characteristics (Spencer, Farmer, and Jury, 1982).

If a chemical exists in the soil and the vapor phase, Equation (6.7) is expanded to include the vapor phase as

$$c_T = \theta c_d + c_p + ac_g = \theta c_d + m_{ss} \times r + ac_g$$
$$= c_d(\theta + \Pi m_{ss}) + ac_g \qquad (6.10)$$

where
a = volumetric air content ($a = p - \theta$)
c_g = vapor density of the chemical (μg/l of soil air)
p = volumetric porosity

The relation between the vapor density and corresponding concentration of the chemical in (pore) water solution is given by Henry's law

$$c_d = K_H c_g \qquad (6.11)$$

where K_H = Henry's constant for the chemical that in this system is dimensionless. Hence Henry's constant K_H is the partitioning coefficient of the chemical between the air and water (U.S. EPA, 1986).

Substituting Equation (6.11) into Equation (6.10) yields the relationship between the total chemical concentrations in the soil and pore water as

$$c_d = \frac{c_T}{\theta + \Pi m_{ss} + (p - \theta)/K_H} \qquad (6.12)$$

When soil dries to a point that only a few monolayers of water remain on the solids, adsorption of chemicals will greatly increase. This is due to the fact that soil particles preferentially adsorb water molecules. Thus in dry regions chemical losses by volatilization are sometimes reduced to insignificant levels. On the other hand, when soil is saturated ($\theta = p$), all transfer of the chemical to the surface of the soil is by diffusion. For this reason volatilization is significant when the soil has an intermediate moisture content (Spencer and Cliath, 1973).

The mathematical description of the vapor flux of the chemical from the soil surface is generally derived from the boundary layer theory, which presumes that at the soil (water)–air interface there is a relatively stagnant layer through which the gaseous chemical must move by molecular diffusion. Using Fick's law, Jury, Spencer, and Farmer (1983) represented this vapor flux away from the soil surface as

$$J = \frac{D_g^{(air)}}{d}[c_g(0) - c_g(d)] \qquad (6.13)$$

where
$D_g^{(air)}$ = molecular diffusion coefficient of the gaseous chemical in air
d = thickness of the soil surface–air boundary layer
$c_g(0)$ = vapor phase concentration of the chemical in the upper layer of the soil defined previously
$c_g(d)$ = gaseous concentration of the chemical in the air (above the stagnant boundary layer)

The stagnant boundary layer is a concept. Its statistical thickness depends primarily on the meteorological factors (wind) and the roughness of the surface.

Biological Degradation and Transformation

Biological degradation of a chemical usually implies a breakdown by the living organism to more simple compounds, ultimately to carbon dioxide, water, and possibly other end products. Ultimate degradation means decomposition into mineral products (mineralization), while partial degradation is a decomposition into less complex compounds. *Transformation* on the other hand, means a change to another form, which may or may not be less environmentally hazardous. An example of such a transformation is the biochemical change of DDT to DDE, which is considered just as toxic as the original compound. Chemicals that are not biodegradable are thus called persistent in the environment, and their removal is only by a transfer to another medium (from soil to water or air) or by burial.

Biotransformation of chemicals in soil is accomplished by soil microorganisms and/or fungi (Valentine and Schnoor, 1986). These organisms are mostly *heterotrophic* and require an energy and organic carbon source, and nutrients for their growth. In contrast *autotrophic* organisms derive their energy from sunlight and *chemotropic* rely on an exothermic chemical reaction as a source of energy. Both autotrophs and chemotrophs use carbon dioxide (alkalinity) as a source of carbon for their growth.

Autotrophic decomposition of chemicals in soil is rare; however, chemotropic transformation of nitrogen in soils is common (see the section titled "Specific Pollutant Interactions" in this chapter). In some cases the chemicals themselves may be the source of organic carbon and energy, in some other instances the chemicals are broken down by enzymatic reactions of microorganisms using a more simple and digestible carbon source (for example, plant residues). To break down some more complex chemicals the organisms require adaptation.

Biodegradation may occur in an *aerobic* (oxygen-rich aerated soils) or *anaerobic* (saturated soils devoid of oxygen) soil environment. Microorganisms that can function in both environments are called *facultative*, while strict aerobes and anaerobes can only function depending on whether oxygen is or is not available. Hence, based on the oxygen availability the biochemical reactions are either oxidation (aerobic) or reduction (anaerobic) processes. As an example, if an organic chemical is made of ring compounds, such decomposition associated with ring cleavage can be accomplished in an anaerobic environment; however, anaerobic biochemical decomposition is generally slower than aerobic breakdown.

Biodegradation kinetics is commonly represented by the Monod's equation derived from the Michaelis–Menton equation for substrate loss in enzyme catalyzed biochemical reactions. The Monod's formula adapted for the substrate variables defined earlier would be

$$\frac{dc_d}{dt} = \frac{-\mu_m X c_d}{K_s + c_d} \qquad (6.14)$$

where
μ_m = is the maximum substrate utilization rate (µg/l-day)
X = microbial biomass per unit volume of liquid (pore water) (µ/l)
K_s = half-saturation constant for the chemical (µg/l)

As was stated previously, only dissolved (pore water) fractions of the chemical are bioavailable.

For a small concentration of the chemical in the soil ($c_d \ll K_s$) and a sufficient and constant microbial population, the biodegradation equation is converted to a first-order reaction such as

$$\frac{dc_d}{dt} = -k_b c_d \qquad (6.15)$$

Major factors affecting the biodegradation of a chemical are (Valentine and Schnoor, 1986) pH, temperature, water content, organic carbon content in the soil, clay content, oxygen availability, nutrients, nature of the microbial population, acclimation of the microbial population to the chemical, and concentration of the chemical.

Although some texts contain magnitudes of the microbiological decay–decomposition rates such as k_b (see Schnoor et al., 1987), the reported ranges of these rate coefficients are very wide, and generally the reported values are not reliable. Furthermore, microorganisms may have the capacity of adapting to a particular chemical. Biodegradation experiments, both in situ and in the laboratory, may have to be performed with soils and sediments to obtain an insight on biodegradability of the chemical and its reaction-rate type and parameters. Aerobic and anaerobic conditions, soil–sediment organic content, sufficient nutrients, and acclimation of the microorganisms are the factors that may affect the decomposition process. In laboratory experiments, radioactive labeled organic chemicals can be used to estimate metabolic degradation (mineralization) by measuring the $^{14}CO_2$ carbon production in the reaction. A more detailed discussion on soil microbial populations and decay of pathogens (disease-causing microorganisms) concludes this chapter.

Plant Uptake

Plant uptake of nutrients and chemicals from soil is a form of immobilization of the pollutants. The nutrients and chemicals uptake process is a part of the overall transpiration process of plants. Nutrients and chemicals are transported into plants only from the dissolved or ionic pool of chemicals in pore water. Consequently, chemicals and nutrients adsorbed on soil particles or precipitated in soils (sediments) are not available to plants; thus, only the mobile components of the soil are of concern, and high uptake rates are observed when the mobile component concentrations are high. The total uptake rates and nutrient requirements also depend on crop type and yield.

The annual crop yields and nutrient content of various plants are presented in Table 6-4. The nutrient content of plants in Table 6-4 also represents the nutrient uptake for annual crops. To obtain annual nutrient uptake for perennial plants, divide the nutrient content by the average age of the plants. Dividing the nutrient content into grains and straw or stubble becomes important when planning certain management practices. For example, no-till planting leaves the stalks (and associated nutrients) on the field, while conventional planting removes the residues from the field; hence, nutrients must be resupplied by fertilizing the field.

Other Transformation Reactions

In addition to biochemical reactions a chemical present in soils and sediments can be degraded or transformed by chemical oxidation-reduction

TABLE 6.4 Approximate Yields and Nutrient Content of Selected Crops

Crop		Yield/ha	kg N/ha	kg P/ha
Alfalfa		9 Tonnes	225	30
Barley	Grains	100 bu	39	7
	Straw	2.2 tonnes	17	2
Beans[a]	(Dry)	75 bu	84	11
Cabbage		45 tonnes	165	18
Clover[a]	Red	4.5 tonnes	89	11
	White	4.5 tonnes	145	11
Corn	Grain	370 bu	151	27
	Stover	10 tonnes	111	18
	Silage	56 tonnes	225	34
Cotton	Lint and seed	2.2 tonnes	67	13
	Stalks	2.2 tonnes	50	7
Lettuce		45 tonnes	100	13
Oats	Grain	22 bu	62	11
	Straw	4.5 tonnes	28	9
Onions		17 tonnes	50	9
Peanuts[a]	Nuts	3.4 tonnes	123	7
Potatoes	Tubers	990 cwt	106	14
	Vines	2.2 tonnes	100	9
Rice	Grains	225 bu	62	13
	Straw	5.6 tonnes	34	4
Rye	Grain	75 bu	39	4
	Straw	3.4 tonnes	17	4
Sorghum	Grain	150 bu	56	11
	Stubble	6.7 tonnes	73	9
Soybean[a]	Grain	111 bu	179	18
	Straw	2.2 tonnes	28	4
Sugar beets	Roots	45 tonnes	95	16
	Tops	27 tonnes	123	11
Sugar cane	Stalks	67 tonnes	112	22
	Tops	29 tonnes	56	11
Tobacco		3.4 tonnes	129	11
Tomatoes	Fruit	56 tonnes	162	22
	Vines	3.4 tonnes	78	11
Wheat	Grain	123 bu	73	16
	Straw	3.4 tonnes	22	2
Bermuda grass[b]			540–670	
Fescue[b]			300	
Medium mature forest[b]				
Deciduous			30–60	
Evergreens			20–30	

Source: From Steward et al. (1975), after U.S. Department of Agriculture.

[a] Legumes that do not require fertilizer nitrogen.
[b] After Powell (1976).

reactions, hydrolysis, and photochemical transformation. The last reaction can only occur in the presence of light in a relatively thin surface layer.

Both hydrolysis and the nature and kinetics of the oxidation-reduction reactions are controlled by the moisture content. Other factors, such as pH and oxygen content, which are also important, are also related to the soil moisture.

Photochemical reactions (photolysis) are important for some organic chemicals, such as some pesticides (for example, DDT and parathion), and aromatic hydrocarbons (Schnoor et al., 1987).

Overall Degradation Rates

Typically due to lack of information it is not possible to quantify specific degradation rates. The only information usually available for many organic chemicals is their half-life, which is the mean number of days required for 50% of the original chemical to degrade in soil.

Using the first-order decay reaction (Eq. (6.15)) as the most accepted model to use to obtain the half-life from, the solution of the equation, by substituting overall pollutant concentration, P, for C_d, is

$$P(t) = P(0)e^{-K_t t} \tag{6.16}$$

where

$P(t), P(0)$ = chemical masses at time t and time 0, respectively
K_t = overall degradation rate coefficient
t = time

The half-life is then

$$t_{1/2} = -\frac{\ln(0.5)}{K_t} = \frac{0.69}{K_t}$$

SPECIFIC POLLUTANT INTERACTIONS

Soil Phosphorus

In natural systems, phosphorus (P) occurs as an orthophosphate anion (PO_4^{3-}), which may exist in inorganic or organic forms. The source of all organic P is plant and organic biomass residues. The origin of all inorganic orthophosphate is the class of minerals known as apatites. These minerals are insoluble calcium phosphates existing in several forms, and the orthophosphate ions are liberated by chemical weathering processes.

Phosphorus is an important nutrient to aquatic ecosystems. In excess it causes accelerated eutrophication (see Chapter 13). Unlike nitrogen (N), phosphorus is not particularly mobile in soils, and phosphate ions do not leach readily. Phosphorus is held tightly by a complex union by the clays and soil particulates, and organic matter. The amount of P in solution is small. Most of the P is removed from soils either by crop uptake or by soil erosion.

Figure 6.5 shows P solubility in soils. At higher pH values—characteristic of calcareous soils—the P precipitates mostly in combination with calcium (Ca). Below a pH of 7, which is characteristic for soils with high clay and organic matter content, Ca rapidly disappears from soils and P reacts predominantly with the Fe and Al ions in soils. From Figure 6.5 it can be seen that the maximum phosphate concentration in the pore water solution is of the order of 10^{-5} mole/l, which corresponds to about 0.3 mg/l. Depending on soil pH, the dissolved P concentrations may decrease to 0.01 mg/l or less.

From the foregoing discussion it can be seen that the P concentrations in the soil (sediment) pore water solution is very low and the excess is fixed by soil particles. Porcella et al. (1974) have reported that almost all P arising from weathering of minerals and/or from P application to soils by atmospheric fallout, commercial fertilizer, plant residues, or animal manure remains near the place of application. The exception is in sandy or peat soils, which exhibit little tendency to react with phosphorus. These authors have also shown that after a normal growing season, fertilizer P applied in the spring is confined to the 5 cm of the surface soil.

The process of fixation of P is controlled by several factors:

1. Al and Fe oxides are responsible for P retention in acid soils (Hsu, 1965; Vijayachandran and Harter, 1975).
2. Calcium compounds control solubility of P in calcareous soils (Hsu and Jackson, 1960; Hsu, 1965).
3. Organic matter contributes to P adsorption.

Figure 6.6 shows typical adsorption isotherms for some selected soils. Several authors have attempted to correlate P sorptivity to various soil parameters. Novotny and Chesters (1981) correlated adsorption Langmuir isotherm parameters to several independent soil parameters for acid soils. The most satisfactory combination of variables found by the authors to assess the adsorption maximum (Q^o) and the energy coefficient (b) was

$$Q^o(\mu g/g) = -3.5 + 10.7(\% \text{ clay}) + 49.5(\% \text{ organic C}) \qquad (6.17)$$

and

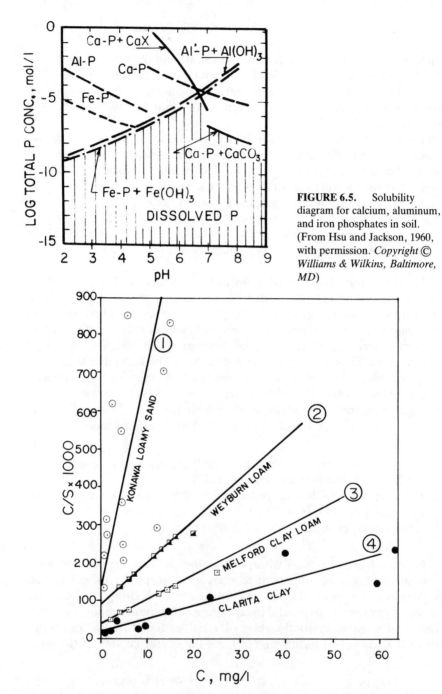

FIGURE 6.5. Solubility diagram for calcium, aluminum, and iron phosphates in soil. (From Hsu and Jackson, 1960, with permission. *Copyright © Williams & Wilkins, Baltimore, MD*)

FIGURE 6.6. Langmuir isotherms for phosphate adsorption in soils. (Data 1 and 4 from Rennie and McKercher, 1959; data 2 and 3 from Enfield, 1974.)

$$b(\text{l/mg}) = 0.061 + 170{,}000 \times 10^{-\text{pH}}$$
$$+ 0.027(\% \text{ clay}) + 0.76(\% \text{ organic C}) \qquad (6.18)$$

The coefficients of multiple correlation for Equations (6.17) and (6.18) were 0.83 and 0.53, respectively. The percentage of organic carbon can be roughly calculated from the content of the soil (sediment) organic matter with the use of conversion factors ranging from 0.5 to 0.6. For calcareous soils the distribution of particulate and as dissolved P is governed by the solubility diagram shown in Figure 6.5.

Example 6.2: Distribution of Sorbed and Dissolved Phosphorus in Soil

Fertilizer was applied to a field at a rate of 100 kg/ha. The fertilizer was plowed in and uniformly distributed to a depth of 0.3 m. The soil was a silt loam and contained 20% clay, 55% silt, and 25% sand. The porosity of the soil was 40%, pH was 6.0, organic C content was 1%, and the specific density of the soil sample was 1500 kg/m^3 (= g/l). Estimate the approximate ground-water contamination by P during a long period of rain with an average intensity of 1 cm/hr. Assume that all water infiltrated into the soil and that the infiltration rate was slow enough that full equilibrium between the adsorbed and dissolved P was established. Assume also that the antecedent adsorbed P was negligible.

Solution Using Equation (6.17), the P adsorption maximum for the Langmuir isotherm (Eq. (6.4)) is

$$Q^o = -3.5 + 10.7 \times 20 + 49.5 \times 1 = 260\,\mu\text{g/g}$$

and the adsorption energy coefficient from Equation (6.18) is

$$b = 0.061 + 170{,}000 \times 10^{-6} + 0.027 \times 20 + 0.76 \times 1$$
$$= 1.53\,\text{l/mg} = 0.00153\,\text{l/}\mu\text{g}$$

The total inorganic P content of the soil becomes

$$c_T = \frac{100\,(\text{kg/ha}) \times 10^9\,(\mu\text{g/kg})}{0.3\,(\text{m}) \times 10^4\,(\text{m}^2/\text{ha}) \times 1000\,(\text{l/m}^3)} = 33{,}300\,\mu\text{g/l of soil}$$

The P concentration is distributed between adsorbed and dissolved (pore water) fractions. Hence

$$c_T = r\rho + c_d\theta = \frac{Q^\circ b c_d}{1 + b c_d}\rho + c_d\theta$$

where
ρ = the specific density of the soil
θ = moisture content (assume θ = porosity)

Substituting into the preceding equation

$$33{,}330\,(\mu g/l) = \frac{260\,(\mu g/g) \times 0.00153\,(l/\mu g) \times c_d}{1 + 0.00153\,(l/\mu g) \times c_d} \times 1500(g/l) + c_d \times 0.4$$

and solving, c_d is 57 µg/l.

Note that in the denominator $b \times c_d \ll 1$; therefore, a linear partitioning concept can be used for which the linear partitioning coefficient, $\Pi = Q^\circ b = 260 \times 0.00153 = 405.6\,l/g$.

Adsorbed P concentration is

$$r = \frac{260 \times 0.00153 \times 57}{1 + 0.00153 \times 57} = 20.85\,\mu g/g$$

The enrichment ratio (ER) for P is generally higher than that for clay or organic matter due to desorption of P from soil into runoff water and possibly due to other causes. Data by Massey and Jackson (1952), Massey, Jackson, and Hays (1953), and Stoltenberg and White (1953) indicate that the ER for P is about 1.5 to 2 times that for clay or soil organic matter.

The adsorption characteristics for P (and by the same reasoning for other chemicals) adsorbed on soil particles, are different for parent soils and sediments, and when solids are suspended in runoff or surface-water body flow (Green, Logan, and Smeck, 1978; Ryden, Syers, and Harris, 1972; Novotny and Chesters, 1981). An example of an adsorption isotherm of P on suspended solids in runoff is shown on Figure 6.7. The ranges for suspended sediment adsorption maxima, Q°, in the watershed of the Maumee River in Ohio reported by Green, Logan, and Smeck (1978) were between 500 and 2000 µg/g (average 988 µg/g), while the ranges of Q° for the parent soils were 189 to 287 µg/g (average 257 µg/g). Similarly, the adsorption energy coefficient b ranged from 0.00011 to 0.00045 l/µg (average 0.00032 l/µg) for the suspended sediment as compared to 0.0009 to 0.0045 l/µg (average 0.0026 l/µg) for the parent soil. However, it can be argued that for the linear partitioning coefficient $\Pi = Q^\circ \times b$ the differences would be much less. Using average values of the Maumee River

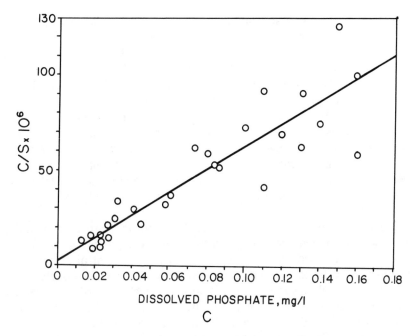

FIGURE 6.7. Isotherm correlation of dissolved and adsorbed phosphates in the suspended solids of a stream. (From Novotny et al., 1978.)

data, the average linear partitioning coefficient for suspended sediment would be $\Pi = 988 \times 0.00032 \approx 0.4\,\text{l/g}$, while that for parent soil is $\Pi = 257 \times 0.0026 \approx 0.66\,\text{l/g}$. Bottom sediments (bedload) exhibited adsorption characteristics between those of suspended sediment and parent soils.

Organic Chemicals

The use of organic chemicals is credited with substantially increasing yields of agricultural crops and assisting in controlling pests and diseases such as malaria. In a situation typical of most modern societies, the availability of agricultural land decreases, giving way to urbanization. On the other hand, severe environmental problems have been created by the use of man-made organic chemicals, which has resulted in ozone depletion, contamination of soils and bodies of water, and chronic damage to animals and humans. The epigraph for this chapter reflects the fears and sometimes realities of the damage done by the improper and excessive use of these chemicals.

Although the use of pesticide chemicals dates back a hundred years, usage did not become substantial until about 1945 following the commercial manufacture of DDT. Rapid growth of the organic pesticide industry continued for about 25 years. Thousands of chemicals were developed and introduced into the market and subsequently discharged into the environment. Pesticide usage revolutionized agricultural production to the point that most agricultural practices formerly used to control weeds, insects, and disease shifted in favor of chemical control. At the same time, however, disposal of organic chemicals became a serious problem, and unsafe dumping of chemicals into the soil, air, and bodies of water has contaminated them to the point that their use has been severely impaired or even negated. Restoration and "repair" of these systems will be the objective of present and future environmental cleanup efforts.

Organic chemicals used throughout the world that are contaminating the environment can be divided into several categories. *Pesticides* include *herbicides* used for weed control (*algicides* for control of unwanted algal growth in lakes), *insecticides* for control of unwanted insects, while *fungicides* control unwanted fungus that may attack orchards or vegetable farms. The use of pesticides is not limited only to the agricultural sector; urban lawn-care chemicals are in the same category. Other chemicals of concern are *cyclic (aromatic) hydrocarbons*, *fluorocarbons* used in sprays and cooling liquids, *polychlorinated bi-phenyls (PCBs)*, which were widely used in transformers, paper, and other products, *volatile organic chemicals* and *solvents*, and several other categories.

Insecticides include organochlorine, organophosphorus, and carbamate chemicals. Organochlorine compounds, such as DDT, dieldrin, aldrin, heptachlor, and lindane, are essentially conservative chemicals. They can persist in soils and aquatic environments for many years. For example, DDT has been frequently detected 10 years after its application. Lindane can be found in some Great Lakes sediments and soils (Green Bay) 20 years after application in cherry orchards. Eventually, burial by new sediment deposits may alleviate the problem.

Use of organophosphorus and carbamate chemicals increased rapidly following restrictions on the use of organochlorine insecticides. These chemicals are less persistent in soils than the "hard" organochlorine form. An extended discussion on the persistence of chemicals in soils and sediments and their gradual disappearance and/or transformation is presented largely in a subsequent section of this chapter.

Herbicides are less ubiquitous in the environment than organochlorine insecticides. However, such compounds as *s*-triazines, picloram, monouron, and related substituted ureas, and 2,4,5-T often persist in soils for as much as a year following application. Atrazine, alone or in combination, and propachlor are the herbicides most commonly used for weed control on corn in the midwestern United States. The herbicide MCPA and 2,4-D amine are used most frequently on small grains. The carbamate herbicides and 2,4-D are short lived in soil. One has to remember, however, that in sandy soil transport of chemicals into ground water may be relatively fast and the pesticide may reach ground water before soil bacteria can degrade it. Since there are few or no bacteria in ground-water-bearing strata, further biochemical decomposition is stopped and ground-water pesticide contamination arises and persists. Persistent ground-water contamination has been found in many agricultural areas throughout the world, including central Wisconsin and the Po River valley of Italy.

Polyaromatic (cyclic) hydrocarbons (PAHs) are mostly of urban and industrial origin. However, their fate and interactions with soils and sediments are similar.

Mobility of Organic Chemicals in Soils and Sediments

Chemicals deposited on soil surfaces follow the general transport path and can be transported into (1) the atmosphere, (2) ground water, and (3) surface runoff.

A general scheme of the distribution and fate of agricultural chemicals in the biosphere is shown in Figure 6.8. Contamination of the atmosphere by chemicals can occur by drift during application and volatilization, and by wind erosion. Drift is that portion of application that does not reach the target area. The extent of drift losses and the subsequent dispersion and transport of chemicals in air is governed primarily by meteorological conditions and methods of application. Chesters and Simsiman (1975) documented that drift losses range from 25% to 75% of aerially applied chemicals.

Contamination of ground water by organic chemicals can occur through leaching. Downward movement of agriculturally applied chemicals is controlled by soil type and chemistry, pesticide composition, and climatic factors (Hern and Melacon, 1986; Adams, 1973; Helling, Kearney, and Alexander, 1971). The leachability of a compound from soils depends primarily on the degree of adsorption of the chemical on soil particles. The U.S. EPA (1984) developed an approach that incorporates the mechanisms of degradation and sorption of constituents and incorporated it into a procedure of assessment mobility of organic chemicals. Using this procedure, a mobility and degradation index (MDI) for ranking the potential leaching of chemicals was proposed by Mahmood and Sims (1986).

For organic chemicals in soils and sediments, Equation (6.10) describes

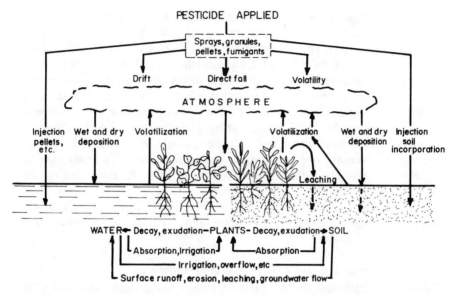

FIGURE 6.8. A scheme showing the possible distribution and fate of agricultural pesticides and their degradation products in the biosphere. (After Foy and Bingham, 1969.)

the distribution of the total chemical concentration between the solid, liquid, and gaseous phases, or

$$c_T = \theta c_d + c_p + a c_g = c_d(\theta + \Pi m_{ss} + a K_H)$$

The adsorptivity of nonpolar organic chemicals is related to their solubility expressed by the octanol partition coefficient K_{ow} and the particulate organic content of the soil (OC) expressed in percent. Then the solid–liquid partition coefficient can be expressed as

$$\Pi \approx K_{oc} \times (\%OC)/100 \qquad (6.19)$$

where K_{oc} = the partitioning coefficient normalized by the organic carbon. The coefficient K_{oc} is a hypothetical partitioning coefficient for a sorbent (sediment or soil) composed solely of particulate organic matter. It has been found that for a given chemical, K_{oc} is relatively constant and varies by a factor of only 2 for a wide range of soils and sediments (Schwarzenbach and Westall, 1981; Weber, McGinley, and Katz, 1991). The coefficient K_{oc} can be correlated to the octanol partitioning coefficient, K_{ow} (Kenaga and Göring, 1980; Schwarzenbach and Westall,

Specific Pollutant Interactions   343

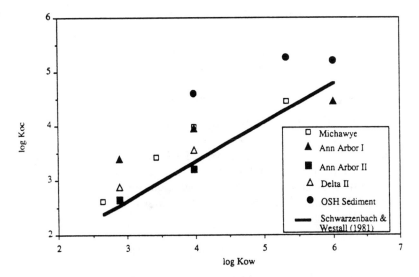

FIGURE 6.9. The relation of organic matter–chemical partitioning coefficient K_{oc} to the chemical's octanol partitioning parameter K_{ow}. (From Weber, McGinley, and Katz, 1991, with kind permission from Pergamon Press, Headington Hill Hall, Oxford, OX3 0BW, UK.)

1981; Weber, McGinley, and Katz, 1991). Several regression equations have been proposed (Karickhoff, 1981; Karickhoff, Brown, and Scott, 1979; Schwarzenbach and Westall, 1981). A relationship of K_{oc} to K_{ow} published by Weber, McGinley, Katz (1991) is shown on Figure 6.9. Values of K_{ow} and K_{oc} for various chemicals are given in the Appendix.

Karickhoff, Brown, and Scott (1979) proposed the following empirical relationships for estimating the partition coefficient for the sorption of aromatic hydrocarbons and chlorinated hydrocarbons on sediments:

$$K_{oc} = 0.63 K_{ow} \quad (6.20a)$$

Almost an identical relation was derived by Rao and Davidson (1982), where

$$K_{oc} = 0.66 K_{ow}^{1.03} \quad (6.20b)$$

This relationship can be expanded to segregate the influence of particle size as follows (Karickhoff, Brown, and Scott, 1979):

$$\Pi \approx K_{oc}[0.2(1-f)x_{oc}^s + fx_{oc}^f] \quad (6.21)$$

where for Π in liters/kilogram
f = mass fraction of fine sediments or soil particles ($d < 50\,\mu m$)
x_{oc}^s = organic carbon content of coarse sediment (soil) fraction
x_{oc}^f = organic carbon content of fine sediment (soil) fraction

Example 6.3: Distribution of a Chemical in Sediment

Determine the fraction of benzo(a)pyrene that is contained in pore water of aqueous sediment with a specific density of $1400\,kg/m_3$ (= g/l). The sediment is 70% fines ($d < 50\,\mu m$) and the weight fraction of organic carbon is 10% of the fines and 5% of the sand fraction. From Table A1 in the Appendix the octanol partitioning coefficient for the compound is 10^6. Porosity of the sediment is 45%, hence the water content is $\theta = 45/100 = 0.45$. Calculate K_{oc}

$$K_{oc} = 0.63 \times 10^6 = 630{,}000$$

Solution From Equation (6.21) the partition coefficient becomes

$$\Pi = 630{,}000[0.2(1 - 0.7)(0.05) + 0.7(0.1)] = 45{,}000\,l/kg$$

From Equation (6.10), neglecting the gaseous phase, the distribution of dissolved (pore water) and total concentration is

$$\frac{c_d}{c_T} = \frac{1}{0.45 + 45{,}000\,(l/kg) \times 1400\,(kg/m^3) \times 0.001\,(m^3/l)}$$
$$= 0.000016$$

Almost all of the chemical will be adsorbed on the sediment, and the pore water concentration will be about 1/1000 of a percent of the total concentration of the contaminant.

A brief description of the behavior of organic chemicals in soils and sediments is taken from Moore and Ramamoorthy (1984) and several other sources.

Aliphatic hydrocarbons. These compounds include a diverse group of open chain compounds of carbon and hydrogen, which may be halogenated by chloride, bromide, iodine, or fluoride ions. Halogenation may occur naturally or by adding halogens (for example, by disinfecting organic wastes or wasting halogenated chemicals into receiving bodies of water and soils). From many possible compounds the following priority pollutants are of environmental concern: carbontetrachloride, dichlorobromomethane, chloroethane, dichloromethane, dichloropropane, vinyl chloride, chloroform, bromoform, tetrachloroethane, trichloroethane, methyl chloride, and methyl bromide. They are in the category of volatile priority pollutants. These priority pollutants have little or no affinity for sorption, and evaporation (volatilization) is the primary mechanisms of their loss. Because of their volatility most aliphatics occur at low concentrations in soils and sediments, except in cases of extreme pollution.

Aromatic hydrocarbons. These chemicals have a cyclic six carbon ring. They can be monocyclic, such as benzene and toluene, or polyaromatic (PAHs). Benzene and toluene are examples of monocyclic compounds, while the family of more complex polyaromatic hydrocarbons includes such priority pollutants as anthracene, benzo(a)anthracene, benzofluoranthene, chrysene, fluoranthene, napthalene, and pyrene. Many PAHs have been found to be carcinogenic.

Benzene, toluene, and some of their derivatives are moderately soluble in water and moderately volatile, so larger scale sorption on soils and sediments does not occur. Mill (1980) proposed the following equation for predicting the sorptivity of monocyclic aromatics:

$$\log K_{oc} = -0.782 \log\{C\} - 0.27 \tag{6.22}$$

where $\{C\}$ is the solubility of the compound in moles/liter.

PAHs found in soils and sediments are generally of a diffuse nature, and their origin can be traced to automobile use, municipal and industrial wastewater effluents, forest fires, and the combustion of coal. The last two sources are the most significant paths for PAHs entering the atmosphere. These contaminants will subsequently reach terrestrial systems by atmospheric deposition. Automobiles, especially those with diesel engines, were in the past a major source of PAHs. Recent restrictions on emissions have significantly reduced their discharge in countries where such environmental regulations were implemented. Typically, urban runoff contains large quantities of PAHs that are mainly incorporated into sediments (see Chapter 9 for a discussion of urban sources).

PAHs have a larger affinity for adsorption on soils and sediments than do the monocyclic compounds. Soil and sediment microorganisms present are capable of degrading PAHs. Photolysis is an important degradation process for some PAHs (for example, anthracene).

Pesticides. Sorption of nonionic (nonpolar) pesticides—organochlorine and organophosphorus insecticides—in soils and sediments is correlated primarily with organic particulate content and to a much lesser degree with clay (Goring and Hamaker, 1972; Moore and Ramamoorthy, 1984). Organochlorine pesticides—priority pollutants—include DDT and its derivative DDE, aldrin, chlordane, dieldrin, endrin, endosulfan, heptachlor, and toxaphene. Lindane and mirex are also of concern and have contaminated parts of the Great Lakes and other bodies of water.

Most developed Western countries have curtailed or banned the use of DDT and several other persistent chemicals, yet these chemicals still may be found in the environment. Retention of acidic and basic pesticides in soils and sediments is affected markedly by pH (Goring and Hamaker, 1972). Soil and sediment pH controls the overall charge of the molecule and its adsorptivity to clay and organic matter. The organic cations—diquat and paraquat (not included in the priority pollutant list)—are held strongly by clay minerals and often are adsorbed irreversibly. Weakly adsorbed, water-soluble compounds are desorbed readily by water, and hence possess a greater potential for leaching. None of the acidic or basic pesticides have been included in the priority pollutant list. Organochlorine insecticides, which have low solubility, are least mobile in soils and sediments, followed in turn by organophosphorus insecticides. The water-soluble acidic herbicides are the most mobile. Other pesticides, including triazines, atrazines, phenyl ureas, and carbamates, have an intermediate degree of mobility.

After initial application losses (by drift and volatilization) the remaining pesticide reaches the soil, where loss is evident for surface-applied and soil-incorporated pesticides. In the soil, pesticides are continuously subjected to dissipation processes and their concentration decreases gradually after application. The mechanisms of dissipation include adsorption on soil particles and subsequent loss by erosion, degradation (photochemical and microbial), volatilization, plant uptake, and leaching. By definition, degradation affects only pesticides that are nonconservative. The general magnitudes of the overall persistence of pesticides in soils and aquatic sediments are shown on Figure 6.10.

Photodecomposition occurs only to pesticides located at the soil surface. The practical significance of photodegradation as a means of pesticide removal from soil and aquatic systems has not been determined quantitatively because of the difficulty of interpolating laboratory data to field conditions. In most cases, photodegradation reactions are comparatively slow, and their rate depends on the physical state of the pesticide (vapor, dissolved, adsorbed), light intensity, photolytic efficiency, and other factors. DDT and parathion are examples of pesticides that may degrade by photolysis (Schnoor et al., 1987).

The rate of volatilization of pesticides depends on such factors as temperature, soil moisture content, pesticide vapor density, soil properties, solubility in water, concentration of pesticides, ambient air humidity, and near-surface wind velocity. Factors controlling and mech-

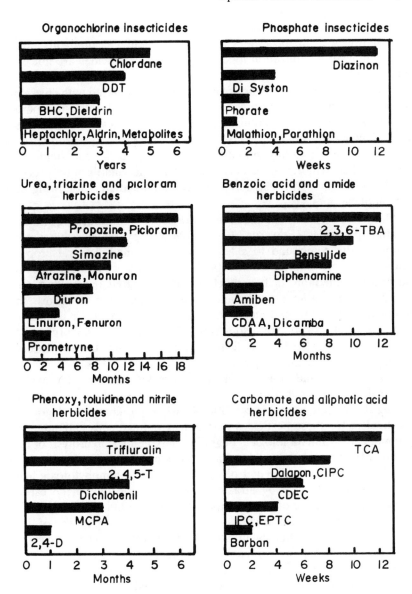

FIGURE 6.10. Persistence of pesticides in soils. (After Kearney et al., 1969.)

anisms of volatilization losses in soils have been discussed in a comprehensive review by Spencer and Cliath (1973). Field measurements indicate that significant volatilization losses may occur if pesticides are not incorporated into the soil.

In a study of pesticide losses from a watershed planted with corn it was found that considerable portions of dieldrin and heptachlor incorporated in the soil were lost by volatilization (Spencer, Farmer, and Cliath, 1973; Cars, Taylor, and Lemon, 1971). Losses in one growing season (from an application of 5.5 kg/ha) amounted to 2.5% to 2.9% for dieldrin and 3.9% for heptachlor. Under field conditions, the volatilization process is continuous, although its highest rate occurs immediately following pesticide application.

Many terrestrial and aquatic plants are capable of adsorbing and translocating pesticides (Edwards, 1970), followed by the possible detoxification of the compound to a less active compound. Several crops are known to adsorb chlorinated insecticides, and evidence of metabolic break down was opined for DDT, heptachlor, endrin, γ-BHC, and aldrin. Corn, which is resistant to atrazine and simazine, was able to adsorb these herbicides from soils and metabolize them to nonphytotoxic compounds. Dissipation of atrazine from soils through uptake by corn, sorghum, and other crops has also been indicated by experiments, but the extent of detoxification by plants is small.

Investigations indicate that chemical breakdown may play a significant role in the dissipation of soil-adsorbed organochlorine insecticides during dry periods (aerobic conditions), while the dominant degradation mechanism for these pesticides in moist or ponded soils and submerged sediments is by microbiological degradation. Chemical hydrolysis of the 2-chloro-s-triazines in soils and sediments has been reported by Armstrong and Chesters (1968). The hydrolysis of atrazine to nonphytotoxic hydroxyatrazine is enhanced by atrazine adsorption possibly to the carboxyl groups present in the organic components of soils and sediments.

Microbial metabolism is considered to be the major pathway of degradation of many pesticides in soils and sediments. The efficiency of this mechanism depends on such environmental factors as temperature, moisture and organic matter content, aeration or saturation, pH, and pesticide concentration. Although intensive studies have been made on the mechanisms by which microorganisms degrade pesticides, the processes are not understood clearly. In general, organochlorine pesticides are the most resistant to microbial attack. As pointed out previously, the partial degradation of DDT results in the formation of TDE (DDD) and DDE. Both components are stable in soils and sediments and have about the same toxicity as the parent compound.

Specific Pollutant Interactions   349

The oxygen status of soils and sediments has a pronounced effect on the microbial breakdown of organochlorine pesticides. In soils and sediments DDT is rapidly converted to TDE (DDD) under anaerobic conditions, and very slowly to DDE in soils and exposed sediments under aerobic conditions (Guenzi and Beard, 1967, 1968). On the other hand, several organochlorine pesticides—heptachlor (Miles, Tu, and Harris, 1969), lindane (Yule and Rosefield, 1964), and endrin (Bouman, Schecter, Carter, 1965)—have been shown to degrade in soils to compounds of lower toxicity and reduced insecticidal activity.

Compared to organochlorine pesticides, other commercial pesticides currently in use are more biodegradable. One of the first extensive reviews of degradability of herbicides was published by Kearney and Kaufman (1964). Paris and Lewis (1973), Paris (1981), Sethunathan (1973), among others have investigated the biodegradability of pesticides in aquatic systems. The newer vinyl phosphate insecticides, such as phosphamidon, chlorfenvinphos, and mevinphos, have half-lives in soils ranging from 1 to 30 weeks. Chlorfenvinphos appears to be the most resistant to microbial degradation.

In spite of the enormous amounts of research on pesticide distribution and transport through the environment, no information is available to accurately estimate enrichment ratios for these chemicals (and by the same reasoning for any other priority organic pollutants). It appears, however, that for those pesticides that are immobile and tightly adsorbed on soil organic matter, the enrichment ratios may be similar to that of organic particulate matter. Conversely, for pesticides that are mobile (mostly dissolved) in soil water the notion of the enrichment ratio is meaningless. An approximate method of computing pesticide loading is shown in the following example.

Example 6.4: Pesticide Load in Runoff

To control weeds, 5.55 kg/ha of atrazine was applied to a field. Estimate how much atrazine will be lost during a 5-cm storm resulting in 1.5 cm of surface runoff and 1.5 tonnes/ha of soil loss due to erosion, which occurred 4 weeks after application. Porosity of the soils is $p = 45\%$, 0.3 bar moisture content is $\theta_{0.3 \text{ bar}} = 35\%$, 15 bar moisture content of the soil is $\theta_{15 \text{ bar}} = 15\%$, and the organic C content is 2.3%. The antecedent soil moisture $\theta = 31\%$ and evapotranspiration are negligible. Specific density of the soil is $\rho = 1500 \text{ kg/m}^3$. Effective moisture $\theta_{\text{eff}} = (31 - 15)/100 = 0.16$

Solution From Figure 6.10 it can be seen that atrazine persists in soils for approximately 10 months, which implies that during the 4 weeks between application and the storm, approximately 10% of the pesticide was dissipated or decomposed.

Table A1 in the Appendix shows that the octanol partitioning number for atrazine $K_{ow} = 10^{2.68} = 478.63 \, l/kg$. Then the organic carbon partition coefficient becomes

$$K_{oc} = 0.63 \times 478.73 = 301 \, l/kg$$

and the soil partitioning coefficient is $\Pi = K_{oc} \times f_{oc} = 301 \times 2.3/100 = 6.02 \, l/kg$.

The depth of penetration of the pesticide in the soil can be approximated as 7.5 cm, with uniform distribution throughout the depth. This is an arbitrarily chosen depth that may only be considered as an order of magnitude; however, it has been seen as an average estimate of pesticide penetration in agricultural soils (Bruce et al., 1975). Thus the pesticide concentration per unit volume of soil is

$$Pe = 0.9 \frac{5.55 \, (kg/ha) \times 10^9 \, (\mu g/kg)}{7.5 \, (cm) \times 10^8 \, (cm^2/ha) \times 0.001 \, (l/cm^3)} = 6670 \, \mu g/l$$

The distribution between the adsorbed and dissolved phases can be computed from the volumetric mass balance equation (Eq. (6.8)) as

Dissolved

$$C_d = \frac{1}{\theta_{eff} + \Pi \rho} Pe = \frac{1}{0.16 + 6.02 \, [l/kg] \times 1500 \, [kg/m^3] \times 0.001 \, [m^3/l])} 6670$$
$$= 726 \, \mu g/l = 0.726 \, g/m^3$$

Adsorbed

$$r = \Pi \rho C_d = 6.02 \, (l/kg) \, 726 \, (\mu g/l) \times 0.001 \, (kg/g)$$
$$= 4.37 \, \mu g/g \, (= g/tonne)$$

The estimated pesticide loss during the storm assumes that desorption from the sorbed phase to solution is negligible. In this case, the adsorbed pesticide loss becomes (assume that the estimated ER for organic matter is ≈ 1.2)

$$Y_{Pe} = r \times ER \times Y_{ss} = 4.37 \, (g/tonnes) \times 1.2 \times 1.5 \, (tonnes/ha)$$
$$= 7.87 \, g/ha$$

If this pesticide was immobile and tightly bound to soil organic matter, the preceding amount would represent the approximate pesticide loss during the storm. However, atrazine has intermediate mobility, and in this case the dissolved fraction must be estimated as well. As a first approximation assume that runoff completely mixes with the pesticide in a 5-mm-thick upper soil surface mixing layer. Then the mass balance equation for the dissolved layer is

$$0.5\,(\text{cm}) \times \theta_{\text{eff}} C_d = C_{\text{runoff}}(0.5 \times \text{porosity} + \text{runoff})$$

from where the pesticide concentration in runoff during and after the storm is

$$C_{\text{runoff}} = \frac{0.5 \times 0.16 \times 726}{0.5 \times 0.45 + 1.5} = 32.72\,\mu\text{g/l} = 0.03272\,\text{g/m}^3$$

and the dissolved pesticide loss in the runoff becomes

$$Y_{PD} = 1.5\,(\text{cm}) \times 0.01\,(\text{m/cm}) \times 0.03272\,(\text{g/m}^3) \times 10^4\,(\text{m}^2/\text{ha})$$
$$= 4.9\,\text{g/ha}.$$

Relating pesticide loss only to erosion for pesticides with medium to high mobility could lead to gross underestimation of the loss.

Since only gravitational water remains in the soil after the storm, the excess water will be leached into the lower soil layer by infiltration. The excess infiltration amounts to

Rainfall depth $-$ runoff $+$ antecedent soil moisture
$-$ gravitational soil moisture $= 5 - 1.5 + \theta \times 7.5 - \theta_{0.3\text{bar}} \times 7.5$
$= 3.5 + 7.5(0.31 - 0.35) = 3.2\,\text{cm}.$

With this amount of excess water the quantity of pesticide leached downward is

$$3.2\,(\text{cm}) \times 0.01\,(\text{m/cm}) \times 0.726\,(\text{g/m}^3) \times 10^4\,(\text{m}^2/\text{ha}) = 232\,\text{g/ha}.$$

For this particular scenario, more pesticide is leached into ground water than is transported from the field by soil erosion. The pesticide has a potential to severely contaminate ground-water resources.

A simple pesticide load model was proposed by Haith and Tubbs

(1981) and was incorporated into the EPA loading model routines for diffuse pollution (Mills et al., 1985).

Polychlorinated Biphenyls

Polychlorinated biphenyls (PCBs) are man-made chemicals that are alien to nature, and as with most of the previous types of chemicals (rare exceptions are some PAHs), no natural-background concentrations in soils and sediments exist. Most of the environmental mass of PCBs is confined to industrial and urban areas; however, PCB contamination of arctic glaciers has been measured. Many freshwater and aquatic sediments have been heavily contaminated by these compounds (the New Bedford, Massachusetts, and Waukeagen, Illinois harbors, and Green Bay of Lake Michigan are examples of such environmental damage). The occurrence and distribution of PCBs in aquatic sediments indicate regional variations (National Academy of Sciences, 1979). The highest level of PCB contamination of sediments is found in heavily industrialized populous areas, particularly in the eastern part of the United States. The sources of these contaminations were mostly traced to past industrial operations. Contamination of rural soils and sediments draining mostly rural watersheds is relatively low and below the detection limit. Lake Ontario and Lake Erie of the Great Lakes system exhibit the highest contamination of sediments by PCBs, while Lake Superior has the lowest. It is suspected that the major source of PCB input to the Great Lakes is atmospheric deposition.

PCBs have very low solubility, consequently, their octanol partition coefficients are high; typically, K_{ow} would range between 10^4 and 10^6 l/kg (see Table A1 in the Appendix). PCBs are almost conservative components that are not readily decomposed by soil and sediment bacteria, and again biodegradability is inversely correlated to the chlorine content of the PCB molecule. Their removal from soils is primarily by volatilization and biomodification of lower PCBs (Hague, Schmedding, and Freed, 1974).

Toxic Metals

Most natural waters, and even more so, sediments, have a natural capacity to reduce the toxicity of added metals that is attributed to the presence of ligands in waters, to adsorption, and precipitation (Salomons and Förstner, 1984). As it is for adsorbable organic chemicals, metallic toxicity of contaminated sediments is affected by the metal activity in the pore water of the sediment.

When toxic metals are added to water both from natural and man-made sources, they undergo complexation with ligands. Ligands are chemical constituents that are both organic and inorganic that combine with the metals in a chemical complex. From the basic physical chemistry it is known that metals precipitate because of changes in pH or when oxidized or otherwise change their chemical composition. However, the process of complexation in natural waters is more complex. Major causes for precipitation and metal complexation are (Salomons and Förstner, 1984):

1. Oxidation of reduced components such as iron, manganese, and sulfides
2. Reduction of higher valency metals by interaction with organic matter (selenium, silver)
3. Reduction of sulfate to sulfide (iron, copper, silver, zinc, mercury, nickel, arsenic, and selenium are precipitated as metal sulfides)
4. Alkaline-type reactions (strontium, manganese, iron, zinc, cadmium, and other elements are precipitated by increased pH, usually caused by interaction with alkaline rocks and sediments or by mixing with alkaline waters)
5. Adsorption or coprecipitation of metallic ions with iron and manganese oxides, clays, and particulate organic matter
6. Ion-exchange reactions, primarily with clays.

In aerobic freshwater sediments the sorption sites are provided by organic carbon, clays, and hydrous oxides of iron and manganese. The Fe and Mn oxides also have limited ion exchange capabilities. Hydrous iron oxides strongly adsorb chromium, while manganese oxides adsorb nickel, and calcium phosphate (also present in sediments) adsorbs cadmium, lead, and other metals. Mercury in sediments (in sediments mercury exists mostly as methyl mercury) is strongly adsorbed by organic matter (Langston, 1985). Oxides of iron and manganese are deemed to be more important than organic matter and clays; however, Combest (1991) documented that the Fe and Mn contents correlate with the clay content. Similarly, Summerhayes, Ellis, and Stoffers (1985) analyzed sediment samples of the New Bedford, Massachusetts, harbor, and found that clay fractions of the sediments were enriched 400 times for silver, 250 times for cadmium, 140 times for copper, 90 times for zinc, 45 times for lead, 40 times for mercury, 32 times for chromium, 7 times for arsenic, and 6 times for nickel when compared to the bulk sediment concentrations. The concentrations of chromium, nickel, lead, and copper correlated strongly with the clay fraction, with the coefficients of correlation ranging from 0.77 for copper to 0.92 for nickel (Moore, 1963). In anaerobic sediments sulfide

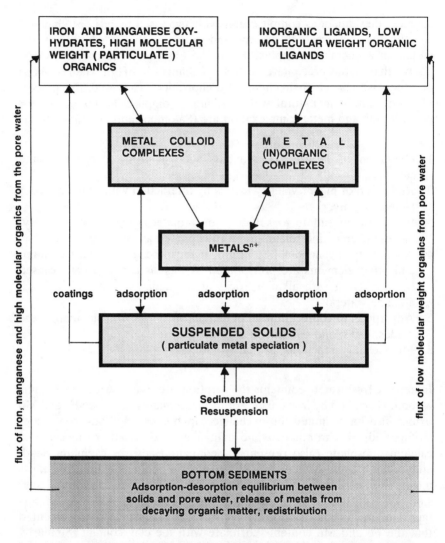

FIGURE 6.11. Major processes and mechanisms in the interaction between dissolved and solid metal species in sediments. (After Salomons and Förstner, 1984, with permission.)

precipitation is important in complexation of toxic metals (DiToro et al., 1989).

It has been found that the free metal ion is the most toxic component for organisms (Gillespie and Vaccaro, 1978; Salomons and Förstner, 1984; DiToro et al., 1989). As a matter of fact Gillespie and Vaccaro and others used a bioassay test with bacteria to detect and measure the complexing capacity and other parameters of the distribution of metals in natural waters. Exchangeable metallic ions (they adsorbed on organic and inorganic complexes, such as humic and clay particles) are far less available due to the ion exchange between the particles and organisms bonding surfaces (Fig. 6.11). Other metal complexes are not bioavailable. Hence, the presence of compounds that will react with the metal ion and cause its precipitation or adsorption on solids (including iron and manganese oxides, sulfides, clay, and organic matter) will reduce the toxicity of the metals and make them less bioavailable for organisms. The colloidal and ionic compounds that combine within a complex with the metal ion (*ligands*) include organic acids and humic substances, dissolved sulfides, chloride, and OH⁻ ions.

Hence, the adsorbing and complexing compounds for toxic metals include

1. Particulates: sulfides, iron and manganese oxyhydrates, particulate organic matter, clays
2. Dissolved: sulfides, humic compounds, organic acids, chloride ion, hydroxyl ion

Dissolved metal–organic (ligand) complexes may also be adsorbed by particulates such as iron and manganese oxyhydrates. On the other hand, some dissolved organic compounds, such as detergents, may reduce the adsorptivity of metals, while chlorides may enhance adsorptivity. From the preceding mechanisms, iron and manganese oxyhydrates provide the strongest adsorption sites, followed by particulate organics and clays.

When metals are present in water the metal ions are distributed with the various complexing ligands (Fig. 6.11) and solids. Under equilibrium conditions the total molar concentration of metals in sediment and (pore) water is (Hart, 1981; Salomon and Förstner, 1984; DiToro et al., 1989)

$$[M_T] = (1 - \phi)\Sigma[M \equiv P_i] + \phi\{[M^{n+}] + \Sigma[M \equiv L_i]\} \qquad (6.23)$$

where
$[M_T]$ = total (molar) concentration of the metal in the sediment volume
$[M^{n+}]$ = concentration of the free metal ion (toxic)
$[M \equiv P_i]$ = inorganic and organic complexes of the metal adsorbed on the sediment (generally biounavailable)
$[M \equiv L_j]$ = organic and inorganic ligand complexes (less toxic or nonbioavailable)
ϕ = porosity

Assuming that in the aerobic soil environment the primary solids interacting with the metals are iron and manganese oxy-hydrates, particulate organic matter, and clays, the following chemical equilibria and isotherms may be introduced:

$$[M \equiv FeO_x] = K_{Fe}[M^{n+}][FeO_x]$$
$$[M \equiv MnO_x] = K_{Mn}[M^{n+}][MnO_x]$$
$$[M \equiv POC] = K_{oc}[M^{n+}]m_{oc}$$
$$[M \equiv clay] = K_{clay}[M^{n+}]m_{clay}$$

where
K = chemical adsorption equilibria or partitioning constants
$[M \equiv FeO_x], [M \equiv MnO_x], [M \equiv POC], [M \equiv clay]$ = metal adsorbed on iron manganese oxyhydrates, organic particulate matter, and clay
m_{oc} = organic carbon portion of the sedimentt
m_{clay} = clay portion of the sediment

DiToro et al. (1989) reported that in sediments the concentration of metal–ligand complexes in pore water is negligible when compared to that adsorbed on the sediments. Then neglecting the ligand–metal concentrations in pore water, the pore water free metal concentration in aerobic sediments or soils becomes

$$M^{n+} \approx \frac{M_s}{K_{Fe}[FeO_x] + K_{Mn}[MnO_x] + m_{ss}(K_{oc}f_{oc} + K_{clay}f_{clay})} \quad (6.24)$$

where
$M_s = M_T/(1 - \phi)$ = total sorbed metal per unit dry weight of sediment
f_{oc} = fraction of the sediment that is organic carbon
f_{clay} = clay fraction of the sediment
m_{ss} = sediment dry weight

In anaerobic sediments, iron and manganese oxides are reduced; however, sulfides become the prime complexing ligands. Hence K_S--$[S^{--}]$ will replace oxide terms in Equation (6.24).

It can be seen that Equation (6.24), which expresses the relationship between the dissociated and precipitated–complexed metal, is similar to the partition relationship between dissolved and adsorbed organic chemicals (Eqs. (6.8) and (6.12)). Hence the denominator of Equation (6.24) could be called a "partition coefficient" for metals, or

$$\Pi_m = K_{\text{Fe}}[\text{FeO}_x] + K_{\text{Mn}}[\text{MnO}_x] + m_{ss}(K_{oc}f_{oc} + K_{\text{clay}}f_{\text{clay}}) \quad (6.25)$$

and similarly to nonpolar organic chemicals $M_s = \Pi_m M^{n+}$

The values of partition constants for various metals and soil/sediment characteristics are given in Table A2 in the Appendix.

From the reactions just listed, the interactions with particulate matter in the water column and, above all, in sediments are most important in determining the toxicity of metals in sediments and water.

The adsorption reactions are strongly pH-dependent, ranging from zero adsorption in low pH to 100% adsorption in higher pH. The change from no adsorption to full adsorption is fairly sharp, usually within two pH units (Salomons and Förstner, 1984). However, at pH > 7 most of metals are complexed, while at a pH of less than 5 the concentrations of free metal ions increase dramatically. This manifests the adverse impact of acid rain and acidity on leaching metals from soils and sediments into the aqueous phase. The leaching of aluminum from acidified soils reduces soil fertility. As pointed out in Stigliani (1991) and Stigliani et al. (1991), in some parts of the world (for example, central Europe) aluminum loss from soil may be a sudden and damaging "time bomb."

The impact of pH can be modeled by including the species MOH^+ in the mass balance equation. The result is that the sorption concentrations are replaced by the following expression (DiToro et al., 1989):

$$K_{\text{Fe}} = K_{\text{Fe}}^{(1)} + K_{\text{Fe}}^{(2)}[\text{H}^+]^{-1} \quad (6.26)$$

In natural waters under oxic condition, iron and manganese oxyhydrates can be present in crystalline forms or as a coating on rocks, sediments, and other solids. Essentially, FE/Mn oxyhydrates are abundant and present in all parts of the hydrological cycle.

Sulfides, mostly particulate, are also common in sediments and have been found even in sediments with sand and gravel texture that do not resemble the anoxic sulfitic sediments (DiToro et al., 1990). A measure of solid or colloidal sulfides that can combine with metals is *acid volatile sulfides (AVS)*, which is sulfide extracted from the sediment by 0.5M HCl in one hour at room temperature. Metals are also extracted by the same treatment of the sediment with acid. These tests provide a quick and inexpensive way of establishing the gross degree by which the sediments are polluted with metals (Salomon and Förstner, 1984; DiToro et al., 1990). The molar concentration of the simultaneously extracted metals (SEM) by hydrochloric acid can then be compared with the molar concentration of the AVS. If the metal toxicity unit (MTU) = SEM/AVS ratio is less than one, then the metals in the sediment are tied in the sulfide complexes and are not bioavailable, and hence are not toxic. Measured AVS concentration ranged from <0.1 mole/g of sediment to >50 mole AVS/g of sediment (literature sources quoted in DiToro et al., 1990). Acid volatile sulfides are typically found in submerged sediments. When the sediments are exposed, the sulfides are oxidized and may not be available for complexation of metals. Iron and manganese oxyhydrates and organic matter then become the major complexing ligands.

Organic adsorbing particulates are also an essential part of terrestrial waters. Research investigations on reducing (anaerobic) sediments have shown that all of the zinc, almost all of the copper, and most of the nickel and cobalt are bound to humic-type materials (Nissenbaum and Swaine, 1976).

Example 6.5: Metal Distribution in Sediments

Sediment was analyzed by the acid extraction method (metals and sulfides were extracted by 0.5 M hydrochloric acid at room temperature in one hour). The results of the test are given in Table 6.5. Make a judgment whether metal is mobile (precipitated–complexed).

Solution From the table we see that $MTU = 2.2473/5 = 0.45 < 1.0$; therefore, the metals are immobile (complexed).

Methylation of metals. Mercury, arsenic and a few other less important metals undergo biological methylation in the anaerobic sediments. The conversion of inorganic mercury by bacteria residing in the sediments into organic methyl or dimethyl compounds is well known and understood. The case has been widely publicized by the famous Minamata Bay disaster

TABLE 6.5 Test Results

Acid Volatile Sulfides		5-μmol AVS/g of (Dry) Sediment	
Metals	Atomic Weight	Dry Weight Concentration (μg/g)	Molar Concentration (μg/g)
Cadmium	112.4	1.0	0.0089
Copper	63.4	50.0	0.7886
Mercury	200.6	0.1	0.0005
Lead	207.0	300.0	1.4493
Totals		351.1	SEM = 2.2473

sickness in Japan, where people were stricken by the symptoms after ingesting fish contaminated by methylmercury. Later incidents occurred in Niagara Falls (New York) with similar debilitating effects (Krenkel, 1973).

Organic complexed compounds are far more toxic than their inorganic counterparts. This is reflected in current water quality standards, although quality criteria for sediments were evolving at the time the manuscript of this book was being prepared. Methylated compounds are still strongly polar and behave in sediments similarly to free metal ions, that is, they are adsorbed by the sediments and the ligands in the sediment.

Specific Metals

Arsenic (As). Inorganic arsenic compounds have been used in agriculture as pesticides and defoliants for many years. Also lake managers have used arsenic salts to control algal and macrophyte growths. Due to the serious environmental problems caused by these applications, their use was banned in the United States in 1967.

Once incorporated into soils or sediments, As reverts to arsenate, which is strongly held by the clay fraction (U.S. EPA, 1976). Arsenate also forms insoluble salts with Mn^{2+}, Ni^{2+}, and other alkaline cations.

Arsenic organic complexation by bacteria is relatively complex; however, Braman and Foreback (1973) found methylated forms of arsenic in a wide range of surface-water bodies. In anaerobic sediments arsenic is complexed by bacteria to methyl- and dimethylarsenic acids, which in interstitial and overlying water are converted to di- and trimethylarsines (Salomons and Förstner, 1984). These acids can be biologically synthesized. Dimethyl arsenic acid is difficult to oxidize and may contribute significantly to dissolved arsenic concentrations in surface waters.

Desorption of complexed arsenic back to solution in water greatly depends on the strength of the reducing environment (typically expressed as the oxydation-reduction potential). Anaerobic sediments release 10 times more toxic arsenite A^{3+} than aerobic sediment layers (Moore and Ramamoorthy, 1984).

Cadmium (Cd). The divalent cadmium ion Cd^{2+} is the predominant species of cadmium in surface waters and interstitial waters of sediments and soils. The mechanisms by which cadmium (Cd^{2+}) is removed from water are precipitation and adsorption and chemosorption on the surface of solids. Partition coefficients for cadmium in soils were measured by Christenson (1984). The partition coefficient for organic river sediments varied with types of solids and concentration of complexing agents. Organic materials such as humates were the main components of the sediment samples responsible for adsorption of cadmium.

DiToro et al. (1990) pointed out that in anaerobic sediments sulfides are responsible for most of the complexation of cadmium, and the acid volatile sulfides test provides a measure of the immobilization of cadmium.

Adsorption increases with pH. At pH > 7, virtually all cadmium is complexed, with humic acid being the primary ligand. Although cadmium is a rare element the median concentrations in sediments of U.S. surface-water bodies is near 2 mg/kg, which is attributed mostly to anthropogenic activities.

Chromium. In a water environment chromium is stable in six different ionic species. Between pH 5 and pH 9 $Cr(OH)^{2-}$ and $Cr(OH)_2^-$ predominate. Cr^{2+} species are rapidly oxidized by manganese oxyhydrates and slowly by oxygen. More toxic Cr^{6+} forms (originating primarily from industries) are easily reduced in natural waters and sediments by ferrous and sulfide compounds. Chromium is transported in water primarily by sediments.

Copper. The dissolved phase of copper in water includes a free ion and complexes with organic and inorganic ligands. Humic materials bind most of the total Cu. Fulvic acid forms insoluble complex with Cu^{2+}. Inorganic complexes include carbonates, nitrates, sulfides, chlorides, ammonia, and hydroxide. Copper hydroxycarbonates are slightly soluble. Coprecipitation with ferric oxyhydroxides and other oxides and adsorption on mineral (clay) surfaces are responsible for the immobilization of copper in soils and sediments. Copper is transported primarily in the solid phase. Unpolluted sediments generally contain less than 20 mg/kg of Cu.

Lead (Pb). Lead is an abundant, naturally occurring metal that exists in four stable isotopes. The accepted average value for the Pb content of the earth's crust is 15 μg/g, but ranges of 0.8 to 500 μg/g have been reported for arable soils.

In addition to natural background concentrations and the use of leaded gasoline by automobiles, sources of Pb include atmospheric deposition, mining and smelting, and some insecticides (for example, lead arsenate, which is now banned in the United States).

The common dissolved inorganic forms of lead are the free ion, Pb^{2+}, hydroxide complexes, and possibly carbonate and sulfide pairs. At pH > 6 lead is almost totally precipitated or sorbed, hence, lead is immobile in soils and sediments; however, Zimdahl and Hasset (1977) reported that the adsorption capacity is related to the type of complexing minerals and pH. The high degree of Pb immobilization may be related directly to the soil cation exchange capacity (CEC) and inorganic matter content and inversely to pH. A regression analysis yielded the following equation for the Langmuir adsorption maximum, Q^o, in micrograms of lead per gram of soil,

$$Q^o = 7100 + 1110 \, (\text{pH}) + 1105 \, (\text{CEC}) \qquad (6.27)$$

Lead forms organic complexes with various ligands, including amino acids, polysacharides, and fulvic and humic acids. Using data compiled by Schnoor et al. (1987), the partitioning coefficient for adsorption of lead by soil organic humic acids (HA) is about $\Pi_{HA} = 10^5 \, \text{l/kg}$, and that for clays ranges from $10^{3.7} \, \text{l/kg}$ for kaolin to $10^{5.5}$ for montmorillonite, respectively.

The transport of Pb from soil to receiving waters occurs almost solely by erosion, with delivery ratios similar to those by clay particles. For all practical purposes, Pb remains only in the few centimeters of soil unless redistributed by plowing.

Mercury (Hg). The background concentrations of mercury in soils range from 0.01 to 0.06 µg/g (Krenkel, 1973; U.S. EPA, 1976). In soils, Hg combines with the exchange complex, forming ionic and covalent bonds. Mercury is strongly adsorbed by clay minerals, and the adsorption process is affected by soil pH. Evidence exists that mercury can be chelated to organic matter. Unlike other metals, mercury may also volatilize.

In sediments, mercury is methylated by bacteria. The final product of microbial methylation of mercury is the dimethyl mercury compound $(CH_3)_2Hg$ (Jernelov, 1972; Krenkel, 1973); however, no methylation has been reported in soils.

Zinc. In natural waters zinc species are limited to zinc carbonates, zinc hydroxides, zinc silicates, and the free ion. Zinc also complexes with ammonia, amines, halides, and cyanides. Zinc is more soluble than other toxic metals such as copper or nickel, but its solubility is affected by pH. Acid rainfall may greatly increase dissolution of zinc from soils and from zinc roofing materials.

Nitrogen

Accumulation of Nitrogen in Soils

Nitrogen (N) is one of the four essential elements (carbon, oxygen, hydrogen, and nitrogen) that form the basic structure of proteins. Nitrogen is also the most abundant gas in the atmosphere, accounting for about 80%. Due to its several valence states it can exist in numerous forms and is abundant in the mineral and organic factions of the soils and sediments. However, not all N forms are related to water pollution or eutrophication of water bodies.

Soils contain 0.07 to about 0.3 of the total N or about 1500 to 6000 kg/ha in the top 15 cm. The source of soil nitrogen include fertilizer applications (46% in the United States), N fixation from the atmosphere by soil bacteria (20%), manure application (7%), plant residues (17%), and precipitation (10%). In suburban areas without sanitary sewers, significant amounts of N enter soils from the seepage of household septic systems.

In the last decades of the twentieth century, nitrogen contamination of surface and ground water resources has reached alarming proportions (see Chapter 1 for a discussion of nitrate trends). Excess nitrogen fertilizer applications in agriculture and from septic tanks have leached into the ground water, with the result that nitrate–nitrogen loads and consequent concentrations of nitrate nitrogen are steadily increasing. The result is an impairment of the use of water for drinking and the acceleration of eutrophication.

Jenny (1941) noticed that the nitrogen content of native soils depends on the average meteorological factors of the region, namely moisture and temperature. For example, soil nitrogen declines as the temperature rises or moisture declines.

Although virgin lands rich with vegetation (forested areas, swamps, prairies) originally contained high amounts of N, on cultivation the natural supply of N is depleted and fertilization may be needed to supplement soil loss of N. Most soil N (>95%) is contained is soil organic matter, or in the case of ammonium ions, it can be sorbed by clays. In these forms, N may be considered immobile and not available to plants. Nitrogen is mostly lost by erosion and through crop harvesting. However, these immobile forms of nitrogen can be converted to nitrate, which is highly mobile. Also part of the ammonia may exist in soil water as free ammonia ions (NH_4^+), which is mobile. The mobile forms of nitrogen are available to plants and can be transported by soil water and infiltrate into ground water. The time of migration of mobile N components in ground-water aquifers can extend to years and even decades. It is also suspected that the heavy contamination of central Europe by nitrates was caused by nitrate fertilizers instead of the less mobile ammoniacal forms.

Nitrogen Cycle

The behavior and transformation of N in soils, sediments of surface waters, and substrates of wetlands is complex, and the pathways from the soil to surface waters are numerous and not well defined. The process of N transformation in these media can be schematically represented by a N cycle similar to that in Figure 6.12.

In soils and sediments, N exists in four basic forms: ammonium, nitrate, organic phytonitrogen in plants and plant residues, and protein nitrogen in living and dead bacteria and small soil inhabitants. The

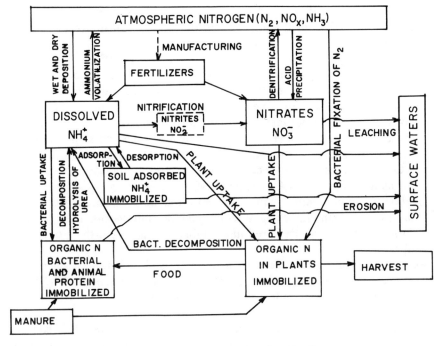

FIGURE 6.12. Nitrogen cycle in soils and sediments.

following definitions are needed for an overall understanding of the N cycle:

Nitrogen fixation is a process by which soil microorganisms in symbiosis with some leguminous plants utilize atmospheric N and change it to an organic form.

Nitrogen accumulation (bacterial uptake) is the conversion of ammonium–N to protein and cell tissue by heterotrophic soil organisms.

Ammonification is a process by which protein and other inorganic forms of N are decomposed to ammonium by a biochemical breakdown of the proteins.

Hydrolysis of urea involves conversion to ammonium ions in the presence of the enzyme urease, which is provided by many heterotrophic organisms.

Nitrification is a complex process by which ammonium–N (NH_4^+) is oxidized to nitrate (NO_3^-) with nitrite (NO_2^-) as an intermediate product. Nitrification is accomplished by two groups of bacteria: namely *Nitrosomonas* (oxidizing NH_4^+–N to NO_2^-–N) and *Nitrobacter* (oxidizing NO_2^-–N to NO_3^-–N). Both groups of bacteria are chemotrophic, utilizing the exothermic reaction as their source of energy and carbon dioxide from the atmosphere or alkalinity of interstitial water as their carbon source. Nitrification is strictly an aerobic process.

Denitrification is a process that occurs under anoxic conditions that usually takes place when water fills most of the available voids and pores in soils. In aquatic sediments and substrates of wetlands anoxic conditions always prevail, so only a very few millimeters thick surface sediment layer can become aerobic. In wetlands rhizomes of some plants have the capability of transferring oxygen down to their roots and creating aerobic pockets near the roots (see Chapter 14). During denitrification $NO_3^- - N$ serves as an electron acceptor and is reduced to gaseous forms including N_2, N_2O, NO, and NO_2.

Fixation of ammonium involves the sorption of NH_4^+ in between the layers of expanding clay minerals such as monmorillonite. In this form, ammonia is considered unavailable for plant growth or bacterial uptake.

Ammonia volatilization may occur at high soil pH values when the ammonium ion (NH_4^+) is converted to gaseous ammonium (NH_3), which volatilizes and is lost to the atmosphere.

Most of the reactions of the N cycle (with the exception of NH_4^+ fixation and NH_3 volatilization) are microbial, and thus their rates are sensitive to temperature, moisture content, and aeration. Warm (32°C) and partially moist soils (80% of voids filled with water) are the optimum conditions for N cycling in soils (Steward et al., 1975).

Unlike phosphorus, most of which is adsorbed on soil particles and may be controlled by erosion control and soil conservation, the available N is mobile in soils and may leach to ground water and reappear with ground-water discharge in the base flow of streams. Seepage represents the major pathway of nitrogen loss from agricultural areas and pervious urban and suburban lands. Erosion control may not be effective for reducing nitrogen contamination of surface and ground-water resources.

Since the mobile components—NH_4^+ and NO_3^-—are carried by soil water, downward or lateral N movement occurs only if the moisture content is above that for gravitational water (0.3 bar tension). Maximum movement occurs when soil moisture content is near saturation, and decreases rapidly with decreasing moisture content. Thus, between rains or irrigation, N movement is slow or nonexistent.

Nitrification and Denitrification

Over 90% of the fertilizer used in the United States is in the form of ammonium salts. Most organic N originating from manure application or from septic tanks can be quickly decomposed to ammonium. In addition, urea is readily hydrolyzed to ammonium. If the ammonium is applied to an aerated, microorganism-rich soil such as farmland, nitrification occurs, resulting in the conversion of NH_4^+ to NO_3^-. In aerated soils or in the top layer of sediments, the nitrification process proceeds as follows:

$$\text{Organic N} \rightarrow NH_4^+ \xrightarrow[O_2]{\text{Nitrosomonas}} NO_2^- \xrightarrow[O_2]{\text{Nitrobacter}} NO_3^-$$

The last reaction, that is, conversion of NO_2^- to NO_3^- is faster than the conversion of NH_4^+ to NO_2^-. Consequently, very little nitrite accumulates in soils and sediments.

Nitrates can readily move with the soil moisture front. The optimum temperature for nitrification is 22°C (Fig. 6.13), and the rate of nitrification decreases rapidly on both sides of the temperature curve. Standford, Free, and Swaininger (1973) stated that nitrification essentially ceases below 10°C and above 45°C. Nitrification also depends on the pH of the soil and its moisture content (Figs. 6.14 and 6.15). Since nitrifying bacteria depend on water as their living environment, the nitrification rate decreases with decreasing moisture content.

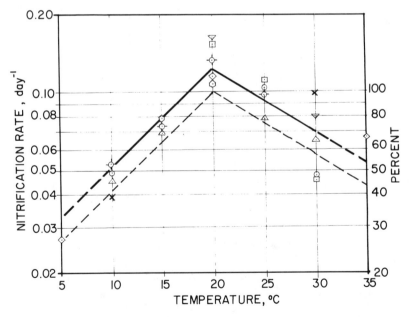

FIGURE 6.13. Dependence of nitrification rate on temperature. (From Zanoni, 1969.)

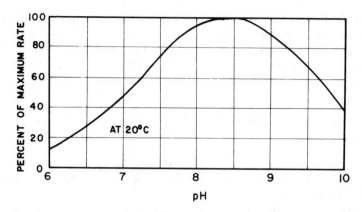

FIGURE 6.14. The effect of pH on the rate of nitrification. (From Wilde, Sawyer, and McMahon, 1971.)

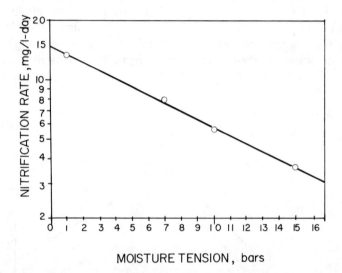

FIGURE 6.15. The dependence of the soil nitrification rate on the moisture tension in soils. For submerged sediments, moisture tension = 0. (After Justice and Smith, 1962.)

By contrast with many heterotrophic microorganisms, the growth of nitrifiers is slow and cell yield per unit of energy source oxidized is low. For practical purposes the NH_4^+ oxidation may be simulated by an equation similar to the Michaelis–Menten–Monod equation (Eq. (6.14)). The value of the rate coefficient for soil at 20°C was suggested as $K_N = \mu X = 0.41$ mg/l-day and the half-saturation constant was $K_s = 1.1$ mg/l, respectively, if the concentration of dissolved ammonia in pore water, C_d, was expressed in milligrams per liter (McLaren, 1969, 1971).

In the absence of oxygen or if the oxygen supply is depleted to a point below the oxygen demand, NO_3^- is reduced mostly to gaseous nitrogenous forms (largely N_2) by denitrification. Denitrifying bacteria are heterotrophic facultative microorganisms, a part of the soil microflora.

The process of denitrification usually occurs in subsoils with low permeability, in soils saturated with water for an extended period, such as in wetlands or after prolonged rainfall. Pratt, Jones, and Hunsaker (1972) reported that denitrification might account for more than 50 percent of N losses in clayey soils. On the other hand, Carter and Allison (1960) concluded that very little, if any, N loss occurred under aerobic conditions, except where higher dosages of dextrose or other carbon-rich compounds were added in the presence of nitrates. Hence they stated that denitrification is of minor importance in soils that are maintained aerobic.

Generally the first-order decay reaction (Eq. (6.15)) is used to describe the reaction where the nitrate concentration in the interstitial water is substituted for C_d. The magnitude of the denitrification coefficient, K_b,

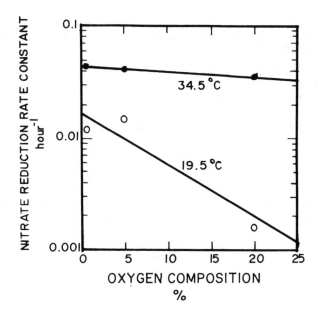

FIGURE 6.16. Dependence of denitrification rate on oxygen composition and temperature. (From Misra et al., 1974. *Reprinted with permission of the publisher, Soil Science Society of America, Madison, WI.*)

measured by Misra, Nielsen, and Biggar (1974) is shown on Figure 6.16. The denitrification rate is a function of easily decomposable carbon that is derived from the organic carbon sources in the soil/sediment. The rate is expected to decrease with the decrease in the available energy source.

Nitrogen Fixation

Legumes (soybean, peas, beans), as well as some aquatic algae, fix nitrogen from the atmosphere by symbiotic microorganisms. It has been assumed that in soils the rate of nitrogen fixation can be related to root growth (Duffy et al., 1975)

$$N_f = K_f r_g \qquad (6.28)$$

where
N_f = rate of fixation of N (mg N/day-cm^2)
K_f = a constant (0.011 mg N/cm^2)
r_g = the rate of root growth (cm/day)

Ammonification of Nitrogen

Painter (1970) described three ways of producing NH_4^+ from organically bound N:

1. From extracellular organic N compounds (e.g., urea) chemically or biochemically.
2. From living bacterial cells during endogenous respiration when cells are becoming smaller.
3. From dead and lysed cells.

Very few data are available on the breakdown of various N containing components to NH_4^+.

Hydrolysis of urea is relatively fast, much faster than the rate of nitrification. Its rate is also temperature dependent. The process of hydrolysis of urea is important where manure or urea-containing sewage or septage is applied to field or land. Soil organic N—other than urea—breaks down at a much slower rate. Most of the organic N is not directly available to plants and must be initially converted to NH_4^+. Straford, Free, and Swaininger (1973) concluded that in the mineralization of soil organic N other than urea, the rate-limiting process is ammonification. Ammonification can be described by the first-order decay equation (Eq. (6.15)); however, with the remaining soil N as substrate, or

$$\frac{dN}{dt} = -K_N N \qquad (6.29)$$

Straford, Free, and Swaininger (1973) found that the coefficient, K_N, is statistically uniform for all investigated soils, that is, $K_N = 0.29 \pm 0.007$ weeks^{-1} at 25°C. As with all biochemical reactions, the coefficient, K_N, is temperature dependent. An Arrhenius' plot yielded the following equation for the overall mineralization of organic N:

$$\log K_N = 6.16 - \frac{2299}{T} \qquad (6.30)$$

where T = temperature in degrees Kelvin (= 273 + °C)

Nitrogen Immobilization

Part of the soil or sediment N is dissolved and can move readily with soil moisture and ground water. The dissolved fractions include NO_3^- and NH_4^+. But a significant portion of n components, especially NH_4^+ and almost all of organic N, can be immobilized. On the other hand, NO_3^- is always dissolved and mobile.

Nitrogen immobilization in soils results from physical–chemical attractions, chemical precipitation, biochemical reactions, and nitrogen uptake by soil organisms and plants. The known nitrogen fixation processes include:

1. Ammonium fixation by clay minerals.
2. Ammonium fixation by lignin-derived substances contained in soil organic matter.
3. Reactions of amino acids derived from plant materials and microbial synthesis with quinones and subsequent polymerization.
4. Biological immobilization, which involves NH_4^+ uptake by heterotrophic bacteria participating in the decay of organic matter in soils.

The immobilized N is not readily available for mineralization and biological uptake, and it appears, as pointed out in the preceding paragraph, that ammonification is much slower than nitrification.

The processes of immobilization and ammonification of soil N depend on the chemical composition of the material undergoing decomposition, primarily its C:N ratio. Plant residues having larger percentage of readily available C stimulate the growth of microbial cells when incorporated in soil under aerobic conditions. Since nitrification is faster than ammonification, the end product of aerobic decomposition is nitrate. Anaerobic degradation of protenaceous materials, such as those occurring in subsurface sediment layers, is slower and results in the accumulation of ammonia.

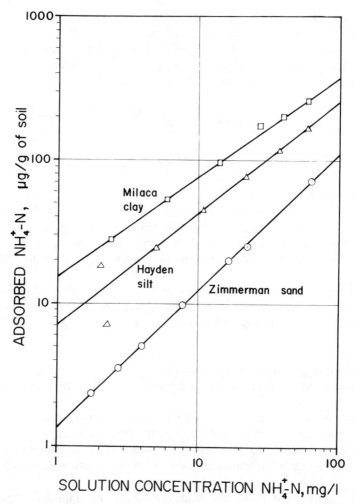

FIGURE 6.17. Freundlich adsorption isotherms for adsorption of ammonia on soils. (From Preul and Schoepfer, 1968.)

Shaffer, Dutt, and Moore (1969) statistically analyzed the rate of NH_4^+ immobilization using the multiple regression technique. The proposed equation was

$$R = \frac{d(NH_4^+)}{dt} = 0.892 - 0.00216T - 0.027 \text{ (organic N)} + 0.392 \log_{10}(NH_4^+) \tag{6.31}$$

where T is the temperature in °C.

Preul and Schoepfer (1968) investigated N adsorption by soils. The Freundlich adsorption isotherms for three soils are shown in Figure 6.17. Bailey (1968) has shown that ammonium is adsorbed by clays in exchangeable form and by organic matter. Much of the NH_4^+ adsorbed by various kinds of organic matter is in a nonexchangeable form and is resistant to decomposition.

Volatilization of Ammonia

For soils of high pH and NH_4^+ content, ammonia (NH_3) can be lost by volatilization. Significant amounts of ammoniacal N (11% to 60%) can be lost from applied sludge, manure, or chemical fertilizers applied to calcareous soils (Ryan and Keeney, 1975; Terry et al., 1978). Losses decrease as clay content increases and more NH_4^+ is immobilized. Since the reaction is pH dependent, all volatilization practically ceases if pH is at or below 7. Note that soil pH is inversely correlated to the soil clay content. Other factors affecting volatilization include moisture content, application rates, and organic content. Due to its effect on pH, liming of soils may enhance ammonium volatilization.

Enrichment and Delivery of Nitrogen during Overland Flow

Due to the fact that N exists in soils in several forms, the notion of enrichment and delivery may be meaningless.

Example 6.6: Distribution between Mobile and Immobile N in Soil

Silt soil contains about 2.5% organic matter and 0.12% total N measured as TKN (total Kjeldahl nitrogen = organic N + NH_4^+ − N), of which about 75% is organic N. Estimate the approximate magnitudes of mobile (available) and immobile N. Effective soil moisture $\theta_{eff} = 20\%$ and the specific density of the soil $\rho = 1500 \, kg/m^3 \, (= g/l)$.

Solution From Figure 6.17, the Freundlich isotherm for silt has the approximate form

$$r = 7.0 \, C_d^{0.8}$$

Organic N constitutes part of the soil organic matter, and for the most part no equilibrium exists between the dissolved and particulate fractions. The dissolved fraction of TKN is a part of the NH_4^+ content, since ammonia can exist both dissolved and adsorbed. Since total NH_4^+ is $(1 - 0.75) \times 0.12\% = 0.03\%$ or $300 \, \mu g/g$ of soil, the volumetric concentration is $300 \, \mu g/g \times 1500 \, g/l \times 0.001 \, mg/\mu g = 450 \, mg/l$. Then

$$450\,(\text{mg/l}) = \rho r + \theta_{\text{eff}} C_d = 1500\,(\text{g/l}) \times 7.0 \times C_d^{0.8}$$
$$\times\, 0.001\,(\text{mg/}\mu\text{g}) + 0.2 \times C_d$$

Solving this equation gives $C_d = 103.5\,\text{mg/l}$. Also the adsorbed ammonia concentration is $r = 7.0 \times 103.5^{0.8} = 286\,\mu\text{g/g}$.

The ratio of available NH_4^+ to TKN is

$$100 \times \{103.5\,[\text{mg/l}] \times 0.2/(0.0012 \times 1500\,[\text{g/l}] \times 1000\,[\text{mg/g}]) = 1.15\%$$

Example 6.7: Nitrogen Loss Estimation during a Storm

A 3-cm storm resulted in 1 cm of surface runoff and 1.3 tonnes/ha of soil loss. Estimate N loss using the N distribution computed in the previous example. Porosity of the soil is $\rho = 45\%$ and $\theta_{15\text{bar}} = 10\%$. Assume that the enrichment ratio for organic nitrogen is the same as that for organic matter. For this particular example assume that $ER_{OR} = 1.2$.

Solution Recall that organic N is 75% of the TKN content and that the adsorbed ammonia is $286\,\mu\text{g/g}$ (= g/tonne). The total TKN content is $0.12\% = 1200\,\text{g/tonne}$, θ_{eff} is 0.2, and dissolved NH_4^+ is $103.5\,\text{mg/l}$. Then the fixed N loss

$$Y_{SN} = (0.75 \times 1200\,[\text{g/tonne}] + 286\,[\text{g/tonne}]) \times 1.3\,[\text{tonnes/ha}] \times 1.2$$
$$= 1850\,\text{g/ha}$$

In order to determine the loss of dissolved N, the depth of mixing of the surface runoff with soil moisture containing the dissolved fraction must be known or assumed. Using the same reasoning as in Example 6.4, assume the mixing depth as 5 mm. Then from the mass balance

$$(\theta_{\text{eff}} + \theta_{15\text{bar}}) \times 0.5\,(\text{cm}) \times C_d = \{\text{porosity} \times 0.5\,(\text{cm}) + 3\,(\text{cm})\}C'$$

from where

$$C' = C_d\{(0.2 + 0.1) \times 0.5\}/\{0.45 \times 0.5 + 3\}$$
$$= 103.5 \times 0.0465 = 4.81\,\text{mg/l}\,(= \text{g/m}^3)$$

and the dissolved N loss in the surface runoff is

$$Y_{DN} = 4.81\,(\text{g/m}^3) \times 1\,(\text{cm}) \times 0.01\,(\text{m/cm}) \times 10{,}000\,(\text{m}^2/\text{ha}) = 481\,\text{g/ha}$$

The total N loss will

$$Y_N = Y_{DN} + Y_{SN} = 481 + 1850 = 2331 \text{ g/ha}.$$

Example 6.8: Mineralization of Ammonia

How much nitrogen will be mineralized during a 2-week dry period if the temperature is 22°C and the soil pH is 6.8? Assume average soil moisture tension close to 0.3 bar. Use the N values estimated in the previous examples.

Solution Only dissolved ammonia can be mineralized. Adsorbed or organic N must first be released into the solution before it can be mineralized.

Since the saturation constant for mineralization of ammonia is about $K_s \approx 1.1 \text{ mg/l}$, and the dissolved ammonia concentration is $C_d = 103.5 \text{ mg/l}$, it is evident that the reaction described by the Michaelis–Menton–Monod equation (Eq. (6.14)) will follow a zero-order pattern and will, for most of the time, not depend on the NH_4^+ concentration in the soil water solution. Then

$$\frac{dC_d}{dt} = \frac{K_N C_d}{K_s + C_d} = \frac{0.41 \times 103.5}{1.1 + 103.5} = 0.41 \text{ mg/l day}$$

Since the temperature is near optimum, no thermal correction is necessary as indicated from Figure 6.13. However, Figure 6.14 shows that at pH 6.8 the reaction will proceed only at about 50% of the optimal rate. Because the moisture tension is close to 0.3 bar, there is no correction for moisture. Then

$$\frac{dC_d}{dt} = 0.2 \text{ mg/l-day}$$

and in 14 days about 2.8 mg/l of the dissolved NH_4^+ will be mineralized to NO_3^-. This form as well as the dissolved NH_4^+ can be taken up by plants and move with the soil water front.

Simple Steady-State Loading Function for Nitrate
Loading to Ground Water
Mills et al. (1985) published a simple steady-state loading function that presumes a constant annual load of nitrogen fertilizer or waste water in the form of organic N. The steady-state loading function is

$$L = 1000N \times [1 - v(1 - F)] - Cu \qquad (6.32)$$

where
- L = annual steady-state nitrate–nitrogen load to ground water (kg/ha)
- X = average annual solids waste–organic fertilizer application rate (tonnes/ha)
- N = nitrogen fraction of applied compound
- F = organic fraction of applied compound
- v = fraction of applied nitrogen that volatilizes (a soil-pH-dependent variable)
- Cu = average crop uptake of nitrogen

Table 6.4 provides the values of nitrogen uptake by crops (note that only nitrogen that is removed with crops is considered).

Soil Microorganisms

Soils contain enormous densities of microorganisms that decompose soil organic materials and/or are an integral part of the soil organic composition. Most of the soil microbial population is contained in the A horizon.

Soil contains five major groups of microorganisms (Alexander, 1977): bacteria, actinomycetes, fungi, algae, and protozoa. These organisms as well as their organic and inorganic food are a part of the soil ecosystem.

Bacteria are the most abundant group of microorganisms. They perform most of the decomposing processes in the soils. As pointed out they can be divided into autotrophic (algal) and heterotrophic groups, and based on their oxygen environment, they may be aerobic, anaerobic, or facultative. Based on their origin they can be an autochthonous species, which are the true residents of soils, or allochthonous species or invaders that have entered soils from precipitation or sewage effluents, manure, and sludge applications and from septic tanks. The allochthonous species are the most important in the soil-decomposition processes. The latter group of bacteria (allochthonous) may persist in soils for some time, but do not contribute significantly to any activity of the soil microflora. However, if soil has a low microbial population, application of organic microorganism-rich compounds, such as manure or sewage, may help restore the soil microbial population.

According to Alexander (1977) plate number counts of bacterial densities in soils may range from a few thousand to up to 200 million bacteria per gram of soil. However, microscopic counts have revealed that the microbial densities can be higher, probably up to 10^{10} of microorganisms/gram of soil. The bacterial population represents between 0.015% and 0.05% of the total mass of fertile soils, or some 300 to 3000 kg live weight of soil bacteria per hectare.

Most of the bacteria in soils are facultative heterotrophs. During dry, well-aerated conditions these bacteria decompose organic matter as aerobes. During flooding or periods of high moisture content, the facultative bacteria or strict anaerobes use other compounds as electron acceptors, including NO_3^-, SO_4^{--}, and Fe^{3+} ions.

The final product of aerobic decomposition is CO_2. Thus the atmosphere of aerated soils and exposed sediments during intensive decomposition processes has depleted O_2 and increased CO_2. This results in increased partial pressure of CO_2 in the soil atmosphere, which in the air above ground level is about 1 mbar, while that in soil may reach 100 mbars. Consequently, by the interaction of the CO_2 with interstitial water the pH may be reduced by 1 pH unit or more when saturated soils or submerged sediments are exposed to the atmosphere and become aerated.

Under unaerobic conditions both CO_2 and methane (CH_4) are produced in various proportions. Also the microorganims in an anaerobic environment may be capable of decomposing such compounds as cellulose and cyclic organics, reduce nitrates to nitrogen gas, and convert DDT to DDE. However, as a rule, aerobic decomposition is more rapid and complete for most other organic and inorganic biodegradable compounds.

Pathogenic Microorganisms in Soils

Various pathogenic (disease-causing) microorganisms (bacteria and viruses) can be found in soils. Their survival in soils depends on many environmental factors and, often, on the availability of an acceptable host organism. Because of the contamination of soils with animal droppings, manure, sewage and sludge, and the tissues of diseased plants, the survival of these organisms in soils has been investigated (Lance, 1978; Loehr et al., 1979; Schaub and Sorber, 1977; Pettygrove and Assano, 1985; Reed, Thomas, and Kowal, 1980).

Pathogenic bacteria. Pathogenic bacteria may survive in soil for a period of from a few hours to several months, depending on the type of organisms, type of soil, the moisture capacity of the soil, moisture and organic content of the soil, pH, temperature, sunlight, rain, and predation by the resident microflora in the soil. In general, enteric bacteria persist in soil for 2 to 3 months; however, under certain favorable conditions pathogenic microorganisms may actually multiply (Loehr et al., 1979). Factors that influence the survival of bacteria in soil are listed in Table 6.6, while the average survival of selected pathogens are presented in

TABLE 6.6 Factors that Affect the Survival of Enteric Bacteria and Viruses in Soil

Factor		Comment
pH	Bacteria	Shorter survival in acid soils (pH 3 to 5) than in neutral and calcareous soils
	Viruses	Insufficient data
Predation by soil microflora	Bacteria	Increased survival in sterile soil
	Viruses	Insufficient data
Moisture content	Bacteria and viruses	Longer survival in moist soils and during periods of higher rainfall
Temperature	Bacteria and viruses	Longer survival at lower temperatures
Sunlight	Bacteria and viruses	Shorter survival at the soil surface
Organic matter	Bacteria and viruses	Longer survival or regrowth of some bacteria when sufficient amounts of organic matter are present

Source: After U.S. EPA (1977).

TABLE 6.7 Survival of Selected Pathogens in Soils

Organism	Range of Survival Time (days)
Salmonella	15 to more than 200
Salmonella typhi	1 to 200
Tubercle bacili	More than 200
Entamoeba histolytica cysts	6 to 8
Enteroviruses	8
Ascaris ova	Up to 7 days
Hookworm larvae	42

Source: After Parson et al. (1975).

Table 6.7. For example, the survival of *E. coli*, *S. typhi*, and *M. avium* is greatly enhanced in moist rather than dry soil. Survival time is less in sandy, permeable soil than in soil with a greater water holding capacity such as clayey soil and peat (Loehr et al., 1979).

Both pathogens and indicator microorganisms (coliforms) survive longer under lower temperatures. Figure 6.18 shows typical die-off curves for fecal coliforms and fecal streptococci under summer and winter conditions. The competition and predation by resident soil bacteria is another important factor. Organisms invading sterilized soils survive longer than they would in unsterilized soil.

Bacteria move in the soil with soil water. However, since fine soil particles can effectively adsorb bacteria, the soil is, in general, a very effective filtering medium. Bacteria are more mobile in sandy soils under

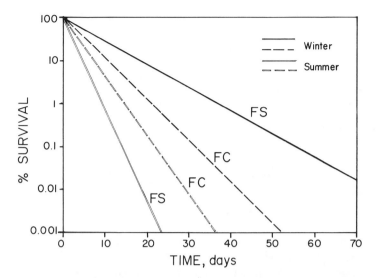

FIGURE 6.18. Die-off curves for fecal coliforms (FC) and fecal streptococci (FS) under winter and summer conditions. (From Von Donsel in Loehr et al., 1979.)

high moisture conditions than in loamy or clayey soils. In fine-textured soils, bacteria can be effectively filtered out by 1 to 2 m of soil. Soil containing clay remove most microorganisms through adsorption. Sandy soils remove them through filtration (Lance, 1978).

Viruses. Viruses in soils are also considered invaders. Loehr et al. (1979) summarized the available information on virus survival as follows:

1. Virus survival in soil depends on the nature of the soil, temperature, pH, moisture, and possibly antagonism from soil microflora.
2. Viruses readily adsorb on soil particles. However, such viruses bound to soil particles are as infectious as free viruses.
3. Viruses survive from as short as 7 days and as long as 6 months in soil. Climatic conditions, particularly temperature, have a major effect on survival time.
4. Enteric viruses can survive from 2 to more than 188 days in fresh water, temperature being the most important factor, with survival greater at lower temperatures.
5. Virus survival in crops is shorter than in soil, because viruses are more exposed to deleterious environmental effects.
6. Contamination of crops most commonly occurs when wastewater comes into contact with the surface of the crop.

TABLE 6.8 Soil Adsorption Characteristics for Viruses: Freundlich Adsorption Isotherms

Soil	pH	Cation Exchange Capacity (meq/100 g)	Specific Surface Area (m^2/g)	Organic Carbon (%)	Freundlich Adsorption Coefficients[a]	
					K_F	$1/n$
Aastad clay loam	6.9	35.2	160	3.02	72.5	0.945
Kranzburg silt loam	6.2	31.4	154	2.38	161	0.908
Palouse silt loam	6.0	22.6	89.2	1.82	45.7	1.24
Parshal silt loam	6.8	13.7	67.6	1.32	4.6	0.92
Quinici loamy sand	7.2	8.9	35.3	0.35		

Source: After Burge and Enkiri (1978), with permission of ASA, CSSA, and SCS.

[a] In number of viruses per gram of soil C_d in number of viruses per milliliter.

7. In rare cases, the translocations of animal viruses from the roots of plants to the aerial parts can occur.
8. Sunlight is believed to be a major factor in killing viruses.
9. Viruses cannot reproduce at all in the soil, and they slowly die off.

Unlike bacteria, for which filtration appears to be the main factor limiting their movement through the soil, viruses are most effectively removed by adsorption on soil particles. The process of adsorption is strongly affected by soil pH. The best adsorption of viruses is achieved at pH 7, but it decreases on both the acid and alkaline sides of the pH scale (U.S. EPA, 1977). Adsorption of viruses does not mean their complete immobilization, since desorption of viruses can occur when pH or other environmental conditions change. Generally, virus adsorption follows the Freunlich isotherm model (Table 6.8). Viruses can be desorbed from soil particles following a heavy rain.

Scheuerman et al. (1979) showed that in organic soils, water-soluble humic substances interfere with the sorptive capacity of soil toward viruses. Wetland soil, such as peat and mulch, were not effective in retaining viruses, and significant viral concentrations were observed in the leachate after land disposal of sewage. One could suspect that similar conclusions can be applied to organic sediments. Sandy soils or sediments with little water-soluble humic substances may retain most or all of the viruses applied.

References

Adams, R. S. (1973). Factors influencing soil adsorption and activity of pesticides, *Residue Rev.* **47**:1–54.

Ahuja, L. R., et al. (1982). Modeling the release of phosphorus and related adsorbed chemicals from soil to overland flow, in *Modeling Component of Hydrologic Cycle*, Water Resources Publ., Littleton, CO.

Alexander, M. (1977). *Introduction to Soil Microbiology*, Wiley, New York.

Armstrong, D. E., and G. Chesters (1968). Adsorption catalyzed chemical hydrolysis of atrazine, *Environ. Sci. Technol.* **2**:683–689.

Bailey, G. W. (1968). *Role of Soils and Sediment in Water Pollution Control. Part I, Reactions of Nitrogenous and Phosphatic Compounds with Soils and Geological Strata*, FWQA (now EPA) South-east Water Laboratory Report, Athens, Georgia.

Bouman, M. C., M. S. Schechter, and R. L. Carter (1965). Behaviour of chlorinated insecticides in a broad spectrum of soil types, *J. Agr. Food Chem.* **13**:360–365.

Braman, R. S., and C. C. Foreback (1973). Methylated forms of arsenic in the environment, *Science* **182**:1247–1249.

Bruce, R. R., et al. (1975). A model for runoff of pesticides from small upland watersheds, *J. Environ. Qual.* **4**:541–548.

Burge, W. D., and N. N. Enkiri (1978). Virus adsorption by five soils, *J. Environ. Qual.* **7**:73–76.

Burns, L. A., et al. (1990). *PIRANHA—Pesticide and Industrial Chemical Risk Analysis and Hazard Assessment-Version 1.0*, Environmental Research Laboratory, U.S. Environmental Protection Agency, Athens, GA.

Cars, J., A. W. Taylor, and E. R. Lemon (1971). Measurement of pesticide concentrations in air overlying field, in *Proceedings, International Symposium on Identification and Measurement of Environmental Pollutants*, Ottawa, Ontario, pp. 72–77.

Carter, J. N., and F. E. Allison (1960). Investigation of denitrification in well aerated soils, *Soil Sci.* **90**:173–177.

Chesters, G. (1986). *Pesticide Transformations and Movement in Soils*, Water Resources Center, University of Wisconsin, Madison, WI.

Chesters, G., and G. V. Simsiman (1975). *Impact of Agricultural Use of Pesticides on Water Quality of the Great Lakes*, Task A, Water Resources Center, University of Wisconsin, Madison, WI, and International Joint Commission, Windsor, Ontario.

Christenson, T. H. (1984). Cadmium soil adsorption at low concentrations: Effect of time, cadmium load, pH, and calcium, *Water, Air, Soil Pollut.* **21**:105–114.

Combest, K. B. (1991). Trace metals in sediments: Spatial trends and sorption process, *Water Resour. Bull.* **27**:19–28.

DiToro, D. M., et al. (1989). *Briefing Report to the EPA Science Advisory Board on the Equilibrium Partitioning Approach to Generating Sediment Quality Criteria*, EPA 440/5-89-002, Office of Water Regulation and Standards, U.S. Environmental Protection Agency, Washington, DC.

DiToro, D. M., et al. (1990). Toxicity of cadmium in sediments: The role of acid volatile sulfide, *Environ. Toxicol. Chem.* **9**:1487–1502.

DiToro, D. M., et al. (1991). Technical basis for establishing sediment quality criteria for non-ionic organic chemicals using equilibrium partitioning, *Environ. Toxicol. Chem.* **10**(12):1541–1583.

Dong, A., G. Chesters, and G. V. Simsiman (1983). Soil dispersibility, *Soil Sci.* **136**(4):208–212.

Dong A., G. Chesters, and G. V. Simsiman (1984). Metal composition of soil, sediment, and urban dust and dirt samples from the Menomonee River watershed, Wisconsin, U.S.A., *Water, Air Soil Pollut.* **22**:257–275.

Dong A., G. V. Simsiman, and G. Chesters (1983). Particle size distribution and phosphorus levels in soil, sediment and urban dust and dirt samples from the Menomonee River Watershed, Wisconsin, *Water Res.* **17**(5):569–577.

Donigian, A. S., D. C. Beyerlein, H. H. Davis, and N. H. Crawford (1977). *Agricultural Runoff Management (ARM) Model Version. II: Refinement and Testing*, EPA 600/3-77/098, U.S. Environmental Protection Agency, Athens, GA.

Duffy, J., C. Chung, C. Boast, and M. Franklin (1974). A simulation model of bio-physicochemical transformation of nitrogen in tile drain corn belt soil, *J. Environ. Quality* **4**:477–485.

Edwards, C. A. (1970). *Persistence of Pesticides in the Environment*, CRC Press, Cleveland, OH.

Enfield, G. G. (1974). Rate of phosphorus sorption by five Oklahoma soils, *Soil Sci. Soc. Am. Proc.* **38**:404–410.

Foy, C. L., and S. W. Bingham (1969). Some research approaches toward minimizing herbicidal residues in the environment, *Residue Rev.* **29**:105–135.

Gillespie, R. A., and R. F. Vaccaro (1978). A bacterial bioassay for measuring the copper chelating capacity of sea water, *Limnol. Oceanogr.* **23**:543–548.

Green, P. B., T. J. Logan, and N. E. Smeck (1978). Phosphate adsorption-desorption characteristics of suspended sediments in the Maumee River Basin of Ohio, *J. Environ. Qual.* **7**:208–212.

Goring, G. A., and J. W. Hamaker (1972). *Organic Chemicals in the Soil Environment*, Marcel Dekker, New York.

Guenzi, W. D., and W. E. Beard (1967). Anaerobic conversion of DDT and DDE in soils, *Science* **156**:1–2.

Guenzi, W. D., and W. E. Beard (1968). Anaerobic conversion of DDT to DDE and aerobic stability of DDT in soils, *Soil Sci. Soc. Am. Proc.* **32**:522–524.

Hague, R., D. W. Schmedding, and V. H. Freed (1974). Aqueous solubility, adsorption, and vapor behavior of polychlorinated biphenyl arochlor 1254, *Environ. Sci. Technol.* **8**:139–142.

Haith, D. A., and L. J. Tubbs (1981). Watershed loading functions for nonpoint sources, *J. Environ. Eng. Div. ASCE* **106**(EE1):121–137.

Hart, B. T. (1981). Trace metal complexing capacity of natural waters: A review, *Environ. Technol. Lett.* **2**:95–110.

Helling, C. S., P. C. Kearney, and M. Alexander (1971). Behavior of pesticides in soils, *Adv. Agron.* **23**:147–240.

Hern, S. C., and S. M. Melacon (1986). *Vadose Zone Modeling of Organic Pollutants*, Lewis Publ., Chelsea, MI.

Hsu, P. H. (1965). Fixation of phosphate by aluminum and iron in acidic soils, *Soil Sci.* **99**:398–400.

Hsu, P. H., and M. L. Jackson (1960). Inorganic phosphorus transformation by chemical weathering in soils as influenced by soil pH, *Soil Sci.* **90**:16–24.

Jenny, H. (1941). *Factors of Soil Formation*, McGraw-Hill, New York.

Jernelov, A. (1972). Factors in the transformation of mercury to methyl mercury, in *Environmental Mercury Contamination*, R. Harrig and B. D. Dinman, eds., Ann Arbor Science Publ., Ann Arbor, MI.

Jury, W. A. (1986). Volatilization from soils, in *Vadose Zone Modeling of Organic Pollutants*, S. C. Hern and S. M. Melacon, eds., Lewis Publ., Chelsea, MI, pp. 159–177.

Jury, W. A., W. F. Spencer, and W. J. Farmer (1983). Use of models for assessing relative volatility, mobility, and persistence of pesticides and other trace organics in soil systems, in *Hazard Assessment of Chemicals*, J. Saxena, ed., Vol. 2, Academic Press, New York.

Justice, J. K., and R. L. Smith (1962). Nitrification of ammonium sulfate in a calcareous soil as influenced by combination of moisture, temperature, and levels of added nitrogen, *Soil Sci. Soc. Am. Proc.* **26**:246–250.

Karickhoff, S. W. (1981). Semiempirical estimation of sorption of hydrophobic pollutants on natural sediments and soils, *Chemosphere* **10**:833–846.

Karickhoff, S. W., D. S. Brown, and T. A. Scott (1979). Sorption of hydrophobic pollutants on natural sediments, *Water Res.* **13**:241–248.

Kearney, P. C., E. A. Woolson, J. R. Plimmer, and A. R. Isensee (1969). Decontamination of pesticides in soils, *Residue Rev.* **44**:137–149.

Kearney, P. C., and D. D. Kaufman, eds. (1964). *Degradation of Herbicides*, Marcel Dekker, New York.

Kenaga, E. E., and C. A. I. Göring (1980). Relationship between water solubility, soil sorption, octanol-water partitioning, and biocencentration of chemicals in biota, in Proceedings, Third ASTM Symposium on Aquatic Toxicology, *ASTM Special Tech. Pub.* **707**:78–115.

Knisel, W. J., ed. (1980). *CREAMS: A Field Scale Model for Chemicals, Runoff and Erosion from Agricultural Management Systems*, Conservation Research Report No. 26., U.S. Department of Agriculture, Washington DC.

Krenkel, P. A. (1973). Mercury in the environment, *CRC Crit. Rev. Environ. Control* **3**(3):303–373.

Lance, J. C. (1985). Fate of bacteria and viruses in sewage applied to soils, *Trans. Am. Soc. Agr. Eng.* **21**:114–117.

Langston, W. J. (1985). Assessment of the distribution and availability of arsenic and mercury in estuaries, in *Estuarine Management and Quality Assessment*, J. G. Wilson and W. Halcrow, eds., Plenum Press, New York, pp. 131–146.

Lindsay, W. L. (1979). *Chemical Equilibria in Soil*, Wiley, New York.

Loehr, R. C., W. J. Jewell, H. D. Novak, W. W. Clarkson, and G. S. Friedman (1979). *Land Application of Wastes*, Van Nostrand Reinhold, New York.

Lyman, W. J. (1982). Octanol/water partitioning coefficient, in *Handbook of Chemical Property Estimation Methods*, W. J. Lyman et al., eds., McGraw-Hill, New York, pp. 2-1–2-52.

McCallister, D. L., and T. J. Logan (1978). Phosphate adsorption-desorption characteristics of soils and bottom sediments in the Maumee River Basin of Ohio, *J. Environ. Qual.* **7**:87–92.

McElroy, A. D., A. D. Chiu, S. Y. Nebgen, J. E. Aleti, and F. W. Bennett (1976). *Loading Functions for Assessment of Water Pollution from Nonpoint Sources*, EPA-600/2-76-151, U.S. Environmental Protection Agency, Washington, DC.

McLaren, A. D. (1969). Steady state studies of nitrification in soil: Theoretical considerations, *Soil Sci. Soc. Am. Proc.* **33**:273–276.

McLaren, A. D. (1971). Kinetics of nitrification: Growth of the nitrifiers, *Soil Sci. Soc. Am. Proc.* **35**:91–95.

Mahmood, R. J., and R. C. Sims (1986). Mobility of organics in land treatment systems, *J. Environ. Eng. ASCE* **112**(2):236–245.

Massey, H. F., and M. L. Jackson (1952). Selective erosion of soil fertility constituents, *Soil Sci. Soc. Proc.* **82**:353–356.

Massey, H. F., M. L. Jackson, and O. E. Hays (1953). Fertility erosion of two Wisconsin soils, *Agron. J.* **45**:543–547.

Miles, J. R. W., C. M. Tu, and C. R. Harris (1969). Metabolism of heptachlor and its degradation products by soil microorganisms, *J. Econ. Entomol.* **69**: 1334–1338.

Mill, T. (1980). Data needed to predict the environmental fate of organic chemicals, in *Dynamic Exposure and Hazard Assessment of Toxic Chemicals*, R. Hague, ed., Ann Arbor Science Publ., Ann Arbor, MI.

Mills, W. B., et al. (1985). *Water Quality Assessment: A Screening Procedure for Toxic and Conventional Pollutants*, EPA-600/6-82-004a and b, Vol. I and II, U.S. Environmental Protection Agency, Athens, GA.

Misra, C., D. R. Nielsen, and J. W. Biggar (1974). Nitrogen transformation in soil during leaching: III—Nitrate reduction in soil, *Soil. Sci. Soc. Am. Proc.* **38**:300–304.

Moore, J. B. (1963). Bottom sediment studies, Buzzards Bay, Massachusetts, *J. Sediment. Pet.* **33**:511–538.

Moore, J. W., and S. Ramamoorthy (1984). *Organic Chemicals in Natural Waters*, Springer-Verlag, New York.

National Academy of Sciences (1979). *Polychlorinated Biphenyls*, National Academy of Sciences, Washington, DC.

Nissenbaum, A., and D. J. Swaine (1976). Organic matter–metal interactions in recent sediments: The role of humic substances, *Geochim. Cosmochim. Acta* **40**:809–816.

Novotny, V., and G. Chesters (1981). *Handbook of Nonpoint Pollution: Sources and Management*, Van Nostrand Reinhold, New York.

Novotny, V., et al. (1978). Mathematical modeling of land runoff contaminated by phosphorus, *J. WPCF* **90**:101–112.

Painter, H. A. (1970). A review of literature on inorganic nitrogen metabolism in microorganisms, *Water Res.* **4**:393–450.

Paris, D. F., and D. L. Lewis (1973). Chemical and microbial degradation of selected pesticide in aquatic systems, *Residue Rev.* **45**:95–124.

Paris D. F., et al. (1981). Second-order model to predict microbial degradation of organic compounds in natural waters, *Appl. Environ. Microbiol.* **41**(3): 603–609.

Parson, D., et al. (1975). *Health Aspects of Sewage Effluent Irrigation*, Pollution Control Branch, British Columbia Water Research Service, Dept. of Lands, Forests, and Water Research, Parliament Bldg., Victoria, B. C.

Pettygrove, G. S., and T. Assano, eds. (1985). *Irrigation with Reclaimed Municipal Wastewater: A Guidance Manual*, Lewis Publ., Chelsea, MI.

Porcella, D. B., et al. (1974). *Comprehensive Management of Phosphorus Water Pollution*, EPA/600/5-74/010, U.S. Environmental Protection Agency, Washington, DC.

Powell, G. M. (1976). *Land Treatment of Municipal Wastewater Effluents. Design Factors II*, U.S. Environmental Protection Agency, Washington DC.

Pratt, P. F., W. W. Jones, and V. E. Hunsaker (1972). Nitrate in deep soil profiles in relation to fertilizer rates and leaching volume, *J. Environ. Qual.* **1**:97–102.

Preul, H. C., and G. J. Schroepfer (1968). Travel of nitrogen in soils, *J. WPCF* **40**:30–48.

Rao, P. S. C., and J. M. Davidson (1982). *Retention and Transformation of Selected Pesticides and Phosphorus in Soil-Water Systems: A Critical Review*, EPA 600/3-82-060, U.S. Environmental Protection Agency, Athens, GA.

Reed, S., R. Thomas, and N. Kowal (1980). Long-term land treatment; Are there health or environmental risks? in *Proceedings, ASCE National Convention*, Portland, OR.

Rennie, D. E., and R. B. McKercher (1959). Adsorption of phosphorus by four Saskatchewan soils, *Can. J. Soil Sci.* **39**:64–75.

Ryan, J. A., and D. R. Keeney (1975). Ammonia volatilization of surface-applied wastewater sludge, *J. WPCF* **50**:386–393.

Ryden, J. C., J. K. Syers, and R. F. Harris (1972). Potential of an eroding urban soil for the phosphorus enrichment of streams, *J. Environ. Qual.* **1**:430–434.

Salomons, W., and U. Förstner (1984). *Metals in the Hydrocycle*, Springer Verlag, Berlin, New York.

Schaub, S. A., and C. A. Sorber (1977). Virus and bacteria removal from wastewater by rapid infiltration through soil, *Appl. Environ. Microbiol.* **33**:609.

Scheuerman, P. R., G. Bitton, A. R. Overman, and G. E. Gifford (1979). Transport of viruses through organic soils and sediments, *J. Environ. Eng. Div. ASCE* **40**:R257–271.

Schnoor, J. L., C. Sato, D. McKechnie, and D. Sahoo (1987). *Processes, Coefficients, and Models for Simulating Toxic Organics and Heavy Metals in Surface Waters*, EPA 600/3-67/015, U.S. Environmental Protection Agency, Athens, GA.

Schwarzenbach, R. P., and J. Westall (1981). Transport of nonpolar organic compounds from surface water to groundwater, laboratory sorption studies, *Environ. Sci. Technol.* **5**:1360–1367.

Sethunathan, N. (1973). Chemical and microbial degradation of ten selected pesticides in aquatic systems, *Residue Rev.* **45**:95–124.

Shacklette, H. T., J. C. Hamilton, J. G. Boergnagen, and J. M. Bowles (1971). *Elemental Composition of Surficial Materials in the Conterminous United States*, U.S. Geological Survey Prof. Paper 574-D, Washington, DC.

Shaffer, M. J., G. R. Dutt, and W. J. Moore (1969). Predicting changes in nitrogen compounds in soil-water systems. Collected papers: Nitrates in agricultural waste waters, FWQA (now EPA) WPC Series, 13030 ELY 12/69, pp. 15–28.

Sharpley, A. N., L. R. Ahuja, and R. G. Menzel (1981). The release of phosphorus to runoff in relation to the kinetics of desorption, *J. Environ. Qual.* **10**(4):386–391.

Spencer, W. E., and M. M. Cliath (1973). Pesticide volatilization as related to water loss from soil, *J. Environ. Qual.* **2**:284–289.

Spencer, W. F., W. J. Farmer, and M. M. Cliath (1973). Pesticide volatilization as related to water loss from soil, *J. Environ. Qual.* **2**:284–289.

Spencer, W. F., W. J. Farmer, and W. A. Jury (1982). Review: Behavior of organic chemicals at soil, air, water interfaces as related to predicting the transport and volatilization of organic pollutants, *Environ. Toxicol. Chem.* **1**:17–26.

Stanford, G., M. H. Free, and D. H. Swaininger (1973). Temperature coefficient of soil nitrogen mineralization, *Soil Sci.* **115**(4):321–328.

Steward, B. A., et al. (1975). *Control of Pollution from Cropland*, EPA 600/2-75/026, U.S. Environmental Protection Agency—U.S. Department of Agriculture, Washington, DC.

Stigliani, W. M., ed. (1991). *Chemical Time Bombs: Definition, Concepts, and Examples*, Ex. Report No 16, International Institute for Applied System Analysis, Laxenburg, Austria.

Stigliani, W. M., et al. (1991). Chemical time bombs—predicting the unpredictable, *Environment* **33**(4):4–9.

Stoltenberg, N. I., and J. L. White (1953). Selective loss of plant nutrients by erosion, *Soil Sci. Soc. Am. Proc.* **27**:406–410.

Summerhayers, C. P., J. P. Ellis, and P. Stoffers (1985). *Estuaries as Sinks for Sediment and Industrial Waste—A Case History from the Massachusetts Coast*, E. Schweizerbart'sche Verlagbuchhandlung, Stuttgart.

Terry, R. E., D. W. Nelson, L. E. Simons, and G. J. Meyer (1978). Ammonia volatilization from wastewater sludge applied to soils, *J. WPCF* **50**:2657–2665.

U.S. Environmental Protection Agency (1976). *Application of Sewage Sludge to Cropland. Appraisal of Potential Hazard of Heavy Metals to Plants and Animals*, EPA 430/9-76-103, U.S. EPA, Washington, DC.

U.S. Environmental Protection Agency (1977). *Process Design Manual for Land Treatment of Municipal Wastewater*, EPA 625/1-77/008, U.S. EPA, Washington, DC.

U.S. Environmental Protection Agency (1984). *Review of In-place Treatment Techniques for Contaminated Surface Soils, Background Information for In Situ Treatment*, Vol. 2, EPA 540/2-84-003b, U.S. EPA, Cincinnati, OH.

U.S. Environmental Protection Agency (1986). *Permit Guidance Manual on*

Hazardous Waste Land Treatment Demonstrations, EPA 530/SW86-032, Office of Solid Waste, U.S. EPA Washington, DC.

Valentine, R. L., and J. L. Schnoor (1986). Biotransformation, in *Vadose Zone Modeling of Organic Pollutants*, S. C. Hern and S. M. Melacon, eds., Lewis Publ., Chelsea, MI, pp. 191–222.

Vijayachandran, P. K., and R. D. Harter (1975). Evaluation of phosphate adsorption by a cross section of soil types, *Soil Sci.* **119**:119–125.

Weber, W. J., P. M. McGinley, and L. E. Katz (1991). Review paper—Sorption phenomena in subsurface systems: Concepts, models and effects on contaminant fate and transport, *Water Res.* **25**(5):499–528.

Wilde, H. E., C. N. Sawyer, and T. C. McMahon (1971). Factors affecting nitrification kinetics, *J. WPCF* **43**:1845–1854.

Young, R. A., and C. A. Onstad (1976). *Predicting Particle Size Composition of Eroded Soil*, Paper No. 76-2052, presented at 1976 American Society of Agricultural Engineers Meeting, Lincoln, NE.

Yule, W. N., and I. Rosefield (1964). Fate of insecticide residues. Decomposition of lindane in soils, *J. Agr. Food. Chem.* **15**:1000–1004.

Zanoni, A. E. (1969) Secondary effluent degradation at different temperatures, *J. WPCF* **41**:640–659.

Zimdahl, R. L., and J. J. Hasset (1977). Lead in soil, in *Lead in the Environment*, W. R. Rodgers and B. G. Wixson, eds., NSF 770214, U.S. Government Printing Office, Washington, DC.

7
Ground-Water and Base-Flow Contamination

You never know the worth of water till the well runs dry.

Benjamin Franklin

GROUND WATER (BASE FLOW) AND DIFFUSE POLLUTION

Extent of the Problem

Approximately one-half of the United States population depends on ground water for its supply of potable water; that is, approximately 36% of all municipal public drinking water supply systems and 95% of the rural population draw potable water from ground-water resources (Conservation Foundation, 1985). As stated in Chapter 3, water recovered from wells is considered hydrologically part of ground-water runoff when it reaches surface drainage systems.

Ground-water and surface-water systems are interconnected by *recharge* of surface-water into ground-water zones and by *discharge* of ground-water into surface systems. In many cases surface-water bodies are an extension of the ground-water systems. Perennial streams may be both recharging ground-water zones or ground water is discharging as base flow into the streams. Ephemeral streams are mostly recharging during the time of flow (see Chapter 3 for a discussion on hydrologic surface-water–ground-water interactions). Hence with ground-water discharge (base flow), diffuse pollution from the land may reach surface-water bodies via ground-water routes.

The quantity of ground water underlying the conterminous United States is very large, and exceeds all freshwater surface storage. The amount of water within the top 800 meters of earth surface layer is equal to about 35 years of all surface-water runoff (Geraghty et al., 1973; Knox, and Fairchild, 1987). However, as far as water use is concerned, more surface freshwater is used than ground water.

Ideally, ground water should be characterized by clarity, bacterial purity, and constant temperature and chemical quality, and should require very little treatment prior to its use. However, increased usage of ground-water resources and a general increase of inputs of surface pollution into ground-water zones have caused contamination and a general deterioration of ground-water quality in many parts to the world. Specifically, the severe contamination by nitrates in Central Europe and in some parts of the United States, and pesticide contamination are of concern, as both pollutants may render ground water unsuitable for human consumption without expensive treatment. Excessive mining of ground-water resources results in more costly pumping from deeper geological zones, which yields water with high salt and ionic content or sea water intrusion into aquifers in coastal areas.

While the basic chemical composition of ground-water reflects its contact with soils, minerals, and rocks, surface diffuse pollution is often the primary source of ground water-contamination. With an increased emphasis on land disposal of sewage and industrial wastes promulgated by the first amendments of the Clean Water Act, the surface-water pollution problem could have been shifted to ground water. Deep-well injection of wastewater, subsurface disposal of toxic wastes, unsanitary landfills, excessive use of chemicals and fertilizers, and suburban and rural septic tank waste disposal systems are the main causes of ground-water contamination. However, Table 7.1 shows that other sources of ground-water contamination can be also considered.

It should be realized that only a relatively small portion of the ground-water problem has been discovered. Ground-water pollution is not visible and is detected only when a water supply or spring is noticeably polluted or the pollutant is discharged into surface waters. The monitoring of ground-water quality is still inadequate, as it is mostly done using observation wells that are located great distances apart.

All ground water results from surface precipitation. However, in northern North America and Europe, most of the present ground water originated from the melting of glaciers at the end of the glacial period approximately 10,000 years ago, but some ground water is even older. Also, ground-water movement is very slow. Plus the practices and land-use activities that have an impact on ground-water quality have been

TABLE 7.1   Principal Sources of Ground-Water Contamination and Their Relative Importance by Region

	Northeast	Northwest	South Central	Southwest
Septic tanks and cesspools	I[a]	I	I	I
Petroleum exploration and development	II	II	I	I
Landfills	I	II	II	II
Irrigation return flows	IV	II	II	II
Surface discharges	II	I	III	I
Surface impoundments	I	I	II	III
Spills	I	II	II	III
Mining activities	II	I	III	II
Agricultural activities				
Fertilizers	III	II	III	II
Pesticides	II	III	II	II
Feedlot and barnyard wastes	III	III	II	III
Highway deicing	I	III	IV	IV
Artificial recharge	III	IV	III	II
River infiltration	II	II	IV	IV
Land disposal of wastewater and sewage	III	IV	III	II
Underground (leaking) storage of chemicals	II	II	II	II

Source: After Miller and Scalf (1974).
[a] I—high; II—moderate; III—low; IV—not significant.

occurring for the past 40 to 50 years. Apparently, this time span is very short in the geological time scale, so most ground-water contamination is still confined near the source.

In industrialized countries, ground-water pollution by toxic chemicals and nitrates has locally and regionally reached alarming levels. For years, industries and municipalities have been disposing of their toxic and other wastes and refuse on land or burying them without regard to the possible contamination of ground water (Fig. 7.1). Examples of widely publicized severe ground-water contamination include the Love Canal site in Niagara Falls, New York, as well as ground-water quality impairment in other parts of the northeastern United States, specifically in the state of New Jersey.

In the Niagara Falls Love Canal incident, a chemical company between 1947 and 1952 used an abandoned river channel to store toxic wastes, mostly solvents. At the time, due to the lack of knowledge on the environmental consequences of such disposal, such containments were common and considered as acceptable. After 1952 the canal was filled

FIGURE 7.1. Illegal and unsanitary dumps and solid waste disposal sites are the primary sources of ground-water contamination.

with soil and, subsequently, the area was developed into a residential subdivision. (It should be noted, however, that the company gave explicit caveats to the authorities when giving them the land for development.) In the 1970s the chemicals began seeping through the basement walls into household drainage systems. Air pollution monitoring in the basements throughout the contaminated area registered pollution that exceeded 10 to 5000 times the ambient standards. The toxic chemicals that created this extreme diffuse pollution problem included mostly aliphatic hydrocarbons (benzene, toluene, chloroform, etc.). The Love Canal site has been cleaned up at great cost, and in the 1990s it appeared safe for reinhabitation.

Cleanup of Contaminated Ground Water

In many cases, a leachate from a localized point source (a storage lagoon, landfill, leaking petroleum and gasoline storage tanks, scrap yard) becomes a regional diffuse pollution problem. Many such unsanitary and contaminating sites have been classified by the U.S. EPA as hazardous sites, often requiring costly remediation.

The tasks facing those concerned and responsible for ground-water quality are enormous, and the solution expensive or nonexistent. There are several approaches that have been used to clean up contaminated ground water (Canter, Knox, and Fairchild, 1987):

1. *Physical control measures.* These controls include surface-water control, the containment of the sources of contamination by capping, and the use of liners and other barriers to isolate contaminated sites and/or to prevent contaminated surface runoff from entering the groundwater zones. The ground-water level can also be lowered by pumping to a safe depth below surface contamination.
2. *In-situ technologies.* Both chemical and biological treatment may be employed. Chemical treatment may be considered only if contaminants are known and the extent of the contaminated zone is defined. Some of the chemical treatment methods involve oxygen injection that will precipitate iron and manganese. Biological in-situ treatment is based on the same principle as that in biological wastewater treatment systems. The method requires the addition of nutrients and oxygen, and allowing microorganisms to degrade the chemicals.
3. *Ground-water treatment.* The most popular ground-water cleanup measure remains removal and treatment and subsequent return of the treated water into underground zones. Water is removed by wells or by injection/extraction systems. Although a wide variety of technologies exists, the treatment options used for ground-water decontamination are usually limited to air stripping, carbon adsorption, or biological treatment for organic removal, and chemical precipitation for inorganic removal.

High nitrate content caused by fertilizer applications and/or septic tank effluents must be remediated at any cost because it is usually widespread. When such waters are to be used for potable water supply in central Europe, biological denitrification of the extracted water has been seriously considered by the authorities. The problem with this methodology is finding a suitable nontoxic source of organic carbon instead of methanol (toxic to humans), which is commonly used in many denitrifying wastewater treatment plants. Because of the slow movement of water in underground zones and the large volumes of water in ground-water systems, the recovery of a contaminated aquifer is slow, usually lasting many years.

Ground-Water Quality Standards

Ground-water quality should comply with standards set forth by environmental control legislation. In the United States the three acts that have the greatest impact on ground-water pollution control are

1. The Safe Drinking Water Act of 1974 (PL-93-523)

2. The Federal Environmental Pesticide Control Act of 1972 (PL-92-516)
3. The Toxic Substance Control Act of 1976 (PL-94-469)

In addition the Clean Water Act also specifies the quality of water that is to be used for drinking purposes.

It should be pointed out that the standards based on the Safe Drinking Water Act are for water quality at the point of use, not at the point of intake into the system. Water quality standards derived from the Clean Water Act (see the Appendix) represent the mandated raw drinking water quality of the source, be it a surface-water body or a ground-water zone. These standards assume that intake water will be sufficiently uncontaminated so that with the application of the most effective treatment method, a public water supply system would be able to protect public health.

GROUND-WATER MOVEMENT

Aquifers and Aquitards

This section describes water movement, chemical composition and contamination, and the fate of contaminants in saturated ground-water zones. In contrast the preceding chapter dealt with the phenomena occurring in the vadose–aerated zone.

The basic chemical content of uncontaminated ground water can be related to the contact of water with rocks and soils. However, it should be noted that while mineralization (enrichment of water by salts and ions) occurs throughout the entire movement of water in the aquifer, ground-water contamination and pollution mostly occur in the recharge area (with the exception of deep wastewater disposal by wells and buried toxic waste disposal sites).

An *aquifer* was defined in Chapter 3 as a saturated permeable geologic underground stratum that can transmit significant quantities of water. An *aquitard* is a less permeable layer that may be significant for consideration in the regional transport of water, but its permeability is not sufficient enough to permit economical development (Freeze and Cherry, 1979). The most common aquifer composition—about 90%—consists of unconsolidated materials (Todd, 1980). Other materials capable of forming aquifers mainly include sedimentary rocks. Crystalline and metamorphic rocks are relatively impermeable, and can form aquifers only when they are fractured and/or weathered near the surface.

Clay and clayey materials can retain large quantities of water, but they are generally considered impermeable and unsuitable as aquifers. Such materials and rocks with poorer water-holding and transmitting capacities (shales and dense crystalline rocks) are classified as aquitards. Completely impermeable geological layers (aquicludes) are rare, so hydrogeologists tend to avoid this classification.

Unconsolidated Deposits
The unconsolidated deposits are composed of gravel, sand, silt, or clay particles that may be bound or hardened by mineral content, pressure, or mineral alteration. The three major types of unconsolidated deposits are

1. Glacial deposits formed during the last glacial age, approximately 15,000 to 100,000 years ago.
2. Alluvial (fluvial) deposits that resulted from deposition of sediments by streams and valleys and floodplains.
3. Aeolian deposits that consist of finer soil materials transported by wind.

Glacial deposits are of particular hydrogeologic importance in Europe and the northern part of the North American continent (Fig. 7.2). Water

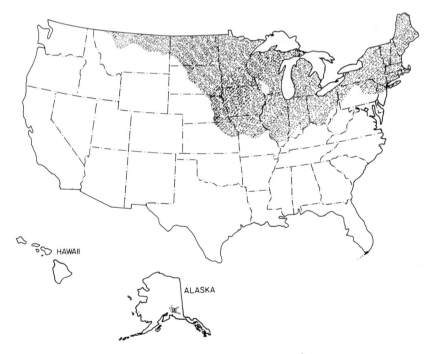

FIGURE 7.2. Extent of glacial deposits in the United States.

in the unconsolidated materials can be divided into four categories: underground water courses, abandoned or buried valleys, plains, and intermontane valleys. Water courses consist of alluvium that forms and underlies stream channels as well as adjacent floodplains. The interconnecting water movement between stream channels depends on the general hydraulic grade. Stream water can be an influent into the ground-water zone (recharging) or vice versa (discharging ground water).

Abandoned or buried valleys are valleys no longer occupied by the streams that formed them. In many areas of the United States, especially the Great Plains and Florida, large regional plain aquifers are located near the surface in unconsolidated deposits. Intermontane aquifers are located in unconsolidated rock deposits, often of considerable thickness, formed by the geological formation of mountains. These are mostly located in the western United States. Water from unconsolidated deposits often has low mineral content; however, due to the higher permeability of these materials and their near-surface location, these aquifers are most commonly affected by surface diffuse pollution.

Sedimentary Rocks

Sedimentary rocks include sandstone limestone or dolomite limestone. Limestone (mostly calcium carbonate) and dolomite (calcium and magnesium dolomite) are also called *carbonate rocks*.

Sandstone. Sandstone is a sedimentary rock with a porosity of 5% to 30%. Its potential as an aquifer is excellent. Sandstone is formed by the binding of sand or gravel by a cementing material such as calcite, dolomite, or clay. These minerals occur as a result of salt precipitation and the filtration of finer clay particles from penetrating water during the geologic times.

Carbonate rocks. Carbonate rocks were formed by the compaction and crystallization of the shells and bones of aquatic animals living in geological oceans and lakes.

Although some of the porosity in carbonate rocks is retained, most of the ground-water movement through them occurs along joints, fractures, and channels formed mostly by the dissolving action of percolating water (recall from Chapter 4 that rainwater is a weak acid that can chemically dissolve calcites and other minerals). The dissolution of limestone formations can reach such proportions that underground streams, caverns, and cavities occur. These limestone formations, called *karst* or *karst limestone*, are especially susceptible to contamination from surface sources. Notable ground-water pollution problems occurred in England (Edworthy, Wilkinson, and Young, 1978) and elsewhere.

Ground-Water Movement

Permeability and Darçy's Law

The flow of water in an aquifer can be expressed by Darçy's law, which relates water velocity in a porous medium to the hydraulic gradient as

$$v = kS = k\frac{\partial h}{\partial l} \tag{7.1}$$

where
v = velocity
S = hydraulic slope of ground water or piezometric table
h = hydraulic head
l = distance in the direction of flow
k = constant of proportionality

In order to compute flow in the aquifer, Equation (7.1) should be multiplied by the cross-sectional area of pores of the media. Hence

$$Q = A_p v = pAkS = KAS \tag{7.2}$$

where
Q = the flow
A_p = the area of voids
A = total cross-sectional area
p = cross-sectional porosity (Table 7.2)

TABLE 7.2 Typical Porosities

	Porosity (%)
Soils	50–60
Clay	45–55
Silt	40–50
Medium to coarse mixed sand	35–40
Uniform sand	30–40
Fine to medium mixed sand	30–35
Gravel	30–40
Gravel and sand	20–35
Sandstone	10–20
Shale	1–10
Limestone	1–10

Source: From Todd (1980).

TABLE 7.3 Typical Hydraulic Conductivities

	Hydraulic Conductivity (cm/sec)
Clay, sand, and gravel mixes (till)	10^{-6}–10^{-4}
Sand and gravel mixes	10^{-3}–10^{-1}
Gravel	0.1–1
Coarse sand	0.01–0.1
Medium sand	0.01
Fine sand	10^{-3}–10^{-2}
Silty sand	10^{-5}–10^{-3}
Loam soils (surface)	10^{-4}–10^{-3}
Glacial outwash	10^{-3}–0.1
Deep clay beds	10^{-11}–10^{-5}
Clay soils (surface)	10^{-2}–10^{-1}
Volcanic rock	Almost 0–1
Fractured or weathered rock (core samples)	Almost 0–1
Fractured or weathered rock (aquifers)	10^{-6}–0.01
Dense solid rock	$<10^{-8}$
Shale	10^{-10}
Carbonate rocks with secondary porosity	10^{-5}–10^{-3}
Sandstone	10^{-6}–10^{-3}

Sources: Based on data from Bouwer (1978) and Fetter (1988).

The coefficient, $K = kp$, is called the *coefficient of permeability* or *hydraulic conductivity*. The coefficient of permeability as defined herein can be visualized as flow velocity in a porous medium under slope that equals unity. Table 7.3 presents coefficients of permeability for various geological materials.

Hydraulic conductivity, K, is a function depending on the porous media and also on the fluid itself. The media characteristics affecting hydraulic conductivity are grain diameter, porosity packing, and the distribution of the material. Fluid characteristics that also affect K are density, viscosity, and its ionic nature. Since K depends on the viscosity of water (the flow is laminar), permeability is affected by temperature.

Example 7.1: Flow Velocity in an Aquifer

An aquifer consisting of sandstone with average slope of the groundwater table of 1% has been contaminated by a highly water-soluble contaminant (it does not adsorb on soil particles). Estimate how fast the contaminated water will move through the aquifer.

Solution This is a typical straightforward application of Darçy's equation. From Equations (7.1) and (7.2)

$$v = \frac{KS}{p} \tag{7.3}$$

and substituting appropriate values for K and p from Tables 7.2 and 7.3, the velocity becomes

$$v = \frac{10^{-4} \times 0.01}{0.15} = 6.67 \times 10^{-6}\,\text{cm/sec} = 210\,\text{cm/yr}$$

which is a very slow advancement of the contaminated water front. If the compound has an affinity for adsorption on soil the advancement would be further retarded, as will be shown later.

Typical velocities of ground water may range from less than 1 cm/yr in tight clays to more than 100 m/yr in permeable sand and gravel. Todd (1980) indicated that the normal range for ground-water velocities is 1.5 m/yr to 1.5 m/day. However, highly permeable glacial outwash deposits, fractured basalts and granites, and cavernous limestone formations may allow much higher velocities.

Homogeneity and Anisotropy of Hydraulic Conductivity

Homogeneity

A formation is homogenous if the hydraulic conductivity is uniform at all points within the aquifer. If the hydraulic conductivity varies with location, the formation is heterogenous. Freeze and Cherry (1979) have defined three types of heterogeneity: (1) layered heterogeneity, (2) discontinuous heterogeneity, and (3) trending heterogeneity. Layered heterogeneity is common in sedimentary deposits where different materials have been deposited throughout the geological ages. Discontinuous heterogeneity is caused by the presence of faults or large-size stratigraphic features. Trending heterogeneity results in the progressive change of hydraulic conductivity over a large spatial extent of a geological formation.

Domenico and Schwartz (1990) have pointed out that since the permeabilities in the individual layers of a heterogenous aquifer can vary by several orders of magnitude, the layer with the highest permeability becomes the main route of flow, while the layers of the lowest permeability become essentially aquitards or confining layers. Heterogeneity may also lead to a situation where some layers become saturated with water and some do not. This is called a *perched water table*.

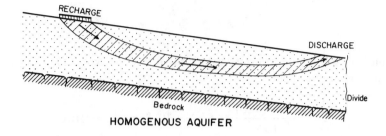

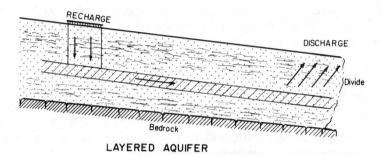

FIGURE 7.3. Flow paths in homogenous and layered aquifers (according to Freeze and Cherry, 1979). *(Copyright © 1979 Prentice Hall, reprinted by permission.)*

Anisotropy

A geologic formation is isotropic at a given point if the hydraulic conductivity is the same in all directions. The formation is anisotropic if the hydraulic conductivity varies with direction. Heterogeneity and anisotrophy can greatly influence flow patterns of contaminants in aquifers (Fig. 7.3).

Dispersion

Average water movement alone does not fully explain the transport and spread of contaminants in ground-water aquifers. The first processes that contribute to forming a concentration field of a contaminant in ground-water zones are advection, dispersion, and diffusion. The second category of processes affecting contaminant movement are adsorption–desorption and chemical precipitation.

Advection is a process in which soluble contaminants are transported by the bulk of the water. The rate of transport can be directly related to the average linear ground-water velocity defined by Equations (7.1) and (7.2). However, the velocity field in ground-water zones is not uniform due to the heterogeneity and anisotropy of the aquifer, which are caused

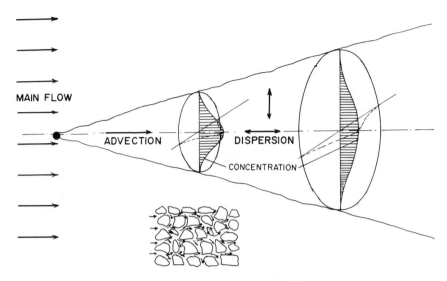

FIGURE 7.4. Contaminant spread in ground-water zones from a continuous source by advection and dispersion.

by differing pore sizes and the branching of pores, plus many obstacles in the path of ground-water movement. These factors cause a spread of mobile contaminants in all directions (longitudinal and transverse), as shown on Figure 7.4. The process is called *hydrodynamical dispersion*. *Molecular diffusion* is a process by which mobile contaminants move as a result of the kinetic activity of molecules and ions in the direction of their concentration gradient. Molecular diffusion can occur independently of ground-water movement. However, normal ground-water velocities and hydrodynamical dispersion are such that molecular diffusion can be neglected (Bouwer, 1978). Hydrodynamical and molecular dispersion cannot be separated and can be included in one dispersion parameter.

The spread of contaminants by dispersion in ground-water zones has a similar appearance to air pollution plumes or discharges of pollutants in surface-water bodies. It is, basically, a mixing process. Nevertheless, it must be remembered that the dispersion of pollutants in air and water environments is caused primarily by turbulent mixing of the air and water masses, while in ground-water zones, mixing occurs in laminar (viscous) flow patterns as a result of the continuous splitting, slowing down, and deflecting of water particles in the pores.

The spread of contaminants in the direction of the ground-water flow is called *longitudinal dispersion*. *Transverse dispersion* is perpendicular to the main direction of flow, and is usually weaker than the longitudinal dispersion. In addition to the movement of water particles, the spread of semimobile contaminants can be caused by adsorption–desorption processes whereby a portion of the contaminant is adsorbed on soil particles at higher concentrations and released back to the solution when the concentration decreases.

Equation for Dispersive Movement

Most of the ground-water contamination problem can be analyzed assuming steady-state flow conditions, which implies that the flow velocity and dispersion characteristics remain constant with time. Furthermore, contamination problems can be limited to the instantaneous injection of

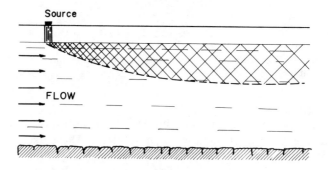

(a) CONTINUOUS SOURCE

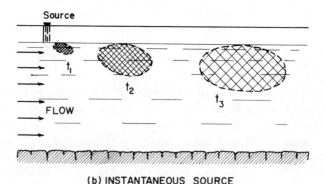

(b) INSTANTANEOUS SOURCE

FIGURE 7.5. Spreading of a pollutant in a two-dimensional uniform field in an isotrophic sand. (a) Continuous source—leaching. (b) Instantaneous source—spill.

a pollutant such as a spill or a continuous release, that is, leaching (Fig. 7.5). General equations of hydrodispersive movement have been published in many references (Anderson, 1979; Freeze and Cherry, 1979; NCASI, 1985; Canter, Knox, and Fairchild, 1987; Fetter, 1988).

The basic differential equation of the hydrodispersive movement is

$$\frac{\partial C}{\partial t} = \frac{\partial}{\partial X_i}\left(D_i \frac{\partial C}{\partial X_i} - V_i C\right) \quad (7.4)$$

where
X_i = the coordinate
t = time
$C(X_i, t)$ = the concentration of the contaminant
V_i = the velocity in the ith direction
D_i = is the coefficient of dispersion in the ith direction

The solution of this equation for a continuous point source in a one-dimensional flow field (all flow is in the direction of the X-axis) will yield (Ogata, 1970; Fetter, 1988)

$$C = \frac{C_0}{2}\left[\text{erfc}\left(\frac{L - V_x t}{2\sqrt{D_t t}}\right) + \exp\left(\frac{V_x L}{D_L}\right)\text{erfc}\left(\frac{L + V_x t}{2\sqrt{D_t t}}\right)\right] \quad (7.5)$$

where
C = concentration at a distance L from the source at time t
C_0 = concentration at the source
D_L = longitudinal dispersion coefficient
D_t = transverse dispersion coefficient

Due to the slow movement of ground water (a few meters per year or less), most of the ground-water contamination problems could be considered as an instantaneous release. There are two solutions available to the problem. One observes the concentration field from a fixed point, for example, the point of discharge. The other solution views the movement of the contaminant cloud as if an observer were located in the center of gravity of the cloud. The former concept yields the following equation for a two-dimensional (longitudinal and transverse) movement of an inert dissolved contaminant (Peaudecerf and Sauty, 1978)

$$C(X, Y, t) = \frac{m}{4\pi p(D_L D_T)^{0.5} t} \exp\left(-\frac{(X - Vt)^2}{4D_L t} - \frac{y^2}{4D_T t}\right) \quad (7.6)$$

where
m = the mass of the released contaminant per unit of aquifer thickness
p = porosity

The point of the coordinate origin for Equation (7.6) ($X = Y = Z = 0$) is the point of release of the contaminant in a two-dimensional aquifer.

Baestlé (1969) published a solution for the latter case, where the coordinate origin is located at the center of gravity of the contaminated cloud. The cloud is carried away from the source with an average velocity equal to the ground-water movement (Eq. (7.1)). The concentration distribution of the contaminant mass is then

$$C(X, Y, Z, t) = \frac{M}{8(\pi t)^{1/2} p \sqrt{D_x D_y D_z}} \exp\left(-\frac{X^2}{4D_x t} - \frac{Y^2}{4D_y t} - \frac{Z^2}{4D_z t}\right) \quad (7.7)$$

where M = the total mass of the contaminant. In the equation the transformed coordinates are

$$X = x - Vt$$
$$Y = y$$
$$Z = z$$

The maximum concentration occurs at the center of the cloud where $X = Y = Z = 0$. Hence

$$C_{max} = \frac{M}{8(\pi t)^{1/2} p \sqrt{D_x D_y D_z}} \quad (7.8)$$

Assuming a true dimensionless point of discharge, Equations (7.6) to (7.8) can be used very rarely. For most cases, such as a surface application of a pesticide, the mass term, M, should be replaced by

$$\frac{M}{p} = C_0 V_0 \quad (7.9)$$

where
V_0 = the volume of ground water below the area of application
C_0 = initial concentration

Then Equations (7.6) to (7.8) are applicable where

$$8(\pi t)^{1/2}\sqrt{D_x D_y D_z} \gg V_0$$

Character and Magnitude of the Dispersion Coefficient

The coefficient of hydrodynamic dispersion is the sum of bulk molecular diffusion and mechanical dispersion. In ground-water zones molecular diffusion is not as fast as that in open water, since molecules have to follow longer pathways because the grains block many possible pathways and the pathways are in the pores. Fetter (1988) pointed out that the actual magnitude of the ground-water diffusion coefficient, D, may be only a few percent of that for water. Molecular diffusion becomes the dominant transport process when ground water is not moving. Perkins and Johnson (1963) have shown that molecular diffusive transport is important when the Peclet number $N_{Pe} = Vd_m/D_m < 1$, where d_m is the grain size of the material and D_m is the molecular diffusion coefficient adjusted for ground-water zones.

There are three basic causes of mechanical dispersion: (1) the fluid in the pores moves faster in the center, while the movement near the wall is slowed down by adhesion of water to solid surfaces, (2) water movement takes different pathways, sometimes compared to a so-called "random walk," (3) fluid traveling through larger pores will move faster than that in smaller pores. Lateral dispersion is caused by the fact that the flow paths can split and branch to the side (Fig. 7.4).

Researchers have found that the longitudinal and transverse dispersivity can be related to the average advective velocity, V, as (Fetter, 1988; Domenico and Schwartz, 1990)

$$D_L = \alpha_1 V \quad \text{and} \quad D_t = \alpha_2 V$$

The total magnitude of the dispersion is then

$$D_i = \alpha_i V + D_m \tag{7.10}$$

The proportionality coefficients, α, are called *coefficients of dynamic dispersivity*. These empirical coefficients are a characteristic property of the porous medium, while the molecular diffusion coefficient, D_m, is a property of the contaminant compound. Anderson (1979) compiled the magnitudes of the dynamic dispersivity coefficients. The ranges are given in Table 7.4.

The magnitude of the dispersion coefficient should be measured rather

TABLE 7.4 Ranges of Dynamic Dispersivity in Aquifers

Type of Aquifer	Longitudinal Dispersivity: a_L (m)	Ratio of Transverse Dispersivity to Longitudinal: a_T/a_L
Alluvial sediments	12–61	0.3
Glacial deposits	21	0.2
Chalk	1–3	
Limestone	7–61	0.1–0.3
Fractured basalt	30–91	0.2–1.5

Source: Based on data from Anderson (1979).

than estimated. The measured dispersion coefficients will have the advantage of accounting for the heterogeneity and anisotropy of the aquifer. Peaudecerf and Sauty (1978) described a method for determining the coefficients of dispersion from a dye pulse test by fitting an observed concentration curve to the dimensionless graphical form of Equation (7.6).

Retardation of Contaminants Moving Through an Aquifer

In a fashion similar to contaminant movement in soils, contaminants can be adsorbed on the solids in the aquifer. In some special cases the contaminants may also biodegrade in ground-water zones; however, the microorganism density in the ground-water zones is far less than that in soil. By definition, ground-water aquifers are saturated, hence volatilization does not occur. Similarly, there are no photochemical reactions therein because of the absence of light.

The concept of adsorption–desorption–coprecipitation reactions and models was introduced in detail in the preceding chapter for vadose–soil zones. The same concepts apply to the transport of solutes in ground-water zones. Adsorption is commonly incorporated into ground-water transport models using a retardation factor that is an empirical term derived from the partitioning coefficient, Π defined in the preceding chapter, and other parameters. The retardation factor is (Anderson, 1979; NCASI, 1985; Fetter, 1988)

$$R = \frac{V_x}{V_c} = 1 + \frac{\rho_b}{p_e}\Pi \qquad (7.11)$$

where
V_x = average convective velocity of water in the aquifer

V_c = velocity of the solute front where the solute concentration is one-half the original value
ρ_b = specific density of the porous medium (kg/m³ = g/l)
p_e = effective porosity (dimensionless)
Π = partitioning coefficient for the contaminant (l/gram)

Recall from Chapter 6 that the partitioning coefficient, Π, for chemical transport is related to the solubility of the chemical and to the properties of the porous medium, such as particulate organic carbon content (nonpolar chemicals), and the pH, clay, metal oxide, or sulfide content for polar chemicals and toxic metals. When a mixture of sorbable and nonsorbable contaminants enters the ground-water zone, each species will travel at a rate that is dependent on its relative velocity, V_c/V_x. It is clear from these concepts that contaminants with higher partitioning coefficients ($>10^3$ l/kg) will move at very slow rates, if at all, from the source of contamination if the aquifer is a composite of materials that have adsorbing capacity. It was pointed out in Chapter 6 that some porous media, such as sand and gravel, possess a very low adsorption capability. The concept of contaminant travel and the impact of dispersion and adsorption–desorption is shown on Figure 7.6.

Example 7.2: Movement of a Chemical Through Aquifer

A pesticide with the octanol partitioning coefficient $K_{ow} = 10^4$ l/kg was applied to a 10-ha orchard overlaying an unconfined aquifer. The average depth of the aquifer is 100 m, and the hydrodynamic slope of the aquifer is 0.5%. The aquifer contains primarily alluvial deposits (sand and gravel)

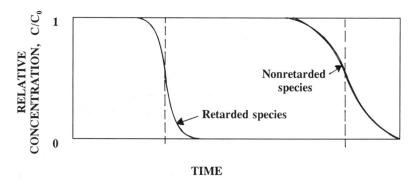

FIGURE 7.6. Influence of retardation on the movement of a solute front in a homogenous aquifer. (After Fetter, 1988.)

with about 0.1% organic matter. The specific volumetric density of the material is 1800 kg/m³. Estimate the spatial spread of the pesticide after 10 years in ground-water zones.

Solution

1. Estimate the advective velocity assuming the hydraulic conductivity of the aquifer (Table 7.3) is $K = 10^{-2}$ cm/sec and the porosity (Table 7.2) is 30%. Following Example 7.1 $V_x = KS/\rho = 10^{-2} \times 0.005/0.3 = 0.000167$ cm/sec $= 0.144$ m/day
2. Estimate the retardation factor. The partitioning coefficient (Eqs. (6.19) and (6.20))

$$K_{oc} = 0.63 \times K_{ow} = 0.63 \times 10^4 = 6300 \, l/kg$$
$$\Pi = K_{oc} \times (\%OC)/100 = 6300 \times 0.1/100 = 6.3 \, l/kg$$

Using Equation (7.11)

$$R = V_x/V_c = 1 + (\rho_b/p_e)K_{oc}$$
$$= 1 + (1800 \, [kg/m_3] \times 0.001 \, [l/kg]/0.3) \times 6.3 = 38.8$$

Hence the velocity of the plume is

$$V_c = V_x/R = 0.144 \, [m/day]/38.8 = 0.0038 \, m/day$$

In 10 years the pesticide will move only 0.0038 [m/day] × 365 [days/year] × 10 [years] = 13.8 meters from the application area (neglecting dispersion).

ORIGIN OF NATURAL GROUND-WATER (BASE-FLOW) QUALITY

The constituents that appear in ground-water can enter the aquifer through the recharge area, as leachate from the upper soil layer and from subsurface disposal of wasterwater, or they can originate from dissolution of minerals during ground-water passage through rocks and geological formations of the aquifer itself. As pointed out in Chapter 3 ground water becomes the base flow of surface-water bodies by ground-water discharge. The relation of geology to ground water and the origin of ground-water quality for the regions of the United States are described in Todd (1983). Also the U.S. Geological Survey publication by Hem (1985) and hydrogeological texts by Freeze and Cherry (1979), Moore and Ramamoothy (1984), Adriano (1986), Fetter (1988), Domenico and Schwartz (1990), and others, discuss this topic in detail.

The dissolved chemical composition of ground water includes *cations*, or positively charged ions, and *anions*, which are negatively charged ions. Calcium (Ca^{2+}), iron (Fe^{2+} or Fe^{3+}), magnesium (Mg^{2+}), and sodium (Na^+) are examples of the most abundant cations. Sulfate (SO_4^{2-}), nitrate (NO_3^-), chloride (Cl^-), and bicarbonate (HCO_3^-) are the most common

TABLE 7.5 Dissolved and Dissociated Constituents in Ground Water Listed According to their Relative Abundance

Major constituents (greater than 5 mg/l)
    Bicarbonate   Calcium
    Chloride   Magnesium
    Sulfate   Silicon
    Carbonic acid   Sodium

Minor constituents (0.01 to 10 mg/l)
    Carbonate
    Fluoride
    Nitrate
    Boron
    Iron
    Potassium
    Strontium

Trace contaminants (less than 0.1 mg/l)
    Aluminum   Nickel
    Antimony   Phosphate
    Arsenic   Radium
    Barium   Radon
    Beryllium   Selenium
    Cadmium   Silver
    Chromium   Thallium
    Cobalt   Thorium
    Copper   Uranium
    Lead   Vanadium
    Manganese   Zinc

Organic compounds (shallow aquifers)
    Humic acids   Tannins
    Fulvic acids   Lignins
    Carbohydrates   Hydrocarbons
    Aminoacids   Total Organic Carbon (TOC)

Organic compounds (deep aquifers)
    Acetate   Propionate

Source: After Davis and DeWiest (1966).

anions. In addition ground water may contain organic compounds, such as fulvic and humic acids, amino acids, tannins, and lignins. Table 7.5 lists some of the typical dissolved constituents.

Processes Controlling Natural Ground-Water (Base-Flow) Quality

Acid–base reactions. Water entering ground-water zones from atmospheric precipitation is generally acidic, that is, it has a pH below neutral. Dissolution of minerals in soil and ground-water zones provides the buffering of acidity.

Under normal conditions of an unpolluted atmosphere, the pH of precipitation is in an equilibrium with the saturated CO_2 concentrations in the atmosphere. As reported in Chapter 4, the partial pressure of CO_2 is 0.0003 bars, which will result in the normal pH of unpolluted precipitation of 5.6. However, as a result of the acid rainfall phenomenon, the pH of precipitation may become as low as 3. Also due to CO_2 production by soil bacteria from decomposing soil organic matter, the partial pressure of CO_2 in soils can reach values up to 0.1 bar. This alone can result in pH values of soil and ground water that are significantly below 5.

After entering into soil and ground-water zones, the acidic water dissolves minerals until its dissolving capacity is exhausted. The salt and ionic content of ground water depends on the type of minerals, their solubility, and time of contact. The evolution of chemical ground-water quality begins in the upper (local) ground-water zone (defined in Chapter 3) by the dissolution of soil and subsoil minerals. Due to its elevated acidity, water entering the aeration zone is a reducing agent with the ability to reduce various substances from their oxidized state. For example, ferric iron (less soluble) can be reduced to far more soluble ferrous iron.

The bicarbonate (HCO_3^-) content of water in the upper zone is a result of the dissolution of limestone ($CaCO_3 \times nH_2O$) and dolomite ($CaMgCO_3 \times nH_2$) carbonate minerals, as well as silicate minerals, by the acidic action of soil water and ground water. The following simple reactions describe the process for carbonate minerals

$$H_2O + CO_2 \rightleftharpoons H_2CO_3 \rightleftharpoons H^+ + HCO_3^-$$

and

$$CaCO_3 + H_2CO_3 \rightleftharpoons Ca^{2+} + 2HCO_{3-}$$

which indicate that carbonic acid is buffered in the reaction. Similarly, stronger acids (pH < 4.5) in acid rainfall will also undergo a neutralization process in contact with limestone or dolomite, for example,

$$CaCO_3 + 2H^+ \rightleftharpoons Ca^{2+} + H_2O + CO_2$$

From the preceding reactions it can be seen that the removal of acidity by limestone and dolomite minerals will increase the hardness of ground water (*hardness* is defined as the content of polyvalent cations, such as Ca^{2+}, Mg^{2+}, Fe^{2+}, and Sr^{2+}, commonly expressed on a $CaCO_3$ equivalent basis).

The bicarbonate (HCO_3^-) and carbonate (CO_3^{2-}) content of ground water represents the basic buffering system for neutralizing acid. This chemical buffering capacity is called *alkalinity*, and is defined as

$$\{Alkalinity\} = [HCO_3^-] + 2[CO_3^{2-}] + [OH^-]$$

and is expressed as $CaCO_3$ equivalent (Sawyer and McCarty, 1978).

Dissolution of minerals. Domenico and Schwartz (1990) have stated that dissolution of minerals is probably the most important process by which ground-water chemistry is controlled, and the recharge water derives almost the entire solute content through the dissolution of minerals along the flow path. The solubility of minerals and solids is determined by the dissolution–precipitation reaction and their equilibria. For example, a carbonate mineral such as limestone will partially dissolve into its ions

$$CaCO_3 \rightleftharpoons Ca^{2+} + CO_3^{2-}$$

which is then expressed by the dissolution–precipitation equilibrium

$$K = \frac{[Ca^{2+}][CO_3^{2-}]}{[CaCO_3]}$$

where K = reaction equilibrium coefficient. The magnitude of the equilibrium coefficient for carbonate minerals is about $K = 10^{-8.3}$, indicating very low solubility in the neutral pH range; however, other minerals, such as salt (NaCl, $K = 34.67$) or gypsum ($CaSO_4$, $K = 10^{-4.62}$), have relatively high solubility, so when ground water encounters these layers, high concentrations of Na^+, Cl^- or Ca^{2+}, and SO_4^{2+} are common.

Of interest is the solubility of minerals containing some priority pol-

TABLE 7.6 Sulfides Containing Minerals and Dissociation Reactions

Sulfide	Reaction	Equilibrium Coefficient	Reference
Sphalerite	$ZnS \rightleftharpoons Zn^{2+} + S^{2-}$	$K = 10^{-23.9}$	Domenico and Schwartz (1990)
Galena	$PbS \rightleftharpoons Pb^{2+} + S^{2-}$	$K = 10^{-27.5}$	Domenico and Schwartz (1990)
Hydroxylapatite	$Ca_5OH(PO_4)_3 \rightleftharpoons 5Ca^{2+} + 3PO_4^{3-} + OH^-$	$K = 10^{-55.6}$	Freeze and Cherry (1979)
Gibsite	$Al_2O_3 + 3H_2O \rightleftharpoons 2Al^{3+} + 6OH^-$	$K = 10^{-34}$	Freeze and Cherry (1979)
Fluorite	$CaF_2 \rightleftharpoons Ca^{2+} + 2F^-$	$K = 10^{-9.8}$	Freeze and Cherry (1979)

lutants (metals). Table 7.6 contains some examples using sulfides. The equilibrium dissolution reaction constant for many minerals and compounds is strongly affected by pH.

Complexation reactions. There are thousands of possible combinations of ions and molecules in water that may produce less soluble compounds. A *complex* is an ion that forms by combining simpler cations, anions, and molecules into a larger, typically less soluble, unit. The cation or central atom is in most cases one of metals. The anions, called ligands, include common inorganic species, such as F^-, Cl^-, SO_4^{2-}, PO_4^{3-}, in reduced conditions S^{2-} or HS^- and CO_3^{2-}, as well as organic molecules, such as humic and fulvic acids. This process of metal complexation was described in the preceding chapter for sediments and soils.

Redox reactions. When oxygen is present in ground water many compounds are present in oxidized and less soluble states. When oxygen is exhausted by the decomposition of organic carbon or other oxygen-consuming reactions redox conditions will convert them to more soluble compounds. However, conversion of sulfate (oxidized) to sulfite (reduced) may cause the more soluble metallic ions to be complexed by sulfides and precipitate as metal-sulfides.

Ion-exchange or surface adsorption–desorption reactions. The ionic species and polar and nonpolar molecules may react with surfaces of solids. *Ion exchange* is a process in which ions in the mineral lattice are replaced by one of the ions in the aqueous solution. In *adsorption–desorption* processes the solid surface attracts and retains a layer of ions or molecules from the solution (Fetter, 1988).

Adsorption–desorption reactions for phosphorus, ammonia, and priority pollutants were described in Chapter 6. Adsorption sites are provided mostly by clays and organic particulates and precipitated metal oxides and sulfide particles.

Ion exchange sites can be found mostly in clay and organic materials; however, Fetter (1988) pointed out that most soil materials have some ion exchange capacity. Both cation and anion exchanges can occur; however, the cation exchange capacity is more dominant. Of concern is the ion exchange process in which sodium from water replaces divalent cations (Ca^{2+} and Mg^{2+}) in soil minerals.

Ground-Water Quality Zones

Mineral ground-water quality is variable. Shallow aquifer water has low mineral content, but often exhibits seasonal or even day-to-day variations. Waters from deep underground zones have high mineral content, but fairly constant quality and temperature. The longer water resides in the underground zones, the higher the measured salt content.

Freeze and Cherry (1979) reported the evolution of mineral ground-water quality known as the Chebotarev or Ignaovich and Souline sequence. According to this concept, ground-water evolution tends to be in the direction of atmospheric to seawater quality. For large sedimentary basins, the sequence can be described in three main zones that correlate well with the depth:

1. The upper zone is characterized by active ground-water flushing through relatively well-leached rocks and soils. Water in this zone has HCO_3^- as the dominant anion and is relatively low in dissolved salts.
2. The intermediate zone has less active ground-water circulation and higher total dissolved solids. Sulfate is normally the dominant anion in this zone.
3. The lower zone has a very sluggish ground-water flow. Highly soluble minerals are commonly present. Very little flushing and leaching by ground water occurs due to the extremely slow movement to water. Dissolved solids are very high and chlorides are common anions.

These three water quality zones are similar to the ground-water zones (local, intermediate, and regional) described in Chapter 3.

Sources of Natural (Background) Ground-Water Quality

Mineral salts and dissolved (ionized) minerals are the most common sources of natural ground-water quality. In terms of contaminant content, natural ground-water contamination may be quite high and sometimes may hinder or even prevent the intended use of water from the source or, after discharge, use of the receiving body of water. For example, streams draining watersheds containing lead minerals (such as galena) in surface layers may exhibit high natural lead content. Klusman and Edwards (1977) measured toxic metals in ground water from the mineral belt of Colorado and noted that the drinking water standards were violated in 14% of the samples for Cd, 1% for Cu, 2% for Hg, and 9% for Zn. It may be noted that present water quality standards for toxic metals are more stringent than those in use when Klusman and Edwards' study was conducted.

Of the 129 compounds and chemicals on the U.S. EPA's Priority Pollutant List, 13 are inorganic elements that have natural (background) occurrence. These priority pollutants include: antimony, arsenic, beryllium, cadmium, chromium, copper, lead, mercury, nickel, selenium, silver, thallium, and zinc. The following list contains the natural sources and minerals from which these elements may originate:

Element	*Natural Source of Minerals*
Antimony	Stibnite (Sb_2S_3), geothermal springs, mine drainage
Arsenic	Metal arsenides and arsenates, sulfide ores (arsenopyrite), arsenite ($HAsO_2$), volcanic gases, geothermal springs
Beryllium	Beryl ($Be_3Al_2Si_6O_{16}$), phenacite (Be_2SiO_4)
Cadmium	Zinc carbonate and sulfide ores, copper carbonate and sulfide ores
Chromium	Chromite ($FeCr_2O_4$), chromic oxide (Cr_2O_3)
Copper	Free metal (Cu^0), copper sulfide (CuS_2), chalcopyrite ($CuFeS_2$), mine drainage
Lead	Galena (PbS)
Mercury	Free metal (Hg^0), cinnabar (HgS)
Nickel	Ferromagnesian minerals, ferrous sulfide ores, pentladite (($Ni,Fe)_9S_8$), nickel oxide (NiO_2), nickel hydroxide ($Ni(OH)_3$)
Selenium	Free element (Se^0), ferroselite ($FeSe_2$), uranium deposits, black shales, chalcopyrite–pentladite–pyrrhotite deposits
Silver	Free metal (Ag^0), silver chloride ($AgCl_2$), argentide (AgS_2), copper, lead, zinc ores
Thallium	Copper, lead, silver residues
Zinc	Zinc blende (ZnS), willemite ($ZnSiO_4$), calamine ($ZnCO_3$), mine drainage

Although nitrate in ground water is primarily of anthropogenic (cultural) origin geologic N, that is, N associated with certain geologic formations of sedimentary origin, is also known (Boyce et al., 1976; Chalk and Keeney, 1971). Chalk and Keeney found widely varying concentrations of NH_4^+ and NO_3^- in Wisconsin limestones and suggested that many limestones are potential sources of NO_3^- in ground water, and hence base-flow contamination.

Kreitler and Jones (1975) reported nitrate levels in the ground water of Runnels County, Texas, that reached values of 250 mg/l, well above the drinking water standard of 10 mg/l. It was found that almost 80% of the nitrate content leached from natural soil nitrogen as a result of cultivation during the last 50 years.

Estimating Base-Flow Quality

The nature and concentrations of dissolved constituents in natural ground water (base-flow contributions) are dependent on the composition of the aquifers through which the ground water flows. The regional characteristics of aquifers are described in a publication by Hem (1985). Depending on the type of geology of the region, water may originate from primarily carbonate formations, crystalline rocks, or sedimentary systems (Freeze and Cherry, 1979). Furthermore, other regional characteristics such as type of surficial vegetation, primarily forestation of the watershed and/or degree of urbanization or agricultural land use, slope and elevation of terrain, soil permeability, and precipitation amount and distribution will affect ground-water and base-flow quality. In agricultural areas with irrigation, ground-water (base-flow) quality is affected by irrigation return flow. The new ecoregional approach to water quality grouping (Omernik and Gallant, 1990; Galant et al., 1989) may provide new relationships for regional base-flow quality determinations.

Mineral loads in base flow can be estimated using an approach developed by Betson and McMaster (1975). These authors correlated the water quality of some undisturbed streams in the Tennessee Valley Authority area using the following logarithmic functional relationship between the quality and flow:

$$C = a(13.66 \times Q/DA)^b \qquad (7.12)$$

where
C = the concentration of a mineral constituent (mg/l)
Q = the stream flow (m³/s)
DA = the drainage area (km²)
a, b = empirically determined coefficients

Ground-Water and Base-Flow Contamination

The two coefficients in Equation (7.12) were related to land use, soils, and geological factors by a linear multiregression formula

$$a, b = N_1 F + N_2 CR + N_3 S + N_4 I + N_5 U \qquad (7.13)$$

where
- a, b = the coefficients just defined
- F = the fraction of the watershed area that is forested
- CR = the fraction of the watershed over carbonate rock
- S = the drainage area fraction over shale–sandstone rock
- I = the drainage area fraction over igneous rock
- U = the drainage area fraction over unconsolidated sediments
- $N1, \ldots, N5$ = regression coefficients

The four independent geological variables simply allocate the drainage area among the rock types present in the watershed and must sum one.

Table 7.7 shows regression coefficients for major mineral constituents obtained by analysis of TVA watersheds. As stated by the authors, the use of Equation (7.13) requires caution since constituent rating curves (Eq. (7.12)) have been found to display a hysteric effect with seasons and with rising and falling stages of the hydrograph (see Chapter 5 for an explanation of the difference in concentrations for suspended solids). Variations among watersheds are also influenced by other factors, as previously mentioned.

Example 7.3: Base Flow Quality

Estimate base background nitrate content in a watershed with the following characteristics:

$$\text{Specific flow} \quad \frac{Q}{DA} = 0.005 \text{ m}^3/\text{sec-km}^2$$

% forest	$F = 25\%$
% carbonate rock	$CR = 5\%$
% sandstone	$S = 35\%$
% igneous rock	$I = 2\%$
% unconsolidated rock	$U = 58\%$

Solution Using the Betson and MacMaster concept, select the coefficients from Table 7.7. Then the coefficients for the rating equation become (Eq. (7.13))

$$a = -1.13 \times 0.25 + 3.52 \times 0.25 + 1.02 \times 0.35$$
$$+ 1.3 \times 0.02 + 0.84 \times 0.58$$
$$= 1.47$$
$$b = -0.70 \times 0.25 + 0.262 \times 0.25 + 0.899 \times 0.35$$
$$+ 1.063 \times 0.02 + 0.297 \times 0.58$$
$$= 0.67$$

Hence, the nitrate concentration becomes (Eq. (7.12))

$$C_{NO_3^-} = a(13.66Q/DA)^b = 1.47(13.66 \times 0.005)^{0.67} = 0.24 \, mg/l$$

IMPACT OF DIFFUSE POLLUTION ON GROUND WATER AND BASE FLOW

There are many potential sources of ground-water contamination, including septic tank systems, solid waste disposal, land disposal of sewage, industrial wastes and sludge, agricultural and suburban irrigation, fertilizer and pesticide application in agricultural and suburban land uses, leaks and spills of chemicals and oils on the surface and from underground leaking storage tanks, mining, and road deicing. Some of these sources can be viewed as point sources of pollution. They are regulated and may result in only a localized aquifer contamination. Other sources, such as septic tanks and large-scale agricultural practices, are typical nonpoint pollution problems still mostly unregulated. It has been estimated that about 2% of aquifers have been contaminated (Lehr, 1981).

Effects of Septic Tank Disposal Systems

The term septic system is commonly used to describe a subsurface, anaerobic sewage disposal that uses soil filtration and adsorption for attenuating the effluent. Detailed information on ground-water impact of septic tank disposal systems is contained in a publication by Canter and Knox (1985).

Approximately 20 million residents, or 29% of the U.S. population, dispose of their sewage by individual on-site systems. The total amount of sewage and wastewater discharged to the subsurface in the United States is $3 \times 10^9 \, m^3/yr$ (U.S. EPA, 1977a, 1977b; Canter, Knox, and Fairchild, 1987). Apparently, septic tanks represent the highest total volume of wastewater discharged directly to ground water and are the most frequently recorded sources of contamination of ground water and surface flow.

The amount of discharge from septic tank systems is commonly es-

TABLE 7.7 TVA Base-Flow Mineral Quality Model: Regression with Forest and Geological Variables

Quality Parameter	Regression Coefficient	N^b	Regression Coefficient Value					Statistics		
			N_1 F	N_2 C	N_3 S	N_4 I	N_5 U	R	Standard Deviation	F
SiO_2	a	64	−1.26	5.42	6.78	10.2	8.95	0.64	1.69	8.11
	b		−0.135	0.051	0.099	0	−0.029	0.40	0.114	2.26
Fe	a	29	0.035	0.020	0.009	−0.008	0.387	0.95	0.004	42.25
	b		−0.173	0.272	0.104	−0.125	0.397	0.38	0.482	0.80^a
Ca	a	66	−8.52	53.9	13.4	8.32	8.41	0.85	9.0	31.42
	b		0.064	−0.116	−0.203	−0.229	−0.005	0.32	0.153	1.32^a
Mg	a	66	−2.81	11.4	3.41	3.05	2.45	0.75	2.62	15.50
	b		−0.148	−0.145	−0.074	−0.104	0.513	0.67	0.197	9.99
Na	a	44	−1.79	2.23	2.84	3.00	3.74	0.74	0.50	9.45
	b		−0.318	0.079	0.122	0.110	−0.007	0.48	0.138	2.33
K	a	44	−1.08	2.51	1.94	1.58	1.80	0.72	0.47	8.12
	b		−0.152	−0.195	−0.061	0.033	−0.158	0.20	0.254	0.32^a
HCO_3	a	66	−22.8	200	35.3	26.8	21.8	0.86	32.1	34.67
	b		0.110	−0.156	−0.294	−0.355	−0.139	0.35	0.132	1.71^a

		N								
SO$_4$	a	66	−7.41	9.15	12.5	7.90	9.56	0.39	5.35	2.19
	b		−0.302	0.103	0.155	0.272	0.592	0.49	0.274	3.69
Cl	a	66	−1.86	3.21	2.95	2.58	3.81	0.60	0.93	6.68
	b		−0.171	0.010	0.088	0.099	0.067	0.24	0.14	0.72[a]
NO$_3$	a	63	−1.13	3.52	1.02	1.30	0.84	0.80	0.71	20.11
	b		−0.70	0.262	0.899	1.063	0.297	0.26	0.671	0.84[a]
TDS	a	66	−39.0	195.6	68.5	55.5	57.7	0.84	30.6	28.42
	b		0.016	−0.094	−0.146	−0.142	0	0.26	0.14	0.88[a]
CaCO$_3$	a	66	−33.4	182.8	48.1	34.3	31.5	0.86	29.3	33.06
	b		0.033	−0.150	−0.176	−0.222	0.131	0.49	0.151	3.78
Specific conductance	a	46	−145	357	180	142	128	0.88	54	26.74
	b		−0.015	−0.078	−0.134	−0.095	0.051	0.43	0.106	1.82[a]
pH	a	65	−0.573	8.37	7.32	7.33	6.86	0.73	0.42	13.53
	b		−0.003	−0.010	−0.003	−0.013	−0.003	0.16	0.021	0.32[a]
Color	a	63	−1.79	2.50	9.17	9.75	10.8	0.23	8.65	0.62[a]
	b		−0.376	0.211	0.339	0.204	0.448	0.32	0.364	1.26[a]

Source: After Betson and McMaster (1975).

[a] Not significant at 0.9 level.

[b] N is number of watershed analyses.

timated as 280 liter/cap-day (75 gpcd). Septic tank effluents contain 40 to 80 mg N/l, 11 to 31 mg P/l, and 200 to 400 mg/l of BOD_5 (Sikora et al., 1976; Canter, Knox, Fairchild, 1987). Based on reported efficiencies of soil absorption systems, Canter and Knox (1985) reported the following typical concentrations entering the ground water: BOD_5—28 to 84 mg/l; COD—57 to 142 mg/l; ammonia nitrogen—10 to 78 mg/l; and total phosphates—6 to 9 mg/l. Other ground-water constituents of concern include bacteria, viruses, nitrates, synthetic organics, and toxic metals. On the other hand, Brown et al. (1979) and Reneau and Petry (1976) reported that organic matter (BOD_5, pathogenic microorganisms) and phosphates were effectively removed by most properly designed and permitted subsurface disposal systems and did not penetrate more than 1.5 m below the level of discharge or beyond the immediate vicinity of the seepage field.

Nitrification is typically completed in seepage fields located in well drained soils and the mobile nitrate-N will move into ground-water zones. An investigation of nitrogen mass balance in Long Island, New York, revealed that more 20% of the total nitrogen contribution from subsurface disposal systems leached as nitrate-N into ground water (Andreoli et al., 1979).

According to Fetter (1988) septic tanks are most likely to contribute to ground-water contamination in areas where (1) there is a high density of homes with septic tanks, (2) the soil layer over permeable bedrock is thin, (3) the soil is extremely permeable such as gravel, or (4) the water table is shallow, one meter or less below the surface. Typically, local ordinances regulating permits for septic tank disposal do not pay enough attention to the preservation of ground-water resources from pollution by septic tank effluents. The standard percolation test simply favors highly permeable, sandy, and gravel soils. Such soils do postpone the failure of the system by hydraulic overloading; nevertheless, adsorption and the purifying capacity of such soils is greatly reduced and pollutants can move downwards and contaminate the ground water. It also has to be realized that most septic tank systems installed in the 1950–1970 period are now exceeding their design life, which is typically 10 to 15 years. Use of manmade organic chemical additives to prolong the life of the systems and/or improve their impaired function resulted in organic contamination of ground water (U.S. EPA, 1980).

Underground Storage Tanks

Underground leaking storage tanks are now considered to be a major source of ground-water contamination by chemicals, especially those from petroleum hydrocarbons. Because there are so many underground tanks (about 2 million) and only a few of them are corrosion resistant (Canter, Knox, and Fairchild, 1987), the problem could be considered to be of a diffuse nature that could reach regional rather than localized scope (for example, New Jersey). The problem has received serious attention and concern of the U.S. Environmental Protection Agency which, in 1985 created the Office of Underground Storage Tanks.

Buried gasoline tanks in service stations are the most common source of the problem. The leaking tanks are a source of a number of aliphatic and aromatic hydrocarbons in ground water (see Chapter 6 for a definition). Many of these hydrocarbons are classified as priority pollutants.

Land Application of Water and Wastewater

Sewage and Wastewater Disposal

The "no pollution" policy of the Clean Water Act advocated in the 1970s caused an increased interest in land disposal of liquid wastes such as conventionally treated sewage effluents, processing plant wastes, animal wastes and feedlot runoff, and sewage sludge. In many arid regions land application of sewage helps alleviate shortages of irrigation water and is even used to replenish the ground-water aquifers. The city of Tucson in Arizona is reusing almost all of its municipal treatment plant effluent for urban irrigation. The long-lasting drought in southern California in the late 1980s and early 1990s increased demand for land disposal of sewage as a means of water reuse.

Although the soil has a great capacity for attenuating contaminants, the reclaimed wastewater from ground-water zones is obviously not of the same quality as the native ground water (Bouwer, 1974). There are three types of land applications of wastewater (Reed, Middlebrooks, and Crites, 1988; U.S. EPA, 1980, 1984):

slow-rate systems
overland flow systems
rapid infiltration

The three systems are shown on Figure 7.7.

The *slow-rate systems* (*SR*) are most common in the United States for treatment of municipal wastewater and effluent reuse in arid areas. In Europe, these systems have been in use for centuries. Their hydraulic loading rate is mostly matched to irrigation and nutrient requirements for crops and soil permeability. In arid regions the hydraulic loading is related to the irrigation requirement and prevention of salt buildup in

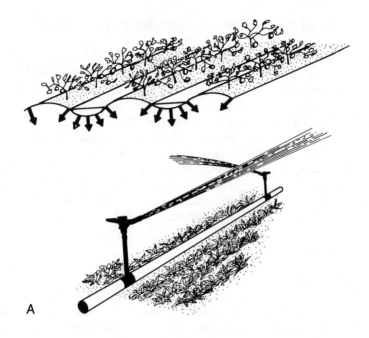

A

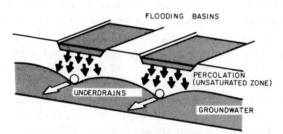

(a) RECOVERY OF RENOVATED WATER BY UNDERDRAINS

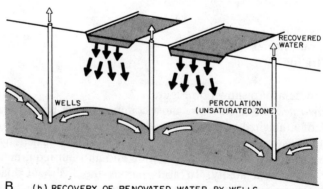

B (b) RECOVERY OF RENOVATED WATER BY WELLS

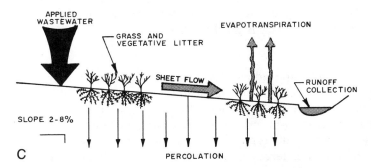

FIGURE 7.7. Land application systems of sewage and wastewater. (*A*) Slow-rate application. (*B*) Rapid infiltration. (*C*) Overland flow systems. (From Novotny et al., 1989.)

soils. These systems are essentially irrigation systems and have similar problems with irrigation return flow and its impact on ground water and base flow (see the next section on irrigation). For the winter nongrowing period, wastewater flows require storage. Of the three systems mentioned the SR exhibits the highest nutrient removal due to the combined effects of nutrient uptake by crops and attenuation by soils.

In *overland flow systems* (*OF*) wastewater is treated as it moves in graded and maintained grassed and vegetated sloped areas and the treated effluent is collected as residual runoff at the bottom of the slope. Percolation of wastewater is not desirable and should be minimized by the selection of slowly permeable soils, soil compaction, and/or locating these systems over an impermeable subsurface stratum. Under these conditions the impact of these systems on ground-water resources should be minimal. These systems are similar to grassed buffer strips used for the treatment of urban and agricultural runoff (see Chapters 10 and 11). Nitrogen removal is accomplished by nitrification–denitrification processes and depends on the BOD/nitrogen ratio. If the nitrogen in the effluent is primarily in nitrate form, then the removal is minimal (Reed, Middlebrooks, and Crites, 1988).

In contrast to OF systems *rapid infiltration* (*RI*) systems rely on the infiltration and filtration of wastewater in permeable soils. If subsoils are permeable, the effluent will reach ground water and, if improperly designed, has the potential of becoming a cause of ground-water contamination. Removal of contaminants in the upper soil layer is accomplished by physical–chemical interaction (adsorption) and biochemical degradation (both aerobic and anaerobic). Vegetation and its nutrient uptake is not considered. Again if most of the nitrogen is in nitrate form, the removal efficiency is greatly reduced.

Application

Sewage can be applied at a low rate in SR systems or a high rate for OF and RI systems. The low application rates—in centimeters week—are based on matching the nutrients from the applied wastewater with their uptake by crops, which reduces nitrate accumulation and pollutant buildup in soils. These systems present the minimal potential for the contamination of ground-water resources and base flow. The disadvantage of low-rate application systems is the large area requirement (approximately 30 ha of land per 1000 m^3/day of sewage).

The high-rate systems require only a fraction of the land needed for low-rate systems. Since water in these systems is applied in excess of evapotranspiration rate, a portion of the applied flow will leach into ground-water zones. To minimize the impact on ground-water resources, the land application systems should be designed and operated (a) to obtain recharge water of the best possible quality (particularly with regard to nitrogen), and (b) to restrict the spread of recharged water into the native ground water (Bouwer, 1974; Reed, Middlebrooks, and Crites, 1988; U.S. EPA, 1981, 1984).

Problems and Restrictions

The problems associated with the land application of wastewater, especially RI systems, are similar to septic systems discussed previously; however, much greater volumes of wastewater are concentrated in a smaller area. Mobile pollutants, such as nitrates are of the greatest concern, since evidence indicates that several other common contaminants (BOD, pathogenic microorganisms, and phosphates) remain near the area of application. Bacteria and viruses die off quite rapidly as wastewater passes through the soil material (Bell and Bole, 1978; U.S. EPA, 1980, 1984).

Freeze and Cherry (1979) state that ground-water contamination by mobile organics may become a serious problem. Treated wastewater contains many dissolved organic compounds, and some of the potentially dangerous components such as chlorinated hydrocarbons, may be created by the treatment process itself. Since many of these compounds are not biodegradable and can be partially mobile, their impact should always be evaluated.

As shown on Figure 7.8, land application systems may contaminate all three ground-water systems, that is, local, intermediate, and regional aquifers. The portion of the aquifer that is recharged by treated waste-

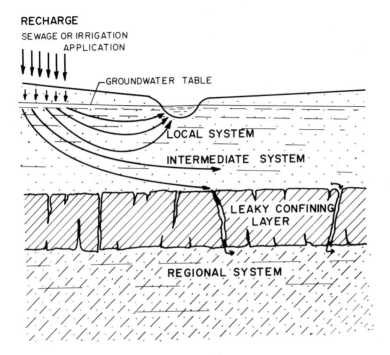

FIGURE 7.8. Possible routes of contaminants in ground-water systems.

water effluents should not be used as a source of drinking water and should be restricted, and water should be taken out at some distance from the recharge area (Fig. 7.9). This distancing could occur naturally as a base flow into a nearby stream or lake, or artificially by drains (shallow aquifer), or by wells (deep aquifer). After collection, water can be reused for irrigation, recreation, aesthetics, or other nonpotable uses, or discharged into receiving streams to augment their flows (Bouwer, 1974). With such aquifer zoning and partitions, the portion of the aquifer between the land application site and the point of discharge is used as a natural filtration system.

Land Application of Sludge
Sludge generated by wastewater treatment facilities is commonly applied to agricultural lands as a fertilizer and soil conditioner. The effect of the land application of sludges on ground-water quality depends on the transformation that occurs within the topsoil horizon and in unsaturated soils. Although most toxic metals will be retained by the topsoil, the toxic

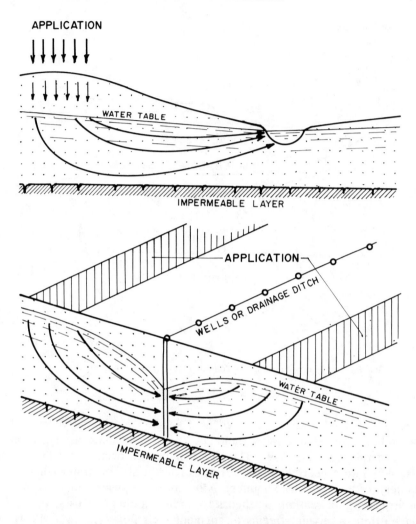

FIGURE 7.9. Restriction of aquifer contamination by sewage application, natural drainage (upper portion), and artificial drainage (lower portion) (according to Bouwer, 1974).

metal content of sludges is of concern. Concentrations of toxic metals in wastewater sludges is much higher that those in raw wastewater.

Irrigation and Irrigation Return Flow

Water applied to land either in the form of a treated effluent or as water withdrawn from a nearby surface-water body contains certain dissolved salts and dissociated ions. A portion of the irrigation water, after application on an irrigated area, is returned to the atmosphere by evapotranspiration. Since evaporated or transpired water has no salt content, there is a subsequent salt and contaminant buildup in soils. The portion returned to the atmosphere may range from less than 20% in high-rate application systems in humid climatic conditions to almost 100% in low-rate application systems in arid or semiarid climatic zones.

In order to maintain an acceptable salt content of soils to sustain crop growth and the fertility of the soils, excess irrigation water must be applied if natural precipitation is not sufficient to control salt buildup. The excess water containing increased salinity and leachate from soils is collected by drainage (natural or man-made) and/or will percolate into ground-water zones. The irrigation tail water collected by drainage systems (Figs. 7.10 and 7.11) or leached into groundwater zone is called *irrigation return flow*, and it represents one of the more serious problems associated with diffuse pollution from agriculture.

The concentration of salts in water percolating through the soil root level zone into irrigation return flow or to the ground water can be computed from the following mass balance:

$$C_i D_i = C_d (D_i - D_e) \tag{7.14}$$

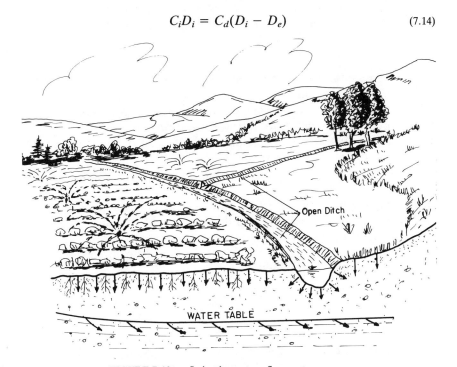

FIGURE 7.10. Irrigation return flow systems.

FIGURE 7.11. Drainage ditch in agricultural fields. Drainage water often contains pollutants leached from soils by percolated water and in irrigation return flows.

where

C_i = salt or conserved contaminant concentration of water or wastewater used for irrigation

D_i = amount of irrigation water, also including effective precipitation (precipitation that is not lost as surface runoff)

C_d = salt or conserved contaminant concentration of water percolating from the root zone downward

D_e = amount of water used by evapotranspiration

The amount of excess irrigation water that has to be applied to control salt or contaminant buildup in soil depends on the tolerance of crops to the compound in soil water, the compound content in irrigation water, evapotranspiration rate, crop uptake, and other losses from the system. The leaching ratio is computed from Equation (7.14) as

$$\frac{D_i}{D_e} = \frac{C_d}{C_d - C_i} \qquad (7.15)$$

The salinity of irrigation water is usually expressed as conductivity in micromhos per centimeter (1000 μmho/cm ≈ 640 mg/l m of total dissolved solids—TDS). The salt tolerance of crops ranges from less than 500 μmho/cm for salt-sensitive crops, such as most fruit trees and some vegetables (celery, strawberries, or beans) to more than 1500 μmho/cm for salt-tolerant crops, such as cotton, beets, barley, and asparagus. Most common grain crops and vegetables have medium tolerance (500 to 1500 μmho/cm) to salts. The leaching requirement is then defined as

$$LR = \frac{EC_i}{EC_d} = \frac{C_i}{C_d} \qquad (7.16)$$

where
EC_i = electric conductivity of irrigation water
EC_d = salt tolerance of crops or conductivity of drainage water

Combining Equations (7.15) and (7.16), the leaching ratio becomes

$$\frac{D_i}{D_e} = \frac{1}{1 - LR} \qquad (7.17)$$

Although the irrigation return flow has been recognized as a water quality problem, the Clean Water Act specifically excluded agricultural runoff and irrigation return flows from the definition of pollution. However, there is no doubt about the pollution effects of these flows. Investigations in central Wisconsin (Saffigna and Keeney, 1977) and elsewhere (Burwell et al., 1976; Brown, 1975) revealed that nitrate concentrations in the subsurface water in agricultural areas receiving irrigation are well above the background level. In the Wisconsin study, nitrate-N concentrations ranged up to 56 mg/l.

However, Kreitler and Jones (1975) and Brown (1975) have documented that the nitrate in the ground water below irrigated fields originated from the natural soil organic nitrogen that was leached from the soil as a result of cultivation and irrigation over a 50-year period. Nitrate levels of the ground water below irrigated fields in Runnels County in Texas reported by Kreitler and Jones have reached average values of 250 mg/l, and over 80% of it was attributed to native nitrogen leaching from the soil. The presence of NH_4^+ in ground water usually indicates incomplete nitrification of nitrogen in sewage. Similar conclusions were reached by Brown for subsurface nitrate content in the irrigated fields of the San Joaquin Valley in California.

The salinity problem is especially troublesome in arid zones of the western United States. It must be realized that water in some streams can be reused for irrigation several times, since downstream users irrigate mostly with irrigation return flows from upstream farms, with increased salinity and pollution content at each reuse.

Example 7.4: Amount and Quality of Irrigation Return Flow

An agricultural field growing crops was irrigated by a treated effluent with the following quality characteristics:

TDS—500 mg/l ($EC = 781\,\mu\text{mho/cm}$)
nitrogen—10 mg/l ($= \text{g/m}^3$)

The effluent was applied at a rate of 10 cm/week. The evapotranspiration rate during the irrigation period (lasting 2 months) was 5 cm/week and there was no appreciable precipitation during the period. The crop yield was about 5 tonnes/ha and the nitrogen content of the crop was 20 kg/tonne (see Table 6.4 for yields and N contents of specific crops). Estimate the amount and concentration of nitrogen leached into the ground water and the salinity of the soil and drainage water (irrigation return flow). The pH of the soil is around 7, which implies that ammonia volatilization is minimal.

Nitrogen uptake assuming 4 months growing period

$$\begin{aligned} UP_N &= \text{crop yield} \times \text{nitrogen content/growing period} \\ &= 5\,(\text{tonnes/ha}) \times 20\,(\text{kg/tonne})/16\,\text{weeks} \\ &= 6.25\,\text{kg/ha-week} = 6250\,\text{g/ha-week} \end{aligned}$$

Nitrogen input from the effluent is

$$0.1\,(\text{m/week}) \times 10{,}000\,(\text{m}^2/\text{ha}) \times 10\,(\text{g/m}^3) = 10{,}000\,\text{g/week-ha}$$

Solution The amount of nitrogen leached can be obtained by subtracting the plant uptake from the nitrogen input. For calcareous soil ammonia volatilization should be also considered (see Eq. (6.31)). Hence

$$\text{Nitrogen leached} = 10{,}000 - 6250 = 3750\,\text{g/week}$$

Nitrogen concentration in the leachate is

$$C_{dN} = \frac{\text{Nitrogen leached}}{\text{Volume of water leached}} = \frac{3750\,(\text{g/week})}{(0.1 - 0.05) \times 10{,}000} = 7.5\,\text{mg/l}$$

Estimate the salinity of the leachate. Using Equation (7.17) the leaching requirement factor is

$$LR = 1 - D_e/D_i = 1 - 5/10 = 0.5$$

and by combining Equations (7.16) and (7.17)

$$C_d = \frac{C_i}{LR} = \frac{300}{0.5} = 1000\,\text{mg/l} = 1562\,\mu\text{mho/cm}$$

Note: This example presumes no leaching of the original nitrogen content of the soil. Examples 6.6 to 6.8 from the preceding chapter may be used to approximately estimate mineralization and leaching of the soil nitrogen.

Ground-Water Pollution from Solid Waste Disposal Sites

In 1970 there were about 20,000 solid waste disposal sites in the United States. However, only 6% were classified as sanitary landfills that do not cause environmental problems and were properly operated. Today there are few licensed and permitted landfills and the suitable landfill site is very difficult to locate. Solid waste disposal sites are now sophisticated engineering operations employing resource recovery (collection of methane and subsequent conversion to energy), collection of liquid wastes (leachate) produced by the landfill with subsequent pretreatment and treatment and daily covering of wastes with soil. A landfill site after ceasing operation should be reclaimed. Figure 7.12 shows an example of a well-operated landfill site in southeastern Wisconsin.

FIGURE 7.12. Omega Hill solid waste (refuse) disposal site in southeastern Wisconsin.

However, for each well-designed and operated landfill there are hundreds of abandoned unsanitary dumps of refuse and toxic chemicals that are now causing ground-water contamination problems. During the decade between 1970 and 1980, a large number of the landfills, including some receiving radioactive wastes, were established. Stored and decomposing wastes and leaching from disintegrating drums left on these sites will represent a serious problem for decades. In the United States, such sites have been inventoried and if severe problems have occurred they were classified by the U.S. Environmental Protection Agency as hazardous waste disposal sites (so-called "Superfund" sites).

Although solid waste disposal sites are considered point sources of pollution, leachate from unsanitary landfills and dumps may have polluted large portions of adjacent aquifers and appear as contaminated base flow in a diffuse manner. Furthermore, some dangerous toxic compounds are commonly a part of the overall composition of the landfill leachate, especially when the landfill is used for disposal of toxic chemicals. Table 7.8 shows the ranges in concentration for various chemical constituents and the physical parameters of typical leachates from municipal solid waste disposal sites. It should be noted that in countries that use coal for household heating, the composition of leachate may be quite different from that typical for U.S. conditions (Johansen and Carlson, 1976).

TABLE 7.8 Leachate Characteristics from Municipal Solid Waste Disposal Sites

Leachate	Median Value (mg/l)	Ranges of All Values (mg/l)
Alkalinity (as $CaCO_3$)	3050	0–20,850
Biochemical oxygen demand (BOD_5)	5700	81–33,360
Chemical oxygen demand (COD)	8100	40–89,520
Copper (Cu)	0.5	0–9.9
Lead (Pb)	0.75	0–2.0
Zinc (Zn)	5.8	3.7–8.5
Chloride (Cl^-)	700	4.7–2500
Sodium (Na^+)	767	0–7700
Total dissolved solids (TDS)	8955	584–44,900
Ammoniacal nitrogen (NH_4^+)	218	0–1106
Total phosphate (PO_4^{3+})	10	0–30
Iron (Fe)	94	0–2820
Manganese (Mn)	0.22	0.05–125
pH (pH units)	5.8	3.7–8.5

Source: After U.S. Environmental Protection Agency (1977b).

Leachate Management and Minimization of Ground-Water Impact

There are several general methods for managing leachate: natural attenuation by soils, prevention of leachate formation, collection and treatment, pretreatment to reduce volume and solubility, and detoxification of hazardous wastes prior to landfilling. Leachate undergoes natural attenuation by various chemical, physical, and biological processes as it migrates through soil. Whether natural attenuation will be adequate to prevent ground-water pollution should be evaluated for each site. The generation of leachate can be minimized by restricting water from infiltrating the landfill. This is accomplished by providing appropriate surface drainage and/or placing an impermeable liner over the daily accumulation of refuse. Another method of controlling leachate is to collect it at the bottom of the landfill and treat it before discharging it into surface water or the land. In most cases collected leachate must be pretreated by an anaerobic biological treatment unit (anaerobic filter of suspended growth reactors) before it is discharged into sewers. The high BOD strength of the leachate makes it difficult to treat it in conventional aerobic treatment units, and without pretreatment conventional biological treatment plants could become overloaded.

Figure 7.13 shows a well barrier, which prevents leachate from reaching the ground water. Newly constructed landfill requires a clay and geomem-

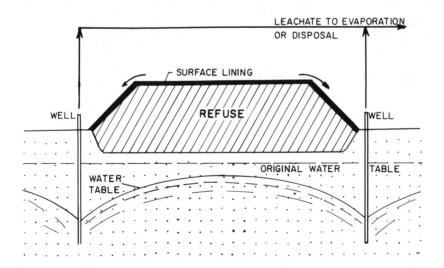

FIGURE 7.13. Control of ground-water pollution from landfills by creating a ground-water depression barrier.

brane lining and a suitable low permeability (clay) substratum to virtually eliminate potential seepage of leachate into ground water.

Most regulations recommend or require that landfill sites are developed on uplands rather than floodplains and in low-permeability soils. Geologically such sites are difficult to find; however, the sites must also be socially and politically acceptable. If the landfill receives hazardous (toxic) waste, a TCLP extraction toxicity analysis must be performed. The solid waste disposal site is then considered toxic (hazardous) if following the application of the TCLP test the leachate (extract) from a representative sample of waste contains any of the 52 listed toxic compounds in concentrations that exceed the limit (see the Appendix for the numerical values of the TCLP criteria). The method and limits have been published by the U.S. EPA (1986).

When leachate from a landfill reaches the saturated ground-water zone it moves as a plume that spreads in the direction of the flowing ground water. This process has been described in the preceding sections of this chapter and shown in Figure 7.5. As the plume slowly moves, the concentrations of contaminants decrease owing to adsorption and retardation, hydrodynamical dispersion, and biochemical degradation.

Tracing contaminated plumes from land disposal sites requires testing the levels of certain water quality parameters. Although the most appropriate parameters may vary somewhat, depending on the types of solid wastes deposited in the landfill, certain parameters have been found to be generally suitable. Key indicators of the presence of leachate that have been suggested by the U.S. EPA (1977c) are specific conductance, pH, temperature, chloride ion, color, turbidity, and COD (TOC).

GROUND-WATER QUALITY MODELS

There are two types of ground-water quality models in present use (Bachmat et al., 1978):

1. Predictive models, which simulate the behavior of the ground-water system and its response to boundary inputs.
2. Resource management models, which integrate prediction with explicit management decision procedures.

Predictive Models

The predictive models are primarily numerical even though electric analogs or even physical modeling can be used for simulation of ground-water quality systems. The numerical models are either *deterministic analytical and computer-simulation models* or *stochastic* (random walk, etc.) *models*. The predictive models represent the vast majority of models developed for ground-water-related problems (Bachmat et al., 1978; Anderson 1979; Canter, Knox, and Fairchild, 1987).

The ground-water movement models, which are similar to models of surface hydrology and water quality, can be either distributed parameter (two- or three-dimensional) models or lumped-parameter models. A distributed parameter is one in which variables are determined at many discrete locations throughout the system dictated by the breakdown of the system into small uniform segments. In the lumped-parameter models aquifer is treated as one uniform element, and only spatially averaged values are considered.

Analytical models are developed by considering highly simplified conditions or by using simplifying assumptions to obtain a solution of the governing differential equations (hydraulic continuity and motion, and mass continuity). The most common analytical solutions are those assuming steady-state and uniform systems. Some examples were presented in the preceding sections of this chapter or in the literature (Bear 1979; Anderson, 1979). Such models typically use the lumped-parameter concept.

The Gelhar–Wilson (1974) model is presented here as an example of a simpler analytical lumped-parameter model of water quality in an aquifer–stream system. The authors of the model pointed out that when long-term basinwide changes in ground-water quality are desired, spatial variations become less important than temporal variations. The fluctuation of the water table is represented mathematically by the following equation:

$$p\frac{dh}{dt} = -q + \varepsilon + q_r - q_p \tag{7.18}$$

where
h = average thickness of the saturated zone
p = average effective porosity
ε = natural discharge rate
q = natural outflow from the aquifer
q_r = artificial recharge/unit area
q_p = pumping rate/unit area
t = time

It can be demonstrated that $q = a(h - h_0)$, where h_0 is the elevation of the river and

$$a = 3T/L^2$$

where
$T = h_0 K$ = transmissivity of the aquifer
K = hydraulic conductivity
L = length of the aquifer

The change in concentration of a nonadsorbable pollutant is represented by an equation of the form

$$ph\frac{dc}{dt} + (\varepsilon + q_r + \alpha ph)c = \varepsilon c_L + q_r c_r \qquad (7.19)$$

where
c = concentration
c_L = concentration of the natural recharge
c_r = concentration of the artificial recharge
α = a first-order rate constant that accounts for degradation of the contaminant

Dispersion is assumed to be negligible, which is a reasonable assumption if only regional average concentrations are sought.

The hydraulic response time and the solute response are measures of the lag observed in the response of the system to a given input. Hydraulic response time (t_h) is defined as follows:

$$t_h = p/a \qquad (7.20)$$

The solute response time, t_c, is then

$$t_c = ph_0/\varepsilon_0 \qquad (7.21)$$

where ε_0 = initial recharge rate. In general, t_c is time dependent, but can be estimated from Equation (7.21). A representation of the model is shown on Figure 7.14.

Gelhar and Wilson (1974) based their model on the concept of a well-mixed linear reservoir. They postulated that aquifer response to a given input would be similar to the response of a well-mixed linear reservoir. They showed that the concentration of water leaving the aquifer is representative of the average concentration within the aquifer. Therefore, such a model is indeed suitable for determining the quality of ground water discharging to surface waters provided that the simulated contaminant is not attenuated by adsorption in the aquifer zone. Such contaminants may include nitrates and chlorides.

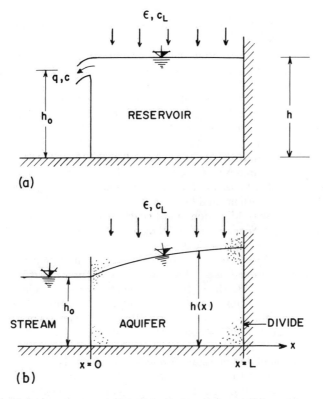

FIGURE 7.14. Representation of the Gelhar–Wilson (1974) model.

Since the Gelhar–Wilson model is based on a lumped-parameter concept, it must be calibrated. For example, historical records of chloride concentrations in wells or base flow can be used for calibration and verification (Anderson, Eisen, and Hoffer, 1978).

Distributed-Parameter Models

The distributed-parameter model differs from the simpler single linear reservoir model by assuming that the aquifer is composited from a number of linear reservoirs that are interconnected. Typically water continuity and mass balance differential equations are written for each cell and simultaneously solved. In almost all cases the equations are solved using finite-difference approximation. A review of such models is included in Canter, Knox, and Fairchild (1987), Anderson (1979), Bachmat et al. (1978), Wang and Anderson (1982), or NCASI (1985), and others.

References

Adriano, D. C. (1986). *Trace Elements in the Terrestrial Environment*, Springer Verlag, New York.

Anderson, M. P. (1979). Using models to simulate the movement of contaminants through groundwater flow systems, *CRC Crit. Rev. Environ. Control* **9**(2):97–156.

Anderson, M. P., C. E. Eisen, and R. N. Hoffer (1978). *Groundwater Hydrology*, Vol. 5, Menomonee River Pilot Watershed Study, International Joint Commission, Windsor, Ontario.

Andreoli, A., N. Bartilucci, R. Forgione, and R. Reynolds (1979). Nitrogen removal in a subsurface disposal system, *J. WPCF* **51**:841–834.

Bachmat, Y., B. Andrews, D. Holtz, and S. Sebastian (1978). *Utilization of Numerical Groundwater Models for Water Resources Management*, EPA 600/8 78/012, U.S. Environmental Protection Agency, Ada, OK.

Baestlé, L. H. (1969). Migration of radionuclides in porous media, in *Progress in Nuclear Energy. Series XII. Health Physics*, A. M. F. Duhamel, ed., Pergamon Press, New York, pp. 707–730.

Bear, J. (1979). Groundwater quality problem (hydrodynamic dispersion), in *Hydraulics of Groundwater*, McGraw-Hill, New York, pp. 263–275.

Bell, R. G., and J. B. Bole (1978). Elimination of fecal coliforms from soil irrigated with municipal sewage lagoon effluent, *J. Environ. Qual.* **7**:193–196.

Betson, R. P., and W. M. McMaster (1975). Nonpoint source mineral water quality model, *J. WPCF* **47**:2461–2473.

Bouwer, H. (1974). Design and operation of land treatment systems for minimum contamination of groundwater, *Groundwater* **12**:140–147.

Bouwer, H. (1978). *Groundwater Hydrology*, McGraw-Hill, New York.

Boyce, J. S., et al. (1976). Geologic nitrogen in Pleistocene loess of Nebraska, *J. Environ. Qual.* **5**:93–96.

Brown, K. W., H. W. Wolf, K. C. Donnelly, and J. F. Slowey (1979). The movement of fecal coliforms and coliphages below septic lines, *J. Environ. Qual.* **5**:121–125.

Brown, R. L. (1975). The occurrence and removal of nitrogen in subsurface agricultural drainage from the San Joaquin Valley, California, *Water Res.* **9**:529–546.

Burwell, R. E., G. E. Schuman, K. E. Saxton, and H. G. Heineman (1976). Nitrogen in subsurface discharge from agricultural watersheds, *J. Environ. Qual.* **3**:325–329.

Canter, L. W., and R. C. Knox (1985). Groundwater pollution from septic tank systems, in *Septic Tank System Effects on Groundwater Quality*, Lewis Publ., Chelsea, MI.

Canter, L. W., R. C. Knox, and D. M. Fairchild (1987). *Ground Water Quality Protection*, Lewis Publ., Chelsea, MI.

Chalk, P. M., and D. R. Keeney (1971). Nitrate and ammonium content of Wisconsin limestone, *Nature* **229**:42–47.

Conservation Foundation (1985). *Groundwater—Saving the Unseen Resource.*

Proposed Conclusions and Recommendations, Conservation Foundation, Washington, DC.

Davis, S. N., and R. J. M. DeWiest (1966). *Hydrogeology*, Wiley, New York.

Domenico, P. A., and F. W. Schwartz (1990). *Physical and Chemical Hydrogeology*, Wiley, New York.

Edworthy, K. J., W. B. Wilkinson, and C. P. Young (1978). The effect of the disposal of effluents and sewage sludge on groundwater quality in the Chalk of the United Kingdom, *Prog. Water Tech.* 10:479–493.

Fetter, C. W. (1988). *Applied Hydrogeology*, 2d ed., Merrill Publ. Columbus, OH.

Freeze, R. A., and J. A. Cherry (1979). *Groundwater*, Prentice-Hall, Englewood Cliffs, NJ.

Gallant, A. L., et al. (1989). *Regionalization as a Tool for Managing Environmental Resources*, NSI Technology Services Corporation, U.S. Environmental Protection Agency, Corvallis, OR.

Gelhar, L. W., and J. L. Wilson (1974). Groundwater quality monitoring, *Groundwater* 12:399–408.

Geraghty, J. J., et al. (1973). *Water Atlas of the United States*, Water Information Center, Syosset, NY.

Hem, J. D. (1985). Study and Interpretation of the Chemical Characteristics of Natural Water, U.S. Geological Survey Water-Supply Paper 2254, Washington, DC.

Johansen, O. J., and D. A. Carlson (1976). Characterization of sanitary landfill leachate, *Water Res.* 10:1129–1134.

Klusman, R. W., and K. W. Edwards (1977). Toxic metals in groundwater of front range, Colorado, *Groundwater* 15:160–169.

Kreitler, C. W., and D. C. Jones (1975). Natural soil nitrate: The cause of the nitrate contamination in groundwater in Runnels County, Texas, *Groundwater* 13:53–61.

Lehr, J. H. (1981). A problem yes—A disaster no, *Groundwater* 19:2–3.

Miller, D. W., and M. R. Scalf (1974). New priorities for groundwater protection, *Groundwater* 12:335–347.

Moore, J. W., and S. Ramamoothy (1984). *Heavy Metals in Natural Waters: Applied Monitoring and Impact Assessment*, Springer Verlag, New York.

National Council of the Paper Industry for Air and Stream Improvement (1985) *A Review and Practical Guide for Analytical Groundwater Models Used to Predict Contaminant Transport in Subsurface Systems*, NCASI, New York.

Novotny, V., K. R. Imhoff, M. Olthof, and P. A. Krenkel (1989). *Karl Imhoff's Handbook of Urban Drainage and Wastewater Disposal*. Wiley Interscience, New York.

Ogata, A. (1970). *Theory of Dispersion in a Granular Medium*, U.S. Geological Survey Prof. Paper 411-1, Washington, DC.

Omernik, J. M., and A. L. Gallant (1990). Defining regions for evaluating environmental resources, in *Global Resources Monitoring and Assessments: Preparing for the 21st Century*, Proceedings, International Conference and

Workshop, American Society for Photogrammetry and Remote Sensing, Bethesda, MD.

Peaudecerf, P., and J. P. Sauty (1978). Application of a mathematical model to the characterization of dispersion effect on groundwater quality, *Prog. Water Technol.* **10**:443–455.

Perkins, T. K., and O. C. Johnson (1963). A review of diffusion and dispersion in porous media, *J. Soc. Petrol. Eng.* **3**:70–83.

Reed, S. C., E. J. Middlebrooks, and R. W. Crites (1988). *Natural Systems for Waste Management and Treatment*, McGraw-Hill, New York.

Reneau, R. B., and D. E. Petry (1976). Phosphorus distribution from septic tank effluents in coastal plain soils, *J. Environ. Qual.* **5**:34–39.

Saffigna, P. G., and D. R. Keeney (1977). Nitrate and chloride in groundwater under irrigated agriculture in Central Wisconsin, *Groundwater* **15**:170–177.

Sawyer, C. N., and P. L. McCarty (1978). *Chemistry for Environmental Engineering*, McGraw-Hill, New York.

Sikora, L. J., M. G. Bent, R. B. Corey, and D. R. Keeney (1976). Septic nitrogen and phosphorus removal test system, *Groundwater* **14**:304–314.

Todd, D. K. (1980). *Groundwater Hydrology*, 2d ed., Wiley, New York.

Todd, D. K. (1983). *Ground-water Resources of the United States*, Premier Press, Berkeley, CA.

U.S. Environmental Protection Agency (1977a). *Environmental Effects of Septic Tank Systems*, EPA 600/3-77/096, U.S. EPA, Washington, DC.

U.S. Environmental Protection Agency (1977b). *The Report to Congress—Waste Disposal Practices and Their Effects on Groundwater*, U.S. EPA, Washington, DC.

U.S. Environmental Protection Agency (1977c). *Procedures Manual for Groundwater Monitoring of Solid Waste Disposal Facilities*, EPA 530/SW-611, U.S. EPA, Washington, DC.

U.S. Environmental Protection Agency (1980). *Design Manual–Onsite Wastewater Treatment and Disposal Systems*, EPA 625/1-80-012, U.S. EPA, Cincinnati, OH.

U.S. Environmental Protection Agency (1981). *Process Design Manual for Land Treatment of Municipal Wastewater*, EPA 625/1-81-013, U.S. EPA, Cincinnati, OH.

U.S. Environmental Protection Agency (1984). Process Design Manual for Land Treatment of Municipal Wastewater: Supplement on Rapid Infiltration and Overland Flow, EPA 624/1-81-013a, U.S. EPA, Cincinnati, OH.

U.S. Environmental Protection Agency (1986). Hazardous waste management system, *Fed. Regist.* **51**(114):21648–21693.

Wang, H. F., and M. P. Anderson (1982). *Introduction to Groundwater Modeling—Finite Difference and Finite Element Models*, W.H. Freeman, San Francisco, CA.

8
Urban and Highway Diffuse Pollution

As a result of increased traffic, congestion, higher pollution levels, littering, reduced green space and the loss of other environmental amenities, living conditions in many urban areas in the OECD region have become worse in the past 20 years.

Organization for Economic Co-operation and Development (1991), The State of the Environment, *Paris*

LAND-USE EFFECTS ON URBAN NONPOINT POLLUTION LOADS

In the first half of this century deterioration of water quality due to urbanization and urban sources was associated with point sources from industrial and commercial operations and with domestic sewage. However, not until 1970 was it realized that a significant portion of pollution from urban and urbanizing areas originated from nonpoint diffuse sources such as construction, washoff of dust and dirt from impervious surfaces, or sewage inputs from unsewered suburban areas.

Urban nonpoint sources have been identified as a major cause of pollution of surface-water bodies by the U.S. EPA (U.S. EPA, 1984; Athayde, Myers, and Tobin, 1986; Myers et al., 1985). In the 1988 Report to Congress (U.S. EPA, 1990a) it was stated that urban stormwater runoff is the fourth most extensive cause of the impairment of the water quality of the nation's rivers and the third most extensive source of water quality impairment of lakes. The combined sewer overflows

440   Urban and Highway Diffuse Pollution

(CSO's) were the tenth most significant source of impairment for both types of surface-water bodies.

Overflows from both separate storm sewers and combined sewer systems are considered diffuse pollution, although in the United States their discharges are regulated. For example, a permit system is used for point sources (see Chapters 1 and 2 for legal classifications of point and nonpoint diffuse sources and effluent discharge permits).

Diffuse-pollution generation in urban areas is quite different from that in rural lands. Several factors cause the difference (Novotny and Chesters, 1981):

1. Large portions of urban areas are impervious (Fig. 8.1), resulting in much higher hydrological activity. The coefficient of runoff (defined as a ratio of runoff volume to that of rainfall) is generally directly proportional to the degree of imperviousness (Fig. 8.2). Hence, urbanization increases the volume of runoff.
2. The hydrological response of the watershed to precipitation is faster, which decreases the time of concentration. As a result the runoff

FIGURE 8.1.   Highly impervious urban areas are one of the largest sources of diffuse pollution.

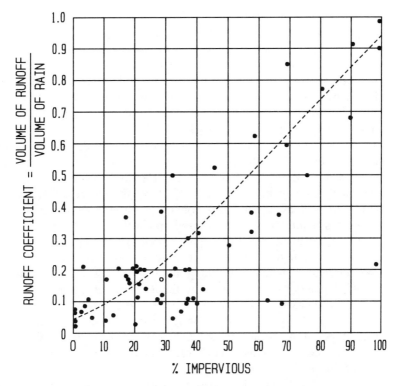

FIGURE 8.2. Relation of the statistical coefficient of runoff for urban areas to imperviousness. (From National Urban Runoff Project, U.S. EPA, 1983.)

peaks are increased (typically two to five times) over those for the predevelopment period. This may greatly increase the flooding potential, hence, many communities have enacted strict runoff control ordinances (for example, metropolitan Chicago).

3. Urbanization commonly decreases ground-water levels. Ground-water levels are lowered by the installation of sewers and consequent sewer infiltration–inflow inputs, and by drainage below the ground levels of buildings. As a result, base flow of many urban streams may decrease to the point where some smaller headwater urban streams may become ephemeral or the flow in them may consist of predominantly undiluted sewage and wastewater effluents (effluent-dominated streams).

4. Except for construction sites, most pervious land surfaces in residential areas and in urban areas east of the Mississippi River are protected by lawns and other vegetation and, consequently, erosion is reduced. However, increased volume and peak flow due to urbanization causes

the velocities in the streams to become faster, resulting in increased stream-bank erosion. Excessive fertilizer and pesticide applications onto urban and suburban lawns also represents a water quality problem. In the urban centers located in the arid regions of the world, including southwestern portions of the United States, erosion of pervious lands can be quite intensive, especially when water use restrictions on lawn irrigation are imposed. Rainfall, although rare, may have devastating erosion potential.

5. Over a longer period of time (for example, a year), in areas with storm sewers, all of the pollution deposited on impervious surfaces that has not been removed by street cleaning, wind, or decay will eventually end up in surface runoff. Thus the street solids accumulation, as well as accumulation on other impervious urban surfaces represents a limited source of predominantly particulate pollutants. On the other hand, soil is an infinite pool of sediments and potential pollutants in nonurban and suburban areas and in construction zones. The removal of sediments and pollutants from pervious lands into runoff depends on the energy of the rainfall (in addition to other factors, such as the soil, slope, and vegetation cover characteristics described in Chapter 5). The contributions of pollutants washed off from impervious surfaces and additional loads of overfertilized and contaminated soils change the character and the type of pollution from that of the predevelopment period.

6. The frequency of pollution-carrying runoff events is greatly increased in developed watersheds with higher imperviousness. On the other hand, nonpoint pollution loads from pervious lands (crop lands, woodlands, urban lawns, and parks) occur only during very large storms or, to a lesser degree, during frozen ground conditions in the spring.

7. Polluted runoff from impervious urban surfaces is generated during rainfall that exceeds a certain minimal threshold value of depression storage, which is about 1 to 2 mm in areas drained by separate storm sewers. The threshold value for overflows from combined sewers is somewhat greater due to the fact that a portion of storm runoff is diverted to the treatment plant. In rural areas or areas with large pervious surfaces, polluting surface runoff events are only generated during large hydrologically rare storms and snowmelt.

Urbanization increases both pollution loadings and flooding; consequently, similar practices may be used to remedy both problems.

Comprehensive reviews of sources of pollution of urban runoff and combined sewer overflows were presented by Lager et al. (1977), Ellis (1986), and by Waller and Hart (1986). Ellis compiled information on both North America and European cases. The edited book by Torno, Marsalek, and Desbordes (1986), which contains the latter two papers, is a comprehensive treatise dealing with the urban runoff problem. Also proceedings of the IAHR-IAWPRC conferences on urban runoff management (Yen, 1982; Balmér, Malmqvist, and Sjöberg, 1984; Gujer and Krejci, 1987; Iwasa and Sueishi, 1990) contain numerous articles dealing with the quality of urban runoff and its characterization.

The pollutant loads from urban areas are strongly affected by drainage. The smallest pollutant loadings are typical for suburban areas with so-called natural surface drainage and sanitary sewers. The highest pollutant loadings are emitted from highly impervious, densely populated (or heavily used) urban centers with separate or combined sewers. As was pointed out in Chapter 1, and will be further elaborated in this chapter, the total pollution impact of overflows from areas served by combined sewers (without CSO abatement) and separate storm sewers may be about the same for many conventional pollutants. This is due to the fact that a substantial portion of storm water in combined sewer systems does not overflow and is conveyed to treatment.

URBAN LAND USE AND MAGNITUDE OF DIFFUSE POLLUTION

As pointed out in Chapter 1, typical urban land uses include

Residential land (low, medium, and high density)
Commercial land use
Industrial land use
Other developed (large parking areas, sport complexes) lands
Open lands (parks, golf courses, idle urban lands)
Transportation (airport, railroad, vehicular traffic)

Since this categorization is insufficient to express the impact of land-use activities on the magnitude of pollution loads emitted from these lands, Marsalek (1978) found it more appropriate to divide the land producing nonpoint pollution into more specific categories, such as:

- *Land-Use Group I: Low Pollution Loads.* This category includes low-and medium-density residential land uses (<125 people/ha) and limited–nuisance industrial activities (wholesale, warehouses).
- *Land-Use Group II: Intermediate Pollution Loads.* Typical land uses in this category include high-density residential (>125 people/ha) and commercial land use.

- *Land-Use Group III: Highest Pollution Loads.* Typical land uses include medium- and high-intensity industrial uses.
- *Land-Use Category IV: Lowest Pollution Potential.* Typical land uses include parks and playgrounds. In many cases pollution loads from these lands are negligible.

The land-use impact applies only to unit loadings of pollutants. A statistical analysis of data collected during the pilot studies of the Nationwide Urban Runoff Program (NURP), sponsored by the U.S. Environmental Protection Agency (1983), found no significant statistical difference between the mean concentrations of pollutants in urban runoff among the typical urban land uses. This finding is elaborated further in the section titled "Nationwide (NURP) Characterization of Urban Runoff Quality."

Sources and Magnitude of Urban Diffuse Pollution

It is not the land itself or land use per se that causes pollution. Pollution is caused by various boundary inputs and polluting processes and activities that occur on the land. The inputs and processes that are the cause of urban diffuse pollution are:

- Pollution contained in precipitation
- Erosion of pervious lands
- Accumulation of dry atmospheric deposits (dust) and street dirt and subsequent washoff from impervious surfaces. The sources of the accumulate pollutants are:

    Dry atmospheric deposition
    Street refuse accumulation, including litter, street dust and dirt, and organic residues from vegetation and animal population
    Traffic emissions

- Solids accumulation and growth in sewers
- Leaching of pollutants from septic systems and other sources, such as landfills, onto surfaces and into ground water and subsequently into storm drainage
- Application, storage, and washoff of deicing and other chemicals
- Application of pesticides and fertilizers onto grassed urban lands
- Discharge of pollutants, such as car oil, detergents, and other household and commercial solvents and chemicals, into the drainage systems
- Cross-connections of sewage and industrial wastes from sanitary sewers, failing septic tanks, and other sources, into storm sewers

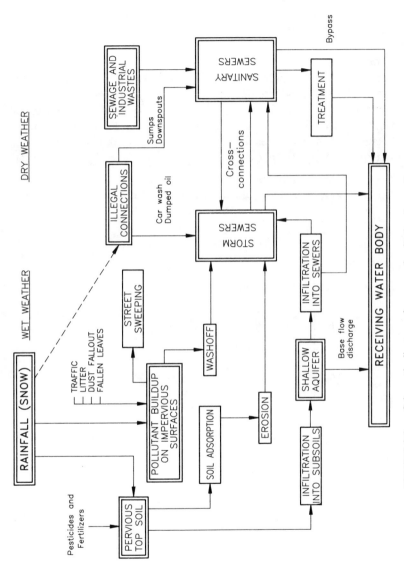

FIGURE 8.3. Schematic diagram of sources of pollution from urban areas.

Figure 8.3 schematically shows the sources of diffuse pollution and the process of accumulation.

Unit Loads of Pollutants

Definition

As first defined in Chapter 1, a unit load is a simple value or function (sometimes termed *export coefficient*) expressing pollutant generation per unit area and unit time for each land use or averaged over the entire contributing basin. The units are usually expressed in kilograms or tonnes per hectare per year or season. In urban areas the unit loads are commonly expressed in mass per unit length of curb (g/m = kg/km or lb/ft).

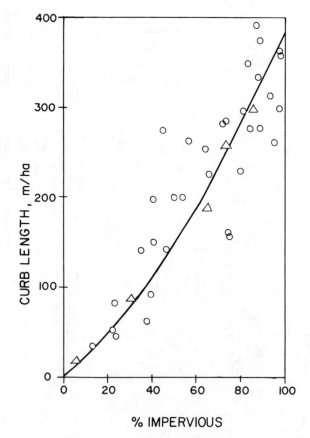

FIGURE 8.4. Relationship of curb length density to total imperviousness of urban areas.

A strong correlation exists between the curb density and the percent of imperviousness of residential areas, as shown on Figure 8.4. The American Public Works Association (1969) developed a regression formula between the curb length of urban areas and population density based on analysis of many American cities. The resulting equation (converted to metric units) was

$$CL = 311.67 - 266.07 \times 0.839^{(2.48\ PD)} \tag{8.1}$$

where
CL = curb length (m/ha)
PD = population density (person/ha)

As pointed out by Novotny and Chesters (1981), unit loads established by measurements or by simulation are variable, often expressing averages from a wide range of measured values. The variations are not only caused by randomness of the process but may also include some cyclic (periodic) and systematic factors and trends, such as seasonal variations, geographical differences, meteorological and land cover factors, soil characteristics, and reliability of measured data due to insufficient frequency of measurements and measurement errors. However, the NURP study of the quality of urban runoff was unable to statistically identify a nationwide effect of any systematic factors on the unit loads except imperviousness, which affects the runoff volume and, consequently, the unit loads.

The unit loads can also be estimated by modeling. The same reservation applies to simulated unit loads; however, some variability factors are usually included in the models and are respected.

Measured Unit Loads

A number of studies provided measured unit loads from various types of urban lands. Several studies compiled the measured data into ranges and averages that have been subsequently related to the type of land use.

The first comprehensive and now classic study of urban pollution loadings by storm water and combined sewer overflows was conducted by the American Public Works Association (1969). This study identified and quantified sources of pollution by urban runoff, including air pollution, street refuse, catch basins, and sewer systems themselves. The data analyzed by the study were mostly from Chicago, Illinois.

Sartor and Boyd (1972) conducted a similar study and analyzed data from several U.S. cities. They noted that the pollutant accumulation in curb storage is not constant, but exhibits a decreasing rate of increase, that is, there is a certain limiting value of pollutant accumulation that is related to land use. The total accumulation of pollutants converted to grams/meter of curb length are presented in the following list (after Sartar and Boyd (1972)):

Land Use	Curb Loading (g/m)
Residential	339
Industrial	790
Commercial	82
Overall	395

Ellis (1986) presented values and ranges of accumulation of street-surface pollutants (Table 8.1). It can be seen that some 20% to 40% of the accumulated material is organic; however, it is not easily biodegradable, being derived from leaf and wood litter, and rubber and bituminous road-surface materials. The high bacterial values recorded in the street-surface solids can largely be ascribed to animal feces. The high metal content of highway solids reflects traffic emissions.

The most comprehensive compilation of unit loads based on measured values to date is the one prepared by the Midwest Research Institute (McElroy et al., 1976). The manual contains estimates and estimation procedures for unit loads of sediment from erosion and impervious urban surfaces, nutrients and organic matter, pesticides, heavy metals, terrestrial disposal, and background emissions of pollutants. For developed urban areas the unit load is related to the curb density factor as

$$Y = CD * L \tag{8.2}$$

TABLE 8.1 Solids Accumulation and Associated Pollutant Concentrations in Urban Areas

Land Use		Residential Low Density	High Density	Commercial	Light Industrial	Highways
Solids accumulation (g/curb m)		10–182	30–210	13–180	80–288	13–1100
Pollutant concentration (μg/g)	BOD$_5$	5260	3370	7190	2920	2300–10,000
	COD	39,300–40,000	40,000–42,000	39,000–61,730	25,100	53,650–80,000
	Tot.N	460–480	530–610	410–420	430	223–1600
	Pb	1570	1980	2330	1390	450–2346
	Cd	3.2	2.7	2.9	3.6	2.1–10.2
Fecal Coliforms (MPN/g)		60,570–82,500	25,621–31,800	36,900	30,700	18,768–38,000

Source: After Ellis (1986).

where
- Y = loading of pollutant in kg/ha per storm event
- CD = curb density in km/ha
- L = measured or estimated pollutant loading–rate curb load of the pollutant in g/m

The manual by McElroy et al. provides loadings of sediments for typical urban areas located in four different geographical locations in the United States. To obtain loads of pollutants associated with sediments, the equation is multiplied by a given concentration of the pollutant on the sediment. These concentrations have sometimes been called *potency factors*.

Marsalek (1978) compiled the data measured by PLUARG (Pollution by Land Use Activities—Reference Group of the International Joint

TABLE 8.2   Unit Loads of Pollutants from PLUARG Studies

Constituent	Land Use Category I	Land Use Category II	Land Use Category III	Land Use Category IV
		Storm Sewers		
BOD	34	90	34	1.12
N	9	11.2	7.8	0.22
P	1.6	3.4	2.2	0.04
SS	390	360	672	11.2
Cd	0.013	0.016	0.024	0.002
Cr	0.026	0.028	0.044	0.003
Cu	0.045	0.049	0.077	0.007
Hg	0.038	0.043	0.065	0.006
Ni	0.029	0.032	0.030	0.004
Pb	0.157	0.174	0.269	0.022
Zn	0.570	0.630	0.980	0.081
		Combined Sewers		
BOD	134.0	293	112	1.6
N	31.5	36.5	34.5	1.1
P	10.2	11.6	10.9	0.34
SS	773	672	740	11.2
Cd	0.016	0.017	0.027	0.002
Cr	0.028	0.031	0.048	0.003
Cu	0.064	0.071	0.109	0.009
Hg	0.043	0.047	0.073	0.006
Ni	0.034	0.037	0.057	0.004
Pb	0.162	0.180	0.277	0.022
Zn	0.640	0.703	1.088	0.090

Source: Based on data from Marsalek (1978).
Note: In kg/ha-year.

Commission) and other estimates (such as the APWA study mentioned previously and measured data from the Canadian Great Lakes) and related them to the four land-use categories defined on the preceding pages. Table 8.2 provides some of the unit loads by Marsalek.

Beaulac and Reckhow (1982) compiled, measured, and published unit loads from various nonurban land uses and from general urban land-use types. The ranges of unit loads of phosphorus for urban lands were 0.5 to 6.25 kg/ha-year (average 1 kg/ha-year), and those for nitrogen were 1 to 38.5 kg/ha-year (average 5 kg/ha-year), respectively.

Similarly, Uttormark, Chapin, and Green (1974) compiled ranges of nutrient export from urban watershed that were 2 to 9 kg/ha-year for nitrogen and 1.1 and 1.2 kg/ha-year (only two values reported) for phos-

TABLE 8.3   Unit Loads of Pollutant from PLUARG Experimental Watersheds

Pollutant	General Urban	Residential	Commercial	Industrial	Developing Urban
Suspended solids	200–4800	620–2300	50–830	450–1700	27,500
Total phosphorus	0.3–4.8	0.4–1.3	0.1–0.9	0.9–4.1	23
Total nitrogen	0.2–18	5–7.3	1.9–11	1.9–14	63
Lead	0.14–0.5	0.06	0.17–1.1	2.2–7	3
Copper	0.02–0.21	0.03	0.07–0.13	0.29–1.3	—
Zinc	0.3–1.0	0.02	0.25–0.43	3.5–12.0	—
Chloride	130–750	1050	10–150	75–160	—

Source: After Sonzogni et al. (1980). *(Copyright © 1980 by the American Chemical Society, reprinted by permission.)*

Note: In kg/ha-year.

TABLE 8.4   Unit Loads of Pollutants from New Jersey Residential Watersheds

	Unit Loads in kg/ha-year	
	Single-Family Housing	Multiple-Family Housing
BOD^5	16.7	73.6
Total P	1.2	8.0
Lead	2.3	1.0
Zinc	1.3	2.0
Copper	0.4	0.5
Nickel	0.3	0.12
Chromium	0.11	0.09

Source: After Whipple et al. (1978).

TABLE 8.5 Unit Loads of Pollutants from Homogenous Urban Watersheds Located in Milwaukee, Wisconsin (excluding winter months December to March)

Land Use	Suspended Solids	Total Phosphorus	Total Lead
Freeways	979	1.04	4.96
Industrial	957	1.49	2.70
Commercial	957	1.49	2.70
Parking lots	453	0.78	0.96
High-density residential	487	1.12	0.90
Medium-density residential	216	0.58	0.24
Low-density residential	11	0.04	0.01
Parks	3	0.03	0.006

Source: After Bannerman et al. (1984).
Note: In kg/ha-year.

phorus, respectively. These ranges were consistent with Beaulac and Reckhow's compilation of a large data base. Sonzogni et al. (1980) summarized the PLUARG watershed data and provided the unit loads of pollutants that are included in Table 8.3. Whipple, Hunter, and Yu (1978) measured the unit loads shown in Table 8.4 in residential watersheds located in New Jersey.

The National Urban Runoff Project (NURP) research conducted in Milwaukee, Wisconsin (Bannerman et al., 1984) measured pollutant loads from very small homogenous urban watersheds. The study also attempted to determine the effect of street sweeping on pollutant loads, which was found to be insignificant (see Chapter 10). The measured pollutant loads are given in Table 8.5.

INDIVIDUAL SOURCES OF POLLUTION

At the beginning of the preceding section on unit loads it was emphasized that it is not the land that produces pollutant loadings. Pollution is caused by boundary pollution inputs and by polluting land uses and land-use activities by man taking place on the land. A number of publications describe individual sources of pollution by urban runoff (Lager et al., 1977; Novotny and Chesters, 1981; Ellis, 1986; James and Boregowda, 1986; U.S. EPA, 1984).

Atmospheric Deposition

Atmospheric deposition of pollutants is generally divided into wet and dry or bulk (combined) surface loading. Atmospheric deposition of pollutants is a boundary input caused by local or distant air-pollution emission sources. Both industrial (urban and transportation) and agricultural activities may contribute to the pollution content of atmospheric deposits. As pointed out in Chapter 4, rain droplets and snowflakes absorb pollutants from the atmosphere, including acid-forming components.

As shown in Chapter 4, in most larger cities the deposition rate of atmospheric particulates in wet and dry fallout range from 7 tonnes/km^2-month (= gm/m^2-month) to more than 30 tonnes/km^2-month. As expected, higher deposition rates occur in congested downtown and industrial zones, and lower rates are typical for residential and other low-density suburban zones.

Halverson et al. (1982) measured the pollutant content of precipitation in a nonindustrial urban area in central Pennsylvania (population about 100,000). The loads presented in Table 8.6 were divided into wet deposition (by rainfalls) and bulk deposition (wet + dry deposition). Similarly, Ng (1987) studied the relationship between rainfall chemistry and the

TABLE 8.6 Atmospheric Deposition in an Urban Area in Pennsylvania

Constituent	Wet Deposition	Bulk Deposition
Total N	0.13	0.20
Total P	0.006	0.006
Lead	T	T
Cadmium	T	T
Copper	T	T
Zinc	T	T

Source: After Halverson et al. (1982).
Note: In kg/ha-year.

TABLE 8.7 Atmospheric Deposition in an Urban Area in Ontario

Constituent	Wet Deposition	Storm Runoff
Total N	0.06	0.06
Total P	0.002	0.005
Lead	0.001	0.004
Cadmium	0.0003	0.0005
Copper	0.002	0.001
Zinc	0.004	0.012

Source: After Ng (1987).
Note: In kg/ha-year.

water quality of urban runoff in an urban catchment with prevailing industrial land use located in Burlington, Ontario, Canada. Rainfall was found to be a significant source of nitrogen (nitrite, nitrate, and ammoniacal N), copper, and nickel. For other measured pollutants (phosphorus, cadmium, chromium, lead, and zinc) rainfall contributions were insignificant. Table 8.7 presents the loads measured by Ng.

Extensive information on concentrations and wet and dry depositions of pollutants related to land use are contained in the report by Pitt and Barron (1989). Novotny and Kincaid (1982), and Randall et al. (1982), reported chemical composition of precipitation for various urban and suburban land-use types.

In residential areas, the fugitive dust (dry deposition) mostly originates from surrounding soils, construction sites, refuse disposal sites, and from biological sources (pollen, spores, and other organic residues). The average composition of dustfall and the deposition rate observed in Milwaukee, Wisconsin, experimental watersheds in the early 1980s is given in the following lists (after Bannerman, 1984):

Average Wet Deposition Characteristics

Suspended solids	4.0 mg/l
Organics (VSS)	1.0 mg/l
Organics (COD)	7.0 mg/l
Total nitrogen	0.9 mg/l
Total phosphorous	0.015 mg/l
Total lead	0.012 mg/l
Ph	4.2

Dry Deposition	*December–May*	*June–November*
Atmospheric fallout rate (solids), kg/ha-day	0.6	0.43
COD, kg/ha-day	0.23	0.16
Total phosphorus, g/ha-day	0.55	0.47
Total lead, g/ha-day	0.52	0.51
Sulphate, g/ha-day	25.0	20.0

Pitt (1979), after an extensive study in the San Francisco Bay area, concluded that a majority of deposited street dust particles are related to local geological conditions, with added fractions from motor vehicle emissions and road wear. Most of the street refuse particles originate from local erosion of soils and are transported by air.

A study of dry atmospheric deposition in an industrial valley located in Milwaukee, Wisconsin (Anon., 1981) showed that fugitive dust emissions and transport are also the major sources of dry deposition from the atmosphere. The origin of dustfall (particles less than 60 μm are considered dust in most urban pollution studies) in the central city— downtown locations are mostly from unpaved roads and parking lots, unpaved railroad rights-of-way, uncovered material storage sites, construction and demolition sites, dirty paved roads, urban refuse (garbage) disposal landfill operations, and industrial park emissions.

Lead, which is associated with the combustion of leaded gasoline (in the United States the use of leaded gasoline was phased out during the 1980s), is transported with atmospheric particulate matter. The influence of traffic using leaded gasoline on lead concentrations is profound and a significant reduction has been achieved after switching to unleaded gasoline. Cadmium, strontium, zinc, nickel, and many organic hazardous chemicals are also transported with atmospheric aerosols, and can be found in the atmospheric fallout.

Street Refuse Deposition

Particles that are larger in size than dust (>60 μm) are considered as street refuse or street dirt. In the NURP studies, these deposits were divided into median-sized deposits (street-dirt particle size ranges 60 μm to 2 mm) and litter (>2 mm). Litter deposits contain items such as cans, broken glass, bottles, pull tabs, papers, building materials, plastic, garbage, parts of vegetation, dead animals and insects, animal excreta, and the like. The sources of street dirt are numerous and often very hard to control. Although originally most of the street refuse deposition occurs in larger sizes (greater than 3 mm), it is reasonable to assume that a part of dust originates from the mechanical breakdown of larger litter particles.

Vegetation Inputs

In residential areas, fallen leaves and vegetation residues including grass clippings dominate street refuse composition during the fall season (Fig. 8.5). During defoliage in the fall a mature tree can produce from 15 to 25 kg of organic leaf residue (dry weight) that contains significant amounts of nutrients (Heaney and Huber, 1973). During the growing season trees can enrich the throughfall (rainwater that penetrates the canopy) with nutrients and organics (Halverson, DeWalle, and Sharpe, 1984). A typical value of foliage and leaf fallout for a forested area in Minnesota with about 420 trees/ha is about 3.8 tons/ha-year or about

Individual Sources of Pollution 455

FIGURE 8.5. Leaf fallout in urban areas and its washoff into storm sewers is a source of large urban loads of biodegradable organics. (Photo: V. Novotny.)

9 kg/tree. Of the yearly values, about 65% of the fallout occurs during the fall. The fallen leaves are about 90% organic and contain about 0.04% to 0.28% of phosphorus.

Only a portion of the vegetation residue that accumulates on impervious surfaces is a pollution threat to surface waters. Vegetation fallout on soils becomes an integral part of the soil composition, and in most cases may even improve soil permeability and erosion resistance.

Traffic

Motor vehicular traffic is directly responsible for the deposition of substantial amounts of pollutants, including toxic hydrocarbons, metals, and asbestos, in addition to oils. The particulates contributed by traffic are primarily inorganic. Most of the traffic exhaust pipe emissions are dust size ($<60\,\mu$m). However, vehicle exhaust pipes are not the only source of traffic-related pollution. Tire wear, solids carried on tires and vehicle bodies, wear and break down of parts, and loss of lubrication fluids add to the pollution inputs contributed by traffic. The subsequent section on sources of toxics presents various sources of toxic materials in urban runoff, including those from traffic.

Only a small portion (<5%) of traffic-related pollution can be directly traced to vehicle emissions. However, the pollutants that motor vehicles emit are among the most important because of their potential toxicity.

In addition to traffic density and vehicle pollutant emissions, pavement conditions and compactions are significant in determining the traffic impact on pollutant loads. Streets paved entirely with asphalt have loadings of about 80% higher than all concrete streets (Sartor and Boyd, 1972). Streets whose conditions were rated as "fair to poor" were found to have total solids loadings 2.5 times greater than those rated "good to excellent." Similar conclusions were also reported by Pitt (1979).

A study by Shaheen (1975), in the Washington, D.C., metropolitan area quantified the traffic contributions to nonpoint pollution. Table 8.8 shows the composition of traffic-related emissions by Shaheen. Shaheen's study estimated that approximately 0.7 g/axle-km of roadway solids can be directly attributed to traffic. In another EPA study, direct traffic emissions were reported to be 0.2 g/vehicle-km from vehicle exhaust and 0.125 g/vehicle-km from tire wear (U.S. EPA, 1977).

In research sponsored by the Federal Highway Administration, Strecker et al. (1987) summarized data on the quality of runoff from highways. Average concentrations of constituents in highway runoff are given in Table 8.9. Bannerman (1991) has found that the average con-

TABLE 8.8 Traffic Emissions

Pollutant	Percent of Total Solids by Weight
Volatile solids	5.1
BOD_5	0.23
COD	5.4
Grease	0.64
Total P	0.06
TKN	0.016
Nitrate	0.008
Asbestos	$3.6 * 10^5$ fibers/g
Lead	1.2
Chromium	0.008
Copper	0.012
Nickel	0.019
Zinc	0.15
Emission rates of total solids	0.671 g/axle-km

Source: Based on data from Shaheen (1975).

TABLE 8.9 Average Concentration of Pollutants in Highway Runoff

Constituent	Mean Concentration (mg/l)	Coefficient of Variation[a]
Suspended solids	143	1.16
Total Kjeldahl N	1.8	0.97
Lead	0.53	2.01
Zinc	0.37	1.37

Source: Based on data from Strecker et al. (1987).

[a] The coefficient of variation is a ratio of standard deviation to the mean.

centration (event mean concentration—see the section titled "Statistical Quality Characteristics of Urban Runoff" for a definition) of metallic toxic pollutants can be correlated to traffic volume (Fig. 8.6).

Particle-Size Distribution of Street Refuse

Table 8.10 and Figure 8.7 indicate that most street refuse is in coarser fractions, roughly in sizes equivalent to sand and gravel. However, most

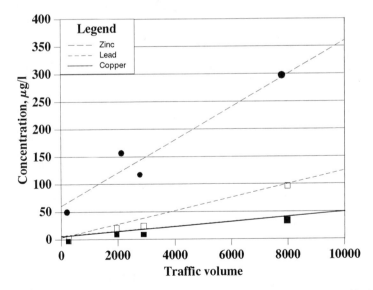

FIGURE 8.6. Relationship of even mean concentrations of some metals to traffic density. (Bannerman, 1991.)

TABLE 8.10 Percent of Street Pollutants in Various Particle-Size Ranges

Pollutant	Particle Size (μm)					
	>2000	840–2000	240–840	104–240	43–104	<43
Total solids	24.4	7.6	24.6	27.8	9.7	5.9
Volatile solids	11.0	17.4	12.0	16.1	17.9	25.6
COD	2.4	4.5	13.0	12.4	45.0	22.7
BOD_5	7.4	21.1	15.7	15.2	17.3	24.3
Phosphates	0	0.9	6.9	6.4	29.6	56.2
All toxic metals	16.3	17.5	14.9	23.5	—	27.5
TKN	9.9	11.6	20.0	20.2	19.6	18.7
All pesticides		27.0			73.0	
PCBs		66.0			34.0	

Source: After Sartor, Boyd, and Agardy (1974).

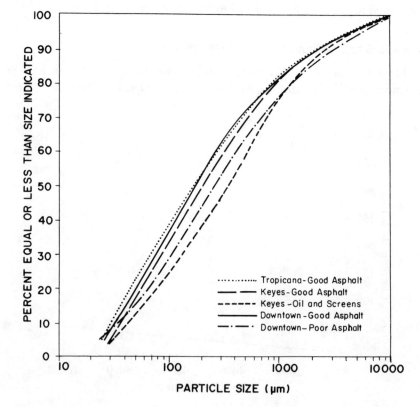

FIGURE 8.7. Particle-size distribution of street refuse in the San Francisco Bay area before sweeping. (From Pitt, 1979.)

pollutants, as they are in soils, are associated with fine fractions. For example, 6% of solids with sizes less than 43 μm (equivalent to clay and silt fraction in soils) contain more than 50% of phosphorus.

Street cleaning practices are selective for coarser particles, while street washing and surface runoff are selective for finer particles. Chapter 10 discusses the efficiency and water quality impact of street sweeping.

Use of Deicing Chemicals

During winter in snowbelt urban areas (approximately the northern half of the United States, all mountain states of the United States, and the northern two-thirds of Europe and the high mountain areas of southern Europe), road deicing salts and sand are applied to road surfaces in addition to snow plowing and removal to provide safe and desirably bare pavement driving conditions. In the United States road deicing salts are applied at rates of 75 to 330 kg/km of highway (street) lane. As an example, in 1984, 40,000 tonnes of salts were applied in Milwaukee, Wisconsin, and during the 1981–1984 period the average seasonal application was 12 tonnes of salt per kilometer of street. Most of the applied

FIGURE 8.8. Urban diffuse pollution during the winter in snowbelt area: Milwaukee, Wisconsin. (Photo: V. Novotny.)

salt in Milwaukee ended up in three relatively small urban streams and ground-water aquifers. Figure 8.8 shows a typical street in Milwaukee after a snowfall and salting.

Salt usage varies among communities. For example, the use of salt in nearby Madison, Wisconsin, is about one-third of that of Milwaukee. The leading states in highway salt consumption in the United States are Minnesota, Michigan, New York, Ohio, and Pennsylvania. Public pressure is directing the municipal and state legislatures to adopt ordinances curtailing salt usage.

Currently, according to information provided by the Salt Institute, the road salt used in most of the northern United States is primarily sodium chloride (about 98%) with added calcium chloride and calcium sulfate. In the past, cyanide compounds and phosphate additives were present in commercial highway deicing salts to control corrosion and to minimize caking; however, due to the severe environmental implications, the use of such toxic and environmentally dangerous additives has been eliminated. Marine salts have proved to be comparable in cost to rock salt and have been used in some New England states (Field et al., 1974).

Extremely high concentrations of salt on street surfaces and in runoff (often more than 20 g/l, which is comparable to or greater than the salt content of sea water) are a major cause of vehicular corrosion and infrastructure deterioration. Most of the metal loss due to corrosion is incorporated in the snowmelt runoff. Other pollution-causing effects include damage to road surfaces and parking lot pavements. High levels of salts are also damaging to soil, vegetation, and ground water. Oberts (1986) noted that significant amounts of phosphorus, lead, and zinc can be found in the sand–salt mixture applied to streets in the Minneapolis–St. Paul area.

Erosion

Urban erosion can be divided into sheet and rill erosion of pervious surfaces or channel erosion. Surface erosion is caused by the energy of rainfall and overland flow, while channel erosion is a hydraulic phenomenon (see Chapter 5 for a discussion). Whipple and DiLouie (1981) discussed the problem of channel erosion, which in urban watersheds, has increased two to three times when compared to predevelopment conditions.

Urban pervious surfaces in humid areas are usually well protected by vegetation and yield pollutant inputs only during large, extreme storm events. In arid areas, recent severe droughts and water shortages have caused a change from urban lawns to a less water-demanding landscape called *xeriscape*, which primarily utilizes native plants and wood or stone mulches (see Chapter 10 for a discussion of diffuse-pollution abatement in arid regions). Rare but intensive storms in arid regions may result in very large sediment losses from unprotected idle lands.

RUNOFF POLLUTION GENERATION PROCESS

Runoff from urban watersheds originates from both pervious and impervious surfaces. The impervious surfaces are almost always hydrologically active because their depression storage represents only a very small subtraction from rainfall. Thus the amount of pollutants washed off by rainfall from impervious surfaces depends on the amount of pollutants that has accumulated during the preceding dry period and on the energy of the runoff. On the other hand, the hydrological subtraction from rainfall on pervious surfaces (soils) is much larger, and usually only severe storms will yield appreciable runoff and sediment load. Although soils become hydrologically active only for larger rainfalls, they do provide an infinite pool of solids. Suspended solids (SS) concentrations in runoff from eroded soils may be quite high. The sources and processes contributing to urban diffuse pollution are shown schematically in Figure 8.3.

Accumulation on Impervious Surfaces (Buildup)

Theoretically, one can assume that deposition of pollutants from the sources described in the preceding section is randomly distributed over the entire street surface. However, wind- and traffic-induced air turbulence is constantly shifting the particles away from the road surface until they become either reentrained into the air, trapped in zones of more quiescent air, or redeposited on a pervious or hydrologically inactive surface. Because a typical curb or median barrier represents a zone of reduced air turbulence or an obstacle in the path of the shifted solids, almost all street refuse can be found within 1 meter of the curb (Fig. 8.9). On streets with a large number of parked cars, the distribution of solids over a street cross section is more spread out. Nevertheless, it is justifiable to consider the pollutant mass balance to be expressed per meter of curb (see also the definition and magnitude of urban pollutant loads discussed in the first section of this chapter).

The pollutant accumulation in near-curb storage follows a simple mass balance: the change of the mass of pollutants is equal to the sum of the inputs minus losses. The concept is depicted on Figure 8.10. The losses are by washoff by rainfall-generated surface runoff, and, during dry

FIGURE 8.9. Most street refuse remains within 1 meter of the curb. (Photo: V. Novotny.)

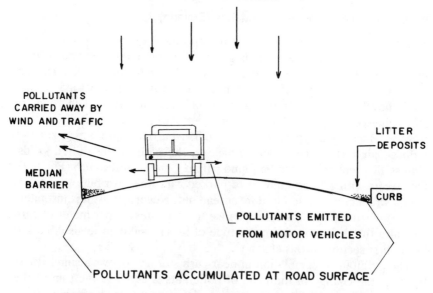

FIGURE 8.10. Schematic of pollutant accumulation on impervious street and highway surfaces.

periods by street sweeping, reentrainment of dust particles, and shifting coarser particles away from the road.

The first urban runoff quantity–quality models, such as the first version of the Stormwater Management Model, SWMM (U.S. EPA, 1971; Heaney and Huber, 1973), and STORM (Hydrologic Engineering Center, 1975) assumed a linear increase in pollutants in curb storage and no losses, whereby each day the amount of solids was increased by a constant increment. This would mean that in the absence of street sweeping, runoff pollution resulting from larger rainfalls would be directly proportional to the number of antecedent dry days. Whipple et al. (1978) criticized the linear accumulation process. Later, theoretical justifications for more complex mass balance models were published by Novotny et al. (1985), James and Boregowda (1986), Huber (1986), and others. The latter two articles contain an extensive discussion of deterministic buildup and washoff models.

The mass balance accumulation function for street solids is

$$\frac{dP}{dt} = I - \xi P \qquad (8.3)$$

where
P = the amount of pollutants in curb storage
I = the sum of all inputs
r = time
ξ = a removal coefficient

Introducing this equation into urban runoff models added another variable that had to be estimated by calibration—the removal coefficient. However, it is unwise to speculate that every urban runoff study and permit site will have the extensive data base available that would be needed to calibrate all model parameters. Hence we present information on the approximate magnitude of the removal parameter, as well as the inputs onto the mass balance models.

Several studies have established that wind- and traffic-induced turbulence can dislocate dust particles that may subsequently become airborne or just be shifted away from the road. Findings in the literature on reentrainment of particles from street surfaces were summarized by Pitt (1979). Another study found that the reentrained portion of the traffic-related emissions (by weight) is an order of magnitude greater than the direct vehicle emissions because of exhaust and tire wear (U.S. EPA, 1977).

The reentrained portion of the traffic-related particulate emissions also contains relatively large particle sizes. The median particle size measured in the EPA study was about 15 μm, with approximately 22% of the particles greater than 30 μm. About 35% of resuspended particles fall out within 30 m of the street. Automobile exhaust and tire wear particulate emissions (by weight) were less than 10% of the total fugitive roadway particulate emissions.

As stated previously, fugitive particulate losses from streets are caused by a combination of wind- and traffic-induced turbulence. Cowherd et al. (1977) stated that wind becomes a factor when the wind speed exceeds a threshold of about 20 km/hr. In most situations, the fugitive particulate losses from the street seem to be due to automobile-induced turbulence.

An equation for the removal coefficient was developed by the author from the statistical analysis of the data published by Shaheen (1975) and NURP study data from Milwaukee. The equation is as follows (Novotny and Chesters, 1981; Novotny et al., 1985):

$$\xi = 0.0116 e^{-0.08H}(TS + WS) \tag{8.4}$$

where
H = the curb height in cm
TS = traffic speed in km/hr
WS = wind speed in km/hr

The unit of ξ is day^{-1}. The coefficient of correlation for this equation is $r = 0.86$. Most of Shaheen's data were collected on roads with a very high traffic density in the Washington, D.C., area (75% of traffic counts exceeded 50,000 axles/day).

Another factor that may affect the overall magnitude of the removal coefficient is the presence of sink surfaces in the vicinity of the road. A sink surface is a hydrologically inactive surface (such as a lawn) on which the reentrained particles can settle and be eliminated from the hydrological transport. The most simple measure of this effect is the fraction of the impervious area directly connected to the drainage system (DC). Then in a model the removal coefficient can be corrected by a correction factor that equals $1 - DC$. This term represents a portion of runoff that overflows from an impervious area on an adjacent hydrologically inactive surface.

Integration of the Accumulation Function

Equation (8.3) can be integrated to

$$P(t) = \frac{I}{\xi}(1 - e^{-\xi t}) + P(0)e^{-\xi t} \tag{8.5}$$

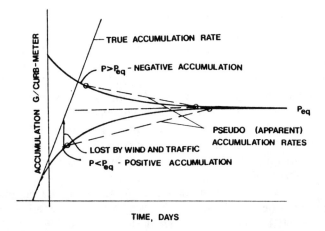

FIGURE 8.11. Street (highway) surface pollution accumulation process.

where in addition to the previously defined variables,
$P(0)$ = initial load of solids
$P(t)$ = the load after time, t

This accumulation function is shown in Figure 8.11. The true input rate, I, is a tangent to the curve at $P = 0$. A variety of apparent pseudoaccumulations can be ascertained by sampling the accumulation pollution loads and by subsequently dividing the difference in loads by the time between the samplings. It also should be noted that in such processes there is always a tendency to attain an equilibrium, whereby

$$\frac{dP}{dt} = 0$$

or

$$I = \xi P_{eq}$$

or

$$P_{eq} = \frac{I}{\xi} \tag{8.6}$$

The accumulation rate then becomes

$$\frac{dP}{dt} = -\xi(P - P_{eq}) \tag{8.7}$$

This means that the apparent accumulation rate of solids in street curb storage can be either positive or negative, depending on whether the initial pollutant load, $P(0)$, is greater or smaller than the equilibrium load P_{eq}. This is also shown on Figure 8.7. A situation where $P(0) > P_{eq}$, which would theoretically result in a negative accumulation (loss) of pollutants from curb storage, is typical for spring, when large amounts of pollutants are left in curb storage after snowmelt. Table 8.1 presented equilibrium curb loads measured by Sartor and Boyd (1972) in their studies of several U.S. cities. Equilibrium curb loadings in residential and commercial sites of the Milwaukee project measured from 1981 to 1984 were about 50% of those presented in Table 8.1.

Statistical evaluation and modeling verifications of the buildup equation were described in Novotny et al. (1985). The results using the Milwaukee NURP study sites showed that the removal coefficient in medium-density residential areas was fairly constant, attaining values around 0.2 to 0.4 day^{-1}, meaning that approximately 20% to 40% of the solids accumulated near the curb on the street surface is removed daily by wind and traffic.

The inputs of solids into curb storage from the three major sources (refuse deposition, atmospheric dry deposition, and traffic) can be combined into one loading estimate

$$I(\text{g/m-day}) = I_r + I_a + I_{tr}$$
$$= \text{LIT} + (\text{ATMFL})(\text{SW})/2 + (\text{TE})(\text{TD})(\text{RCC})/2 \quad (8.8)$$

where
I_r = little or street refuse deposition
I_a = dry atmospheric deposition
I_{tr} = deposition due to traffic emissions and wear
LIT = litter and street refuse deposition, g/m-day
ATMFL = dry atmospheric deposition, g/m$_2$-day (= tonnes/km$_2$-day)
SW = street width in meters
TE = traffic emission rate, g/axle-m
TP = traffic density, axles/day
RCC = road condition factor

Example 8.1

Estimate equilibrium solids load in curb storage of a medium-density residential area. The following information is given:

Traffic count   TC = 240 cars/day
Traffic velocity   TS = 50 km/hr

Wind velocity   WS = 15 km/hr
Curb height   H = 10 cm
Atmospheric dry deposition rate   100 kg/km^2-day (mg/m^2-day)
Daily street refuse deposition I_r = 6 g/m-day
Street width   W = 15 meters
Road Condition Index   RCC = 1.0 (average conditions)

Solution   Express street surface load per 1 meter of curb length.

Atmospheric contribution

$$I_{\text{atm}} = 100\,[\text{mg/m}^2\text{-day}] * 0.001\,[\text{g/mg}] * 15\,[\text{m}]/2 = 1.5\,\text{g/m-day}$$

Traffic contribution (using information from Table 8.8)

$$I_t = 0.671\,[\text{g/axle-km}] * 240\,[\text{cars/day}] * 2\,[\text{axles/car}] * 0.001\,[\text{km/m}]/2$$
$$= 0.32\,\text{g/m-day}$$

(Division by 2 in the preceding estimates reflects two curbs on each side of the road.)

Total input

$$I = 6 + 1.5 + 0.32 = 7.82\,\text{g/m-day}$$

Removal coefficient from Equation (8.4)

$$\xi = 0.0116 e^{-0.08*10}(15 + 50) = 0.34\,\text{day}^{-1}$$

Equilibrium curb load by Equation (8.6)

$$P_{\text{eq}} = I/\xi = 7.82/0.34 = 23\,\text{g/m}$$

Effect of Curb Height on Accumulation

The fact that the rate of pollutant accumulation in curb (median barrier) storage is caused by wind- and traffic-induced air turbulence, leads to the logical conclusion that the height of the curb will affect the accumulation rate, that is, the higher the curb, the more pollutants will accumulate. This effect will be more profound for fine dust particles than for coarser litter particles that are less amenable to becoming airborne. Figure 8.12 shows such a relationship measured in the Washington, D.C., area. The graph indeed indicates that the curb height may have a profound effect on the accumulation of finer particles (<3.2 mm in diameter), while larger

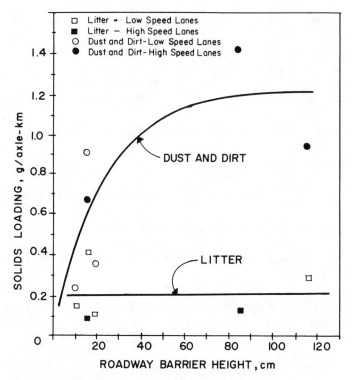

FIGURE 8.12. Effect of curb height on particle loading of street surface. (From Shaheen, 1975.)

particle accumulation will be unaffected. The curve for the finer particles accumulation proves the classic "first-order removal" concept expressed by Equation (8.3), which also implies that the amount of pollutants deposited on the surface or in curb storage will have a decreasing rate of increase approaching an equilibrium.

Example 8.2

What will be the equilibrium of solids accumulation if the curb height in the previous example is lowered to 5 cm?

Solution For $H = 5$ cm the removal coefficient becomes (Eq. (8.4))

$$\xi = 0.0116 e^{-0.088*5}(15 + 50) = 0.51 \, \text{day}^{-1}$$

and the equilibrium accumulation is

$$P_{eq} = 7.82/0.51 = 15.33 \, g/m$$

which represents a 33% reduction.

Winter Accumulation of Pollutants

The concept of translocation and removal of pollutants from curb storage by wind and traffic that determines the magnitude of the curb loading cannot be applied during the winter period in snowbelt urban areas. In addition, snow-removal practices and application of deicing chemicals further complicate the process.

Typically, even after a minor snowfall in most northern U.S. cities snow-removal activities are initiated by the highway and sanitation crews. The snow from the streets and highways is plowed toward the curb and large quantities of salt are then applied to provide bare pavement for traffic. The accumulated snow piles near the curb (Fig. 8.8) may persist for months in northern cities, including Chicago, Milwaukee, Minneapolis, Cleveland, and Buffalo. The snow piles are then effective traps of street pollutants, including large amounts of salt. Also, due to increased heating and energy use, atmospheric deposition from industrial sources is increased. On the other hand, dust from soil erosion is reduced due to the protective snow cover and the freezing of soils. From data by Bannerman et al. (1984) presented in the list on page 453, it can be documented that dry atmospheric deposition in the winter months in Milwaukee (December–May) was 1.5 times higher for COD and solids than in the nonwinter period. Deposition rates for phosphorus, lead, and sulphates were statistically indistinguishable from nonwinter rates.

Because the solids and pollutants are incorporated into snow and are not removed from the snowpack by wind and traffic the accumulation rate is almost linear and, hence, much higher than in nonwinter periods. For this reason the quantity of accumulated pollutants near the curb at the end of the snow period is very high, as shown on Figure 8.13. Bannerman et al. (1984) documented that the street loads of sediment and toxic metals are at their highest level from the onset of the first significant spring melt through the first spring rainfall event. In the Milwaukee research Bannerman et al. reported that 20% to 33% of the annual load of these substances can occur during this period. It should also be pointed out that the possibility of reducing the quantity of street pollutants during the winter is very limited, since the use of sweeping equipment is not possible when snow piles and frozen ice are located on the sides of the streets.

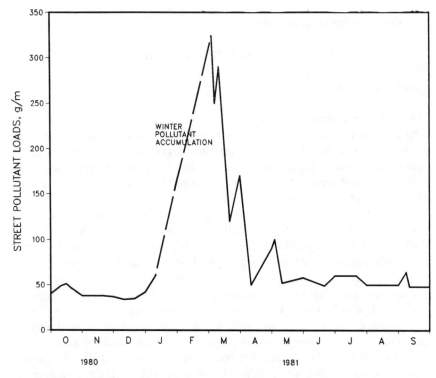

FIGURE 8.13. Annual variation of street accumulation of particulate pollutants in snowbelt urban areas. (Data from Wisconsin Department of Natural Resources.)

Removal of Pollutants from Curb Storage by Precipitation (Washoff)

When surface runoff occurs on impervious surfaces (Fig. 8.14) as a result of either natural storms of street flushing practices, the splashing effect of rain droplets and drag forces of the flow put particles in motion. Many hydraulic models have appeared in the literature on sedimentation that could potentially be applicable to the problem of particle pickup and transport.

From the numerous equations published in the literature, the Yalin equation (Yalin, 1963) (see Chapter 5 for a detailed discussion of the Yalin equation) is viewed by many as the one that best describes the pickup and transport of particles by shallow flows typical of street gutters (Sutherland and McCuen, 1978). Huber (1986) quoted a research investigation by the University of Florida and reviewed several theoretical

FIGURE 8.14. Washoff of street refuse during a rainstorm. (Photo: V. Novotny.)

approaches to modeling sediment pickup and transport. The study at the University of Florida concluded that although the theoretical sediment transport theory is certainly attractive and worth studying, it is often insufficient in practice because of lack of data for parameter evaluation, sensitivity to time period and breakdown into small discrete computational elements, and uncertain cross-sectional geometry of street gutters and surfaces, and besides which, the simpler semiempirical models work as well or better. Huber pointed out quite correctly that urban runoff modeling is generally crude and always requires calibration regardless of the level of sophistication of individual models and submodels.

The simplest, most widely used but empirical model of washoff was introduced by Sartor et al. (Sartor and Boyd, 1972; Sartor, Boyd, and Agardy, 1974). The authors analyzed the solids washoff data using a simple first-order removal concept

$$\frac{dP}{dt} = -K_u r P \tag{8.9}$$

where
r = rainfall intensity

k_u = a constant called "urban washoff coefficient" that depends on street surface characteristics
P = amount of solids remaining on the surface
t = time

The constant, k_u, was found to be almost independent of particle size within the studied range of 10 μm to 1 mm.

Equation (8.9) integrates to

$$P_t = P_0(1 - \exp(-k_u rt)) \tag{8.10}$$

where
P_0 = the initial mass (weight) of solids in the curb storage
P_t = mass (weight) of material removed by rain with intensity r and duration t

The value of the urban washoff coefficient was almost arbitrarily chosen as 0.19 if the rain intensity is in millimeters/hour. Note that this value has been recommended by most subsequent urban runoff models that utilize this concept. The authors of STORM (Hydrologic Engineering Center, 1975) modified Equation (8.10) by assuming that not all the solids are available for transport. Then

$$P_t = AP_0(1 - \exp(-k_u rt)) \tag{8.11}$$

where

$$A = 0.057 + 0.04(r^{1.1}) \tag{8.12}$$

is the so-called availability factor that accounts for the nonheterogenous makeup of particles and the variability in the travel distances of the dust and dirt particles. The maximum value for A is 1.0.

Huber (1986) pointed out that the washoff model expressed by Equation (8.11) is not completely without physical basis since the theoretical sediment transport theory predicts dependence of scour upon flow rate (see the Yalin equation in Chapter 5). The Sartor et al. equation and its modifications have been incorporated into most of the widely used urban runoff models presented in the previous chapter.

Example 8.3: Washoff Estimation by the Sartor et al. Model

Estimate solids removal from the curb storage by the Sartor et al. concept for a 15-mm/hr rain lasting 1.5 hours with $P_0 = 50$ g/m.

Solution From Equation (8.12) the availability factor is

$$A = 0.057 + 0.04(15^{1.1}) = 0.84$$

Then the amount removed by rain is

$$P_t = 0.84 * 50(1 - \exp(-0.19 * 15 * 1.5)) = 41.42 \, g/m$$

The particle removal by rain is 83%.

To obtain the loading of pollutants, the sediment loads are commonly multiplied by multipliers expressing the pollution content of the sediment. In modeling terminology these multipliers are called *potency factors*. The magnitudes of potency factors for urban and rural areas were investigated by Zison (1980).

Pollutant Removal by Winter Snowmelt

The hydrology and pollutant loads of urban areas located in snowbelt areas is different from that typical for nonwinter periods. The melting of snowpack is a result of heat inputs and heat balance that bring the temperature of the snowpack above the melting point. The melting point of ice or snow with small quantities of pollutants is around 0°C. However, the addition of salt and the exothermic nature of the salt dissolution decreases the melting point of the saline melt below that for clean water.

The heat inputs into the snow pack are solar and atmospheric radiation, sensible heat, and heat gained by condensation of moisture above the snowpack; heat losses include reflectivity (albedo) of the snow, back radiation, and sublimation. Reflectivity (albedo) for urban snow may vary between 20% for dirty saturated snow to about 85% of reflected radiation for fresh fallen snow.

Observation and modeling of winter snowmelt rates in urban areas (Novotny, 1987, 1988; Bengtsson, 1982) showed that melting rates, even during days with higher temperatures, are relatively small, much less than those for a typical nonwinter storm event of the same magnitude (meaning that storm depth is the same as the water content of the snowpack). For example, assuming a clear sunny day in late February and considering the average incoming solar radiation at the 46 degree latitude, dry conditions and a 10°C temperature of the ambient air the maximum runoff from melting snow is only about 0.15 cm/hr. One has to also consider the fact that pervious urban areas in spite of freezing are still permeable enough to absorb the snowmelt from above. Therefore, only impervious urban areas can be considered in the hydrological models of snowmelt generation from urban areas.

Enrichment of Urban Snowmelt

In Chapter 5 the term *enrichment* referred to the difference between the pollution content of the soil and that of the runoff. For winter snowmelt processes, the enrichment ratio represents the ratio of the pollutant concentration in the snowmelt and that in the snowpack or

$$ER = \frac{\text{Meltwater concentration}}{\text{Bulk snow concentration of prior snowmelt}} \tag{8.13}$$

The enrichment ratio of unity would mean that the pollutant concentrations in the snowmelt (runoff) and snowpack are similar. However, this is rarely the case. During freezing and refreezing, the ice crystals that form the snowpack effectively reject some dissolved pollutants that may then become easily available for transport by subsequent snowmelt water. This phenomenon is especially evident when dealing with pollution and modeling caused by salt applications (salinity and chloride contents). In studies in Milwaukee (Novotny, 1987), enrichment ratios of 6 to 10 had to be used for modeling salt loads to receiving waters. An enrichment ratio of greater than one will result in a "first flush" effect. A similar study with similar results was also performed in Sweden by Westerström (1990), who measured peak enrichment ratios of 4 to 6 for chlorides and nitrates in the first 25% of snowmelt, and up to 8 for sulphates. All snowmelt concentration curves showed a strong peak, followed by a fast decline (Fig. 8.15). For pH the difference between that of snowpack and the beginning of snowmelt was more than one order of magnitude (snowpack pH was one pH logarithmic unit higher than pH in the beginning of snowmelt).

On the other hand, due to the fact that snowmelt must percolate through the snowpack, particulate pollutants are filtered out and remain mostly in the snowpack, and are subsequently deposited into the curbside deposits, as shown on Figure 8.8. Thus for these pollutants the enrichment ratio is less than one. Such pollutants include suspended solids, both volatile and mineral lead, many metals, and phosphates. This was confirmed by, among others, McComas, Cooke, and Kennedy (1976), who compared winter snowmelt and rainfall runoff events in northeast Ohio. The results indicated that while snow was a significant source of water, the snowmelt load of phosphorus was much smaller, insignificant, when compared to a similar rainfall-runoff event. Most of the phosphorus is associated with fine sediments that remain on site and do not move with snowmelt.

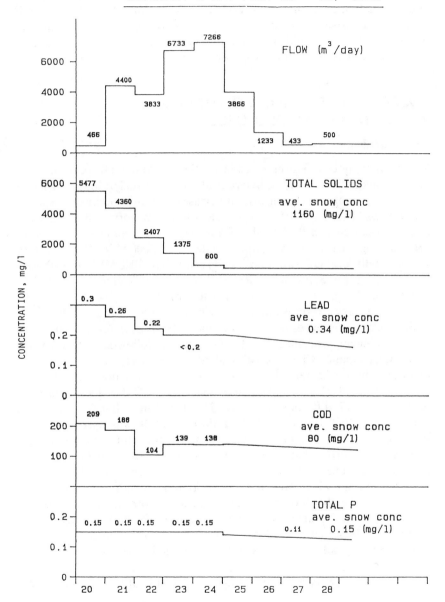

FIGURE 8.15. Snowmelt enrichment by pollutants in urban snowmelt. Data from an urban experimental watershed in Milwaukee. (After Novotny, 1987.)

Oberts (1986, 1990) noted that high concentrations of solids, phosphorus, lead, and zinc in urban snowmelt are due in part to sand and salt spread to improve winter driving conditions. Oberts (1982a) also found that snowmelt is a significant pollution source of urban lakes located in the Minneapolis–St. Paul metropolitan area.

SOURCES OF SPECIFIC POLLUTANTS IN URBAN RUNOFF AND CSOs

Microorganisms

Microbial analyses of storm water in Baltimore, Maryland, performed by Johns Hopkins University (Olivieri et al., 1977) revealed the presence of coliforms, pathogenic organisms, and viruses in both combined and storm sewer flows. The ranges of fecal coliforms found in storm runoff were 200 to more than 2000 MPN/100 ml, with a majority of samples—123 out of 136—having fecal coliform counts greater than 2000 MPN/100 ml. Storm samples with fecal coliform densities greater than 2000 MPN/100 ml were 95% positive for *Salmonella*. Ranges of *Salmonella* densities in urban runoff from Baltimore were less than 1 to more than 11,000 MPN/10 l. Six storm water flows were examined for viruses and all tested positive.

Significant differences have been found at the NURP test sites between the event mean concentration (EMCs; see the following section, titled "Statistical Quality Characteristics of Urban Runoff," for a definition) of fecal coliform bacteria between winter (cold) and nonwinter periods (U.S. EPA, 1983). The average value for EMC for fecal coliforms measured at 17 different sites was 21,000/100 ml, while that for cold periods was 1000/100 ml, a 21-fold difference. It was speculated that these significant seasonal differences cannot be related to any comparable magnitude of urban land-use activities taking place in cold and warm periods. The levels of fecal coliform bacteria, especially those in the warm period, are high and could be considered a health risk. However, it was pointed out that indicators such as fecal coliform counts may not be useful in identifying health risks from urban runoff pollution.

Glenne (1984) correlated the density of total coliforms in surface runoff to the population density in the watershed. The resulting equation

$$T.\ coli\ (\text{MPN}/100\,\text{ml}) = 5700\ (\text{PDE})^{1.35}$$

where PDE is the population density equivalent in persons per acre, had a coefficient of 0.95. This indicates that the coliform bacteria in urban runoff may be caused by the cross-connection of sewage into storm drainage. In view of the results of the NURP statistical analysis of a large number of samples from a number of sites throughout the United States the correlation by Glenne may be only site specific.

Literature values refer to the fecal coliforms to fecal streptococci ratio (FC/FS) as an indicator of the origin of fecal pollution. A value of four or greater indicates human source of fecal pollution, while a value of less than one suggest nonhuman origin. In the Baltimore study just quoted, 86% of all storm water samples had a FC/FS ratio of less than one, and 94% of samples had a FC/FS ratio of less than four, giving an indication that most of the fecal contamination is of nonhuman origin. On the other hand, combined sewer overflows had 58% of the samples with a FC/FS ratio of less than one, and 15% with a FC/FS ratio of greater than four. Raw sewage had 50% of samples with a FC/FS ratio greater than four. The authors therefore stated that the FC/FS ratio should not be employed as a magic number to evaluate the source of contamination in a complex system.

Hydrocarbons (PAHs) and Other Toxic (Priority) Pollutants

Based on the analyses of urban storm runoff effluents, Whipple and Hunter (1979) speculated that after implementation of point source abatement, urban runoff will become the major source of petroleum hydrocarbons. The sources of toxics are numerous (Table 8.11).

Inorganic pollutants. As a group, the toxic metals were by far the most prevalent priority pollutant constituent found in urban runoff by the NURP studies (U.S. EPA, 1983). Fourteen inorganic, potentially hazardous constituents (13 toxic metals plus asbestos) were detected in the NURP study sites. Most often detected among the metals were copper, lead, and zinc, all of which were found at least in 91% of the samples. Other frequently detected inorganic pollutants included arsenic, chromium, cadmium, nickel, and cyanide. Pitt (1988) found asbestos in urban runoff in a watershed in the San Francisco Bay area where no natural sources were known; however, no specific single source was identified.

By statistically analyzing the event mean concentrations (EMCs) (see the next section for a definition and methods of estimating EMCs) of urban runoff in Milwaukee, Bannerman (1991) found that metallic toxicity of urban runoff varies with land use. For commercial areas with a high traffic count, acute toxicity criteria for warm-water fish were violated in 73% of the samples for lead and more than 90% of the samples for silver, copper, and zinc. On the other hand, event mean concentrations of urban runoff from a medium-density residential area with low traffic density the

TABLE 8.11 Sources of Toxic and Hazardous Substances in Urban Runoff

	Automobile Use	Pesticide Use	Industrial/Other Use
Heavy Metals			
Copper	Metal Corrosion	Algicide	Paint, wood preservative, electroplating
Lead	Gasoline, batteries		Paint, lead pipe
Zinc	Metal corrosion Tires, road salt	Wood preservative	Paint, metal corrosion
Chromium	Metal corrosion		Paint, metal corrosion, electroplating
Halogenated Alphatics			
Methylene chloride		Fumigant	Plastics, paint remover, solvent
Methyl chloride	Gasoline	Fumigant	Refrigerant, solvent
Phthalate Esters			
Bis (2-ethylexy) phthalate			Plasticizer
Butylbenzyl phthalate			Plasticizer
D-*N*-butyl phthalate		Insecticide	Plasticizer, printing inks, paper, stain, adhesive
Polycyclic Aromatic Hydrocarbons			
Chrysene	Gasoline, oil, grease		
Phenanthrene	Gasoline		Wood/coal combustion
Pyrene	Gasoline, oil, asphalt	Wood preservative	Wood/coal combustion
Other Volatiles			
Benzene	Gasoline		Solvent
Chloroform	Formed from salt	Insecticide	Solvent, formed from chlorination
	Gasoline, asphalt		Solvent
Toluene	Gasoline, asphalt		Solvent
Pesticides and Phenols			
Lindane (gamma BHC)		Mosquito control	
		Seed pretreatment	
Chlordane		Termite control	
Dieldrin		Insecticide	Wood processing
Pentachlorophenol			Wood preservative, paint
PCBs			Electrical, insulation, several other Industrial applications
Asbestos	Break and clutch lining Tire additives		Insulations

Source: After U.S. EPA (1990b).

toxic criteria were exceeded in only 0% to 4% for lead, 0% for silver, and 20% to 30% for zinc and copper. High concentrations of mercury, cadmium, and lead have also been observed in urban soils (Carey, 1979).

Organic pollutants. In the NURP study by the U.S. EPA (1983) organic pollutants were detected less frequently and at lower concentrations than the inorganic pollutants. The most commonly found organic pollutant was the plasticizer bis(2-ethylhexyl) phthalate (22%), followed by the pesticide α-hexachlorocyclohexane (α-BHC) (20%). An additional 11 organic pollutants were reported with detection frequencies between 10 and 20%. Among the PCB group there was only one detected among all the samples.

Marsalek (1986) reported data on priority (toxic) pollutants in urban runoff gathered in Canadian watersheds of the Great Lakes. The highest concentrations in storm runoff were found for trace elements as a group. Among the individual elements the highest mean concentrations were found for zinc (EMC = 400 μg/l in water), copper (19 μg/l), lead (90 μg/l), and nickel (16 μg/l). Much higher concentrations were measured in the sediment. Significant (greater than the detection limit) concentrations were found for PAHs (poly-aromatic hydrocarbons). The highest concentrations were found for 1,2-dichlorobenzene (0.04 μg/l), α-BHC (0.02 μg/l), and for the total PCBs (0.014 μg/l). Again, higher concentrations were found in the sediments, and it was speculated that most of the PAHs transported by runoff occurred most often with the sediment (Marsalek, 1990). The probability distribution of PAH concentrations in runoff was approximated by a log-normal distribution (see the following section titled "Statistical Quality Characteristics of Urban Runoff").

Urban runoff was found to be the source of 48% of the petroleum hydrocarbons, 3% of the lower molecular weight polycyclic aromatic hydrocarbons, 44% of the higher molecular weight polycyclic aromatic hydrocarbons, 65% of lead, 56% of zinc, and 5% of copper entering the Narragansett Bay annually (Hoffman et al., 1984; Hoffman, 1985). Annual loadings of petroleum hydrocarbons measured by Hoffman (1985) were the highest from heavy industrial land uses (14,000 kg/km^2), followed by busy highways (7800 kg/km^2). Loadings from commercial and residential land uses were one order of magnitude less (580 kg/km^2 for commercial and 180 kg/km^2 for residential land uses, respectively). The highest loadings of PAHs were found on the highway (18.1 kg/km^2), followed by heavy industrial lands (8.4 kg/km^2); similarly, PAHs loadings of one order of magnitude lower were measured on commercial (0.59 kg/km^2) and residential (0.27 kg/km^2) lands.

Fam, Stenstrom, and Silverman (1987) found in the San Francisco Bay area watersheds that land uses with high commercial/industrial activity have much greater aliphatic hydrocarbon emissions than noncommercial ones. The ratio of total extractable organics to total organic carbons was found to vary with land use, with a ratio of six or more indicating significant commercial activity.

Bannerman (1991) found that the EMCs of PAHs in Wisconsin urban runoff (the Milwaukee and Madison areas) violated the human cancer criteria in 60% of samples for chrysene and phenatherene, 82% for pyrene, and more than 95% of samples for fluoranthene and benzo (GHI) perylene. EMCs of aldrin, toxaphene, and DDT in Wisconsin urban runoff almost always exceeded human cancer criteria. Other pesticides found less frequently in such runoff included endrin, chlordane, malathion, diazinon, and lindane. Annual monitoring of urban soils between 1969 and 1979 has demonstrated that they generally have higher pesticide residue concentrations than agricultural soils in the same (nearby) location (Carey, 1979).

A comprehensive study on toxic contamination of urban runoff and CSOs was prepared for the U.S. EPA by Pitt and Barron (1989). They analyzed about 150 samples of surface runoff (sheet flow) and CSO, and found that about 15% of the samples analyzed were extremely toxic. The remaining samples were evenly split between being moderately toxic and nontoxic. The greatest detection frequencies were for 1,3-dichlorobenzene and fluoranthene, which had detection frequencies of 23%.

A higher percentage of the CSO samples were considered extremely toxic than any of the others followed by parking and storage area surface runoff samples. Roof runoff, urban creeks, and CSO samples had the greatest detection frequencies, while vehicle service and parking areas had the highest observed concentrations of organic toxicants. Roof runoff had the greatest concentrations of zinc, while parking areas had the greatest concentrations of nickel. Traffic- and parking-associated areas had the greatest concentrations of lead, and urban creeks were highest in copper content.

Effect of Acidity of Urban Precipitation on Pollutant Loads

As pointed out in Chapter 4, the acidity of urban precipitation caused by local and regional sources of sulphur oxides and mostly local (traffic) sources of nitric oxides is quickly buffered by overland flow (Novotny and Kincaid, 1982). However, this "buffering" capacity in urban areas is provided primarily by the infrastructure, including buildings, pavements, and sewers. During buffering, the excess hydrogen ions from acid precipitation is buffered by carbonates, which releases the cations, such as

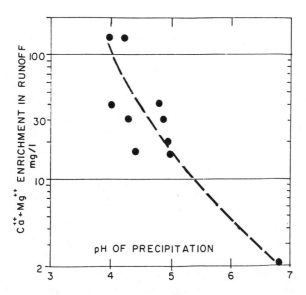

FIGURE 8.16. Effect of the acidity of urban precipitation on the elutriation of minerals and cations from urban infrastructure. (From Novotny and Kincaid, 1982.)

calcium and magnesium (primary components in concrete, mortars, and soil), into the runoff. Figure 8.16 shows the relationship between the acidity of precipitation and the calcium–magnesium content of runoff from a large concrete parking lot basin located in Milwaukee.

Acids in precipitation also dramatically increase the solubility of metals used in cars, zinc gutters, and other places. Förster (1990) showed that urban metal roofs and roof-and-gutter materials (such as zinc) are good buffers for acid rainfall, but metals and other pollutants (for example, PAHs from roofs with tar shingles or cover) are elevated in the runoff from these surfaces. Although zinc roof materials provided very high buffering of the pH of the acid rainfall (pH of runoff was near neutral), dissolved zinc content in the runoff was 450 times greater than that in the rainfall (0.005 mg of dissolved Zn/l in rainfall compared to 2.2 mg dissolved Zn/l in roof runoff from zinc-covered roofs and gutters) when the pH of the rainfall was less than 4.

Growth and Accumulation of Sewer Solids

In the 1970s Pisano et al. (1979) investigated solids accumulation in combined sewers in Boston during dry periods. Research in Switzerland

(Krejci et al., 1987) showed that solids accumulation in sewers has two components: (1) settling of solids in places with low velocity during low-flow, dry-weather flow, and (2) slime growth on the walls of the sewers. A part of the slime growth component is less affected by the hydraulics of the flow and is not easily washable. However, during high-flow storm events scouring of both components will contribute to the solids load of the overflows and may lead to the "first flush" effect.

By statistical analysis, Pisano et al. developed the following equation for estimating solids accumulation (converted and modified for metric units)

$$TS = 11.14 S^{-0.4375} q^{-0.51} \qquad (8.14)$$

where
TS = solids deposition and growth rate in combined sewers in grams per meter of sewer per day
S = average slope of the sewer, dimensionless (m/m)
q = per capita dry weather waste rate, liters/cap-day

and the coefficient of correlation was 0.85.

The inverse proportionality of the solids deposition rate to the slope reflects the effect of flow velocity that controls the deposition. The per capita waste flow rate may reflect the concentration of solids in the flow. The higher the per capita flow, the more diluted solids may become in the sewer flow, resulting in less deposition.

Research in Switzerland by Krejci et al. (1987) revealed that solids deposition and slime growths in combined sewers represent a significant portion of the total pollution load. They found that the growth of the washable slime layer varied between 10 to 30 g of TSS per square meter of sewer surface per day, and depended on the local dry-weather flow velocity. The sewer slime contained by weight 80% COD, 5% TKN, and 1% phosphate. In the overall balance of pollution carried by the combined wet-weather flow during storm events, the sewer slime–sediment layer contribution represented 55% to 80% of the total load for TSS, COD, TOC, TKN, and total P. This contribution exceeded that by surface runoff and sewage flows during the wet-weather period.

Krejci et al. also found that the length of the dry period preceding the wet-weather flow event had no noticeable effect on the quantity of deposits in sewers, which appears to contradict the results in Boston by Pisano et al. Apparently, the slime-deposit layer is in equilibrium with the scouring energy of the dry-weather flow, and this equilibrium is attained after a relatively short period of time.

Cross-Connections and Illicit Discharges into Storm Sewers

Research and investigations of the impact of illicit and cross-connection on the pollution content of storm water discharges has revealed that these inputs can contribute significant pollutant loads (Schmidt and Spencer, 1986; Pitt et al., 1990a; Pitt, Lalor, and Driscoll, 1990).

Non-storm water discharges in storm sewers originate primarily from sewage and industrial wastewater leaking from sanitary sewers, failing septic systems in storm-sewered areas, and from vehicle maintenance activities. However, almost any type of pollution may find its way into urban storm sewers by illicit discharges and accidental spills. Deliberate dumping into storm sewers and catch basins of used oil or waste paint are especially common and troublesome. Leaking underground storage tanks, leachate from sanitary landfills, and hazardous waste treatment, storage, and disposal sites can also contribute to urban storm water and combined sewer overflow pollution. In the United States the detection, identification, and elimination of such discharges is a major focus of the NPDES Stormwater Permit program (see Chapter 2 for a discussion). The manual by Pitt, Lalor, and Driscoll (1990) provides procedures for identification of sources and magnitudes of non-storm water discharges into storm sewers.

Mobile pollutants such as nitrates may also contaminate the shallow aquifer and discharge into urban drainage systems in the form of base flow in streams and infiltration inflows into sewers. The pollutional impact of these contributions is generally unknown.

Suburban Nonpoint Pollution Loads

Septic systems and construction erosion are the most common and significant sources of pollution in suburban areas. Septic system pollution is due to two pathways: (1) subsurface transport of mobile pollutants (primarily nitrate) via shallow discharging aquifers toward the receiving bodies of water, occurring primarily during base flow (during high-flow runoff events most urban streams may be recharging the aquifer), and (2) from effluent surfacing from failing septic systems.

Suburban nonpoint pollution loads are also affected by the presence or absence of wetlands (Oberts, 1982b; Brown, 1988) within the watershed. Wetlands are known to attenuate pollutants and runoff rates.

STATISTICAL QUALITY CHARACTERISTICS OF URBAN RUNOFF

During the Nationwide Urban Runoff Program (U.S. EPA, 1983) 28 urban sites were monitored throughout the United States. This research effort provided a large data base on the quality and loads by urban runoff (not including the combined sewer overflows). Major conclusions are presented herein. Prior to the presentation of the results and their interpretations, some fundamental statistics have to be reviewed.

Parameters and Statistics of the Urban Runoff Water Quality Evaluation

Event mean concentration. The water quality parameters exhibit large variations throughout each runoff event. Some events exhibit the first flush effect, for some other events the first flush effect has not been noticeable (Ellis, 1986). Thus, the NURP studies focused on evaluating the event mean concentrations, defined as

$$\text{EMC} = \frac{\text{Mass of pollutant contained in the runoff event}}{\text{Total volume of flow in the event}} = \frac{\Sigma Q_i C_i}{\Sigma Q_i} \quad (8.15)$$

where
Q_i = discrete flow ordinates on the event hydrograph
C_i = corresponding concentrations on the pollutograph

In most cases the total load by the runoff event is more important than the individual concentrations within the event due to the fact that runoff events are relatively short, the receiving water body provides some mixing, and the concentration in the receiving body of water is a response to the total load rather than to the concentration variability within each particular event. For some pollutants, such as nutrients, the total load is the most important decision variable in determining water quality impact. For this reason the EMC parameter has been found to be the most appropriate variable for evaluating the impact of urban runoff.

Mean, coefficient of variation, and probability distribution. The evaluation of the NURP data base has revealed that the probability distribution of EMCs follows a so-called *log-normal* probability distribution. A probability density or frequency distribution (pdf) expresses the probability that a concentration with a given magnitude falls within a specified interval, or

$$\int_{x_1}^{x_2} \text{pdf}(x) = p(x_1 > x > x_2)$$

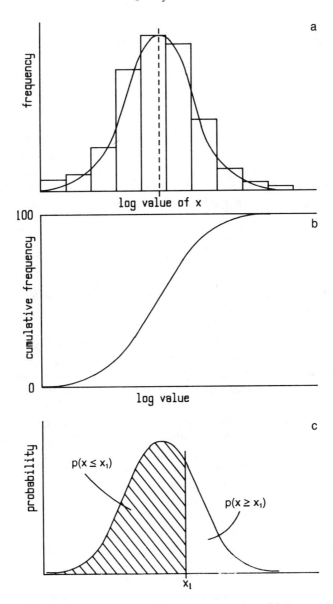

FIGURE 8.17. Frequency and probability distribution functions. (a) Frequency distribution of logarithmic values. (b) Cumulative probability distribution. (c) Probability of exceedence.

where x is the value of the parameter and x_1 and x_2 are the limits of the interval. The frequency distribution of a parameter can be simply found by plotting or tabulating the number or percent of cases for which the value of the parameter falls within each interval (at this point one may select an arbitrary number of intervals between the smallest and the largest value of the record) versus the magnitude of the interval, as shown in Figure 8.17(a).

A cumulative frequency (probability) function (cpf) is thus a summation of all individual probabilities of all x's that are either less or equal (or greater or equal) than a given value of x (Fig. 8.17(b)). The cumulative frequency curve can also be obtained by integration of the probability density (frequency) distribution function, as shown on Figure 8.17(c). Therein

$$\text{cpf}(x \leq) = p(x \leq X_1) \quad \text{or} \quad \text{cpf}(x \geq) = p(x \geq X_1)$$

Note that $p(x \leq X_1) = 1 - p(x > X_1)$ or a probability of a value being less or equal is one minus probability of the same value being exceeded. This is shown on Figure 8.17c.

If the interval becomes very small, then the probability (frequency) density function becomes continuous and may assume several typical shapes, such as are shown in Figure 8.17(b). There are several mathematical formulations that may describe the shape of these curves, of which Gaussian normal distribution and Pearson Type III distribution are the most common probability density functions used in hydrology. The reader is referred to the several standard texts on hydrology referenced in Chapter 3, or to standard texts on statistics, for more detailed discussion. Probability paper that has one ordinate scaled according to the Gaussian cpf is very convenient for representation of a series of observations such as a sequence of EMCs of surface runoff.

To plot EMCs on probability paper (arithmetic or logarithmic) the concentrations (or any other parameter values subject to statistical analysis, including flows and rainfalls) are arranged according to their order of magnitude, which may be ascending or descending. For pollution studies the ascending (from the smallest to the greatest) order of magnitude is commonly selected; for flood studies the descending order of magnitude is common. The plotting position on the probability paper is then assigned to each parameter value as follows:

$$p(\text{in } \%) = \frac{m}{n+1} \tag{8.16}$$

where
m = ascending or descending order of magnitude
n = total number of data in the time series

The *mean*, μ, of the series is the arithmetic average of all values, X_i, in the series, or

$$\mu = \frac{\Sigma X_i}{n} \tag{8.17}$$

To fit the series of EMCs or any other series to the log-normal distribution, a logarithmic transformation of the values must be performed, or $X_i = \ln Y_i$, where Y is the original series and X is the transformed series of natural logarithms of the values of the series.

The *median* of the series is the central value for which one-half of the data values in the series has a magnitude that is less than or equal. If the series fits normal distribution (original or transformed), the median approximately equals the mean. However, if the original series has been transformed into a series of logarithms exhibiting normal (meaning lognormal) distribution, then the median and the mean of the original nontransformed series are different. On a probability plot the median corresponds to a 50 percentile value on the plot.

The *standard deviation* is a measure of the scatter of the values of the series around its mean.

$$s = \sqrt{\frac{\Sigma(X_i - \mu)^2}{n - 1}} \tag{8.18}$$

The *coefficient of variation* is a ratio or the standard deviation and the mean, or

$$CV = \frac{s}{\mu} \tag{8.19}$$

The expected value of the series at any probability or frequency of occurrence (X_α) of a series that has been logarithmically transformed can be determined by

$$X_\alpha = \exp(\mu_{\ln x} + Z_\alpha s_{\ln x}) \tag{8.20}$$

where
Z_α = the standard normal probability

$\mu_{\ln x}$ = mean of log-transformed data
$s_{\ln x}$ = standard deviation of log-transformed data

The standard normal probability value, Z_α, can be obtained from probability tables or calculated from the Gaussian cumulative probability distribution function. For the 90 percentile value (10% exceedence) Z_α = 1.2817, for 95 percentile value (5% exceedence) Z_α = 1.645, and so forth.

Also X_α can be expressed as a ratio of the median value of the original series by the following equation, which defines the ratio in terms of the coefficient of variation

$$\frac{X_\alpha}{\text{Median}} = \exp(Z_\alpha \sqrt{\ln(1 + CV_{\ln x}^2)}) \quad (8.21)$$

Example 8.4: Statistics and Probability Distribution of Event Mean Concentrations

Nine urban runoff events were measured at a storm sewer outlet location. The event mean concentrations for each event were calculated using Equation (8.15) and are given in column 2 of Table 8.12. The resulting probability plot is in Figure 8.18.

Solution Using the results reported in Table 8.12, the arithmetic mean is

$$\mu_x = 13.08/9 = 1.45 \, \text{mg/l}$$

TABLE 8.12 Statistical Evaluation of EMCs (Lead) in Storm Water Runoff

Date of Sampling	X[EMC mg Pb/l]	$\ln X$	Order of Magnitude (m)	Plotting Position $p = 100 m/(n + 1)$
3/5/1989	1.41	0.3423	6	60
4/15/1989	2.24	0.8065	8	80
5/2/1989	1.10	0.0953	4	40
5/25/1989	0.90	−0.1054	3	30
6/14/1989	1.27	0.2390	5	50
7/12/1989	1.72	0.5423	7	70
7/20/1989	3.20	1.1632	9	90
9/17/1989	0.50	−0.6931	1	10
10/4/1989	0.74	−0.3011	2	20
	$\Sigma X = 13.08$	$\Sigma \ln x = 2.0890$		

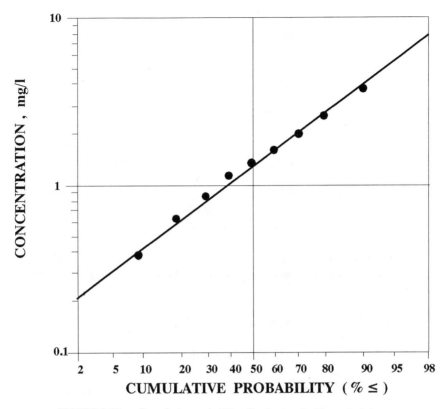

FIGURE 8.18. Cumulative probability distribution for Example 8.4.

The arithmetic median is 1.27 mg/l. Hence, since the mean does not equal the median, the frequency distribution is asymmetrical (skewed).

The logarithmic mean is

$$\mu_{\ln x} = 2.0890/9 = 0.2321$$

The median of the logarithmic transformed series is 0.239. Hence, the transformed series appears to be symmetrical and normally distributed.

Example 8.5: Probability of Exceedence

For the series in the previous example determine a value that would have a 95% probability of not being exceeded.

TABLE 8.13 The Transformation Used to Calculate the Standard Deviation

Data	Value of $\ln x$	$\ln x - \mu_{\ln x}$	$(\ln x - \mu_{\ln x})^2$
1	0.3423	0.1102	0.0121
2	0.8065	0.5744	0.3299
3	0.0953	−0.1368	0.0187
4	−0.1054	−0.3375	0.1139
5	0.2390	0.0069	0.00005
6	0.5423	0.3102	0.0962
7	1.1632	0.9311	0.8669
8	−0.6931	−0.9252	0.8560
9	−0.3011	−0.5332	0.2843
		$\Sigma = 2.5781$	

Solution Calculate the standard deviation of the transformed series (see Table 8.13). From the preceding example the mean of the transformed series is $\mu_{\ln x} = 0.2321$.

The standard deviation of the transformed series is

$$s_{\ln x} = [2.5781/(9-1)]^{0.5} = 0.5677$$

The coefficient of variation is then

$$CV = 0.5677/0.2321 = 2.44$$

The standard normal probability value for the 95% nonexceedence is $Z_\alpha = 1.645$. Thus the EMC for lead that will have this probability on nonexceedence (and hence $1 - 0.95 = 0.05$, 5% probability of exceedence) is, by Equation (8.20),

$$X_\alpha = \exp(0.2321 + 1.645 \times 0.5677) = 3.21 \, \text{mg/l}$$

This checks with the probability graph on Figure 8.18.

The establishment of the fundamental probability distribution of EMCs as log-normal has a number of benefits (U.S. EPA, 1983):

- Concise summaries of highly variable data can be developed.
- Comparisons of results from different sites, events, etc., are convenient and are more easily understood.

- Statements can be made concerning frequency of occurrence. One can express how often values exceed various magnitudes of interest.
- A more useful method of reporting data than the use of ranges is provided; one that is less subject to misinterpretations.
- A framework is provided for examining "transferability" of data in a qualitative manner.

The log-normal distribution is completely specified by the mean and standard deviation of the transformed logarithmic series.

Nationwide (NURP) Characterization of Urban Runoff Quality

The NURP study established that even mean concentrations for total suspended solids (TSS), total phosphorus, total Kjeldahl nitrogen (organic nitrogen plus ammonia), total lead, and total zinc are extremely well represented by the log-normal distribution. For COD and nitrate–nitrite the log-normal distribution fitted the data quite well. For other constituents, including BOD, soluble phosphorus, and total copper, the log-normal distribution cannot be rejected. These were the pollutants of interest analyzed in the NURP study.

Land use effect on EMCs. The nationwide analysis did not find significant statistical correlations of the EMCs to the geographical locations of the site. Also three typical urban land uses, residential, mixed, and commercial, were not statistically different. Only open/nonurban lands were statistically significantly different from the previous three land-use types. Hence the NURP report concluded that

> regardless of the analytical approach taken, we are forced to conclude that, if land use category effects are present, they are eclipsed by the storm to storm variabilities and that, therefore, land use category is of little general use to aid in predicting urban runoff quality at *unmonitored* (emphasized by the author) sites or in explaining site to site differences where monitoring data exists.

Correlation between EMCs and runoff volume. A total of 67 sites from 20 of the NURP projects were examined for possible correlation for nine constituents. Of the 517 linear correlation coefficients calculated 116 (22%) were significant at the 95% confidence level and 154 (30%) were significant at the 90% confidence level. Both positive and negative correlations were detected. Hence, the NURP study concluded that there is no significant linear correlation between EMCs and runoff volume.

This finding does have some importance in designing storm water monitoring programs. First, there is natural and appropriate bias that favors emphasizing resource allocation to larger storm events. Since no significant linear correlation was found, such biases and differences are not expected to influence EMCs comparisons to any appreciable extent. Second, the probabilistic methodologies for examining the receiving water impact of storm water runoff assumes that concentration and runoff volume are independent (i.e., that there is no significant correlation).

Other factors' effects. In deterministic concepts, factors such as slope, soil types, and rainfall characteristics, are all potentially important. However, in a statistical sense derived from a large number of observations at various sites throughout the United States, these factors did not appear to have any real consistent significance in explaining observed similarities or differences among individual sites. However, the sites that were investigated all had storm sewers, hence, the effect of, for example, natural (swale) drainage on pollutant loads cannot be determined. Also no abatement, such as ponds or infiltration, were included in the test sites.

Overall urban runoff quality characteristics. Having determined that geographic location, land-use category, runoff volume, or other factors appear to be of little utility in explaining overall site-to-site, event-to-event variability to predicting the characterization of urban runoff at unmonitored sites, the best general characterization may be obtained by pooling the site data for all sites (other than the open/nonurban ones). The results for the single overall urban land-use category are given in Table 8.14. In the absence of better information, the NURP study recommended these EMC characteristics for planning purposes as the best description of the quality of urban runoff.

TABLE 8.14 Overall Water Quality Characteristics of Urban Runoff

Constituent	Typical Coefficient of Variation	Site Median EMC	
		For Median Urban Site	For 90 Percentile Urban Site
TSS (mg/l)	1–2	100	300
BOD (mg/l)	0.5–1	9	15
COD (mg/l)	0.5–1	65	140
Total P (mg/l)	0.5–1	0.33	0.70
Soluble P (mg/l)	0.5–1	0.12	0.21
TKN (mg/l)	0.5–1	1.50	3.30
NO_{2+3}-N (mg/l)	0.5–1	0.68	1.75
Total Cu (µg/l)	0.5–1	34	93
Total Pb (µg/l)	0.5–1	144	350
Total Zn (µg/l)	0.5–1	160	500

Source: From NURP studies, U.S. EPA (1983).

TABLE 8.15 EMC Mean Values for Pollutant Load Estimates

	Site	Mean EMC
Constituent	Median Urban Site	90 Percentile Urban Site
TSS (mg/l)	141–224	424–671
BOD_5 (mg/l)	10–13	17–21
COD (mg/l)	73–92	157–198
Tot. P (mg/l)	0.37–0.47	0.78–0.99
Sol. P (mg/l)	0.13–0.17	0.23–0.30
TKN (mg/l)	1.68–2.12	3.69–4.67
NO_{2+3}-N (mg/l)	0.76–0.96	1.96–2.47
Tot. Cu (µg/l)	38–48	104–132
Tot. Pb (µg/l)	161–204	391–495
Tot. Zn (µg/l)	179–226	559–707

Source: U.S. EPA (1983).

It should be remembered that the values in Table 8.14 are derived from log-normal distribution of EMCs. For estimating loads of pollutants to receiving bodies of water these values should be converted to the mean values, which are given in Table 8.15. The ranges reflect the variability of the coefficient of variation of the EMCs.

Reconciliation of NURP Statistical Results with the Deterministic Urban Runoff Quality Concepts

This somewhat surprising conclusion of the NURP statistical analysis appears, at first, to contradict the deterministic findings presented in the first sections of this chapter. For example, does the poor correlation of EMCs to land use exclude the use of unit loads or more complex deterministic models? The first conclusion would be to throw out these earlier concepts. However, more detailed and thorough analysis leads to a different conclusion.

First, most of the unit loads related to land use and/or other parameters, as well as more complex deterministic models, are site specific, while the NURP statistical analysis encompasses the entire United States.

Second, the NURP conclusions are related to event mean concentrations, while at the same time a relationship between the runoff volume and the degree of imperviousness has been found in the NURP studies (Fig. 8.2). Since the unit load is a product of runoff volume and EMC, there may be a relationship of the unit loads to land use, while a relation to EMCs may not be profound.

Third, site specificity of deterministic hydrologic models is so dominant that conclusions obtained on a continental scale are not applicable to sites with a basin area ranging from a few hectares to hundreds square kilometers. Deterministic models are mostly applicable to site specific designs of abatement for which nationwide statistical averages would not be appropriate.

In the discussion of deterministic buildup and washoff concepts, it was pointed out that the curb solids accumulation (and by the same reasoning curb-side pollutant loads) stay more or less constant. The washoff rate is proportional to rainfall intensity and, in a cumulative manner, to the rainfall depth. This may lead to the conclusion that the EMCs under these circumstances may not vary and, in a statistical sense, may be considered as constant.

Hence the NURP findings do not a priori disprove the deterministic concepts. Furthermore, even though the street loads of pollutants tend to be inversely proportional to the imperviousness of the area, total loads from less impervious areas also include pollutants eroded from pervious surfaces, which again tends to equalize the EMCs and pollutant loads among the land use areas.

The following example documents that there is no major discrepancy between the unit loads presented in the first portion of this chapter and units loads obtained from the statistical nationwide characteristics of urban runoff.

Example 8.6: Unit Loads Based on NURP EMC Data and their Comparison with Other Estimates

Estimate annual unit loads and 90 percentile exceedence event mean concentrations (EMC) of total phosphorus and lead from an urban area. The following information is given:

Land use: residential
Percent imperviousness: 50%
Annual rainfall: 76 cm (30 in.)

From Table 8.14 the median and 90 percentile EMCs for total phosphorus are 0.33 mg/l and 0.70 mg/l, respectively. EMCs for lead are 144 µg/l (median) and 350 µg/l, respectively.

Solution The annual load can be estimated using the relation of the coefficient of runoff (defined as a ratio of runoff depth to the total rainfall depth) as determined by the NURP study, which is shown on Figure 8.2.

From the figure for the 50% impervious area, CR = 0.45 (recall from the discussion in Chapter 3 that for urban areas the runoff coefficient is approximately a measure of the degree of imperviousness).

For estimating the loads for median and 90 percentile urban sites, the mean EMCs are taken from Table 8.15.

For a median urban site:

Average Tot. P EMC = 0.42 mg/l

Tot. P_{mean} (kg/ha) = CV * rainfall volume * EMC = 0.45 *
76 (cm) * 0.01 (cm/m) * 10,000 (m²/ha) *
0.42 (g/m³) * 0.001 (kg/g) = 1.44 kg/ha

Average Tot. Pb EMC = 0.182 mg/l

Tot. Pb_{mean} (kg/ha) = 0.45 * 76 * 0.01 * 10,000 * 0.182 * 0.001
= 0.62 kg/ha

For a 90 percentile urban site

Tot. P EMC (90%) = 0.9 mg/l

Tot. $P_{90\%}$ (kg/ha) = 0.45 * 76 * 0.01 * 10,000 * 0.9 * 0.001
= 3.08 kg/ha

Tot. Pb (90%) = 0.443 mg/l

Tot. $Pb_{90\%}$ (kg/ha) = 0.46 * 76 * 0.01 * 10,000 * 0.443 * 0.001
= 1.55 kg/ha

By comparison, the PLUARG data measured in the Great Lakes region given in Table 8.8 reported ranges of annual unit loads for urban areas of 0.3 to 4.8 kg/ha for phosphorus and 0.14 to 0.5 kg/ha for lead. The range of Tot. P unit loadings compiled by Beaulac and Reckhow (1982) for urban lands were 0.5 to 6.25 kg/ha-year (average 1 kg/ha-year). No conclusions can be drawn from this comparison.

COMBINED SEWER OVERFLOWS

Unlike the separate storm sewers that mostly contain "cleaner" surface runoff and infiltrations during wet weather, combined sewers carry a mixture of storm runoff, infiltration, and wastewater (dry-weather) flows. During wet weather, combined sewer systems may overflow, until recently commonly without any treatment, directly to the receiving waters.

Combined sewers were originally built during the last century by connecting sewage discharges to subsurface and surface storm drainage. By the end of the nineteenth century, the need for wastewater treatment became increasingly apparent, but urban storm water was generally considered to be "clean" and suitable for the dilution of "dirty" wastewater. However, after the turn of the twentieth century most of the urban sewer systems were built separate.

In Europe, the practice of building separate systems was not as widely accepted as in the United States, and many urban sewer systems are combined. Typically in European urban areas, the city centers have combined sewers, to which separate suburban sanitary systems are connected. Because of the period in which the combined sewers were built in the United States, they tend to be located in areas that experienced major growth during the period from approximately 1850 to 1900. Major combined sewer service areas are located along the upper East Coast, in the upper Midwest, and in the far east. There are at least 1 million hectares of combined sewer service area in the United States today, located in 1100 to 1300 distinct collection systems with an average population density of 40 people per hectare and total population served of about 37,000,000 (U.S. EPA, 1978). In total, there are between 15,000 to 20,000 CSOs currently in operation (Association of Metropolitan Sewerage Agencies, 1988).

Sources and Characterization of Combined Sewer Overflows

An overflow from combined sewers is initiated when the mixed flow in the sewer exceeds the sewer capacity. To prevent backwatering into the sewers and for safe conveyance the separation of the flows into an overflow and downstream conveyance occurs in so-called *flow regulators* or *flow dividers* whose schematic is shown on Figure 8.19. There are many types of flow regulators, some of which use simple static weir controls to separate flows and some that use moving parts. Many of the existing regulators became inoperable in the past (Moffa, 1990).

Overflow frequency and duration. Combined sewer overflows (CSOs) occur only during runoff-producing rainfall events which, in general, in an average year range from 200 hours to more than 1300 hours in the United States (U.S. EPA, 1978). In Germany (Ruhr area) there are on average about 700 precipitation hours, and 1000 hours are typical for Switzerland (Novotny et al., 1989). However, during these runoff-producing precipi-

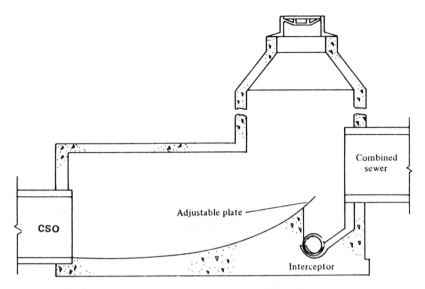

FIGURE 8.19. Leaping weir CSO flow divider (regulator).

tation hours pollutant loads may be extremely large. Combined sewer overflows containing raw wastewater with pathogenic (disease-causing) microorganisms, are repugnant and typically result in unpleasant odors. They are suspected of being a source of objectionable debris, including those illegally dumped into sewers (i.e., medical waste found on the East Coast beaches that is suspected of originating with CSOs).

Typically, urban combined sewage systems are designed to carry flow that is about four to eight times the average dry-weather (sewage) flows. Treatment plants serving these systems built based on past U.S. and European practices, were designed to handle mixed flows that were four to six times the average dry-weather flow. Simply said, any flow in excess of the treatment plant capacity must overflow either in CSOs or in a treatment plant bypass.

The rainfall that initiates overflow can then be related to the hydrological characteristics of the drainage basin, population density (per capita sewage flow), and the excess treatment plant capacity as follows. Assume that the peak surface runoff flow is adequately represented by the rational formula (see Chapter 3) or $Q_{s\text{-peak}} = CAI$, were C is the coefficient of runoff defined in Table 3.13, A is the contributing area in square meters, and I is the rainfall intensity in millimeters/hour, and $Q_{s\text{-peak}}$ is in liters/hour. However, in this concept one has to realize that the minimal CSOs causing rainfalls have relatively low intensity and, in most cases, pervious areas do not contribute to the runoff. Thus the term $C*A$ should be replaced by the directly connected impervious area $A_{connected} = DC*$ (Percent imperviousness/100) $*A$. Then if Q_{dw} is the dry-weather sewage flow and the treatment system capacity is $m*Q_{dw}$ (where m is typically four to six), then the rainfall intensity at which the overflows occur is

$$I_{crit} = (m - 1)\frac{Q_{dw}}{A_{connected}} \qquad (8.22)$$

where I_{crit} is an average rainfall intensity at which CSOs begin in liters/m²-hr = millimeters/hour.

The typical magnitude of the critical rainfall intensity for initiation of CSOs is evaluated in the following Example 8.7.

Example 8.7: Calculation of the Critical Rainfall Intensity for CSOs

Calculate the critical rainfall intensity for a typical urban area with a population density of 40 persons per hectare and an average dry-weather sewage flow of 300 liters per capita and day. Assume that infiltration inflow equals the dry weather flow. The capacity of the treatment plant is four times the dry-weather flow (including infiltration), average runoff coefficient for the area is 0.7, and the drainage area is 1000 ha.

Solution The dry-weather sewage flow is

$$\begin{aligned}Q_{dw} &= 40 \text{ [people/ha]} * 300 \text{ [liters/capita-day]} * 1000 \text{ [ha]} \\ &= 12*10^6 \text{ liters/day} \\ &= 12*10^6/24 \text{ (hr/day)} \\ &= 5*10^5 \text{ l/hr}.\end{aligned}$$

The total dry-weather flow (including infiltration) is thus

$$Q_{dw} = 2*5*10^5 = 10^6 \text{ l/hr}$$

Hence, the capacity of the treatment plant is

$$Q_{cap} = 4*10^6 \text{ l/hr}$$

Assuming $DC = 0.5$ (see Chapter 2 for the magnitude of the directly connected impervious area factor) and 70% average imperviousness, the directly connected area is

$$A_{con} = 0.5 * (70\%/100) * 1000 \,(\text{ha}) * 10^4 \,(\text{m}^2/\text{ha}) = 3.5 * 10^6 \,\text{m}^2$$

The critical rainfall intensity is then (Eq. (8.22))

$$I_{crit} = (4-1)\frac{10^6\,[\text{l/hr}]}{3.5*10^6} = 0.85\,\text{mm/hr}\ (=0.033\,\text{in./hr})$$

The critical CSO-initiating rainfall intensity of about 1 mm/hr may be typical for U.S. conditions. In Europe population density and imperviousness of urban areas are higher. Consequently, the critical rainfall intensity, I_{crit}, is about 3 to 6 mm/hr in Germany and 6 to 12 mm/hr in Switzerland (Novotny et al., 1989). Typically, without CSO control, overflows from combined systems in most urban areas occur on average between 10 and 50 times per year. In Milwaukee a typical number of overflows from the combined sewer area is about 40 per year.

Water Quality Characteristics of CSOs. Although CSOs have received scrutiny by researchers and the EPA, and a large data base of quality data has been assembled, no statistical analysis comparable to NURP studies is available. Hence, only ranges and arithmetic averages have been reported. There is a great likelihood that the statistical characteristics of EMCs from combined sewer overflows may also follow the log-normal distribution, and similar conclusions could be drawn as to the variability as were for storm water quality characteristics. This is especially true for the constituents that almost solely originate from storm water runoff, such as toxic metals and organics. Table 8.16 shows some average quality characteristics of CSOs.

There is ample evidence indicating that the first flush effect is more pronounced in the CSO discharges. Organic solids accumulate and sewer slime grows during dry periods; hence, far more pollutants accumulate in combined sewers than in storm sewers, which are idle and mostly dry between rainfall events.

TABLE 8.16 Nationwide Average Characteristics of CSOs

Parameter	Average Concentration
BOD_5 (mg/l)	115
Suspended solids (mg/l)	370
Total nitrogen (mg/l)	9–10
Phosphate (mg/l)	1.9
Lead (mg/l)	0.37
Total coliforms (MPN/100 ml)	10^2–10^4

Source: From U.S. EPA (1978).

Nationwide assessment has also shown that the annual load from CSOs for BOD are about the same as those for the entire year from the secondary treatment plants treating wastewater from the same area. The annual discharges of suspended solids and lead are approximately 15 times higher from combined sewer overflow than from secondary treatment plant effluents. On the other hand, annual loads of total N and phosphates from secondary treatment plant effluents are 4 and 7 times the discharge from combined sewer overflows, respectively.

References

American Public Works Association (1969). *Water Pollution Aspects of Urban Runoff*, WP-30-15, U.S. Dept. of Interior, FWPCA (now the EPA), Washington, DC.

Anonymous (1981). *Fugitive Dust Emissions: Their Sources and Their Control in Milwaukee's Menomonee River Valley*, Department of City Development, City of Milwaukee, WI.

Association of Metropolitan Sewerage Agencies (1988). *Draft CSO Permitting Strategy, August 1988*, Bulletin No. GB88-22, AMSA, Washington, DC.

Athayde, D. N., C. F. Myers, and P. Tobin (1986). EPA's perspective of urban nonpoint sources, in *Urban Runoff Quality—Impact and Quality Enhancement Technology*, Proceedings, Engineering Foundation Conference, American Society of Chemical Engineers, pp. 217–225.

Balmér, P., P. A. Malmqvist, and A. Sjöberg, eds. (1984). *Proceedings, Third International Conference on Urban Storm Drainage*, Chalmers University of Technology, Göteborg, Sweden.

Bannerman, R. (1991). *Pollutants in Wisconsin Stormwater*, unpublished report, Wisconsin Department of Natural Resources, Madison, WI.

Bannerman, R., K. Baun, M. Bohm, P. E. Hughes, and D. A. Graczyk (1984). *Evaluation of Urban Nonpoint Source Pollution Management in Milwaukee County, Wisconsin*, Report No. PB84-114164, U.S. Environmental Protection Agency, Region V, Chicago, IL.

Beaulac, M. N., and K. H. Reckhow (1982). An examination of land use—nutrient export relationships, *Water Resour. Bull.* **18**(6):1013–1024.

Bengtsson, L. (1982). Snowmelt-generated runoff in urban areas, in *Urban Stormwater Hydraulics and Hydrology*, B. C. Yen, ed. Water Resources Publ., Littleton, CO, pp. 444–451.

Brown, R. G. (1988). Effects of precipitation and land use on storm runoff, *Water Resour. Bull.* **24**(2):421–426.

Carey, A. E. (1979). Monitoring pesticides in agricultural and urban soils of the United States, *Pesticides Monit. J.* **13**(1):23–27.

Cowherd, C., Jr., et al. (1977). *Quantification of Dust Entrainment from Paved Roadways*, EPA 450/3-77-027, U.S. Environmental Protection Agency, Research Triangle Park, NC.

Ellis, J. B. (1986). Pollutional aspects of urban runoff, in *Urban Runoff Pollution*,

H. C. Torno, J. Marsalek, and M. Desbordes, eds., Springer Verlag, Berlin, New York, pp. 1–38.

Fam, S., M. K. Stenstrom, and G. Silverman (1987). Hydrocarbons in urban runoff, *J. Environ. Eng. Div. ASCE* **113**(5):1032–1046.

Field, R., et al. (1974). Water pollution and associated effects from street salting, *J. Environ. Eng. Div. ASCE* **100**(EE2):459–477.

Förster, J. (1990). Roof runoff: A source of pollutants in urban storm drainage systems, in *Proceedings, Fifth International Conference on Urban Storm Drainage*, Y. Iwasa and T. Sueishi, eds., Osaka University, Japan, pp. 469–474.

Glenne, B. (1984). Simulation of water pollution generation and abatement on suburban watersheds, *Water Resour. Bull.* **20**(2):211–217.

Gujer, W., and V. Krejci (1987) *Topics in Urban Storm Water Quality, Planning and Management*, in Proceedings, Fourth International Conference on Urban Storm Drainage. IAHR-IAWPRC École Polytechnique Fédérale, Lausanne, Switzerland, pp. 1–20.

Halverson, H. G., D. R. DeWalle, and W. E. Sharpe (1984). Contribution of precipitation to quality of urban storm runoff, *Water Resour. Bull.* **20**(6): 859–864.

Halverson, H. G., D. R. DeWalle, W. E. Sharpe, and D. L. Wirries (1982). Runoff contaminants from natural and manmade surfaces in a nonindustrial urban area, in *Proceedings, 1982 International Symposium on Urban Hydrology, Hydraulics and Sediment Control*, University of Kentucky, Lexington, pp. 233–238.

Heaney, J. P., and W. C. Huber (1973). *Stormwater Management Model. Refinements, Testing and Decision-making*, Department of Environmental Engineering & Science, University of Florida, Gainesville, FL.

Hoffman, E. J. (1985). Urban runoff pollutant inputs to Narragansett Bay: Comparison to point sources, in *Perspectives on Nonpoint Pollution*, Proceedings, National Conference, EPA 440/5-85-001, U.S. Environmental Protection Agency, Washington, DC, pp. 159–164.

Hoffman, E. J., et al. (1984). Urban runoff as a source of polycyclic aromatics to coastal waters, *Environ. Sci. Technol.* **18**:580–587.

Huber, W. C. (1986). Deterministic modeling of urban runoff quality, in *Urban Runoff Pollution*, H. C. Torno, J. Marsalek, and M. Desbordes, eds., Springer Verlag, Berlin, pp. 167–242.

Hydrologic Engineering Center (1975). *Urban Stormwater Runoff—STORM*, U.S. Army Corps of Engineers, Davis, CA.

Iwasa, Y., and T. Sueishi, eds. (1990). *Proceedings, Fifth International Conference on Urban Storm Drainage*, IAHR-IAWPRC, Suita-Osaka, University of Osaka, Japan.

James, W., and S. Boregowda (1986). Continuous mass balance of pollutant build-up processes, in *Urban Runoff Pollution*, H. Torno, J. Marsalek, and M. Desbordes, eds., Springer Verlag, Berlin, pp. 243–271.

Krejci, V., L. Dauber, B. Novak, and W. Gujer (1987). Contribution of different sources to pollutant loads in combined sewers, in *Proceedings, Fourth International Conference on Urban Storm Drainage, W. Guejer and V. Krejci, eds., IAHR-IAWPRC, École Polytechnique Fédérale, Lausanne, Switzerland, pp. 34–39.

Lager, J. A., W. G. Smith, W. G. Lynard, R. M. Finn, and E. J. Finnemore (1977). *Urban Stormwater Management and Technology: Update and User's Guide*, EPA-600/8-77-014, U.S. Environmental Protection Agency, Cincinnati, OH.

McComas, M. R., G. D. Cooke, and R. H. Kennedy (1976). A comparison of phosphorus and water contributions by snowfall and rain in northern Ohio, *Water Resour. Bull.* **12**(3):519–527.

McElroy, A. D., S. Y. Chiu, J. W. Nebgen, A. Aleti, and F. W. Bennett (1976). *Loading Functions for Assessment of Water Pollution from Nonpoint Sources*, Office of Research and Development, U.S. Environmental Protection Agency, EPA-600/2-76-151, Washington, DC.

Moffa, P. E. (1990). *Control and Treatment of Combined Sewer Overflows*, Van Nostrand Reinhold, New York.

Myers, C. F., J. Meek, J. S. Tuller, and A. Weinberg (1985). Nonpoint sources of water pollution, *J. Soil Water Conserv.* **40**(1):14–18.

Marsalek, J. (1978). *Pollution Due to Urban Runoff: Unit Loads and Abatement Measures*, Pollution from Land Use Activities Reference Group, International Joint Commission, Windsor, Ontario.

Marsalek, J. (1986). Toxic contaminants in urban runoff, in *Urban Runoff Pollution*, H. Torno, J. Marsalek, and M. Desbordes, eds., Springer Verlag, Berlin, pp. 39–57.

Marsalek, J. (1990). PAH transport by urban runoff from an industrial city, in *Proceedings, Fifth International, Conference on Urban Storm Drainage*, Y. Iwasa and T. Sueishi, eds., University of Osaka, Japan, pp. 481–486.

Ng, H. Y. F. (1987). Rainwater contribution to the dissolved chemistry of storm runoff, in *Urban Storm Water Quality, Planning and Management*, W. Gujer and V. Krejci, eds., IAHR-IAWPRC, École Polytechnique Fédérale, Lausanne, Switzerland, pp. 21–26.

Novotny, V. (1987). *Effect of Pollutants from Snow and Ice on Quality of Water from Urban Drainage Basins*, PB87-162509, Dept. of Civil Eng., Marquette University, Milwaukee, WI (available from NTIS).

Novotny, V. (1988). Modeling urban runoff pollution during winter and off-winter periods, in *Advances in Environmental Modelling*, A. Marani, ed., Elsevier, Amsterdam, pp. 43–58.

Novotny, V., and G. Chesters (1981). *Handbook of Nonpoint Pollution: Sources and Management*, Van Nostrand Reinhold, New York.

Novotny, V., and G. W. Kincaid (1982). Acidity of urban precipitation and its buffering during overland flow, in *Urban Stormwater Quality, Management and Planning*, B. C. Yen, ed., Water Resources Publ., Littleton, CO, pp. 1–9.

Novotny, V., K. R. Imhoff, M. Othof, and P. Krenkel (1989). *Karl Imhoff's Handbook of Urban Drainage and Wastewater Disposal*, Wiley Interscience, New York.

Novotny, V., H. M. Sung, R. Bannerman, and K. Baum (1985). Estimating nonpoint pollution from small urban watersheds, *J. WPCF* **57**(4):339–348.

Oberts, G. L. (1982a). *Water Resources Management: Nonpoint Source Pollution Technical Report*, PUB-10-82-016, Metropolitan Council, St. Paul, MN, Environmental Protection Agency, Chicago, IL, Region V (available from NTIS), 260p.

Oberts, G. L. (1982b). Impact of wetlands on nonpoint source pollution. *Proceedings, International Symposium on Urban Hydrology, Hydraulics and Sediment Control*, University of Kentucky, Lexington, KY, pp. 225–232.

Oberts, G. L. (1986). Pollutants associated with sand and salt applied to roads in Minnesota, *Water Resour. Bull.* **22**(3):479–484.

Oberts, G. L. (1990). Design considerations for management of urban runoff in wintry conditions, *Proceedings, International Conference on Urban Hydrology under Wintry Conditions*, Narvik, Norway, March 19–21.

Olivieri, V. P., C. W. Kruse, K. Kawata, and J. E. Smith (1977). *Microorganisms in Urban Stormwater*, EPA-600/2-77-087, Municipal Environmental Research Laboratory, U.S. Environmental Protection Agency, Cincinnati, OH.

Pisano, W. C., G. L. Aronson, C. S. Queiroz, F. C. Blanc, and J. C. O'Shaughnessy (1979). *Dry-weather Deposition and Flushing for Combined Sewer Overflow Pollution Control*, EPA-600/2-79-133, U.S. Environmental Protection Agency, Cincinnati, OH.

Pitt, R. (1979). *Demonstration of Nonpoint Pollution Abatement through Improved Street Cleaning Practices*, EPA 600/2-79-161, U.S. Environmental Protection Agency, Cincinnati, OH.

Pitt, R. (1988). Asbestos as an urban area pollutant, *J. WPCF* **60**(11):1993–2001.

Pitt, R., and P. Barron (1989). *Assessment of Urban and Industrial Stormwater Runoff Toxicity and the Evaluation/Development of Treatment for Runoff Toxicity Abatement—Phase I*, Storm and Combined Sewer Pollution Program, U.S. Environmental Protection Agency, Edison, NJ.

Pitt, R., M. Lalor, and G. Driscoll (1990). *Assessment of Non-stormwater Discharges into Separate Storm Drainage Networks—Phase I: Manual of Practice*, Storm and Combined Sewer Pollution Program, U.S. Environmental Protection Agency, Edison, NJ.

Pitt, R., R. Field, M. Lalor, and G. Driscoll (1990). Analysis of cross-connections and storm drainage, in *Proceedings, Urban Stormwater Enhancement Source Control, Retrofitting and Combined Sewer Technology*, Davos, Switzerland, Engineering Foundation of ASCE, New York.

Randall, C. W., T. J. Grizzard, D. R. Helsel, and D. M. Griffin (1982). Comparison of pollutant mass loads in precipitation and runoff in urban area, in *Urban Stormwater Quality, Management and Planning*, B. C. Yen, ed., Water Resources Publ., Littleton, CO, pp. 29–38.

Sartor, J. D., and G. B. Boyd (1972). *Water Pollution Aspects of Street Surface Contamination*, EPA R2-72-081, U.S. Environmental Protection Agency, Washington, DC.

Sartor, J. D., G. B. Boyd, and F. J. Agardy (1974). Water pollution aspects of street surface contamination, *J. WPCF* **46**:458–465.

Schmidt, S. D., and D. R. Spencer (1986). The magnitude of improper waste discharges in an urban stormwater system, *J. WPCF* **58**(11):744–748.

Shaheen, D. G. (1975). *Contributions of Urban Roadway Usage to Water Pollution*, EPA 600/2-75-004, Office of Research and Development, U.S. Environmental Protection Agency, Washington, DC.

Sonzogni, W. C., et al. (1980). Pollution from land runoff, *Environ. Sci. Technol.* **14**(2):148–153.

Strecker, E. W., E. D. Driscoll, P. E. Shelley, and D. R. Gaboury (1987). Characterization of pollutant loadings from highway runoff in the USA, in *Urban Storm Water Quality, Planning and Management*, W. Gujer and V. Krejci, eds., IAHR-IAWPRC, École Polytechnique Fédérale, Lausanne, Switzerland, pp. 85–90.

Sutherland, R. C., and R. H. McCuen (1978). Simulation of urban nonpoint source pollution, *Water Resour. Bull.* **14**:409–428.

Torno, H. C., J. Marsalek, and M. Desbordes (1986). *Urban Runoff Pollution*, Springer Verlag, Berlin, New York.

U.S. Environmental Protection Agency (1971). *Stormwater Management Model*, EPA 11024D0OC 07/71 to 1102D0C10/71, U.S. EPA, Washington, DC.

U.S. Environmental Protection Agency (1977). *Control of Reentrained Dust from Paved Streets*, EPA 905/9-77-007, U.S. EPA, Kansas City, MO.

U.S. Environmental Protection Agency (1978). *Report to Congress on Control of Combined Sewer Overflow in the United States*, EPA 430/9-78-006, U.S. EPA, Washington, DC.

U.S. Environmental Protection Agency (1983). *Results of the Nationwide Urban Runoff Program*. Vol. I. *Final Report, Water Planning Division*, U.S. EPA, Washington, DC.

U.S. Environmental Protection Agency (1984). *Report to Congress: Nonpoint Source Pollution in the U.S.*, Office of Water Program Operations, Synectics Group, Inc., U.S. EPA, Washington, DC (available from NTIS).

U.S. Environmental Protection Agency (1990a). *National Water Quality Inventory—1988 Report to Congress*, EPA 440-4-90-003, U.S. EPA, Office of Water, Washington DC.

U.S. Environmental Protection Agency (1990b). *Urban Targeting and BMP Selection. An Information and Guidance Manual for State NPS Program Staff Engineers and Managers*, EPA 68-C8-0034, U.S. EPA, Washington, DC.

Uttormark, P. D., J. D. Chapin, and K. M. Green (1974). *Estimating Nutrient Loadings of Lakes from Nonpoint Sources*, EPA 660/3-74-02C, Office of Research and Monitoring, U.S. Environmental Protection Agency, Washington, DC.

Waller, D. H., and W. C. Hart (1986). Solids, nutrients, and chlorides in urban runoff, in *Urban Runoff Pollution*, H. C. Torno, J. Marsalek, and M. Desbordes, eds., Springer Verlag, Berlin, New York, pp. 59–85.

Westerström, G. (1990). Pilot study of the chemical characteristics of urban snow meltwater, in *Proceedings, Fifth International Conference on Urban Storm Drainage*, University of Osaka, Japan, pp. 305–310.

Whipple, W., et al. (1978). Effect of storm frequency on pollution from urban runoff, *J. Water Pollut. Control Fed.* **50**:974–980.

Whipple, W., Jr., and J. V. Hunter (1979). Petroleum by hydrocarbons in urban runoff, *Water Resour. Bull.* **15**(4):1096–1105.

Whipple, W., Jr., and J. DiLouie (1981). Coping with increased stream erosion in urbanizing areas, *Water Resour. Res.* **17**(5):1561–1564.

Whipple, W., Jr., J. V. Hunter, and S. L. Yu (1978). Runoff pollution from multiple family housing, *Water Resour. Bull.* **14**(2):288–301.

Yalin, M. S. (1963). An expression for bed load transportation, *J. Hydraul. Div. ASCE* **89**:221–250.

Yen, B. C. (1982). *Urban Stormwater Quality, Management and Planning*, Water Resources Publ., Littleton, CO.

Zison, S. W. (1980). *Sediment-Pollutant Relationships in Runoff from Selected Agricultural, Suburban, and Urban Watersheds. A Statistical Correlation Study* (Final rept. Sep. 77–Sep. 78), EPA 600/3-80-022, Tetra Tech, Inc., Lafayette, CA, Environmental Research Lab., Athens, GA (available from NTIS), 150p.

9
Modeling and Monitoring Diffuse Pollution

Most natural hydrologic phenomena are so complex that they are beyond comprehension, or exact laws governing such phenomena have not been fully discovered. Before such laws can ever be found, complicated hydrologic phenomena (the prototype) can only be approximated by modeling.

Ven Te Chow

GENERAL CONCEPTS

As a result of research in the 1970s and early 1980s, a large number of models have been developed by various agencies, universities, and researchers throughout the world that could be used for modeling diffuse pollution and for assessment of remedial measures. Such models range from simple applications of basic hydrological procedures with added unit pollutant loads (see Chapter 1 for a definition and discussion of the unit load concept), to highly complex hydrological surface and ground-water runoff quantity and quality models. Models that in the past could be run only on large mainframe computers are now available for personal minicomputers. Several are available from agencies (for example, the U.S. EPA Research Laboratory in Athens, Georgia) and software vendors. As a matter of fact, the availability of personal computers that now have the same capability, or even surpass, medium mainframe computers of ten years ago, have meant that almost every engineer or planner can now possess a computer capable of running very sophisticated hydrological models. On-line interactions of user and computer graphics are being incorporated into the available models. Use of geographic information systems (GIS) and CAD programs for diffuse pollution modeling is now evolving.

Diffuse pollution hydrological models are a part of *loading models* that represent and simulate the generation and movement of water and its pollution content from the point of origin (source area) to a place of treatment and/or disposal (discharge) into receiving waters. These models may be used for planning and design alone, or may interface with receiving water quality models that assess the impact of nonpoint pollution on the aquatic biota and beneficial downstream uses. In most cases, the loading models provide concentrations and flow rates (pollutographs and hydrographs) and/or unit and total loads of pollutants. This chapter briefly describes the concepts and basic approaches of loading models. The processes included in the design of these models are described in pertinent chapters throughout this book.

Stream and receiving water quality models simulate the movement of materials through streams, river impoundments, estuaries, or near-shore ocean dispersion. The concepts of these models are introduced in Chapter 12. Reviews of the available runoff-quality models applicable to diffuse-pollution modeling of urban and agricultural watersheds have been prepared by Haan, Johnson, and Brakensiek (1982), Giorgini and Zingales (1986), Novotny (1986, 1988) and for the U.S. EPA by Donigian and Huber (1990) and Huber (1990).

Types of Models

Every model can be represented by the "black box" concept (Fig. 9.1). The model, just like the real system, produces output to various inputs. The structure of the model is always a very simplified version of the interactions and reactions taking place in the real system. The variables describing the physical state of the system are called *system parameters*; the variables affecting the state of the system are *state variables*. Some input variables may be considered state variables and vice versa. Watershed size, slope and roughness characteristics, erodibility, and texture of soils are examples of system parameters, while temperature and solar radiation and vegetation cover may be considered state variables. Rain, atmospheric fallout, and daily litter inputs on impervious areas can be considered inputs to most diffuse-pollution–water quality models. Note that a time series of unknown random perturbations can also be considered an input series, which is a fundamental premise of *stochastic water* quality models.

In his now classic paper on hydrologic modeling, Chow (1972) clas-

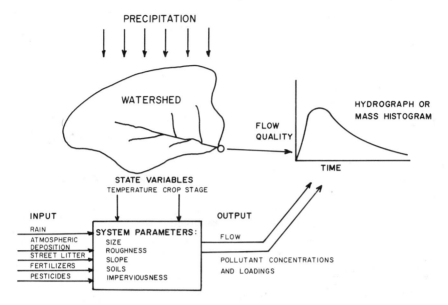

FIGURE 9.1. "Black box" concept of the watershed.

sified models into a number of categories, including physical, deterministic, and stochastic models. Physical models are of little importance to diffuse pollution modeling although they are used extensively for studying individual processes such as infiltration, soil water movement, and adsorption–desorption. Abstract models attempt to represent the prototype theoretically in mathematical form. These models replace the features of the system by relevant mathematical relationships.

Following and expanding Chow's definitions, the diffuse-pollution models can be broadly divided into the following basic groups (Fig. 9.2):

- Simple statistical routines and screening models
- Deterministic hydrologic models
- Stochastic models

A *deterministic model* ignores the input of random perturbations and random variations of system parameters and state variables. Such models for a given set of inputs provide only one set of outputs, while the output from stochastic models is often expressed in terms of mean and probabilistic ranges.

There are basically two approaches to modeling diffuse pollution. The more widely used are *lumped-parameter* models, while some more complex deterministic models are based on the *distributed-parameter* concept.

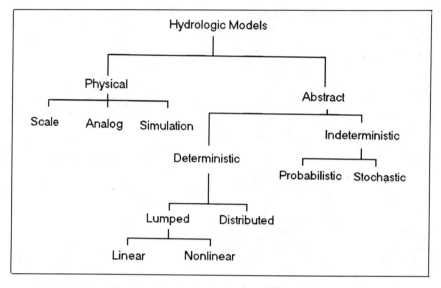

FIGURE 9.2. Classification of watershed hydrologic models. (After Chow, 1972.)

Lumped-parameter models can be both stochastic and deterministic (Fig. 9.3). The lumped-parameter models treat the watershed or a significant portion of it as one unit. The various characteristics of the watershed are often lumped together by an empirical equation, and the final form and magnitude of the parameters are simplified to represent the modeled unit as a uniform system. The coefficients and system parameters for each homogenous unit are often determined by calibration by comparing the response of the model with extensive field data. For such models the input–output relationship can be represented as

$$Y = \phi X \tag{9.1}$$

where
X = input vector (single or multivariate)
Y = output vector (single or multivariate)
ϕ = simple or complex transfer function

According to this definition, a mathematical model is itself a computerized complex transfer function. Note also that Equation (9.1) does not mean that the output vector, Y, is obtained by multiplication of the input vector, X, by a transfer function, ϕ. This simplified concept may involve complex

General Concepts 511

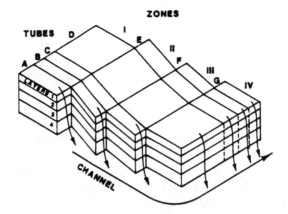

Lumped Parameter Model Concept

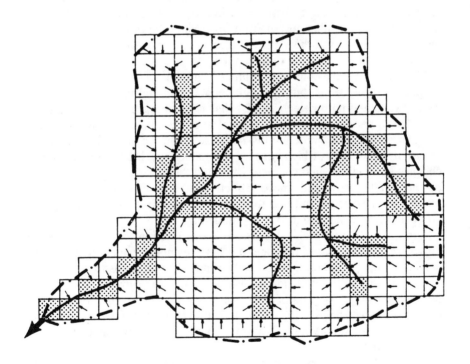

Distributed Parameter Model Concept

FIGURE 9.3. Modeling concepts.

mathematical operations, such as integrations, finite-difference approximations, and matrix inversions.

The models, once they have been calibrated and verified, can produce long time-series of outputs reflecting different hydrologic and meteorologic conditions (input variables). However, it has to be realized that a true time series of input and output parameters also contains random fluctuations that have been ignored by the deterministic nature of the model. Therefore, using statistical routines to estimate probabilistic characteristics of a time series produced by a deterministic model is inappropriate and may provide erroneous information about the statistical nature of the time series or of the modeled phenomenon. This is explained further in the section on stochastic modeling.

The distributed-parameter approach involves dividing the watershed into smaller homogenous units—areal elements—with uniform characteristics (soil, imperviousness, crop, slope). Each areal unit is described individually by a set of differential mass-balance equations. The input to each unit consists of the distributed inputs, such as rainfall and atmospheric deposition, plus output from upstream adjacent units. The mass-balance equations for the entire system are then solved simultaneously over a small computational time element Δt.

Theoretically, the lumped-parameter model can provide only one output location, while outputs can be obtained throughout the system from distributed-parameter models, that is, from each modeled subunit. Distributed-parameter models require larger computer storage for performing comparable modeling tasks and an extensive detailed description of system parameters, which must be provided for each element. However, changes in the watershed characteristics and their effect on the output can be modeled easily and more effectively. The distributed-parameter models are also more suitable for geographic information systems (GIS) and computer-aided design (CAD) modeling.

Some larger hydrological models allow division of the watershed into a smaller number of homogenous land segments, which then become essentially lumped-parameter subwatersheds. The output is then obtained by summing up the individual outputs from the subwatersheds, and not by simultaneous mass-balance solutions of a large number of areal elements. Nevertheless, the overall character of these more complex, spatially variant models are still more lumped in character than distributed-parameter models.

In size, the watershed models may range from small uniform segments of less than 1 ha to entire watersheds with the area of several hundred km^2. The U.S. Environmental Protection Agency (Sauders, 1976) has divided the loading models into basin-scale models ($>500 \, km^2$), areawide models (55 to 500 km^2), and small watersheds or fields (<55 km^2). Even the smallest (55 km^2) scale may be too large for some more detailed hydrological models, and introduction of a subbasin scale of less than 5 km^2 would be appropriate. Recall from Chapter 3 that watersheds with an area of less than 5 km^2 can be represented by a single-reservoir hydrological model, while for larger scale modeling more complex watershed representations are needed. A trade-off always exists between the size of the modeled area and the detail and reliability of the model.

Models can be designed or run on an *event* or *continuous* basis. Discrete-event models simulate the response of a watershed to a major rainfall or snowfall event. The principal advantage of event modeling over continuous simulation is that it requires little meteorological data and can be operated with shorter computer run time. The principal disadvantage of event modeling is that it requires specification of the design storm and antecedent moisture conditions, thereby assuming equivalence between the recurrence interval of the design storm with the recurrence interval of the runoff. This disadvantage is more pronounced when modeling more pervious watersheds than in modeling impervious urban basins.

Continuous process modeling sequentially simulates processes such as precipitation, available surface storage, snow accumulation and melt, evapotranspiration, soil moisture, surface runoff, infiltration, soil water movement, pollutant accumulation, erosion, and possibly other processes. Such models typically operate on a time interval ranging from a day to a fraction of an hour, and continuously balance water and pollutant mass in the system.

The principal advantage of continuous modeling is that it provides long time-series of water and pollutant loadings. Some users of such models attempted to statistically analyze the output time series as to the frequency and occurrence of runoff and pollutant load events. As was pointed out in the preceding discussion and will be further elaborated in the subsequent section on stochastic models, such statistical analyses using outputs from deterministic models are not proper and may provide erroneous statistical characteristics. A principal disadvantage of continuous modeling is that it requires long simulation runs, thus imposing restrictions on the number of alternatives that can be investigated. It also requires historical data on precipitation, often in less than hourly intervals, which is not always available.

Reliability and Usefulness of the Models

The epigraph at the beginning of this chapter introduced the dilemma of hydrologic modeling. It must be understood that as with any simulation of "real-world" systems, mathematical models are only a rough approximation. A computer model is the formulation into a computer language of the modeler's concept of the physical system and processes. Models are simplifications of the real system, and the degree of simplification may be the result of the modeler's understanding of the processes involved in the system or the desired accuracy or the purpose of the model itself. For example, a scientist conducting research on infiltration of rainfall into the soil might very precisely describe mathematically the infiltration process and generation of runoff volume, but might grossly oversimplify the movement of surface runoff rates and the erosion process (Knisel, 1985).

Hence, the accuracy and reliability of models are limited. Although some models represent the best available technology for analysis of environmental systems, a common error made by many planners is that they accept simulation results as true and absolute for unknown conditions. In order to avoid disappointments and court challenges, users should be aware of the limited accuracy of the models. Furthermore, as stated previously, deterministic models that are prevalent in today's practice inherently neglect random variations in the input and output series. Such variations can sometimes overwhelm deterministic interrelationships. Also the measured data contain inherent (systematic) and random measurement errors. Often measured data are not available at the location of interest and must be obtained by extrapolation from nearby observations.

The most accurate deterministic models (± few percent) are hydrologic models simulating runoff from small, uniform, impervious areas; the least reliable (an order of magnitude or more) are water quality models for large watersheds. Figure 9.4 shows the approximate accuracy of modeling water and pollution loads from diffuse sources using deterministic models. It should be noted that for simulating pollutant transport, hydrology must be calibrated and determined first, followed by sediment, and finally pollutant transport. Any errors that appear in the hydrologic or erosion component will be transferred and magnified in all dependent follow-up components. The highest error associated with bacterial estimations is caused primarily by analytical techniques and to a lesser degree by modeling reliability.

Since stochastic models provide a measure of error (a time series of errors between predicted and calculated output series), the reliability of the models can be accurately estimated (an opposite of reliability is the risk of failure of the model, or $p_r = 1 - p_f$, where p_r is reliability and p_f is probability of failure). A similar but inferior series of errors can be obtained by comparing the resulting output time series of deterministic models with the measured data. However, such series may still have

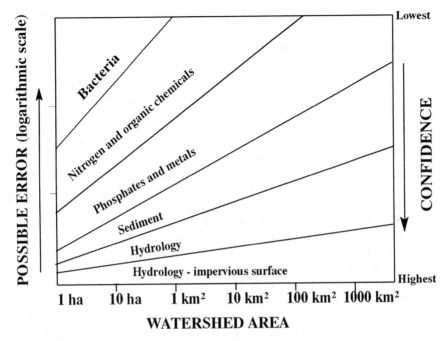

FIGURE 9.4. Relative accuracy and reliability of hydrologic models of diffuse pollution. Accuracy and reliability decreases with the increased complexity and size of the modelled system.

"hidden" deterministic and stochastic subprocesses that have not been included in the model. This will be further elaborated in the section titled "Stochastic and Neural Network Models."

In spite of the errors involved in modeling such complex environmental systems as diffuse pollution, the model as a planning tool cannot be replaced by any "rule-of-thumb" approach. The use of models is beneficial and greatly enhances the planning process for the following reasons:

1. Models can provide a forecast of impact of planned actions on water quality and pollution loadings.
2. Models provide an understanding of the processes involved in pollution generation from nonpoint sources.
3. The data base necessary to construct and calibrate the model is useful for other planning activities. Many problems will be answered or become clearer just by evaluating the data and compiling them into an appropriate input format.

4. Critical processes and areas of concern can be delineated and detected by modeling.
5. Regulators almost always require proof of water quality impact during conditions for which monitoring data may not be available, especially in cases when some action is in the planning stage and has not been implemented. Such impact can only be established by modeling.
6. Models can be updated continuously according to the state of the art of modeling technology and understanding of the modeled process.
7. Models can generate numerous alternatives and their impact on the environment according to the specifications. Various strategies can be investigated, and the impact of remedial measures can be evaluated.
8. Although the absolute accuracy of the outputs from the model is limited and sometimes even small, a comparison and ranking of outputs for various alternative remedial measures are commonly reliable and in most cases adequate.
9. Models can estimate and analyze trade-offs between planning objectives. A system providing the lowest pollutant load may not be optional for other objectives such as the enhancement or even viability of agricultural production (e.g., limiting or eliminating pesticide and fertilizer use may result in significant yield reduction; urban development versus no future development). If the environmental objective is known, the alternative to achieve it can be measured in terms of economic efficiency by considering the willingness of those involved by the measures to pay for the consequences. If there is a financial limit, it must be treated as a constraint.
10. Water quality loading models are now a required and integral part of the permit application for discharging urban and some agricultural runoff into receiving water bodies, as specified in the United States by the Clean Water Act.

Selection of the Models

As stated before a large number of models were developed in the 1970s and early 1980s. For a review of models developed in this "model creation period" the reader is referred to papers by Renard, Rawls, and Fogel (1982), Huber (1986, 1990), and other reviews by EPA and other agencies. Many mainframe computer models have been converted for use on small personal computers and are available from agencies, universities, and software vendors. However, as pointed out by Knisel (1985), who is one of the most prominent model creators, "the burden is upon the user to decide which model is appropriate for the problem on hand. The user must learn the model concepts, assumptions, and limitations and whether or not it will adequately treat the problem of concern." Knisel also correctly stated that "a person really cannot learn a model until they actually run it, get familiar with it, and feel comfortable in what they are doing."

Potential users should be well aware of and know by reputation the model creators, and should obtain references just as they would when purchasing any other goods. Purchasing and using anonymous models from software vendors without learning about the model creators and their reputation or acquiring models that have not been extensively tested and used by others is not wise. Model selection is an important part of the modeling process. Barnwell and Krenkel (1982) delineated the following level-of-use for the models: screening, planning, and design. Sutherland (1980) and Novotny (1986) divided diffuse pollution models into five categories:

I. *Simple statistical procedures and unit loads* with no interactions among the processes
II. *Simplified procedures* with some interactions among the processes
III. *Simplified deterministic models*, either event oriented or continuous
IV. *Sophisticated (detailed) event simulation models*
V. *Sophisticated continuous models*

Hence, several models may be used in the modeling process, as shown on Figure 9.5. In the process:

1. Overview statistical and simplified screening procedures (Levels I or II) will identify the problem areas.
2. Detailed continuous models (Level III or V) can be used for obtaining time series of simulated water and pollutant loads and for the screening of various management practices and scenarios. Such modeling activities must consider both the source strength (pollutant emission rates) and delivery.
3. A detailed event-oriented model (Level IV) can be used to finalize the design of some technically complex structural and nonstructural management practices and measures selected and evaluated in Phase 2. Event-oriented models should not be used for evaluating the impact of diffuse sources on receiving water bodies since the worst water quality impact may not occur during the high-flow "design" storms.

Models have been designed for specific applications, such as evaluation of pollutant loads and management practices for agricultural operations,

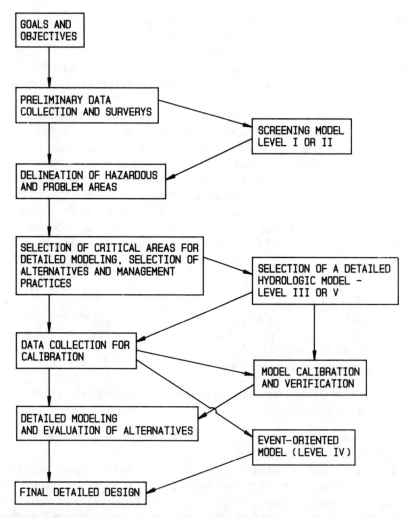

FIGURE 9.5. Schematics of the selection process for using models in diffuse-pollution simulation and abatement design.

urban and highway drainage-water quality models, mining pollution models, watershed models for the evaluation of the effects of atmospheric deposition, subsurface drainage and pollutant transport models, and specific designs of various structural and nonstructural measures, such as detention–retention ponds. There is no "universal" reliable diffuse-pollution model available, and the user must match the selected model with the modeling task.

Calibration and verification of the models with field data may be necessary for Phase 2 modeling activities if accurate loading of pollutants is desired. Calibration is less important if only the relative effect of various management practices is compared. One important consideration in model selection is the experience of others with its use. Few models have adequate documentation, have been extensively tested and applied by others than the model creators, and are continuously supported and maintained.

SIMPLE STATISTICAL ROUTINES AND SCREENING MODELS

Screening models (Levels I and II) evolved primarily from established traditional design procedures and/or the statistical results of monitoring programs. The concepts are the same for both urban and rural (agricultural) land uses, and the land-use type and distribution is generally the most dominant independent input parameter. The hydrologic fundamentals of these models are not complex or may not even be considered. Simple water quality estimation procedures for diffuse-source areas rely primarily on estimating pollutant accumulation in curb storage or use directly established unit loads related to land use for each land use within the watershed.

Diffuse (nonpoint)-pollution unit loads are expressed in mass of pollutant per unit of area per time. Unit loads of pollutants from dry-weather flow in combined sewers are often expressed in so-called population equivalents (pollutant emission rate expressed as mass per capita per day). The pollution load in this simple way is then found by multiplying the unit load by partial area of the land-use category and for sewage by the population number of the contributing basin. Publications by McElroy et al. (1976) and by Chesters et al. (1978) summarized unit loads and functions for nonpoint pollution, including that from urban drainage basins.

The basic characteristic of simplified screening models is that there is very little interaction, if any, among the processes generating the flow and quality. The statistical equation used to derive the unit loads and parameters of the models were based on elemental statistical routines, such as mean, standard deviation, and multiple-regression analyses.

Unit Loads of Pollutants from Impervious Urban Surfaces

Buildup and Washoff. In the late 1960s and early 1970s, investigators (APWA, 1969; Sartor and Boyd, 1972) analyzed distribution of pollutants on street surfaces and found that the majority of solids accumulated within a 1-m-wide strip on the street side of the curb. Sartor and Boyd (1972) then provided accumulation plots for three major land uses (residential, commercial, and industrial). These findings were then confirmed by a number of studies throughout the United States (for example, Shaheen, 1975, and Bannerman et al., 1984). In such cases, it was convenient to express the accumulated pollutants in grams per unit length of curb. Such curb pollutant loads were provided by McElroy et al. (1976).

Instead of using a constant curb or surface load of "dust and dirt" and associated pollutants, it has been demonstrated that there is a build up of particular pollutants on the impervious surfaces, particularly in the curb storage of urban basins. Originally, it was assumed that this buildup is linear; however, Sartor and Boyd (1972) demonstrated a gradual decrease in the accumulation rates until a certain pseudoequilibrium is achieved. Novotny et al. (1985) showed that this steady-state pollutant load is a result of a quasi equilibrium between the pollutant inputs (atmospheric and litter depositions and traffic emissions) and translocation of pollutants from the curb storage on adjacent pervious and hydrologically inactive surfaces. The steady-state curb pollutant loading may be maintained and has been observed in some experimental urban basins. Typically, these loads were measured between 20 to 100 g/m of curb, depending on the character and "cleanness" of the area. A linear steady buildup of accumulated pollutants on impervious surfaces and curb storage is typical only for urban areas that are nearly 100% impervious. The *buildup* then refers to a complex process occurring during the dry period preceding a rainfall that includes deposition, wind erosion, and street cleaning effects.

Washoff is thus the process by which the accumulated particles are removed from the surface by rainfall-generated runoff. Hence, the quantity of the pollutant in the sewer flow could be computed from:

$$P = CD * L * A * r \qquad (9.2)$$

where
P = pollutant load in the sewer or at the basin outlet (kg/storm)
L = accumulated curb load of the pollutant during an antecedent dry period (g/m)
CD = curb density (km/ha)
A = subwatershed area (ha)
r = washoff factor that is related to runoff energy

The magnitudes of the unit loads and simple procedures were compiled by McElroy et al. (1976). The pollutant loads computed by these simple procedures (using either the literature or local monitoring statistical values) can be complemented by simple flow determination in order to crudely estimate the average concentrations of the pollutant in urban runoff.

The buildup and washoff modeling concept has been used in simple screening models, as well as in more complex hydrological urban runoff pollution models. A more detailed presentation of buildup and washoff processes and their implication on urban runoff quality management was presented in Chapter 8.

Unit Loads of Pollutants from Pervious Surfaces

The annual or seasonal form of the universal soil loss equation (USLE) proposed by Wischmeier and Smith (see Chapter 5 for a detailed description of the USLE with quotations from the authors and developers) is the most widely used and respected model for estimating soil loss from pervious areas. In order to obtain solids load (sediment yield) at the watershed outlet, the potential soil loss estimated by the USLE must be multiplied by the delivery ratio factor, DR. Generally, DR equates the measured sediment yield to the potential soil loss estimated by the USLE. As pointed out in Chapter 5, the DR is a highly variable hydrological parameter that normally is estimated by calibration. For sewered areas, DR is close to unity. The magnitudes of DR for urban areas and different land uses are not known and should be researched. In most cases only simplified morphological estimates of the delivery ratio are considered and included.

To obtain loading of particulates other than sediment, the sediment loads obtained by the USLE are multiplied by *potency factors*, which relate the concentration of the pollutant to that of the sediment. These factors have also been summarized by McElroy et al. (1976) and investigated by Zison (1980). The enrichment is either not considered or considered only superficially.

A number of excess rainfall estimating procedures can be used for calculating flow and, hence, average event concentrations of pollutants. The best known hydrological procedure in the United States was developed by the Soil Conservation Service (SCS) and is called the runoff curve number method, which was introduced in Chapter 3 (Soil Conservation Service, 1972). The SCS method provides a means of estimating surface runoff volume and hydrograph using a semiempirical relationship between rainfall and runoff based on a runoff curve number that depends on land use, soil type, vegetation, and antecedent soil moisture conditions.

For combined sewer systems, the simplified procedures mentioned earlier provide only the wet-weather water and pollutant contributions by surface runoff. In order to obtain the total pollution load, dry-weather (wastewater) contributions must be added. These contributions are estimated using population equivalent loads such as those reported in Novotny et al. (1989).

Screening Models

The EPA screening procedures (Mills et al., 1982, 1985) evolved from the revisions and expansions of the water quality assessment procedures initially developed for 208 planning areas. The procedures are essentially a compilation of functions and unit loads developed previously by the Midwest Research Institute (McElroy et al., 1976) and provide annual estimates of loads. The nonurban loads are based on the annual form of the universal soil loss equation (see Chapter 5), while urban loads are estimated using the buildup–washoff concept presented in the preceding paragraph. The revised edition of 1985 contains a comprehensive compilation of procedures for estimating loadings of toxic pollutants, including their partitioning between solid and dissolved phases.

The data requirement is minimal and the procedures have excellent user documentation and guidance, including workshops sponsored by the EPA Water Quality Modeling Center, Athens, Georgia. The presentation of loading functions is supplemented with additional parameters and sample calculations.

Model-enhanced unit loads. A large hydrologic-pollution simulation model, calibrated by extensive monitoring data from several experimental watersheds located in southeastern Wisconsin, was used to generate unit loads of pollutants for typical land uses (Novotny et al., 1979; Novotny and Bannerman, 1980). The method called MEUL (model enhanced unit loads) and the generated unit loads were later used by local planning agencies to delineate hazardous lands and watersheds requiring abatement in the Wisconsin Priority Watershed Program and 208 local studies. Figure 9.6 shows the loading diagrams generated by the MEUL method. These unit loads can then be used along with the land-use distribution within the watershed to generate total annual or seasonal watershed loadings.

Cornell University researches (Haith and Tubbs, 1981; Dickerhoff and Haith, 1983; Haith and Shoemaker, 1987) presented generalized loading functions for estimating nutrient loadings. These simple functions describe both rural and urban sources. For urban sources the buildup and washoff concept is used. For rural sources the loading functions have the following form:

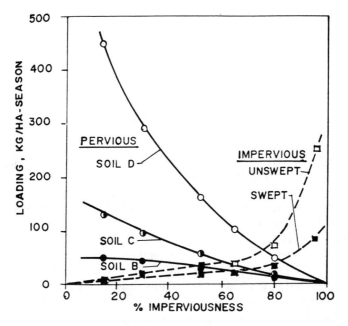

FIGURE 9.6. Unit loads of sediment (suspended solids) obtained by hydrologic simulation (model-enhanced unit loads).

$$LD_{kt} = 0.1 CD_{kt} Q_{kt} TD_k$$
$$LS_{kt} = 0.001 CS_{kt} X_{kt} TS_k \qquad (9.3)$$

where
LD_{kt} = unit load of dissolved pollutant from land-use area k during the event on day t, kg/ha
LS_{kt} = unit load of suspended pollutant from land-use area k during event t, kg/ha
CD_{kt} = dissolved pollutant concentration of pollutant in runoff, mg/l
CS_{kt} = solid-phase (adsorbed) pollutant concentration, µg/g of sediment
Q_{kt} = runoff for the day of the event on day t from land-use area k, cm
X_{kt} = soil loss from land-use area k due to the event on day t, tonnes/ha
TD_k, TS_k = transport factors that indicate the fractions of dissolved and solid-phase pollutants that move from the edge of the source area (field) to the watershed outlet.

To obtain the total watershed seasonal or annual (nonwinter) loadings the unit loads expressed by Equation (9.3) are integrated over the entire watershed area and time span of the simulation period as

$$LD \text{ (total)} = \sum_k \sum_t LD_{kt} A_k \qquad (9.4a)$$

$$LS \text{ (total)} = \sum_k \sum_t LS_{kt} A_k \qquad (9.4b)$$

where A_k = watershed area associated with land use k, ha. Implementation of these loading functions requires estimates of runoff, soil loss, pollutant concentrations, and attenuation and transport from each unit land-use source area within the watershed. Simplified procedures are provided by the authors.

Regional regression models. USGS developed regional regression models for estimating urban runoff quantity and quality (Driver and Tasker, 1988; Driver and Lystrom, 1987; Driver, 1990) in which the nonpoint pollution loads were correlated to the percentage of impervious area, industrial land use, commercial land use, nonurban land use, mean annual rainfall, and mean annual nitrogen-precipitation load. The required input data can easily and inexpensively be obtained from local rain gages, land-use maps, annual weather summaries, and census reports.

The model has the form (Driver and Lystrom, 1987):

$$Y = (B_0 \times X_1^{B1} \times X_2^{B2}, \ldots, X_n^{Bn}) BCF \qquad (9.5)$$

where
Y = estimated storm-runoff pollutant load or volume *dependent variable
$X_1, X_2, \ldots, X_n$ = basin morphological or climatic characteristics (independent variables)
$B1, B2, \ldots, B_n$ = regression coefficients
n = number of basins or climatic characteristics in the models
BCF = bias correction factor correcting the impact that the logarithmic transformation of the variables has on the mean

Independent variables included in the model were:

Basin characteristics
1. Total contributing drainage area
2. Impervious area as a percentage of the total area

3. Industrial land use
4. Commercial land use
5. Residential land use
6. Nonurban land use

Climatic characteristics

1. Total rainfall for each storm
2. Mean annual rainfall
3. Mean annual nitrogen precipitation load

In the analysis the United States was divided into three regions based on mean annual rainfall. The USGS data base from which the models were developed included 1123 storms for 98 urban stations, which were mostly stations of the National Urban Runoff Project, in 21 metropolitan areas. The dependent (modeled) parameters for which adequate regression models were obtained were runoff volume and chemical oxygen demand and nitrogen loads.

Caveats. In spite of their questionable accuracy and reliability, the simple screening procedures have found wide applications. It may be surprising to note that such simple quality estimations have been attached to several complicated hydrological models currently in use. One explanation for this popularity is that these procedures provide a simple mechanism and quick answers to pollution problems of larger watersheds where more complicated models might fail because of the enormous amount of information (not commonly available) required to set up the input data bases. The screening procedures enable identification of hydrologically active areas that generate nonpoint pollution that should then be subjected to further studies and subsequent management.

The effect of various abatement measures on nonpoint pollution loads estimated by the simplified screening procedures can be only approximately determined by using simple rule-of-thumb factors and pollution-removal efficiencies. For example, it can be assumed that an effective street sweeping program will reduce solids loads from impervious urban surfaces by 15% to 25%, or that a properly designed and operated wet detention pond can attenuate particulate pollution by about 50%, or that a certain soil-conservation practice will reduce soil loss as delineated in the guidelines for determining the cover factor, C, and the erosion control practice factor, P, of the universal soil loss equation.

If statistical regression models are selected, the user should be aware that all statistical models reflect and correspond to the data sample from which the model was derived. Using the models for conditions and locales outside the original data is always improper.

Geographical information systems. The unit load models and simpler screening models along with basin attributes can be incorporated in computer-aided design (CAD) and the geographic information systems (GIS) to generate graphic maps of source areas, transport routes, and other components and parameters of diffuse-pollution generation and transport processes, as well as the contaminant (pollutant) loads from these areas. The idea of using GIS for estimating diffuse pollution is not new and was used in the 1970s for calculating basinwide phosphate loads and developing source maps in the Great Lakes basin (Johnson et al. 1978). The GIS have been used in many disciplines, including mapping, military, and water resources–watershed descriptions.

Geographical information systems have a lot in common with the computer-aided design systems used for drafting a wide range of technical objects. Both GIS and CAD need to be able to relate objects to a frame of reference (coordinates), both need to handle graphic attributes and both need to be able to describe topological relations (Burrough, 1986). GIS can be linked to remote sensing, surveying data, and can include functional relationships and unit loads. Geographic information systems can be thought as being both the means of storing and retrieving data about aspects of the earth's surface, and systems by which the data can be transformed and manipulated interactively for studying environmental processes and the impact of planning decisions. They are ideally suited for studying the processes and impacts of diffuse pollution (Connors-Sasowski and Gardner, 1991; DeRoo, Hazelhoff, and Burrough, 1989; Evans and Miller, 1988; Potter, Gilliand, and Long, 1986).

Cline, Molinas, and Julien (1989) described an auto-CAD system that was used in connection with the watershed hydrologic model HEC-1 developed by the U.S. Army Corps of Engineers in the 1970s. They have pointed out that the commercially available auto-CAD system can organize and display spatial watershed data and also has some analytical capability. In a similar application, Stuebe and Johnston (1990) used a public domain GIS system GRASS (available from the U.S. Army Engineering Research Laboratory) for estimating runoff volumes. The GRASS system is the designated GIS of the U.S. Department of Agriculture, Soil Conservation Service.

In a GIS system, information of the spatial characteristics of a geographic area (such as a watershed) can be stored in a square grid system called a *raster* or a *vector–polygon system*. In the raster system, each square may correspond to a dot (pixel) on the computer screen. Information can be stored in different layers (for example, soil information in one layer, elevation in the second layer, and land cover in the third layer. This concept of information storage is shown in Figures 9.7 and 9.8. As

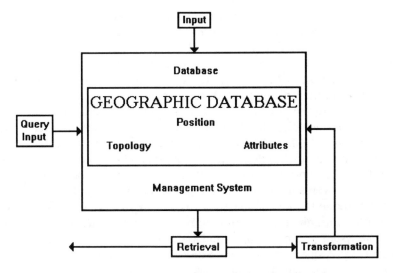

FIGURE 9.7. Schematic of the geographical information systems (GIS). (After Burrough, 1986, by permission of Oxford University Press.)

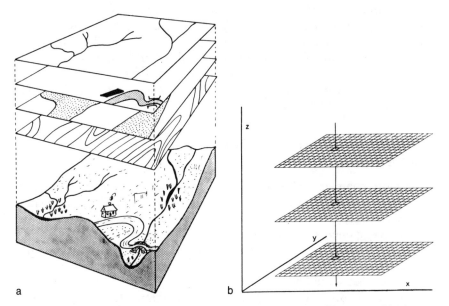

FIGURE 9.8. Overlay and storage of geographic information in a GIS system. (*a*) The "overlay" concept: the real world is portrayed by a series of overlays, in each of which one aspect of reality has been recorded (e.g., topography, soil type, roads, rivers). (*b*) Three-dimensional arrays used for coding map overlays in raster data base structures. (After Burrough, 1986, by permission of Oxford University Press.)

soon as the entire watershed information is coded and entered into the GIS system it can be processed using, for example, the universal soil loss equation, to produce source maps of sediment and pollutants. The GIS can also be used for spatial water quality and pollution analyses, as well as for evaluation of the impacts of various land-use changes within the watershed on the water quality of receiving water bodies.

The GIS are used by the agencies to store and manipulate geographical information, including that related to diffuse pollution and water quality. For example, the Soil Conservation Service (SCS) of the U.S. Department of Agriculture has developed the State Soil Geographic (STATSGO) data base (Reybold and TeSelle, 1986; Bliss and Reinbold, 1989). This data base overcame many of the problems associated with previous tedious methods of mapping soils over a larger area. STATSGO data have polygon units that incorporate several components of the soil information system, such as slopes and water capacity. The U.S. Geological Survey was cooperating with the SCS in an effort how to use the soil GIS data base in watershedwide analyses.

Probabilistic statistical methods. The so-called *EPA Probabilistic Method* for analyzing water quality effects of urban runoff is based on the derived frequency distribution of the flows and quality, which is presumed to be log-normal (Driscoll, 1986). When coupled together the distribution of runoff loads can be derived (Driscoll, DiToro, and Thomann, 1979; DiToro, 1984; and Athayde, Meyers, and Tobin, 1986). Further analytical treatment has been developed to account for storage (DiToro and Small, 1979). These models are not identical to the stochastic models discussed in the next section.

DETERMINISTIC HYDROLOGIC SIMULATION MODELS

The more complex hydrologic simulation models in the second category represent a description of the hydrologic rainfall–runoff transformation process with associated erosion, pollution buildup and washoff, and other quality components. In combined sewer systems, sewage flow contributions, infiltration, quality contributions by infiltration (for example, nitrate), and pollutant buildup in sewers are additional sources and processes that should be included.

The basic premise of hydrologic simulation models, in contrast to the previous simplified computational procedures, is an interaction among hydrological and pollution-generation and -transport processes. Obviously, the complexity of the processes that are considered by the models and their feedback and interactions vary from model to model.

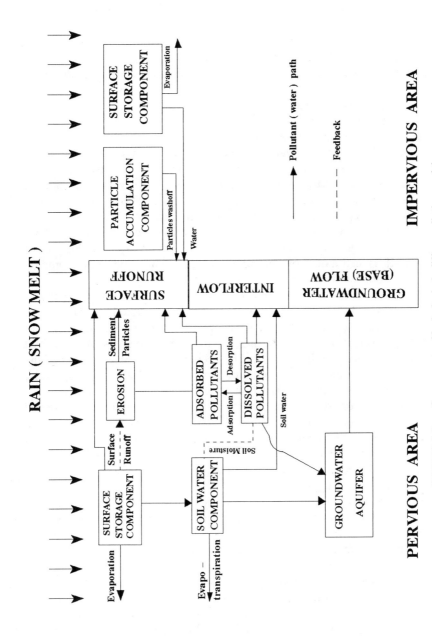

FIGURE 9.9. Components of hydrological–diffuse-pollution models.

Diffuse-pollution hydrological flow-quality models have the following basic components (Novotny, 1986, 1988), which are also depicted in Figure 9.9:

1. A surface-runoff-generation component that computes the transformation into excess (net) rainfall. Modeling surface runoff includes the following processes that were described in Chapter 3:
   a. Exhaustion of surface storage
   b. Evapotranspiration
   c. Snow accumulation and melt
2. The soil and ground-water component (not common to all models) that describes the movement of water through the unsaturated soil zone into the saturated ground-water zone. These submodels balance soil moisture with infiltration rate, evapotranspiration, and water loss into the deep ground-water zone. Since infiltration is a function of soil moisture content, an iterative procedure may be employed. If the soil component is not included in the model, the infiltration rate of surface runoff is estimated by an empirical equation. The most common infiltration models were presented in Chapter 3.
3. Accumulation, removal, and washoff of pollutants from impervious surfaces. This component, typical for urban models, balances the particulate pollutants that deposit onto the impervious surfaces near the curb of paved roads during dry days and estimates their washoff into the runoff during rainfall. The accumulated mass is a result of the deposition of particulate pollutants from various sources (atmospheric dry deposition, litter deposition, traffic emissions, salt application), and their removal from curb storage by wind and traffic-induced turbulence and street sweeping (Novotny et al., 1985).
4. Soil-erosion components, by which soil loss from impervious areas can be estimated. The universal soil loss equation (USLE) is the most common model to represent this process. Soil-erosion models were introduced in Chapter 5.
5. Adsorption of pollutants on soils and urban dust. This component, which in agricultural models is represented by adsorption–desorption relationships presented in Chapter 6, is in most urban models replaced by potency factors that relate the particulate pollutant to that of total suspended solids.
6. Runoff and pollutant routing components transform the excess rainfall into an inlet hydrograph and quality histogram that is then routed through the sewer network. Common techniques for excess rainfall–surface-runoff overland routing (convolution) rely on a synthetic unit hydrograph, in most cases. Routing in sewers and channels is most

often accomplished by kinematic wave formulas. In the absence of a detailed routine for sediment redeposition and trapping during the overland flow, sediment delivery and enrichment factors are used to relate the measured sediment and pollutant yields to their generation at the source.
7. In the combined sewer systems, flow generation and deposition of solids in sewers, as well as the growth of biomass therein, should be considered. The contribution of solids deposited from sewage can be substantial. Dry-weather fraction of the pollution carried by the combined sewers can be estimated by population equivalents, while sewer solids accumulation, growth of biological mass, and first flush effects due to scouring of these materials can only be estimated using highly empirical and unreliable procedures. One can expect that the parameters affecting these processes will include the hydraulic conditions in the sewers and the length of the antecedent dry-weather period.

Discrete event modeling simulates the response of a watershed to a major rainfall. Continuous dynamic models simulate flow and pollutant loading over an extended period of time. However, it is erroneous to expect that continuous models can provide adequate information on the variability of the flow or its quality. By their nature and design, the deterministic dynamic models could be called short-term moving average models without a noise component (this loose classification should not be confused with the moving average ARMA models that are introduced later in this chapter). Therefore, attempts to apply statistical frequency and probabilistic excedence routines to the outputs from the deterministic models are improper.

Urban Models

All urban runoff and CSO pollution models presented below are lumped-parameter models, although they do allow segmentation of the watershed into uniform subbasins according to land use and drainage patterns. As to their complexity, the urban hydrologic models can be classified into three groups:

Level III. Simplified models, either event oriented or continuous. In these models some hydrologic or water quality components are missing, some others are replaced by a simple relationship. There is some interaction between the processes, but this is usually done in an approximate way. Water and pollutant routing may not be included.

Level IV. Sophisticated single-event, simulation models. These models provide for extensive interactions among the various processes that are important in the simulation of storm-water quantity, including infiltration, overland and gutter flow, and flow routing. The quality components may be less sophisticated, relying on simpler pollutant buildup and washoff concepts. Street sweeping effects can be included only during a predetermined relatively short antecedent dry period and, in general, such models may not be suitable for evaluation of typical long-term best management practices for controlling urban runoff pollution. Their primary use is in the design and evaluation of urban drainage systems.

Level V. Sophisticated continuous models. These models provide for extensive interactions among the various processes that are important in the simulation of both storm-water quantity and quality in separate and combined sewer systems. Quality and routing of flow are included. Modeling intermittent dry periods accounts for soil moisture, pollutant buildup and removal, and other important processes occurring between rainfalls.

Level III Models

Examples of simplified urban models in Level III include the Corps of Engineers' (Hydrologic Engineering Center, 1975) Storage-Treatment-Overflow-Runoff Model (STORM), Model of Urban Runoff and Sewer Flow (MOURSEF) (formerly the Wisconsin Urban Runoff Model) (Novotny et al., 1985; Novotny and Capodaglio, 1991), the Areawide Stormwater Model (ABMAC) (Litwin, Lager, and Smith, 1981), TVA's model HYSIM (Milligan, Wallace, and Betson, 1984), and a number of proprietary models. Most of these models simulate runoff either by the modified rational formula (STORM) or by the SCS Runoff Curve Number Model (Wisconsin Urban Runoff Model). The models simulate street drainage and storage-treatment systems. The TVA-HYSIM model also incorporated surface–ground-water interrelation due to specific characteristics of the NURP watersheds in Knoxville, Tennessee.

The STORM model is available from the Hydraulic Engineering Center of the U.S. Army Corps of Engineers in Davis, California, but the center does not provide user's support nor further development and maintenance of the program, which was left with private software vendors. The MOURSEF model by the senior author and his coworkers can be provided as supplementary software to the users of this book. Figure 9.10 shows a schematic of the MOURSEF model.

Deterministic Hydrologic Simulation Models   533

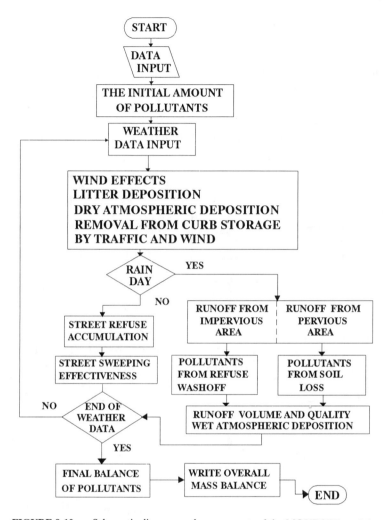

FIGURE 9.10.   Schematic diagram and components of the MOURSEF model.

Level IV Urban Models

The best known Level IV models (event-oriented models) are the EPA's original Stormwater Management Model (SWMM) (Huber and Dickinson, 1988) and the Illinois Urban Drainage Area Simulator (ILLUDAS) (Terstriep and Stall, 1974). In Great Britain the WASSP-QUAL model (a proprietary but widely used model) was developed and sold throughout Europe (Henderson and Moys, 1987). Several "synthetic" design storms have been developed and are available (Soil Conservation Service, 1972, 1985; Chow, Maidment, and Mays, 1988). These design storms can be used as a substitute for measured storm hyetographs in event modeling.

The Storm Water Management Model (SWMM) is a widely used stormwater and combined sewer overflows quantity–quality model that is readily available from the U.S. EPA Laboratory in Athens, Georgia. SWMM, developed in the 1970s (U.S. EPA, 1971), has been applied to urban hydrologic quantity–quality problems in many locations world-

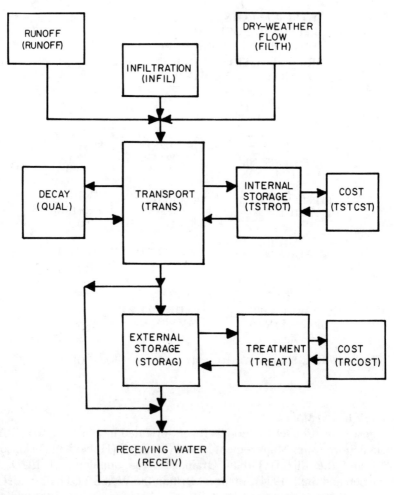

FIGURE 9.11. Components of the Storm Water Management Model (SWMM). Subroutine names are shown in parentheses. (After U.S. EPA, 1971.)

wide. The model is in the public domain, has excellent documentation, and is continuously maintained and updated. The model can analyze single storm events on multiple catchments. Figure 9.11 shows a schematic of the basic components of the program.

Version 4 (Huber and Dickinson, 1988) of the SWMM model performs both continuous and single-event simulation throughout the whole model. It can simulate backwater, surcharge, pressure flows, and looped connections, and has a variety of options for quality simulations, including traditional buildup and washoff functions, as well as rating curves and regression techniques. Subsurface flow routing (constant quality) can be performed in addition to surface quantity and quality routing, and treatment devices may be simulated using removal functions and sedimentation theory. A hydraulic design routine is included for sizing of pipes, and a variety of regulator devices can be simulated, including orifices, weirs, pumps, and storage. This version is essentially a Level III model (Maahel and Huber, 1984).

The volume of data needed to run SWMM varies with the scope of the application, and can be very extensive in some complex situations. At a minimum it requires information on area, imperviousness, slope, roughness, depression storage, and infiltration characteristics. Information on buildup of pollutants is required for quality generation. Precipitation input is in the form of hyetographs for individual storm events (hourly or 15 minute intervals). Channel-pipe data include shapes, dimensions, slopes or invert elevations, and roughness.

Calibration of the model is needed, especially for quality simulation, but it is tedious and an expert computerized system is needed. Even with calibration the performance of the models for quality is relatively weak (Donigian and Huber, 1990; Delleur and Baffaut, 1990). The PC version of the model is not "user friendly;" however, an expert system for calibration of the model is available (Delleur and Baffaut, 1990).

The Highway Drainage Model (Dever and Roesner, 1982, 1983; Roesner, Dever, and Aldrich, 1983; Dever, Aldrich, and Williams, 1983) developed for the U.S. Federal Highway Administration consists of a package of six computer programs. The model has used some components of the SWMM as basic blocs. The general capabilities of the model include statistical analysis of long-term rainfall records, preliminary highway drainage system design, hydraulic analysis of drainage systems during extreme storm events, and simulation of buildup and washoff of pollutants. Its computational capabilities are similar to those of SWMM (Donigian and Huber, 1990).

DR3 QUAL is a version of the USGS Distributed Routing Rainfall Runoff Model that includes quality simulations (Alley and Smith, 1982).

The model is available for general use from the U.S. Geological Survey. Runoff generation is followed by routing with the kinematic wave formula. Quality is simulated using buildup and washoff functions with settling of solids in storage units dependent on particle size distribution. The model has been used in some of the NURP studies that were conducted by the USGS. No microcomputer version is available.

Level V Urban Hydrologic Models

Hydrologic Simulation Program—FORTRAN (HSP-F). The FORTRAN version of the Hydrologic Simulation Program (HSP-F) is an example of a Level V model (Johanson et al., 1982; Barnwell and Kittle, 1984). This model is capable of simulating a hydrologic time series of runoff quantity–quality events, including flows (hydrographs) and conventional pollutants and toxics. The HSP-F incorporates the watershed scale Agricultural Runoff Model (ARM) and urban Nonpoint Simulator (NPS) into a basin-scale analysis framework that includes fate and transport in one-dimensional stream channels. It is the only comprehensive model of watershed hydrology and water quality that allows the integrated simulation of land and soil-contaminant runoff processes with in-stream hydraulic, water temperature, sediment transport, nutrient, and sediment–chemical interactions. The runoff quality capabilities include both simple relationships, such as empirical buildup and washoff functions and constant concentration pollutant inputs, as well as detailed soil process options (including, leaching, soil attenuation, and soil nutrient transformation). The model was originally developed from the Stanford Watershed Model. It is a large model and requires considerable effort when applied to a watershed. The use of this model for combined sewer systems is very limited.

The HSP-F contains three applications moduli—PERLND, IMPLND, and RCHES—and five utility moduli—COPY, PLTGEN, DISPLY, DURANL, and GENER (Fig. 9.12). Basically, the HSP-F performs the simulation in a lumped-parameter mode, whereby the magnitudes of the parameters and coefficients must be determined by calibration.

The PERLND simulates a pervious land segment with homogenous hydrologic and climatic characteristics. Water movement is modeled along three flow paths, overland flows, interflow, and ground-water flow in the manner conceived and incorporated in the Stanford Watershed Model, which is the predecessor of the HSP-F. Erosion is modeled by the Negev model (see Chapter 5). Water quality constituents can be simulated in a simple way by relating them to sediment loads with potency factors or by a more complex adsorption–desorption concept. The former pollutant transport concept is incorporated in the Nonpoint

APPLICATION MODULES		
PERLN	**IMPLBD**	**RCHRES**
SNOW	SNOW	HYDRAULICS
WATER	WATER	CONSERVATIVE
SEDIMENT	SOLIDS	TEMPERATURE
QUALITY	QUALITY	SEDIMENT
PESTICIDE		NONCONSERVATIVE
NITROGEN		BOD/DO
PHOSPHORUS		NITROGEN
PHOSPHORUS		PHOSPHORUS
TRACER		PLANKTON

UTILITY MODULES		
COPY	**PLTGEN**	**DISPLY**
DATA TRANSFER	PLOT DATA	TABULATE AND SUMMARIZE
	DURANI	**GENER**
	DURATION ANALYSIS	TRANSFORM OR COMBINE2

FIGURE 9.12. Components (modules) of the HSP-F watershed model. (After Barnwell and Kittle, 1984.)

Pollution Source (NPS) model, typically used for urban areas, while the latter was incorporated in the Agricultural Runoff Management (ARM) model. The NPS model has been in public use since 1980 and is available for PC computers from the U.S. EPA Environmental Research Laboratory in Athens, Georgia.

Module INPLND is designed to simulate impervious land segments with no infiltration. Water yield and movement is similar to PERLND, except that no water movement occurs by subsurface flows. Solids are simulated using the buildup–washoff concept in a manner similar to other urban models. Water quality constituents are simulated using empirical relationships with solids and water yields.

Module RCHRES simulates the processes that occur in a single reach of an open channel or a completely mixed lake. Hydraulic behavior is modeled by a kinematic wave routing concept (Chapter 3). Water quality algorithms are similar to many other stream and lake models that have evolved in the past 50 years (see Chapter 12).

The result of the modeling is a time history of runoff flow rate, sediment load and nutrient and pesticide concentrations, along with a time history of water quantity and quality at any point in a watershed. HSP-F simulates three sediment types (sand, silt, and clay) in addition to a single organic chemical and nutrients.

The WASSP-QUAL model was developed in Great Britain by Hydraulic Research, Ltd. The model is a result of a joint research effort with the Water Research Centre and other British institutions. The model can be

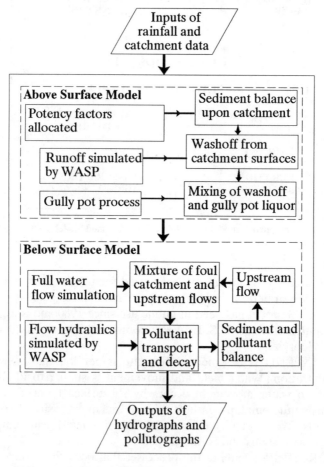

FIGURE 9.13. Overview of the WASSP-QUAL model structure and modeled processes. (After Henderson and Moys, 1987.)

run either in a continuous or single-event mode, and it models the following quality parameters (Henderson and Moys, 1987):

- Suspended solids, both total (TSS) and volatile (VSS)
- Dissolved oxygen
- Oxygen demand (BOD or COD)
- Ammoniacal nitrogen
- Hydrogen sulphide
- Large grain-size sediments

A schematic diagram of the components of the model is shown on Figure 9.13. The model requires calibration, and in some aspects is more complex than the SWMM, but not necessarily more reliable. This model should not be mistaken for the WASP model of the U.S. EPA (see Chapter 12 for a description of WASP).

Operational Urban Models

Huber (1986) listed the six operational urban water quality models that are included in Table 9.1. By *operational* he implied that (1) a user's manual and documentation are available, (2) the model had been used by others than just the model developer, and (3) continuous support was available. Two additional proprietary operational models that fit this definition were also included.

The Proceedings of the International Conferences on Urban Storm Drainage (Yen, 1982; Balmér, Malmqvist, and Sjöberg, 1984; Gujer and Krejci, 1987; Iwasa and Sueishi, 1990) reported numerous modeling efforts throughout the world. Also the NATO Advanced Seminar pro-

TABLE 9.1 Operational Urban Runoff Models

Model	Sponsoring Agency	Year of Origin	No. of Pollutants	Simulation Type
DR3M-QUAL	USGS	1982	4	C, SE
FHWA	FHWA	1979	13	SE
HSP-F	EPA	1976	10	C, SE
STORM	HRC	1974	6	C, SE
SWMM	EPA	1971	10	C, SE
WASSP-QUAL	Proprietary[a]	1987	7	C, SE
MOURSEF	Proprietary[b]	1983	6	C

Source: Adapted from Huber 1986.
Note: C = continuous simulation; SE = single-event simulation.
[a] Water Resources Center Engineering, Swindon, England.
[b] Available from the senior author.

ceedings (Torno, Marsalek, and Desbordes, 1986) has several overview chapters dealing with modeling. Also see Moffa (1990) for chapters on models dealing specifically with the CSO and their impact on receiving waters.

Rural (Agricultural Models)

Some of the previous hydrological models (for example, HSP-F, MOURSEF) can also be used for simpler modeling of agricultural watersheds, primarily erosion and movement of particulate pollutants. Agricultural models may also be divided into the three levels of sophistication; however, the models presented herein have mostly Level IV or V sophistication. Unlike urban models, which typically use the lumped-parameter concept, both distributed- and lumped-parameter concepts have been used for agricultural models. Their level of sophistication is based on the number of modeled processes and constituents. The agricultural models were reviewed by a number of authors; for example, models for the transport of chemicals were reviewed by Donigian and Dean (1985). The following discussion was adapted from reviews by Novotny (1986) and Donigian and Huber (1990).

Areal, Nonpoint Source Watershed Environment Response Simulation (ANSWERS) was developed by the Agricultural Engineering Department of Purdue University (Beasley and Huggins, 1981; Beasley, 1986). It is a distributed parameter model, primarily event oriented. Currently the model is maintained and distributed by the Agricultural Engineering Department, University of Georgia, Tifton, Georgia.

In order to use the ANSWERS model, the watershed is divided into uniform square elements, as shown in Figure 9.3(b). The elements range from one to four hectares. Within each element the model simulates the process of interception, infiltration, surface storage, surface flow, subsurface drainage, sediment detachment, and movement across the element. The output from one element then becomes a source of input into an adjacent element, which can be either area of the channel segment.

Nutrients (nitrogen and phosphorus) are simulated using correlation relationships between chemical concentrations, sediment yield, and runoff volume. Snowmelt or pesticide movement cannot be simulated. Data needs compose a detailed description of the watershed topography, drainage network, soils, and land use. A single storm hyetograph drives the model. A PC version is available for small watershed applications. For larger watersheds, a mainframe–large memory capacity computer is required.

Agricultural Nonpoint Source Pollution Model (AGNPS) was developed by the U.S. Department of Agriculture, Agriculture Research Service (USDA-ARS) (Young et al., 1986). The primary emphasis of the model is on nutrients and sediment, and on comparing the effects of various best management practices on the pollutant loadings.

The AGNPS can simulate sediment and nutrient loads from agricultural watersheds for a single storm event or for a continuous simulation. The watershed must be divided into uniform square cells. Grouping of the cells is by division of the basin into subwatersheds. However, flow and pollutant routing is accomplished by a function of the unit hydrograph type, which is a lumped-parameter approach. The model does not simulate pesticides.

AGNPS is also capable of accepting and handling point inputs such as feedlots, wastewater discharges, and stream bank and gully erosion. In the model, pollutants are routed from the top of the watershed to the watershed outlet in a series of steps. The modified universal soil loss equation is used for predicting soil loss in five different particle sizes (clay, silt, sand, small aggregates, and large aggregates). The pollutant transport portion is subdivided into one part handling soluble pollutants and another part handling sediment absorbed pollutants.

The input data requirements are extensive, but most of the data can be retrieved from topographic and soil maps, local meteorological information, field observations, and various publications, tables, and graphs provided in the manual or references. Both mainframe and PC versions of the model are available from the USDA-ARS in Morris, Minnesota.

The Agricultural Runoff Management Model (ARM). This model is a version of the HSP-F model that can be run independently or included in HSP-F. The model simulates runoff (including snow accumulation and melt), sediment, pesticides, and nutrient loadings from surface and subsurface sources (Donigian and Davis, 1978). The ARM model (as the more complex HSP-F model that incorporates ARM) requires extensive calibration. The PC version is only available as a part of the HSP-F.

Chemicals, Runoff, and Erosion from Agricultural Management Systems (CREAMS). This model was developed by the U.S. Department of Agriculture, Agricultural Research Service (Knisel, 1980, 1985; Leonard and Knisel, 1986). CREAMS is a field scale model that uses separate hydrology, erosion, and chemistry submodels connected together by pass files.

The hydrology component has two options, depending upon availability of rainfall data. Option one estimates storm runoff when only daily rainfall data are available. This is accomplished by the SCS Runoff Curve Number model. When hourly rainfall data are available, option two estimates runoff by the Green-Ampt equation. Both submodels were described in Chapter 3.

The erosion component of the model considers the basic processes of soil detachment, transport, and deposition. Detachment of soil particles is modeled by the modified universal soil loss equation for a single storm event. The transport capacity of the overland and channel flow is derived from the Yalin sediment movement model (see Chapter 5). The basic concepts for nutrient modeling treat their transport as proceeding separately in adsorbed (with sediment) and dissolved (with runoff) phases. Soil nitrate is lost both with surface and subsurface flows. Soil nitrogen is modified by nitrification–denitrification processes and by plant uptake.

The pesticide component estimates concentrations of pesticides in runoff (water and sediment phases) and the total mass carried from the field for each storm during the period of interest. Pesticides in runoff are partitioned between the solution and sediment phases using a simplified isotherm model (Chapter 6). The model has the capability of simulating up to 20 quality components at one time.

CREAMS can simulate user-defined management practices. These activities include aerial spraying (foliar or soil directed) or soil incorporation of pesticides, animal waste management, and agricultural best management practices (minimum or no tillage, terracing, etc.; see Chapter 11).

Groundwater Loading Effects of Agricultural Management Systems (GLEAMS) was developed by the USDA-ARS to utilize the management-oriented CREAMS model. GLEAMS is essentially a vadose zone component for CREAMS (Leonard, Knisel, and Still, 1987). The soil column is divided into three to twelve layers of variable thickness in which pesticide and nutrient mass balance and routing are executed. The input data requirement for CREAMS–GLEAMS simulations are extensive and quite detailed. The maximum size of the watershed is limited to a field plot consisting of a maximum of three segments.

The CREAMS–GLEAMS model can be obtained from the USDA-ARS laboratory in Tifton, Georgia. It is available for PC computers, and there is a very active users group comprised of several hundred users throughout the world. Several other models were developed by adaptation of the CREAMS model, including *Simulator for Water Resources in Rural Basins (SWRRB)* and *Pesticide Runoff Simulator (PRS)*.

The SWRRB model was developed for evaluating basin-scale quality in rural watersheds (Williams, Nicks, and Arnold, 1985). SWRRB operates on a daily time step and simulates meteorology, hydrology, crop growth, sedimentation, flood plain degradation and aggradation, and nitrogen, phosphorus, and pesticide movement. The model was developed by modifying the CREAMS daily rainfall hydrology model for applications to large, complex rural basins. The PRS objective is to simulate pesticide runoff and adsorption onto the soil in small agricultural watershed.

Pesticide Root Zone Model (PRZM) was developed at the U.S. EPA's Environmental Research Laboratory in Athens, Georgia (Carsel et al., 1984; Muelkey, Carsel, and Smith, 1986; Muelkey, Ambrose, and Barnwell, 1986). It is a one-dimensional, dynamic, compartmental model that can be used for simulation of the movement of chemicals in an unsaturated zone within and immediately below the plant root zone. The model is divided into two major components, namely hydrology (and hydraulics) and chemical transport.

The hydrology and erosion components use the SCS runoff curve number and the universal soil loss equation, respectively. Evapotranspiration is estimated directly from the pan evaporation data and/or by empirical equations. Soil-water capacity terms, including field capacity, wilting point, and water saturation content, are used for simulating water movement through the soil column within the unsaturated zone. The model can also consider the application of irrigation.

Pesticide application onto plant foliage or into soils is also considered. Dissolved, adsorbed, and vapor-phase concentrations in the soil are estimated by simultaneously considering the processes of pesticide uptake by plants, surface runoff losses, erosion, decay, volatilization, foliar washoff, advection, dispersion, and retardation. The model is dynamic, hence pulse (instantaneous) pesticide loads can be simulated.

PRZM is also an integral part of a unsaturated/saturated zone model RUSTIC (Dean et al., 1989). RUSTIC is an acronym for risk of unsaturated/saturated transport and transformation of chemicals concentrations. This shell model links three subordinate models (including PRZM) in order to predict pesticide fate and transport through the crop root zone, and saturate zone to drinking water wells.

Data requirements to run the model are extensive. Predictions are made on a daily basis. The model only simulates organic chemicals in a one-dimensional downward transport. The model is available for PC computers and can be obtained from the Center for Exposure Assessment Modeling of the U.S. Environmental Protection Agency, Athens, Georgia.

Water Erosion Prediction Project (WEPP)
Hillslope Profile Model

A new generation water erosion model that was developed by the USDA-ARS National Soil Erosion Research Laboratory (Foster and Lane, 1987;

Lane and Nearing, 1989; Laflen, Lane, and Foster, 1991). It is a continuous simulation model, although it can be run on a single storm basis. By continuous simulation the model "mimics" the processes that are important to erosion prediction as a function of time, and as are affected by management decisions and the climatic environment. The output of the continuous simulation are time-integrated estimates of erosion. The model calculates both detachment and deposition, hence, the delivery process is considered. The output includes both on-site and off-site erosion effects. The on-site effects of erosion include the time-integrated (average annual) soil loss of the net soil loss over the area on the hillslope, which is analogous (but not identical) to USLE estimates. The output also includes deposition of the net deposition over the area on the hillslope. The output describing off-site effects includes sediment loads and particle size information. The output options also include the potential for obtaining monthly or daily (storm-by-storm) estimates of the on-site and off-site effects of erosion.

The model has six components (Fig. 9.14): climate generation

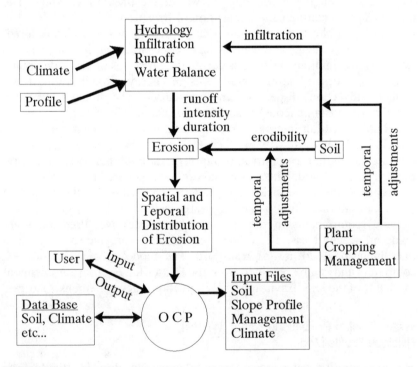

FIGURE 9.14. Flow chart of the Water Erosion Prediction Project (WEPP) model. (After Nicks et al., 1989.)

(CLIGEN), hydrology, plant growth, soils, irrigation, and erosion. The climate generator is run separately from the WEPP model. It generates rainfall amount, duration, maximum intensity, time to peak intensity, maximum and minimum temperature, and solar radiation for the location. The hydrology component calculates infiltration, the daily water balance, including runoff, evapotranspiration, and deep percolation. Infiltration is calculated by the Green–Ampt infiltration equation. Runoff is calculated using the kinematic wave equation or its approximation. The water balance routines are a modification of the SWRRB water balance (Williams, Nicks, and Arnold, 1985) and account for snowmelt, percolation below the root zone, movement of water downward between soil layers within the root zone, and both bare soil evaporation and plant transpiration. The plant growth component calculates the growth, senescence, and decomposition of plant material. This component also calculates leaf area index for transpiration calculations. Grazing and harvest removes material at user-specified intervals. After harvest, decomposition of vegetative residue, if present, is simulated.

Many of the soil parameters that are used in hydrology and erosion calculations change with time as a result of tillage operations, freezing and thawing, compaction, weathering, or history of precipitation. The soil component makes adjustments to soil properties on a daily time step. The irrigation component simulates the effects of irrigation, which are simulated as a rainfall event of uniform density. Several irrigation schemes and options can be accommodated.

The erosion component uses the steady-state sediment continuity equation as a basis for erosion computations. Soil detachment in the interrill areas is calculated from the rainfall intensity. Soil detachment in the rills occurs if the hydraulic shear stress is greater than critical shear and the flow is at less than transport capacity. Deposition occurs when the sediment load is greater than the transport capacity of the flow. The calculated sediment sizes are a function of the original soil material and the preferential deposition of certain sediment-size fractions along the hillside profile.

The erosion predictions from the WEPP profile model are applicable to "field size" areas or conservation treatment units. The maximum modeling unit size is about 270 ha (640 acres). The profile model is also applicable to areas that are farmed over and known as concentrated flow or "cropland ephemeral gullies." The model can also be used for modeling range and forest lands with large concentrated flow channels, and for estimating erosion in terrace channels or grassed waterways on cropland.

The models just discussed are operational models that can be obtained

TABLE 9.2 Summary Overview of Agricultural Models

Model	Modeling Tasks	Event or Continuous Simulation	Lumped or Distributed Parameter	Number and Type of Pollutants Modeled	Agency, Support, Source
AGNPS	SR	C, SE	DP[a]	S, N	USDA-ARS, Morris, Minnesota
ANSWERS	SR	SE	DP	S	University of Georgia, Tifton
ARM[b]	SR	C	L	S, N, 1 OP	
CREAMS-GLEAMS	SR, SW, GW	C, SE	L	S, N, 20 OP	ASDA-ARS, Tifton, Georgia
HSP-F	SR	C	L	S, N, 1 OP	EPA, Athens, Georgia
PRZM	SW, GW	C	L, one-dimensional	S, OP	EPA, Athens, Georgia
SWRRB	SR	C	L	S, N, OP	USDA-ARS, Temple, Texas
WEPP	CL, SR	C, (E)	L	S	USDA-ARS, West Lafayette, Indiana

Note: SR = surface runoff; SW = soil water; GW = ground water; CL = meteorology; C = continuous; SE = single event; L = lumped parameter; DP = distributed parameter; S = sediment; N = nutrients; OP = organic pesticides.
[a] Distributed parameter watershed description, lumped parameter routing.
[b] ARM is a part of the HSP-F model.

either free of charge or for a nominal charge from the agencies that developed and maintain the model. Table 9.2 summarizes some of the most important features of these operational models.

STOCHASTIC AND NEURAL NETWORK MODELS

Stochastic Autoregressive, Moving Average, Transfer Function Models

More than 15 years experience with the development of deterministic urban sewerage flow and quality models, along with their calibration and verification with numerous data, have revealed that:

- Deterministic hydrological models of flow and quality are highly variable and often difficult to reproduce even after extensive calibration of the model

- Increased complexity of the models does not a priori imply a better representation of the process
- The best models can only reproduce general trends in data, leaving a significant variability component aside

The reason for the general failure of deterministic models to represent the variability of the data is, simply said, the deterministic nature of the models themselves, which inherently neglects random and other variations. This defect has been known to researchers for a long time and they have tried to alleviate the problem by Monte Carlo simulations (Brutsaert, 1975). A serious drawback of using deterministic steady-state models and Monte Carlo simulations is the failure to detect the non-deterministic variation of the output. Only the uncorrelated variations of the input and system parameters can be adequately simulated. If these input and system parameters are cross-correlated, the Monte Carlo simulation must be modified to incorporate the cross-correlation, which is tedious and sometimes impossible. The second drawback is the use of steady-state deterministic models. Essentially, the variations of the input and system parameters are filtered by a deterministic model that is not sufficient to identify the entire variability of the output.

Time series analysis techniques have been successfully used in hydrology and water resources for over 20 years. In the earlier applications, one method for analyzing time series was to decompose the series (say, for example, daily or hourly BOD or suspended solids loads in combined sewers) into three components: (1) trend, (2) periodic component, and (3) random fluctuations. A trend was estimated by fitting a straight line or a polynomial to the series, the periodic component was extracted by Fourier analysis, and forecasting was made using the two extracted functions. However, in one of the first papers on the use of stochastic models for wastewater flow analyses, Berthouex et al. (1975) stated that such decompositions can be misleading, even for a deterministic process. Treating a stochastic process (the flow in combined and separate sewers is undoubtedly a stochastic process) the same way is even more dangerous, because it could give highly misleading results.

There are literally dozens of articles in the literature on the application of stochastic models to stream flow or rainfall and other hydrological events; however, only a few articles and reports were published in the last 12 years on applications of stochastic models to sewer flow quantity and quality (Berthouex et al., 1975, 1976, 1978; Barnes and Rowe, 1978; Novotny and Zheng, 1989; Capodaglio et al., 1990; and Novotny et al., 1990a, 1990b). In spite of the apparent lack of interest by researchers, there is no doubt that diffuse flow and quality and, hence, diffuse pollution are stochastic processes. Stochastic models are highly appropriate and can be successfully used, especially for forecasting and real time control of urban sewerage systems.

The term *stochastic* does not mean "completely random" or "unpredictable." Box and Jenkins (1976) the pioneers of stochastic modeling, have emphasized that such systems contain both the dynamic (deterministic) and stochastic nature of the underlying processes. There are basically two types of stochastic models:

1. Univariate ARMA
2. Transfer function
   a. Single input–single output (SISO)
   b. Multiple input–single output (MISO)

Figure 9.15 presents the concepts of stochastic modeling.

The univariate ARMA models relate the present or predicted value of the parameter to the series of its immediate past values. As shown on Figure 9.15 an uncorrelated random input called "white noise" is the primary driving force of the univariate stochastic model. This random input is "filtered" by the system transfer function to produce the output autocorrelated series. The nomenclature of ARMA modeling makes use of the so-called backshift operator B such as $X_{t-(\Delta t)} = B \times X_t$, $X_{t-2(\Delta t)} = B^2 \times X_t$, and so forth.

The univariate ARMA model is

$$Y(t) = \frac{\theta(B)}{\Phi(B)} a_t = G(B) a_t \tag{9.6}$$

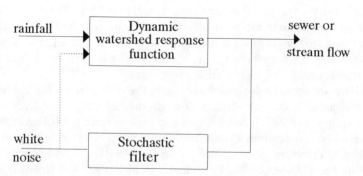

FIGURE 9.15. Representation of a stochastic modeling system. In addition to rainfall, other inputs may also be considered.

where
- Y = the values of the time series
- $\Phi(B)$ = polynomial of autoregession coefficients
- $\theta(B)$ = polynomial of moving average coefficients
- $G(B)$ = Green's function describing the behavior of the series
- a_t = uncorrelated, zero-mean random-shock series

Univariate models were used by Barnes and Rowe (1978) for analyzing combined sewer flow series of Springfield, Ohio, and Anderson, Indiana. For example, the ARMA model of Barnes and Rowe for flow in a combined sewer in Springfield after simplification (Novotny, 1988) is

$$(1 - 0.71B)Q_t = (1 + 0.17B)a_t \tag{9.7}$$

which is a first-order univariate ARMA(1,1) model. The number in parentheses denotes the order of autoregressive (left side of the equation) and moving average (right side) polynomials. Berthouex et al. (1975) found that a first-order univariate ARMA model fits the influent BOD time series of sewer flow in Madison, Wisconsin. Higher order models are needed for simulating quality parameters. Methods for estimating the coefficients of univariate ARMA models were presented by Box and Jenkins (1976), Pandit and Wu (1982), Capodaglio et al. (1990), and others.

The transfer function ARMA models are

$$Y(t) = \sum_j \omega^j(B)X_t^j + N_t \tag{9.8}$$

where
- $\omega^j(B)$ = transfer function polynomial for the jth input series
- X^j = input series
- N_t = noise series described by a univariate ARMA model (Equation 9.6)

Novotny et al. (1990a) documented that stochastic transfer function models can successfully separate dry-weather- (sewage) and wet-weather-driven component flow and quality components in combined sewer systems. Capodaglio et al. (1990) provided a methodology for univariate, MISO, and SISO model identification. Novotny and Zheng (1989) also showed that the stochastic rainfall-runoff transfer function is theoretically identical to the deterministic unit hydrograph, and that the deterministic transfer function could and should be incorporated into the stochastic models.

Figure 9.16 shows a comparison of measured time series of BOD loads in combined sewers and one-step-ahead forecasts by stochastic ARMA-transfer function models. Stochastic models are more crude and can incorporate only a few input and system parameters. In a simplified way, they can have a *deterministic* core model as documented by Novotny and Zheng (1989) and Novotny et al. (1990a, 1990b). Stochastic models can only detect and incorporate the input–output relationships that are not overwhelmed by noise. However, the fact that the stochastic models can differentiate between the deterministic relationships and noise (which can be correlated or uncorrelated) makes them an attractive, unbiased tool for forecasting and control. Stochastic transfer function models are very useful for separation of dry-weather and wet-weather (rainfall-driven) quantity and quality components in combined sewers or storm sewers and surface channels with strong infiltration and/or base-flow inputs. Figure 9.17 shows a rainfall-runoff transfer function for a combined sewer outlet identified by stochastic modeling. The shape of the hydrograph contains faster (presumably surface runoff) and slower (possibly infiltration) components.

The stochastic time series models require for identification an uninterrupted time series in which data are collected at uniform time intervals. Techniques are available for substitutions of missing data.

Neural Network Models

Neural network models represent a new and very powerful modeling technology that, to our knowledge, has not yet been used for modeling urban (or nonurban) nonpoint pollution. They are basically multiple-input–multiple-output complex models that use parallel processing and learning to find a relationship between inputs and outputs (Novotny et al., 1990a). A neural network model consists of a large number of simple *processing elements* that are connected by one directional channels called *connections*. Each processing element can have multiple input signals, but produce only one output. The input signals in each processing unit are weighed by a transfer function that operates on the input signals, weighs them, and modifies the weights according to the strength of the stimulus by the input signal.

There are two operational modes of the neural network—*learning* and *recall*. During learning the system (neural network) adopts connection weights between processing units in response to stimuli being processed at the input and output. These stimuli, in the case of quality and a sewer flow, are the measured time series of, for example, treatment plant influent data or a time series measured by specialized surveys. Such time

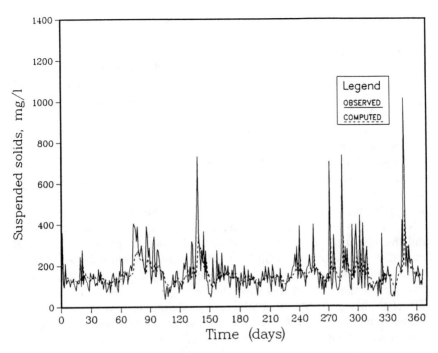

FIGURE 9.16. Comparison of measured BOD concentrations and corresponding one-step-ahead forecasts by an ARMA-transfer function model of a combined sewer system.

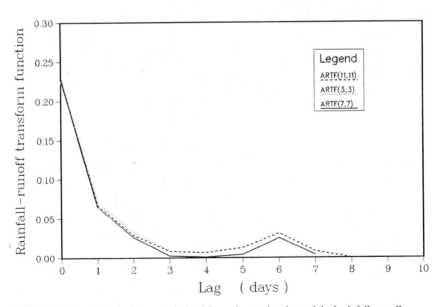

FIGURE 9.17. Unit hydrograph derived from the stochastic model of rainfall-runoff transformation for a combined sewer basin.

series do not have to be continuous as they would be for ARMA models. The input stimuli can be, for example, rainfall, atmospheric deposition (air-pollution levels), traffic density, and similar other parameters used for deterministic modeling. The network learns by implementing a learning rule that determines how the weights will adapt in response to the learning example. The learning process usually involves showing the network learning patterns (examples) many times. The network is said to have *learned* when the error between the measured output series and the output estimated by the network reaches some prescribed value. The type of learning described herein is called *supervised learning*, in which the input and output stimuli are presented to the system.

In the *recall* mode, weights inside the neural network model are frozen and the system produces output from input time series (such as rainfall and other typical inputs mentioned before). The system may be asked to periodically *relearn* the process any time the system parameters or the process has changed. In this way, the neural network is highly adaptive. Although being highly adaptable to modeling any system, including diffuse pollution, neural network models have not been used in this domain at the time of writing this book.

REAL-TIME CONTROL MODELS FOR URBAN SEWERAGE SYSTEMS

Real-time control (RTC) systems actually represent a set of models and concepts that are used for optimal and flexible control of urban sewerage systems (Béron et al., 1984; Schilling, 1989). In a typical sewerage system, especially in those involving combined sewers, there are numerous points that are "bottlenecks." The downstream capacity of the sewer system or the capacity of the treatment plant are potential bottlenecks.

The invariant steady-state model of operation typical for most sewerage systems dictates that when the bottleneck capacity is exceeded, the excess flow is discharged into the receiving body of water, and hence, overflows. The steady-state solution to the problem is to enlarge the bottleneck. This mode of operation is highly inefficient and results, on the one hand, in large quantities of sewage being discharged without treatment into the receiving bodies of water while, on the other hand, there may be unused capacity downstream. A conversion of the system into a dynamic RTC operation may be more economical than conventional structural enlargement of the bottleneck.

In a dynamic RTC scheme, the excesses are temporarily stored in an auxiliary or in-line storage, and the stored flow is then manipulated either by later release for treatment when the flow subsides below the capacity of the bottleneck, or by a side treatment and release into the receiving body of water if it is permitted by the receiving water standards. Commonly, the waste-assimilative capacity of receiving bodies of water may be greater during wet-weather flows due to higher dilution.

In a real-time control system (RTCS) the following elements apply (Schilling, 1989):

- A (measurement) sensor that is used to monitor an on-going process, for example, water-level gauge, turbidity meter
- A (corrective) regulator that manipulates the process, for example, gates and valves
- A controller than causes the regulator to bring the process back to its desired value (set point)
- A communication system that carries the measured data from the sensor to the controller and the signals of the controller back to the regulator, for example, a telemetry system

The RTC systems do not a priori imply automation. As a matter of fact, significant and substantial operator interface and override are desirable. In either case, computers equipped with RTC software are an integral part of the RTC system.

In the context of RTC, the meaning of "system" is the controllable part of the urban drainage system. A real-time control can be planned and implemented if numerical simulation models (deterministic, stochastic, neural network, etc.) are available and included in the RTC system. The following processes have to be modeled in order to obtain a comprehensive overview of the performance of the system:

- The input to the system: deterministic and/or stochastic models of rainfall-runoff sewer flow and quality models will provide short-term forecasts of the inputs (Labadie, Lazaro, and Morrow, 1981)
- The system response to the input: transfer function models will provide pollutant loads, flows, and water levels at the key bottleneck points of the system
- The total output to the environment: all partial outputs will be summarized in order to minimize, in an optimal way, the total (pollution) output from the system
- The response of the environment to the output from the system: the total output is compared with the waste-assimilative capacity; further restrictions are imposed on the system if the waste-assimilative capacity is exceeded

In a typical application, the input to an urban drainage (nonpoint) system is measured or predicted rainfall. Radar measurements have been considered for rainfall forecasting. Other inputs, such as traffic density and atmospheric deposition, may also be considered. The input is then transformed by a deterministic or stochastic model (similar to those discussed in the preceding sections) into input parameters, such as predicted flows and pollutant loads. Chapter 10 discusses various strategies employed in RTC systems.

CALIBRATION AND VERIFICATION OF DETERMINISTIC MODELS

Due to the fact that stochastic models are identified from measured time series by an unbiased multiple-regression analysis, further calibration and verification of the models is not needed as long as the series of residuals (a series of one-step-ahead forecasts minus corresponding measured values) is an uncorrelated zero-mean series—the "white noise"—which is the fundamental condition in the identification and design of stochastic models. Similarly, the neural network models are developed from measured input–output multiple time series; however, the white noise residual series is not inherently guaranteed.

Calibration and verification do not make sense for simple unit load models and procedures. A majority of deterministic hydrologic models require calibration and verification of the model parameters during the design and model identification process. Many mathematical equations and formulas used in the deterministic flow-quality models are of an empirical or semiempirical nature, requiring knowledge of a large number of coefficients and reaction rates. Although the model manuals often provide guidelines for a rough estimation of the most important parameters, the ranges are commonly quite wide. On average, twenty or more coefficients must be input for each subunit of a hydrological diffuse-pollution loading model.

As an example consider soil permeability and surface storage, two very simple hydrological parameters. Although the magnitude of surface runoff is very sensitive to these parameters, their exact measurement is tedious and they are not uniform even on very small (less than one hectare) watershed segments. The permeability ranges reported in the U.S. Soil Conservation Service soil maps are broad; for example, silt loam soils are listed as having permeability between 1.6 and 5 cm/hr. In addition, no adequate physical method is available to measure surface storage. Under these circumstances, the only way to arrive at a set of coefficients adequately describing the watershed is by calibrating the model against measured data. Usually, for calibration and verification of event-oriented models (level IV) hydrographs and pollutographs of a few representative storms should be available. For calibration and verification of dynamic continuous models (Level III and especially, Level V, models), continuous long-term series may be also needed.

Calibration means varying the coefficients of the designed model within acceptable ranges until a satisfactory agreement between measured and computed output values is achieved. Trying to adjust all twenty or more coefficients for each land-use segment at the same time is tedious and often impossible. For diffuse-pollution hydrologic models, the hydrologic components must be calibrated first, followed by erosion, and finally the pollutant component. Also, all models are more sensitive to a few important variables and less to others. As a rule of thumb, the variables to which the model is most sensitive should be calibrated first, with the other coefficients kept at their optimal or average values. The hydrologic components of diffuse-pollution models are sensitive to the magnitude of surface storage and soil permeability, which determine the runoff volume. Surface roughness affects the magnitude and time location of runoff peaks. Other variables—slope, soil moisture characteristics—are less important.

In the erosion component, the factor to which soil loss is most responsive is the vegetative factor, C. For example, a typical range of the cover factor for grassed areas is between 0.01 and 0.03, a 300% difference. Uncertainty about the sediment delivery will complicate the calibration. Adsorption characteristics and attenuation rates are most important for the pollutant submodel.

Since calibration is a subjective process requiring experience, Delleur and Baffaut (1990) developed an expert system for calibrating the Storm Water Management Model, and reported improved adequacy of the model.

Once the model has been calibrated, that is, once a satisfactory fit of computed and simulated data has been achieved for one calibration storm or time series, the model must be verified. *Verification* is accomplished by running the model with the coefficients established during calibration and with inputs corresponding to another (verification) storm and/or time series. If a satisfactory fit of computed and measured data is obtained, calibration and verification are accomplished. Very often this is not the case, and calibration and verification must be repeated until a satisfactory fit is found. For some locations, data for calibration and verification of hydrological urban runoff models can be retrieved from data banks such as the EPA's STORET (Huber et al., 1979, 1982).

Caveats. Often the model's structure may have an excessive number of "degrees of freedom." For example, sediment load can be changed by varying three parameters of the universal soil loss equation, slope, and the delivery ratio. There is no methodology available to distinguish between the parameters, yet selection of the wrong magnitude of the parameter can lead to erroneous conclusions on management alternatives. Therefore, the modeler must be experienced and must be familiar with the watershed. A reconnaissance tour of the watershed by the modeler, who should take photographs and notes, should always be a part of the model design, calibration, and verification process.

MONITORING AND DATA ACQUISITION

Data are collected routinely by agencies for a multitude of purposes. Very often these data bases are suitable for the design and identification of models. For example, sewerage agencies and industrial wastewater sources must routinely and periodically, often daily, collect information on flow and various quality parameters in sewers, including storm sewers. Soil data have been collected throughout the United States for the preparation of soil maps, meteorological data are collected by the National Oceanic and Atmospheric Administration (NOAA), and flow quality data in streams is available from the U.S. Geological Survey, which has offices in every state. Special surveys and monitoring, however, are almost always needed for calibration, and verification of deterministic complex models. This type of data acquisition is different from the routinely performed monitoring by agencies that require preliminary design and a carefully executed monitoring program. Specially designed monitoring programs are also needed for establishing unit loads, quantity and quality characteristics of diffuse pollution, identification of stochastic and regression (statistical) models, and for research of the processes involved in the generation and transmission of diffuse pollution.

The data needed for design (identification), calibration, and verification of hydrologic models can be divided into the following groups, which includes an approximate list of monitored parameters and variables:

1. System parameters
   a. watershed size
   b. subdivision of the watershed into homogenous subsegments
   c. imperviousness of each subsegment
   d. slopes
   e. fraction of impervious areas directly connected to a drainage channel or sewer
   f. maximum surface storage (depression plus interception storage)
   g. soil characteristics, including texture, permeability, erodibility, and composition

h. crop and vegetation cover
i. curb density or street gutter length
j. sewer system or natural drainage characteristics
2. State variables
    a. ambient temperature
    b. reaction rate coefficients
    c. adsorption–desorption coefficients
    d. daily accumulation rates of litter
    e. traffic density and speed
    f. potency factors for pollutants (concentration of pollutants on street dust and dirt and on soil)
    g. solar radiation (for some models)
    h. growth stage of crops
3. Initialization parameters
    a. length of antecedent dry period
    b. initial soil moisture
    c. depth of snowpack
    d. initial concentration of pollutants in soil
4. Input variables
    a. precipitation
    b. atmospheric fallout
    c. evaporation rates
5. Calibration and verification parameters
    a. storm-water and receiving water flows (hydrographs)
    b. water quality parameters (quality histograms)
    c. biological assessment of receiving waters (toxicity impact on biota)

Sources of Data

Walesh (1989) divided the data for planning studies into three groups: (1) completed or ongoing studies; (2) natural resources data; and (3) infrastructure data. In the first category, data for the design of diffuse-pollution models and for planning can be found in the past studies carried out under the mandate of the Section 208 of the 1972 Clean Water Act (208 studies), data obtained by the National Urban Runoff Project by the U.S. Environmental Protection Agency, Flood Plain Management studies by various planning agencies, and numerous other studies. Such data may contain calibration and verification surveys for runoff quantity–quality modeling.

Data related to land and land use can be obtained from maps and/or aerial photographs and remote sensing. The data obtainable by high plane or satellite imagery can be digitized in order to provide more specific information on the degree of imperviousness, vegetation cover, surface roughness, and soil moisture. Soil data are available from U.S. Soil Conservation Service (SCS) maps. These maps provide information on soil type, texture, slope, approximate ground-water table, permeability, erodibility, soil profile, and other valuable information. An example of the map was shown in Figure 6.2. Land-use data can be obtained from the U.S. Bureau of Land Management, U.S. Census Bureau, and from local and regional planning agencies.

Hydrologic and basic water quality information is collected and published regularly by the U.S. Environmental Protection Agency and by the U.S. Geological Survey. Both agencies have computerized systems for storing and retrieving information on water quality and quantity. Information on dust and dirt accumulation and traffic densities in some urban areas is usually available from the local city engineer's office, sanitation department, and air-pollution control agencies. Meteorological data are routinely measured by NOAA.

It must be emphasized that the format and frequency of data obtainable from various agencies (with the exception of meteorological data) usually do not conform to modeling requirements, and often significant time and financial resources must be spent to get available information into conformity with the requirements of the models. Additional in situ surveys and monitoring are necessary to gather missing information. Expenses associated with data-collection activities often exceed the expenses for setting up and running the model.

Monitoring programs must also be set up to comply with the regulatory requirements of the National Pollution Discharge Elimination Systems (NPDES) permit system. Both combined and separate storm sewer systems in larger urban areas are covered by the NPDES permitting system, which requires monitoring both flows and quality and long-term load estimations by modeling.

Field Monitoring

A typical monitoring station has the following components:

1. Rain gauge
2. Wet- and dry-atmospheric deposition collector
3. Flow monitoring device
4. Quality monitoring device
5. Power source
6. Telecommunication link or data recorder

Rain Gauges and Atmospheric Deposition Monitoring

Rain gauges measure rainfall. Typically the rain gauge is activated when rainfall depth is more than 0.25 mm (0.01 in.). Smaller rainfalls are recorded as *trace precipitation*. Many rain gauges signal each 0.25-mm occurrence of rainfall.

Wet and dry deposition samplers are essentially two collecting containers. One is always open and collects the bulk atmospheric deposition (wet and dry); the other is covered, but the lid opens when a moisture sensor is activated by precipitation. This sampler collects only wet precipitation. The dry deposition is thus the difference between the bulk and wet depositions.

Flow Measurements

Typical monitoring programs include monitoring of flow and its quality. In most cases, measuring flow involves measuring the depth (stage) in the channel or sewer that is then converted to flow using the so-called rating curve for the section where the flow stage is measured.

In the most simple case, the flow stage is measured manually or by a recording float device. In order to avoid backwatering effects, the flow should be impounded. This is best accomplished by standard sharp-crested weirs (triangular, rectangular, trapezoidal), as shown on Figure 9.18, or by narrowing the cross section, as is done in a standard Parshall flume. Rating curves (flow-stage relationships) are available for most standard weirs and flow-measuring flumes. These rating curves must be checked by several in situ flow measurements using simple devices, such as a bucket and stopwatch, or more complex flow-measuring devices. Periodic field checks and manual flow measurements are always necessary and are integral to any monitoring programs.

Velocity sensors can help to alleviate the problem with backwatering effects that could greatly affect the flow-stage relationship of the rating curve. Many newer monitors incorporate both depth and velocity measuring in a single microprocessor that can be telemetrically interrogated. Such sensors must be calibrated by manual measurements at the time of installation. It is important that the point velocity sensor is placed at the right depth, that is, one that corresponds to the average cross-sectional velocity.

Telemetry rather than recording is becoming more common and economical today. Any site that requires permanent NPDES monitoring should consider telemetered monitors. Once installed and calibrated, a telemetered flow monitor should require much less labor to maintain than a portable monitor. On the other hand, a portable manually operated

560   Modeling and Monitoring Diffuse Pollution

FIGURE 9.18. Typical monitoring station for a small watershed used for gathering calibration and verification data and/or for storm-water monitoring. (Photo: V. Novotny.)

monitor is more economical for occasional monitoring surveys, such as those required for the calibration of event-oriented models.

In storm-water monitoring stations the flow meters are commonly interconnected with the sampler, hence, they activate sampling during episodes of runoff. They also send a signal to the quality sampler when a sample is to be collected. For example, the flow meter can signal the sampler each time $50\,m^3$ of flow has passed by the flowmeter. In this way a flow-weighted composite sample can be collected.

Quality Monitoring

Quality monitors are devices that periodically withdraw a small volume of water (or wastewater) from the stream at the monitoring site. This withdrawn volume is then analyzed for various constituents. On-site analyses are typically done for only a few constituents, generally those that can be measured by electrochemical probes such as dissolved oxygen, turbidity, temperature, pH probes, and a few others. This information can be telemetrically transmitted. For most other constituents, the samples collected by the sampler must be transported to a laboratory and analyzed therein. Sample preservation between the time of the withdrawal from the channel and the time of analysis must be implemented. Many automatic samplers use refrigeration to preserve the sample.

There are basically two types of sampling. *Grab samples*, collected periodically at predetermined intervals and analyzed separately, can document the extent, frequency, and general variability of quality. The sampling interval should be selected according to the rate of quality change or frequency of quality variation. It may range from intervals as short as 5 minutes for some flush storm runoff events and CSOs to as long as 1 hour or more. Daily or even less frequent grab sampling is unreliable and can be only used for monitoring slowly changing streams, such as river flows. When the source composition varies in space, as it is in a typical urban drainage system, sampling should be done at several appropriate locations. Most of the runoff of storm sewer monitoring stations are flow activated, hence, no sampling is conducted when the flow in the conduit is below a prespecified minimum level.

A *composite sample* is a mixture of grab samples or continuously withdrawn flow. The mixing of the volumes withdrawn from the flow is either based on a constant volume–constant time (time composite) or is flow proportional. *Time composite* samples provide the average concentration of the constituent during the time of composition irrespective of flow, or

$$\hat{C} = \Sigma C_i / n$$

where
$\hat{C}$ = time-averaged composite concentration
C_i = individual concentrations of grab samples
n = number of samples

A *flow proportional composite sample* is generated by specifically designed automatic monitoring equipment that measures flow and activates grab sampling when prespecified flow volume passes through the flow-measuring device. Then the concentration of the composite sample can be represented by

$$\hat{C} = \Sigma C_i \times \Delta Q / (n \times \Delta Q)$$

The flow proportional composite sample is representative of the average mass concentration for the sampled runoff event.

There are certain common sense rules related to monitoring. Inadequate frequency of data acquisition and incomplete monitoring may be useless, and high-frequency monitoring and sampling for many constituents may be costly and will create a backlog of unusable data.

High-frequency grab sampling is used for monitoring flush runoff flow events. Often the period of data acquisition is in minutes; however, high-frequency sampling precludes an extensive variety of monitored constituents. Low-frequency composite sampling is typical for between the event sampling that is to characterize infiltration and base-flow contributions. Low-frequency grab sampling (for example, daily, weekly, or monthly intervals) may not provide representative results and should be avoided (essentially, such sampling may be a waste of funds).

High-frequency or continuous monitoring is also feasible for constituents that can be reliably monitored by electronic or electrochemical probes. Such constituents include pH, temperature, dissolved oxygen concentrations, conductivity (which is related to salt or dissolved solids contents), and turbidity (which can be correlated to suspended solids content). Relatively inexpensive in situ or in-laboratory analyses include organic contents expressed as the chemical oxygen demand (COD), total organic carbon (TOC), total suspended and total dissolved solids (TSS, TDS), volatile suspended and dissolved solids, nitrates, and metals. These analyses can still be performed on grab samples, both filtered and unfiltered. A filtered sample represents dissolved contamination, while the difference between the total and filtered concentrations of the constituent reflects the contamination of the suspended sediment. Analyses for organic chemicals (pesticides, polyaromatic hydrocarbons (PAHs), PCBs, etc.), bacteria, oil and grease, nitrogen compounds other than nitrate (total Kjeldahl nitrogen, ammonia), phosphates, and asbestos are more demanding and costly and are typically performed only on composite samples. In many cases, a few surrogate constituents are selected instead of analyzing a complete spectrum of constituents. For example, in a group of toxic metals, lead and/or zinc and/or cadmium are often selected to indicate pollution by toxic metals. Instead of analyzing for every possible pathogenic microorganism, total and fecal coliform bacteria analyses are commonly used to indicate bacterial and viral contamination. COD or TOC are used in high-frequency grab sampling programs as indicators of pollution by organics. Volatile suspended solids indicate organic contamination of sediments, as does a difference between the total and filtered COD.

Event-oriented models are calibrated and verified using detailed grab sampling over several storm runoff hydrographs. Care should be taken that the grab samples are distributed more-or-less evenly over the rising and receding portions of the hydrograph. Continuous long-term models can be calibrated using daily (wet-weather) composite sampling. However, long-term sampling must include both work days and weekend (holiday) data.

The guidelines for the NPDES permit for urban storm water require that monitoring data programs provide information on the pollution content of urban runoff, including the first flush portion of runoff. This may require grab sampling programs at various locations of the urban storm drainage system. The new sampling devices available on the market today allow for programmed sample collection in order to characterize separately the more polluted first flush and less polluted tail portion of a storm event hydrograph.

Computer Software. The monitoring station can also be equipped with internal memory and a connection to a computer. The data of a storm event can be printed on the internal chart during the storm event, or stored in the internal memory of the flowmeter for later retrieval by a computer. An internal modem enables convenient retrieval by a remote computer over telephone lines, or a laptop computer can be used for on-site collection of data.

Log-Normal Distribution of Data

Event mean concentration (EMC) and site mean concentration (SMC) are calculated or graphically estimated from monitored grab (within event) or composite (long-term monitoring) data by fitting the data to a log-normal probability distribution function. The mathematical form of the function has been presented in many statistical handbooks, and the concept was also shown in the preceding chapter (Figs. 8.17 and 8.18).

Log-normal distribution has been used and suggested to characterize diffuse-pollution runoff at a sampling site for a large number of samples (Driscoll, 1986). This distribution has a number of important benefits (Woodward-Clyde Consultants, 1989):

- Concise summaries of highly variable data can be developed, and the variability can be identified and dealt with appropriately.
- Comparison of results from different sites, events, etc., are convenient and more easily understood.
- Statements can be made about frequency of occurrence, that is, one can express how often values will exceed various magnitudes of interest.
- A more useful and informative method for reporting data than the use of ranges is provided, one that is less subject to misinterpretations.
- A framework is provided for examining the transferability of data in quantitative manner.

Since most of the deterministic models inherently neglect the statistical variability of the data, statistical evaluation of the monitored data must precede the calibration and verification process in order to minimize subjectivity.

References

Alley, W. M., and P. E. Smith (1982). *Multi-event Urban Runoff Quality Model*, U.S. Geological Survey Open File Rep. B2-764, U.S. Geological Survey, Reston, VA.

American Public Works Association (1969). *Water Pollution Aspects of Urban Runoff*, WP-20-15, U.S. Department of Interior, FWPCA (present EPA). Washington, DC.

Athayde, D. N., C. F. Myers, and P. Tobin (1986). EPA's perspectives of urban nonpoint pollution, in *Proceedings, Urban Runoff Quality-Impact and Quality Enhancement Technology*, Eng. Foundation Conf., American Society of Civil Engineers, pp. 217-225.

Balmér, P., P. A. Malmqvist, and A. Sjöberg, eds. (1984). *Proceedings, Third International Conference on Urban Storm Drainage*, 4 vol., Chalmers University of Technology, Göteborg, Sweden, 1669p.

Bannerman, R., et al. (1984). *Evaluation of Urban Nonpoint Source Pollution Management in Milwaukee County, Wisconsin. Vol. I. Urban Stormwater Characteristics, Pollutant Sources and Management by Street Sweeping*, U.S. Environmental Protection Agency, Region V, Chicago, IL.

Barnes, J. W., and F. A. Rowe (1978). Modeling sewer flows using time series analysis, *J. Environ. Eng. Div. ASCE* **109**(EE4):639-646.

Barnwell, T. O., and J. J. Kittle (1984). Hydrologic simulation program—FORTRAN: Development, maintenance and applications, in *Proceedings, Third International Conference on Urban Storm Drainage*, Vol. 2, Chalmers University of Technology, Göteborg, Sweden, pp. 493-502.

Barnwell, T. O., Jr., and P. A. Krenkel (1982). The use of water quality models in management decision making, *Water Sci. Tech.* **14**:1095-1107.

Beasley, D. B. (1986). Distributed parameter hydrologic and water quality modeling, in *Agricultural Nonpoint Source Pollution: Model Selection and Application*, A. Giorgini and F. Zingales, eds., Elsevier, Amsterdam.

Beasley, D. B., and L. F. Huggins (1981). *ANSWERS Users Manual*, EPA 905/9-82-001, U.S. Environmental Protection Agency, Chicago, IL.

Béron, P., F. Brière, J. Rouselle, and J. P. Riley (1984). An evaluation of some real-time techniques for controlling combined sewer overflows, in *Proceedings, Third International Conference on Urban Storm Drainage*, Vol. 3, Chalmers University of Technology, Göteborg, Sweden, pp. 1093-1097.

Berthouex, P. M., et al. (1975). Modeling sewage treatment plant input BOD data, *J. Environ. Eng. Div. ASCE* **101**(EE1):127-138.

Berthouex, P. M., et al. (1976). The use of stochastic models in the interpretation of historical data from sewage treatment plants, *Water Res.* **10**:689-698.

Berthouex, P. M., et al. (1978). Dynamic behavior of an activated sludge plant, *Water Res.* **12**:957-972.

Bliss, N. B., and W. U. Reybold (1989). Small-scale digital soil maps for interpreting natural resources, *J. Soil Water Conserv.* **47**(1):30–34.
Box, G. E. P., and G. M. Jenkins (1976). *Time Series Analysis, Forecasting and Control*, Holden Day, Oakland, CA.
Brutsaert, W. F. (1975). Water quality modeling by Monte Carlo simulation, *Water Resour. Bull.* **11**(2):229–236.
Burrough, P. A. (1986). *Principles of Geographical Information Systems for Land Resources Assessment*, Clarendon Press, Oxford.
Capodaglio, A. G., S. Zheng, V. Novotny, and X. Feng (1990). Stochastic system identification of sewer flow models, *J. Environ. Eng. ASCE* **116**(2):284–298.
Carsel, R. F., et al. (1984). *User's Manual for the Pesticide Root Zone Model (PRZM): Release 1*, EPA 600/3-84-109, U.S. Environmental Protection Agency, Athens, GA.
Chesters, G., et al. (1978). *Pilot Watershed Studies. Summary Report*, International Joint Commission, Windsor, Ontario.
Chow, V. T. (1972). Hydrologic modeling, *J. Boston Soc. Civ. Eng.* **60**:1–27.
Chow, V. T., D. R. Maidment, and L. W. Mays (1988). *Applied Hydrology*, McGraw-Hill, New York.
Cline, T. J., A. Molinas, and P. Y. Julien (1989). An auto-CAD-based watershed information system for the hydrologic model HEC-1, *Water Resour. Bull.* **25**(3):641–652.
Connors-Sasowsky, K., and T. W. Gardner (1991). Watershed configuration and Geographic Information System parametrization for SPUR model hydrologic simulation, *Water Resour. Bull.* **27**(1):7–18.
Dean, J. D., et al. (1989). *Risk of Unsaturated/Saturated Transport and Transformation of Chemical Concentrations (RUSTIC)*, EPA 600/3-89/048a, U.S. Environmental Protection Agency, Athens, GA.
Delleur, J. W., and C. Baffaut (1990). An expert system for urban runoff quality modeling, in *Proceedings, Fifth International Conference on Urban Storm Drainage, Suita-Osaka*, Y. Iwasa and T. Sueishi, eds., University of Osaka, Japan, pp. 1323–1328.
DeRoo, A. P. J., L. Hazelhoff, and P. A. Burrough (1989). Soil erosion modeling using 'ANSWERS' and Geographical Information Systems, *Earth Surf. Process. Landf.* **14**:517–532.
Dever, R. J., Jr., and L. A. Roesner (1982). Development and application of a dynamic urban highway drainage model, in *Proceedings, Second International Conference on Urban Storm Drainage*, Vol. I. *Hydraulics and Hydrology*, B. C. Yen, ed., Water Resources Publ., Littleton, CO, pp. 229–235.
Dever, R. J., Jr., and L. A. Roesner (1983). *Urban Highway Storm Drainage Model*. Vol. 5. *Drainage Design Program User's Manual and Documentations* (Final Report Sep. 76–Mar. 81), Camp, Dresser, and McKee, Inc., Annandale, VA, FHWA/RD-83-045, Federal Highway Administration, Washington, DC, 117p.
Dever, R. J., Jr., J. A. Aldrich, and W. M. Williams (1983). *Urban Highway Storm Drainage Model*. Vol. 1. *Model Development and Test Application* (Final Report Sep.–Mar. 81), Camp, Dresser, and McKee, Inc., Annandale,

VA, FHWA/RD-83/041, Federal Highway Administration, Washington, DC, 71p.

Dickerhoff, L. L., and D. A. Haith (1983). Loading functions for predicting nutrient losses from complex watersheds, *Water Resour. Bull.* **19**(6):951–959.

DiToro, D. M. (1984). Probability model of stream quality due to runoff, *J. Environ. Eng. Div. ASCE* **110**(3):617–628.

DiToro, D. M., and M. J. Small (1979). Stormwater interception and storage, *J. Environ. Eng. Div. ASCE* **105**(EE1):43–54.

Donigian, A. S., and H. H. Davis (1978). *User's Manual for Agricultural Runoff Management (ARM) Model*, EPA 600/3-78-080, U.S. Environmental Protection Agency, Athens, GA.

Donigian, A. S., and J. D. Dean (1985). Nonpoint source pollution models for chemicals, in *Environmental Exposure from Chemicals*, Vol. II, CRC Press, Boca Raton, FL, pp. 75–105.

Donigian, A. S., and W. C. Huber (1990). *Modeling Nonpoint Source Water Quality in Urban and Non-urban Areas*, EPA 68-03-3513, Environmental Research Laboratory, U.S. Environmental Protection Agency, Athens, GA.

Driscoll, E. D. (1986). Lognormality of point and nonpoint source pollutant concentrations, in *Urban Runoff Quality—Impact and Quality Enhancement Technology*, Proceedings, Engineering Foundation Conference, American Society of Civil Engineers, pp. 438–458.

Driscoll, E. D., D. M. DiToro, and R. V. Thomann (1979). *A Statistical Method for Assessment of Urban Runoff*, EPA 440/3-79-023, U.S. Environmental Protection Agency, Washington, DC.

Driver, N. E. (1990). Summary of nationwide analyses of storm-runoff quality and quantity in urban watersheds, in *Proceedings, Fifth International Conference on Urban Drainage, Suita-Osaka*, Y. Iwasa and T. Sueishi, eds., University of Osaka, Japan.

Driver, N. E., and D. J. Lystrom (1987). Estimation of urban storm-runoff loads and volumes in the United States, in *Urban Storm Water Quality, Planning and Management*, Proceedings, Fourth International Conference on Urban Storm Drainage, W. Gujer and V. Krejci, eds., Ecole Polytechnique Federale, Lausanne, Switzerland, pp. 214–220.

Driver, N. E., and G. D. Tasker (1988). *Techniques for Estimation of Storm-Runoff Loads, Volumes, and Selected Constituents in Urban Watersheds in the United States*, U.S. Geological Survey Open File Rep. 88-191, Denver, CO.

Evans, B. M., and D. A. Miller (1988). Modeling non-point pollution at the watershed level with the aid of a Geographic Information System, in *Nonpoint Pollution 1988—Policy, Economy, Management, and Appropriate Technology*, American Water Resources Association, Bethesda, MD, pp. 283–291.

Foster, G. R., and L. J. Lane (1987). *User Requirements: USDA—Water Erosion Prediction Project (WEPP)*, NSERL Rep. No. 1, USDA-ARS, W. Lafayette, IN.

Giorgini, A., and F. Zingales (1986). *Agricultural Nonpoint Source Pollution: Model Selection and Application*, Elsevier, Amsterdam.

Gujer, W., and V. Krejci (1987). *Topics in Urban Storm Water Quality, Planning and Management*, in Proceedings, Fourth International Conference on Urban Storm Drainage, W. Gujer and V. Krejci, eds., IAHR-IAWPRC, Ecole Polytechnique Federale, Lausanne, Switzerland.

Haan, C. T., H. P. Johnson, and D. L. Brakensiek, eds. (1982). *Hydrologic Modeling of Small Watersheds*, Monogr. No. 5, American Society of Agricultural Engineers, St. Joseph, MI.

Haith, D. A., and L. L. Shoemaker (1987). Generalized watershed loading functions for stream flow nutrients, *Water Resour. Bull.* **23**(3):471–478.

Haith, D. A., and L. J. Tubbs (1981). Watershed loading functions for nonpoint sources, *J. Environ. Eng. Div. ASCE* **107**(EE1):121–137.

Henderson, R. J., and G. D. Moys (1987). Development of a sewer flow quality model for the United Kingdom, in *Urban Storm Water Quality, Planning and Management*, Proceedings, Fourth International Conference on Urban Storm Drainage, Ecole Polytechnique Federale, Lausanne, Switzerland, pp. 201–207.

Huber, W. C. (1986). Deterministic modeling of urban runoff quality, in *Urban Runoff Pollution*, H. C. Torno, J. Marsalek, and M. Desbordes, eds., Springer Verlag, Berlin, New York, pp. 167–244.

Huber, W. C. (1990). Current methods for modeling of nonpoint source water quality, in *Proceedings, Seminar on Urban Storm Water Quality Management*, American Society of Civil Engineers, Orlando, FL.

Huber, W. C., and R. C. Dickinson (1988). *Storm Water Management Model. User's Manual*, Version 4, EPA 600/3-88/001a (NTIS PB88-236641/AS), U.S. Environmental Protection Agency, Athens, GA.

Huber, W. C., et al. (1979). *Urban Rainfall-runoff Quality Data Base: Update with Statistical Analysis*, EPA 600/8-79-004, U.S. Environmental Protection Agency, Cincinnati, OH.

Huber, W. C., et al. (1982). *Urban Rainfall-runoff-Quality Data Base*, EPA 600/S2-81-238, U.S. Environmental Protection Agency, Cincinnati, OH.

Hydrologic Engineering Center (1975). *Storage, Treatment, Overflow, Runoff Model, STORM, Users Manual*, Generalized Computer Program 723-S8-L7520, Hydr. Eng. Center, U.S. Army Corps of Engineers, Davis, CA.

Iwasa, Y., and T. Sueishi (1990). *Drainage Models and Quality Issues*, in Proceedings, Fifth International Conference on Urban Storm Drainage, IAHR-IAWPRC, Suita-Osaka, University of Osaka, Japan.

Johanson, R. C., J. C. Imhoff, J. L. Kittle, and A. S. Donigian (1984). *Hydrologic Simulation Program—Fortran (HSPF): User's Manual*, Release 8, EPA 600/3-84-006, U.S. Environmental Protection Agency, Athens, GA.

Johnson, M. G., et al. (1978). *Management Information Base and Overview Modeling*, International Joint Commission, Windsor, Ontario.

Knisel, W. G. (1980). *CREAMS: A Field Scale Model for Chemicals, Runoff, and Erosion from Agricultural Management Systems*, Conservation Research Rep. No. 26, U.S. Department of Agriculture, Washington, DC.

Knisel, W. G. (1985). Use of computer models in managing nonpoint pollution from agriculture, in *Proceedings, Non-point Pollution Abatement Symposium*, V. Novotny, ed., Marquette University, Milwaukee, WI.

Labadie, J. W., R. C. Lazaro, and D. M. Morrow (1981). Worth of short-term rainfall forecasting for combined sewer overflow control, *Water Resour. Res.* **17**(5):1489–1497.

Laflen, J. M., L. J. Lane, and G. R. Foster (1991). WEPP—A new generation of erosion prediction technology, *J. Soil Water Conserv.* **46**(1):34–38.

Lane, L. J., and M. A. Nearing, eds. (1989). *USDA—Water Erosion Prediction Project: Hillslope Profile Model Documentation*, NSERL Rep. No. 2, USDA-ARS National Soil Erosion Research Laboratory, West Lafayette, IN.

Leonard, R. A., and W. G. Knisel (1986). Selection and application of models for nonpoint source pollution and resource conservation, in *Agricultural Nonpoint Source Pollution: Model Selection and Application*, A. Giorgini and F. Zingales, eds., Elsevier, Amsterdam.

Leonard, R. A., W. G. Knisel, and D. A. Still (1987). GLEAMS: Groundwater Loading Effects of Agricultural Management Systems, *Transact. ASAE* **30**(5): 1403–1418.

Litwin Y., J. A. Lager, and W. G. Smith (1981). *Areawide Stormwater Pollution Analysis with the Macroscopic Planning (ABMAC) Model*, EPA 600/S2-81-223, U.S. Environmental Protection Agency, Cincinnati, OH.

Maahel, K., and W. C. Huber (1984). SWMM calibration using continuous and multiple event simulation, in *Proceedings, Third International Conference on Urban Storm Drainage*, Vol. 2, Chalmers University of Technology, Göteborg, Sweden, pp. 595–604.

McElroy, A. D., et al. (1976). *Loading Functions for Assessment of Water Pollution from Nonpoint Sources*, EPA 600/2-76-151, U.S. Environmental Protection Agency, Washington, DC.

Milligan, J. D., I. E. Wallace, and R. P. Betson (1984). *The Relationship of Urban Runoff to Land Use and Groundwater Resource, Knoxville, Tennessee 1981–1982*, Office of Natural Resources and Economic Development, Tennessee Valley Authority, Chattanooga, TN.

Mills, W. B., et al. (1982). *Water Quality Assessment: A Screening Procedure for Toxic and Conventional Pollutants*, EPA 600/6-82-004a and b, U.S. Environmental Protection Agency, Athens, GA.

Mills, W. B., et al. (1985). *Water Quality Assessment: A Screening Procedure for Toxic and Conventional Pollutants* (revised 1985), EPA 600/6-85-002a and b, U.S. Environmental Protection Agency, Athens, GA.

Moffa, P. E., ed. (1990). *Control and Treatment of Combined-Sewer Overflows*, Van Nostrand Reinhold, New York.

Muelkey, L. A., R. B. Amrose, Jr., and T. O. Barnwell, Jr. (1986). Aquatic fate and transport modeling techniques for predicting environmental exposure to organic pesticide and other toxicants—A comparative study, in *Urban Runoff Pollution*, H. C. Torno, J. Marsalek, and M. Desbordes, eds., Springer Verlag, Berlin, New York, pp. 463–496.

Muelkey, L. A., R. F. Carsel, and C. N. Smith (1986). Development, testing and application of nonpoint source models for evaluation of pesticide risk to the environment, in *Agricultural Nonpoint Source Pollution: Model Selection and*

Application, A. Giorgini and F. Zingales, eds., Elsevier, Amsterdam, pp. 383–398.

Nicks, A. D., V. L. Lopes, M. A. Nearing, and L. J. Lane (1989). Overview of WEPP hillslope profile erosion model, in *USDA-Water Erosion Prediction Project: Hillslope Profile Model Documentation*, L. J. Lane and M. A. Nearing, eds., NSERL Rep. No. 2, USDA-ARS National Soil Erosion Research Laboratory, W. Lafayette, IN.

Novotny, V. (1985). Role of mathematical models in design and selection of best management practices for control of pollution from urban and urbanizing areas, in *Proceedings, Non-point Pollution Abatement Symposium*, V. Novotny, ed., Marquette University, Milwaukee, WI.

Novotny, V. (1986). A review of hydrologic and water quality models used for simulation of agricultural pollution, in *Agricultural Nonpoint Source Pollution: Model Selection and Application*, A. Giorgini and F. Zingales, eds., Elsevier, Amsterdam, pp. 9–35.

Novotny, V. (1988). Modeling of sewer flow quality, *Proceedings, Bilateral Conference US-Italy Urban Stormwater Management*, University of Cagliary, Italy, Water Resources Publications, Littleton, CO.

Novotny, V., and R. Bannerman (1980). Model enhanced unit loadings of pollutants from nonpoint sources, in *Proceedings, Hydraulic Transport Modeling Symposium*, Publ. No. 4–80, American Society of Agricultural Engineers, St. Joseph, MI.

Novotny, V., and A. Capodaglio (1991). *Model of Urban Runoff and Sewer Flow (MOURSEF)*, AquaNova International, Mequon, WI.

Novotny, V., and S. Zheng (1989). Rainfall-runoff transfer function by ARMA modeling, *J. Hydraul. Eng. ASCE* **115**(10):1386–1400.

Novotny, V., S. Zheng, and X. Feng (1990) Wet weather-dry weather flow and pollutant load separation by ARMA modeling, *Proceedings, Fifth International Conference on Urban Storm Drainage*, IAWPRC-IAHR, Suita-Osaka, University of Osaka, Japan, pp. 403–408.

Novotny, V., et al. (1979). *Simulation of Pollutant Loads and Runoff Quality*, EPA 905/4-79-029E, U.S. Environmental Protection Agency, Chicago, IL.

Novotny, V., H. M. Sung, R. Bannerman, and K. Baum (1985). Estimating nonpoint pollution from small urban watersheds, *J. WPCF* **57**(4):339–348.

Novotny, V., K. Imhoff, M. Olthof, and P. Krenkel (1989). *Karl Imhoff's Handbook of Urban Drainage and Wastewater Disposal*, Wiley, New York.

Novotny, V., H. Jones, X. Feng, and A. Capodaglio (1990). Time series analysis models of activated sludge plant, *Water Sci. Technol.* **23**:1107–1116.

Pandit S. M., and S. M. Wu (1983). *Time Series and System Analysis with Applications*, Wiley, New York.

Potter, W. B., M. W. Gilliand, and M. D. Long (1986). A Geographic Information System for predicting of runoff and non-point source pollution potential, in *Hydrologic Applications of Space Technology*, Publ. No. 160, International Association of Hydrological Sciences, Wallingford, United Kingdom, pp. 437–446.

Renard, K. G., W. J. Rawls, and M. M. Fogel (1982). Currently available models, in *Hydrologic Modeling of Small Watersheds*, C. T. Haan, H. P. Johnson, and D. L. Brakensiek, eds., Monog. No. 5, American Society of Agricultural Engineers, St. Joseph, MI.

Reybold, W. U., and G. W. TeSelle (1986). Soil geographic data base, *J. Soil Water Conserv.* **44**(1):28–29.

Roesner, L. A., R. J. Dever, Jr., and J. A. Aldrich (1983). *Urban Highway Storm Drainage Model. Vol. 6. Analysis Module User's Manual and Documentation* (Final Rep. Sep. 76–Mar. 81), Camp, Dresser, and McKee, Inc., Annandale, VA, FHWA/RD-83-046, Federal Highway Administration, Washington, DC, 253p.

Sanders, W. M. (1976). Non-point source modeling for Section 208 planning, in *Best Management Practices for Non-point Source Pollution Control*, EPA 905/9-76-005, U.S. Environmental Protection Agency, Washington, DC.

Sartor, J. D., and G. B. Boyd (1972). *Water Pollution Aspects of Street Surface Contaminants*, EPA-R2-72-081, U.S. Environmental Protection Agency, Washington, DC.

Schilling, W., ed. (1989). *Real-time Control of Urban Drainage Systems—The State-of-the-Art*, Sci. Tech. Rep. No. 2, IAWPRC, Pergamon Press, Oxford.

Shaheen, D. G. (1975). *Contributions of Urban Roadway Usage to Water Pollution*, EPA 600/2-75-004, Office of Research and Development, U.S. Environmental Protection Agency, Washington, DC, 346p.

Soil Conservation Service (1972). *National Engineering Handbook 1964*, Section 4, Hydrology, rev. ed., U.S. Dept. of Agriculture, Washington, DC.

Soil Conservation Service (1985). *Urban Hydrology for Small Watersheds*, Release No. 55, U.S. Department of Agriculture, Washington, DC.

Stuebe, M. M., and D. M. Johnston (1990). Runoff volume estimation using GIS techniques, *Water Resour. Bull.* **26**(4):611–620.

Sutherland, R. C. (1980). An overview of stormwater quality modeling, in *Proceedings, International Symposium on Urban Storm Runoff*, University of Kentucky, Lexington, KY.

Terstriep, M. L., and J. B. Stall (1974). *The Illinois Urban Drainage Area Simulator, ILLUDAS*, Bull. 58, Illinois State Water Survey, Urbana, IL.

Torno, H. C., J. Marsalek, and M. Desbordes, eds. (1986). *Urban Runoff Pollution*, Springer Verlag, Berlin, New York.

U.S. Environmental Protection Agency (1971). *Storm Water Management Model*, 4 Vol., EPA No. 11 024DOC07/71 to 11 024DOC10/71, U.S. EPA, Washington, DC.

Walesh, S. G. (1989). *Urban Surface Water Management*, Wiley, New York.

Williams, J. R., A. D. Nicks, and J. G. Arnold (1985). Simulator for Water Resources in Rural Basins, *J. Hydraul. Eng. ASCE* **111**(6):970–986.

Woodward-Clyde Consultants (1989). *A Probabilistic Methodology for Analyzing Water Quality Effects of Urban Runoff on Rivers and Streams*, U.S. Environmental Protection Agency, Washington, DC.

Yen, B. C., ed. (1982). *Urban Stormwater Quality, Management and Planning*, in Proceedings, Second International Conference on Urban Storm Drainage, Water Resources Publ. Littleton, CO.

Young, R. A., et al. (1986). *Agricultural Nonpoint Source Pollution Model: A Watershed Analysis Tool*, Agricultural Research Service, U.S. Department of Agriculture, Morris, MN.

Zison, S. W. (1980). *Sediment Pollutant Interrelationships in Runoff from Selected Agricultural, Suburban, and Urban Watersheds. A Statistical Correlation Study*, EPA 600/3-80-022, U.S. Environmental Protection Agency, Athens, GA.

10
Control of Urban Diffuse Pollution

When a city takes a shower, what do you do with the dirty water?

Paraphrased from Lager et al. (1977)

POLLUTION-CONTROL MEASURES

Urban runoff from separate sewers and combined sewer overflows can be controlled by a variety of measures that can be structurally intensive and expensive or nonstructural and relatively inexpensive. The term *best management practices* usually refers to less structurally intensive modifications of the drainage system aimed at the reduction of pollution loads.

Several treatises and manuals deal extensively with control of urban runoff. The following manuals were reviewed and are recommended for further reference:

DeGroot (1982): Stormwater Detention Facilities
Dorman et al. (1988): Retention, Detention, and Overland Flow for Pollutant Removal from Highway Stormwater Runoff
Lager and Smith (1974): Urban Stormwater Management and Technology: An Assessment.
Lager, Smith, and Tchobanoglous (1977): Urban Stormwater Management and Technology: Update and Users' Guide.
Minnesota Pollution Control Agency (1989): Protecting Water Quality in Urban Areas
Moffa (ed.) (1990): Control and Treatment of Combined Sewer Overflows
Novotny and Chesters (1981): Handbook of Nonpoint Pollution: Sources and Management

Schueler, Kumble, and Heraty (1991): A Current Assessment of Urban Best Management Practices. Techniques for Reducing Non-point Pollution in the Coastal Zone.
Stahre and Urbonas (1990): Stormwater Detention
Torno, Marsalek, and Desbordes (1986): Urban Runoff Pollution
Walesh (1989): Urban Surface Water Management
Whipple et al. (1983): Stormwater Management in Urbanizing Areas

In addition, the *Proceedings of the International Conference on Urban Storm Drainage* (Yen, 1982; Balmér, Malmqvist, and Sjöberg, 1984; Gujer and Krejci, 1987), *Proceedings of an Engineering Foundation Conference* (Urbonas and Roesner, 1986; Roesner, Urbonas, and Sonnen, 1989), and *Proceedings of the National Conference—Perspectives on Nonpoint Pollution*, held in Kansas City in 1985, were also quoted by individual papers and are recommended for further reading. Manuals by the Water Pollution Control Federation (1990), the American Public Works Association (1981), and numerous research reports sponsored by the U.S. Environmental Protection Agency published since the 1970s provide a comprehensive review of various practices and structural measures for control of the quantity and quality of urban runoff.

The National Urban Runoff Project was the most comprehensive urban runoff research effort to date anywhere in the world. Research that addressed issues of the characterization of the quality of urban runoff and the effectiveness of various measures of abatement was carried out at 28 localities throughout the United States (for example, Bannerman, Baun, and Bohn, 1983; Bannerman et al., 1983; Metropolitan Washington Council of Governments, 1983; Pitt and Shawley, 1981; Terstriep, Bender, and Noel, 1982, and others). A summary of the NURP research was published in a U.S. EPA (1983) report.

Review papers on urban nonpoint pollution abatement measures include articles and reports by Field (1984, 1985a, 1985b, 1986, 1990) and Novotny (1984). A report by the U.S. EPA (1990) provides an information and guidance manual for state NPS program staff engineers and planners. Livingston and Roesner (1991) pointed out that runoff quality control (and by the same reasoning, CSO control) is not yet a technical science; rather it is an engineering art, with few design criteria for pollution removal having been established to date. Nevertheless, some empirical rules have been suggested by the authors:

1. The most effective runoff quality controls reduce the runoff peak and volume (these are generally infiltration controls).

2. The next most effective controls reduce the runoff peak. (These controls generally involve storage.)
3. For small storms (those with return intervals of less than two years), the peak rate of runoff should not exceed the peak rate of runoff from a two-year storm in a *preurbanized* condition, in order to control stream erosion.
4. Most obnoxious pollutants in urban runoff can be settled out; however, appreciable amounts of nutrients and some heavy metals are dissolved pollutants and require further treatment.

Urban storm-water management, which has an objective of flood control, must consider storms that have a relatively long recurrence interval. Typically urban storm sewers are designed to carry flows generated by a storm with a recurrence interval of 5 to 10 years, while urban floodplains and other flood conveyance systems should be able to convey without major damage 50- to 100-year floods (Walesh, 1989). On the other hand, storm-water management for *water quality control* is most efficient and economical if *small frequent storms* (smaller than the 1-year storm) events are considered (Livingston and Roesner, 1991).

After an extensive review of urban best management practices, Schueler, Kumble, and Heraty (1991) pointed out that

- Not all urban best management practices (BMPs) to control the quality of urban runoff can reliably provide high levels for both particulate and soluble pollutants.
- The longevity of some BMPs is limited to such a degree that their widespread use is not encouraged. Of particular concern are such infiltration practices as basins, trenches, and porous pavements.
- BMP options are adaptable to most regions of the United States, with the exception of extremely arid regions of the west and the colder climates of the north. In these regions, conventional BMP designs may need to be refined to account for high evaporation rates or subfreezing snowmelt conditions, respectively.
- No single BMP option can be applied to all development situations, and all BMP options require careful site assessment prior to design.
- Several BMPs can have a significant secondary environmental impact, although the extent and nature of these impacts is uncertain and site specific. These impacts can be both positive (such as the creation of a wetland wildlife habitat) or negative (accumulation of toxic compounds in ponds).
- Relatively limited cost data exists to aid in the assessment of the comparative cost-effectiveness of urban BMPs.

- Many conventional urban BMPs need to be enhanced to provide more reliable pollutant removal and greater longevity.

URBAN DRAINAGE AND RUNOFF POLLUTION CONTROL

Urban drainage systems are composed of two systems. One, the *convenience* or *minor* system, contains components that accommodate the smaller more frequent events. This system includes sewers, either combined or separated, their appurtenances (manholes, inlets, flow divides, overflow fixtures, etc.), and/or natural (swale) drainage. These systems are designed to accommodate the design flows (usually with a 5–10 yr recurrence interval) without ponding on the streets and parking lots.

The emergency, *major* or *overflow*, system considers rare storm runoff events, typically up to a 100-year storm. Larger storms are generally considered to be an "Act of God," and abatement for such large storms is commonly not required unless loss of human lives is a concern. Accordingly, the urban drainage systems for these storms should minimize or eliminate major flood damages, but this may be accomplished by nonstructural means, such as restricting development in floodplains, allowing for temporary storage on street surfaces, parking lots, within open public space areas, by establishing building grades above the 100-year flood levels, etc. In many urban areas, especially where unrestricted development has been allowed, major systems have not been explicitly designed; however, the emergency system will function by default when major runoff events occur, sometimes with catastrophic damage and disruption (Walesh, 1989).

Urban runoff water quality control, even though designed for smaller, more frequent runoff events, can be made a part of both the major and minor systems. As pointed out previously, urban runoff quality control may also have the side benefit of flood control. The urban drainage systems as they evolved in the last several hundred years can now be separated into two distinct methodologies and drainage objectives. The first methodology has the objective of *fast conveyance* of high flows from the premises. This is the more traditional approach still favored by many city engineers and developers. Systems designed in accordance with this approach provide for the collection of urban runoff, followed by the immediate and fast conveyance of storm-water from the collection area. The principal components of these systems are sewers with inlets, catch basins, overflow points, and lined open or covered channels. This system may result in more damaging floods downstream below the discharge point from the system.

It is more difficult to consider pollution abatement in the fast conveyance systems. The delivery ratio for pollutants in these systems is close to one (all pollutants delivered to the systems from the surface will be conveyed to the receiving body of water). Runoff and CSO quality management must then rely on expensive structural measures and treatment.

A potentially more effective system for pollution control is the *storage-oriented* (retardance) system. These systems provide for temporary storage at many points throughout the system, beginning with the source areas, and subsequent release after the flows subside. Pollution can be attenuated better when the flow is slowed down and temporarily stored. Therefore, pollution mitigation is more effective and inexpensive in the storage-oriented systems.

Walesh (1989) pointed out that fast conveyance systems can be implemented in both existing (developed) and newly developing urban areas, while the storage-oriented system, which uses landscape to a great extent, is mostly applicable to new developments. The principal advantages of the storage-oriented approach are possible cost reductions in newly developing areas, far better prevention of downstream adverse flooding and pollution associated with storm-water runoff, and potential for multiple-purpose reuse.

Urban runoff pollution-control measures can be divided into several categories. Such categorization separates the runoff and combined sewer overflow control measures according to where the measure is implemented, namely, into on-site land pollution-control measures, hydrologic modification and land management, reduction of delivery of pollutants in the collection (drainage) system, and end-of-pipe storage and treatment. Typically in a study or a plan, hydrological models or design routines are used for the design and evaluation of the effectiveness of the proposed measures. Many measures have multiple benefits in addition to water quality control, such as flood control, aesthetic enhancement and recreation, remedy of sewer backwatering problems, aquifer recharge, wetland restoration and protection, and water conservation.

SOURCE-CONTROL MEASURES

Source-control measures are most effective for control of urban runoff pollution delivered to receiving waters from separate storm sewers, and the efficiency of such measures may be minimal for controlling pollution by combined sewer overflows. These measures commonly involve reduction of pollutant accumulation on the impervious surfaces of the contributing watersheds or reduction of the erosion of pervious lands.

Source controls are generally difficult to implement because the sources are diffuse, originating from the atmosphere, streets, parking lots, and so forth, as documented in Chapter 8. Many of these controls can be incorporated into the landscape and, in fact, they can often enhance the aesthetics of the urban landscape (Roesner, 1988). Generally, however, planners have relatively little control over the policies that would implement source controls. Homeowners are not restricted in using chemicals on their properties as long they are approved for sale. Street cleaning practices are mostly for aesthetic purposes, and their frequency depends on the financial resources of each individual community and on the willingness of public officials to implement these practices.

Implementing source-control practices requires extensive public education and grass-roots public effort. Local environmental organizations and schools must be involved in planning, providing educational materials outlining sound environmental uses of chemicals, and promoting clean and well-kept neighborhoods. These organizations and citizen groups must keep pressure on elected officials to implement and enforce controls. Without these efforts implementation of these practices is impossible.

Control of Atmospheric Deposition

After implementation of stricter industrial air-pollution control measures, switching from coal to natural gas as a primary energy source for household heating, and by implementing limitations on leaded gasoline, dry deposition of pollutants has dropped substantially in the United States during the last 30 years. However, other problems, such as acidity of rainfall and deposition of nitric oxides from traffic have not improved. As shown by Novotny and Kincaid (1982), wet-atmospheric deposition (precipitation) in the midwestern U.S. cities is acidic, with pH around 4, and contains appreciable amounts of nitrogen.

Control of atmospheric deposition originating from air pollution by industrial and traffic sources is regulated in the United States by the Clean Air Act, which also includes provisions to control sources of acid rainfall. It was pointed out in Chapter 8 that the acidity of urban precipitation elevates loads of many pollutants that are elutriated during the buffering of acid rainfall from the urban infrastructure, including metal and tar shingle roofs, from vehicular corrosion, and from other sources.

Removal of Solids from Street Surfaces

Litter-Control Programs

Novotny (1984) described the effects of litter-control programs. Litter includes larger items and particulates deposited on street surfaces, such as paper, vegetation residues, animal feces, bottles and broken glass, and plastics. In the fall, fallen leaves become the most dominant component of street litter. It has been shown that litter-control programs can reduce the amount of deposition of pollutants by as much as 50%, and litter control, especially in watersheds that are highly impervious, may be an effective measure of controlling pollution by storm runoff. During the fall leaf fallout period the organic and nutrient pollution inputs in areas with trees require an effective leaf pickup program to minimize the impact on the receiving waters. As stated in the preceding section, implementation of litter-control programs requires public education and acceptable enforcement.

Street Cleaning—Sweeping

Street cleaning practices include the sweeping of streets and parking lot surfaces by mechanical vehicles or flushing from tanker trucks (Figs. 10.1 and 10.2). Sweeping is more common in the United States, while street flushing is practiced more in Europe. Both practices are used primarily for aesthetics and removal of unsightly litter. The choice whether street sweeping or flushing is to be used for water quality control depends above all on the type of sewer system.

Street flushing. Pravoshinsky (1975) has shown that the quality of runoff from street flushing is very poor and is apparently inferior to storm runoff. However, street flushing is an advantageous and efficient means of control in areas served by combined sewers. Flushing cleans a larger street area (not just a narrow strip near the curb) and is more efficient for picking up fine particles. Because the capacity of flow separators and sewer capacity in the combined sewer system is selected to be about six times the peak dry-weather flow, street runoff generated by this practice during the dry period is generally far below the critical runoff rate that would initiate an overflow, hence, all polluted flow is conveyed to the treatment plant. In addition, frequent flushing can dislocate solids and slime that accumulate in sewers that otherwise would be discharged untreated with the overflow into the receiving waters.

Street sweeping. This method is appropriate for urban areas served by storm sewers. Application of street flushing in these areas would defy the purpose of this activity, that is, all solids washed into the sewer system would be carried directly into the receiving waters, while sweepers physically remove them from the surface, thereby making them unavailable for pickup by subsequent runoff-generating rainfall. Two types of sweepers are currently used to remove solids from impervious surfaces. The most common design (mechanical street cleaners) uses a rotating gutter broom to remove the particles from the street gutter area and place them in the

FIGURE 10.1. Street sweeping vehicle. (Photo: University of Wisconsin.)

FIGURE 10.2. Street flushing vehicle. (Photo: University of Wisconsin.)

path of a large cylindrical broom that rotates to carry the material onto a conveyor belt and into the hopper. Vacuum-assisted street cleaners use gutter brooms to loosen the deposits and move street refuse into the path of a vacuum intake. The vacuum places the debris in the hopper.

Both types of sweepers are relatively ineffective for removing fine particles (American Public Works Association, 1969; Sartor and Boyd, 1972; Sartor, Boyd, and Agarty, 1974; Bannerman et al., 1983). Broom mechanical sweepers are ineffective for particulates in the dust and dirt range (<3.2 mm in diameter), and their overall efficiency is only about 50%. Vacuum sweeper efficiency is higher, but still ineffective for silt and clay size particles (American Public Works Association, 1969; Sartor and Boyd, 1972). Tables 10.1 and 10.2 show typical street cleaner efficiencies for removal of particles of various sizes. If the efficiency is combined with

TABLE 10.1 Mechanical Street Cleaner Efficiencies for Various Equipment Passes (%)

	Curb Loading (50–500 g/curb m)		
Size Range (μm)	1 Pass	2 Passes	3 Passes
<43	15	28	59
43–104	20	36	49
104–246	50	75	88
246–840	60	84	94
840–2000	65	86	96
2000–6370	80	96	99

Sources: Data from Sartor and Boyd (1972) and Pitt (1979).

TABLE 10.2 Removal Efficiencies for Vacuum Street Cleaners at Different Initial Particulate Loadings and for Various Equipment Passes

	Street Surface Loading and Number of Passes								
	5–50 (g/curb m)			20–200 (g/curb m)			280–2800 (g/curb m)		
Size range (μm)	1	2	3	1	2	3	1	2	3
44–74	3	6	9	20	36	49	70	91	97
74–177	25	40	50	40	60	72	75	94	99
177–300	50	75	88	60	84	94	80	96	99
300–750	60	84	94	65	88	96	70	91	94
750–1000	50	75	88	60	84	94	70	91	97

Sources: Data from Clark and Cobbins (1963) and Pitt (1979).

the pollutant distribution on particles of various sizes (Table 8.11), the sweeper removal efficiency for pollutants can be estimated.

A normal sweeping effort was defined by Sartor and Boyd (1972) as 2.56 equipment minutes per 1000 m² of cleaned area, which can be translated to the average sweeper vehicle velocity of about 10 km/hr (6 mph). The information on street sweeper efficiencies in Table 10.2 refers to the normal operating effort. Increased street sweeping efficiency can be achieved by operating the vehicle at a lower speed or by conducting multiple passes. From the work by Sartor and Boyd (1972) it follows that the removal of particles by street sweepers can be approximated by the equation

$$P = P^* + (P_0 - P^*)e^{-kE} \tag{10.1}$$

where
P = the amount of street surface particulates in a given size range remaining after sweeping
P_0 = the initial amount of particulates in the size range
E = the amount of sweeping effort involving (min/1000 m²) or as a relative effort (i.e., actual effort divided by the standard effort)
P^*, k = empirical coefficients depending on sweeper characteristics and design, particle size of particulates, and street surface characteristics

If E is normalized by expressing it as a relative effort, the coefficient k becomes dimensionless. From Sartor and Boyd's data it appears that k can be approximated best by the following equation

$$k = \alpha d^\beta \tag{10.2}$$

where α and β are empirical coefficients and d is particle size. The approximate magnitudes of α and β are around 0.027 and 0.35, respectively, if d is in micrometers. The actual magnitudes of α and β should be determined for each type of equipment by the manufacturer or by a testing laboratory.

Example 10.1: Removal of Pollutants by Sweeping

A street dust and dirt sample had the following particle-size distribution

Particle Size (μm)	% Distribution
<43	5
43–104	10

Particle Size (μm)	% Distribution
104–246	20
246–840	25
840–2000	25
>2000	15

The sweeper efficiency coefficients, α and β, were given by the manufacturer as α = 0.03 and β = 0.55. Determine the overall removal efficiency of the sweeper if it moves at a sweeping speed of 5 km/hr.

Solution The removal efficiency for each particle-size fraction can be computed using Equation (10.1). Then approximately (assuming that $P^* \approx 0$)

$$\text{Eff}(\%) = 100(1 - e^{-kE})$$

where $k = 0.03 d^{0.55}$ The normalized sweeper effort variable is

$$E = \frac{\text{Actual effort (km/hr)}}{\text{Standard effort (km/hr)}} = \frac{(1/5 \text{ km/hr})}{(1/10 \text{ km/hr})} = 2.0$$

The solution can be obtained either graphically or numerically as shown in Table 10.3. The calculation may also be incorporated into a computer

TABLE 10.3 Computation of Overall Sweeper Efficiency

Particle Size (μm) (1)	Initial Distribution (%) (2)	k (3)	e^{-kE} (4)	Average Removal $1 - e^{-kE}$ (5)	% Removal (6) = (2) * (5)	% Remaining (7) = (2) − (6)
0		0	1.0			
	5			0.19	0.95	4.05
43		0.24	0.62			
	10			0.46	4.6	5.4
104		0.39	0.46			
	20			0.625	12.5	7.5
246		0.62	0.29			
	25			0.81	20.25	5.75
840		1.22	0.09			
	25			0.945	23.62	1.38
2000		1.96	0.02			
	15			0.98	14.7	0.3
Total	100				76.62	23.38

model (see the Wisconsin Urban Runoff Model in Chapter 8). From Table 10.3, we see that the overall efficiency is 76.6%.

Example 10.2: Lead Removal by Sweeping

By analyzing the lead distribution with particle sizes of the street dust and dirt sample the same as in the previous example, it was found that most lead is associated with fine fractions is as follows:

Particle Size (μm)	% Lead
0–43	48
43–104	25
104–246	20
246–840	16
840–2000	1

Estimate how much lead will be removed by the sweeper.

Solution As seen in Table 10.4, the computation can be arranged in the same fashion as in the previous example. Thus we get that only 48% of lead will be removed, as compared to 76% of solids.

Water quality benefits of the removal of pollutants by street sweeping. Current street sweeping practices are primarily for aesthetic purposes. The results of the National Urban Runoff Project documented the relatively low impact of street sweeping on improvement of water quality on midwestern and eastern U.S. conditions (Bannerman et al., 1983; Terstriep, Bender, and Noel, 1982; Bender and Rice, 1982; Novotny et al., 1985; Field, 1986).

TABLE 10.4   Compution of Percentage of Lead Removed by the Sweeper

Particle Size Range (μm) (1)	Initial Distribution (2)	Average Removal (3)	% Removed (4) = (2) * (3)
0–43	48	0.19	9.12
43–104	25	0.46	11.50
104–246	20	0.625	13.50
246–840	16	0.81	12.96
840–2000	1	0.94	0.94
2000	0		0
Total	100		48.02

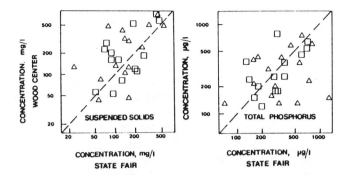

FIGURE 10.3. A comparison of runoff concentrations of suspended solids from paired events from two almost identical small and uniform urban experimental watersheds with predominantly commercial land use. State Fair test site was swept, while Wood Center was kept unswept. (Data from the Wisconsin Department of Natural Resources.)

In the Milwaukee, Wisconsin, NURP project six pairs of small uniform watersheds were investigated. The watersheds in each pair were either identical or similar in land use and surface characteristics. One watershed was swept regularly with relatively high frequency (at least once per week); the other watershed was unswept. The measured results were disapppointing. Street sweeper efficiency was generally low and the effect of street sweeping on runoff pollutant concentrations was minimal, as shown on Figure 10.3, in which corresponding concentrations of swept and unswept watersheds in the pair with commercial land use were plotted against each other. Modeling results for these experimental paired watersheds were quite similar (Novotny et al., 1985).

Several hypotheses have been given to explain the reasons for there results (Novotny et al., 1985). First, street sweeper efficiency is relatively low, and sometimes, when initial street loads were low, it was negative. Second, simulations by a mathematical model showed that in areas of lower imperviousness, more particles are blown away by traffic and wind onto adjacent pervious and hydrologically inactive areas than can be removed even by a very vigorous sweeping program. Third, erosion of pervious areas, although not as frequent as the washoff of pollutants from impervious areas, yields unit loadings of pollutants that are often of the same order of magnitude or greater than the corresponding loads from impervious areas affected by street sweeping.

Tests under real conditions in San Jose, California, showed that street cleaning can remove up to 50% of the total solids and heavy metals in urban storm water with very frequent (once or twice daily) cleaning.

Typical cleaning programs of once or twice a month proved ineffective (Pitt, 1979; Field, 1986).

Control of Pervious Areas

Soil loss from unprotected bare soils may be considerable—up to 100 tonnes/ha-year. Temporary or permanent seeding of grass, sodding, and mulching is used to reduce erosion of pollutants. Such measures are important and mandated in some states for control of pollution caused by construction activities. Ports (1975) presented procedures and criteria for the design of urban sediment-control practices. More extensive design procedures for erosion control were included in a publication by Goldman, Jackson, and Bursztynsky (1986).

Chemical stabilization of soils. This is a temporary measure employed on bare soils until permanent vegetation is established or other long-term erosion-control measures are implemented. The chemical emulsions that were previously used for erosion control included polyvinyl acetate emulsions, vinyl acrylic copolymer emulsions, or methacrylates and actylates. The use of organic chemicals and oil derivatives may not be possible due to suspected surface- and ground-water contamination by carcinogenic priority organic pollutants (vinyl chloride and poly-cyclic aromatic hydrocarbons (PAHs)). Hydrate lime and cement is also used for stabilization of clayey soils.

Mulching and protective covers. Covering an exposed area with any of a number of available mulches generally increases surface roughness and hydrological surface storage, protects the surface against rainfall impact, and subsequently reduces erosion. The effect of land-cover materials is reflected in a reduction of the land-cover factor (C) of the universal soil loss equation (Table 5.5). Straw mulch (small grain) has proved effective on 12% slopes at application rates of 5 tonnes/ha and on 15% slope with an application rate of 10 tonnes/ha. These application rates should reduce erosion by 75% to 80%. Straw mulch application can be combined with grass seeding for more permanent surface protection. As a matter of fact, the most economical, effective, and practical surface protection in construction sites is hydromulching, which is an application of a slurry mixture of seed, mulch, fertilizer, and lime (Fig. 10.4).

Figure 10.5 shows an example of straw mulch application protected by plastic or paper (biodegradable) fibers. Straw mulching loses its effectiveness on steep slopes because of rill formation and its tendency to be washed away by more concentrated and erosive overland flow. Alternate materials include woodchips, crushed stone, and blankets and mats from textile materials (Fig. 10.6). Chapter 5 contains an extensive discussion of erosion-control measures and design parameters with examples.

FIGURE 10.4. Hydroseeding of grasses. (Photo: University of Wisconsin.)

FIGURE 10.5. Mulch net applied over straw mulch, seed, and fertilizer. The net is of paper fiber with plastic strands that are pinned down over the straw. (Photo: USDA, Soil Conservation Service.)

FIGURE 10.6. Control of erosion by burlap netting. (Photo: USDA, Soil Conservation Service.)

Control of surface application of chemicals. Measures to control the surface application of chemicals include control of herbicide use on pervious grassed areas (lawns and golf courses) and road deicing salt storage. Studies by the EPA and the Federal Highway Administration quoted in Field (1986) prompted several states to enact legislation controlling salt application and storage. For example, chemical loading/unloading and storage areas should be covered or diked to capture about 1.25 cm (1/2 inch) of runoff, with a closeable outlet so that, if spillage occurs, it will not be washed into the drainage system before it can be cleaned up. In areas served by separate sewers, runoff from washing should be discharged into sanitary sewers (Livingston and Roesner, 1991).

As was pointed out previously, control of pollution caused by chemical use by individual homeowners on their lands is difficult due to a lack of legal instruments to enact regulation. Public education is currently the only possible mechanism to implement some partially effective measures. A switch to *xeriscape*, that is, landscape that incorporates native plants and shrubs that do not require chemicals and excessive irrigation, should be encouraged and demonstrated to the public. Such changes may even be mandated in the future in some areas that are short of water because of their water-conservation benefits in addition reducing the pollution of the surface- and ground-water resources.

Hydrologic Modifications

Hydrologic modifications of urban watersheds include measures and practices that reduce the volume and intensity of urban runoff entering the separate storm or combined sewer systems. In separate systems it is expected that the pollution load will be proportional to the volume of runoff, and with the volumetric reduction of flow, one may also expect a reduction in pollution load. In combined sewers, reduction of peak flow and volume of flow reduces the frequency of untreated overflows from the sewer systems. Thus hydrological modification practices are effective for the control of pollution from both systems (storm sewers and CSOs) and both have pollution-control and flood-control benefits. A special report by the EPA (1977) deals with practices involving hydrologic modifications of urban watersheds.

The hydrologic measures can be divided into:

1. Practices that increase permeability and enhance infiltration, such as the use of pervious pavements or vegetation filtration strips;
2. Practices that increase hydrological storage;
3. Practices that reduce the size of impervious areas that are directly connected to the sewer system.

Schueler, Kumble, and Heraty (1991) cautioned against the indiscriminate use of infiltration practices in an urban setting. The poor longevity of these BMPs is attributable to lack of pretreatment, poor construction practices, application at unsuitable sites, lack of regular maintenance, and faulty design. The life span of such practices may be increased if local communities adopt enhanced maintenance and inspection programs.

Porous (Pervious) Pavements

Porous pavement is an alternative to conventional pavement. Its use allows rainfall to percolate through it into the subbase. The water stored in the subbase then gradually infiltrates the subsoil.

Porous pavement provides storage, enhancing soil infiltration that can be used to reduce runoff and combined sewer overflows. These pavements are either made from asphalt, in which fine filler fractions are missing, or

590   Control of Urban Diffuse Pollution

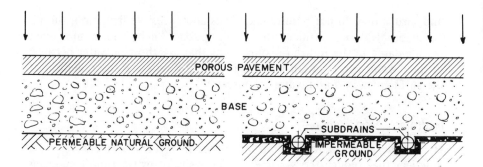

FIGURE 10.7.   Porous pavement installed over permeable and impermeable ground.

are modular or poured-in concrete pavements (Fig. 10.7). The primary benefit of porous pavements is a significant reduction or even complete elimination of surface runoff rate and volume from an otherwise impervious area. If the pavement is designed properly, all or most of the runoff can be stored and subsequently allowed to infiltrate into the natural ground. Aquifer recharge by infiltrated water is the second important benefit. The third benefit is the reduced need for storm drainage. As a matter of fact, if subsoils are very permeable, there may be no need for installing storm drainage. However, porous pavement is most feasible when subsoils are permeable and the slopes are relatively flat. If the soils are permeable, porous pavements are feasible even in cold snowbelt areas (see Chapter 3 for a discussion of infiltration into frozen soils).

In areas with poorly draining subsoils or if the porous pavement is installed over an existing impervious base, a drainage system can be installed. The pavement and its base in this case enhances hydrological storage, reduces the peak runoff rate, and provides filtration of pollutants from runoff. Field (1986) summarized the results of several U.S. EPA studies on the experimental applications of porous pavements. Results from a study in Rochester, New York, indicated that peak runoff rates were reduced by as much as 83% where porous pavement was used. The structural integrity of the pavement was not impaired by a heavy traffic load and, if properly installed, porous pavements have load-bearing strength and longevity similar to conventional pavement. Although clogging may occur during construction and during operation, it can be remedied by flushing and sweeping.

Typically, hydraulic conductivity (permeability) of porous pavements is much greater than runoff rates. Hydraulic conductivity measured by Jackson and Ragan (1974) was about 250 cm/hr, which is an order of magnitude higher than a typical catastrophic design storm. This means

FIGURE 10.8. Example of porous pavement installation in an experimental basin in the Tokyo metropolitan area. Before (*top*) and after (*bottom*) installation. (Photo: S. Fujita.)

that infiltration into the base should occur without ponding. An extensive report summarizing EPA research on the hydraulic characteristics of pervious pavements was prepared by Goforth, Diniz, and Rauhut (1984). Porous pavement has an excellent potential for use in parking areas and on side streets. These pavements have been installed in many localities throughout the world (Fig. 10.8) and a design manual has become available in Florida (Florida Concrete Products Association, 1989) and other states (Maryland Water Resources Association, 1984; Schueler, Kumble, and Heraty, 1991).

Longevity of the porous pavement function is a problem (Schueler, Kumble, and Heraty, 1991) since it has a high failure rate. Failure is due to partial or total clogging that occurs during or immediately after construction or over time, when porous pavement is clogged by sediment and oil. For this reason, porous pavements should not be installed where high solids loads from wind erosion or other sources (heavy truck traffic, use of sand for ice skidding control) are expected. The Japanese experience (Fujita, 1984; Fujita and Koyama, 1990) has shown that frequent and proper sweeping by vacuum sweepers resolves the problem of clogging.

Contamination of shallow aquifers by toxic materials attributed to asphalt, vehicular traffic, and road usage, including salt application for deicing, represents a slight to moderate environmental risk that depends on soil conditions and aquifer susceptibility.

The construction cost of porous pavement is about the same or even less than that for conventional pavement when savings on storm drainage are included. A laboratory and economic study on feasibility of porous pavements was undertaken by Thelen et al. (1972). Haak and Oberts (1983) showed that the cost of conventional pavement on streets and highways with drainage can be as much as two to three times higher than the cost of a pervious pavement alternative.

Increasing Surface Storage

Rooftop storage on flat roofs, temporary ponding, and restriction of storm-water inlets are used to control combined sewer overflows and to reduce flooding (American Public Works Association, 1981). As a water quality control measure in areas served by combined sewers, these practices may be effective and may reduce the requirements for in-line or off-line storage. In areas served by storm sewers, they can control flooding and reduce the requirements for detention–retention basin volumes, but alone they do not have a significant water quality benefit.

Decreasing Connected Impervious Area

Directly connected impervious area (DCIA) is defined as the impermeable area that drains directly to an improved drainage system, that is, paved gutter and, subsequently, sewer. Livingston and Roesner (1991) suggest that the minimization of DCIA is by far the most effective method of runoff quality control because it delays the concentration of flows into the sewers and maximizes infiltration.

Practices used for minimization of DCIA include:

1. Disconnecting roof drains from storm sewers. This practice has a low direct water quality benefit since only wet- and dry-atmospheric contributions are controlled. It may have some impact on the necessary detention volume. Since this practice represents an infringement on private property, it may be difficult to implement.
2. Permitting surface runoff to overflow on adjacent pervious surfaces.
3. Use of dry wells, infiltration basins, and ditches into which storm water is directed.

Porous pavements obviously also reduce the directly connected pervious area.

Filter strips. Filter strips are grassed strips situated along the roads or parking areas between the impervious road surface and drainage sewers or ditches (Fig. 10.9). The filter strips remove the pollutants from runoff by filtering, provide some infiltration, and slow down the runoff flow. Their efficiency is related to the velocity and depth of flow. The higher the flow rate and flow depth, the lower the efficiency. Growing grasses almost guarantee laminar flow conditions of the shallow flow. Typically grass height should be maintained between 15 and 30 cm. If grass is submerged such, as in grassed swales and ditches, the flow is turbulent and efficiency drops significantly. Thus, low velocity and shallow non-submerged flow depth are key design criteria. The Maryland design guidelines recommend that the slope of the strips be less than 5% and overland flow velocities less than 0.75 m/s. Failures will occur if rill erosion creates concentrated flows. Filter strips can effectively reduce particulate contaminants; however, their ability to remove soluble pollutants is variable and depends on the fraction of infiltrated runoff in the strip and adsorption of contaminants on organic matter and soils.

Kao, Barfield, and Lyons (1975) showed by experiments on artificial grass strips that grass filter strips provide excellent trapping efficiency, especially for construction and surface mining sites. Grass strips are also widely used as borders to control soil and pollutant loss from agricultural fields and barnyards (see Chapter 11). Grassed areas are also effective traps for pollutants associated with particulate matter, such as phosphates and pesticides (Asmussen et al., 1977).

The distance at which close to 100% removal of solids is achieved (both by filtering and infiltration) is called the *critical distance*. Wilson

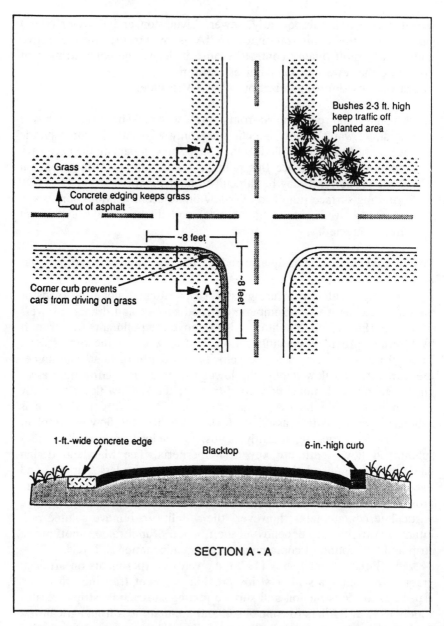

FIGURE 10.9. Simple installation of grass filters along roads. (After Livingston and Roesner, 1991, by permission of ASCE.)

(1967), in an empirical study on Bermuda grass, found that the maximum percentages of sand, silt, and clay were removed at about 3, 15, and 122 meters distances along the test strip. A minimum of 85% sediment removal can be achieved with a 2.5-m-wide grass strip during shallow (nonsubmerged) flow (Kao, Barfield, and Lyons, 1975; Barfield, Kao, and Toller, 1975). This efficiency is increased when alternating grass and bare land are used.

Experimental research at the University of Kentucky yielded an empirical model for particle trapping efficiency of grass filters (Barfield, Kao, and Toller, 1975; Toller et al., 1977). This efficiency was found to be a function of two dimensionless parameters, namely, the flow Reynolds number, Re^t, and the particle fall number, N_f (a determination of the probability of how many times a particle can reach the bottom during the flow period). These two variables were defined as

$$\text{Re}_t = \frac{v_s R_s}{\nu} \tag{10.3}$$

and

$$N_f = \frac{L_T w}{v_s} \tag{10.4}$$

where
v_s = flow velocity through the grassed media
L_T = overland flow length
R_s = spacing parameter defined as

$$R_s = \frac{s D_f}{2 D_f + s} \tag{10.5}$$

where
s = spacing of grass blades
D_f = depth of flow
ν = kinematic viscosity
w = settling velocity of particles from Stokes law

$$w = \frac{gD^2}{18\nu}(\rho_s - 1) \tag{10.6}$$

where
D = particle diameter

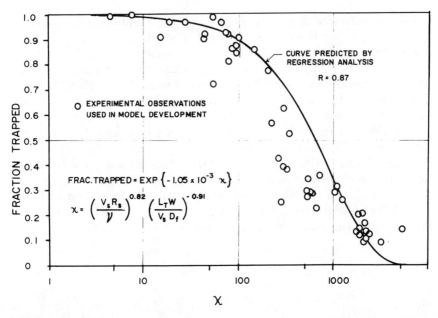

FIGURE 10.10. Relationship of particle removal efficiency of grass filters to Re_t and N_f. (From Kao, Barfield, and Lyons, 1975.)

g = gravity acceleration
ρ_s = specific gravity of the grain with respect to water

For cohesive sediments (clays) use Table 5.9 for settling velocity of aggregate particles rather than that for settling individual particles according to Stokes' law.

The relationship of the removal efficiency to Re_s and N_f is shown on Figure 10.10. From the figure a removal efficiency approaching 100% can be achieved when

$$\chi = \left(\frac{v_s R_s}{\nu}\right)^{0.82} = \left(\frac{L_T W}{v_s D_f}\right)^{-0.81} \approx 5 \tag{10.7}$$

Example 10.3: Grass Strip Estimation

What should be the minimum width of a grass buffer strip dividing a construction site from the nearest drainage that would reduce clay loads to the receiving body of water by 90% during a storm that yielded an average surface runoff of $q = 0.003 \, \text{m}^3/\text{s-m}$ width? The slope of the strip, S, is 5%.

Solution Assume:

Grass blades spacing $s = 3\,\text{mm} = 0.003\,\text{m}$
Clay particle size $d = 2\,\mu\text{m} = 2 \times 10^{-6}\,\text{m}$
Clay settling velocity from Table 5.9 $w = 0.17\,\text{mm/sec} = 0.00017\,\text{m/sec}$
Manning roughness factor for grass (from Table 3.14) $n = 0.35$

The flow depth can be estimated from the Manning equation

$$v_s = \frac{1}{n} D_f^{2/3} S^{1/2} = \frac{q}{D_f} \tag{10.8}$$

and assuming shallow flow in a wide channel

$$D_f = \frac{(qn)^{0.6}}{S^{0.3}} = \frac{(0.003 \times 0.35)^{0.6}}{0.05^{0.3}} = 0.0365\,\text{m}$$

Assume that the grass height is greater than the flow depth (nonsubmerged flow conditions). Then the flow velocity becomes

$$v_s = \frac{q}{D_f} = \frac{0.003}{0.0365} = 0.082\,\text{m/sec}$$

The spacing parameter, R_s, is

$$R_s = \frac{sD_f}{2D_f + s} = \frac{0.003 \times 0.0365}{2 \times 0.0365 + 0.003} = 0.00144\,\text{m}$$

Ninety percent removal occurs when the removal parameter, χ, from Figure 10.10 is ≤ 100. Hence

$$\chi = \left(\frac{v_s R_s}{v}\right)^{0.82} \times \left(\frac{L_m w}{v_s D_f}\right)^{-0.91} \leq 100$$

Solving for L_m, the critical distance for which 90% removal is achieved is

$$L_m = v_s^{1.9} R_s^{0.9} D_f v^{-0.9} w^{-1} 100^{-1.1}$$
$$= 0.082^{1.9} \times 0.00144^{0.9} \times 0.0365 \times (10^{-6})^{-0.9} \times 0.00017^{-1} \times 100^{-1.1}$$
$$= 7.93\,\text{m}$$

or about 10 m.

Example 10.4: Efficiency of grass strip

For the critical distance, L_m, calculated in Example 10.3, estimate removal efficiencies for silt and sand fractions.

Solution For silt, assume particle size $d = 0.02$ mm the settling velocity $v = 0.03$ cm/sec $= 0.0003$ m/s (Table 5.9). Then the removal factor χ is

$$\chi = \left(\frac{0.082 \times 0.00144}{10^{-6}}\right)^{0.82} \times \left(\frac{10 \times 0.0003}{0.082 \times 0.0365}\right)^{-0.91} = 6.15$$

According to Figure 10.10, for $\chi = 6.15$ about 99% or more of the silt will be removed in the 10-m-wide buffer strip. The same is true for sand that will yield lower values of χ.

It is possible to speculate that this estimation of solids removal efficiency is on the conservative side since it does not include the removal of solids by infiltration and the effect of increased infiltration.

Infiltration

Perforated concrete street gutters (Fig. 10.11) and infiltration pipes laid in trenches have been proposed and tested in Japan (Fujita, 1984; Fujita and Koyama, 1990). The infiltration pipes are combined with an infiltra-

FIGURE 10.11. Example of street gutter modification for enhanced infiltration. Thorough maintenance and frequent solids removal are needed to maintain efficiency of infiltration. (Photo: S. Fujita.)

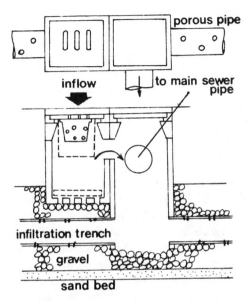

FIGURE 10.12. Example of manhole modification for enhanced infiltration. The manhole has an impermeable bottom and overflow area, not a perforated infiltration pipe laid in an infiltration trench. (From Fujita, 1984.)

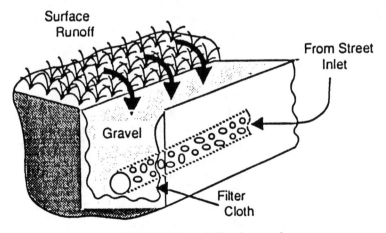

FIGURE 10.13. Infiltration trench.

tion manhole (Fig. 10.12) that has a pervious bottom. The soils in the Tokyo metropolitan area, where the experimental watersheds are located, are mostly composed of volcanic highly permeable turfs.

In the United States, excavated infiltration trenches, basins, and wells are more common (Fig. 10.13). These devices are especially advantageous in areas where ground-water recharge benefits are desired. Properly designed trenches are believed to be highly capable of removing particulate pollutants and moderately capable of removing soluble contaminants. Concerns persist about the possibility of ground-water contamination; however, sources in the literature to date do not indicate a major risk, but have noted migration of chlorides and to a lesser degree of nitrates (Schueler, Kumble, and Heraty, 1991). Generally, the nitrate content of urban runoff is lower and below the human health protection criterion.

An *infiltration trench* is an excavated trench that is backfilled with stone aggregates, gravel, or sand. An *infiltration basin* is made by constructing an embankment or by excavating it down to relatively permeable soils. The basin stores storm water until it infiltrates through the bottom and the sides of the system. The basins that are mostly dry can be incorporated into the landscape design of open areas or even recreational areas, such as sport fields (Roesner, 1988; Livingston and Roesner, 1991). Infiltration trenches can be located along highways and parking lots, in the median strip of divided highways, as a part of residential drainage, and several other possible locations (Harrington, 1989).

Both infiltration trenches and basins are prone to clogging by deposited solids. To increase their life span when sediment-laden storm-water flows are anticipated, sediment traps can be employed (Haak and Oberts, 1983). The Maryland experience (Schueler, Kumble, and Heraty, 1991; Galli, 1992) has shown that about one in five conventional trenches installed in the 1970s and 1980s failed to operate as designed immediately after construction. Furthermore, barely half of all conventional infiltration trenches operated after five years (Galli, 1992). The best pretreatment device is a grassed filter or buffer strip, which should be at least 6.6 m wide, along the trench or basin (Harrington, 1989). If trenches become clogged, the gravel and sand backfill will need to be removed and the trenches washed before they can be reused (Haak and Oberts, 1983), which could be expensive. To minimize this possibility it is necessary that the entire contributing area to the infiltration device be stabilized before construction of the trench or basin begins.

Oil and grease should also be removed before they enter these devices since these contaminants are difficult to remove and pose a danger to ground water. Treatment of runoff contaminated with these pollutants is accomplished through specially designed three-chamber storm-water inlets. Grit and sand are removed in the first chamber, and oil and grease are removed in the second chamber, from which runoff water is siphoned into the third chamber for final storage and polishing. These devices must also be regularly cleaned and the accumulated sediment removed.

Infiltration devices obviously require permeable soils or subsoils. The minimum infiltration rate of soils underlying the device should be more than 1.25 cm/hr, which corresponds to the Soil Conservation Service's soil permeability classification of C or better (Table 3.2). After taking moderate clogging into consideration, the permeability rates should call for the diverted volume to infiltrate within 72 hours, or within 24-36 hours for infiltration basins that are planted with grasses. The seasonal high water table should be at least 1.2 meters below the bottom of the infiltration device to assure that pollutants present in storm water are removed by vegetation, soil, and soil bacteria before they reach ground water.

Infiltration devices should not be installed in areas with shallow bedrock and in areas with Karst limestone formations, where sinkholes, caverns, and large rock fractures are common. They should also be located at least 6.6 m from buildings to avoid cross-connections with basements.

Both infiltration trenches and basins should be carefully designed. Infiltration trenches can handle runoff from watersheds ranging in size up to 5 hectares. Maryland's design guidelines for sizing infiltration trenches recommend the size to be sufficient to capture the first 1.25 cm (1/2 in.) of rainfall from connected impervious areas (Maryland Water Resources Administration, 1986). The U.S. Environmental Protection Agency suggests that infiltration devices be designed to store and treat flow from an average storm, which typically may have a rainfall depth lower than 1.25 cm. Stahre and Urbonas (1989) presented a methodology for site selection and design of infiltration basins, which for site selection, assigns points to various site, watershed, and soil characteristics.

Dry (French) *wells* are smaller borings that are filled with gravel to permit infiltration of accumulated storm water. Their best use is for infiltration of rooftop runoff. Again it is important that the well is located several meters from the foundation wall of the house to prevent back entry of water into foundation drain pipes.

Environmental Corridors and Buffer Zones

The corridors—usually a park along a stream, lake, or adjacent to the drainage system—act as buffers between the polluting urban area and the receiving body of water. In most cases the buffer zones also provide storage for flood and pollution control (Wiesner, Kassem, and Cheung, 1982). The corridors lose their efficiency if the storm drainage outlet bypasses the grassed and vegetated areas and discharges directly into the receiving body of water or into a channel with concentrated flows that is directly connected with the body of water. The "treatment" processes for storm runoff, such as vegetated filters, infiltration basins, detention–retention ponds (dry or wet), and wetlands, are incorporated into the landscape of the corridor.

Buffer strips made of uneven shoreline vegetation may also be used to attenuate runoff pollutants that would otherwise reach the body of water. Woodard (1989) measured the efficiency of buffer strips that had vegetation typical of a Maine lakeshore (mixed growth, uneven age stand, predominantly hardwood, moderate ground cover of shrubs, ferns, etc.). Similar measurements were made by Potts and Bai (1989) in Florida. They found that the critical distance of a grass strip, used for the control of suspended sediment and phosphates, from residential developments was 22.5 meters; however, they concluded that the efficiency of the buffer strips is highly dependent on a sufficient cover of organic matter (natural vegetation) and on the initial concentration of the pollutants or the density of shoreline development. A 30-m-wide buffer strip is recommended for the protection of surface-water reservoirs that are part of the drinking water supply, in states where residential development is permitted in such watersheds (Nieswand et al., 1990). In some countries, urban and agricultural land-use practices are greatly restricted or not permitted at all in watersheds of water supply reservoirs.

Buffer strips are ineffective on steep slopes with loose soils. Also their effectiveness is reduced by exposed soil on any part of the buffer strip, which can actually erode and contribute suspended solids and other pollutants instead of attenuating them. Woodard (1989) recommends that a porous organic "duff" layer and/or a dense growth of underbrush cover the mineral soil if buffer strips are to be effective. Environmental corridors in urban areas are typically a part of the major drainage system.

COLLECTION SYSTEM CONTROL AND REDUCTION OF DELIVERY OF POLLUTANTS

The second group of management practices for urban diffuse-pollution control involves methods and structures for removing pollutants from runoff after they leave the source area. Collection system controls refer to management alternatives for storm-water interception and transport. These include (Field, 1986) improved maintenance and design of catch basins, sewers, in-sewer (in-pipe) and in-channel storage, and elimination

Collection System Control and Reduction of Delivery of Pollutants 603

of sanitary and industrial cross-connections from separate storm sewers. Most of these facilities are a part of the minor drainage system.

Urban areas with good separate storm sewer drainage have a delivery ratio (the ratio of a mass of pollutants delivered to a stream divided by the mass of pollutants generated at the source) close to one, while residential areas with natural (swale) drainage have a delivery ratio ranging from several percent to about 50% (Novotny and Chesters, 1989). The objective of management practices is thus reduction of the delivery ratio. In addition to grass filters, buffer strips, and other practices mentioned in the preceding section, such practices in urban basins also include grassed waterways (natural drainage) and catch basins.

In separate sewer systems, sanitary and industrial wastewater cross-connections into separate sewers are a nationwide problem (Pitt et al., 1990a). A manual of practice for identification and control of cross-connections has been prepared (Pitt et al., 1990b). Most of the methods in this category require some engineering structures and design, and are more expensive than source-control measures. In the disturbed areas the polluted waters should be segregated as much as possible from the rest of the runoff. Segregation can be achieved by diversion dikes, culverts, and other drainageways constructed to divert upslope cleaner runoff from the source of diffuse pollution (disturbed area). Diversion structures can be temporary or permanent.

Grassed Waterways and Channel Stabilization

If runoff is allowed to leave the source areas without a conveyance system, rill and gully erosion will create an uncontrolled channel drainage network. Channel erosion may also occur if natural and man-made channels are damaged by the increased flows generally caused by urbanization or by construction activities (Fig. 10.14). To minimize or prevent this occurrence, collection systems must be designed in such a way that the shear stress caused by the flow on the channel bottom and banks is less than the resistance of the channel to erosion. Example 5.6 presented a method of estimating flow erosivity in evaluating the stability of the channels.

Several engineering means are available to mitigate channel erosion. The shear stress caused by flow is proportional to flow velocity, channel slope, and depth. Thus, collection systems should be designed to keep the velocity of the flow or the slope of the channel below a certain critical value representing the resistance of the channel to erosion. The resistance of unprotected earth-dug channels against erosion is very low. Therefore excavated channels and newly created roadside ditches should always be

FIGURE 10.14. Erosion of unprotected channels in construction zones. (Photo: University of Wisconsin.)

protected. Materials used for dry channel protection (such as roadside ditches) are straw mulch, matting (Fig. 10.15), grass (sod), wire mesh with stone, gabions, rip rap, and other less used and more expensive protective measures. These erosion-control measures should be implemented immediately after the channel is excavated.

In extreme, high-slope situations, channels can be lined with concrete or covered concrete pipes can be used. Lining of natural perennial streams with concrete for control of erosion or flooding is not recommended because such methods destroy aquatic habitat and would meet public resistance.

Grassed Waterways

Grassed waterways are probably the most inexpensive but most effective means of conveying water (Figs. 10.16 and 10.17). In a simple case, a grassed roadside swale will perform as well or better than a far more expensive buried storm sewer. If the grassed waterways are designed properly, in-channel erosion should be minimal, and the grass lining may even serve as a trap and treatment for pollutants.

FIGURE 10.15. Adding jute matting to give additional protection in a channel where grass cover has failed during excessive flush runoff events. (Photo: USDA, Soil Conservation Service.)

FIGURE 10.16. Grassed waterway in a residential zone. (Photo: USDA, Soil Conservation Service.)

FIGURE 10.17. Urban swale connection to an underground storm sewer. Straw bale provides detention and filtration.

FIGURE 10.18. Temporary sediment traps made of straw bales. Sod is laid immediately after excavation to prevent excessive channel erosion. (Photo: V. Novotny.)

Grassed waterways and roadside swale drainage are man-made channels of parabolic, triangular, or trapezoidal cross section. A typical swale is a shallow trench with side slopes of one (vertical) to three (horizontal). Swales are used solely for the conveyance of surface runoff, hence they are mostly dry and contain water only following a rain. Planted vegetation (primarily grass) is for soil stabilization and stormwater treatment. Grassed waterways are primarily used for the conveyance and treatment of runoff from surface mining and similar operations, and from agricultural and silvicultural lands (Chapter 11). As with grass filters, swales and grassed waterways remove pollutants by slowing down the flow, filtering by grasses, infiltration, and by nutrient uptake by vegetation. In contrast to grass filters, however, the grasses in swales are submerged, flow in the waterways is turbulent, and are therefore less effective in removing particulates than grass filters. The efficiency of grassed waterways (swales) in removing pollutants is about 30% (Oakland, 1983). Generally, the lower the slope and the velocity, the better the treatment performance of swales. Higher slopes can be reduced by check

TABLE 10.5 Maximum Permissible Design Velocities for Waterways

Cover	Range of Channel Gradient (%)	Permissible Velocity (m/sec)
Vegetative[a,b]		
1. Tufcote, Midland and Coastal Bermuda grass[c]	0–5.0	1.8
	5.1–10.0	1.5
	Over 10	1.2
2. Reed canary grass, Kentucky 31 tall fescue, Kentucky bluegrass	0–5.0	1.5
	5.1–10.0	1.2
	Over 10	0.9
3. Red fescue	0–5.0	0.75
4. Annuals[d]—ryegrass	0–5.0	0.75
Unlined Channels[e,f]		
5. Fine sand		0.5–0.75
6. Silt loam		0.6–0.9
7. Alluvial silts, colloidal		1.15–1.5
8. Fine gravel		0.75–1.5
9. Coarse gravel		1.2–1.8

[a] After Ree and Palmer (1949).
[b] To be used only below stabilized or protected area.
[c] Common Bermuda grass is considered to be a restricted noxious weed in Maryland.
[d] Annuals—use only as temporary protection until permanent vegetation is established.
[e] After Goldman, Jackson, and Bursztynsky (1986).
[f] Lower velocity is recommended for clean water, higher is allowed for silty water.

dams and/or by temporary barriers made of straw bails or cloth (Fig. 10.18).

The basic design criteria for grassed waterways are given in Tables 10.5 and 10.6 and in Figure 10.19. Table 10.5 shows the maximum flow velocities that would cause minimum erosion for different types of vegetation linings. Table 10.6 classifies vegetative linings into retardance groups, which basically express the hydraulic roughness of the vegetal cover. For design velocity of less than 1 m/sec, seeding and mulching are needed to establish vegetation. For design velocities over 1 m/sec, the waterway or swale should be stabilized with sod (bottom), the seeded sides protected by jute or excelsior matting, or with seeding and mulching, along with diversion of runoff until the vegetation is established.

Channels can be designed using the common Manning's formula

$$Q = A\frac{1}{n}R^{2/3}S^{1/2} = Av \tag{10.8}$$

where
Q = flow rate in the channel (m^3/sec)

TABLE 10.6 Classification of Vegetative Cover in Waterways Based on Degree of Flow Retardance by the Vegetation

Cover	Stand	Condition and Height	Retardance
Reed canary grass	Excellent	Tall (average 1 m)	A
Kentucky 31 tall fescue	Excellent	Tall (average 1 m)	
Tufcote, Midland and Coastal Bermuda grass	Good	Tall (average 30 cm)	B
Reed canary grass	Good	Mowed (30–40 cm)	
Kentucky 31 tall fescue	Good	Unmowed (average 50 cm)	
Red fescue	Good	Unmowed (average 40 cm)	
Kentucky bluegrass	Good	Unmowed (average 40 cm)	
Redtop	Good	Average	
Kentucky bluegrass	Good	Headed (15–30 cm)	C
Red fescue	Good	Headed (15–30 cm)	
Tufcote, Midland and Coastal Bermuda grass	Good	Mowed (average 15 cm)	
Redtop	Good	Headed (40–60 cm)	
Tufcote, Midland and Coastal Bermuda grass	Good	Mowed (6 cm)	D
Red fescue	Good	Mowed (6 cm)	
Kentucky bluegrass	Good	Mowed (5–12 cm)	

Source: After Ree and Palmer (1949).

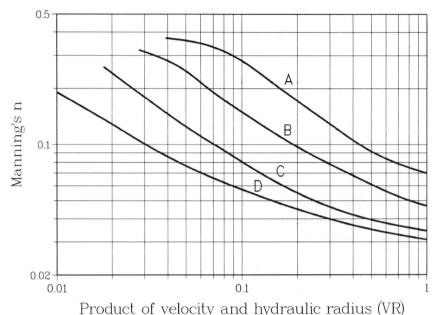

FIGURE 10.19. Manning's n for grassed waterways related to velocity v (m/sec), hydraulic radius R (m), and vegetal retardance. (After Ree and Palmer, 1949.)

A = cross-sectional area of the channel (m²)
n = Manning's roughness coefficient for the channel (Fig. 10.19)
R = hydraulic radius defined as cross-sectional area divided by the wetted perimeter (m)
S = slope of the energy line of flow, which under steady conditions equals approximately the slope of the channel bottom (dimensionless—m/m)
v = mean velocity (m/sec)

The roughness coefficient for grassed waterways depends on flow conditions and vegetative cover. Figure 10.19 shows the magnitude of the roughness coefficient for the four retardance classes defined in Table 10.6. In the figure Manning's roughness coefficient is related to the product of velocity and the hydraulic radius.

Example 10.4: Grassed Waterway Design

Determine the nonerosive velocity and dimensions of a trapezoidal waterway given that $Q = 2\,\text{m}^3/\text{sec}$, the slope of the channel is 5%, the bank slope is 1 (vertical) to 3 (horizontal), and the vegetation cover is a good stand of headed Kentucky bluegrass (C-retardance curve (see Table 10.6)).

From Table 10.5 the permissible velocity $v_{max} = 1.5$ m/sec. The solution is a trial-and-error process, since the Manning roughness coefficient in Figure 10.19 depends on the product of the velocity and hydraulic radius.

Solution First, make a first estimate of the cross-sectional area:

$$A = \frac{Q}{v_{max}} = \frac{2}{1.5} = 1.33 \, m^2$$

Then try depth of flow $H = 0.3$ m and calculate the width of the channel (note that the channel banks have slope of $1:Z$, where $Z = 3$. Then the cross-sectional area is

$$A = BH + ZH^2$$

where B is the bottom width, B is computed from

$$1.33 = B \times 0.3 + 3 \times 0.3^2$$

or

$$B = 3.53 \, m$$

The wetted perimeter (P) is then calculated

$$P = B + 2H\sqrt{Z^2 + 1} = 3.53 + 2\sqrt{3^2 + 1} = 5.43 \, m$$

and the hydraulic radius is

$$R = \frac{A}{P} = \frac{1.33}{5.43} = 0.24 \, m$$

Estimate the Manning's roughness coefficient using the product of velocity and hydraulic radius

$$vR = 1.5 \times 0.24 = 0.33 \, m^2/sec$$

From Figure 10.19, for a value of $vR = 0.33 \, m^2$/sec and C-retardance, the Manning factor $n = 0.043$.

Next compute v and compare with the initial assumption:

$$v = \frac{1}{n}R^{2/3}S^{1/2} = \frac{1}{0.043} \times 0.24^{2/3} \times 0.05^{1/2} = 2.03 \text{ m/sec}$$

which is not acceptable since $v > v_{\max}$. Therefore try $H = 0.2$ m.

$$B = \frac{A - ZH^2}{H} = \frac{1.33 - 3 \times 0.2^2}{0.2} = 6.05 \text{ m}$$

$$P = 6.05 + 2 \times 0.2\sqrt{3^2 + 1} = 7.31 \text{ m}$$

$$R = \frac{A}{P} = \frac{1.33}{7.31} = 0.18 \text{ m}$$

$$vR = 1.5 \times 0.18 = 0.27 \text{ m}^2/\text{sec}$$

From Figure 10.19, $n = 0.048$ and

$$v = \frac{1}{n}R^{2/3}S^{1/2} = \frac{1}{0.048} \times 0.18^{2/3} \times 0.05^{1/2} = 1.5 \text{ m/sec}$$

which is acceptable since the computed v agrees with the initial estimate.

Therefore, for $Q = 2.0 \text{ m}^3/\text{sec}$, design a grassed waterway about 6.1 meters wide at the bottom, which will result in a design flow depth of roughly 0.2 meter and a nonerosive flow velocity of $v \leq 1.5 \text{ m/sec}$.

Ripraps and Gabions

Riprap is a layer of loose rock or concrete blocks placed over an erodible soil surface. *Gabions* are blocks made of wire mesh filled with smaller rocks or gravel. Both ripraps and gabions are primarily used for channel stabilization in high erosion zones such as sharp bends, channel drops, and flow-energy dissipators (stilling basins) below the outlets from sewers, narrow bridges, and near connections with lined high-velocity channels. Note that energy dissipators are permanent structure that require hydraulic design. The Manning roughness coefficient, n, used for determining velocities and flows of riprap- (gabion-) lined channels can be obtained from Figure 10.20.

Sediment Barriers and Silt Fences

Sediment barriers and silt fences are small temporary structures used at various points within, and at the periphery of a disturbed area to detain runoff for a short period of time and trap heavier sediment particles. Sediment barriers are built from many materials (Goldman, Jackson, and Bursztynsky, 1986), such as straw bales (Fig. 10.18), filter fabric attached

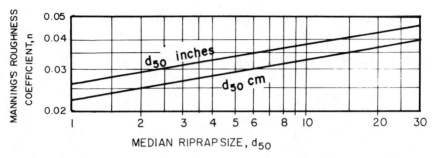

FIGURE 10.20. Manning's n for riprap surfaces. (After U.S. EPA, 1976.)

to a wire or wood fence, filter fabric on straw bales, and gravel and earth berms. These barriers are placed in the path of sediment-laden runoff, commonly from construction sites and surface mining. A sediment barrier should not be placed across a drainageway that carries a large volume of runoff.

Collection Sewer Control

Combined sewer overflows (CSOs) have been recognized as a significant source of diffuse urban pollution since the late 1960s and early 1970s (Sullivan, 1968; Graham, 1978). Initially, sewer separation was considered a solution; however, since then discussions about whether "to separate or not to separate" have continued (see Chapter 1). Nevertheless, combined sewer overflow pollution-control practices use unit processes and units that are either unique to CSO control or common to both types of sewer systems (for example, detention, hydrologic modifications, infiltration).

Catch Basins

A *catch basin* is a chamber or well, usually built at the curbline of a street, through which storm water is admitted into the sewer system. In contrast to a simple inlet chamber, catch basins are equipped to retain grit, detritus, and other sediment. Historically, the role of catch basins has been to minimize sewer clogging by trapping coarse debris and to reduce odor emanation from low-velocity sewers by providing a water seal (Field, 1990). Typically, a catch basin has a sump at its base. This sump should be large enough to provide storage for trapped debris. A Japanese version of catch basins in the Experimental Sewer System (ESS) described by Fujita (1984) includes a perforated removable bucket above the sump for trapping larger sediment and debris and a permeable bottom to the sump for enhanced infiltration.

Lager, Smith, and Tchobanoglous (1977) claimed that catch basins installed in sewer systems were ineffective, as they do not remove appreciable amounts of pollutants and the decaying organics trapped in them are actually a source of pollution. However, in a project conducted in Boston, Massachusetts, catch basins were shown to be quite effective for solids reduction (60–97%). Removal of associated pollutants, such as COD and BOD, were also significant (10–56% and 54–88%, respectively). To maintain the effectiveness of catch basins, they must be cleaned about twice a year, depending on local conditions (Aronson, Watson, and Pisano, 1983; Field, 1986, 1990).

Sewer Flushing

The solids that are deposited on the bottom of sewers and the biological slime that grows on the walls of combined sewers during the dry, low-flow period have been recognized as a problem for a long time. As a matter of fact, as documented by Krejci et al. (1987) this pollutant source may represent a major portion of the total pollution load in CSOs. Estimates of the magnitude of solids accumulation in sewers were presented in the preceding chapter.

Sewer flushing during dry weather is designed to periodically remove the accumulated material and convey it to the treatment plant downstream. Flushing is especially necessary in sewer systems that have low grades, as a result of which velocities during low flow periods fall below those needed for self-cleaning (typically, self-cleaning velocities are above 0.6 to 1 m/sec). It may be convenient to combine sewer flushing with street flushing, although street flushing from tanker trucks (Fig. 10.2) may not provide enough flow and thus should be used for solids removal in smaller-diameter laterals and trunk sewers (Field, 1990). Flow for flushing can also be provided by gates installed in some strategically located manholes (Novotny et al., 1989). Internal automatic flushing devices have also been developed for sewer systems. An inflatable bag is used to stop flow in upstream reaches until a volume capable of generating a flushing wave is accumulated, at which point the bag is deflated with the assistance of a vacuum pump, and the released sewage cleans the sewer segment (Field, 1990).

Regulators, Concentrators, and Separators

Regulators, concentrators, and separators are capable of separating solids from the flow of storm water. The dual-purpose swirl-flow regulator–solids concentrator has shown a potential for simultaneous quality and quantity control (Field 1986, 1990). A helical-type regulator–separator has also been developed. These devices have been primarily applied to CSO; however, they can also be installed as storm-runoff pollution-control devices. The concentrated flows that may amount only to a few percent of the total runoff flow can be stored and subsequently directed toward sanitary sewers for treatment during low-flow periods. The swirl concentrator was included in a CSO abatement program in Michigan (Pisano, Connick, and Aronson, 1985).

The *vortex solids separator* is a compact solids separation device. As early as in 1932, the idea of separating solids from CSO in a vortex chamber was conceived in England (Brombach, 1987, 1989; Pisano, 1989, 1991). The idea was pursued in the United States in the 1970s, resulting in a device known today as a *swirl concentrator*. Similar devices known as *fluidsep*, developed in Germany, or *Storm King*, developed in the United Kingdom, have also been implemented throughout the United States and Europe.

Vortex separation devices have no moving parts (Fig. 10.21). Dry-weather sewage passes unimpeded through the unit, if used as a combined sewer, in-line regulator. If the device is intended to operate as an

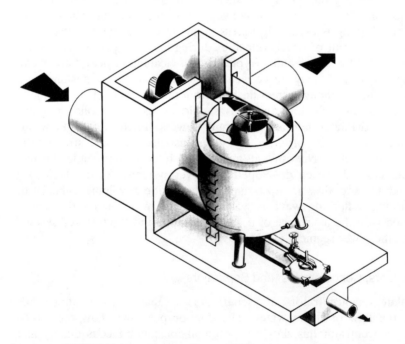

FIGURE 10.21. Fluidsep (vortex) separator. (Courtesy of W. Pisano.)

off-line treatment unit only, then storm flows are deflected into the unit by gravity or by pumps. During wet weather the unit's outflow is throttled (typically only 3% to 10% of the flow passes through the foul sewer outlet toward the treatment plant), causing the unit to fill and to self-induce a "swirling" vortexlike operation. Settleable grit and floatable matter are rapidly removed. Concentrated foul matter is sent to the treatment plant (or sent to temporary storage), while the cleaner, treated flow discharges into the receiving waters.

The three primary determinants governing the process performance of these devices are the hydraulic throughput, amount of internal fluid turbulence generated in the device, and the relative preponderance of particles whose settling velocities exceed 1 mm/sec (Pisano, 1991). Solids removal decreases with an increasing flow rate, and increases with more coarse (gritty) particle grain distribution. Any internal vessel protrusions that disrupt smooth vortex patterns cause decreased performance.

Vortex solids separators remove settleable matter by two mechanisms (Pisano, 1990, 1991; Field, 1990):

1. The sweeping action of solids by secondary vortex flow currents toward the centroidal axis of rotation
2. The transport of particles by gravity in the laminar sheet flow regime on the floor of the unit toward the same axis

There are some differences, none major, between the U.S. "swirl concentrator" and the German "fluidsep" (Pisano, 1990), but both work on the same principle. A design handbook for vortex separators was prepared by Sullivan et al. (1982). A partial list of U.S. installations with experience in their use was presented by Pisano (1989, 1991).

The design overflow rate for vortex separators installed in the United States ranges from 10 to 30 l/sec-m^2 (swirl concentrator) and 18 to 140 l/sec-m^2 (fluidsep), respectively. Up to 60% of suspended solids removal can be achieved, but generally the performance of such units is less than that for primary treatment.

A *helical bend regulator/concentrator* induces helical motion in a curved separator with a bend angle of about 60° and a radius of the curvature equal to 16 times the inlet pipe diameter. Dry-weather flow passes through the lower portion of the device to the intercepting sewer. As the flow increases during a wet weather period, the helical motion begins and the particles are drawn to the inner wall and drop to the lower channel leading to the treatment plant. The excess cleaner flow overflows over a weir into a CSO. The removal efficiency of helical bend separators is about the same as that of the swirl concentrator (Field, 1990). The

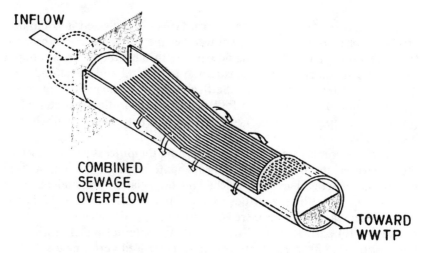

FIGURE 10.22. Self-cleaning screens for the control of pollution in combined sewer overflows and urban storm-water runoff. (After Krejci, 1988.)

handbook by Sullivan et al. (1982) also contains design parameters for helical bends.

Screening
Unlike screens in treatment plants that can be supervised, manually or mechanically cleaned, and where the screenings (material collected on the screens) can be removed and transported away, screens used for removal of solids form CSOs or urban runoff must be self-cleaning and reliable without supervision. An example of a self-cleaning structure developed and implemented in Switzerland is shown on Figure 10.22. In this simple fixture solids that are captured by the screens are washed by the flow that is diverted to the treatment plant. Pilot investigations indicated a treatment capacity of 0.2 to 0.4 m^3/sec-m^2 for screens without clogging (Novak, 1983; Krejci, 1988; Krejci and Baer, 1990).

DETENTION-RETENTION FACILITIES

Urban runoff entering a drainage system is a highly variable, intermittent phenomenon. In this case, storage and flow equalization, which can also be connected with pollution removal, is necessary and storage facilities must be included in the master plan for pollution abatement. Storage facilities also provide for maximum use of existing dry-weather flow treatment facilities, minimize their overload, and allow for the subsequent treatment of stored excess flow. Storage facilities range from inexpensive multipurpose terrain depressions with restricted outlets to expensive underground tunnels drilled in underlying bedrock (Milwaukee, Chicago). Wetlands also provide a combination of storage and treatment and are becoming an attractive alternative for storm-water and CSO pollution control. Today, ponds and storage basins are the backbone of urban storm-water quantity–quality management.

Detention Volume for Quality Control

Before the release of current storm-water quality control regulations ponds and detention basins were primarily used for peak attenuation from storm runoff rather than for quality control. As a matter of fact, many dry storage facilities designed previously for large design storms (5-year or greater recurrence interval) are ineffective for quality control. For example, Roesner (1988) has shown that in Cincinnati, Ohio, 74% of the total annual rainfall volume is contained in storms with a rainfall depth of less than 2.54 cm (1 in.). Considering the capture of the first 2.5 cm of rainfall would not only capture the total volume of a majority of storms (94% of storms in Cincinnati are less than 2.5 cm) but also two-thirds of the volume of larger than 2.5-cm storms. Hence, 91% of the runoff that falls on the watershed would be captured and subsequently treated. A 2.5-cm rainfall is substantially less than a 1-year storm. Similarly, Vitale and Spray (1975) reported that an 85% decrease in total BOD load can be realized by capturing the first 0.8 to 2.5 cm of the runoff. Hence, most guidelines for storm-water quality control call for the capture of rainfall depths of less than 2.54 cm.

As a rule-of-thumb calculation, consider a 2.5-cm rainfall falling on a residential watershed that is 50% impervious. Only flow from impervious areas will be considered because in most cases surface runoff is not generated from pervious areas by smaller storms. If surface storage of impervious surfaces is assumed to be about 0.15 cm (see Chapter 3), the flow generated from a 1-ha $(10,000\,\text{m}^2)$ catchment will be

$$0.5(2.5\,\text{cm} - 0.15\,\text{cm}) \times 0.01\,\text{m/cm} \times 10,000 = 117\,\text{m}^3$$

If one assumes further that the typical depth of the water in the pond is about 1 meter, then the planimetric extent of the pond surface area is $117\,\text{m}^2$, or slightly above 1% of the catchment area, or 2.2% of the connected impervious area. Typically, about 2% to 4% of the connected impervious area should be devoted to management of urban diffuse pollution by detention–retention facilities. The 90% capture and treatment of the total load from sewered urban areas is mandated in Germany. However, this volume, designed to capture small storms and the first flush of larger storms, is not sufficient to remove compounds such as nutrients that require longer detention times.

Sizing the CSO and Storm-Water Storage–Quality Control Basins

Kuo and Zhu (1989) presented designs for diversion systems for first flush overflow control. Griffin, Randall, and Grizzard (1980) developed a method by which it is possible to size the detention basin based on a predetermined portion of runoff (first flush) to be stored in the basin.

The sizing of storage basins can best be accomplished by continuous simulation models, since it involves the optimization (minimization of cost) of two components of the system, that is, the storage volume and the excess treatment plant capacity (the rate at which the storage basin is gradually emptied during the subsequent dry period). Few dynamic models are available for modeling the transient phenomenon of filling and emptying the storage tanks. Of the models discussed in Chapter 9, the Storm Water Management Model (SWMM) is capable of computing storage volumes; however, in most applications the SWMM is an event-oriented model. Lessard and Beck (1991) presented a conceptual model for dynamic simulation of the dynamic performance of storm retention tanks. Four modes of behavior have been identified: fill, draw, dynamic sedimentation, and quiescent settling.

Using popular so-called "design storms" to size the CSO storage facilities may not be appropriate, since in the United States the constraining parameter is the number of overflows allowed per year, while in Europe total stored and treated volume is the controlling variable These control parameters do not render themselves to the "excess design runoff" storage capacity determination that is used for flood control. The difference in storage capacity determination between flood control and water quality control is shown on Figure 10.23. However, these two constraints are similar, and their application may lead to a unified methodology, as pointed out by Roesner, Burges, and Aldrich (1991).

By continuous simulation modeling Roesner, Burges, and Aldrich have shown that the basin storage volume needed to capture a certain portion of runoff and the number of overflows are interrelated, as shown in Figure 10.24 and Table 10.7. To a lesser degree the storage volume is related to the excess treatment plant capacity, which regulates the outflow from the storage. The pumping or release rate of stored runoff (CSO) from the basin should provide a minimum of 24 hours retention (Grizzard et al., 1986; Schueler, Kumble, and Heraty, 1991). The storage volume

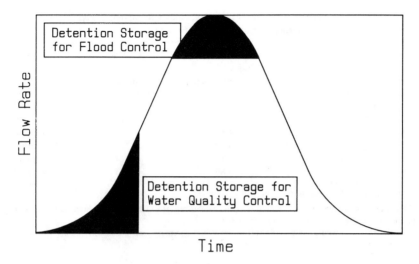

FIGURE 10.23. Storage strategies for urban flood control and pollution capture. (After Roesner, Burges, and Aldrich, 1991.)

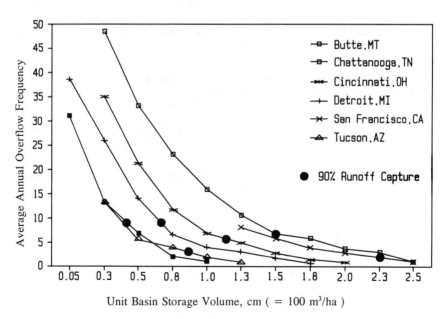

FIGURE 10.24. Relation of overflow frequency to unit storage volume. (After Roesner, Burges, and Aldrich, 1991.)

TABLE 10.7 Relation of Overflow Frequency, Storm Recurrence Interval, and Storage Volume Needed to Capture 90% of the Flow

City	Storage Volume Required for 90% Runoff Capture (cm)[a]	Overflow Frequency (times/year)	Recurrence Interval for Design Storm[b]
Butte, Montana	0.46	6	2 month
Chattanooga, Tennessee	1.52	10	1.2 month
Cincinnati, Ohio	1.14	8	1.5 month
Detroit, Michigan	0.69	12	1 month
San Francisco, California	2.29	4	3 month
Tucson, Arizona	0.89	3	4 month

Source: After Roesner, Burges, and Aldrich (1991).

[a] Depth of rainfall uniformly distributed over the catchment area. To convert to cubic meters of storage per hectare of catchment area, multiply by 100.

[b] The design storm would be a single storm that would completely fill the initially empty storage basin without an overflow.

of the sediment-control basins must also be sized to accomplish two functions, namely, to effectively remove a certain percentage of the suspended sediment, and to provide sufficient storage capacity for the removed sediment during a period of about five years or more between cleaning.

The planning and design of storm-water detention basins were covered extensively in the following publications:

American Association of State Highway and Transportation Officials (1980): Design of sedimentation basins

DeGroot (1982): Stormwater detention facilities

Ormsbee (1984): Systematic planning of dual purpose detention basins

Urbonas (1984): Summary of findings by ASCE Task Committee on detention basins

Stahre and Urbonas (1990): Stormwater detention for drainage, water quality, and CSO management

Schueler, Kumble, and Heraty (1991) A current assessment of urban best management practices. Techniques for reducing non-point source pollution in the coastal zone

Ponds and Detention Basins

Two types of detention basins are used for quality control of urban runoff. The first type is the *wet detention pond*, which maintains a permanent pool of water with additional storage designated to capture

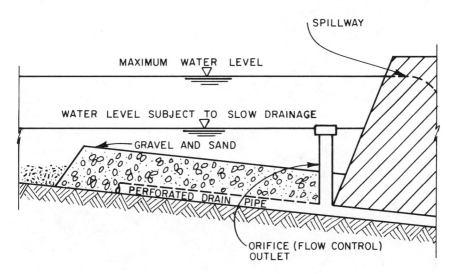

FIGURE 10.25. Modified (extended) dry pond.

transient storm runoff. The second type is the *extended* or *modified dry pond*, which provides a part of its storage capacity for enhanced settling of solids and auxiliary removal of pollutants by filtering.

The *dry pond* is a storm-water detention facility that is normally empty and is designed to hold storm water temporarily during high peak flows. The outlet is restricted and the storage volume is filled only during flows that exceed the outlet capacity. The pond also has a safety overflow spillway for the conveyance of very high flows when the storage capacity of the pond is exhausted. Since the outlet of dry ponds used for flood control is typically meant for large storms, smaller but polluting runoff events will pass through them without appreciable attenuation of the pollution load. Hence, such dry ponds are ineffective for urban runoff quality control. However, by combining the dry detention pond with an infiltration system located at the bottom of the pond, its pollution-control capability is enhanced. Modified dry ponds are effective pollution-control devices in both sewer systems (Fig. 10.25).

A *wet detention pond* has a permanent pool of water. This type of pond acts as a settling facility with relatively low efficiency (Fig. 10.26). Accumulated solids must be removed (dredged out) in order to maintain the removal efficiency and aesthetics of the pond. Improper design and maintenance can make such facilities an eyesore and a mosquito-breeding mudhole. On Figure 10.26 one can see that the pond located in a park is already partially filled with solids after about 10 years of operation and

FIGURE 10.26. Storm-water wet detention pond in Wauwatosa, Wisconsin. Note large sediment deposits in the middle of the basin, which should be periodically dredged. (Photo: V. Novotny.)

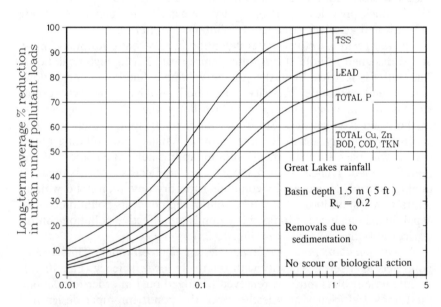

Basin area as % of contributing catchment area

FIGURE 10.27. Approximate removal efficiencies of conventional wet detention ponds. (After Driscoll, 1988.)

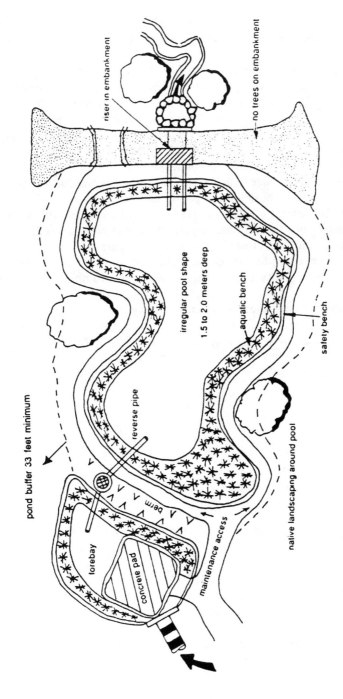

FIGURE 10.28. Enhanced high-efficiency wet pond system. (After Schueler, Kumble, and Heraty, 1992.)

should be cleaned out. The removal efficiency of wet ponds for constituents, obtained by a statistical analysis of NURP study sites is shown on Figure 10.27.

A well-designed wet pond consists of (1) a permanent water pool, (2) an overlying zone in which the design runoff volume temporarily increases the depth of the pool while it is stored and released at the allowed peak discharge rate, and (3) a shallow littoral zone acting as a biological filter (Fig. 10.28). Wanielista et al. (1982) described a system consisting of a detention pond with effluent infiltration, which increased the otherwise low efficiency of the detention pond. Similarly, Urbonas and Ruzzo (1986) stated that in order to achieve a 50% removal rate for phosphorus, a properly designed wet detention pond must be followed by filtration of infiltration. Schueler and Helfrich (1988) described an improved design for detention ponds that includes a permanent wet pool, extended detention storage, and storm-water storage. The perimeter wetland area created by the extended detention and stormwater storage provides additional water quality improvement (Schueler, Kumble, and Heraty, 1991).

The simplest design method for determining the size of sedimentation basins is based on the classic "overflow rate" theory of settling-tank design. The overflow rate is defined as the flow rate divided by the surface area of the pond. A particle is removed if

$$OR = \frac{Q}{A_s} \leq w \qquad (10.9)$$

where
OR = overflow rate (m/day)
Q = inflow in the basin (m³/day)
A_s = surface area of the basin
w = particle settling velocity

The particle settling velocity is estimated from Stokes' equation, or

$$w = \frac{gD^2}{18v}(\rho_s - 1) \qquad (10.10)$$

where
D = particle diameter (m)
ρ_s = specific gravity of the particle with respect to water
g = gravity acceleration (m/sec²)
v = kinematic viscosity (m²/sec)

Under the "ideal settling basin" assumption, all particles with settling velocities greater than OR will be trapped in the basin. The depth is determined from the required trap volume of the basin, which should be enlarged (at least doubled) to provide storage for the trapped sediment.

Example 10.5: Sedimentation basin design by overflow method

Design a settling basin that would remove sediment with particle sizes up to 0.03 mm. The basin receives flow from a 60% impervious watershed and should be large enough to store flow from a 2.5-cm rainfall that will reach the basin within one hour. Excess rainfall over 2.5-cm depth might be allowed to overflow. The average residence time of runoff in the basin is one day. The catchment area is 10 hectares. Kinematic viscosity is $v = 10^{-6}$ m²/sec, and the specific gravity of the particle is $\rho = 2.5$.

Solution   Estimate flow in the basin

$$Q = 0.6 \times 10\,[\text{ha}] \times 10,000\,[\text{m}^2/\text{ha}] \times 0.025\,[\text{m of rainfall}]$$
$$= 1500\,\text{m}^3/\text{hr}$$

From Stokes' law (Eq. (10.10)) estimate the settling velocity of the 0.03-mm- (0.00003 m) diameter particles:

$$w = \frac{gD^2}{18v}(\rho_s - 1) = \frac{9.81 \times 0.00003^2}{18 \times 10^{-6}}(2.5 - 1) = 0.000736\,\text{m/sec}$$

$$w = 0.000736\,\text{m/sec} \times 3600\,\text{sec/hr} = 2.65\,\text{m/hr}$$

The basin's surface area size is then (Eq. (10.9))

$$A_s = \frac{Q}{w} = \frac{1500}{2.65} = 566\,\text{m}^2$$

To account for nonideal settling, increase the calculated area by 20%, or

$$A_{\text{adjusted}} = 1.2 \times A_s = 1.2 \times 566 = 679\,\text{m}^2$$

Then the depth of the basin is $H = 1500/679 = 2.2$ m. This depth may be too large if safety considerations are included. Therefore the depth should be reduced and the surface area increased proportionally, which may improve the treatment efficiency of the basin.

The design of the settling basin based on the overflow rate is simple and has been used for some time for the design of sedimentation basins in sewage and wastewater treatment plants and for the control of sediment problems from strip mines (U.S. EPA, 1976). This method presumes uniform flow conditions at the outflow, so it may be inappropriate for the design of settling basins for storm water when discharge is not uniform and the surface area of the basin varies considerably during the storm event.

A better method of sizing sedimentation basins is based on investigations by Chen (1975). Figure 10.29 shows trap efficiency curves for a settling reservoir related to the reciprocal of the overflow rate. Note that the overflow rate is based on the outflow discharge, which is commonly less variable than the inflow rate. The plot also indicates that the earlier overflow models overestimated the trap efficiency for coarse materials.

Example 10.6: Efficiency of Sedimentation Basin for Solids Removal

Using Figure 10.29, estimate trap efficiency of a sedimentation basin with surface area $A_s = 680\,\text{m}^2$ and average inflow rate of $1500\,\text{m}^3/\text{hr}$. The particle size of the sediment is 0.03 mm:

$$A_s/Q = 680\,[\text{m}^2]/(1500\,[\text{m}^3/\text{hr}]) = 0.4533\,[\text{m}^2/(\text{m}^3/\text{hr})] \times 3600\,[\text{sec/hr}]$$
$$= 1632\,\text{m}^2/(\text{m}^3/\text{sec})$$

From Figure 10.29 the removal efficiency of particles with 0.03 mm diameter is approximately 60%. Complete removal is roughly achieved

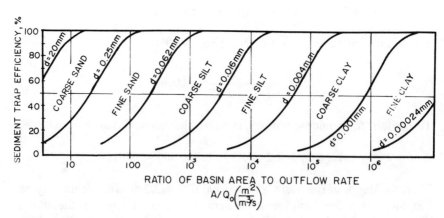

FIGURE 10.29. Settling-trap efficiency curves for detention basins related to top overflow rate. (From Chen, 1975.)

for particle sizes >0.062 mm. It thus appears that the 'ideal settling basin" concept overestimated the removal efficiency.

Whipple and Hunter (1980), reported settleability of some pollutants associated with urban runoff. By laboratory settling experiments they found that lead and hydrocarbons settled out at 60% to 65% in 32 hours. Slightly higher settleability was observed for suspended solids. The removal rates for BOD, zinc, and copper were smaller. Coliform bacteria were reduced by an order of magnitude. Driscoll (1988) evaluated and summarized pollutant trap efficiencies of wet ponds measured by the NURP studies (Fig. 10.27). Trap efficiency of storm-water management basins for pollutant attenuation was reviewed by McCuen (1980). Data collected in Maryland showed that a detention basin can trap as much as 98% of pollutants associated with the sediment. This high efficiency observed by McCuen contradicts the evaluation of detention–retention pond performance by Driscoll (1988). Efficiencies of detention ponds for quality improvement and settleability of various pollutants were also extensively covered by Ellis (1985).

Hartigan (1989) compared the removal efficiencies of modified (extended) dry ponds and wet ponds. He reported that removal of total phosphorus in wet ponds is 2 to 3 times greater than in modified dry ponds (50–60% versus 20–30%) and 1.3 to 2 times greater for total nitrogen (30–40% versus 20–30%). For other pollutants, the average removal rates for wet detention basins and extended dry detention basins were very similar: 80–90% for all the dissolved solids, 70–80% for lead, 40–50% for zinc, and 20–40% for BOD and COD. Negative removal efficiencies caused by resuspension of sediment during runoff events were observed by Dally and Lettenmaier (1984). Similarly, Mulhern and Steele (1988) measured negative removals for phosphorus in wet ponds during base-flow conditions.

A small lake in the Chicago area (a wet pond type) removed 91–95% of suspended sediment and 76–94% of copper, iron, lead, and zinc. The accumulated sediment in the lake was progressively reducing the available storage space and interfered with the aquatic habitat (Striegl, 1987). Nix and Tsay (1988) conducted an experiment in which stormwater flows were partially diverted at various levels into an off-line detention basin. The experiment showed that off-line storage basins are an effective means of controlling peaks and pollution by urban runoff. DiToro and Small (1979), Ferrara and Hildick-Smith (1982), Kuo and Ni (1984), Nix, Heaney, and Huber (1988), and Loganthan, Delleur, and Segarra (1985) presented mathematical models used for estimating the trap efficiencies of detention basins.

The "ideal solids settling" models are inappropriate if removal of pollutants that are partially dissolved is required. Such is the case of nutrients (nitrogen and phosphorus), which exist in urban runoff in both dissolved and particulate forms. For such designs Hartigan (1989) recommends using "a small lake eutrophication" model (for example, a model by Walker (1987)). The solids settling model is appropriate for simulation of removal efficiency for total suspended solids, a majority of toxic metals and organic toxic compounds that have a strong affinity for adsorption on suspended solids.

Typical Sizes of Ponds Combining Storage and Treatment

Hartigan (1989) pointed out that the size of wet detention ponds (which are apparently more efficient for removing nutrients) is 2 to 5 times the storage area of an extended dry pond. The storage volume of wet detention ponds is based on an average hydraulic detention time (typically two weeks), while the dry pond is designed to capture small rains and a first flush of larger rainfalls, as delineated in the previous section. Wet detention ponds are therefore recommended and cost-effective primarily for nutrient control. As an example, based on a typical average 2-week detention period for wet detention ponds and first flush capture for extended dry detention ponds located in Northern Virginia (Washington, D.C., metropolitan area), the sizes of the ponds are given in Table 10.8 (Hartigan, 1989). The depth of wet ponds should range between 1 and 3 meters. It is important that the side slope of the basin be very mild (5 to 10 horizontal to 1 vertical) to minimize the danger of drowning.

TABLE 10.8   Sizes of Wet and Dry Detention Ponds in Northern Virginia

Land Use	Percent of Imperviousness	Wet Detention (cm)[a]	Extended Dry Detention (cm)[a]
Low-density residential	20	1.8	0.25
Medium-density residential	50	2.5	1.0
Industrial/office	70	3.0	1.25
Commercial	80–90	3.3	2.0
Forest/park/undeveloped	0	1.25	0

[a] One cm of captured runoff for a watershed is equivalent to 100 m^3 of pond volume per one hectare of the basin area.

Dual Use of Detention Basins for Flood and Quality Control

From the foregoing discussion it follows that the design of storage facilities for flood and quality controls use different objectives and design criteria. Hence, facilities designed solely for flood control using statistically rare design storms may be ineffective for quality control. It is, however, possible to retrofit existing flood-control storage facilities, for example, by installing an additional small orifice and implementing a dual-level control strategy. This was shown in Figure 10.25.

Storage Basins for CSO Control

For CSO control, retention basins are often used to catch the first, most polluted portion of the overflow (the first flush). Storage facilities that have been employed for this purpose can be either *in-line* or *off-line* (Fig. 10.30) and include ponds or surface basins, underground tunnels, excess sewer storage, and flexible or collapsible tanks located in the receiving body of water. The in-line storage basins do not need pumps and the outflow is controlled by restricting the outlet. From off-line basins wastewater is pumped at a rate determined by the capacity of the next treatment facility for disposing of the stored overflows.

Off-line storage facilities are dry between rainfall events and typically collect the "first flush," more polluted portion of the overflow during large rainfalls or the entire overflow during smaller rains. The overflow occurs when the downstream capacity of the interceptor is exceeded. The flow is regulated in the flow divider (FD). After the basin is filled during longer rainfalls, the safety overflow of the basin (BO), located upstream or on the inflow is turned on and the rest of the overflow is then discharged into the adjacent receiving body of water (disinfection of the overflow may be required). Such basins are advantageous if the sewer system is not overloaded and the time of concentration is less than 15 minutes.

In-line flowthrough basins can have an additional clarifying overflow weir (OW) over which, after the basin is full, the mechanically clarified mixture of runoff and sewage is discharged into the receiving water. When incoming flow exceeds a predetermined critical flow (by regulating agencies and/or by design), the excess flow can be diverted in an upstream flow divider (BO).

Most of the sewers in a typical combined sewer system are designed with a capacity to handle 5- to 10-year storms; however, interceptors that carry water toward the treatment plant have a capacity of 4 to 6 times the

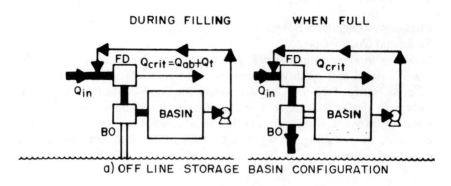

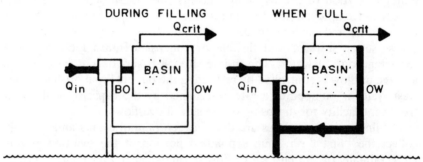

FIGURE 10.30. In-line and off-line storage basins for first flush control of CSOs. (After Novotny et al., 1989.)

maximal dry-weather flow. Hence sewers may have some unused capacity that can be used as in-line or in-pipe storage.

Ponds and surface storage are used to collect and store primarily "cleaner" and less malodorous storm water before it enters the collection system, as described in the preceding section. Once the runoff enters the combined sewer system and mixes with sewage, open air storage may not be recommended unless sufficient distance from inhabited areas is available for locating the storage basins or the basins are aerated and a part of the treatment plant. The storage basin must be surrounded by fences. Covered or underground storage basins are more appropriate but, consequently, far more expensive than open surface storage basins.

When geological underground formations are composed of soft but

FIGURE 10.31. Deep tunnel pumping station in Chicago, Illinois. (Photo: V. Novotny.)

FIGURE 10.32. Deep tunnel construction in Milwaukee, Wisconsin. (Photo: V. Novotny.)

solid rocks (such as limestone or dolomite), the drilling of underground storage tunnels and caverns may be economical. However, cavernous (Karst) limestone formations should not be considered for underground storage of CSOs. A system of underground tunnels for storage of CSOs has been implemented in Chicago, Illinois. Located 100 meters below the surface, the TARP (Tunnel and Reservoir Project) system, when fully completed, will have 190 km of 10-m-diameter underground tunnels for storage of CSO. The CSO can be subsequently pumped to one of the largest treatment plants in the world (Fig. 10.31). An almost identical but smaller tunnel system is in Milwaukee, Wisconsin (Fig. 10.32), and similar systems are now being considered in other U.S. cities (Cincinnati, Ohio; Boston, Massachusetts; Cleveland, Ohio). Instead of conventional blasting, these tunnels have been drilled by a special giant drilling machine and grouted for minimization of clean ground-water inflows. Up to 30 meters of tunnel can be drilled in one day by the machine.

Tunnels in fractured rocks or purely consolidated geological materials may require lining by concrete, which will escalate the cost. When constructing the so-called "deep tunnel" in Milwaukee, soft rock formations were encountered, which resulted in ground water flooding the construction site (the tunnel was 100 meters below the ground-water table in most locations, resulting in high static pressures) and large remedial expense. Pockets of underground methane are also a problem, as they were in Milwaukee and Chicago, and all mining safety precautions must be taken to prevent explosions.

Sizing of First Flush CSO Control Retention Basins

In a typical routine application, the retention basin is designed to capture the more polluted, first flush portion of the combined sewer overflow. Thousands of such first flush capture basins have been installed in Germany and other parts of Europe. U.S. CSO control regulations require larger capture volume, and the captured flow must be subsequently treated in the main or satellite treatment plant.

A simple method of determining the size of the capture basins was developed in Germany (Novotny et al., 1989). In this method first calculate the specific capacity, r_{ab}, of the treatment plant or of the downstream interceptor

$$r_{ab} = \frac{Q_{ds}}{DC \times A_{\text{imp}}} \text{(l/sec-ha)} \qquad (10.11)$$

where

Q_{ds} = capacity of the flow that is allowed downstream from the storage basin, l/sec

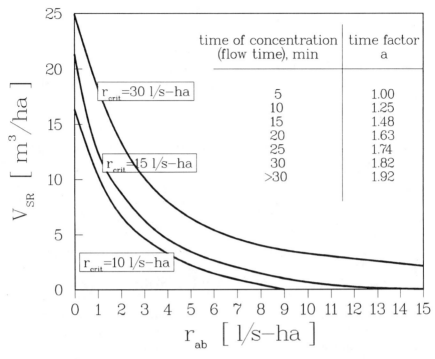

FIGURE 10.33. Dimensioning diagram for storage basins for first flush control of CSOs. (After Novotny et al., 1989.)

DC = fraction of directly connected impervious area of the drainage basin

A_{imp} = impervious area of the drainage basin

If r_{crit} is the minimum rainfall intensity at which overflow is allowed in l/sec-ha (to convert from mm/hr to l/sec-ha, divide by 0.36), then the specific detention volume V_{SR} can be obtained from Figure 10.33. The volume of the first flush retention basin is then obtained from

$$V = V_{SR}a(DC)A_{imp} \qquad (10.12)$$

where the coefficient, a, which is related to the time of concentration, is read from the table in Figure 10.33. Dimensioning of the first flush basin is shown in Example 10.7. This design provides 10 to 30 minutes detention in the basin at the critical flow $Q_{crit} = r_{crit} \times DC \times A_{imp}$. As shown on Figure 10.24 taken from Roesner (1988), r_{crit} can be related to the frequency of overflows.

Example 10.7: First Flush In-Line or Off-Line CSO Control Storage Basin

Estimate the size of the CSO control retention basin that would capture overflow resulting from rainfalls with an intensity of up to 10 mm/hr. The drainage basin of 200 ha is 60% impervious, the directly connected impervious area is about 65% of the impervious surface, the time of concentration (overland and sewer) is 20 minutes. Downstream interceptor capacity is $0.3\,\text{m}^3/\text{sec} = 300\,\text{l/sec}$.

Solution The impervious area $A_{imp} = (60\%/100) \times 200\,(\text{ha}) = 120\,\text{ha}$. Calculate the specific allowable flow, r_{ds}

$$r_{ds} = \frac{300}{0.65 \times 120} = 3.84\,\text{l/sec-ha}$$

Calculate the critical rainfall intensity at which overflow is permissible

$$r_{crit} = 10\,[\text{mm/hr}]/0.36\,[(\text{mm/hr})/(\text{l/sec-ha})] = 27.78\,\text{l/sec-ha}$$

By interpolation in Figure 10.33 the specific volume is obtained as $V_{SR} = 8\,\text{m}^3/\text{ha}$. The time factor for the concentration of 20 minutes is $a = 1.63$. Then the basin volume becomes

$$V = V_{SR}aDCA_{imp} = 8\,(\text{m}^3/\text{ha}) \times 1.63 \times 0.65 \times 120 = 1017.12\,\text{m}^3$$

Estimate the detention time at the critical rainfall rate

$$Q_{crit} = r_{crit} \times DC \times A_{imp} = 10 \times 0.65 \times 120 = 780\,\text{l/sec}$$
$$= 0.78\,\text{m}^3/\text{sec}$$

Detention time $t = V/Q_{crit} = 1017.12/0.78 = 1304\,\text{sec}/60 = 21.7\,\text{min}$.

Flow Balancing Retention Basins

The flow-balancing method (FBM) was developed in Sweden by Karl Dunkers (Field, Dunkers, and Forndran, 1990; Field, 1990). The features that make the flow-balancing method attractive are (1) this type of facility uses a portion of an existing receiving body of water, thus reducing the need for land space for siting, and (2) the materials used offer low structurally intensive and low-cost alternatives compared with more costly conventional construction on land.

Typically, an FBM detention facility is built of pontoons and curtains that form a series of interconnected rectangular tanks (Fig. 10.34). The

Detention–Retention Facilities

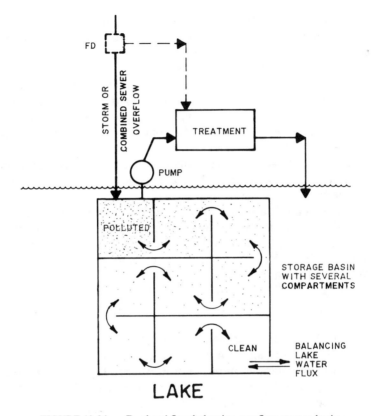

FIGURE 10.34. Dunkers' flow-balancing overflow storage basin.

arrangement of the tanks in series simulates so-called "plug flow" conditions, which minimizes mixing between the compartments and short-cutting. Typically, nine tank compartments suffice. If the facility is located in a body of water with brackish or saline waters, an interface between the less saline combined sewer overflow or storm-water will be formed in the tanks. For example, in a saline water environment, because CSO water is lighter than the saline water from the receiving body of water, the CSO will begin to fill the tank from the top.

Before the first overflow arrives after the detention facility is installed the entire volume of all tanks is filled with water from the receiving body of water (clean water). During the overflow event, polluted water from the overflow is directed into the first compartment. Polluted water will then begin to push clean water from the compartments, beginning with the first compartment, into which the overflow is directed. During storms that exceed the design storm volume, the process continues until the entire facility is filled. Only at this point can polluted water enter the receiving body of water. For all other smaller storms, the polluted water should be contained by the FBM facility.

After the CSO or storm-water discharge event, a floating pump located in the first receiving compartment starts to pump the polluted water back into the sewer interceptor for conveyance to the treatment plant. As clean water enters the facility from the opposite side, it pushes the polluted water back toward the first compartment. After clean water refills the system, pumping stops.

Solids and floatables that accumulate in the facility must periodically be pumped out. A majority of the solids typically settle out in the first two compartments.

The same methods are used to determine the size of the facility as are used for detention facilities located on land. An experimental–demonstration FBM detention facility was built in a tributary of Jamaica Bay in Brooklyn, New York (Forndran, Field, and Dunkers, 1989; Field, Dunkers, and Forndran, 1990).

Wetlands

Oberts (1982), studied 17 urban and suburban basins in the Minneapolis-St. Paul area and found that wetlands can have a significant impact on pollution attenuation of urban watersheds. His research found a significant inverse relation between the pollutant loads and the percent of the watershed area that is wetland.

Incorporation of wetlands into a comprehensive storm-water management system achieves several objectives: reduced operation/maintenance, aesthetic buffering, development amenities, wetland preservation, and restoration, and, hence, enhanced wetland value (Livingston and Roesner, 1991). Wetlands combine both sedimentation and biological utilization effects to remove pollutants from runoff. The largest pollutant reduction can be achieved during the "active" wetland growing season, that is, May to September in northern climatic conditions. During the dormant (winter) condition, wetlands can become a source of pollution that is leached from dead vegetation. In southern (Florida) climatic conditions wetlands efficiency remains more-or-less constant throughout the year.

Wetlands can be used in combination with other treatment, such as detention ponds (Wotzka and Oberts, 1988; Schueler, Kumble, and Heraty, 1991). A more detailed discussion and presentation of the design of wetlands for quality control of runoff and in-stream water quality restoration, as well as the impact of diffuse-pollution loads on wetlands, is presented in Chapter 15.

Both natural and man-made wetlands have been used for runoff pollution control; however, natural wetlands are considered to be receiving waters and are subjected to effluent restrictions. On the other hand, man-made wetlands are considered to be treatment facilities and the standards mostly apply to the outlet from the wetland (Hey et al., 1982; Stockdale and Horner, 1987; Linker, 1989; Hammer, 1989). Therefore, currently only man-made or restored wetlands can be used for the treatment of runoff and CSOs.

However, state-of-the-art use of wetlands for urban storm-water and CSO control is still evolving, and many issues have not yet been resolved or answered by long-term research (Livingston and Roesner, 1991). For example, how long can a wetland continue to accumulate pollutants without impairing its biological functions? How long can a wetland remove pollutants effectively? What type of maintenance (harvesting, sediment banking, and others) should be planned for a given wetland area?

TREATMENT

In areas that are highly developed and have existing combined or separate sewer systems, a large-scale structural treatment method may be the only feasible way to eliminate pollutants from storm sewer discharges and CSOs. With the exception of vortex separators, use of treatment unit processes requires storage, mostly underground (tunnels). Most large-scale treatment facilities used for the control of pollution by CSOs and urban runoff are a part of the dry-weather treatment facilities that are enlarged to accommodate the stored wet-weather flows. However, the Emscher River treatment plant in the Ruhr industrial district of Germany treats the entire flow of a river that is heavily polluted by municipal and industrial point and nonpoint sources before its confluence with the Rhine River.

Wetland systems and modified dry and wet ponds provide inexpensive treatment of urban runoff, and have been used extensively and successfully. Traditional treatment methods that are used for point sources are difficult to implement for non-CSO urban nonpoint sources due to their variable and intermittent nature. This is especially true for the microorganism-dependent (biological) treatment processes, although a report by Agnew et al. (1975) describes a relatively successful combined treatment of municipal wastewaters and stored overflows.

There are three types of treatment used for treating wastewater discharges, including those occurring during wet-weather conditions, namely, physical, physical–chemical, and biological. Some systems use a

TABLE 10.9 Efficiency of Various Storm-Water and CSO Treatment Processes

Process	Efficiency (%)				
	Suspended Solids	BOD$_5$	COD	Total P	TKN[a]
Physical–chemical					
Sedimentation					
Without chemicals	20–60	50	34	20	38
With chemicals	68	68	45		
Vortex separation	40–60	25–60	50–60		
Screening					
Microstrainers	50–95	10–50	35	20	30
Rotary screens	20–35	1–30	15	12	10
Sand–peat filters[b]	90	90	NA	70	50
Biological[c]					
Contact stabilization	75–95	70–90		50	50
Biodiscs	40–80	40–80	33		
Oxidation ponds	20–57	10–17		22–40	57
Aerated lagoons	92	91			
Facultative lagoons	50	50–90			

Source: After Lager et al. (1977).
[a] Total Kjeldahl nitrogen.
[b] After Galli (1990), the peat–sand filter are similar to biological anaerobic-aerobic slow filters. They are applicable for treatment of urban runoff.
[c] Biological treatment is feasible only for CSOs.

combination of two or all three types. For example, even simple systems such as wetlands provide a multitude of treatment processes to separate pollutants (see Chapter 14 for a description of pollutant separation and removal processes in wetlands and for wetland applications). The treatment efficiency of some processes is given in Table 10.9. Design parameters for treatment units are given in Novotny et al. (1989) and Field (1990). Field's summary also includes pilot and prototype performance and cost information.

Physical–chemical treatment processes have shown some promise in overcoming the shock loadings (Field, 1986). Chemical additives used to enhance separation of particles from liquid include chemical coagulants, such as lime, alum, ferric chloride, and various polyelectrolytes. During the process of chemical clarification, which consists of chemical addition, flocculation, and particle separation (clarification and/or filtration), the flocs are formed from finer particles, mainly by reduction of surface charge and/or by adsorption on surfaces provided by the added chemicals. Chemical addition is also effective for removal of phosphorus, toxic metals, and some organic colloids. Handling and disposal of produced sludge is a costly problem.

The physical–chemical processes that have been successfully demonstrated and/or installed include fine-mesh screening, fine-mesh screening/high-rate infiltration, sedimentation, sand and peat–sand filtration, fine mesh-screening/dissolved-air flotation, and swirl separator. A pilot plan study (Innerfeld et al., 1979) using high-rate filtration of CSOs was conducted in New York.

Sand filters represent a new application of an old technology for treatment of urban storm water (Schueler, Kumble, and Heraty, 1992). These filters can be used effectively for the control of runoff from small sites, or the treatment of runoff prior to entering another storm-water management structure (for example, before a vegetative filter), or as a retrofit strategy in an urban storm-water system. The design includes both sedimentation and filtration components (Fig. 10.35). Sedimentation removes the larger particles (grit and sand), and filtration removes silt and clay-size particles. The drainage area should be less than 2 ha of impervious surface, and the sedimentation chamber volume should be about $50 \, m^3$ per hectare of connected impervious area. The sedimentation chamber is followed by a filtration chamber that has the same volume.

Since sand is inert and has little or no adsorbing capacity, a dissolved fraction of the dissolved priority pollutants is not removed (unless an organic–microbiological population develops in the filter). However the top layer of the filter, which contains the organic matter, has to be periodically removed and replaced by clean sand.

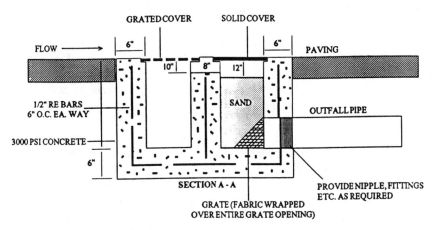

FIGURE 10.35. Sand filter used for storm-water pollution control. (From Schueler, Kumble, and Heraty, 1992.)

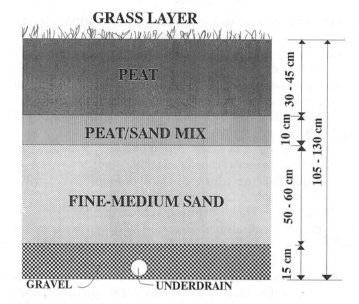

FIGURE 10.36. Peat–sand filter for storm-water treatment. (After Galli, 1990.)

Slow-rate filters, in which development of mostly anaerobic microbiological layer occurs, remove priority pollutants by microbiological action and by adsorption on organic particulates. Carlo et al. (1992) showed that a sand filter, when operated at higher loading rates, did not remove selenium, while when operated as an anaerobic slow-rate filter, its removal efficiencies for selenium were 74–97%. Selenium was retained in the organic layer of the filter.

Enhanced (peat–sand) filters utilize layers of peat, limestone, and/or top soil, and may also have a grass cover top (Fig. 10.36). Sand filters can be used in areas with thin soils, high evaporation rates, low soil infiltration, and limited space. Sand filters remove mostly particulate pollutants. To minimize clogging both sand and enhanced filters should be preceded by a solids-removing unit, such as a pond or a filter strip.

Peat–sand filters provide high phosphorus, BOD, and nitrogen removals, in addition to the removal of solids. For sizing the filters, Galli (1990) recommends that the filter area should be about 0.5% of the contributing watershed area, and the annual hydraulic loading should equal 75 m/year or less. As with most biological filtration systems, the peat–sand filter works best during the growing season, since a part of the nutrient load is taken up by grass. Also the filter should remain aerobic. The peat has a very high affinity for adsorbing and removing toxic compounds (see Chapter 14 for a discussion), hence peat-containing filters are effective for removing priority pollutants.

Biological treatment processes are effective for flows, such as CSOs, where organics are concentrated. Storm-water discharges have low BOD concentrations (on the order of biologically treated effluents), and are not amenable to biological treatment. Biological treatment processes do not perform satisfactorily or not at all when highly irregular flows and quality are treated. Because it is very difficult to keep the biota alive between events, conventional activated sludge can be used for treatment of CSO only in combination with dry-weather sewage treatment (Milwaukee and Chicago) and storage for CSOs to equalize the transient loads.

Other biological treatment processes that have been tested and/or implemented for treatment to CSOs include contact stabilization, attached media units (trickling filters and rotating biological contractors), and lagoons. The peat–sand filter described earlier also includes biological removal and uptake of biodegradable organics and nutrients.

Due to the permanent water content in lagoons, it is less difficult to keep active biota alive in them between events. Many storage ponds can be, and in fact are, used as biological lagoons. Lagoons can be categorized according to type and the amount of aeration they receive. Lagoons relying mostly on atmospheric aeration and photosynthesis for their oxygen supply are called *oxidation ponds*. Aerated lagoons rely on mechanical floating or mounted aerators for their oxygen supply. These more complicated unit operations are less suitable for CSO control.

Oxidation ponds are too shallow (0.2 to 0.5 m plus additional volume for sludge storage) to allow for light penetration. Their treatment efficiency depends on temperature and climatic conditions, and detention time is between 3 days (warmer climatic conditions) and 20 days (colder zones of the United States, Canada, and central and northern Europe). Lagoons and ponds can be used in combination with pretreatment or posttreatment. Typical inexpensive units include grass filters and/or wetlands and settling ponds. Without posttreatment the effect of lagoons on organic (BOD, COD) loads may be negative due to possible high algal growths.

Disinfection

Disinfection has been used primarily to control CSOs. Disinfectant and contact times are usually long, and research has concentrated on the use of high-rate applications by static and mechanical mixing, higher disinfection dosages, and more rapid disinfectants, such as chlorine dioxide, ozone, and ultraviolet light (Field, 1986). Adding chlorine to storm water and CSOs is questionable in today's systems. Chlorine, in addition to being toxic to bacteria and viruses, is also a strong oxidizing agent for other organics and for some reduced chemical components, such as iron, manganese, sulphides, and ammonia, that may be present in water. Since these oxidation processes must be accomplished before chlorine acts as a disinfectant, from 20 to 40 mg of Cl_2/l may be required to kill bacteria. Some chlorinated organic matter of natural (for example, humates) and man-made origin found in urban runoff and wastewater are carcinogenic and subject to regulation.

EFFICIENCY OF BEST MANAGEMENT PRACTICES FOR CONTROL OF PRIORITY POLLUTANTS

Dissolved versus Total Concentration of Priority Pollutants in Runoff

In Chapter 6 and throughout the book it was pointed out that priority pollutants can exist in the dissolved and adsorbed–precipitate phases. The proportion between these two phases is basically given by the solubility of the pollutant, which for nonpolar organic chemicals is given by the octanol partitioning coefficient, K_{ow}. A similar partitioning concept can also be applied to metals and some polar compounds, for which the partitioning coefficient would be pH dependent.

Few investigators have measured both soluble and particulate fractions of priority pollutants in urban runoff and other similar discharges. Some of the findings were presented in Chapter 8. Whipple and Hunter (1980) found that petroleum hydrocarbons in urban runoff are about 4.5% soluble. The flow-weighted total hydrocarbon concentrations measured in North Philadelphia, Pennsylvania, and Trenton, New Jersey, urban storm discharges ranged from 2.16 to 5.9 mg/l. The following hydrocarbons were detected: Di-benzo-thiopene and PAHs, such as napthalene, anthracene, pyrenem fluoranthene, chrysene, and benzo(a)pyrene.

Herrmann and Kari (1990) measured particle-size distribution of PAHs and PCBs in urban runoff, and found that most of these contaminants can be found in particle-size fractions between 2 to 63 µm (silt size), while the clay fraction ($d < 2\,\mu m$) was less contaminated. It can be suspected that the organic particulates to which these contaminants are most attracted are in these size fractions. The authors also found that the priority pollutants with high partition coefficients ($K_{ow} > 10^5\,l/kg$) exhibited a more pronounced "first flush effect" than those that are more dissolved. The first flush phenomenon indicates that the load of the pollutants is limited by the "supply" on the street surface. Most of the PAHs are associated with street dust (see also Herrmann, 1984).

Morrison et al. (1984) found that in urban runoff zinc and cadmium exhibited preference for the dissolved phase, whereas lead predominaned in the suspended phase. Similarly, Hewitt and Rashed (1992) found that the particulate phase (>0.45 μm) in the contents of highway runoff in England contained more than 90% lead, 70% copper, and 50% cadmium.

Data on urban runoff quality collected by Pitt and Baron (1989) showed that most of the organics and metals were associated with non-filterable (particulate) sample fractions. However, some organic pollutants and zinc had significant filterable fractions. The organic pollutants with high filterable portions included bis(2-chloroethyl) ether (19–50% filterable), 1,3-dichlorobenzene (31–100% filterable), hexachloroethane (93%), napthalene (2–79%), anthracene (25%), benzyl butyl phthalate (33%), fluoranthene (5–100%), and pyrene (15–100%). These compounds have an octanol partitioning coefficient between 10^3 and 10^5.

Most of the reports and manuals on best management practices report the removal efficiencies for a few conventional pollutants, such as suspended sediment. However, such information may be misleading when attempting to estimate the efficiencies of BMPs for priority pollutants. The pollutants exist in water–sediment systems either in a dissolved–dissociated state in water or adsorbed onto fine sediments and particulate organic matter. Sand particles do not adsorb most priority pollutants, and BMPs that effectively remove coarser particles only may be ineffective for removal of priority pollutants. A swirl concentrator is an example of a control unit that removes mostly coarser particles from runoff, but may be ineffective for the removal of priority pollutants.

The proportion between the dissolved and adsorbed portions of a priority pollutant in water depends on its adsorption characteristics—typically its octanol partitioning number—and the concentration of fine particles—primarily particulate organic matter. Figure 10.37 shows a simplified proportion of an adsorbed fraction versus the total (particulate and dissolved) content of a priority pollutant related to its octanol partitioning characteristics and content of suspended organic particles indicated as volatile suspended solids.

Particulate organic carbon (POC) is the best surrogate parameter for estimating the removal efficiencies of nonionic toxic chemicals (DDT, PCBs, PAHs, and a large number of organic pesticides) with a higher octanol partitioning coefficient ($K_{ow} > 10^5$ l/kg). The POC measurements are relatively simple and should be made a routine part of chemical analyses of waste loads from point and nonpoint urban sources. In the absence of the POC measurements, volatile suspended solids (VSS) parameters may be substituted. The efficiency of various abatement

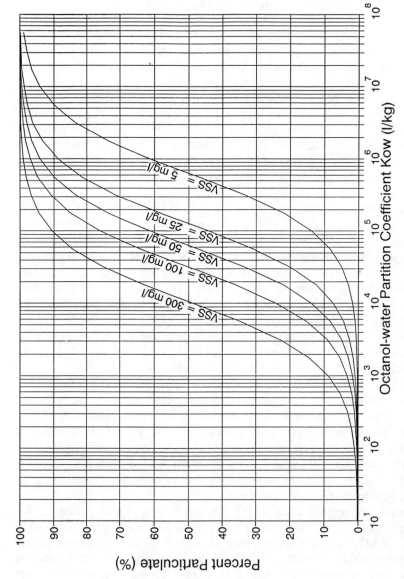

FIGURE 10.37. Relationship of dissolved and total concentrations of organic priority pollutants related to the octanol partitioning number and volatile suspended soilds content of runoff.

practices for the removal of nonionic organic priority pollutants may be correlated to the removal or fate of organic particulate matter, although the fate models may not always be simple (for example, for processes that produce organic matter such as eutrophic ponds). Compounds with a low octanol partitioning cannot be removed with particulates, and the removal efficiency is related to their biodegradability and volatilization.

Toxic Metals
When toxic metals are added to water, both from natural and man-made sources, they undergo complexation with ligands. Ligands are chemical constituents, both organic and inorganic, that combine with the metals in a chemical complex. From the basic physical chemistry it is known that metals precipitate because of changes in pH, when oxidized, or when they change their chemical composition. However, the process of complexation in natural waters is more complex.

Of the metal complexing mechanisms that were described in Chapter 6, iron and manganese oxyhydrates provide the strongest adsorption sites, followed by particulate organics and clays. The adsorption reactions are strongly pH dependent, ranging from zero adsorption in low pH to 100% adsorption in higher pH. The change from no adsorption to full adsorption is fairly sharp, usually over two pH units (Salomons and Förstner, 1984). However, at pH > 7 most of metals are complexed, while at pH 5 the concentrations of free metal ions increase dramatically. The methylated compounds are still strongly polar and in sediments behave similarly to free metal ions, that is, they are adsorbed by the sediments and ligands in the sediment.

The best surrogate parameter for estimating efficiency of runoff and waste pollution-control facilities for metals would be suspended sediment without its sand fractions and particulate organic carbon. The efficiency of removal of metals can be correlated to the efficiency of removing clay and silt fractions of the sediment and POC (VSS) parameters. However, only lead appears to be strongly associated with particulates. Other toxic metals are typically found both in dissolved and particulate phases.

Removal of Priority Pollutants or Toxicity by Best Management

Practices and Treatment
The objective of best management practices (BMPs) for control of priority pollutants has to be specified. There are two criteria by which the efficiency of BMPs can be judged; one is the removal of the total mass of a pollutant or pollutants, and the other is the removal of toxicity.

Typically, storm-water permits rely on two sets of standards. One involves numerical limits on concentrations of priority pollutants; the second requires testing for the toxicity of the runoff. The methods used for toxicity testing must conform to EPA-recommended guidelines for conducting static acute toxicity tests (see Chapter 12 for a discussion on toxicity).

The EPA has also published guidelines pertaining to the performance of toxicity reduction evaluation (TREs) for discharges from municipal and industrial wastewater treatment plants. It is certain that most of the major elements of TREs are also applicable to BMPs for storm-water discharges (Collins, Roller, and Walton, 1992), although very little information is available on the performance of various BMPs regarding toxicity reduction. According to these guidelines the efficiency of BMPs is tested using the cumulative acute toxicity mesurements.

The partitioning theory of priority pollutants specifies that a majority of the priority pollutants in an aqueous environment exist as either dissolved or adsorbed precipitates and are associated with particulates. Numerous studies have proved that only dissolved pollutants are bioavailable, and therefore, toxic. This is true of both metals, where the divalent ion Me^{++} has been identified as a toxic compound, and for organics. In a simplified way, Equation (6.8) provides the key to understanding the relation of toxicity removal to total pollutant removal in a BMP process designed primarily to remove solids. This is explained in the following example.

Example 10.8: Removal of Toxicity by Best Management Practices

Suppose that storm-water runoff contains 1 mg/l (1000 µg/l) of a priority pollutant that has the octanol partitioning coefficient $K_{ow} = 10^5$ l/kg. The runoff also contains 300 mg/l (0.3 g/l) of suspended solids, of which 10% are organic carbon—$f_{OC} = 0.1$. A BMP facility provides 80% removal of total solids and 65% removal of OC.

Solution Calculate total and toxic (dissolved) concentrations of the priority pollutant before and after treatment.

BMP Influent From Equation (6.20a)

$$K_{OC} = 0.63 K_{ow}/1000 = 0.63 \times 10^5/1000 = 63 \, \text{l/g}$$

The partitioning coefficient is then (Eq. (6.19))

$$\Pi = K_{OC} \times (OC)/100 = 63 \times 10/100 = 6.3$$

The dissolved (toxic) priority pollutant concentration is then calculated from Equation (6.8) as

$$c_{d\text{-in}} = \frac{c_{Tin}}{1 + \Pi m_{ss}} = \frac{1000}{1 + 6.3 \times 0.3} = 340 \, \mu g/l$$

The adsorbed (nontoxic) fraction of the pollutant in the influent is then $1000 - 340 = 660 \, \mu g/l$.

BMP Effluent The BMP facility removed 85% of solids and 65% of OC. Hence the effluent solids concentrations is $(1 - 0.85) \times 300 = 45 \, mg/l$, and the OC concentration is $(1 - 0.65) \times 0.1 \times 300 = 10.5 \, mg/l$ and the organic carbon fraction $f_{OC} = 10.5/45 = 0.23$. Since the particulate fraction of the pollutants is associated with organic carbon, the same removal rate (65%) applies to the removal of adsorbed pollutant. The dissolved fraction is not removed by the facility. Hence after the treatment the total concentration of the pollutant in the effluent is

$$C_{T\text{-ef}} = (1 - 0.65) \times 660 + 340 = 571 \, \mu g/l$$

The facility removed $100 \times (1000 - 571)/1000 = 43\%$ of the total mass of the contaminant.

A new equilibrium is reached in the effluent between the dissolved (toxic) and adsorbed (unavailable) fractions of the contaminant. The new partition coefficient is

$$\Pi = K_{OC} \times f_{OC} = 63 \times 9.23 = 14.5 \, l/g$$

and the dissolved concentration is

$$C_{d\text{-ef}} = \frac{C_{T\text{-ef}}}{1 + \Pi m_{ss\text{-ef}}} = \frac{571}{1 + 14.5 \times 0.045} = 345 \, \mu g/l$$

which indicates that no reduction of toxicity was achieved by the facility.

This example illustrates the dilemma planners of BMPs for storm-water toxicity reduction are facing. Apparently, toxicity cannot be reduced by BMPs relying on settling only, so other mechanisms such as biodegradation, volatilization and adsorption on additional organic matter must be considered.

Conclusions on the Efficiencies of BMPs

A detailed evaluation of the efficiencies of BMPs for removing priority pollutants was prepared by Novotny et al. (1992), and a summary paper was published by Scholze, Novotny, and Schonter (1993). The U.S. EPA Risk Reduction Engineering Laboratory (RREL) in Cincinnati, Ohio, has compiled an extensive computerized "Treatability Data Base" of many chemicals, including all priority pollutants. The information contained in the data base was retrieved from the literature. The data base also considers a large number of treatment processes, such as lagoons (both aerobic and anaerobic), suspended and attached growth reactors (both aerobic and anaerobic), physical chemical coagulation–flocculation processes, dissolved air flotation, ion exchanges, filtration and ultrafiltration, reverse osmosis, and electrodialysis. Altogether 35 different treatment unit processes are considered. However, only a few can be considered as BMPs for the control of runoff pollution.

The following conclusions are pertinent when BMPs are proposed for control of priority pollutants.

1. Priority pollutants are a category of pollutants that in many aspects differ from such traditional pollutants as biodegradable organic matter and nutrients. Hence not all of the BMPs designed for the removal of traditional pollutants may be effective for the removal of priority pollutants.
2. Although a large number of pollutants have been classified as *priority*, the pollutants themselves and their treatability and treatment-removal process selection may be characterized by a few parameters, including octanol partitioning number for polar organics, biodegradability, and volatility.
3. Because most of the priority pollutants are effectively immobilized by organic matter, the most effective removal processes include organic matter in some form. The organic matter is provided in several effective forms, such as vegetation and its residues, peat and other wetland vegetation and its residues, or particulate organic solids contained in runoff.
4. BMPs designed to remove solids, such as settling ponds, sand filters, and microstrainers, can remove the mass of priority pollutants associated with particulates, but are not effective for removing toxicity that is attributed to the dissolved components of the priority pollutants.
5. Most effective BMPs are pond enhanced with wetlands, wetlands themselves, water hyacinth ponds, overland flow systems, and sand–

peat filters. Typically these BMPs should provide 80% to 99% removal of priority pollutants.
6. Some priority pollutants, such as toxic metals and several organics, are best immobilized in an anaerobic environment by binding to sulfides and other complexing ligands, or are best biodegraded in an anoxic or anaerobic environment. Wetlands, sand–peat filters, and facultative lagoons provide such a reducing environment.
7. Priority pollutants with high volatility can be removed in systems that provide a high degree of aeration and exposure to the atmosphere. Such systems include aerated ponds (followed by another unit to remove the remaining dissolved and particulate fractions of the pollutant), and overland (sheet flow) flow systems.
8. BMPs with moderate removal efficiencies include most ponds, grassed swales, sand filters, porous pavements, and microstrainers. Since such systems primarily remove solids, their efficiency for removal of pollutants with low octanol partitioning and toxicity reduction may be diminished.
9. BMPs using infiltration may be moderately effective for the removal of pollutants with a higher octanol partitioning number and for some metals, but these methods are unsuitable for the control of pollutants with a low octanol partitioning number and for some metals (for example, zinc) that have a high dissolved fraction in the runoff.
10. Street sweeping and similar surface control of pollutants may be ineffective, although they may be required for aesthetic enhancement and for maintaining the cleanliness of the area.
11. Due to the multiplicity of priority pollutants present in runoff, there is no single BMP that would be effective for all priority pollutants. Typically a combination of units is required. Proper and effective combinations of unit processes are still being researched.

REAL-TIME CONTROL OF CSOs AND POLLUTION LOAD TRADE-OFFS

An urban sewerage system, including all of its components (sewers, storage, main and satellite treatment, and disposal) is a time-variable–dynamic system. Input into the system—rainfall, sewage and industrial wastewater flows, infiltration—and the treatment processes themselves are highly variable, with ranges of parameters often over one order of magnitude. Process upsets, malfunctions, monitoring errors, labor force behaviors, and management communications are additional factors contributing to the variability of the processes and parameters.

Traditionally however, these systems have been designed and operated using time-invariant (fixed), rather simple models and rules and criteria. For example, as specified previously, a combined sewer system typically has the design capacity of four to eight times the dry-weather flow contributions, which is also a fixed parameter. Use of storage facilities and treatment of overflows make the system more dynamic; however, the operation can still be based on static invariant control. Instead of the hydraulic capacity of the interceptor, which determines whether the overflow is initiated, the capacity of the storage basin controls the overflows.

It is apparent that an urban sewerage system has a number of so-called "bottlenecks" or control–decision points. These bottlenecks include the interceptor capacity, storage capacities, and the capacity of the treatment plant. If the flow or pollutant load exceeds the capacity of the bottleneck, an overflow results. In the treatment process there are a number of points where a bottleneck situation may occur. The most important control point is the secondary clarifier of the biological treatment units. If the hydraulic or solids loading of the final clarifiers is greater than its cap-

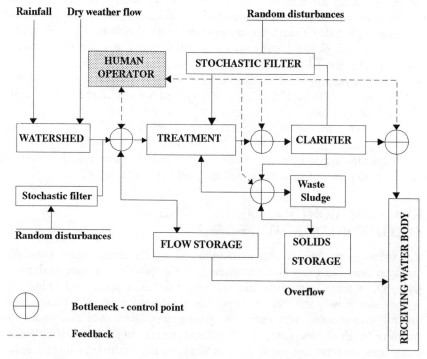

FIGURE 10.38. Schematic of an urban sewerage system with control (bottleneck) points. (After Novotny and Capodaglio, 1992.)

acity, a massive spill of solids into the effluent can occur. The effluent standards and the waste-assimilative capacity of the receiving body of water are the final bottlenecks of the system. Figure 10.38 shows a typical schematic of an urban sewerage system with the locations of possible bottleneck points.

The invariant steady-state mode of operation typical of a majority of sewerage and wastewater disposal facilities dictates that when the input to a bottleneck-control point exceeds its capacity, the excess is bypassed and discharged without treatment. The steady-state solution is to enlarge the bottleneck, which is costly. This mode of operation is inefficient and results in large quantities of untreated sewage being discharged into receiving waters on the one hand, but on the other hand, there may be unused or idle capacity in some other places throughout the system. Furthermore, funds for structural enlargement of existing capacities of sewers and treatment facilities to accommodate transient wet-weather loads may be limited. Thus a conversion of existing facilities to a dynamic operational-control scheme or a conversion of existing facilities to such control may be an economical and more efficient alternative to a massive structural enlargement of sewer and treatment facilities.

Dynamic control of CSOs implies a computerized monitoring, forecasting, and optimization system by which the drainage system, including pipes, overflows, and storage can be controlled and optimized. Most of the research has been conducted in Europe and reported in the Proceedings of the International Conferences on Urban Storm Drainage (Balmér, Malmqvist, and Sjöberg, 1984; Gujer and Krejci, 1987; Iwasa and Sueishi, 1990) or in a special International Association for Water Pollution Research and Control (IAWPRC) publication edited by Schilling (1989).

Objectives and Strategy of RTC Systems for Control of Urban Sewerage

The objective of RTC control is to provide optimization and minimization of the total pollution load of a joint conveyance and disposal of dry-weather (sewage and wastewater) and wet-weather (storm runoff) flows. However, the RTC systems have been researched and proposed mostly for the control of CSOs (Schilling, 1989). Only recently has their use been extended to consider the treatment facility in a pollution trade-off (Novotny, Capodaglio, and Feng, 1990; Durchschlag and Schilling, 1990). Note that control and minimization of overflows require treatment and storage, and the operation of the two components should be optimized in a coordinated fashion. The alternative objectives that have been presented for the control of urban sewerage systems are (Novotny and Capodaglio, 1992):

1. *Minimization of discharges of untreated overflows.* This objective was suggested in several RTC systems dealing with combined sewer overflows (CSOs). In this type of control, storage and release from storage, and overflow within the system are manipulated to minimize the volume of the untreated mixture of rain and wastewater from CSOs (Labadie, Morrow, and Chen, 1980). Several modifications of this control strategy have been proposed by German researchers (Weynard and Dohman, 1990; Kollatsch and Schilling, 1990).
2. *Stability of the treatment process and meeting the effluent standard.* This is a typical control objective and strategy of manual control for PTW. In this system the operator maintains stable MLSS and dissolved oxygen throughout the treatment process, and maintains the effluent BOD and SS below the effluent standards. The control variables in this type of RTC are the return sludge flow, excess sludge, sludge waste, and the air flow into the aeration basins. Avoidance of sludge bulking is another objective in this category.

    This type of control mostly responds to the present situation occurring in the plant, unless the operator is capable of anticipating and predicting the plant overloads and upsets. Although bypasses are not allowed by law (with exceptions stipulated by the U.S. EPA), the control strategy itself does not pay much attention to what happens before treatment. The excess flow over the capacity of the treatment plant is automatically bypassed.
3. *Minimization of total pollution load.* This objective considers both the effluent discharges and bypasses–overflows from the system, and minimizes the total pollution load from the system. The system recognizes the excess capacity of the sewer system to temporarily store a part of the wet-weather flow and the full dynamic capacity of the treatment plant. Auxiliary off-line storage for excess influent flow and excess solids enhances the capability of the system to handle and treat wet-weather flows (Novotny, Capodaglio, and Feng, 1990). Since this control focuses on the treatment facility, overflows from distant CSOs within the sewer system cannot be considered. The control variables are a manipulation of the off-line storage of flow and solids, return sludge flow, and sludge waste. Nelen (1990) proposed a control objective based on the water quality impact of the overflows, which depends on the total pollution load and conditions of the receiving water body.

4. *Avoidance or minimization of "bottleneck" situations.* This control objective was formulated by Capodaglio (1990). A bottleneck occurs when:
   a. the capacity of a process unit within the system is exceeded due to
      (1) flow exceeding the hydraulic capacity of the unit
      (2) mass load exceeding the mass loading limit of the unit
   Exceeding these limits would lead to a drastic reduction of treatment efficiency and/or overflow. Such limitations are caused by the hydraulic capacity of the units, limited aeration capacity, settling characteristics of solids, and other factors.
   b. effluent or stream quality limitations would be exceeded
      This factor represents a failure of the total system similar to the partial failure specified previously.
   c. capacity of residue (sludge) disposal units would be exceeded
      Temporary auxiliary sludge storage may be needed to handle this bottleneck situation. Many plants employ digesters and/or sludge lagoons for the storage of sludge. Including these facilities in the RTC system enhances the dynamic efficiency of the treatment process.
5. *Control strategy by Tan et al. (1990).* A group of Australian RTC researchers specified the following for an RTC urban sewerage system:
   a. Main goals
      (1) minimize overflows
      (2) reduce peak energy consumption of pumping (lift) stations
      (3) reduce treatment cost at the downstream treatment plant
   b. Constraints
      (1) the capacity of the treatment plant
      (2) pumping capacity of lift stations
      (3) in-line storage capacity

Additional alternative control objectives were presented and discussed by Béron et al. (1984), Schilling (1986, 1989), and several other authors.

Strategy

In a typical RTC system, pumps, valves, gates, etc., have to be operated to store wastewater or solids in-line or off-line before the bottleneck and route them to subsequent treatment and receiving waters. It is highly undesirable to create flooding, overflows, and/or violation of standards if the system has an unused storage capacity upstream or downstream from the bottleneck at the same time.

The time sequence of the set points of all the regulators in an RTC system is termed *control strategy*. The determination of a control strategy can be either automatic (built-in) or manual. The strategy can be found through mathematical modeling, optimization, search, decision matrices, control scenarios, trial-and-error (heuristically), or through a self-learning expert system. This can be done either during the on-going process (online) or beforehand, after predictions (off-line). After the control strategy has been defined, it is distributed throughout the system for implementation.

Components of RTC Systems

The RTC system represents a set of models and concepts that is used for optimal and flexible control of urban sewerage systems. In an RTC system the following elements are used (Schilling, 1989):

- A (measurement) sensor that is used to monitor the on-going process, for example, water level gauge or turbidity meter
- A (corrective) regulator that manipulates the process, for example, gates, valves, and so forth
- A controller that causes the regulator to bring the process back to its desired value (set or goal point)
- A communication system that carries the measured data from the sensor to the controller and the signals of the controller back to the regulator, for example, a telemetry system

The RTC systems do not have to a priori imply automation. As a matter of fact, significant and substantial operator interface, control, and override are desirable. In either case, computers equipped with RTC software are an integral part of the system.

In the context of the RTC, the meaning of "system" is the controllable part of the urban drainage and treatment system. A real-time control can be planned and implemented if numerical simulation models (deterministic, statistical, stochastic, neural network, etc.) are available and included in the RTC system.

As pointed out in Chapter 8, the following processes have to be modeled in order to obtain a comprehensive overview of the performance of the system:

- *Input to the system*: deterministic and/or stochastic rainfall-runoff-sewer flow and quality models will provide short-term forecasts of the inputs
- *System response to the input*: transfer function models will provide pollutant loads, flows, water levels, mixed liquor solids, etc., at key bottleneck points of the system
- *Total output to the environment*: all partial outputs (CSO, bypasses, effluent) will be summarized in order to minimize in an optimal way the total output from the system

- *Optimization of the system*: optimization routines are applied to achieve the goal
- *Response of the receiving environment to the total output*: the total output after system optimization can be compared with the waste-assimilative capacity of the receiving body of water, and further restrictions may be imposed on the system if the waste-assimilative capacity is exceeded. In such case the system optimization is repeated with the modified goals

Modeling Technology

The operational models to be used in an RTC system for control of discharges and pollution loads from urban sewerage and industrial wastewater processes ought to be adaptive both in response to any changes of the input waste loads and to the variations in the system parameters. Such models should be capable of on-line installation in an operational mode of a typical medium to large wastewater disposal facility. An automated operational mode is an option, but not a requirement, nor constraint imposed on the RTC operational systems.

A new idea advanced in research and implementation of the RTC systems is that all wastewater treatment and disposal organizations routinely and periodically collect data on input waste loads, effluent quality, and various operational parameters (temperature, sludge volume index, mixed liquor solids, dissolved oxygen in the basins, and others). The period of acquisition for most operations varies from hours to days, while some variables are recorded continuously. Currently, most of the data-collecting activities are for administrative and statistical purposes, and for recordkeeping and/or are required by the discharge permit. However, the same data may also contain needed and valuable information on the parameters and interrelationships between the parameters that can assist the operator in selecting the most optimal mode of operation.

A model of the system and of the processes occurring therein is a fundamental component of the RTC system. Since daily or even hourly series of data are accepted and managed by the RTC system, the models are in the category of a time series analysis. The adaptive modeling technology available to carry out the tasks according to the stated objectives of RTC control (i.e., adaptiveness, predictiveness, and efficiency) can be divided into two categories (Novotny, Capodaglio, and Feng, 1990):

1. *Auto-regressive moving-average models (ARMA)*, introduced by Box and Jenkins, 1976). Both univariate and transfer function models and their modifications can be considered.

2. *Artificial Neural Network Models* (ANN), proposed by Hopfield (1982) and Kohonen (1972).

The fundamentals of these models were introduced in the preceding chapter.

The "traditional" nonadaptive deterministic models can also be used alone, or the most important known deterministic relationships can be incorporated into the stochastic ARMA–transfer function models.

Existing RTC Systems for Wet-Weather– Dry-Weather Wastewater–Urban Runoff Treatment and Disposal Operation

CSO control. Most of the existing RTC systems in prototype use today are for the operation of the combined sewer overflows. In the United States such systems are in operation, for example, in Milwaukee, Wisconsin, Seattle, Washington, and Lima, Ohio. The objective of these systems is to minimize the overflows from combined sewers. The municipality of Seattle has been using computers for the control of their sewerage system since 1973. Currently, the city is developing an advanced system for real-time control of pump stations and regulator control in the combined sewer system to minimize overflows. The system is called CATAD (computer-augmented treatment and disposal). It is a centralized control system that has been implemented on a large mainframe computer. Of note is the AM analysis of the economy of the application of the real time control performed for the municipality by CH2M-Hill Consultants, which estimated that incorporating the real-time software would result in savings of $18 million annually in Seattle alone (Vitasovic, Swarner, and Speer, 1988).

The new CATAD system employs hydrologic and hydraulic routines to estimate dynamic flows at various points of the sewer system. The models are essentially deterministic. A number of CSO control RTC systems have been implemented in Europe (Durschlag and Schilling, 1990; Kolatsch and Schilling, 1990; Bennerstedt, Arnel, and Svenson, 1990; Weyand and Dohman, 1990).

Wastewater Treatment Operation

CS-FOCUS. The senior author and coworkers (Novotny, 1989; Novotny, Capodaglio and Feng, 1990) developed a computerized real-time control system with the acronym CS-FOCUS (computerized system for overflow control of urban sewerage). The overall schematic block diagram is shown on Figure 10.39.

Real-Time Control of CSOs and Pollution Load Trade-offs   657

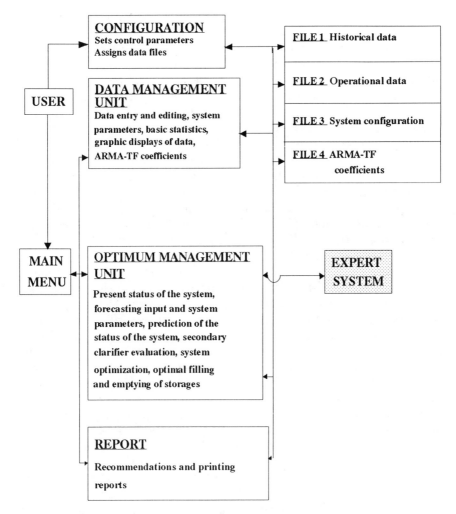

FIGURE 10.39. Real-time control (RTC) system for wet-weather overflow control (CS-FOCUS). (After Novotny, 1989, and Novotny and Capodaglio, 1992.)

Other RTC Systems. Berthouex, Lai, and Darjatmoko (1989) of the University of Wisconsin in Madison, developed a statistics-based expert system for the treatment plant operation. The system has three basic units: the data entry system, the data analysis system, and the expert advisory system. The data entry and data analysis systems are similar in their function to the data management unit of the CS-FOCUS (data entry and basic and intermediate statistics). The expert system incorporated in the Berthouex, Lai, and Dayatmoko system is rule based. The rules were taken from domain experts. At the time of the writing of their paper there were only four rules tested for activated sludge plant operation: however, up to thirteen rules were anticipated by the creators. The system remembers the variables that are periodically tested and the limits within which they should be maintained. The computer software of the University of Wisconsin RTC system was written for a larger mainframe computer and relies on some specific software packages.

An advanced computerized system for control operation of wastewater treatment plants is being developed in Japan (Hiraoka and Tsumura, 1988; Hiraoka et al., 1990) by a team from the University in Kyoto. This approach uses a computer statistical package SACCES (Statistical Analysis and Computer Control of Environmental Systems). The statistical package provides AR-TF functional models from the observed time series of the input and output data.

The Kyoto group selected the multivariate AR models as the most appropriate models for operational control of activated sludge systems. The control process and methodology proposed by the Japanese team is very complex; nevertheless, the approach is theoretically very sound. Of note also is the hierarchical concept of treatment plant operational control suggested by the Kyoto group. An upper level control system controls long-term (seasonal) variations, an intermediate system is aimed at an overall optimization of the process, and a lower level system deals with diurnal or hourly fluctuations.

Research conducted in Australia by Tan et al. (1990) was mentioned previously. An ARMAX model (similar to the AR-TF model developed and used by the senior author and coworkers), recursively updated by Wittenmark's (1974) self-tuning predictor, has been used for predicting flows in the sewer network. Excellent predictive capability has been achieved. The strategy of the RTC system worked and led to improved economy and reduction of overflows and pumping energy requirements, and to the improved performance of the treatment plant during wet-weather flow condition.

The LOCUS system being developed by Delft University in Holland (Nelen, 1990) included treatment in the consideration of the RTC sewerage system. The assumption incorporated in the LOCUS concept is that the efficiency of most treatment units is not affected by an increase in the hydraulic load as long as the critical hydraulic capacity is not exceeded. The simplified treatment mode of clarifier control uses four parameters: undisturbed clarifier treatment efficiency; period of time with maximum load, after which sludge overflow occurs; disturbed treatment efficiency;

and the period of time it takes for the system to recover when the hydraulic load is reduced below the critical value.

Practical Applications

The RTC systems for wastewater treatment plant operations discussed herein have not yet been fully implemented in a real prototype situation. However, following the impetus by the latest amendments of the Federal Water Pollution Control Act (Clean Water Act), the Environmental Protection Agency will be implementing tougher standards in order to minimize pollution from overflows and bypasses of existing wastewater treatment facilities. As has been documented in Seattle in a real prototype situation, implementing on-line RTC can save a municipality substantial amounts of financial resources.

Considering the fact that Congress is unlikely to authorize any substantial grants for upgrading and enlarging the sewerage systems that anticipate unacceptable overflows or violations of effluent standards, applications of alternative, less costly solutions are needed. The presented computerized RTC system will provide such remedial alternatives. The operators of the treatment system will also benefit from the interpretive expert system and graphical display features that may be incorporated into the RTC system. The modern RTC systems are "smart" control systems that enable the operators to forecast and anticipate the problems caused by transient wet-weather loads and, with the aid of the system, suggest optimal solutions to what otherwise could be a crisis situation in a treatment process. Such systems also provide for recordkeeping of data, graphical displays, and, possibly, reporting required by regulating and permitting authorities.

References

Agnew, R. W., et al. (1975). *Biological Treatment of Combined Sewer Overflow at Kenosha, Wisconsin*, EPA 670/2-75-019, National Environmental Research Center, Office of Research and Development, U.S. Environmental Protection Agency, Washington, DC, 247p.

American Association of State Highway and Transportation Officials (1980). *Design of Sedimenntation Basins*, Transportation Research Board, National Cooperative Highway Research Program Synthesis of Highway Practice, 53p.

American Public Works Association (1969). *Water Pollution Aspects of Urban Runoff*, Res. Project No. 120, Federal Water Pollution Control Administration, Washington, DC.

American Public Works Association (1981). *Urban Stormwater Management*, APWA, Special Rep. No. 49, 285p.

Aronson, G. L., D. S. Watson, and W. C. Pisano (1983). Evaluation of catch-basin performance for urban stormwater pollution control: Project summary, EPA 600/S2-83-043, Municipal Environmental Research Laboratory, U.S. Environmental Protection Agency, Cincinnati, OH.

Asmussen, L. E., A. W. White, Jr., E. E. Hauser, and J. M. Shendon (1977). Reduction of 2,4-D load in surface runoff down a grassed waterway, *J. Environ. Qual.* **6**(2):159–162.

Balmér, P., P. A. Malmqvist, and A. Sjöberg (1984). *Proceedings, Third International Conference on Urban Storm Drainage*, 4 vol., Chalmers University of Technology, Göteborg, Sweden, 1669p.

Bannerman, R., K. Baun, and M. Bohn (1983). *Nationwide Urban Runoff Program, Milwaukee, Wisconsin. Evaluation of Urban Nonpoint Source Pollution Management in Milwaukee County, Wisconsin. Executive Summary*, EPA P005432-01-5, Wisconsin Dept. of Natural Resources, Madison, Environmental Protection Agency, Chicago, IL, Region V (available from NTIS), 11p.

Bannerman, R., K. Baun, P. Hughes, and D. Graczyk (1983). *Nationwide Urban Runoff Program, Milwaukee, Wisconsin. Evaluation of Urban Nonpoint Source Pollution Management in Milwaukee County, Wisconsin.* Vol. 1. *Urban Stormwater Characteristics, Sources and Pollutant Management by Street Sweeping.* Vol. 2. *Feasibility and Application of Urban Nonpoint Source Water Pollution Abatement Measures*, EPA P005432-01-5, Wisconsin Dept. of Natural Resources, Madison, Environmental Protection Agency, Chicago, IL, Region V (available from NTIS).

Barfield, B. J., D. T. Y Kao, and E. W. Toller (1975). *Estimation of the Sediment Filtering Action of Grasses Media*, Res. Paper No. 90, Water Research Institute, University of Kentucky, Lexington, KY.

Bender, G. M., and W. W. Rice (1982). Effects of municipal street sweeping on urban storm runoff quality, in *Proceedings, 1982 International Symposium on Urban Hydrology, Hydraulics and Sediment Control*, University of Kentucky, Lexington, KY, pp. 217–223.

Bennerstedt, K., V. Arnel, and G. Svenson (1990). Real time control of a combined sewer system, in *Proceedings, Fifth International Conference on Urban Drainage, Suita-Osaka*, Y. Iwasa and T. Sueishi, eds., University of Osaka, Japan, July 23–27, pp. 1255–1260.

Béron, P., F. Brière, J. Rouselle, and J. P. Riley (1984). An evaluation of some real-time techniques for controlling combined sewer overflows, in *Proceedings, Third International Conference on Urban Storm Drainage*, Vol. 3, Chalmers University of Technology, Göteborg, Sweden, pp. 1093–1097.

Berthouex, P. M., W. Lai, and A. Dayatmoko (1989). Statistics-based approach to wastewater treatment plant operations, *J. Environ. Eng. Div. ASCE* **115**(3): 650–671.

Box, G. E., and G. M. Jenkins (1976). *Time Series Analysis: Forecasting and Control*, Holden-Day, Oakland, CA.

Brombach, H. (1987). Liquid-solid separation on vortex-storm-overflows, in *Topics in Urban Storm Water Quality, Planning and Management*, W. Gujer and V. Krejci, eds., *Proceedings, Fourth International Conference on Urban*

Storm Drainage, IAWPRC-IAHR, École Polytechnique Fédérale, Lausanne, Switzerland, pp. 103–108.

Brombach, H. (1989) Combined-sewer overflow control in West Germany—history, practice, and experience, in *Design of Urban Runoff Quality Controls*, L. A. Roesner, B. Urbonas, and M. B. Sonnen, eds., American Society of Civil Engineers, New York, NY, pp. 359–374.

Capodaglio, A.G. (1990). Identification of a Stochastic Modeling Procedure for Wastewater Treatment Operation and Control, Ph.D. thesis, Department of Civil Engineering, Marquette University, Milwaukee, WI.

Carlo, P. L., L. P. Owens, G. P. Hanna, Jr., and K. E. Longley (1992). The removal of selenium from water by slow sand filtration, *Water Sci. Technol.* **26**(9–11):2137–2140.

Chen, C. (1975). Design of sediment retention basins, in *Proceedings, National Symposium on Urban Hydraulics and Sediment Control*. University of Kentucky, Lexington, KY, July 28–31.

Clark, D. E., and W. C. Cobbins (1963). *Removal Effectiveness of Simulated Dry Fallout from Paved Areas by Motorized and Vacuumized Street Sweepers*, U.S. Naval Radiological Defense Laboratory, San Francisco, CA.

Collins, M. A., K. Roller, and G. Walton (1992). Managing toxic pollutants in stormwater runoff, *Water Environ. Technol.* **4**(5):60–64.

Dally, L. K., and D. P. Lettenmaier (1984). Urban storm drainage detention facility for runoff and water quality control, *Proceedings, Third International Conference on Urban Storm Drainage*, Vol. 1, Chalmers University of Technology, Göteborg, Sweden, pp. 47–56.

DeGroot, W., ed. (1982). *Proceedings, Conference on Stormwater Detention Facilities: Planning, Design, Operation and Maintenance*, American Society of Civil Engineers, New York, 431p.

DiToro, D. M., and M. J. Small (1979). Stormwater interception and storage, *J. Environ. Eng. Div. ASCE* **105**(EE1):43–54.

Dorman, M. E., et al. (1988). *Retention, Detention, and Overland Flow for Pollutant Removal from Highway Stormwater Runoff: Interim Guidelines for Management Measures*, FHWA/RD-87/056, U.S. Department of Transportation, Federal Highway Administration, McLean, VA.

Driscoll, E. D. (1988). Long term performance of water quality ponds, in *Design of Urban Runoff Quality Controls*, Proceedings, Eng. Foundation Conference, L. A. Roesner, et al., eds., American Society of Civil Engineers, New York, pp. 145–162.

Durchschlag, A., and W. Schilling (1990). The total pollution discharge as a design criterion for combined sewer systems, in *Proceedings, Fifth International Conference on Urban Drainage, Suita-Osaka*, Y. Iwasa and T. Sueishi, eds., University of Osaka, Japan, July 23–27, pp. 1077–1082.

Ellis, J. B. (1985). Design of urban detention basins for water quality control, in *Proceedings, Non-point Pollution Abatement Symposium*, V. Novotny, ed., Marquette University, Milwaukee, WI.

Ferrara, R. A., and A. Hildick-Smith (1982). A modeling approach for storm water quantity and quality control via detention basins, *Water Resour. Bull.* **18**(6):975–982.

Field, R. (1984). The USEPA Office of Research and Development's view of combined sewer overflow control, in *Proceedings, Third International Conference on Urban Storm Drainage*, Vol. 4, P. Balmér, P. A. Malmqvist, and A. Sjöberg, eds., Chalmers University of Technology, Göteborg, Sweden, pp. 1233–1356.

Field, R. (1985a). Urban runoff: Pollution sources, control, and treatment, *Water Resour. Bull.* **21**(2):197–206.

Field, R. (1985b). Management and control of pollution by urban runoff, in *Proceedings, Non-point Pollution Abatement Symposium*, V. Novotny, ed., Marquette University, Milwaukee, WI.

Field, R. (1986). Urban stormwater runoff quality management: Low structurally intensive measures and treatment, in *Urban Runoff Pollution*, H. C. Torno, J Marsalek, and M. Desbordes, eds., New York, pp. 677–699.

Field, R. (1990). Combined sewer overflows: Control and treatment, in *Control and Treatment of Combined-Sewer Overflows*, P. E. Moffa, ed., Van Nostrand Reinhold, New York, pp. 119–191.

Field, R., K. Dunkers, and A. Forndran (1990). Demonstration of in-receiving water storage of combined sewer overflows in a marine/estuarine environment by the Floe Balance Method, in *Proceedings, Fifth International Conference on Urban Storm Drainage, Suita-Osaka*, Y. Iwasa and T. Sueishi, eds., University of Osaka, Japan, July 23–27, pp. 759–764.

Florida Concrete Products Association (1989). *Pervious Pavement Manual*, FCPA, Tallahassee, FL.

Forndran, A., R. Field, and K. Dunkers (1989). *Demonstration of the Flow Balancing Method for Combined Sewer Overflow Abatement in Estuarine Waters*, Water Pollution Control Federation 62nd Annual Conference, San Francisco, CA, 30p.

Fujita, S. (1984). Experimental sewer system for reduction of urban stormwater runoff, in *Proceedings, Third International Conference on Urban Storm Drainage*, P. Balmér, P. A. Malmqvist, and A. Sjöberg, eds., Chalmers University of Technology, Göteborg, Sweden, pp. 1211–1220.

Fujita, S., and T. Koyama (1990). Pollution abatement and the experimental sewer system, *Proceedings, Fifth International Conference on Urban Storm Drainage*, IAWPRC-IAHR, *Suita-Osaka*, University of Osaka, Japan, pp. 799–904.

Galli, J. (1990). *Peat-sand Filters: A Proposed Stormwater Management Practice for Urban Areas*, Department of Environmental Programs, Metropolitan Washington Council of Governments, Washington, DC.

Galli, J. (1992). *Preliminary Analysis of the Performance and Longevity of Urban BMPs Installed in Prince George County, Maryland*, prepared for Department of Environmental Resources, Prince George's County, MD.

Goforth, G. F., E. V. Diniz, and J. B. Rauhut (1984). Stormwater hydrological characteristics of porous and conventional paving systems: Project summary,

EPA 600/S2-83-106, Municipal Environmental Research Laboratory, U.S. Environmental Protection Agency, Cincinnati, OH.
Goldman, S. J., K. Jackson, and T. A. Bursztynsky (1986). *Erosion and Sediment Control Handbook*, McGraw-Hill, New York.
Graham, P. H. (1978). *Report to Congress on Control of Combined Sewer Overflow in the United States*, EPA 430/9-78-006, Facility Requirements Division, Office of Water Program Operations, U.S. Environmental Protection Agency, Washington, DC.
Griffin, D. M., Jr., C. W. Randall, and T. J. Grizzard (1980). Efficient design of stormwater holding basins used for water quality protection, *Water Res.* **14**:1549–1554.
Grizzard, T. L. et al. (1986). Effectiveness of extended detention ponds, in *Urban Runoff Quality*, American Society of Civil Engineers, New York.
Gujer, W., and V. Krejci, eds. (1987). Topics in urban storm water quality, planning and management, in *Proceedings, Fourth International Conference on Urban Storm Drainage*, IAHR-IAWPRC, Lausanne, Switzerland.
Haak, A., and G. Oberts (1983), *Surface Water Management: Management Practices Evaluation*, Metropolitan Council of the Twin Cities Area, St. Paul, MN, 215p.
Hammer, D. A. (1989). *Constructed Wetlands for Wastewater Treatment: Municipal, Industrial and Agricultural*, in Proceedings, First International Conference on Constructed Wetlands for Wastewater Treatment, Chattanooga, TN, June 13–17, 1988, Lewis Publ., Chelsea, MI.
Harrington, B. W. (1989). Design and construction of infiltration trenches, in *Design of Urban Runoff Quality Controls*, L. A. Roesner, B. Urbonas, and M. B. Sonnen, eds., American Society of Civil Engineers, New York, pp. 290–304.
Hartigan, J. P. (1989). Basis for design of wet detention basin BMP's, in *Design of Urban Runoff Quality Controls*, L. A. Roesner, B. Urbonas, and M. B. Sonnen, eds., American Society of Civil Engineers, New York, pp. 122–143.
Herrmann, R. (1984). 3,4-benzopyrene washoff efficiency in different urban environment, in *Proceedings, Third International Conference on Urban Drainage*, P. Balmér, P. A. Malmqvist, and A. Sjöberg, eds., Chalmers Technical University, Göteborg, Sweden, pp. 1001–1008.
Herrmann, R., and F. G. Kari (1990). Grain size dependent transport of nonpolar trace pollutants (PAH, PCB) by suspended solids during urban runoff storm runoff, in *Proceedings, Fifth International Conference on Urban Storm Drainage, Suita-Osaka*, Y. Iwasa and T. Sueishi, eds., University of Osaka, Japan, July 23–27, pp. 499–503.
Hewitt, C. N., and M. B. Rashed (1992). Removal rates of selected pollutants in the runoff waters from a major rural highway, *Water Res.* **26**(3):311–319.
Hey, D. L., J. M. Stockdale, D. Kropp, and G. Wilhelm (1982). *Creation of Wetland Habitats in Northeastern Illinois*, Illinois Department of Energy and Natural Resources Document No. 82/09, May 1982, 117p.
Hiraoka, M., and K. Tsumura (1988). System identification and control of ac-

tivated sludge process by use of a statistical model, *Water Sci. Tech.* **21**:1161–1173.
Hiraoka, M., K. Tsumura, I. Fujita, and T. Kanaya (1990). System Identification of Activated Sludge Process by Use of Autoregressive Model, in *Advances in Water Pollution Control*, Proceedings, Fifth IAWPRC Workshop on Instrumentation, Control and Automation of Water and Wastewater Treatment and Transport Systems, R. Briggs, ed., Pergamon Press, Oxford, pp. 121–128.
Hopfield, J. J. (1982). Neural networks and physical systems with emergent collective computational abilities, *Proc. Nat. Acad. Sci.* **79**:2554–2558.
Innerfeld, H., A. Forndran, D. D. Ruggiero, and T. J. Hartman (1979). Dual process high-rate filtration of raw sanitary sewage and combined sewer overflows, EPA 600/2-79-015, Municipal Environmental Research Laboratory, Office of Research and Development, U.S. Environmental Protection Agency, 105p.
Iwasa, Y., and T. Sueishi, eds. (1990). *Proceedings, Fifth International Conference on Urban Storm Drainage, Suita-Osaka*, University of Osaka, Japan, July 23–27.
Jackson, T. J., and R. M. Ragan (1974). Hydrology of porous pavement parking lots, *J. Hydraul. Div. ASCE* **12**:1739–1752.
Kao, D. T. Y., B. J. Barfield, and A. E. Lyons (1975). On site sediment filtration using grass strips, in *Proceedings, National Symposium on Urban Hydrology and Sedimentation Control*, University of Kentucky, Lexington, KY.
Kohonen, T. (1972). Correlation matrix memories, *IEE Trans. Comput.* C-21: 353–359.
Kollatsch, D., and W. Schilling (1990). Control strategy of sanitary sewage detention tanks to reduce combine sewer overflow pollution loads, in *Proceedings, Fifth International Conference on Urban Drainage, Suita-Osaka*, Y. Iwasa and T. Sueishi, eds., University of Osaka, Japan, July 23–27, pp. 1365–1370.
Krejci V. (1988) New strategies in urban drainage and stormwater pollution control in Switzerland, in *Proceedings of the Symposium on Nonpoint Pollution: 1988—Policy, Economy, Management and Appropriate Technology*, V. Novotny, ed., AWRA, Bethesda, MD, pp. 203–213.
Krejci, V., and E. Baer (1990). Screening structures for combined sewer overflow treatment in Switzerland, in *Proceedings, Fifth International Conference on Urban Storm Drainage, Suita-Osaka*, Y. Iwasa and T. Sueishi, eds., University of Osaka, Japan, July 23–27, pp. 927–932.
Krejci, V., L. Dauber, B. Novak, and W. Gujer (1987). Contribution of different sources to pollutant loads in combined sewers, in *Topics in Urban Storm Water Quality, Planning and Management*, W. Gujer and V. Krejci, eds., Proceedings, Fourth International Conference on Urban Storm Drainage, IAWPRC-IAHR, École Polytechnique Fédérale, Lausanne, Switzerland, pp. 34–39.
Kuo, C. Y., and W. Y. Ni (1984). Pollutant trap efficiency in a detention basin, in *Proceedings, Third International Conference on Urban Storm Drainage*, Vol. 1,

P. Balmér, P. A. Malmqvist, and A. Sjöberg, eds., Chalmers University of Technology, Göteborg, Sweden, pp. 21–28.

Kuo, C. Y., and J. L. Zhu (1989). Design of a diversion system to manage the first flush, *Water Resour. Bull.* **25**(3):517–525.

Labadie, J. W., D. M. Morrow, and Y. H. Chen (1980). Optimal control of unsteady combined sewer flow, *J. Water Resour. Planning Manage. ASCE*, **106**(WR1):205–223.

Lager, J. A., and W. G. Smith (1974). *Urban Stormwater Management and Technology: An Assessment*, EPA 670/2-74-040, National Environmental Research Center, Office of Research and Development, U.S. Environmental Protection Agency, 447p.

Lager, J. A., W. G. Smith, W. G. Lynard, R. M. Finn, and E. J. Finnemore (1977). *Urban Stormwater Management and Technology: Update and Users' Guide*, EPA 600/8-77-014, Municipal Environmental Research Laboratory, Office of Research and Development, U.S. Environmental Protection Agency, 313p.

Lessard, P., and M. B. Beck (1991). Dynamic simulation of storm tanks, *Water Res.* **25**(4):375–391.

Linker, L. C. (1989). Creation of wetlands for the improvement of water quality: A proposal for the joint use of highway right-of-way, in *Constructed Wetlands for Wastewater Treatment: Municipal, Industrial and Agricultural*, D. A. Hammer, ed., Lewis Publ., Chelsea, MI, pp. 695–701.

Livingston, E. H., and L. A. Roesner (1991). Stormwater management practices for water quality enhancement, in *Manual of Practice Urban Stormwater Management*, American Society of Civil Engineers, New York.

Loganthan, V. G., J. Delleur, and R. I. Segarra (1985). Planning detention storage for stormwater management, *J. Water Resour. Planning Manage. ASCE* **111**(4):382–398.

McCuen, R. H. (1980). Water quality trap efficiency of storm water management basins, *Water Resour. Bull.* **16**(1):15–21.

Maryland Water Resources Administration (1984). *Standards and Specifications for Infiltration Practices*, MWRA, Annapolis, MD.

Maryland Water Resources Administration (1986). *Minimum Water Quality Objectives and Planning Guidelines for Infiltration Practices*, MWRA, Annapolis, MD.

Metropolitan Washington Council of Governments (1983). *Nationwide Urban Runoff Program, Washington Metropolitan Area Urban Runoff Demonstration Project: Final Report, 1979–1982*, Metropolitan Washington Council of Governments, Washington, DC; Northern Virginia Planning District Commission, Annandale, VA; Water Planning Div., Environmental Protection Agency, Washington, DC, 120p. (available from NTIS).

Minnesota Pollution Control Agency (1989). *Protecting Water Quality in Urban Areas*, Division of Water Quality, St. Paul, MN.

Moffa, P. E., ed. (1990). *Control and Treatment of Combined-Sewer Overflows*, Van Nostrand Reinhold, New York.

Morrison, G. M. P., D. M. Revitt, J. B. Ellis, G. Svensson, and P. Balmér

(1984). The physico-chemical speciation of zinc, cadmium, lead and copper in urban stormwater, in *Proceedings, Third International Conference on Urban Storm Drainage*, P. Balmér, P. A. Malmqvist, and A. Sjöberg, eds., Chalmers University of Technology, Göteborg, Sweden, pp. 989–1000.

Mulhern, P. F., and T. D. Steele (1988). Water-quality ponds—are they the answer? in *Design of Urban Runoff Quality Controls*, Proceedings, Engineering Foundation Conference, L. A. Roesner, B. Urbonas, and B. Sonneu, eds., American Society of Civil Engineers, New York, pp. 203–213.

Nelen, A. J. M. (1990). Control strategies based on water quality aspects, in *Proceedings, Fifth International Conference on Urban Drainage, Suita-Osaka*, Y. Iwasa and T. Sueishi, eds., University of Osaka, Japan, July 23–27, pp. 1311–1316.

Nieswand, G. H., et al. (1990). Buffer strips to protect water supply reservoirs: A model and recommendations, *Water Resour. Bull.* 26(6):959–966.

Nix, S. J., and T. K. Tsay (1988). Alternative strategies for stormwater detention, *Water Resour. Bull.* 24(3):609–614.

Nix, S. J., J. P. Heaney, and W. C. Huber (1988). Suspended solids removal in detention basins, *J. Environ. Eng. ASCE* 114(6):1331–1343.

Novak, B. (1983). Schutz des Vorfluters vor Schmutzstoffen aus Regenentlastungen, in *Jahresbereicht des EAWAG*, Dübendorf/Zurich, Switzerland.

Novotny, V. (1984) Efficiency of low cost practices for controlling pollution by urban runoff, in *Proceedings, Third International Conference on Urban Storm Drainage*, P. Balmér, P. A. Malmqvist, and A. Sjöberg, eds., Chalmers University of Technology, Göteborg, Sweden, pp. 1241–1250.

Novotny, V. (1989). *CS-FOCUS-Computerized System for Overflow Control of Urban Sewerage: Concepts, Theory and System Description*, AquaNova International, Mequon, WI.

Novotny, V., and A. Capodaglio (1992). Strategy of stochastic Real-Time Control of wastewater treatment plants, *ISA Trans.* 32(1):73–86.

Novotny, V., and G. Chesters (1981). *Handbook of Nonpoint Pollution: Sources and Management*, Van Nostrand Reinhold, New York.

Novotny, V., and G. Chesters (1989). Delivery of sediment and pollutants from nonpoint sources: A water quality perspective, *J. Soil Water Conserv.* 44(6): 568–576.

Novotny, V., and G. Kincaid (1982). Acidity of urban precipitation and its buffering during overland flow, in *Urban Stormwater Quality, Management and Planning*, B. C. Yen, ed., Water Resources Publ., Littleton, CO, pp. 1–9.

Novotny, V., A. Capodaglio, and X. Feng (1990). Stochastic Real Time Control of wastewater treatment plant operation, in *Advances in Water Pollution Control*, Proceedings, Fifth IAWPRC Workshop on Instrumentation, Control and Automation of Water and Wastewater Treatment and Transport Systems, R. Briggs, ed., Pergamon Press, Oxford, pp. 539–544.

Novotny, V., R. Schonter, and L. Novotny (1992). *Efficiency of Best Management Practices for Control of Pollution by Priority Pollutants with Emphasis on Army Operations*, US Army Corps of Engineers Civil Eng. Res. Lab., Proj. No. DACA88-92-M0594, Champaign, IL.

Novotny, V., H. M. Sung, R. Bannerman, and K. Baum (1985). Estimating nonpoint pollution from small urban watersheds, *J. WPCF* **57**(4):339–348.

Novotny, V., K. R. Imhoff, M. Othoff, and P. A. Krenkel (1989). *Karl Imhoff's Handbook of Urban Drainage and Wastewater Disposal*, Wiley Interscience, New York.

Oakland, P. H. (1983). An evaluation of urban stormwater pollution removal through grassed swale treatment, in *Proceedings, International Symposium on Urban Hydrology, Hydraulics and Sediment Control*, University of Kentucky, Lexington, KY.

Oberts, G. (1982). Impact of wetlands on nonpoint source pollution, in *Proceedings, International Symposium on Urban Hydrology, Hydraulics and Sediment Control*, University of Kentucky, Lexington, KY.

Ormsbee, L. E. (1984). Systematic planning of dual purpose detention basins, University of Kentucky, Office of Engineering Services, *(Bulletin) UKY BU* **135**:143–151.

Pisano, W. C. (1989). Swirl concentrators revisited: The American experience and new German technology, in *Design of Urban Runoff Quality Controls*, L. A. Roesner, B. Urbonas, and M. B. Sonnen, eds., American Society of Civil Engineers, New York, pp. 390–402.

Pisano, W. C. (1990). Recent United States experience with designs of German vortex solids separators for CSO control, in *Proceedings, Fifth International Conference on Urban Storm Drainage*, IAWPRC-IAHR, Suita-Osaka, University of Osaka, Japan, pp. 933–938.

Pisano, W. C. (1991). *Design of CSO Treatment Facilities*, paper presented at Combined Sewer Overflow Conference, Technical University of Nova Scotia Centre for Water Resources, May 10, 1991.

Pisano, W. C., D. J. Connick, and G. L. Aronson (1985). *In-system Storage Controls for Reduction of Combined Sewer Overflow, Saginaw, Michigan*, EPA 905/2-85-001-A, Great Lakes National Program Office, U.S. Environmental Protection Agency, Chicago, IL.

Pitt, R. (1979). *Demonstration of Nonpoint Pollution Abatement through Improved Street Cleaning Practices*, EPA 600/2-79/161, U.S. Environmental Protection Agency, Cincinnati, OH.

Pitt, R., and P. Baron (1989). *Assessment of Urban and Industrial Stormwater Runoff Toxicity and the Evaluation/Development of Treatment for Runoff Toxicity Abatement—Phase I*. Storm and Combined Sewer Pollution Program, U.S. Environmental Protection Agency, Edison, NJ.

Pitt, R., and G. Shawley (1981). *San Francisco Bay Area National Urban Runoff Project. Demonstration of Non-point Source Pollution Management on Castro Valley Creek: Executive Summary* (final rept.), Water Planning Div., Environmental Protection Agency, Washington, DC (available from NTIS).

Pitt, R., R. Field, M. Lalor, and G. Driscoll (1990a). Analysis of cross-connections and storm drainage, in *Proceedings, Urban Stormwater Enhancement—Source Control, and Combined Sewer Technology*, Davos, Switzerland, Eng. Foundation, American Society of Civil Engineers, New York.

Pitt, R., M. Lalor, N. Miller, and G. Driscoll (1990b). *Assessment of Non-*

Stormwater Discharges into Separate Storm Drainage Networks—Phase I: A Manual of Practice, Storm and Combined Sewer Section, U.S. Environmental Protection Agency, Edison, NJ.

Ports, M. A. (1975). *Urban Sediment Control Design Criteria and Procedures*, Technical Paper No. 75-2567, American Society of Agricultural Engineers, 28p.

Potts, R. R., and J. L. Bai (1989). Establishing variable width buffer zones based upon site characteristics and development type, in *Proceedings, Symposium on Water Laws and Management*, American Water Resources Association, Bethesda, MD.

Pravoshinsky, N. A. (1975). Basic principles for determining regulating structures to prevent rain sewers receivers from contamination, *Prog. Water Tech.* **7**(2): 301–307.

Ree, W. O., and V. J. Palmer (1949). *Flow of Water in Channels Protected by Vegetative Lining*, U.S. Soil Conservation Serv. Bull. No. 967, Washington, DC.

Roesner, L. A. (1988). Aesthetic implementation of nonpoint source controls, in *Proceedings, Symposium on Nonpoint Pollution: 1988—Policy, Economy, Management, and Appropriate Technology*, V. Novotny, ed., American Water Resources Association, Bethesda, MD, pp. 213–223.

Roesner, L. A., E. H. Burges, and J. A. Aldrich (1991). *The Hydrology of Urban Runoff Quality Management*, paper presented at WPCF-ASCE Conference on Urban Runoff, New Orleans, LA, May 21–22.

Roesner, L. A., B. Urbonas, and B. Sonnen, eds. (1989). *Design of Urban Runoff Quality Controls*, Proceedings, Engineering Foundation Conference, American Society of Civil Engineers, New York.

Salomons, W., and U. Förstner (1984). *Metals in the Hydrocycle*, Springer Verlag, Berlin, New York.

Sartor, J. D., and G. B. Boyd (1972). *Water Pollution Aspects of Street Surface Contamination*, EPA R2-72-081, U.S. Environmental Protection Agency, Washington, DC.

Sartor, J. D., G. B. Boyd, and F. J. Agardy (1974). Water pollution aspects of street surface contamination, *J. WPCF* **46**:458–463.

Schilling, W. (1986). Urban runoff quality management by Real-Time Control, in *Urban Runoff Pollution*, H. C. Torno, J. Marsalek, and M. Desbordes, eds., Springer Verlag, New York, pp. 765–817.

Schilling, W. (1989). *Real-Time Control of Urban Drainage Systems. The State-of-the-Art*, Sci. Tech. Rep. No. 2, IAWPRC, London.

Scholze, R., V. Novotny, and R. Schonter (1993). *Efficiency of Best Management Practices for Control of Pollution by Priority Pollutants*, Paper presented at the First IAWQ Conf. on Diffuse Pollution—Chicago, IL, Sept. 1993, *Water Sci. Technol.* **28**(3–5):215–224.

Schueler, T., and M. Helfrich (1988). Design of extended detention wet pond systems, in *Design of Urban Runoff Quality Controls*, L. A. Roesner, B. Urbonas, and B. Sonnen, eds., Proceedings, Engineering Foundation Conference, American Society of Civil Engineers, New York, pp. 180–200.

Schueler, T. R., P. A. Kumble, and M. A. Heraty (1992). *A Current Assessment of Urban Best Management Practices. Techniques for Reducing Non-point Source Pollution in the Coastal Zone*, Technical Guidance Manual, Metropolitan Washington Council of Governments, Office of Wetlands, Oceans, and Watersheds, U.S. Environmental Protection Agency, Washington, DC.

Stahre, P., and B. Urbonas (1989). Swedish approach to infiltration and percolation design, in *Design of Urban Runoff Quality Controls*, L. A. Roesner, B. Urbonas, and M. B. Sonnen, eds., American Society of Civil Engineers, New York, pp. 307–322.

Stahre, P., and B. Urbonas (1990). *Stormwater Detention For Drainage, Water Quality, and CSO Management*, Prentice Hall, Englewood Cliffs, NJ.

Stockdale, E. C., and R. R. Horner (1987). Prospects for wetlands use in stormwater management, *Proceedings, Fifth Symposium on Coastal and Ocean Management*, Vol. 4, American Society of Civil Engineers, pp. 3701–3714.

Striegl, R. G. (1987). Suspended sediment and metals removal from urban runoff by a small lake, *Water Resour. Bull.* **23**(6):985–996.

Sullivan, R. H. (1968). Problems of combined sewer facilities and overflows, *J. WPCF* **41**(1):113–121.

Sullivan, R. H., et al. (1982). *Design Manual: Swirl and Helical Bend Pollution Control Devices*, EPA 600/8-82-013, Storm and Combined Sewer Section, U.S. Environmental Protection Agency, Edison, NJ.

Tan, P. C., K. P. Dabke, R. G. Mein, and C. S. Berger (1990). Real-Time Control of sewer networks, in *Proceedings, Fifth International Conference on Urban Drainage, Suita-Osaka*, Y. Iwasa and T. Sueishi, eds., University of Osaka, Japan, July 23–27, pp. 1231–1237.

Terstriep, M. L., M. G. Bender, and D. C. Noel (1982). *Nationwide Urban Runoff Project, Champaign, Illinois: Evaluation of the Effectiveness of Municipal Street Sweeping in the Control of Urban Storm Runoff Pollution*, Final report, Illinois State Water Survey Div., Champaign; Illinois State Environmental Protection Agency, Springfield; Environmental Protection Agency, Chicago, IL, Region V, 246p. (available from NTIS).

Thelen, E., W. C. Grover, A. J. Hoiberg, and T. I. Haigh (1972). *Investigation of Porous Pavements for Urban Runoff Control*, Office of Research and Monitoring, U.S. Environmental Protection Agency, Washington, DC, 142p.

Toller, E. W., B. J. Barfield, C. T. Haan, and D. T. Y. Kao (1977). Suspended sediment filtration capacity of simulated vegetation, *Trans. ASAE* **19**(5): 678–682.

Torno, H. C., J. Marsalek, and M. Desbordes, eds. (1986). *Urban Runoff Pollution*, Springer Verlag, Heidelberg, New York.

Urbonas, B. (1984). Summary of findings by ASCE Task Committee on Detention Basins, in *Proceedings, Third International Conference on Urban Storm Drainage*, Vol. 2, Chalmers University of Technology, Göteborg, Sweden; pp. 733–742.

U.S. Environmental Protection Agency (1976). *Erosion and Sediment Control, Surface Mining in Eastern U.S.A.*, EPA 625/3-76/006, U.S. EPA, Washington, DC.

U.S. Environmental Protection Agency (1977). *Nonpoint Source Control Guidance Hydrologic Modifications*, Water Planning Division, Nonpoint Sources Branch, U.S. EPA, Washington, DC.

U.S. Environmental Protection Agency (1983). *Results of the Nationwide Urban Runoff Program*, Water Planning Division, U.S. EPA, Washington, DC.

Urbonas, B., and L. A. Roesner, eds. (1986). *Urban Runoff Quality—Impact and Quality Enhancement Technology, Urban Runoff Quality Enhancement Technology*, Proceedings, Engineering Foundation Conference, ASCE, New York, 477p.

Urbonas, B., and W. P. Ruzzo (1986). Standardization of detention pond design for phosphorus removal, in *Urban Runoff Pollution*, H. C. Torno, J. Marsalek, and M. Desbordes, eds., New York, pp. 739–760.

Vitale, A. M., and P. M. Spray (1975). *Total Urban Water Pollution Loads: The Impact of Storm Water*, U.S. Council on Environmental Quality, Washington, DC.

Vitasovic, Z., R. Swarner, and E. Speer (1988). *Real Time Control for CSO Reduction*, paper presented at WPCF Annual Conference, Dallas, TX.

Walesh, S. (1989). *Urban Surface Water Management*, Wiley-Interscience, New York, 518p.

Walker, W. W. (1987). Phosphorus removal by urban runoff detention basins, in *Lake and Reservoir Management*, Vol. III, North American Lake Management Society, Washington, DC, pp. 314–326.

Wanielista, M. F., Y. A. Yousef, H. H. Harper, and C. L. Cassagnol (1982). Detention with effluent filtration for stormwater management, in *Urban Stormwater Quality, Management and Planning*, B. C. Yen, ed., Water Resources Publ., Littleton, CO, pp. 314–330.

Water Pollution Control Federation (1990). *Manual of Practice for the Design and Construction of Urban Stormwater Management Systems*, WPCF, Alexandria, VA.

Weyand, M., and M. Dohman (1990). Experience and results of Real-Time Control within a combined sewers system, in *Proceedings, Fifth International Conference on Urban Drainage, Suita-Osaka*, Y. Iwasa and T. Sueishi, eds., University of Osaka, Japan, July 23–27, pp. 1237–1248.

Whipple, W., Jr., et al. (1983). *Stormwater Management in Urbanizing Areas*, Prentice Hall, Englewood Cliffs, NJ.

Whipple, W., Jr., and J. V. Hunter (1980). *Settleability of Urban Runoff Pollution*, PB80-182017, National Technical Information Service, Springfield, VA.

Wiesner, P. E., A. M. Kassem, and P. W. Cheung (1982). Parks against storms, in *Urban Stormwater Quality, Management and Planning*, B. C. Yen, ed., Water Resources Publ., Littleton, CO, pp. 322–330.

Wilson, L. G. (1967). Sediment removal from flood water, *Trans. ASAE* **10**(1): 35–37.

Wittenmark, B. (1974). A self-tuning predictor, *IEEE Trans. Autom. Control* **A-19**:848–851.

Woodard, S. E. (1989). The Effectiveness of Buffer Strips to Protect Water

Quality, Master of Science thesis, Department of Civil Eng., University of Maine, Orono, ME.

Wotzka, P., and G. Oberts (1988). The water quality performance of a detention basin-wetland treatment system in an urban area, in *Nonpoint Pollution: 1988- Policy, Economy, Management, and Appropriate Technology Symposium Proceedings*, American Water Resources Association, Bethesda, MD, pp. 237–247.

Yen, B. C. (1982). *Urban Stormwater Quality, Management and Planning*, Water Resources Publ., Littleton, CO.

11
Agricultural Issues

Soil and water are critical components of the resource base upon which agriculture depends. To move towards sustainability, agriculture and natural resource management interests must recognize that they are equal partners in the effort. The challenge is to adapt and extend our knowledge about soil and water to develop economically productive, culturally appropriate, and environmentally sound agricultural systems. A flexible, ongoing process is necessary to set research priorities to support inherently dynamic agricultural systems.

Committee on International Soil and Water Research
and Development, National Research Council

American agriculture is a large and complex industry. The extent of land devoted to agricultural production, the unusual pricing system in place, the intensity and efficiency of modern farming, ranching, and other animal production systems all contribute to the complexity and diversity of the system. Agricultural production has a great potential to adversely affect the environment, most particularly from nonpoint source runoff, hazardous waste disposal, habitat destruction, and in localized areas, nuisance odors. In other developed countries agriculture can occupy a

---

Susan Alexander was the primary writer of this chapter. A literature review by Smolen, Jennings, and Huffman (1990) was also a major source of materials.

similar niche in a nation's economy and social structure depending upon the extent of arable land and population pressures. In developing and undeveloped countries agriculture operates at more of a subsistence level status. Subsistence farming practices have an even greater potential for causing environmental damage than modern methods used in developed countries due to the high erosion rates associated with slash and burn or other types of subsistence farming. Additionally, many small or poor nations do not have the regulatory framework in place to prohibit the use of highly toxic or persistent pesticides, nor do farmers in these countries have access to the educational materials and programs offered in developed countries.

This chapter examines how current agricultural management techniques contribute to the diffuse-pollution problem; explores which best management practices (BMPs) are most effective in protecting water quality; explains some of the many programs or methods state and federal governments use to get BMPs installed; discusses a few of the reasons that agricultural nonpoint source (NPS) controls have not been more widely adopted; and recommends or suggests some method that might speed up the installation of agricultural BMPs. For the most part, it focuses on the agricultural systems in place in the United States, because the data base is most extensive.

The following manuals may be studied for more detailed discussions of the management of agricultural and silvicultural pollution:

Stewart et al. (1975): Control of Water Pollution from Cropland
Loehr et al. (1979): Best Management Practices for Agriculture and Silviculture
U.S. EPA (1992c): Managing Nonpoint Source Pollution—Final Report to the Congress
U.S. Forest Service (1989): Best Management Practices Guide for Integrated Resource Planning on National Forest Lands in the Southwestern Region
Humenik et al. (1982): Best Management Practices for Agricultural Nonpoint Source Control—Animal Waste
Thompson et al. (1989): Poison Runoff: A Guide to State and Local Control of Nonpoint Source Water Pollution
U.S. EPA (1993): Guidance Specifying Management Measures for Sources of Nonpoint Pollution in Coastal Waters

Some concepts to look for in this chapter that will provide a more thorough understanding of agricultural issues include:

- Agricultural nonpoint source abatement is implemented mostly through voluntary programs.
- BMPs may not have a direct measurable benefit to the land or the water immediately adjacent to the land to which they are applied (benefits and impacts are off-site).
- Agricultural producers (in general) have a particular outlook on their profession.
- Agricultural producers "buy at retail and sell at wholesale" (they do not set the prices), thus a large network of commodity programs has been established in most developed countries).
- Variability in BMP effectiveness can be extreme.
- Diffuse-pollution control on agricultural lands can turn into a private lands rights issue, quickly becoming an emotional and political "hot potato" that no one wants to touch.
- Most of the technology to control agricultural nonpoint pollution is currently available.

EXTENT OF AMERICAN AGRICULTURE

According to the USDA Economic Research Service (USDA, National Agricultural Statistics Service, 1992a–1992g), sales of agricultural commodities account for 16% of the gross national product of the United States. Agricultural land is generally divided into the following categories for inventory purposes:

Dryland cropland
Irrigated cropland
Pasture land
Range land
Forest land
Confined animal feeding operations
Specialty areas (such as aquaculture, orchard crops, and wildlife land)

These divisions are useful to water quality planning efforts and nonpoint source pollution control since each type of land area generally has a somewhat distinct set of pollutants of most concern associated with that land use, and because most current BMP reference guides are divided by these categories rather than by pollutant.

Row and field crops, such as corn or hay, are generally produced on privately owned land in the United States, but beef cattle, sheep and goats, and timber are often produced on publicly held lands that are leased to private individuals for production. To be effective in controlling

TABLE 11.1 Extent of American Agriculture Production for Selected Commodities in 1991

Commodity	Amount	Cash Receipts or Value of Product ($)	Acres Planted
Corn	7,474,480,000 bu	18,063,205,000	75,951,000
Cotton	17,541,500 bales[a]	5,322,413,000	14,143,000
Grain sorghum	579,490,000 bu	1,347,366,000	11,014,000
Soybeans	1,985,564,000 bu	11,078,422,000	59,060,000
Small grains[b]	2,697,486,000 bu	7,117,902,000	89,172,000
Rice	154,457,000 cwt	1,166,077,000	2,851,000
Vegetables[c]	—	5,277,606	—
Fruits and nuts	—	9,574,134,000	—
Floricultural crops[d]	—	2,802,510,000	—
Hay	153,485,000 tons	9,800,766,000	62,575,000
Beef cattle	100,110,000 head	39,632,086,000	—
Sheep	10,850,000 head	399,097,000	—
Hogs	57,684,000 head	11,064,101,000	—
Broilers[e]	6,138,350,000 birds	8,385,284,000	—
Turkeys	285,000,000 birds	2,344,016,000	—
Eggs	68,958,000,000 eggs	3,886,810,000	—
Spent layers	197,518,000 birds	67,548,000	—
Catfish[f]	390,870,000 lb	346,638,970	—
Dairy cows	9,990,000 head produced 148,526,000 pounds of milk	18,075,191,000 milk and cream	—

Sources: From USDA NASS (1992a–g).
[a] One bale of cotton = 500 lb.
[b] Small grains = wheat, rye, barley, oats.
[c] Not all vegetable crops reported upon.
[d] Estimate, consists of flowering plants and cut flowers.
[e] Excludes states that produced less than 500,000 broilers.
[f] Four states only reporting, estimate.

water pollution from agriculture it is important to understand the issues and control options of agricultural production on both types of land.

Table 11.1 summarizes the extent of American agriculture production of selected food and fiber crops and their associated cash receipts or value during 1991.

Agricultural Philosophy and Perspective

The Family Farm. The average farm size in North America is about 200 ha in developed European countries, such as Italy, it averages about 10 ha. Although many farms are operated by a single family, that family often forms a corporation to protect its financial assets. A growing number of farms and ranches are owned by absentee landlords and operated on a cash-lease or crop-share agreement. About 30% of the agricultural land in the United States is owned by an absentee landlord, although much of this land may be farmed by a neighbor. An undetermined number of acres are owned by large corporations that may or may not specialize in agricultural products. Insurance companies, banks, chemical companies, petroleum conglomerations, etc., often hold agricultural land for the tax benefits it provides (American Farm Bureau Federation, 1989). Agricultural land is taxed at a lower rate than land zoned for development. Use of land for agricultural production at a net loss, when held in conjunction with other interests, can also provide tax advantages to the land owner (Levi, 1978). The small 15–40 ha family farm typical of the 1940s has basically disappeared due to economic pressures and population migration. It is very difficult for the small farmer or rancher to produce the volume of product required to be competitive and to pay for the capital investment farming requires. Regardless of the love of farming or the commitment to produce cheap food and fiber, agricultural producers must be good businesspeople. They must make their land-use decisions based on the economic conditions under which they operate (Brade and Uchtmann, 1985). This economic reality must also be recognized by persons wishing to control nonpoint source pollution from agricultural lands.

Commodity Programs and Price Supports: Short-Term Economic Gain versus Long-Term Environmental Stability. In the United States most farmers operate as independent small businesses, buying production inputs locally (and at retail cost), but selling products on the world market at wholesale prices. The profit margin for most of the commodity crops grown in the United States has steadily declined since the 1940s. Agricultural producers have compensated by increasing production. Farm policies of the USDA encouraged this trend in the 1970s and tried to cushion its effects in the 1980s and 1990s. Technological advances in equipment, hybrid varieties of crops and livestock, fertilizers, and pest control have supported improved efficiency and increased production (Tankersley, 1981). For example, during the 1920s wheat sold for $3.10 per bushel and the tractor used on that farm cost $6500; however, in the 1990s while wheat still sells for $3.10 per bushel, tractor prices have risen to $72,000 (Andrielenas and Eichers, 1982).

A combination of factors, including fluctuating land values, low market prices for commodity crops, high input costs, and a high debt ratio, have caused most producers to operate on an annual credit basis. Many utilize systems like the Commodity Credit Corporation (CCC—a federally managed corporate commodity brokerage and lending firm) to obtain annual operating loans to purchase feed, seed, fertilizer, gasoline, and other annual inputs for the upcoming year, using as collateral their anticipated crop yield. This dependence on "upfront" financing limits farmer flexibility since many lending institutions will only make loans on certain high-value/commodity crops. Thus most producers manage and plan for short-term economic returns and specialize in one or two continuously grown crops or livestock. Little cash is available and little incentive provided for conservation work. Long-term planning and management for resource protection do not provide the short-term profits most large specialized producers need to stay solvent. Diversified operations that produce a variety of grain, cash, grass, animal, and specialty crops are often not as dependent upon the short-term cash flow from one particular crop of animal, and thus have more ability to operate in a resource-conserving manner. This diversification provides an economic cushion in case other crops fail or prices are low (Hosapple, 1992).

To compensate for the low sales price (in relation to the high input costs) for most commodity crops, the surplus of most crops produced each year by American farmers, and to provide consumers with somewhat stable low prices for food and fiber products and farmers with a more competitive edge in world agricultural markets (which are even more heavily subsidized than those in the United States), the USDA has developed a variety of price-support and other commodity or subsidy programs. These programs attempt to balance the crop to be planted with expected market needs and assist producers with deficiency payments that pay the difference between average sales price and average production price (U.S. General Accounting Office, 1990).

For many years a conflict existed within the USDA between conservation programs (designed to protect the resource base) and commodity programs (designed to stabilize farm income, etc.). In many cases commodity programs provided much more financial incentive for farms to overproduce and utilize land inappropriately than conservation programs provided incentives to protect the resource base (Sampson, 1981). In recent years these programs have been made more flexible, more careful of their environmental effects, and have decreased the actual cash or in kind (commodity) payments provided to each farmer. Recent farm legislation has also placed limits on total revenues collected. The newer environmental "consciousness" of recent American agricultural policy can be seen in the Food, Agriculture, Conservation and Trade Act of 1990, commonly referred to as the Farm Bill, which contain "cross compliance" provisions that make eligibility for participation in commodity programs dependent upon use of sound conservation practices (U.S. General Accounting Office, 1990).

Role of the European Community and Other World Market Systems

For national security and other reasons every country wishes to be food self-sufficient. In some European countries, Japan, and other countries where land available for agriculture is limited, subsidies and market price guarantees are used extensively to ensure national production of important food crops that are not possible to produce profitably. These heavily subsidized crops, combined with restrictive trade barriers and tariffs, make international crop and livestock marketing difficult (Harrington, Krupnich, and Peskin, 1985).

Agricultural Land-Use and Private Property Rights in the United States

There is a long tradition of personal freedom associated with private property land-use decisions in the United States. In the past, since there was plenty of land still available, land-use decisions that degraded soil and water were not considered by many to be a problem. Once one farm was depleted, settlers move on to more fertile, less worn out land. Thomas Jefferson wrote and spoke extensively on this trend at the end of the 1700s, saying that land stewardship was essential for maintenance of private property rights. The U.S. Constitution provides that the federal government cannot "take" land from an individual without just compensation. A number of court cases (mostly involving wetland issues) have been filed claiming that land-use control regulations (in those cases denial of dredge and fill permits for wetland conversion to shopping complexes) constitute a "taking" without just compensation since the economic value of the land for development far exceeds its agricultural or wetlands value. Americans in general are extremely resistant to being told what to do on "their" property. In Germany where land is at a premium and population densities high, citizens routinely recycle garbage and abide by garbage bulk limits for the public benefit without making it a private land right issue. More than any other factor, the concept of balance in private property rights with the good of society is at the heart of any agricultural diffuse-pollution dispute.

Private Use of Public Lands by Ranching

In the western United States a large percentage of the land is owned or controlled by the federal government. These lands are generally part of the National Forest System, the Bureau of Land Management System, or a part of a public works project such as a Bureau of Reclamation irrigation and water supply reservoir and canal system. Under the multiple-use, sustained-yield, and greatest benefit management mandates of many federal land management agencies these lands are harvested for timber and leased for cattle grazing. A large percentage of these rangelands are in poor to fair condition, their natural productivity in decline, and their contribution to poor water quality significant (New Mexico Environmental Improvement Division, 1989). The rental rates for this property are usually below the comparable rates for leasing of private lands. Stocking rates allowed on this range have traditionally been more than the resource could support; however, in recent years livestock numbers have come more into balance. The traditional practices of continuous yearlong grazing in one undivided pasture continues to be a barrier to water quality protection. Ranchers on public lands are reluctant to install costly structural BMPs on property that they do not own, yet they have also resisted low-cost management BMPs (U.S. Forest Service, 1989).

Sustainable Agricultural Practice

As the epigraph for this chapter indicates, farm operators must realize that the production process must be in a harmony with the natural resources (land and soil, water), must not overburden them, and provide for sustainable agricultural production of future generations and not destroy the resources. Agriculture that destroys forests, drains wetlands, overuses fertilizers, and causes excessive soil loss has been and still is practiced in some parts of the world.

While based on concepts of stewardship, biodiversity, and mixed farming systems, the concept of sustainable agriculture should not advocate return to the past, but rather it focuses on an integrated approach that embraces biotechnology, engineering, and system studies for individual farming operations also using ecoregional concepts and knowledge of diversity. Sustainable agriculture employs a multiplicity of approaches, such as hybrids, sustainable pest and weed management, and soil fertility practices (Francis, Flora, and King, 1992; Committee on International Soil and Water Research and Development, 1991). Because agricultural diffuse pollution is primarily caused by the unwise use or overuse of lands, monocultural farming, and excessive use of fertilizers and herbicides, it is expected that sustainable agricultural practices should bring about a significant reduction if not even outright elimination of diffuse pollution.

AGRICULTURE AND ITS EFFECT ON THE ENVIRONMENT

Sources of Pollution

As pointed out in Chapter 1 diffuse pollution is a landscape phenomenon that is strongly affected by land-use activities. Chapter 1 also introduced the types of land uses in a dynamic concept and how they are related to the generation of diffuse pollution. Many factors affect pollution loads from agricultural operations. Among those are the types of land use, type of crops or animals, crop rotation, soils on which crop is grown, climatic conditions, farming technology, and irrigation and drainage. Proximity of polluting agricultural operations, such as feedlots and barnyards, to watercourses (Fig. 11.1) is one of the major causes of agricultural pollution. On the other hand, pollutant loads are reduced if buffer strips and measures that prevent cattle and cattle waste from entering the watercourses are installed.

Ritter (1988) reviewed the literature on the topic and compared loads

FIGURE 11.1. Proximity of polluting agricultural operations to a watercourse is a cause of agricultural pollution. (Photo: University of Wisconsin.)

TABLE 11.2 Comparative Magnitude of Diffuse Sources from Agriculture

Source	Total N		Total P	
	ml/l	kg/ha-yr	mg/l	kg/ha-yr
Precipitation (U.S.)[a]	0.7–1.3	5.6–10	—	0.05–01
Precipitation (Minn.)[a]	—	—	0.01	0.1
Precipitation (Ohio)[a]	2.0–2.8	12.8	—	—
Precipitation (Del.)[a]	—	45	—	1.5
Forest (Minn.)[a]	—	—	0.04–1.2	0.1
Forest (Ohio)[a]	0.05–0.9	2.1	0.01–0.01	0.4
Silviculture (Va.)[a]	1.1–1.8	2.7	0.01–0.2	0.3
Upland native prairie (Minn.)[a]	—	1	—	—
Grassland—rotate graze (Okla.)[a]	1.5–1.6	1.5	0.06–0.8	0.9
Grassland—continuous graze (Tex.)[a]	0.6	—	1.3	3.2
Grassland—44 kg/N/ha (N.C.)[a]		8.4		
Farmland (Ohio)[a]	0.09–3.1	5.1	0.02	0.06
Agricultural (Va.)[a]	1.1–1.8	2.7	0.02–0.3	0.3
Poorly drained coastal plain (Va.)[a]	1.7–2.3	1.6	0.4–0.7	0.2
Well-drained coastal plain (Va.)[a]	1.5–4.1	4.9	9–40	0.9
Corn—240 kg N/ha (Ga.)[a]	0.2–0.4	—	0.1–0.2	—
Feedlot runoff (Great Plains)[a]	3000–18,000	100–1600	47–300	10–620
Land applied dairy manure (Minn.)[a]	13–62	2.8–3.7	7.5–8.9	0.5–0.6
Land applied dairy manure (Wis.)[a]	—	2.8–8	1.8–4.9	0.1–17
Land applied dairy manure (S.C.)[a]	10–12	—	—	8.2–14
Irrigation return flow[b]	—	5–30		1–4
Subsurface tile drainage[b]	—	5–20	—	3–10

[a] Compiled by Ritter (1988).
[b] Compiled by Loehr (1974).

and concentrations from a variety of sources. Table 11.2 shows that nitrogen and phosphorus contamination may vary over a wide scale.

The major pollutants associated with agriculture include sediment, nutrients (especially N and P), pesticides and other toxins, bacteria or pathogens, and salts or salinity. Different types of agricultural land use are more likely to contribute certain pollutants than others. Runoff and subsurface water from agricultural and silvicultural operations is a source of the following pollution:

1. Dryland cropland: most often contains sediment, adsorbed nutrients, and pesticides
2. Irrigated cropland: irrigation return flow most often contains sediment, both adsorbed and dissolved nutrients and pesticides, traces of certain metals, salts and sometimes bacteria, viruses, and other microorganisms

3. Pasture land: contains bacteria, nutrients, sediment, and sometimes pesticides
4. Range land: can contain sediment, bacteria, nutrients, and occasionally metals or pesticides
5. Forest land: most often contains sediment, organic materials, and adsorbed nutrients due to logging operations
6. Confined animal feeding operations: contain bacteria, viruses, and other microorganisms; both dissolved and adsorbed nutrients, sediment, organic material, salts, metals
7. Specialty areas
   a. aquaculture: most often contain dissolved nutrients, bacteria, and other pathogens
   b. orchards or nurseries: contain nutrients (generally dissolved), pesticides, salts, bacteria, organic material, and trace metals
8. Wildlife land: can contribute bacteria and nutrients if wildlife populations become unbalanced

Just as significant as pollutants in the water column are the physical changes that occur in or adjacent to riparian areas as a result of a variety of agricultural land-use activities. These changes are responsible for much of the nonattainment of water quality standards and associated designated beneficial uses reported by state water quality agencies in their Nonpoint Source Assessments Reports and biennial Water Quality Inventories (U.S. EPA, 1990; U.S. EPA, 1992a). As states increase the use of biologic criteria in their water quality standards programs, these physical effects on biological communities will become more widely reported and used as measurements of maintenance of the biologic integrity of the nation's waters (U.S. EPA, 1990).

Some specific factors and activities that affect the diffuse loadings of pollutants from agricultural and silvicultural areas are:

- Livestock permitted uncontrolled access to riparian areas cause shear and sloughing of stream-bank soils and eliminate stream-bank vegetation, which results in changes to the surface and subsurface hydrology and morphology of the stream area, subsequent changes in stream flow, increases in water temperature, and reductions in dissolved oxygen content (Platts 1990; U.S. EPA, 1992a, 1992b; New Mexico Environmental Improvement Division, 1989).
- Crop production systems that plow fields right up to the edge of the streambank destabilize the banks causing them to collapse or for gully headcuts to form. The material from these caved-in streambanks and active gullies smothers remaining riparian vegetation and silts in the

adjacent stream channel. Aquatic organism species composition and richness changes as a result of loss of cover, food source, and changes in flow, temperature, and dissolved oxygen (DO) (Robillard et al., 1982; U.S. EPA, 1992a).
- Draining low-lying areas (generally wetlands) for crop production has had a profound effect on the physical state of the area. Loss of riparian vegetation through clearing, loss of a natural high water table from drainage, and loss of the flood buffering capacity of these areas results in a stream system that is more flashy when wet and more drought prone when dry. Loss of consistent stream base flow is one of the most detrimental changes affecting the reproduction habits of most aquatic organisms (U.S. EPA, 1992a; Platts, 1990).

Crop Production

On croplands pollutants arise from surface runoff (erosion of topsoil particles and irrigation return flows), interflow (mostly tile drainage and leachate of excess irrigation), and ground-water base flow (high nitrate content due to overfertilization). Often the reduction of pollution in one component may cause an increase in another component. Figures 11.2 and 11.3 show a typical example of the water quality consequences of diffuse pollution from farm runoff.

Erosion and soil loss by surface runoff and nitrate leaching into ground water are considered predominant sources of pollution from cropland. The disturbing activity associated with tillage increases the erosion potential of croplands. Conversely, increased hydrologic surface storage and permeability of tilled fields reduce hydrologic activity, which sometimes balances the increased erosion potential.

Alberts, Schuman, and Burwell (1978) and McDowell et al. (1989) reported that of the nutrient (N and P) losses (reported in Table 11.2), about 90% was associated with soil loss. Alberts et al. also noted that the losses of nutrients from croplands represent a relatively small portion of the applied fertilizer; nevertheless, the concentrations in the runoff almost always exceeds the criteria for preventing accelerated eutrophication of receiving of bodies of water.

Dudley and Karr (1979) noted that bacterial contamination from agricultural drainage in the Black Creek Watershed of Indiana exceeded accepted standards for recreation. Organic pollutants and fecal contaminants originate from manure application, unconfined livestock, and septic tank drainage fields. Applying manure to frozen fields is a practice that is most damaging to surface-water bodies.

The contribution of pesticides of diffuse pollution is also of great concern. Atrazine and small amounts of other pesticides have been

FIGURE 11.2. Algal infestation of a drainage canal caused by agricultural drainage in the watershed of the Lagoon of Venice in Italy. (Photo: G. Bendoricchio.)

FIGURE 11.3. Poor water quality of a sluggish stream in an agricultural watershed. (Photo: University of Wisconsin.)

detected in drinking water supplies in, among other places, Tennessee (Klaine et al., 1988) and in the Po River Valley in Italy (Capodaglio, 1988). Measurements by Klaine et al. have shown that just over 1% of the total atrazine applied to a small agricultural watershed was lost with runoff; however, concentration in runoff reached as high as 250 µg/l.

Pasture- and Rangeland

Pasture- and rangeland account for the largest proportion of total land use in the United States and includes about 40% of all nonfederal land. Range- and pastureland is used directly for livestock production. The grazing practices include continuous and seasonal or rotational grazing. *Rangeland* typically refers to land covered with native or adapted grasses or shrub vegetation used for large-scale, but low-density animal operations in arid or semiarid western states. In the past, rangeland livestock grazing had adverse impact on fragile watersheds of the west and southwest United States.

Smolen, Jennings, and Huffman (1990) have pointed out that pastureland becomes a source of diffuse pollution when proper erosion control practices are not in place or when grazing livestock is allowed to approach or enter surface waters (Fig. 11.4). They also reported that sediment yield from pastures in the state of Washington was minimized when vegetation cover remained greater than 50%, regardless of animal trampling disturbances. Owens et al. (1989) investigated nutrient and sediment losses from unimproved pastures, that is, pastures not treated with fertilizer, weed control, or selected forage species, and found insignificant differences in nutrient losses from grazed as compared to ungrazed areas, although sediment losses were greater in grazed than in ungrazed areas.

From these findings it becomes apparent that overgrazing and permitting livestock to approach and enter watercourses are the major pollution activities on pastures and rangelands. If such activities are controlled, pollution from pastures and rangelands may be minimal.

Woodlands (Silviculture)

Undisturbed forests or woodlands represent the best protection of lands from sediment and pollutant losses. Woodlands and forests have low hydrologic activity due to high surface storage in leaves (interception), ground, mulch, and terrain roughness. Even lowland forests with a high ground-water table (containing wooded wetlands) absorb large amounts of precipitation and actively retain water and contaminants.

Uncontrolled logging operations (clear-cutting), however, disturb the forest's resistance to erosion (Fig. 11.5). Observations and records indicate that almost all sediment reaching waterways from forestlands originate from construction of logging roads (Haupt, 1959; Beschta, 1978)

Agriculture and Its Effect on the Environment 687

FIGURE 11.4. Pollution of bodies of water located in rangeland and pastures is due to the practice of allowing cattle to enter the water body. (Photo: University of Wisconsin.)

FIGURE 11.5. The highest pollutant loads from silviculture (forestry) are caused by clear-cutting and road building.

and from clear-cuts (Beschta, 1978; Cheng, 1988). The chief sources are roads that disturb the natural drainage channels.

The potential for substantial water quality impact from increased aluminum concentrations resulting from whole tree harvesting was demonstrated by Lawrence and Driscoll (1988). The aluminum content of runoff is greatly increased when deforestation is caused by acidic deposition. (See Chapters 1 and 6 for a discussion of the so-called "chemical time bomb," which is an excessive loss of soil aluminum and its consequences caused by acidic precipitation.) Streams draining lowland forests with wetlands may have elevated organic and nutrient levels caused by leaching from soils. Also the dissolved oxygen may be depressed due to intensive decay processes taking place in the wetland portions of the forests. Despite these effects, pollutant loads from woodlands are near background levels.

Intensive Animal Production

Feedlots and barnyards can be the most intensive land uses in rural areas. With the advent of improved feeding methods and handling of ensiled materials, cattle are no longer put out to pasture but are held in relatively small areas. An obvious potential problem exists in feedlots, barnyards, and exercise areas where herds are confined on a year-round or seasonal basis (Fig. 11.6).

Commonly, feedlots themselves (beef, dairy, swine, sheep, and poultry) are classified as point sources, requiring in the United States a permit for discharge of pollution. Indirectly, however, the solid waste disposal methods and methods of management cause this subcategory to have a major diffuse-pollution impact. The majority of feedlot wastes reaching surface waters are transported by surface runoff. Loehr (1972) has pointed out that the quantity of runoff and the hydrologic activity of feedlot areas depend on the degree of perviousness of the lot and the permeability of the soils (which is greatly reduced by cattle trampling and high compacted organic content), antecedent soil moisture conditions, the number of cattle on the lot, surface storage characteristics, and rainfall intensity. In fact, permeability and infiltration characteristics in many feedlots are only remotely related to the native soil. The high organic content of the soil surface crust in a feedlot protects against erosion, and consequently sediment yields from feedlots are lower. Loehr (1972) conducted one of the first analyses of agricultural runoff pollution potential, reporting feedlot runoff concentrations of 1000 to 12,000 mg BOD_5/l, 2400 to 38,000 mg COD/l, 6 to 800 mg/l of organic N, and 4 to 15 mg/l of P. Runoff from feedlots is turbid and represents a high nutrient and organic shock loading to receiving waters.

Land application of feedlot waste has been traditionally a source of

FIGURE 11.6. Hog and beef feeding lot adjacent to a body of water. Bacterial contamination in the pond was 10 times the maximum for safe swimming and recreation use. (Photo: USDA, Soil Conservation Service.)

fertilizer as well as means of disposal. However, rates must be controlled to ensure stand persistence if perennial crops are grown as well as to protect water quality (Burns et al., 1987). In the follow-up study, Westerman et al. (1987) concluded that pollution hazards from disposal of manure exists for any rate, especially when rainfall occurs soon after application.

BEST MANAGEMENT PRACTICES: THEIR IMPLEMENTATION AND EFFECTIVENESS

Best management practices (BMPs) are methods and practices or combinations of practices for preventing or reducing nonpoint source pollution to a *level compatible with water quality goals*. Effective BMPs must also be economically and technically feasible (see Chapter 1 for a definition of best management practices). BMPs can be classed as structural, vegetative, or management, and each class is somewhat more effective in controlling certain types of diffuse pollution than others. As discussed previously throughout this book diffuse water pollution occurs after a pollutant becomes available, detached, and is transported to the body of water. When selecting BMPs it is important that the pollutants and the forms in which they are transported is known. It is also important to realize that BMPs are not necessarily synonymous with resource management systems or conservation treatments. Many practices listed in the technical handbooks of land resource agencies were designed to protect soil or grass resources or to provide economic stability for farmers by improving productivity. For example, while conservation tillage is extremely effective in erosion control, it is not very effective in removing or controlling soluble pollutants from runoff or leached water (USDA, Soil Conservation Service 1988; Robillard et al., 1982; Haith and Loehr, 1979).

BMPs can be selected in basically two ways: to control a known or suspected type of pollution (for example, phosphorus or bacteria) from a particular source (runoff from a corn field or a diary feedlot), or to prevent pollution from a category of land-use activity (such as agriculture row crop farming or containerized nursery irrigation return flow). (Robillard et al., 1982; Alexander, 1992). When selecting BMPs to solve a water quality problem the following process can be used (USDA, Soil Conservation Service 1988; USDA, Soil Conservation Service, 1980; Gordon and Hansen, 1989; Brach, 1990);

- Identify the water quality problem (e.g., annual summer algal bloom in lake)
- Identify the pollutants contributing to the problem and their probable sources (e.g., nutrients from septic systems adjoining the lake and runoff from a nearby horse pasture)
- Determine how each pollutant is delivered to the water (e.g., soluble nutrients from septic tank drain fields rise to the surface when the systems are overloaded and are carried to the lake by overland flow during rain storms or snow melt)
- Set a reasonable water quality goal for the resource and determine the level of treatment needed to meet that goal
- Evaluate feasible BMPs for water quality effectiveness, effect on ground water, economic feasibility, and suitability of the practice to the site

When selecting BMPs to prevent a problem a more technology-based approach can be followed. Such an approach divides agricultural land into use categories such as irrigated cropland and rangeland, and specifies a minimum level of treatment necessary to protect the resource base. This type of planning has been used for years by the USDA in developing conservation plans for soil conservation. Technical guidance for the new Coastal Zone Management Act employs this basic "minimum treatment level" approach for diffuse-pollution prevention. Technically and economically achievable performance standards for each type of agricultural land use (and its associated pollutants) are used in this approach, which does not require that a cause-and-effect linkage between site-specific land-use activities and site-specific water quality problems be established. This approach is based on the rationale that neither the money nor the time is available to establish the complex monitoring programs that would be needed for such site-specific problem and source identification (U.S. EPA, 1993).

The following section discusses some individual BMPs and their effectiveness. In a fashion similar to urban practices, agricultural BMPs can be divided into source controls, hydrologic modifications, reduction of delivery, and storage and "treatment." Efficiencies of various practices have been summarized by Smolen, Jennings, and Huffman (1990); Brach (1990); U.S. EPA (1992a, 1992b) and others. Table 11.3 lists the factors for the selection of BMPs.

Cropping Practices: Conservation Tillage, Cover Crops, and Crop Rotation (Source Control Measures)

Definitions. These three types of cropping systems stress maintenance of vegetative cover during critical times, and their primary objective is to reduce erosion and hence soil loss. Conservation tillage is any tillage method that leaves at least 30% of the soil surface covered with crop residue after planting. The soil is only tilled to the extent needed to prepare a seedbed. Cover crops are close-growing grasses, legumes, or small grain crops that cover the soil during the critical erosion period for the area. Crop rotation is a system of periodically changing the crops grown on a particular field. The most effective crop rotations for water quality protection involve at least two years of grass or legumes in a four-year rotation.

No-till planting (Fig. 11.7) has been considered the most effective erosion-control practice applicable to agricultural lands. Planting is accomplished by placing seeds in the soil without tillage and retaining previous plant residues. The previous planting can be killed and weed control accomplished by the use of chemical herbicides. The plant and root residues provide the necessary surface protection. The magnitude of

TABLE 11.3 Selecting BMPs by Pollutant: Rules of Thumb

Pollutant	Methods of Control	Structural	Vegetative	Management
Sediment (TSS, cobble embeddedness, turbidity)	1. Control erosion on land and streambank	Terraces; diversions; grade stabilization structures; streambank protection and stabilization	Cover crops and rotations; conservation tillage; critical area planting	Contour farming; riparian area protection; proper grazing use and range management
	2. Route runoff through BMPs that capture sediment	Sediment basins	Filter strip; grassed waterway; stripcropping; field borders	
	3. Dispose of sediment properly			Beneficial use of sediment—wetland enhancement
Nutrients: N, P (nuisance algae, low dissolved oxygen, odor)	1. Minimize sources	Animal waste system (lagoon, storage area); fences (livestock exclusion); diversions; terraces	Range management; crop rotations	Range and pasture management; proper stocking rate; waste composting; nutrient management
	2(a). Uptake all that is applied to the land or contain and recycle/reuse (dissolved form control—commercial nutrients)	Terrace; tailwater pit; runoff retention pond; wetland development	Cover crop; strip cropping; riparian buffer zone; change crop or grass species to one that is more nutrient demanding	Recycle/reuse irrigation return flow and runoff water; nutrient management; irrigation water management
	2(b). Contain animal waste, process and land apply, or export to a different watershed (dissolved form control—animal waste)	Diversion; pit/pond/lagoon; compost facility	See 2(a)	Lagoon pump out; proper irrigation management

	3. Minimize soil erosion and sediment delivery (adsorbed form control)	Terrace; diversion; stream-bank protection and stabilization; sediment pond; critical area treatment	Conservation tillage; filter strip; riparian buffer zone; cover crop	Nutrient management
	4. Intercept, treat runoff before it reaches the water (suspended form control)	See 1–3; water treatment (filtration or flocculation) for high-value crops	See 1–3	
Pathogens (bacteria, viruses, etc.)	Minimize source	Fences	Riparian buffer zone	Animal waste management, especially proper application rate and timing
	Minimize movement so bacteria dies	Animal waste storage; detention pond	Filter strips; riparian buffer zones	Proper site selection for animal feeding facility; proper application rate of waste
	Treat water	Waste treatment lagoon; filtration	Artificial wetland/rock reed microbial filter	Recycle and reuse
Metals	Control soil sources		Crop/plant selection	Avoid adding materials containing trace metals
	Control added sources	Tailwater pit; reuse/recycle system	Crop selection	Irrigation water management; integrated pest control
	Treat water	Filtration	Artificial wetland/rock reed microbial filter system	
Salts/salinity	Limit availability	Evaporation basins; tailwater recovery pits; ditch lining; replace ditches with pipe	Crop selection; saline wetland buffer; land-use conversion	Drip irrigation
	Control loss			Irrigation water management

TABLE 11.3 (*Continued*)

Pollutant	Methods of Control	Structural	Vegetative	Management
Pesticides and other toxins	Minimize sources		Plant variety/crop selection	IPM; change planting dates; proper container disposal
	Minimize movement and discharge	Terrace; sediment control basin; retention pond with water reuse/recycle system	Buffer zone; conservation tillage; filter strips (adsorbed control only); wetland enhancement	Irrigation water management; IPM
	Treat discharge water	Carbon filter system (high-value crops)	Rock–reed microbial filter system/artificial wetland	
Physical habitat alteration	Minimize disturbance within 100 feet of water	Road and turnrow realignment; fencing/livestock water crossing facility	Buffer strips; riparian buffer zones	Proper grazing management, including limiting livestock access
	Control erosion on land	See sediment BMPs		
	Maintain or restore natural riparian area vegetation and hydrology	Streambank stabilization; channel integrity repair	Wetland enhancement	Proper grazing use and range management; limit livestock access

Sources: U.S. EPA (1993); Brach (1990); Alexander (1993a); USDA, Soil Conservation Service (1988).

FIGURE 11.7. No-till planting. Seeds are placed in the soil and vegetation residues are left on the ground to minimize erosion. Weeds are killed by herbicide application. (Photo: University of Wisconsin.)

FIGURE 11.8. Strip-cropping. (Photo: USDA, Soil Conservation Service.)

TABLE 11.4 Effect of the No-Till Practice on the Magnitude of the Cover Factor (C) in the USLE

Tilling Practice	C Factor Production Level	
	High	Moderate
No-till		
Corn after sod[a]	0.017	0.053
Corn after corn	0.18	0.18
Soybean after corn	0.18	0.22
Grain	0.11	0.18
Corn after small grain[a]	0.062	0.14
Conservation tillage[b]		
Corn—chisel plowing	0.19	0.26
Corn—fall chisel—spring disk	0.24	0.30
Corn—strip till—row zones	0.16	0.24

Source: Adapted from Stewart et al. (1975).
[a] Removed by chemical treatment.
[b] Residue not removed.

the cover factor (C) of the universal soil loss equation (see Chapter 5) for no-till and conservation tillage practices are given in Table 11.4. However, a greater use of herbicides and sometimes lower yields are experienced on some soils.

Contour plowing is a widely recommended supporting practice in which plowing and crop rows follow field contours across the slope. It provides excellent erosion control for moderate rainstorms. The effects of contouring on the erosion-control practice factor (P) of the universal soil loss equation are given in Chapter 5.

Strip-cropping (Fig. 11.8) alternates contour-plowed strips of row crops and close-grown crops.

How the practices work to control pollution. Conservation tillage controls erosion and sediment by decreasing soil detachment, which also helps control the loss of adsorbed pesticides and nutrients. Cover crops also reduce erosion and sedimentation by decreasing detachment. The growing vegetation also retards movement of any detached particles and their adsorbed pollutants. When legumes are used as a cover crop or as part of a crop rotation, they provide a nitrogen source for the subsequent crop, thus minimizing the addition of commercial fertilizers. The cover crops also use nutrients that are otherwise lost through leaching during fallow periods. Crop rotations reduce erosion and associated adsorbed pollutant loads by improving soil structure, which decreases detachment, but they also reduce the need for pesticide and nutrient applications since insect reproductive cycles are disrupted by changing host crops and nutrient needs more correctly balanced for a moderate level of production.

Effectiveness. Conservation tillage has been found to be highly effective in sediment reduction but of very little effect in controlling soluble nutrients and pesticides. Sediment reductions of 30–90%, total phosphorus reductions of 35–85%, and total nitrogen reductions of 50–80% have been found; however pollution potential could increase (and reduction efficiencies decline) if fertilizers are not soil incorporated in a CT system. Cover crops have been found to be 40–60% effective in reducing sediment, 30–50% in removing total phosphorus. The effectiveness of a crop rotation system is difficult to measure. Often the best estimate is an average annual load reduction over the entire rotation, so a higher reduction in one year coupled with a low reduction in another gives a moderate level for the entire rotation. Some rotations have been estimated to reduce nitrogen 50% and phosphorus 30% annually.

Berg et al. (1988) reported in a 10-year study that no-till management controlled erosion within acceptable limits on highly erodible graze-out wheat land. Conservation of nutrients in soil water was shown to be another benefit from no-tillage systems (Stinner et al., 1988). Parkin and Meisinger (1989) reported that no-till did not significantly affect subsurface biological activity and denitrification.

Strip-cropping systems can provide up to a 75% load reduction in sediment and up to a 50% load reduction in total phosphorus.

Integrated Pest Management

Definition. Integrated pest management (IPM) is a combination of practices to control crop pests (insects, weeds, diseases) while minimizing pollution. It uses traditional practices such as selection of resistant crop varieties and crop rotations, modified planting dates along with sophisticated pesticide application management, which includes pest scouting, minimal application rates, most judicious timing, and selection of the least toxic and least persistent chemical.

How the practice works to control pollution. IPM works mainly by decreasing the amount of pesticide or crop-protection chemical available and by selecting the least toxic, least mobile, and/or least persistent material, it decreases the potential for pollution.

Effectiveness. Complete studies on the effectiveness of IPM show varying results depending on the exact type of pesticide, soil, moisture, and crop conditions. Some estimates are extremely high and others low.

Nutrient Management

Definition. Nutrient management is a series of practices designed to decrease the availability of excess nutrients through improvements in timing, application rates, and location selection for fertilizer placement. It also involves fertilizer-type selection, crop-variety selection, and conservation cropping systems (such as rotations, cover crops, and strip-cropping). Nutrient management is currently based on the limiting nutrient concept (that fertilizer application rate should be based upon the nutrient most needed by the plant for optimum growth—most often nitrogen). New research is showing that in some cases, phosphorus, thought to be relatively immobile and strongly sorbed to the soil can become available and contribute to water quality problems. When animal waste is used as a fertilizer in areas with phosphorus water quality problems (nitrogen to phosphorus ratios cannot be manipulated with animal waste), the EPA recommends that phosphorus be considered as the limiting nutrient.

How the practice works to control pollution. Nutrient management most effectively limits the availability of nutrients through a lowered or more precise application rate.

Effectiveness. Nitrogen and phosphorus loss can be reduced by 20–90%. This practice is especially effective in controlling the soluble phases of these nutrients.

Terraces and Diversions

Definition. A *terrace* is an earthen embankment, channel, or a combination of ridge and channel constructed across a slope to intercept runoff. Terraces reduce the slope effect on erosion by dividing the fields into segments with lesser or even near-horizontal slopes (Fig. 11.9). This practice therefore reduces the slope-length factor of the USLE. Terraces can be level or graded. Level terraces store or pond water for infiltration or outlet through an underground conduit. Graded terraces divert water to a suitable outlet. *Diversions* are a form of graded terrace, but are defined as a channel constructed across the slope with a supporting ridge on the lower side.

How the practice works to control pollution. Terraces change the effective slope of the land, which decreases runoff velocity. Soil particles and adsorbed pollutants are thus not transported from the field. Terraces that retain all water applied to the field are also effective in reducing the loss of dissolved pollutants. Graded terraces or those with outlets allow runoff to leave the field and therefore do not control dissolved or suspended pollutants well. Diversions transport excess water from areas needing protection to sites suitable for disposal. They can prevent water from flowing across areas where pollutants are available, such as highly ero-

FIGURE 11.9. Parallel terraces. (Photo: USDA, Soil Conservation Service.)

dible soils, feedlots, or areas where fertilizers or pesticides are mixed, stored, or applied at high rates. Diversions must be provided with a stable outlet, such as a grassed waterway or buffer zone.

Effectiveness. Level terraces can remove up to 95% of the sediment, up to 90% of its associated adsorbed nutrients, and between 30% and 70% of dissolved nutrients. Diversions can reduce sediment movement by 30–60% and adsorbed nutrients by 20–45%.

Example 11.1: Effect of Conservation Tillage

A corn-growing field on a 10% slope, plowed up and down, had estimated erosion losses of soil, organic matter, and P of 20 tonnes/ha-yr, 1.0 tonne/ha-yr, and 30 kg/ha-yr, respectively. Estimate the effect of: (1) No-till planting, (2) sod-based rotation, and (3) terracing on the erosion losses.

Solution Most of the loss (80%) occurred during the critical erosion period (i.e., a period of 2 months after spring plowing). The overland flow length (downslope) is 200 m.

For reference practice of conventional plowing up and down the slope, the average cover factor of USLE for the critical period is $C = 0.8$ and afterward $C = 0.15$ (see Table 5.4 for appropriate magnitudes). The erosion-control factor $P = 1.0$. In the no-till practice, crop residues and root systems from the previous harvest remain on the soil as a protective cover. This can reduce significantly erosion during the critical period. If the preceding crop was small grains, the C factor would approximate 1.0. Hence, the effect of this practice on the soil and pollutant loss is

$$\text{Percent loss reduction} = \left[1 - \frac{\text{weighted } C \text{ with no-till}}{\text{weighted } C \text{ without no-till}} \times 100\right]$$

$$= \left[1 - \frac{0.8 \times 0.1 + 0.2 \times 0.1}{0.8 \times 0.8 + 0.2 \times 0.15}\right] 100 = 87.8\%$$

If all other variables remain the same (a crude and often unrealistic assumption), the soil loss with no-till practice should be

$$A_s = (1 - 0.878) \times 20 = 2.44 \text{ tonnes/ha-yr}$$

and similarly, organic matter and P losses become

$$A_{OM} = (1 - 0.878) \times 1.00 = 0.122 \text{ tonnes/ha-yr}$$

and

$$A_P = (1 - 0.878) \times 35 = 4.27 \text{ kg/ha-yr}$$

With sod-based rotation, sod is planted once during the rotation cycle. For the purpose of this example, assume that the rotation cycle is 2 years and that the sod is plowed during the spring season by conventional plowing methods. The C factor for sod during the off-year is $C = 0.01$. In the following year when corn is planted, the sod residues may reduce C to a value of 0.6. Average soil loss reduction over the 2-year cycle is

$$\text{Percent loss reduction} = \left[\frac{1 - (0.01)/(0.8 \times 0.8 + 0.2 \times 0.15) + 1}{2}\right.$$

$$\left. - \frac{(0.8 \times 0.6 + 0.2 \times 0.15)/(0.8 \times 0.8 + 0.2 \times 0.15)}{2}\right] \times 100$$

$$= 61.1\%$$

If slopes are reduced by terracing, the slope-length factor (LS) is decreased. If terraces are built 1 meter high on a 10% slope, the overland flow length is reduced to 10 m. From Figure 5.14, the original LS factor for a 200 m overland flow distance and 10% slope is 3.7. Having terraces 10-m wide with an average slope of 1% reduces the LS factor to 0.1. Hence the soil and pollution loss reduction becomes

$$\text{Percent loss reduction} = \left[1 - \frac{(LS) \text{ factor after terracing}}{(LS) \text{ factor before terracing}}\right] \times 100$$

$$= \left[1 - \frac{0.1}{3.5}\right] 100 = 97.1\%$$

Although the cover factor or slope-length factor seem to be the predominant factor for the situation discussed in this example, other factors should be included in a more exact analysis. For example, reduction of the slope by terracing increases surface storage (Fig. 3.3); no-till practices, on the other hand, reduce surface storage and permeability. For a more complex analysis, use of mathematical models is the best approach to evaluate the combined effects of several factors. Such models are described in Chapter 9.

Critical Area Treatment (Grade-Stabilization Structure and Critical Area Planting—Pollutant Delivery Control)

Definition. A grade-stabilization structure is used to control the grade and gully-head cutting in natural or artificial channels. It can provide an outlet for other conservation practices. Structures can be constructed of metal, wood, rocks, concrete, or earth. The area surrounding the structure must be suitably stabilized. Critical-area planting involves planting suitable vegetation (such as grass, trees, or shrubs) on highly unstable sites such as gullies, denuded streambanks, dams, and embankments where vegetation is difficult to establish by the usual planting methods. Intensive methods, such as use of erosion blankets, tiedowns, gabions, mulches, hydromulching, increased seeding rate, and hand planting, are often needed.

How the practice works to control pollution. Grade-stabilization structures reduce water velocity, thus preventing additional sediment detachment and decreasing the transport capacity of the water. They also often act as traps, reducing additional sediment transport. Sediment adsorbed nutrients are thus controlled much more effectively than dissolved pollutants. Planting vegetation on other critical areas works in the same manner by slowing velocity and decreasing detachment and transport.

Effectiveness. Grade-stabilization structures can result in a 5–75% reduction in suspended solids leaving the unstable area, but data on the area below the structures are unavailable. Data on the vegetation methods are also lacking, but soil-erosion estimates using soil-loss models could be used as a rough approximation of effectiveness.

Sediment Basins, Water and Sediment-Control Basins, and Detention–Retention Ponds

Definition. All earthen embankments are generally designed to trap and store sediment or other pollutants. Sediment basins are used to preserve the storage capacity of downstream reservoirs, canals, etc., and often serve an entire small watershed (Fig. 11.10). Water- and sediment-control basins are used to reduce on-site water, control gully erosion, and protect downstream water quality. Detention ponds are used to collect runoff from agricultural fields for storage and pollution control.

How the practice works to control pollution. These structures stop water flow, allowing heavier particulate to settle out. Because the sediment basins and water- and sediment-control basin structures are usually drained and the water released downstream or across land, soluble pollutants are not controlled. Detention ponds are not drained and often support aquatic vegetation. Available phosphorus and nitrogen can be taken up by these plants, and it is believed that nonpersistent pesticides will have time to degrade. Design methods for sediment detention basins were presented in Chapter 10.

FIGURE 11.10. Retention pond–sedimentation basin. (Photo: S. Alexander.)

Effectiveness. Sediment basins can remove 40–87% of the incoming sediment, up to 30% of the adsorbed nitrogen, and 40% of the total phosphorus. Effectiveness of detention–retention ponds, especially those with aquatic vegetation, is generally higher. As is pointed out in Chapter 13, sedimentation basins are not effective for the removal of toxicity or dissolved toxic compounds.

Animal Waste Storage and Treatment Facilities and Methods (Lagoons, Storage Bins and Ponds, Composters), Also Known as Animal Waste Management System

Definitions. A combination of practices for storing manure and other wastes, controlling runoff from feedlots, managing removed waste and runoff, and disposing of animal carcasses, or other waste. Waste storage ponds and structures hold waste until it can be land applied without causing a water-pollution problem. Waste treatment lagoons biologically treat liquid waste to reduce the nutrient and BOD content. Composters are structures to biologically degrade dry waste or dead animals (especially poultry carcasses). Storage areas, lagoons, and composters all must be emptied and their contents disposed of properly. Management BMPs are then applicable. The waste can be land applied at the proper rate, time, and in the proper way, or removed from the farm for proper use or disposal elsewhere.

Generally, animal wastes are classified as point sources and require a permit in many states. However, pollution is generated by runoff from the contaminated premises and typically requires runoff control BMPs in addition to treatment and/or safe disposal.

How the practices work to control pollution. Runoff-control systems for feedlots and manure storage are governed by two basic principles. First, all clean water originating outside the feedlot or storage area should be diverted so that it does not come in contact with the contaminated soil. Second, the water originating inside the feedlot should be disposed of in a way that minimizes its pollution potential. To meet these objectives, four components should be considered in the design of runoff pollution-control systems: clear water diversion, runoff collection, runoff containment, and controlled disposal (Fig. 11.11).

Clear water diversion systems include terraces, which direct water from upland watersheds away from the feedlot or barnyard, and gutters with downspouts on buildings, which prevent roof drainage from entering the area. When a natural waterway (creek) crosses a feedlot (Fig. 11.11), it may be necessary to pass it through a culvert or relocate the feedlot.

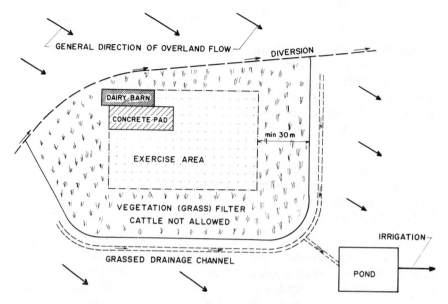

FIGURE 11.11. Best management practices for containment of feedlot runoff.

FIGURE 11.12. Large two-stage lagoon for containment of dairy animal waste. The first stage provides solids separation. (Photo: S. Alexander.)

Collecting runoff with a system of curbs, gutters, or terraces prevents it from directly entering defined watercourses and concentrates polluted runoff for treatment before disposal. Runoff can be contained for short periods of time in settling channels or basins, vegetative filters, or infiltration areas. Holding ponds (Fig. 11.12) provide long-term storage and they also act as facultative (aerobic–anaerobic) lagoons (see Chapter 10) and provide sedimentation and biological treatment (Loehr, 1972).

Often waste storage ponds or structures contain all pollutants, thus allowing none to enter the receiving bodies of water (of course, they must be pumped out later and disposed of). With waste treatment lagoons the detention time allows bacteria to die off and nitrogen to volatilize, and the biological process within the lagoon decrease the oxygen demand of the effluent. Composting of solid wastes from feedlots allows some slight mineralization of nitrogen. The heat and enzymatic activity generated during the composting process usually kills most bacteria. Composting is mainly used to reduce the volume and odor of the waste and improve its handling and management. Proper management is crucial to the effectiveness of each component of the system because it limits the availability of nutrients and pathogens and the detachment potential of the waste pollutants.

Effectiveness. Animal waste systems (all practices combined) have been estimated to reduce pollutant loadings as follows (averages): bacteria 74%, sediment 64%, total nitrogen 62%, total phosphorus 21%. Individual practices in the system vary widely. For example, a waste treatment lagoon (and proper land application) reduces bacteria by 65–100%, dissolved phosphorus 10–69%, and total nitrogen 32–91%. Most BMP recommendations suggest applying manure to satisfy crop nitrogen needs. This practice often results in excess phosphorus being applied.

Livestock Exclusion (Fences)

Definition. Excluding livestock from areas where grazing or trampling will cause erosion of streambanks and lowering of water quality by livestock activity in or adjacent to the water. This is generally accomplished by permanent or temporary fences or herding.

How the practice works to control pollution. By keeping animals out of streams, deposition (availability) of fecal material (nutrient and bacterial source) will be reduced, turbidity from in-stream trampling will be eliminated (particles will not be detached), and streambanks will not be denuded (sediment will not detach), but will instead provide a filter for runoff (transport and delivery of sediment and adsorbed materials will be reduced), and if the riparian area is large enough, uptake some of the nutrient load (availability of dissolved pollutants will be decreased).

Effectiveness. Reductions of 50–90% of suspended solids and total phosphorus have been reported.

Filter Strips and Field Borders

Definitions. Filter strips and field borders utilize strips of closely growing vegetation, such as sod or bunch grasses or small grain crops, with the primary purpose of water quality protection. Filter strips are generally placed between the agricultural land being used (for example, cropland fields, animal waste application areas, or forests) and the body of water to be protected (Fig. 11.13). They are designed to remove sediment and other pollutants from sheet runoff. Field borders normally consist of perennial vegetation planted at the edge of fields to control erosion, regardless of their proximity to water. In contrast, strip-cropping, discussed earlier, involves planting straight or contoured strips of grass or closely growing crops in alternating bands between strips of row crops, and its primary purpose is to reduce soil losses.

How the practices work to control pollution. Both practices interfere with the transport of sediment, slowing water velocity, and allowing the material and any adsorbed pollutants to drop out. If filter strips are wide enough (detention time is adequate) and planted with appropriate

FIGURE 11.13. Vegetated buffer strips between plowed field and drainage ways. (Photo: University of Wisconsin.)

vegetation (high-nutrient-demanding species), they can uptake some nutrients, thus decreasing their availability. Field borders are effective in preventing detachment of soil particles in the areas covered by the border, but have little effect on controlling erosion or pollutant detachment or transport from the fields they surround. Designs and design considerations for grass filter strips were discussed in Chapter 10.

Effectiveness. Filter strips are very effective in removing sediment and sediment-bound nitrogen (about 35–90%), but much less effective in removing phosphorus, fine sediment, and soluble nutrients, such as nitrate (14%) or orthophosphate (5–50%). During the active growing season some filter strips have not been effective in removing phosphorus. Field borders are most effective in sediment removal and erosion control. Pollutant reductions for borders alone have not been fully evaluated. The most effective systems rely upon perennial grasses. New research appears to indicate that deep-rooted native bunch grasses in strip-cropping systems are also effective in "recovering" leached nitrogen that has escaped below the roots of conventional row crops.

Wetland Rehabilitation and Development, and the Beneficial Use of Sediment

Definitions. Rehabilitation and development involve restoring, rehabilitating, or enhancing existing wetlands to function as self-sustaining ecosystems that process, remove, transfer, and store pollutants. Beneficial use of sediment is a newer technology designed to utilize clean sediment recovered from dredging operations to mitigate damage to wetlands in other places by enhancing or rebuilding degraded wetlands. Wetland management and the types of constructed and restored wetlands are presented in Chapter 14.

How the practices work to control pollution. The low slopes of wetlands serve to slow the velocity of water and allow sediment to precipitate out (decrease transport capacity of water), the long retention time promotes bacterial die-off, and the wetland vegetation uptakes and uses nitrogen (decreasing availability) and provides a source of carbon for soil microbial action. This carbon fuels the immobilization of some phosphorus and nitrogen by microbial action and the transference of nitrogen by denitrification. Utilization of dredged materials to mitigate damage to degraded wetlands solves two environmental problems: (1) it provides a safe and effective disposal site for materials that previously have been dumped in off-shore ocean disposal sites or inland in poorly functioning sediment containment pits, and (2) improves wetlands that have been depleted of their needed annual sediment replenishment/nourishment by changes in hydrology.

Effectiveness. Wetlands are 80–90% effective in sediment removal, 40–80% effective in nitrogen removal (depending upon location of riparian vegetation in relation to the systems hydrology; those wetlands located in the upper reaches of the watershed appear to be more effective), and 10–70% effective in phosphorus removal, also depending upon location. See Chapter 14 for a more detailed discussion.

Riparian Buffer Zones

Definition. Vegetated area along a body of water containing a complex assemblage of organisms and their environment, typically part of a riparian system (Fig. 11.14). It can be a complete ecosystem or function as an ecotone between aquatic and terrestrial ecosystems, but maintains a distinct set of vegetational and soil characteristics. The area is maintained in its native vegetation state at widths sufficient for pollution-control functions.

How the practice works to control pollution. The grasses and low vegetation in the zone filters both surface and subsurface flow, while the roots of taller vegetation take up and transform pollutants and nutrients from shallow ground water.

FIGURE 11.14. Riparian buffer zone separates farm operations from a watercourse. (Photo: University of Wisconsin.)

Effectiveness. Sediment removal efficiencies of 80-90%, total phosphorus removal efficiencies of 50-75%, and total nitrogen removal efficiencies of 80-90% have been measured. Pesticide removal has been documented but not adequately quantified.

Irrigation Water Management

Definition. Irrigation water management (IWM) is a combination of practices that control irrigation water to prevent pollution and reduce water loss. It also applies to chemigation application of chemicals (pesticides or fertilizers) through the irrigation system. IWM includes proper scheduling, efficient application, efficient transport systems, utilization and reuse of tailwater and runoff, and management of drainage water.

How the practice works to control pollution. Seepage-control practices (lining ditches with concrete or converting to pipe conveyances) greatly reduces leaching losses and decreases the availability of salts in canals for application (and runoff) from fields. Tailwater pits (which are pits, usually lined, to catch rainwater runoff or the water that runs out of the furrows at the end of a field) retain any pollutants and keep them from entering any bodies of water. Improvements in management include the efficiency of water application and prevention of excess runoff that carries with it dissolved nutrients and pesticides.

Effectiveness. IWM is particularly effective in reducing nitrogen and pesticide loadings to ground water and salt loading to surface waters. Tailwater pits are about 50% effective in sediment removal and can be highly effective in nutrient and pesticide pollution control as long as the collected water is reused and not discharged.

Stream-bank Stabilization

Definition. Structural or vegetative methods to protect the streambank from erosion. Riprap, concrete, wood, or rock gabions can be used, but vegetative stabilization is the most effective for pollution control as long as it is capable of withstanding the hydrology. See Chapter 15 for more information on stream-bank stabilization techniques.

How the practice works to control pollution. Eroding streambanks not only contribute sediment to the water and streambed but also cause increased water temperatures (due to the loss of the riparian vegetation that shades the stream). Loss of floodwater storage or hydrologic assimilative capacity to regulate flow is also an effect. Stabilization methods work to eliminate the detachment of soil particles and to restore one part of the hydrology of the system.

Effectiveness. Effective stabilization can reduce sediment loading by 90%, but is highly dependent upon the type of vegetation used and the stability and width of the reclaimed area. When combined with riparian area restoration, removal rates of 50–70% for phosphorus and 80–90% for nitrogen have been found. When planted in conjunction with a grass filter strip, load reductions of 80% TSS, 60–90% total nitrogen, and 30–90% total phosphorus are possible.

Rock–Reed Microbial Filters

Definition. A long shallow hydroponic plant/rock filter system that treats polluted waste and waste water. Combines horizontal and vertical flow of water through the filter, which is filled with aquatic and semiaquatic plants, microorganisms, and provides a high surface area of support media, such as rocks or crushed stone (see Chapter 14 for a description of submerged bed systems; also, see Jones, 1990).

How the practice works to control pollution. Symbiotic relationship between the microbes on the plant roots and the surface media synergistically effect the degradation of pollutants. Rocks and roots also provide filters and plants further contribute by translocating oxygen and by adsorbing metabolites from the microbes. Thus each system uses the other's waste in a cyclic and sustainable system. If the filters are used for removal of heavy metals and certain other toxins, harvesting of plant materials may be periodically necessary. These materials must be degraded (composted) and the residue tested before proper disposal.

Effectiveness. Biological oxygen demand has been decreased from 110 mg/l to 10 mg/l in a 24-hour retention time, and toxic organic compounds from 9 mg/l to 0.05 mg/l in a 24-hour retention time. The systems have been used for toxic heavy metals and on some radioactive elements, but efficiency figures are not available. The systems provide complete disinfection.

Range and Pasture Management

Definition. Systems of practices to protect the vegetative cover on improved pasture and native rangelands. It includes such practices as seeding or reseeding, brush management (mechanical, chemical, physical, or biological), proper stocking rates and proper grazing use, and deferred rotational systems. Rangeland is land generally in native grass and managed as range. Pastures are normally seeded with an improved grass variety and managed with agronomic practices (liming, fertilization, irrigation).

How the practices work to control pollution. Keeping land permanently covered in high-quality (closely spaced) vegetation decreases soil loss to negligible amounts (detachment and transport are interrupted), and sediment and adsorbed pollutants are not lost from the land surface. Since rangeland is not fertilized or irrigated, it can represent a low input system if properly managed.

Effectiveness. If correctly managed, range and pastures function like filters to reduce the load of nitrogen, phosphorus, and sediment up to 80%.

ROLES, RESPONSIBILITIES, AND PROGRAMS OF FEDERAL, STATE, AND LOCAL AGENCIES AND GROUPS

Within the United States there are more than 31 federal agencies, and numerous state and local agencies with programs that can be used for the control of diffuse pollution from agriculture (U.S. General Accounting Office, 1990). It is important to understand the responsibilities and opportunities available from each of these agencies. Effective control of diffuse pollution from agriculture requires that these programs be coordinated, nonduplicative, and focused on the goal of water quality protection. Table 11.5 lists programs by various agencies. The large number of agencies with overlapping responsibilities indicate fragmentation of efforts and often a lack of coordination.

BALANCING AGRICULTURAL PRODUCTION WITH ENVIRONMENTAL PROTECTION

Maintaining an adequate and available supply of food through domestic production is a goal of almost every country. Such a goal helps ensure national self-sufficiency and provides some degree of internal stability. Maintaining a clean and safe environment through proper resource management is also becoming a priority for most countries. It is possible to protect the environment while maintaining and sometimes enhancing agricultural production. Most consumers want plentiful, inexpensive, and safe food and fiber, as well as clean water and air, all without making personal sacrifices in time, money, or quality of life. This is not possible. What is possible is a series of changes in values, economics, and methods of agricultural production and pollution control. Needed are a modification in the attitudes of citizens, agricultural producers, and legislators, a strengthening of the pollution-control laws and programs, advances in agriculture production technologies, and changes to agricultural programs

(Text continues on page 723.)

TABLE 11.5 Summary of Programs in the United States that Could Be Used for Control of Diffuse Pollution from Agriculture

Agency and Program	Program Descriptions and Agency Responsibilities	Resources Available and Possible Roles
U.S. Environmental Protection Agency (EPA)	Administers educational and regulatory programs designed to protect the environment (prevent and control pollution). Provides environmental assessments, water quality monitoring, regulations and regulatory oversight, education, planning, technical assistance, grants and loans for pollution control. *Works mainly with state, federal, regional, and local agencies on pollution-control efforts*	Staff, information and data, laboratories and research facilities, grants and loans for pollution control, educational materials, monitoring equipment. *Offices located in 10 regional centers and Washington, D.C.*
EPA—permits	NPDES permits for confined animal feeding operations, enforcement for noncompliance	Staff for technical assistance with modeling and permit drafting, site inspections and compliance monitoring. Funds for special studies or projects
EPA—pesticides	Regulation of pesticide labeling and registration, which includes application rates, allowable crops and pests, environmental and human health cautions, disposal procedures. Licensing of restricted use pesticide applicators	Staff for review of research results, assistance with strategic planning, education and training, oversight of enforcement procedures of states. Funds for special projects and studies
EPA—water quality	Overall water quality planning and management through the following programs: 1. Nonpoint Source Control: program that oversees and approves state development of water quality assessments and implantation of management programs designed to control NPS. Directs funds to high-priority watershed or projects 2. Clean Lakes: program provides funds to restore or enhance publicly owned lakes	Staff for technical assistance to state and local agencies, review and approval of state programs, research and special studies. Grants to states for most water quality protection activities, educational materials, and programs. Funds for special studies or projects

	3. Water Quality Standards: program provides technical assistance in developing numeric, narrative, and biological limits (standards) to protect water quality and its use 4. Coastal programs: a number of different programs and initiatives designed to assess coastal resources and study ways to protect coastal waters, including the National Estuary program. Administers the new CZMA	
EPA—ground water	Administers the Sole Source Aquifer Protection program, provides technical and programmatic assistance to state wellhead protection programs	Staff for technical assistance, funds for special studies
EPA—wetlands	Oversight of the Corps of Engineers on wetlands dredge and fill permits, takes enforcement actions for illegal wetlands filling, technical support for wetlands delineations	Staff for oversight and enforcement activities, monitoring of wetland status, health and trends, and funds for special studies, educational materials, and programs, data
EPA—monitoring and surveillance	Environmental assessment, data analysis, oversight of state monitoring programs, special studies and agency research, EPA lab and Office of Research and Development coordination	Staff for technical assistance to states and citizens on monitoring programs and projects, special studies and data analysis upon request, water quality monitoring at select locations
EPA—drinking water	Regulates public drinking water supplies and suppliers, special studies on human health and risk, develops drinking water criteria and MCLs (maximum contaminant levels). Administers a special program that allows watershed treatment work to be done to decrease pollution loads to drinking water supplies if installation of the BMPs is cheaper than the water treatment method needed	Staff for technical assistance in setting drinking water standards, special studies, oversight and compliance monitoring of public water supplies and suppliers

TABLE 11.5 (*Continued*)

Agency and Program	Program Descriptions and Agency Responsibilities	Resources Available and Possible Roles
EPA—Office Research and Development (ORD)	Conducts basic and applied research to support EPA mission, including biological and physical studies on fate and transport of environmental contaminants and ecosystems at large	Reports, data, maps, monitoring equipment, study, and demonstration sites, staff for technical assistance in interpreting research results. *Laboratories and research stations located throughout the country*
U.S. Department of Agriculture (USDA)	Stabilize and support the efficient production, marketing, and distribution of food and fiber. In addition to commodity and public welfare programs, administers a number of conservation programs designed to assist private and federal land owners or managers in natural resource conservation and multiple-use management. *Works mainly with private individuals on improving resource management*	Staff, technical assistance, information and data, educational materials, cost-share funds, engineering equipment. *Unless otherwise indicated each agency has field offices located in almost every county or parish, state offices in each state, and a Washington, D.C. office*
USDA—Multiple agency administration of 1985 and 1990 "Farm Bill" programs: 1. Conservation Reserve Program, 2. Wetlands Reserve Program, 3. Sustainable Agricultural Research and Education Program, 4. Conservation cross-compliance (sodbuster and swampbuster) 5. Water Quality Incentives Program	1. Program to conserve/protect highly erodible or other environmentally sensitive land from production by putting it in permanent vegetative cover through 10-year easements and annual rental payments 2. Program available only in pilot states to return drained wetlands to wetland status and protect existing wetlands. Uses same easement/payment method as CRP 3. A practical research and education and grant program to promote lower input methods of farming	In most cases responsibilities within these programs are divided between departments of USDA as follows: SCS: technical assistance in planning, design, and implementation of BMPs ASCS: Administrative oversight of program and cost-share funding disbursement CES: Education and information about the variety of conservation and economic choices available CSRS: Research, data, and the results of

	4. A quasi-regulatory program that denies subsidy payments to farmers who plow out highly erodible land or drain wetlands	demonstration field trials of new technologies
	5. A watershed treatment program designed to improve or protect soil and water resources in watersheds impacted or threatened by NPS pollution	
USDA—Soil Conservation Service (SCS)	Technical assistance on the planning, site specific design and installation and management of soil and range conservation, animal waste, and water quality management systems and special land and water resource assessments and inventories. Cost-share funds for installation of BMPs on private lands are available from some of the programs listed below.	Staff and equipment in field offices for technical assistance including engineering designs, survey work, and planning for water resource protection
USDA, SCS—Small Watershed Program (PL-566)	Evaluation and treatment of small agricultural watersheds with multiple resources to protect; includes land and natural resource inventories and assessments, basinwide planning and targeting of resources, technical assistance, and educational programs	Staff for technical assistance to landowners and decision-makers in the basin, funds for demonstration projects, reconnaissance, and intensive inventories of resources
USDA, SCS—Great Plains Conservation Program (GPCP)	Intensive conservation treatment for individual farms located within the Great Plains ecoregion through long-term agreements (3–10 year contract) with farmers	Technical assistance, cost-share funds up to 75% of the average cost of selected high-priority conservation practices
SDA, SCS—Resource Conservation and Development Program (RC&D)	Voluntary program to promote the economic development and to intensify resource protection in priority areas through the use of public participation in RC&D councils	Planning assistance for small communities for communitywide resource protection
USDA, SCS—Natural Resource Assessment programs: Soil Survey, Natural Resources Inventory, River Basin Studies	Various programs to map and assess the condition of natural resources (generally soil, water, vegetation, and wildlife) and conservation treatments	Maps, reports, data information, statistical analysis

TABLE 11.5 (*Continued*)

Agency and Program	Program Descriptions and Agency Responsibilities	Resources Available and Possible Roles
USDA—Agricultural Stabilization and Conservation Service (ASCS)	Provides administrative oversight and cost-sharing for approved conservation practices from ASCS and other USDA administered programs. Tracks crop production and other statistics. Distributes crop subsidy and deficiency payments	Maps, conservation practice status information, cost-share funds
USDA, ASCS—Agricultural Conservation Program (ACP)	Cost-sharing on an annual basis for a number of soil-conserving, production efficiency improving, and water quality practices	Funds for cost-share, generally limited to $3500 per farm per year
USDA, ASCS—Emergency Conservation Program (ECP)	Cost-sharing on an annual basis to replace conservation treatments (mainly structural) that were destroyed in areas designated as disaster areas due to an act of nature	Funds for cost-share of high-priority conservation practices
USDA, ASCS—Water Bank Program	Designed to improve and restore wetland areas through financial compensation for 10-year easements on private property	Funds for easement compensation on eligible lands in participating states
USDA, ASCS—Colorado River Salinity Control Program (CRSCP)	Financial assistance for on-farm projects that seek to control salinity levels delivered to the basin, primarily irrigation water management	Funds, reports, data on level of conservation treatment, demonstration sites, funds for cost-share, monitoring, and education
USDA, ASCS—Forestry Incentives Program (FIP)	Cost-share to revegetate and improve timber stands on private lands	Cost-share funds
USDA, Cooperative Extension Service (CES)	Educational programs and information to aid individuals in the selection, operation, and maintenance of the most beneficial conservation treatments. Economic analysis and data for each farm or ranch. Provides technical assistance in integrated pest management. Programs generally carried out in cooperation with state land grant universities	Staff for educational programs and technical assistance, personalized economic analysis, and to coordinate small-scale demonstrations on local farms. Educational materials

Agency	Description	Resources/Outputs
USDA, Cooperative State Research Service (CSRS)	Applied research, usually at state experiment stations on agricultural production and soil and water conservation generally using demonstration plots. Conducts the Sustainable Agriculture Research and Education program (SARE). Many projects in cooperation with state land grant universities	Reports, data, equipment. Occasionally funds for joint/special projects outside the normal research agenda. Grants for Agriculture in Concert with the Environment (ACE) program
USDA—Forest Service (USFS)	Management of national forests and grasslands for sustained production and multiple use. *Works with individuals, industries and other agencies*	Staff, maps, reports, equipment for construction and monitoring, educational materials, occasionally funds for special projects. *Field offices located in each national forest, regional offices located in 9 areas, Washington, D.C. office*
USDA, USFS—Permit Program	Oversight of timber sales and harvest contracts, grazing leases, minerals development on USFS property. Provides technical assistance to permittee on proper resource use	Staff for technical assistance and compliance monitoring
USDA, USFS—Air and Watershed Programs	Overall environmental planning and technical support for forest management decisions. Special studies and watershed demonstration projects in certain areas	Funds for special studies and watershed demonstration projects. Natural resource inventories and reports, water quality/habitat monitoring, environmental analysis of resource trends and conditions
USDA, USFS—Forest Stewardship Initiative	Technical assistance and cost-share to private holdings or lands adjacent to national forest lands for installing BMPs	Funds and technical assistance to individuals
USDA—Farmers Home Administration (FmHA)	Loans and loan guarantees to eligible producers for operating expenses, land purchase, and conservation measures	Funds and loans for property improvement and conservation treatment installation and water-conservation practices
USDA—Agricultural Research Service (ARS)	Basic and applied research on agricultural production and conservation measures, including fertilizers, pesticides, and BMP effectiveness	Reports, BMP effectiveness and environmental fate and transport data, demonstration sites, occasionally funds for joint-sponsored projects. *Research stations located throughout each state; most specialize in particular types of investigations*

TABLE 11.5 (*Continued*)

Agency and Program	Program Descriptions and Agency Responsibilities	Resources Available and Possible Roles
U.S. Department of the Interior (USDOI)	Oversight, management or monitoring of national natural resources, including land, water, and wildlife.	Staff, maps, reports, demonstration sites, educational materials, monitoring equipment. *Offices located in regional centers, field offices in each management area, Washington, D.C. office*
USDOI—Geological Survey (USGS)	Long-term baseline monitoring of water resources (quantity, flow, and quality), hydrologic and geologic investigations and data, special intensive short-term studies	Maps, data, and information on hydrology and water quality status and trends. Staff for technical assistance in designing a monitoring plan
USDOI—Fish and Wildlife Service (USF & WLS)	Oversight and regulation of the nation's wildlife resources. Management of national wildlife reserves, enforcement of federal game and fish laws, cooperative administration of national wetlands program with COE and EPA. Cooperative projects to enhance wildlife habitat, special studies, especially fisheries investigations	Staff for enforcement of Endangered Species Act and other laws on public and private agricultural land, research reports and data on habitat, populations and management of wildlife. Funds for cooperative projects. Educational materials, teacher training, curricula, and maps
USDOI—Bureau of Land Management (BLM)	Administration and management of federal lands. Oversight of grazing leases, mineral exploration, and extraction bids and leases on BLM lands. Technical assistance to permitees on BLM land in proper resource use. Oversight of recreational users of BLM land	Staff for environmental analysis and trend evaluation on BLM land, technical assistance, and oversight. Funds for special studies and cost-share for permitees for certain conservation practices (generally grazing/range management). Funds for range improvement, riparian area management, and recreational area development projects. Maps

Agency	Role	Resources
USDOI—Bureau of Indian Affairs (BIA)	Technical assistance to tribes on tribal lands, mainly for social services. Some assistance for conservation work and educational programs. Natural resource inventories and monitoring of ground and surface water	Maps, natural resource inventories of Indian and tribal lands. Funds for special projects. Staff for technical assistance to tribes
USDOI—Bureau of Reclamation	Administers, constructs, and oversees water supply facilities in western states. Regulates discharge from these facilities. Joint administration of the Colorado River Salinity Control Program with many agencies to set consistent salinity standards and manage public and private lands within the basin. New initiative to reclaim lands damaged by federal irrigation projects	Staff for oversight and management of federal property and facilities, assessment of water quality around reservoirs as part of the national irrigation water quality program. Maps and reports, some data
USDOI—National Park Service	Administers and manages national parks for preservation of natural resources	Staff for oversight and administration. Funds for special studies and occasionally cooperative projects on land adjoining park boundaries
USDOI—Office of Surface Mines (OSM)	Regulates the removal and reclamation of surface mined minerals, mostly coal on private lands	Staff for oversight and technical assistance in mining operations and reclamation efforts, for engineering studies, and for vegetative site inspections and monitoring of resources. Educational materials, data, and reports
U.S. Department of Defense—Army Corps of Engineers (COE)	Oversees construction and operation of large flood-control and public water supply reservoirs. Conducts water quality monitoring on lakes within their jurisdiction. Regulates in-lake activities and shoreline development. Cooperatively administers the wetlands dredge and fill permit program with EPA and USF&WLS. Can enforce permit requirements for BMPs or other mitigation	Maps, special studies, water quality monitoring data. Staff and funds for improvement of existing projects. Staff for review and oversight of 404 (wetlands) permits. *Field offices located in various districts throughout states. Washington, D.C. office*

TABLE 11.5 (*Continued*)

Agency and Program	Program Descriptions and Agency Responsibilities	Resources Available and Possible Roles
U.S. Department of Commerce—National Oceanic and Atmospheric Administration (NOAA)	Administers programs in cooperation with states to inventory and manage coastal resources. Funds and performs basic research and assessments relating to coastal eutrophication. Maintains data base for agricultural pesticides and nutrient loadings	Funds to state coastal programs. Staff for technical assistance. Data, reports, educational materials. Occasionally funds for special demonstration projects
USDOC, NOAA—Coastal Zone Management Act programs (CZMA)	In cooperation with EPA, administers a quasi-regulatory coastal protection program that sets performance-based management measures for control and prevention of NPS pollution in coastal areas for all land-use activities	Staff for technical assistance. Funds for plan development
State Water Quality Agencies	Administers many programs (similar to EPAs) for protection of water quality in ground and surface water, including the NPDES permit program, water quality standards regulations, the NPS program, ambient statewide monitoring programs	Staff for technical assistance to local governments and individuals in BMPS application. Water quality monitoring, data, and reports. Funds for pollution-control projects, educational materials, and programs
State Natural Resource Agencies	Administer programs for wetlands and coastal protection programs	Staff for technical assistance to local governments. Monitoring of natural resource trends. Reports, data. Educational materials and programs
State Departments of Agriculture	Regulate pesticide registration and use, administers marketing and rural development programs. Sometimes issues permits for fertilizer or feedlots	Staff for oversight of applicators and other regulatory functions

Agency	Responsibilities	Resources
State Departments of Health	Administer septic tank and public drinking water regulatory programs. Monitors water supplies. Provides technical assistance to local governments	Staff for technical assistance to local governments, monitoring, and educational programs. Data, reports, and educational materials
State Soil and Water Conservation Commissions	Administer cooperative programs with the USDA SCS to conserve soil and water resources on private lands. Provide technical assistance to individuals	Staff for technical assistance to individuals, engineering, or construction equipment, services and supplies that support BMP implementation. Some states have state cost-share funds for BMPs
State Fish and Game Agencies	Regulate the harvest of fish and wildlife resources by individuals and commercial operations. Responsible for cost recovery to state of lost fish and wildlife due to environmental contamination	Staff for enforcement of state fish and game laws and for technical assistance in wildlife and fisheries managment for private individuals. Educational materials, natural resource inventory data
State Water Rights Agencies	Responsible for allocation of water rights (mostly in western states). Regulates consumptive use of water resources	Staff for permit writing and oversight. Data and reports on water flow
Local Planning and Zoning boards	Specify land-use zoning and boundary determinations, general community planning, oversight of program operation	Maps, long-range plans, inventory of local resources, special reports, budget information, staff for technical assistance
Local County Commissioners	Manage, construct, and maintain county roads and bridges, oversight and approval of county budgets for all county programs. Taxing authority	Information on county conditions, equipment for construction and maintenance, budget reports, occasionally funds for special projects
Local SWCDs	Local field office of state agency. See earlier	
Local School Boards and School Administration	Oversee public education within jurisdictional boundaries. Can set local curricula requirements and priorities. Taxing authority, bond-issuing authority	Information on status of current educational programs, assistance in developing new initiatives
Local Municipal Utilities Districts	Oversees construction and maintenance of public works projects for water and sewer (occasionally energy). Taxing and bond-issuing authority	Information and special reports on water issues. Funds for special projects to enhance system operation and reduce costs

TABLE 11.5 (*Continued*)

Agency and Program	Program Descriptions and Agency Responsibilities	Resources Available and Possible Roles
Regional River Authorities	Manage and coordinate activities within their basin for flood control, water quality protection, energy development. Taxing authority	Data, reports, maps, water quality monitoring. Staff for technical assistance to local government and other agencies or groups. Funds for special projects
Regional Councils of Government	Assist in the coordination of activities of all governments within the councils area. Provides technical assistance, information, and promotes special projects of benefit to all	Staff for technical assistance to local governments, occasionally water quality monitoring, reports, and data about local conditions. Funds for special projects
Others—Commodity Groups	Various groups usually formed to improve marketing and lobbying capabilities for specific crops or livestock interests. Almost every major crop has at least one such group	Staff for data gathering and analysis, public education campaigns, technical support to growers, legislative and market analysis. Funds from members for special projects
Environmental Organizations	Various groups formed to protect, conserve, or preserve the environment in general or to address a specific issue. Lobby for environmental laws and programs as well as funding. Many perform volunteer services such as water quality monitoring, natural resource rehabilitation work, cost-share, or other funds for special projects Too numerous to list, consult a directory	Staff and volunteers for assistance with local projects, occasionally funding for cooperative work. Educational materials and programs. Reports and data on environmental conditions and trends
Social and Service Clubs	Formed for reasons other than resource protection, most do have local projects that enhance or beautify the community. Staffed with volunteers, these organizations can provide labor, supplies, and equipment on mutually beneficial projects, as well as insight into the community	Volunteers for special projects

Sources: Alexander (1993); U.S. EPA (1992b); Mass et al. (1987).

and pricing to support sustainable, low-*impact* farming systems. Most water quality professionals agree that the largest (and most difficult) change must be to attitudes (Ehorn, 1989).

It is clear, however, from a review of the water quality data as well as visual inspection of many streams, rivers, and lakes, that in the United States and elsewhere, diffuse or nonpoint source pollution from agriculture has not been adequately controlled. As more data are gathered by state and federal agencies, and as citizens become more educated about its causes and effects, and thus more active in looking for it, the extent of nonpoint source pollution reported will likely increase. Most of the technology to control diffuse pollution from agriculture is currently available, but to ensure that this technology is implemented a number of changes must occur and a number of political, institutional, and financial barriers must be overcome.

Before change will occur an individual or an institution (government, business, social group, etc.) must (1) know that a problem exists, (2) understand its significance, and (3) have some type of incentive (either internal or external) to change. Experts have noted that diffuse pollution from agriculture has not been adequately controlled because:

1. Farmers, agribusinesspeople, agricultural agencies, agricultural advocacy groups, lawmakers, and the general public have not been made aware (and many do not believe) of the problem or its extent.
   a. The effects on the environment of improper agricultural land management are generally not evident or visible to the land user/owner since they often occur downstream or off-site.
   b. The effects of NPS pollution are cumulative. While the "discharge"/runoff from a single site may not be great (dirty) enough to cause a use impairment, the minor contributions from many other sites add up to a larger and more significant problem.
   c. Many agricultural producers have no historical, highly visible basis of comparison between water quality now and that of 100 years or more ago. The impression that the water has "always looked this way" is often not dispelled, even when data are presented.
   d. Data establishing a discrete cause–effect relationship is not always clear. While adequate and defensible research may exist for some problems, the nature of NPS is so site specific that an argument can always be made that "This area is different; its soils, its hydrology and climate, its crop varieties and farming practices...etc." because in natural systems research this is almost always true.
   e. When data documenting changes to water chemistry or fisheries habitat are available, often they are unclear if the effect is great

enough to cause a use impairment now or in the future. It is especially difficult (and expensive) to obtain data sets extensive enough to establish the sources of each pollutant found, and even more difficult to show the causes (sources) of biological changes.
2. These individuals and institutions do not fully understand (or believe) the potentially serious short- and long-term effects of diffuse pollution on human and environmental health.
   a. Inconclusive data and conflicting reports about the effect of various contaminants on human health and the inexact science of risk analysis breeds skepticism and doubt.
   b. Long-term studies that accurately determine the risks associated with the many overlapping conditions found in biological systems research are lacking and are often cost prohibitive.
   c. The type of contamination from diffuse pollution is insidious and often results in small changes in species composition that go unnoticed.
   d. Fortunately there have been few major "catastrophes" from diffuse pollution, like oil spills or nuclear reactor meltdowns, but lack of such events often allows a problem to go unresolved until one does occur.
3. The incentives currently available are not sufficient to control the problem. A number of political, institutional, and financial barriers exist that must be removed before existing incentives can work effectively, and in some cases new or additional incentives are needed. Barriers to control agricultural diffuse pollution include:
   a. The lack of education of all sectors of society about the problem, effective controls available, and methods that can be used to get BMPs applied.
      (1) Previous educational programs have concentrated on educating the agricultural producer, an important step but an incomplete one. Effective education for legislators, citizens, agricultural agencies, marketing associations, and all other facets affected by agriculture need water quality education targeted to their particular role/responsibility in solving the problem.
   b. The lack of trained professionals specializing in agricultural water quality issues at local, state, regional, and national levels.
      (1) The "lag-time" or built-in institutional inertia that exists has hampered agricultural agencies and interest groups and water quality agencies from taking a strong leadership role in diffuse pollution control and prevention. A lack of vision and grasp of

the changing reality of American politics kept some of the agricultural "institutions" (agencies and special interest and commodity groups) from taking a leadership role in diffuse-pollution research, education, and technical training and assistance. Instead these agencies and groups worked in a counterproductive fashion to "disprove" the claims of water quality agencies that agriculture was and could cause a problem in an effort to "protect" the farmer. The farmer, hearing conflicting information from many sources and trying to adjust production practices to a changing market, does not know where to turn for unbiased, factual, and useful information (Babcock, 1985).

c. The lack of consistent regulatory "back-up" from the state or federal level to support local control efforts.

   (1) The Clean Water Act explicitly exempts agricultural irrigation return flow from the federal NPDES system. For many years people maintained that diffuse pollution from agriculture was not regulatable. They were only partially correct. Diffuse pollution from agriculture cannot be regulated under the NPDES permit system by the federal government, but it can be regulated by state NPDES permit systems and by other state regulations. Some states do have laws that deal with diffuse pollution, most notably erosion and sediment controls from construction activities, but some states are writing laws for agriculture, too. In Wisconsin, a number of farmers were frustrated with their neighbors. Some producers had installed the needed BMPs, but others would not "comply," even when substantial incentives (a high cost-share rate) were offered. As a local initiative Wisconsin passed their "Bad Actors Law" that provides regulatory backup in the form of the state NPDES permit requirements for those farmers not voluntarily installing BMPs (U.S. EPA, 1993).

   (2) When considering legislation to "regulate" agricultural NPS pollution at the federal level, the issues of private land-use rights and "taking" of private property arise. However, without consistent federal regulation states and local governments, which are subject to a great deal of public and political pressure, risk losing industry, jobs, or business income from their state if they pass more stringent environmental laws than their neighbors. A considerable water quality problem from dairy waste in Texas is the result of Arizona and California passing

tough pollution-control laws causing large dairies to move to a state with a more "favorable" environmental climate (Gordon and Hansen, 1989).
d. The lack of targeting of resources to the most threatened or impacted watersheds.
   (1) For the most part agricultural NPS control programs are voluntary and the cash costs of BMP installation, maintenance, and management are often not recoverable as cash, although they are sometimes recoverable through conserving natural resources (soil, grass, timber) for long-term sustainability of the farming system. Thus each individual land owner/user decides which, if any, BMPs will be installed on his/her property. In making this decision agricultural producers must factor the economic benefit and cost of any BMP into the farm's profit margin. For effective NPS control, those farms causing the most problem must be targeted for implementation first. In a purely voluntary program it is very difficult to overcome the economic forces to achieve this targeting. In addition, USDA programs are generally designed to "share the wealth" and provide equal opportunities for cost-share funds to all producers, regardless of the environmental priority (Duda and Johnson, 1985; Mass et al., 1987).
e. Conflicting laws and programs of various agencies that work in a counterproductive fashion.
   (1) Programs within many agencies often have conflicting requirements. Sometimes even within the same agency or department laws and programs can be counterproductive. For example, within the USDA many programs exist that provide farmers price supports and other financial benefits to produce certain crops at the highest yield rate regardless of the impacts of these intensive production systems on the soil, water, or energy resources, yet other programs provide other financial incentives to install conservation treatments. Producers cannot follow the guidelines for conservation programs without jeopardizing their benefits from the commodity programs (U.S. General Accounting Office, 1990).
f. Lack of coordination between existing programs resulting in overlap or duplication of some efforts and scant attention to other needed areas.
   (1) The laws governing natural resource management (land management) and environmental quality have not been developed or implemented in concert with one another. Agricultural laws

Balancing Agricultural Production with Environmental Protection   727

have traditionally been written to stabilize production and ensure that consumers have a safe, reliable source of inexpensive food and fiber. Even today economic, supply, and demand concerns dominate agricultural laws and policies. Environmental laws have concentrated on a one-problem, one-solution type of approach rather than on a more holistic way of managing natural and man-made systems. Current laws have practically mandated this approach by being narrowly focused on quick fixes where "nobody gets hurt too much" (a concession to economic realities). These laws have often turned out not to be fixes at all, but rather redirection of the problem (U.S. EPA, 1992; Ribaudo, 1985).

g. Lack of support services built into the economy that perpetuate the BMPs once the incentives have expired.
  (1) In areas or watershed where projects to control diffuse pollution from agriculture have been successfully implemented, those with long-term lasting results had a number of private enterprises support the implementation and operation and maintenance of the recommended BMPs. These companies supplied services and equipment that individuals could not afford to own or acquire. Without these services or equipment there is a tendency to neglect BMP maintenance once the incentive (cost-share, land easement rental payment) expires. Some examples include firms specializing in animal waste lagoon pump-out (cleaning) and land application, companies that specialize in prescribed burning for brush control and range management, or professional associations such as the Association of Independent Crop Consultants, which are skilled in the use of integrated pest management techniques (American Farm Bureau Federation, 1992; Gordon and Hansen, 1989; Alexander, 1993b).

h. The lack of funds for all facets of control programs (education, technical assistance, cost-share, regulatory aspects).
  (1) The extent of the diffuse-pollution problem is such that there will never be enough state or federal funds available to completely control the problem in all impacted watersheds. Many policy analysts maintain that NPS pollution problems should be treated and funded no differently than point source pollution problems in which the polluter pays for the costs of control and adsorbs it as the cost of doing business. Others believe that because agricultural producers have no way of passing these costs on to consumers, and since diffuse pol-

lution is so much more difficult to pinpoint to a particular location (a single farm) than point source discharges, society should compensate farmers for their pollution-control efforts (Harrington, Krupnick, and Peskin, 1985). In most parts of the United States this second option is most widespread; however, it has not proved to be fully effective in controlling the problem.

Once some of the barriers have been lessened the addition of new incentives and the modification of existing incentives can be used much more effectively to encourage adoption of BMPs by individuals. Not all are equally effective, and some are not the politically astute choice in certain areas, but all should be considered as options. These modified or expanded incentives can include the following.

1. *Education*. Programs that target key audiences and tailor both the message and the method of presentation to that audience are most effective in eliciting a behavior change.
2. *Technical assistance*. Involves one-on-one interaction between the professional water quality person and the farmer, and includes recommendations about environmental conditions and BMPs appropriate for the specific site in question. It also includes assistance of on-site engineering or agronomic work during the installation of BMPs.
3. *Tax advantages*. Can be provided through state and local taxing authorities or by a change in the federal taxing system that rewards those producers who install BMPs.
4. *Price supports or subsidies*. Crops or agricultural products that have a distinct water quality benefit and are not currently in the commodity program could be added to encourage planting or production.
5. *Cost-share to individuals*. Direct payment to individuals for the installation of specific BMPs (for example, terraces) has been effective in many areas if the cost-share rate is high enough to compensate the farmer for the perceived and real risk associated with the use of the BMP.
6. *Cross-compliance legislation built into existing programs*. Currently in effect in the 1985 and 1990 U.S. Farm Bills. Generally a type of quasi-regulatory incentive/disincentive that conditions benefits received on meeting certain requirements or performing in a certain way.

7. *Direct purchase of lands contributing the greatest problem or of riparian corridors for mitigation.* Direct purchase of special areas for preservation has been used extensively by groups such as the Nature Conservancy. Use of community-owned greenbelts in urban areas is another variation. Costs of direct purchase are generally high, but effectiveness can also be exceptional. The practice could be used by any government or group to obtain control of a certain piece of land whose owner is not willing to install needed BMPs through the existing programs.
8. *"Oversight/site inspections" in a nonregulatory program.* Any individual, group, agency, etc., will perform better with personal attention. A powerful incentive for voluntary installation of BMPs is a site visit by the local, state, or federal regulatory agency to encourage BMP installation on the visited site.
9. *Peer pressure.* Social acceptance by ones peers can be a motivational factor for installation of BMPS by some individuals. If a community values a clean environment or the use of certain agricultural BMPs, producers in those communities are more likely to install them.
10. *Direct regulation of land-use and production activities.* Often considered to be the option/incentive of last resort, regulatory programs that are simple, direct, and easy to enforce are quite effective. Such programs can regulate the type of land use allowed (like zoning ordinances), or the kind and extent of activity allowed on that type of land (like pesticide application rates), or set performance standards for the level of environmental damage allowed from that land activity (such as retention and reuse of the first inch of runoff from the property).
11. *Consumer demand.* Demand for a particular product or for a commodity produced in a certain way (organically raised wheat or vegetables) can create a market that may pay a premium for items in high demand.

CONCLUSIONS

Control and prevention of diffuse pollution from agriculture is and will continue to be a challenge. A number of experimental programs have documented varying degrees of success. The current programs for nonpoint source control from agriculture are just the beginning of a long process that will require a long-term commitment of time, resources, and funds from every sector. Obtaining these commitments from agencies whose missions may not be water quality oriented, and working together with diverse groups, a great deal can be done.

References

Alberts, E. E., G. E. Schuman, and R. E. Burwell (1978). Seasonal runoff losses of nitrogen and phosphorus from Missouri Valley loess watersheds, *J. Environ. Qual.* **7**:203–208.

Alexander, S. (1992). *Managing Containerized Nurseries for Pollution Prevention*, Water Quality Branch, U.S. Environmental Protection Agency, Dallas, TX, and Terrene Institute, Washington, DC.

Alexander, S. (1993a). *Clean Water in Your Watershed—A Citizen's Guide to Watershed Protection*, U.S. Environmental Protection Agency, Dallas, TX (in press).

Alexander, S. (1993b). Effective education programs change behavior. Lessons Learned from the Nonpoint Source Program, in *Proceedings, Riparian Biosystems in the Humid Southeast*, Atlanta Ga, March, 1993, National Association of Conservation Districts and U.S. Environmental Protection Agency, Washington, DC.

American Farm Bureau Federation (1989). *Innovation and Technology Transfer—America's Farmers: Profitable Environmental Managers*, AFBF, Park Ridge, IL.

Andridenas, P., and T. Eichers (1982). What farmers buy makes production fly, in *Food from Farm to Table—1982 Yearbook of Agriculture*, J. Hays, ed., Science and Education Administration, Washington, DC.

Babcock, H. (1985). Compelling on-the-ground implementation of measures to control nonpoint source pollution, in *Perspectives on Nonpoint Source Pollution*, EPA 440/5-85-001, Office of Water, U.S. Environmental Protection Agency, Washington, DC.

Berg, H., et al. (1988). Management effects on runoff, soil, and nutrient losses from highly erodible soils in the Southern Plains, *J. Soil. Water Conserv.* **43**:407.

Beschta, R. L. (1978). Long-term patterns of sediment prediction following road construction and logging in Oregon coast range, *Water Resour. Res.* **14**:1011–1016.

Brach, J. (1990). *Agriculture and Water Quality—Best Management Practices for Minnesota*, Division of Water, Minnesota Pollution Control Agency, St. Paul, MN.

Brade, J., and D. Uchtmann (1985). Agricultural nonpoint pollution control: An assessment, *J. Soil Water Conserv.* **40**(1):23–26.

Burns, J. C., et al. (1987). Swine manure and lagoon effluent applied to a temperate forage mixture: Persistence, yield, quality, and elemental removal, *J. Environ. Qual.* **16**:99.

Capodaglio, A. (1988). Diffuse groundwater contamination by herbicides in an agricultural area of Northern Italy, in *Proceedings, Nonpoint Pollution 1988—Policy, Economy, Management and Appropriate Technology*, American Water Resources Association, Bethesda, MD, pp. 303–313.

Cheng, J. D. (1988). Streamflow changes after clear-cut logging of a pine beetle-infested watershed in southern British Columbia, Canada, *Water Resour. Res.* **25**:449.

Committee on International Soil and Water Research and Development, National Research Council (1991). *Toward Sustainability. Soil and Water Research Priorities for Developing Countries*, National Academy Press, Washington, DC.

Duda, A., and R. J. Johnson (1985). Cost-effective targeting of agricultural nonpoint-source pollution controls, *J. Soil. Water Conserv.* **40**(1):108–111.

Dudley, R. R., and J. R. Karr (1979). Concentrations and sources of fecal and organic pollution in an agricultural watershed, *Water Resour. Bull.* **15**:911–923.

Ehorn, D. A. (1989). *Offsite Assessment: A National Workshop*, EPA 440/5-89-001, Office of Water, U.S. Environmental Protection Agency, Washington, DC.

Francis, C. A., C. B. Flora, and L. D. King (1992). *Sustainable Agriculture in Temperate Zones*, Van Nostrand Reinhold, New York.

Gordon, D. G., and N. Hansen (1989). *Managing Nonpoint Source Pollution: An Action Plan Handbook for Puget Sound Watersheds*, Puget Sound Water Quality Authority, Seattle, WA.

Haith, D. A., and R. C. Loehr, eds. (1979). *Effectiveness of Soil and Water Conservation Practices for Pollution Control*, EPA 600/3-79-106, Environmental Research Laboratory, U.S. Environmental Protection Agency, Athens, GA.

Harington, W., A. J. Krupnick, and H. M. Peskin (1985). Policies for nonpoint source water pollution control, *J. Soil Water Conserv.* **40**(1):27–32.

Haupt, H. F. (1959). Road and slope characteristics affecting sediment movement from logging roads, *J. Forestry* **61**:329–332.

Hosapple, T. (1992). Diversification and Stability of Hosapple farms, in Growing into the 21st Century, 1992 Sustainable Agriculture Conference, National Association of Conservation Districts, Washington, DC.

Humenik, F. A., F. A. Koehler, R. P. Mass, S. A. Dressing, J. M. Kreglow, and D. D. Johnson (1982). *Best Management Practices for Agricultural Nonpoint Source Control—Animal Waste*, NC Agricultural Extension Service, NCSU, Raleigh, NC.

Jones, A. (1990). *Technology Designed for Outer Space an Outstanding Success on Earth—Microbial Rock Reed Filter an Emerging and Promising Low Cost Low O&M Wastewater Treatment Bio-Technology*, Sixth Annual Regional Municipal Technology Forum, Louisiana State University Ag Center, U.S. Environmental Protection Agency, Dallas, TX.

Klaine, S. J., et al. (1988). Characterization of agricultural nonpoint pollution: Pesticide migration in a West Tennessee Watershed, *Environ. Toxicol. Chem.* **7**:609.

Lawrence, G. B., and C. T. Driscoll (1988). Aluminum chemistry downstream of a whole-tree harvested watershed, *Environ. Sci. Technol.* **22**:1293.

Levi, D. A. (1978). *Agricultural Law*, Department of Agricultural Economics, Texas A&M University, Lucas Brothers Publ. Columbia, MO.

Loehr, R. C. (1972). Agricultural runoff characteristics and control, *J. Sanitary Eng. Div.* **98**:909–925.

Loehr, R. C. (1974). Characteristics and comparative magnitude of nonpoint sources, *J. WPCF* **46**:1849–1872.

Loehr, R. C., et al., eds. (1979). *Best Management Practices for Agriculture and Silviculture*, Ann Arbor Sci. Publ., Ann Arbor, MI.

McDowell, L. L., et al. (1989). Nitrogen and phosphorus yields in run-off from silty soils in the Mississippi delta, U.S.A., *Agric. Ecosyst. Environ.* **25**:119.

Mass, R. P., M. D. Smolen, C. A. Jamieson, and A. C. Weinberg (1987). *Setting Priorities: The Key to Nonpoint Source Control*, Office of Water, U.S. Environmental Protection Agency, Washington, DC.

New Mexico Environmental Improvement Division (1989). *Priority Waterbodies for Water Quality Action*, New Mexico Environmental Improvement Division, Santa Fe, NM.

Owens, L. B., et al. (1989). Sediment and nutrient losses from an unimproved, all-year grazed watershed, *J. Environ. Qual.* **18**:232.

Parkin, T. B., and S. E. Meisinger (1989). Denitrification below the crop rotating zone as influenced by tillage, *J. Environ. Qual.* **18**:12.

Platts, W. (1990). *Managing Fisheries and Wildlife on Rangelands Grazed by Livestock—A Guidance and Reference Document for Biologists*, Nevada Department of Wildlife, Reno, NV.

Ribaudo, M. O. (1992). Options for agricultural nonpoint-source pollution control, *J. Soil Water Conserv.* **47**(1):42–46.

Ritter, W. F. (1988). Reducing impact of nonpoint pollution from agriculture: A review, *J. Environ. Sci. Health* **25**:821.

Robillard, P., et al. (1982). *Planning Guide for Evaluating Agricultural Nonpoint Source Water Quality Controls*, EPA 600/3-82-021, Environmental Research Laboratory, U.S. Enviromental Protection Agency, Athens, GA.

Sampson, N. (1981). *Farmland or Wasteland: A Time to Choose*, Rodale Press, Emmaus, PA.

Smolen, M. D., G. D. Jennings, and R. L. Huffman (1990). Impact of nonpoint sources of pollution on aquatic systems: Agricultural land uses, in *Nonpoint Source Impact Assessment*, Rep. No. 90-5, CH$_2$M-Hill, Water Environment Research Foundation, Alexandria, VA.

Stewart, B. A., et al. (1975). *Control of Water Pollution from Cropland*, Vol. I, EPA 600/2-75/026a, U.S. Environmental Protection Agency, Washington, DC.

Stinner, B. R., et al. (1988). Phosphorus and cation dynamics of components and processes in conventional and no-tillage soybean ecosystem, *Agric. Ecosystems. Environ.* **20**:81.

Tankersley, H. C. (1981). Whoa There! Let's Smarten Up on Land Use, in *Will There be Enough Food? The 1981 Yearbook of Agriculture*, USDA Science and Education Administration, Washington, DC.

Thompson, P., et al. (1989). *Poison Runoff: A Guide to State and Local Control of Nonpoint Source Water Pollution*, National Resources Defense Council, New York.

U.S. Department of Agriculture, National Agricultural Statistics Service (1992a). *Crop Values 1991 Summary*, Agricultural Statistics Board, Washington, DC.

References   733

U.S. Department of Agriculture, National Agricultural Statistics Service (1992b). *Crop Production 1991 Summary*, Agricultural Statistics Board, Washington, DC.

U.S. Department of Agriculture, National Agricultural Statistics Service (1992c). *Poultry Production and Value 1991 Summary*, Agricultural Statistics Board, Washington, DC.

U.S. Department of Agriculture, National Agricultural Statistics Service (1992d). *Milk Production, Disposition and Income 1991 Summary*, Agricultural Statistics Board, Washington, DC.

U.S. Department of Agriculture, National Agricultural Statistics Service (1992e). *Meat Animals Production, Disposition and Income 1991 Summary*, Agricultural Statistics Board, Washington, DC.

U.S. Department of Agriculture, National Agricultural Statistics Service (1992f). *Floricultural Crops 1991 Summary*, Agricultural Statistics Board, Washington, DC.

U.S. Department of Agriculture, National Agricultural Statistics Service (1992g). *Catfish Production*, Agricultural Statistics Board, Washington, DC.

U.S. Department of Agriculture, Soil Conservation Service (1980). *Field Office Technical Guide—Section III*, USDA SCS, Washington, DC.

U.S. Department of Agriculture, Soil Conservation Service (1988). *National Handbook of Conservation Practices*, Washington, DC.

U.S. Environmental Protection Agency (1990). *Biological Criteria, National Program Guidance for Surface Waters*, EPA 440/5-90-004, Office of Water Regulations and Standards, U.S. EPA, Washington, DC.

U.S. Environmental Protection Agency (1992a). *National Water Quality Inventory—1990 Report to Congress*, EPA 503/9-92/006, Office of Water, U.S. EPA, Washington, DC.

U.S. Environmental Protection Agency (1992b). *Managing Nonpoint Source Pollution—Final Report to Congress on Section 319 of the Clean Water Act*, EPA 506/9-90, Office of Water, U.S. EPA, Washington, DC.

U.S. Environmental Protection Agency (1993). *Guidance Specifying Management Measures for Sources of Nonpoint Pollution in Coastal Waters*, EPA 840-B-92-002, Office of Water, U.S. EPA, Washington, DC.

U.S. Forest Service (1989). *Best Management Practices Guide for Integrated Resource Planning on National Forest Lands in the Southwestern Region*, U.S. Department of Agriculture, USFS, Albuquerque, NM.

U.S. General Accounting Office (1990). *Water Pollution—Greater EPA Leadership Needed to Reduce Nonpoint Source Pollution*, GA/RCED-91-10, Resources, Community and Economic Development Division, Washington, DC.

Westerman, P. W., et al. (1987). Swine manure and lagoon effluent applied to a temperate forage mixture. II. Rainfall runoff and soil chemical properties, *J. Environ. Qual.* **16**:106.

12
Receiving Water Impacts

Problems that are harder to assess and control, such as sedimentation, nutrient enrichment, runoff from farmlands, and toxic contamination of fish tissues and sediments, are becoming more evident. . . . By some extent, it is certainly true that the more we look, the more we find.

EPA's National Water Quality Inventory: 1988 Report to Congress

ASSESSMENT OF WATER QUALITY PROBLEMS

Throughout the late 1970s and early 1980s when the results of the 208 studies were published, using the traditional point source assessment approaches, researchers and pollution abatement administrators struggled with the answers to the question "What is the water quality problem of diffuse pollution?" The following quote taken from Sonnen (1980) illustrates the dilemma of this post-208 period:

> We still have not designed let alone implemented even one monitoring program based on postulated mechanisms of fundamental physics and chemistry to demonstrate in one urban area, much less all of them, whether urban runoff poses a (water) quality problem or not, and to the degree that it does, what could be done about it.

However, ten years later in 1989, the National Resources Defense Council published *Poison Runoff* (Thompson, Adler, and Landman, 1989) claiming that, based on the studies of the Resources for the Future, the

National Oceanographic and Atmospheric Administration, and the U.S. EPA, diffuse pollution (i.e., *poison runoff*) contributed to the nation's receiving waters the following proportions of pollutants:

90% of total nitrogen
90% of fecal coliform bacteria
70% of chemical oxygen demand
70% of oil
70% of zinc
66% of phosphorus
57% of lead
50% of chromium

According to *Poison Runoff*, the high levels of nitrogen in urban and agricultural runoff promotes excess algal growth in streams, lakes, and estuaries, and the toxic components also diminish the beneficial uses of bodies of water.

The 1988 U.S. EPA Report to Congress on the state of water quality of the nation's waters (U.S. EPA, 1990) provided the statistics on the impact of various sources of pollution on the nation's streams and lakes (excluding Alaska and islands). A quote from the report opens this chapter. Of the total river kilometers of U.S. streams, 70% were fully supporting the designated uses (that is, they were considered as unpolluted), 20% were partially supporting the uses of the rivers (moderately polluted) and 10% were not supporting the uses (severely polluted). Referring to the nation's inland lakes (excluding the Great Lakes and Alaska), 74% of lake area fully supported the designated uses, 17% were partially supporting, and 10% were nonsupporting. The designated uses were specified by the Clean Water Act (see Chapter 1) and generally implied attainment of water quality that would support aquatic life, contact recreation (swimming), and in some special cases, potable use of water. Tables 12.1 and 12.2 provide a summary of the assessment. The values reported in the tables are percentages of the impaired river kilometers or lake areas. Hence to obtain a percentage of the total river kilometers, the values would have to be multiplied by 0.3 (fraction of total river kilometers where uses are not being met) and by 0.26 for lakes.

The Great Lakes are the largest freshwater bodies in the world (the largest lake in the world, the Caspian Sea in the former USSR, is a saltwater body). They are so large that they should be considered freshwater seas. However, only 8% of the total shoreline kilometers in the United States were supporting the designated uses, 18% of the shoreline may only partially support the designated uses, while 73% were not supporting.

TABLE 12.1   Causes of Pollution of U.S. Surface Waters

Pollution Cause	Percent of River Kilometers Affected by Source		Percent of Lake Surface Affected by Source	
	Moderately	Severely	Moderately	Severely
Siltation	33	9	18	8
Nutrients	22	4	33	13
Pathogens	15	7	8	2
Organic enrichment	11	3	16	4
Pesticides	8	1	5.5	2
Metals	9	5	5	1
Suspended solids	5	2.5	7.5	3
Salinity	5.5	2	4	2
Habitat modification	5	2.5	12	0.5

Source: After U.S. EPA (1990).

TABLE 12.2   Adverse Water Quality (Pollution) Impact of Sources

Pollution Source	Percent of River Kilometers Affected by Source		Percent of Lake Area Affected by Source	
	Moderately	Severely	Moderately	Severely
Agriculture	40	15	27	14
Municipal (POTW)	13	5	5	2
Mining	8	3	2	0
Habitat modification	12	4	16	2
Urban storm runoff	7	3	3	0.5
Silviculture	4	0.5	1.5	0
Industrial	7	3	3	0.5
Construction	5	1.5	2	0
Land disposal of waste	4	1	4	1
Combined sewers	3	1	<0.5	0

Source: After U.S. EPA (1990).

Priority (toxic) organics are by far the most extensive cause of impairment of Great Lakes water and biota quality. Metals were also reported by the states as a source of impairment, with nutrients and organic enrichment/low dissolved oxygen also cited as contributors to the impairment of use.

Based on the 1988 EPA Report to Congress, of the U.S. coastal waters in late the 1980s, 72% were fully supporting the designated uses, 23% were partially supporting, and 6% were not supporting. Nutrients, pathogens, and organic enrichment were listed as the three major causes of impairment of the coastal shore waters. Unlike the inland waters, for which agriculture was the major polluting land-use activity, urban point sources (POTW effluents), mining, urban storm runoff, CSOs, and land disposal of wastes were the four major sources responsible for about 90% of impairment (pollution) of coastal water, all of which are related to urban land use and urbanization.

As pointed out in Chapter 1 diffuse-pollution problems are not limited only to the United States. The North Sea and shorelines of the Adriatic and Mediterranean seas are grossly polluted by the impact of nutrient discharges, to the point that beaches are closed for swimming and fish is contaminated and unfit for human consumption. Many rivers in Europe, Latin America, and Asia are extremely polluted by waste discharges from a variety of sources. Due to inadequate sanitation in undeveloped countries, urban runoff may contain sewage and gray wastewater (domestic wastewater without fecal matter), which is still being discharged into drainage systems.

A number of researchers have tried to define the impact of diffuse pollution on receiving waters. The impact is different depending on the state of the drainage system development, degree of urbanization expressed by the percent imperviousness of the urban area, land-use distribution, hydrologic and hydraulic characteristics of the body of water, use and type of chemicals for fertilization of soil and pest control, and other factors. For example, Novotny (1985) related the water quality impact to the degree of watershed development, which was discussed previously (Chapter 1).

Myers et al. (1985) summarized the EPA's data and reports on the magnitudes of nonpoint pollution and their water quality impact. The authors emphasized that substantial water quality benefits can be achieved by targeting the resource to lands and land uses that are most polluting. Such land uses include urban and highway construction. Agricultural activities were found to be the main contributors of nonpoint pollution, followed by urban runoff.

Urban runoff has been identified as a major source of toxics, including toxic metals and petroleum hydrocarbons; however very little was known about the long-term effects of these substances on the biota of the receiving bodies of water (U.S. EPA, 1983; Torno, 1984; Niedzialkowski and Athayde, 1985). The NURP study identified receiving water quality problems associated with urban runoff. Frequently, discharges of urban runoff are the cause of streams and rivers exceeding heavy metal ambient water quality criteria for freshwater aquatic life. Copper, lead, and zinc appear to pose a significant threat to aquatic life in some urban areas of the United States. On the other hand, organic priority pollutants in urban runoff do not appear to pose a general threat to freshwater aquatic life.

Assessment of Water Quality Problems   739

Several NURP projects identified possible problems in sediments because of the buildup of priority pollutants contributed wholly or in part by urban runoff. However, there are only a few studies on this topic, and these are limited in scope, so the findings must be considered only indicative of the problem. Further studies are therefore needed, particularly regarding the long-term impact.

Nutrients in urban runoff may accelerate eutrophication problems and severely limit recreational uses, especially in lakes. However, the NURP lake studies indicated that the degree of beneficial use impairment varied widely, as did the significance of the urban component. Coliform bacteria discharges in urban runoff can have a significant negative impact on the recreational use of lakes.

Adverse effects of urban runoff in marine waters were found to be a highly specific local situation. Though estuaries and embayments were studied by NURP to a very limited extent, they were not believed to be threatened by urban runoff, though specific instances where use was impaired or denied could be of significant local or regional importance. Coliform bacteria present in urban runoff could have a direct impact on shellfish harvesting and beach closing. The significant impact of urban runoff on shellfish harvesting was identified in the Long Island, New York, NURP project. The specific and detrimental impact of diffuse pollution on Chesapeake Bay was covered in Chapter 1. It should be pointed out, however, that the NURP studies focused on urban runoff only, primarily on its surface flow component, and did not include mixed overflows from the combined sewers or subsurface flows from septic systems.

In the conclusion of their nationwide assessment, Heaney and Huber (1984) stated that:

1. Documented studies of the impact of urban runoff and combined sewer overflows on receiving waters are scarce.
2. Numerous definitions of the terms "impact" and/or "problem" exist.
3. Receiving waters that were included in the evaluations ranged from small creeks and ponds to large estuaries and oceans. No clear line of demarcation exists to distinguish the urban drainage systems from the receiving bodies of water.
4. Some evidence was gathered that linked fish kill and beach closing to discharges of urban runoff and overflows. Such evidence was weak.
5. The studies of dissolved oxygen downstream of urban areas have produced the most definitive information on the impact of urban runoff on water quality. However, the impact is primarily on dampen-

TABLE 12.3 BOD and Nitrogen Concentration in Border Points of Some Czech Rivers

River Border Crossing	Concentration (mg/l)		
	BOD_5	NO_3^-	NH_4^+
Labe (Elbe)	6.6	19.4	2.67
Morava	9.7	17.4	0.91
Dyje (Thaya)	8.8	37.5	1.7
Odra (Oder)	9.8	19.5	7.58

Source: Data from Ministry of Environment (1991).

ing of diurnal fluctuations of dissolved oxygen concentrations and less on actual reduction of the average daily concentrations.

PLUARG (Pollution from Land Use Activities—Reference Group) activities were reviewed by the Nonpoint Source Task Force of the Water Quality Board of the International Joint Committee (IJC) (1983), which stated that the major sources of water quality degradation are loading of phosphorus and toxics, but that urban runoff pollution has only local impact and does not constitute a significant problem on a lakewide basis.

Although the information on the impact of diffuse pollution in North America is most extensive, in the world context it is not the most severe one. For, example, almost 56% of all major streams in the Czech Republic have been classified as unfit for most uses (Ministry of the Environment, 1991). Average concentration of pollutants in major rivers leaving the Czech Republic measured at the border sections were as documented in Table 12.3. The water quality of these rivers is comparable to treated sewage without nutrient removal. The rivers are also heavily contaminated by toxic organic chemicals and metals. The extremely high nitrate content of the Czech rivers is attributed primarily to excessive use of chemical fertilizers and discharges of untreated or marginally treated urban and industrial wastewater.

THE AQUATIC ECOSYSTEM

Basic Interrelationships

The goal of all our water quality protection, restoration, and pollution abatement activities is to restore, protect, and enhance the ecosystems or ecology. The ecosystems can be divided into two components, the biotic, or living, and abiotic, or nonliving. Ecosystems are considered to be in a dynamic equilibrium with their inputs and surroundings. When one of the inputs is changed, the ecosystem will readjust to a new equilibrium. Typical aquatic ecosystems are rivers, lakes, wetlands, estuaries, or soils.

The hydrodynamical, biological, and chemical processes in aquatic systems cannot be separated and must be treated simultaneously. Studying water quality changes resulting from diffuse inputs requires multidisciplinary approaches. It is beyond the scope of this book to cover the details of these processes, and the reader is referred to monographs (Stumm, 1985; Krenkel and Novotny, 1980) for further review.

Sun energy reaching the aquatic ecosystems is the primary input. However, other energy inputs, such as organics in waste discharges, or waste heat from cooling-water discharges, can contribute significantly to the overall behavior and composition of these systems. The equilibrium behavior of the components of the ecosystem is also affected by the presence or absence of various elements, chemicals, or other compounds that stimulate the growth of biotic components, or compounds that retard the growth, damage, or kill the organisms. The former category of chemicals is called *nutrients* or *food*, the latter include *toxic* (carcinogenic or noncarcinogenic) compounds. The biotic components of aquatic ecosystems and their role can be divided into three groups of organisms.

Producers and productivity. Producers can accept inorganic materials and energy from the sun to produce new organic matter. The process is called *photosynthesis*, and the group of participating organisms are known as *autotrophic*. These include green aquatic plants attached to the bottom or vascular aquatic plants and microscopic floating planktonic organisms known as *phytoplankton*. A special group of microorganisms utilizing energy obtained from the oxidation of reduced chemical compounds (such as the oxidation of ammonia to nitrite and nitrate by *Nitrosomonas* and *Nitrobacter*) are called *chemotrophs*.

The photoautotrophs and chemotrophs use carbon dioxide or alkalinity as their sole carbon source and form new organic matter. The organisms also require nutrients, such as nitrogen, phosphorus, and trace elements. Typically in aquatic ecosystems the quantities of trace elements are in abundance; however, nitrogen, phosphorus, or sometimes alkalinity may be in a short supply. The nutrient that is in the shortest supply is called the *limiting nutrient* because its quantity affects the rate of production (growth) of organic matter by the producers. Water bodies may be carbon (alkalinity) limited if alkalinity is very low—few mg $CaCO_3/l$; however, most of surface-water bodies are either phosphorus (most of the inland lakes) or nitrogen (some estuaries and rivers) limited.

To control the eutrophication process (see the subsequent section) the control should be focused on the limiting nutrient, but this should not be considered as a strictly to obey rule. For example, if nitrogen is the limiting nutrient, it still may be more efficient to control phosphorus loads if nitrogen originates from difficult to control sources, such as nitrates in ground water. For most inland lakes phosphorus is the limiting nutrient. To determine whether nitrogen or phosphorus is limiting the productivity (eutrophication) of a particular body of water, one can plot the nitrogen concentrations versus the consequent phosphorus concentrations on an arithmetic plot. The straight line of the best fit will intercept either the nitrogen or phosphorus abscissa. The nutrient that at low concentrations is exhausted first (zero or negative intercept on the abscissa) is then the limiting nutrient.

The total nitrogen to total phosphorus ratio may also determine what algae may prevail. For example, nitrogen-fixing blue-green algae might be favored during periods of low nitrogen content ($N:P < 10:1$). The nitrogen-fixing capability of some algae is another reason why limiting phosphorus inputs is usually the only practical solution to the excess productivity of water bodies, which results in accelerated eutrophication (U.S. EPA, 1990).

In the photosynthetic (chemosynthetic) reactions organic matter and oxygen are produced. Stumm and Morgan (1962) in a classic paper proposed the following stoichiometric representation of the production of organic matter by aquatic autotrophs. Another, slightly different modification of the equation was presented by Schindler (1985) and others.

$$106CO_2 + 90H_2O + 16NO_3^- + PO_4^{3-} + \text{light energy}$$
$$= C_{106}H_{180}O_{45}N_{16}P + 154O_2 \qquad (12.1)$$

It should be pointed out that phytoplankton can utilize both ammonia (NH_4^+) and nitrate (NO_3^-) forms of nitrogen in the photosynthetic process.

This equation indicates that under abundant light and alkalinity conditions, one atom of phosphorus or 16 atoms of nitrogen (whichever nutrient is limiting) will produce 154 moles of oxygen. Less oxygen (about 106 moles of O_2) is produced if ammonia is substituted for nitrate in Equation (12.1). The productivity relation can be illustrated by the following example.

Example 12.1: Oxygen and Organic Matter Production by Photosynthesis

Lake concentration of phosphate-P was measured as 1 mg P/l. Estimate the amount of organic matter and oxygen produced by photosynthesis stimulated by the phosphate nutrient, assuming that other nutrients (inorganic carbon and nitrogen) are not limiting.

Solution Equation (12.1) states that 1 mole (or millimole) of phosphate will approximately stimulate growth of one (pseudo-) mole (millimole) of organic matter and 154 moles (millimoles) of oxygen. The atomic weights are hydrogen, $H = 1$, carbon, $C = 12$, nitrogen, $N = 14$, oxygen, $O = 16$, and phosphorus, $P = 31$.

This calculation follows the following stoichiometric relationship derived from Equation (12.1):

$$\frac{\text{P concentration (mg/l)}}{[P]} = \frac{\text{organic matter production (mg/l)}}{[C_{106}H_{180}O_{45}N_{16}P]}$$
$$= \frac{\text{oxygen production (mg/l)}}{154\,[O_2]}$$

Then $1\,\text{mg P/l}/[P] = 1/31 = 0.0323\,\text{m mole/l}$.

The molar weight of organic matter is

$$[C_{106}H_{180}O_{45}N_{16}P] = 106 \times 12 + 180 \times 1 + 45 \times 16 + 16 \times 14 + 1 \times 31$$
$$= 2427$$

Hence 1 mg P/l will stimulate production of

$$0.0323 \times 2427 = 78.4\,\text{mg/l of organic matter}$$

The molar weight of produced oxygen is

$$154\,[O_2] = 154 \times 2 \times 16 = 4928$$

Oxygen production is then

$$0.0323 \times 4928 = 159\,\text{mg/l}$$

Such production of oxygen would result in the supersaturation of water by oxygen, and the excess oxygen would eventually be lost by deaeration.

The producers are essential building blocks in the construction of basic primary organic matter and the only organisms that can covert inorganic materials into organic tissue. Since these organisms need sunlight, they are productive in the well-lighted upper zones of bodies of water called the *euphotic layer*, which represents the depth of effective light penetra-

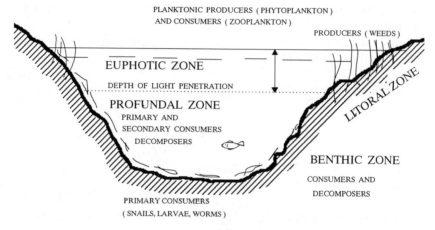

FIGURE 12.1. Biological zones in aquatic ecosystems.

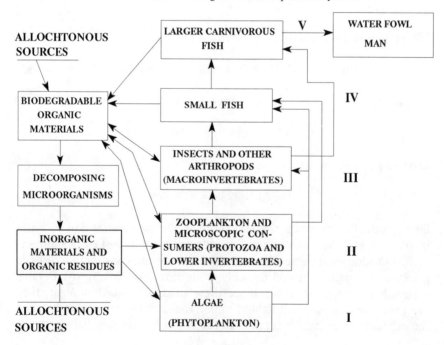

FIGURE 12.2. Food web schematic.

tion for photosynthesis. This depth depends on turbidity of the water (Fig. 12.1). In the absence of light energy (during night hours or in the deeper, profundal zones of bodies of water) the process is reversed and the organisms use oxygen to sustain their living process. This is called *endogenous respiration*.

Consumers and transformation of produced organic matter. Consumers are organisms that utilize organic carbon as their source of energy. Apparently these organisms depend on the producers for their nourishment. The primary consumers can use the produced algal and plant tissues and other biomass directly, while secondary consumers receive their nourishment by feeding on primary or smaller secondary consumers. Organisms relying on organic carbon as their source of energy and nourishment are called *heterotrophic*.

The producers and consumers form a *food web*. This chain has several *trophic levels*, starting with phytoplankton as level I, zooplankton feeding on algae (herbivous zooplankton) as level II, followed by plankton-eating microorganisms (carnivous zooplankton) as level III, and continuing to higher levels of invertebrates, insects, fish, and ending with fish- and shellfish-eating fowl and man (Fig. 12.2). Of interest to water quality is bioaccumulation and biomagnification of some compounds as they are transferred throughout the food web from the lowest trophic levels to the highest with consequent increase of body tissue concentrations of the compounds in each trophic level. The phenomenon of bioaccumulation and biomagnification is discussed in the next chapter.

Decomposers and decomposition of organic matter. Decomposers are the organisms, primarily bacteria and fungi, that can decompose organic matter to basic minerals and organic residues that may then become available again to the producers. Organic matter decomposed by the organisms originate from dead bodies and cellular tissues of all organisms and from excreta and organic byproducts of living processes. Decomposing organisms are also heterotrophic.

The character and intensity of the decompositon process depends on the availability of dissolved oxygen. If oxygen is in abundance, then decomposition is *aerobic*. The stoichiometric representation of the aerobic decomposition is a reverse of Equation (12.1), or

$$C_aH_bO_cN_dP_e + \left(a + \frac{1}{4}b + \frac{3}{2}c - \frac{1}{2}d + 2e\right)O_2$$

$$= aCO_2 + \frac{b}{2}H_2O + dNO_3^- + ePO_4^{3-} \qquad (12.2)$$

Decomposition in the water column is usually aerobic. In the absence of oxygen decomposition, it may be *anoxic* or *anaerobic*. In anoxic decomposition oxygen is supplied by the reduction of oxidized compounds, such as nitrates (NO_3^-) and sulphates (SO_4^{--}). The end products are the same as during aerobic decomposition.

In anaerobic decomposition, which is typical for sediments, organic matter is broken down by bacteria into methane, carbon dioxide, and ammonia. This process is called *diagenesis*. The anaerobic decomposition in sediments is a complex process that progresses in several stages. The intermediate products of the decomposition are fatty and other organic acids. *Methanogenesis*, that is, the formation of methane in addition to carbon dioxide and other byproducts is the final process of the anaerobic decomposition of organics in sediments. In freshwater sediments those products are formed from methyl carbon of acetate (Rudd and Taylor, 1980). Methane is released from the sediments into the overlying water by diffusion and ebullition.

Using Stumm and Morgan's representation of chemical composition of the algal organic biomass the stoichiometric balance of anaerobic decomposition can be represented as follows:

$$C_{106}H_{180}O_{45}N_{16}P + 53\tfrac{1}{4}H_2O \rightarrow 47\tfrac{1}{8}CO_2 + 58\tfrac{7}{8}CH_4 + 16NH_3 + H_3PO_4$$

(12.3)

This equation is only an approximation of stoichiometric proportions and does not represent the biochemical process itself, which is far more complex and progresses in several stages (acidic and methane fermentation). However, using this approximation for aquatic organic sediments one could deduce that after the decomposition is completed the molar proportions of methane and carbon dioxide evolution are about the same and that the methane carbon is about one-half of the organic carbon load.

From sludge digestion studies it is known that not all organic carbon is converted to methane or other mineral products. The proof is in coal deposits that resulted mostly from the anaerobic decomposition of plants millions of years ago. A significant fraction of sedimentary organic matter is either inert or refractory (decays at a very slow to negligible rate). Foree, Jewell, and McCarty (1971) measured the inert–refractory fraction of decomposition algae and found that on average 50% of organic carbon, nitrogen, and phosphorus incorporated in algal biomass is refractory under aerobic conditions, and 40% of organic carbon and phosphorus and 60% of algal nitrogen is refractory under anaerobic conditions. Wide ranges of refractory–inert fraction values were measured in the experiments.

The fraction of inert organic solids also depends on the type of solids. For example, bottom sediments in the Milwaukee harbor contain organic solids from combined sewer overflow, urban runoff, and upstream algae and organic detritus. It was found that most of the inert organic matter originated from the CSOs and urban runoff, while most solids from upstream algae development were biodegradable (Anon., 1987).

Robertson (1979) and Rudd and Taylor (1980) found that for lake sediments (mostly of algal origin) the range of methane carbon release to carbon input varied from 0.36 to 0.6. They also concluded that methane production is correlated to the rate of organic carbon input to the sediments, and the rate is not strongly affected by temperature.

The dissolved methane released from the anaerobic zones is oxidized in the oxic–anoxic sediment–water interface layer. In freshwater sediments most methane oxidation is aerobic and hence exerts *sediment oxygen demand (SOD)*. In marine sediments, oxidation may be anoxic, using oxygen from sulphates and nitrates. Since methane has a relatively low solubility in water, most of it may escape into overlying water by ebullition, where part of it may dissolve and become a part of the pool of the dissolved organic carbon that may be oxidized by heterotrophic bacteria and exert an oxygen demand. This addition to the dissolved carbon pool of the water column should not be omitted. It should be noted that the present standard methods for biochemical oxygen demand

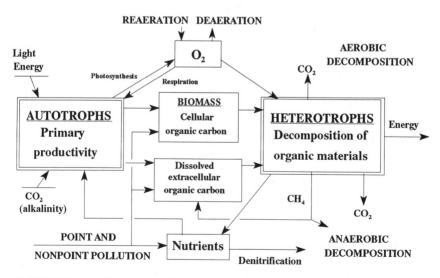

FIGURE 12.3. Carbon and nutrient cycles in water–sediment systems and their impact on the dissolved oxygen balance.

analysis require aeration of the samples, which may strip out the dissolved methane. Figure 12.3 shows the interrelationships in carbon and nutrient transport and the participating organisms in water–sediment systems. Such aquatic systems include lakes, wetlands, estuaries, and slow moving (impounded) rivers.

The relative distribution of producers, consumers, and decomposers depends on the type and amount of energy stored or entering the system. Different types of organisms will be emerging, growing, and dying at rates affected by ambient conditions. The overall growth rate for each group of organisms is a summation of their growth rate, death rate, and grazing by other organisms. If a body of water initially possesses a small number of consumers and decomposers, producers will develop, sometimes in very large numbers, depending on the optimal energy of the sun and the amount of mineral nutrients. As the overall organic matter concentration increases, consumers and decomposers develop, and when the energy inputs diminish, the decomposers liquidate the remaining biodegradable organic matter.

Several other definitions should be mentioned (Krenkel and Novotny, 1980):

The overall rate of organic matter production of a body of water in mg/m^2-day, or a similar unit, is called *productivity*, while *primary productivity* refers to organic matter production by producers only.

Plankton are organisms of small size that drift on the water and are subject to the action of the water. They may be defined further as *phytoplankton*, or photosynthetic planktonic organisms, and *zooplankton*, or animal plankton. As pointed out, phytoplankton represents the first trophic level, while zooplankton, which graze on phytoplankton, are the second trophic level.

Nekton are organisms of larger size, such as fish and insects, that swim freely and determine their own distribution in water, while *seston* are nonliving and living bodies of plants and animals that float or swim in the water.

Peryphyton are attached organisms, as are *sessile* organisms.

Benthos is the living and nonliving parts of the bottom sediments.

Invertebrates are a group of small multicellular organisms that are more complex than plankton. They can range in size from microinvertebrates (such as protozoa) to macroinvertebrates (such as larvae and worms). The benthic macroinvertebrates are of great importance in studying the pollution effects and impact of various remedial measures. These organisms are typically sedentary and live for an extended period of time (months to years). Hence, their occurrence and abundance reflects the biochemical–ecological status of the body of water, and their presence is commonly used by limnologists for determining the health (integrity) and/or the degree of pollution of a body of water.

Mason (1991) has described in detail the effects of pollution and other water quality alterations on the biota of impacted aquatic systems. Such impacts may be reversible or irreversible.

Integrity of Surface-Water Bodies: Ecoregional Approaches

Integrity of Aquatic Ecosystems

The objective of the water pollution control policies in the United States expressed in the Clean Water Act is to *restore and maintain the chemical, physical and biological integrity of the Nation's waters* (Clean Water Act, Sec. 101(a)). When this definition is tied to ecological systems, the term *integrity* has been defined as the ability to support and maintain "a balanced, integrated, adaptive community of organisms having a species composition, diversity, and functional organization comparable to that of natural biota of the region" (Karr and Dudley, 1981).

Recent policy changes of the U.S. Environmental Protection Agency have recognized the fact that surface-water bodies differ among themselves and that there is a regional clustering of their ecological characteristics. Aquatic ecologists have recognized that there is a natural physical–chemical–biological variability in and among the bodies of water. Swamps and low pH bogs, for example, exhibit low dissolved oxygen levels that are unrelated to human activities (Karr et al., 1986). When metallic ores reach the land surface, higher metal concentrations can be found in surface waters. If the water quality standards that are in force do not recognize this natural variability, water quality regulators may find themselves enforcing the unenforceable. Karr et al. (1986) have stated that regulations are still being governed by point source (NPDES) permits that rely on monitoring of chemical and toxic concentrations and overlook perturbations whether they are caused by natural factors or by humans. When diffuse pollution is considered, in addition to or in a combination with point pollution, the integrity of the body of water has to be considered, and the water quality goals of abatement must be formulated in a bioregional context, including natural (background) loads.

An aquatic system is in a quasi-equilibrium with the boundary stresses on the system and its internal perturbations, which change as a result of natural and cultural factors. In Chapters 1 and 2 it was shown that

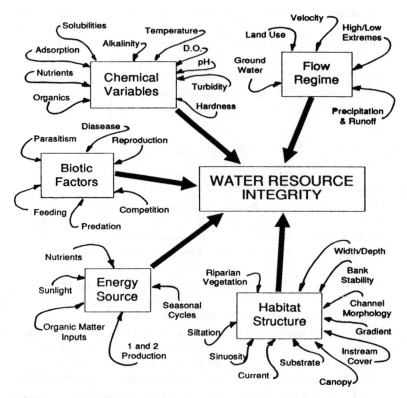

FIGURE 12.4. Five principal factors and their physical, chemical, and biological components that influence the resultant integrity of surface waters. (After Ohio EPA, 1987, modified from Karr et al., 1986.)

ecosystems have changed throughout the history of man (thousands of years) and even more so throughout geological history (millions of years). Although many environmental factors have been instrumental in the evolution of an aquatic ecosystem, Karr et al. (1986) have documented that these factors can be grouped into five major classes (Fig. 12.4). Altering any of these factors and parameters will have an impact on the ecosystem and the biota, which is its integral component. As is pointed out in Chapter 15 on system restoration, to achieve an improvement in the integrity of an aquatic ecosystem the perturbed processes and factors must be identified in all five classes. This obviously leads to an *integrated approach* to point and nonpoint pollution abatement.

Figure 12.4 also indicates why a single-factor, simplistic approach may fail in an attempt to remedy water quality and the ecological integrity of the aquatic system. For example, take a riverine system that is stressed by point source discharges of typical wastewater and nonpoint source discharges of nitrate and phosphate. Without an abatement the river will exhibit low dissolved oxygen levels and a shift to excessive heterotrophic microbial populations (decomposers and grazing zooplankton). These microorganisms typically keep the algal population in check. Focusing on point source control may improve the dissolved oxygen problem and reduce the undesirable excessive heterotrophic microbial and zooplankton population; however, the nutrient levels in the stream may cause nuisance development of algal-dominated biota and an excessive primary production of organic matter, creating algal blooms, and hence negating the benefits of the point source cleanup effort. In spite of large expenditures for point source cleanup of biodegradable organics (BOD) such streams, besides being unsightly, may not be fit for many mandated uses, including swimming (swimmers may get a rash from enzymes produced by algae), fish (fish-kills may occur as a result of low dissolved oxygen caused by algal respiration), and drinking (the taste and color of the water may be severely impaired). In this case, the cleanup effort must include a reduction of nutrient inputs and development of riparian (stream-bordering) and in-stream habitats that are suited to enriched biota. The effects of nutrients can also be reduced by planting shade trees along the river, which restricts the light and limits the growth of algae.

Ecoregions. The work at the U.S. Environmental Protection Agency by Omernik (1987) and others (Gallant et al., 1989) resulted in a division of the conterminous United States into a number of geographical–ecological subunits that exhibit relative homogeneity in ecological systems (Fig. 12.5). These *ecoregions* are typically different from other geographical units such as physiographic units that are based strictly on land-surface forms or hydrological watershed units. The ecoregional map and ecoregional delineation are based on the premise that regional patterns of environmental factors would be reflected in regional patterns of surface quality. The factors that affect the extent of an ecoregion include, among others, land-surface forms, potential natural vegetation, soils, and land use. The boundaries between the ecoregions should not be viewed as abrupt changes, since they are basically subareas of change from one ecoregion to the next one. The homogeneity within the ecoregion is only relative. There is less variability within an ecoregion than between ecoregions.

The ecoregional concept is used for many cartographic purposes dealing with natural resources, including water quality. For example, it is assumed that attainable biological and chemical quality within the ecoregion can be approximated by measuring the physical, chemical,

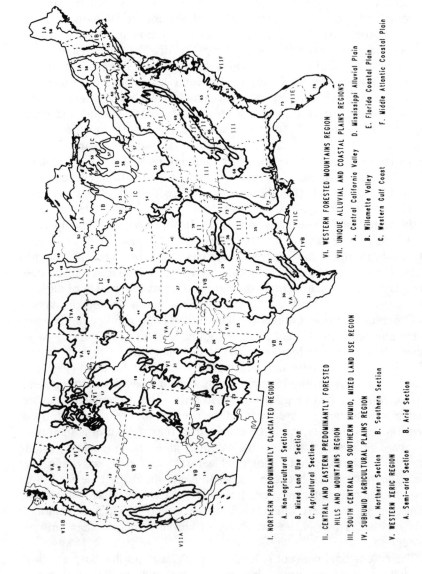

FIGURE 12.5. Ecoregions of the conterminous United States. (From Omernik, 1987.)

and biological quality of the streams draining watersheds that are representative of the natural environmental characteristics typifying the region and subject to the least possible human interference. Such streams and lakes constitute *reference bodies of water*. The degree of water quality deterioration can be ascertained by comparing the biology and chemistry of the water body of interest with the reference body of water of the same character located within the ecoregion. Preserving and protecting the undisturbed "reference reaches" should receive the highest priority by federal and state pollution-control agencies (Committee on Restoration of Aquatic Ecosystems, 1992).

Measures of Biotic Integrity

A healthy surface-water body provides a habitat to many species. However, specific species distribution exhibits geographic variability that reflects the ecosystem equilibrium with boundary inputs. The natural inputs that control the species distribution are meteorology, geography, latitude, elevation (both elevation and latitude affect some meteorological parameters), stream or lake morphology, stream habitat alterations by man, and water chemistry.

Water contamination will alter the species distribution and their abundance. Some organisms are more tolerant to pollution and, hence, they may develop in larger numbers, while the less tolerant will diminish. This is the principle behind several species diversity indices that have been used for water quality assessment (Krenkel and Novotny, 1980; Mason, 1991; Karr et al., 1986).

Use of biological indices for assessing water quality and the health (integrity) of the biota dates back almost 100 years. The earliest proposed biotic index was the saprobien system of Kolkwitz and Marson (1908). In this system indicator organism species are assigned a saprobien value ranging from 1 to 4–7 (depending on the classification) and a weighted average (depending on the abundance of the organisms), then expressing the saprobien index. The original system recognized four saprobien classes of quality related to the state of decomposition of organic matter. An *oligosaprobic* classification reflects the cleanest water quality, corresponding to unpolluted natural waters, while a *polysaprobic* characterization reflects highly polluted waters; α and β-mezosaprobic are intermediate classifications. Sládeček (1979) reviewed this system and its modifications. The saprobien system is widely used in Europe. In fact, water quality standards in most European countries have been based on and related to the four classes of the saprobien system.

Species diversity indices have been developed in the United States and are used to measure the stress of an ecosystem. It is considered that in unpolluted systems a large number of organism species coexist, with no single species dominating. When the ecosystem becomes stressed by contamination, species sensitive to that particular stress will be eliminated, thus reducing the variety of the biota. In addition, certain species may be favored, thus increasing in numbers. This system appears logical, however, Karr et al. (1986) pointed out that combining species richness with species abundance can yield ambiguous results and may prove difficult to interpret. Species diversity indices were reviewed by Krebs (1989). A relationship between water quality and diversity indices is difficult to establish.

Two biotic integrity indices have recently been developed and introduced in the United States. These indices have the advantage of detecting impacts that do not reflect on the chemistry of the receiving body of water, such as channelization of the stream. These systems measure the response of fish or benthic macroinvertebrate communities. The first system by Karr et al. (1986) uses fish as biological monitors. The second system uses macroinvertebrates.

The *Index of Biotic Integrity* (IBI) by Karr et al. (1986) uses measurements of the distribution and abundance or absence of several fish species types listed in Table 12.4. The fish samples are obtained by electrically shocking the sampling reach (fish is not killed). The Index of Biotic Integrity is a comparative index in which the measurements in a test reach of the particular body of water of interest are compared with fish distribution in reference to unimpacted, undisturbed reaches within the same ecoregion (Karr et al., 1986; Rankin, Yoder, and Mishne, 1990). Presumably, these reference bodies of water should exhibit no or minimal human impact. Table 12.5 and Figure 12.6 show the conceptual response of fish communities to various stresses and the ranges of the IBI for the five integrity classes. IBI of less than 12 implies that no fish were found by repeated sampling.

IBI relies on multiparameters, since the fact that the ecosystems to be evaluated are complex. It requires professional judgment and appropriate sampling methodology. As with any other biological tool, IBI must be used appropriately. IBI does not replace chemical indices and evaluation of chemical water quality, but it does provide a long-term picture of impacts, including those that are undetectable by chemical methods. Ideally, IBI can be correlated to chemical water quality, as indirectly indicated in Table 12.5.

It has to be pointed out that in addition to the variation among ecoregions, the fish population and number of fish species are a function of the stream order (headwater streams with no tributaries have order 1, streams with one or more first-order tributaries have order 2, streams

TABLE 12.4 Metrics Used to Assess Fish Communities in the Midwestern United States

Category	Metric	Scoring Criteria[a]		
		5	3	1
Species richness and composition	1. Total number of fish species 2. Number and identity of darter species 3. Number and identity of sunfish species 4. Number and identity of sucker species 5. Number and identity of intolerant species	Expectations for metrics 1–5 vary with stream size and region and are discussed in the text		
	6. Proportions of individuals as green sunfish	5%	5–20%	>20%
Trophic composition	7. Proportion of individuals as omnivores[b]	<20%	20–45%	>45%
	8. Proportions of individuals as insectivorous cyprinids	<45%	45–20%	<20%
	9. Proportion of individuals as piscivores (top carnivores)	>5%	5–1%	<1%
Fish abundance and condition	10. Number of individual in sample	Expectations for metric 10 vary with stream size and other factors and are discussed in the text		
	11. Proportion of individuals as hybrids	0%	>0–1%	>1%
	12. Proportion of individuals with diseases, tumors, fin damage, and skeletal anomalies	0–2%	>2–5%	>5%

Source: After Karr et al. (1986).

[a] Ratings of 5, 3, and 1 are assigned to each metric according to whether its value approximates, deviates somewhat from, or deviates strongly from the value expected at a comparable undisturbed reference body of water.
[b] Omnivores are defined here as fish species with diets composed of >25% plant material and >25% animal material.

TABLE 12.5 Conceptual Model of the Response of Fish Community as Portrayed by the Index of Biotic Integrity

Integrity Class Total IBI Score	Severely Degraded (Very Poor) 12–22	Degraded (Poor) 28–34	Moderate Impact (Fair) 40–44	Enrichment (Good) 48–52	Least Impacted (Excellent) 58–60
Community conditions, characteristics	No community organization—few or no species, very low numbers most tolerant only, high percentage of anomalies, diseases	Poorly organized community—few species, low numbers tolerant species only, many anomalies, fish dominated by omnivores	Reorganized community—tolerant species, intolerant in very few numbers, omnivores predominate, older age classes of top predators rare	Good community organization—good numbers of sensitive species, some intolerant, trophic structure shows some signs of stress	Highly organized community—insectivores, top carnivores, intolerants predominate, high diversity
Chemical conditions	Acutely toxic chemical conditions	Low dissolved oxygen with chronic toxicity	Low dissolved oxygen, nutrient enriched, no recurrent toxicity	Adequate dissolved oxygen, no acute/chronic toxicity effects, elevated nutrients	No effects evident, background conditions within the ecoregion, good dissolved oxygen
AND/OR Physical conditions	Total habitat loss, extremely contaminated sediments	Severe habitat degradation, severe sediment contamination	Modified stream channel, heavy siltation, canopy removal	Good habitat, no significant channel modifications	Excellent habitat, no modification evident
Example of perturbation	Toxic discharges, acid mine drainage, severe thermal stress, anoxic hypolimnetic discharges	Municipal and industrial discharges, intermittent acute impacts	Municipal sewage, combined sewers, heavy agricultural use, nonacid mine drainage, moderate thermal stresses	Minor sewage inputs, most agricultural and urban stormwater diffuse point affected areas	No or minimal human impact, no perturbation evident, reference assessment reaches

Sources: After Karr et al. (1986) and Rankin, Yoder, and Mishne (1990).

The Aquatic Ecosystem 757

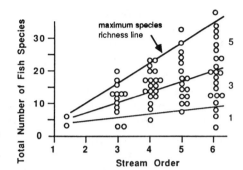

FIGURE 12.6. Total number of fish species versus stream order for least disturbed sites along the Embarras River in Illinois. (After Fausch, Karr, and Yant, 1984.)

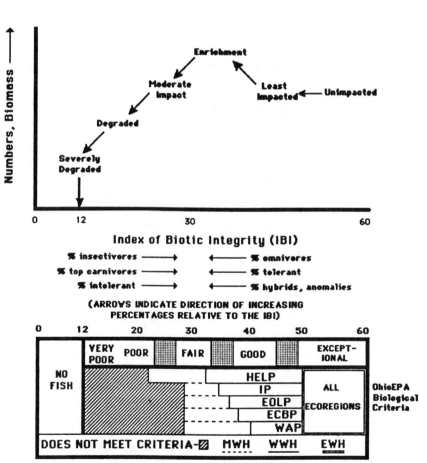

FIGURE 12.7. Biotic integrity (IBI) response to different levels of stress. (After Ohio EPA, 1987.)

with one or more second-order tributaries have order 3, and so forth) as shown in Figure 12.7. It is known that the total number of fish and the species richness vary with the stream size within the ecoregion. Unfortunately, most of the undisturbed reference reaches in most populated ecoregions are small headwater streams. Therefore, when assessing the streams a judgment must be made on the richness of the most probable species of higher order streams as indicated in Figure 12.7.

The *Invertebrate Community Index* (ICI), proposed by the Ohio Environmental Protection Agency (Ohio EPA, 1987), is a modification of the IBI index. This report contains extensive guidelines on the use of biotic indices and criteria. Use of macroinvertebrates has some advantages over fish sampling, as shown in the following list (Mason, 1991; Ohio EPA, 1987):

1. Macroinvertebrates form permanent, relatively immobile communities.
2. Microinvertebrates can be easily collected and are present in large numbers in even the smallest of streams.
3. They can be easily sampled at a low cost.
4. They occupy all stream habitats and, even within family and generic grouping, display a wide range of functional feeding preferences (i.e., predators, collectors, shredders, scrapers).
5. They are quick to react to environmental change and are incapable of escape.
6. They inhabit the middle of the aquatic food web and are a source of food for fish and other aquatic and terrestrial animals.
7. Taxonomy has developed to the point where species-level identification of many larval forms are available along with much environmental and pollution tolerance information.

It should be pointed out that the saprobien system of Kolkwitz and Marson uses extensively benthic macroinvertebrates as test organisms. Publications by Sládeček (1979) and Mason (1991) list pollution tolerance levels (saprobien numbers) for a number of benthic invertebrates. The Ohio EPA's system uses the following organisms as indicators of pollution: dipterans, mayflies, caddisflies, midges, plus other location-specific tolerant organisms, such as worms, snails, and mollusks.

The toxicity impact of various constituents on aquatic biota was incorporated into the formulation of chemical water quality criteria (U.S. EPA, 1987; also, see Chapter 13). The IBI and ICI indices also enable formulation of the ecoregional water quality criteria once the relationship between the water quality and habitat degradation and pollution loads is established. Following this concept a proper and attainable use of the body of water can be determined by use attainability analysis, which is mandated by the Clean Water Act. This concept is introduced in Chapter 16.

Biological Criteria

Until recently the methods used to achieve water quality goals were based primarily on chemical and toxicological criteria (see the Appendix). Recent emphasis on the biological criteria in the United States (note that the biological criteria based on the saprobien concept have been used in Europe for a long time) make aquatic life protection the focus. Several states have adopted the biotic criteria in their water quality assessment and plans. At the time of writing this book Ohio's approach appeared as the most progressive and advanced (Ohio EPA, 1987).

The use of biological communities, particularly fish and macroinvertebrates, offers a holistic system approach to surface-water quality assessment and management. Aquatic organisms not only integrate a variety of environmental influences, but complete their life cycles in the body of water and by so doing are continuous monitors of environmental quality.

Biological criteria can be developed using the biosurvey/ecoregional approach, which may be outlined as follows:

1. Classification and selection of reference sites
2. Conduct water quality and biological surveys at the reference sites
3. Develop a predictive model
4. Conduct biological and chemical surveys at the test site, a habitat assessment, and a survey of existing and potential sources of contaminations
5. Apply the model to the test site (estimation of unimpacted water quality and biological composition at the test site)
6. Determination of impairment. Ascertain toxicity levels at the test site by bioassays with the test organisms obtained from reference reaches
7. Determine acceptable water quality goals and the definitions of site-specific criteria (standards) for water and sediments (site specificity will be more profound for sediments)

The reasons for establishment of site-specific criteria based on the reference water quality and habitat integrity are the shift from "end of the pipe" control to an integrated point and nonpoint source abatement, and aquatic habitat restoration.

The biotic integrity indices can provide the measure of the water quality goals. For example, water quality goals can be met if the biotic

IBI and/or ICI index is classified as "good" to "excellent." The biotic indices and water body assessments described herein were incorporated into Rapid Bioassessment Protocols by the U.S. EPA (Plafkin et al., 1989).

Nitrogen Cycle: Nitrification and Denitrification

Nitrogen transformation and cycle in aquatic bodies of water, which is affected by discharges of waste loads from point and nonpoint sources, is almost as important as the organic matter-dissolved oxygen cycle. Nitrogen transformation may also affect the dissolved oxygen balance. Nitrogen in aquatic water bodies may exist in several forms: (a) dissolved nitrogen gas (N_2); (b) organic nitrogen incorporated into proteinaceous organic matter; (c) ionized and nonionized ammonia (NH_4^+ and NH_3); (d) nitrite ion (NO_2^-); and (e) nitrate ion (NO_3^-).

The nitrogen cycle in aquatic environments is shown on Figure 12.8. The reactions take place in water under aerobic and anaerobic conditions, in sediments mostly under anaerobic conditions, and in the sediment–water interface, where the transformation may be both aerobic and anaerobic. The processes in the sediments and water are similar to nitrogen transformation in soils.

Ammonium. In aquatic environments, decomposers break down organic proteinaceous matter and, consequently, ammonia is released. This process of biological degradation and release of ammonia is called *deamination*. Deamination may occur in sediments, water, soils, and also in biological treatment processes. Three ways of producing ammonia from organically bound nitrogen can be distinguished (Painter, 1970):

1. From extracellular organic nitrogen containing compounds, chemically or biochemically (for example, from urea)
2. From living bacterial cells during endogenous respiration
3. From dead and lysed cells

Depending on the pH of water the dissolved ammonium gas and ammonia ions will exist in the equilibrium given by the following reaction

$$NH_4^+ + OH^- \leftrightarrow NH_3 + H_2O \tag{12.4}$$

At a pH of 7 or below most of the ammonia will be ionized. At higher pH levels the proportions of deionized ammonia will increase.

Deionized ammonium is toxic to fish, while ionized ammonia is a nutrient to algae and aquatic plants and also exerts dissolved oxygen demand (see the nitrification process below). The un-ionized ammonium

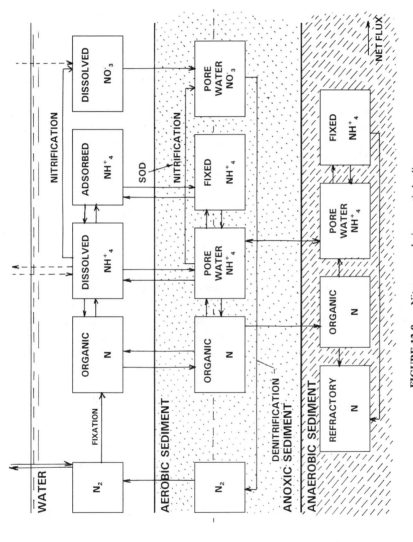

FIGURE 12.8. Nitrogen cycle in aquatic bodies.

is a gas that will mostly volatilize from water. The amount of un-ionized ammonia dissolved in water, which is toxic, depends on its solubility, which is affected by temperature and pH. Consequently, the toxicity of deionized ammonium depends on the same parameters and on the type of biological species being affected. The toxicity criteria for deionized and total ammonia concentrations are included in the Appendix.

As shown in Chapter 6, ammonia ions have an affinity for adsorption on soil particles (primarily clay and organic fractions) and by the same mechanism on sediments. Absorbed (fixed) ammonia is not available as a nutrient in the primary organic matter production process and is not toxic.

Nitrification. Nitrification is a two-stage process involving two types of chemotrophic bacteria (bacteria deriving energy from exothermic nitrification reactions). In the first step an ammonia ion is converted to nitrite by *Nitrosomonas*; in the second step nitrite is converted to nitrate by *Nitrobacter*. In simplified form the reactions may be written as

$$NH_4^+ + \tfrac{3}{2}O_2 \xrightarrow{\text{Nitrosomonas}} NO_2^- + 2H^+ + H_2O \qquad (12.5a)$$

$$NO_2^- + \tfrac{1}{2}O_2 \xrightarrow{\text{Nitrobacter}} NO_3^- \qquad (12.5b)$$

Stoichiometrically, the oxygen requirement for the overall nitrification reaction is $4.56\,\text{mg}\;O_2/\text{mg}$ of NH_4^+. However, since the reaction is basically autotrophic the bacterial growth essentially follows Equation (12.1), that is, oxygen is produced as a result of bacterial growth. Consequently, the overall oxygen requirement for nitrification is less than the stoichiometrical value. Wezernak and Gannon (1968) measured the O_2 requirement in nitrification as about $4.3\,\text{mg}\;O_2/\text{mg}$ of NH_4^+.

Several environmental factors affect the nitrification rate (Krenkel and Novotny, 1980):

1. The reaction is strictly aerobic. If the oxygen concentration is depleted below 2 mg/l, the reaction rate decreases rapidly.
2. Its optimum pH range is between 8 and 9. Below a pH of 6, the reaction essentially ceases.
3. The nitrifying organisms tend to be attached either to sediments or solid surfaces.
4. The growth rate of nitrifying organisms is much less than the growth rate of heterotrophic decomposers. Consequently, if higher concentrations of biodegradable organics are present, the heterotrophs will limit the growth of nitrifiers and nitrification will be suppressed.

5. The optimal temperature range is between 20° and 25°C. The rate decreases if the temperature is less or more than the optimum.

Denitrification. Under anoxic conditions, the nitrate–nitrogen ion becomes the electron acceptor in the essentially organic matter oxidation reaction. The reaction can be represented as (Schindler, 1985)

$$\text{Organic matter } (nC_6H_{12}O_6) + 4.8n(NO_3^- + H^+)$$
$$\rightarrow 6nCO_2 + 2.4nN_2 \uparrow + 8.4nH_2O \quad (12.6)$$

Other kinds of organic compounds may serve as the source of energy in this reaction. For example, dissolved methane that evolves from the anaerobic decomposition in the sediments can be used by the denitrifying organism in the interstitial layer.

This reaction represents a loss of nitrogen from water since the produced nitrogen gas volatilizes because it is in excess of the equilibrium between the gaseous nitrogen in the air and that dissolved in water. Only a few planktonic algal organisms (for example, blue-green algae) are capable of fixing dissolved N_2-nitrogen in the photosynthetic reaction.

In the anoxic sediments, the decomposers break down the organic matter to methane and carbon dioxide (methanogenesis) and nonbiodegradable residues. Ammonia is released in this process. Conditions at the sediment–water interface can then have a major impact on the N balance of the system. Unless the water at the bottom is completely anoxic, an aerobic thin sediment upper layer will form. Nitrification should occur in the oxic sediment layer. Because the gradient of NO_3^- is downward to the sediment (zero nitrate concentration) and not upward, the NO_3^- thus formed diffuses back into the anaerobic sediment where denitrification is certain (Keeney, 1973). This is referred to as a *simultaneous nitrification–denitrification process* by which nitrogen is lost from the system. By combining Equations (12.5) and (12.6), the following simplified overall chemical balance can be obtained

Nitrification
$$NH_4^+ + 2O_2 \rightarrow NO_3^- + 2H^+ + H_2O$$

Denitrification
$$NO_3^- + H^+ \rightarrow \tfrac{1}{2}(H_2O + N_2) + \tfrac{5}{4}O_2$$

Overall reaction
$$NH_4^+ + \tfrac{3}{4}O_2 \rightarrow H^+ 1\tfrac{1}{2}H_2O + \tfrac{1}{2}N_2$$

If these reactions proceed simultaneously three-quarters of a mole of O_2 is used per each mole of ammonia converted to nitrogen gas (1.7 g of O_2/g of $NH_4^+ - N$). This O_2 consumption will then be accounted for in the sediment oxygen demand, which is shown in the next section.

Experiments by Keeney, Schmidt, and Wilkinson (1975) have shown that about one-third of nitrogen input to a lake may be lost from the body of water by the simultaneous nitrification–denitrification process. Kadlec (1988) and others documented similar nitrogen losses for wetlands.

Not all of the ammonia released by decomposition in the sediments will be returned to water. Part of it will remain adsorbed on the sediments and part will be recycled into rooted macrophytes growing in the sediments located in the shallow euphotic (light abundant) zone, provided that the hydraulic conditions are favorable for the macrophyte growth. Simultaneous nitrification–denitrification can only occur in the sediment–water interface. Nitrification in free-flowing water is relatively rare (Tuffey, Hunter, Matulewich, 1974).

DISSOLVED OXYGEN PROBLEM

As pointed out in the first section of this chapter the waste discharges from diffuse sources embody rarely a direct thread to the dissolved oxygen regime of receiving bodies of water. The exceptions to this statement include runoff with a high concentration of biodegradable organics from concentrated animal operations, spring runoff from fields with manure spread on still frozen soils, and some instances of CSO discharges. However, indirect and secondary effects of diffuse-pollution discharges on the dissolved oxygen regime caused by nutrient enrichment and consequent sediment oxygen demand may be profound and detrimental to a receiving body of water and its uses (Novotny and Bendoricchio, 1989).

Dissolved Oxygen Standard

Water quality investigations and toxicity studies indicate that the dissolved oxygen content is the most important parameter for protecting fish and aquatic biota. The level of protection can be classified into three groups (Krenkel and Novotny, 1980). The first level of oxygen concentration would just permit the fish to live (prevention of fish-kills), the second level would permit the fish or aquatic organisms to be active to a specified degree, the third level would allow the organisms to live, grow, and reproduce in a given area. Typically, fish-kills occur when fish are exposed for a certain period of time (a few hours) to dissolved oxygen concentrations of less than 3 mg/l. Present standards provide, however, the third level of protection.

The levels of fish tolerance to low dissolved oxygen stresses vary. Cold-water fish (salmon, trout) and biota require higher dissolved oxygen concentrations than warm-water fish and biota. Therefore the standards distinguish between cold- and warm-water biota protection. Further considerations are given to whether the body of water or a portion thereof represents a breeding area or a migratory route. Appendix B contains the values of water quality criteria for U.S. bodies of water.

For the general well-being of trout, salmon, and associated cold-water biota, the dissolved oxygen concentration should not be below 6.5 mg/l. Under extreme conditions, the concentration may range between 5 and 6 mg/l (second level of protection), provided that all other water quality parameters are within the acceptable ranges. The 4-mg/l dissolved oxygen standard for cold-water biota and 3 mg/l for warm-water biota are essentially set to prevent fish-kills, and should not be included in standards aimed at protecting the biological integrity of receiving waters. Both biodegradable organics (five-day biochemical oxygen demand) and dissolved oxygen standards are used in the Economic Community and other countries of Europe.

Dissolved Oxygen Balance of Streams

The content of biodegradable organics in waste discharges (from both point and nonpoint sources) is commonly expressed as the biochemical oxygen demand (BOD). The reader is referred to standard environmental and water quality texts for the explanation of the fundamentals of the BOD reaction and its course (for example, see Krenkel and Novotny, 1980). Other tests expressing the amount of matter that will then complement the BOD data include chemical oxygen demand (COD), total organic carbon (TOC), and volatile suspended solids (VSS).

When waste discharge containing biodegradable organics is introduced into receiving waters, decomposers begin immediately to decompose the biodegradable organic matter and then convert it, ultimately, to carbon dioxide (alkalinity), water, and mineral and organic nonbiodegradable residues. According to Equation (12.2), dissolved oxygen is used in this biochemical reaction.

The dissolved oxygen balance is an important water quality consideration for streams and estuaries. Reservoirs and lakes can also be affected; however, the autochthonous productivity of these water bodies may increase the dissolved oxygen balance therein and overwhelm the effect of allochthonous inputs of oxygen-demanding wastes. This is especially true for deep reservoirs and lakes that stratify during the productive season. Shallow river mixed impoundments can be considered as streams.

In streams, the dissolved oxygen concentration is a response to various oxygen sinks and sources. The sinks of oxygen, that is the biochemical and biological processes that use oxygen, include:

1. Deoxygenation of biodegradable organics whereby bacteria and fungi (decomposers) utilize oxygen in the bioxidation–decomposition process.
2. Sediment oxygen demand (SOD), where oxygen is utilized by the upper layers of the bottom sediment deposits.
3. Nitrification, in which oxygen is utilized during oxidation of ammonia and organic nitrogen to nitrates.
4. Respiration by algae and aquatic vascular plants that use oxygen during night hours to sustain their living processes.

Major oxygen sources are:

1. Atmospheric reaeration, where oxygen is transported from the air into the water turbulence at the air–water interface.
2. Photosynthesis, where chlorophyll-containing organisms (producers such as algae and aquatic plants) convert CO_2 (or alkalinity of water) to organic matter with a consequent production of oxygen.

The basic concept of the dissolved oxygen balance in streams was proposed by Streeter and Phelps (1925) and was later summarized by Phelps (1944). This concept, which is primarily used for evaluation of the point source impact on water quality, is still used with some modifications. However, it will be subsequently shown that in present state-of-the-art models (WASP or QUAL) the original dissolved oxygen balance concept proposed by Streeter and Phelps constitutes only a small part of the overall water quality picture. Furthermore, the original model presumes a point discharge of waste.

The basic assumptions of more recent dissolved oxygen balance models applicable to diffuse sources are:

1. The biochemical oxygen demand (BOD) load to the stream can take place either as a point source (combined or storm sewer outfall or a major tributary) or as a distributed source.
2. The flow conditions do not change considerably during the time of analysis—an assumption that is often violated, but as shown by Meadows, Weeter, and Green (1978), no significant error will result as

long as $X \leq 0.05 \, Q_0/q_r$, where X is the length of the reach, Q_0 is the flow at the head of the reach, and q_r is the internal inflow per unit length of the reach.
3. Effects of nitrification and photosynthesis are negligible (an assumption that is often violated in streams where most pollution loads originate from diffuse sources).
4. Effects of dispersion, such as in tidal estuaries, is negligible.
5. The reach is uniform with no appreciable changes in cross-sectional dimensions, flow velocity, and depth.

Under these assumptions, the equation for the BOD and dissolved oxygen concentration becomes (Meadows, Weeter, and Green, 1978; Thomann and Mueller, 1987):

BOD variations

$$L = L_0 e^{-K_r(X/u)} + \frac{L_r}{K_r}(1 - e^{-K_r(X/u)}) \tag{12.7}$$

Dissolved oxygen variations

$$D = D_0 e^{-K_a(X/u)} + \frac{K_r}{K_a - K_r}(e^{-K_d(X/u)} - e^{-K_a(X/u)})L_0$$

$$+ \left[\frac{K_d}{K_d K_r}(1 - e^{-K_a(X/u)}) - \frac{K_d}{(K_a - K_r)K_r}(e^{-K_r(X/u)} - e^{-K_a(X/u)})\right]L_r$$

$$+ (1 - e^{-K_a(X/u)})\frac{S_B}{K_a H} \tag{12.8}$$

where
$D = c_s - c$ is the oxygen deficit at the end of the reach (mg/l)
D_0 = initial oxygen deficit (mg/l)
c_s = oxygen saturation (mg/l)
c = oxygen concentration at the end of the reach (mg/l)
L_0 = initial ultimate BOD concentration (mg/l)
L_r = ultimate BOD concentration input due to the lateral distributed sources (mg/l-day)
K_r = overall BOD removal coefficient (day^{-1})
K_a = reaeration coefficient (day^{-1})
K_d = BOD deoxygenation coefficient (does not include BOD removal by sedimentation of particulate BOD component) (day^{-1})
X = length of the reach (m)
u = average stream flow velocity (m/day)

S_B = sediment oxygen demand (SOD) (g/m²-day)
H = average depth of flow (m)

Equation (12.7) does not include the effect of nutrients and photosynthesis.

Coefficient of deoxygenation. The coefficient of deoxygenation measured in streams is a composite of three processes that may or may not occur simultaneously. These processes are deoxygenation of BOD in free-flowing water, effect of benthic slimes on BOD absorption and removal, and sedimentation of particulate organics. Hence (Novotny and Krenkel, 1975)

$$K_r = K_1 + K_3 + B \tag{12.9}$$

and

$$K_d = K_1 + B$$

where
K_1 = deoxygenation rate constant for free-flowing water approximately equaling the laboratory BOD bottle (respirometer) rate
K_3 = BOD removal by sedimentation, which does not result in oxygen demand
B = effect of absorption of BOD by benthic slimes, which results in oxygen demand

The magnitudes of the laboratory BOD removal coefficient, K_1, have been reported in ranges of 0.1 to 0.6 day^{-1} (K_1 base e), while the overall ranges of the deoxygenation rate, K_r, can be between 0.1 and 5 day^{-1}, depending on the hydraulic and biological character of the stream and biodegradability of the organic pollution.

Atmospheric reaeration. The gas transfer theory demonstrated that stream reaeration by atmospheric oxygen is proportional to the turbulence intensity at the water surface and the ratio of the surface area to water volume. The magnitude of the reaeration coefficient has been subjected to intensive investigations ranging from mostly theoretical, such as that of O'Connor and Dobbins (1958) to experimental field investigations, such as those by Churchill, Elmore, and Buckingham (1962). The bulk of the research on stream reaeration was conducted and published 30 to 40 years ago. These studies usually resulted in a relationship of the following type

$$K_a = CU^m H^n S_e^p \tag{12.10}$$

TABLE 12.6   Summary of Coefficients for the Reaeration Formula

Investigators	C	m	n	p
O'Connor and Dobbins (1958)	3.92	0.5	−1.5	0
Churchill, Elmore, and Buckingham (1962)	5.05	0.962	−1.673	0
Owens, Edwards, and Gibbs (1964)	3.0	0.67	−1.65	0
Krenkel and Orlob (1962)	173.6	0.408	−0.66	0.408

Note: K_a is in day^{-1}; U = stream velocity in m/sec; H = flow depth in m; S_e slope is in m/m (dimensionless).

where
U = stream velocity (typically m/sec)
H = flow depth (m)
S_e = energy slope or slope of water surface (dimensionless)
C, m, n, p = coefficients

The coefficients C, m, n, p for some most common estimating equations are given in Table 12.6.

Sediment oxygen demand (SOD). Stream bottoms are commonly covered by sediments. It is the stream morphology that determines what type of sediments will deposit on the bottom and what role, if any, the sediments will play in the transport and storage of pollutants. The sediments stored on the bottom are called *bedload*, while those in suspension are a part of *washload* (see Chapter 5 for a more detailed discussion on sediment transport).

Generally, if the stream velocity is high and the bed is composed mostly of sand and gravel, which have a very low affinity for adsorbing pollutants, most of the fine (cohesive and flocculent) sediments remain in suspension, the organic sediment content will be very low, and the bed will exhibit very low to no demand for oxygen. Consequently, for such streams the BOD removal rate due to sedimentation of particulate organics will be negligible. Fine sediments will deposit if the bottom shear stress drops below a certain threshold. As a rule of thumb, if flow velocity is below 0.3 m/sec, then organic particulates and other finer particles may settle to the bottom and become a part of the organic bottom sediments. The aerobic and anaerobic decomposition processes in the sediment layer are then responsible for the sediment oxygen demand (SOD).

The first scientific measurements of SOD by Fair, Moore, and Thomas (1941) related the oxygen demand of river sludges (primarily sewage solids) to the rate of deposition of organic sludge solids on the bottom and to the organic content of the sediments (expressed as Volatile Solids). Others related the SOD to the chemical oxygen demand (COD) of the sediment (Gardiner, Auer, and Canale, 1984). The measurements revealed that SOD is proportional to the square root of the organic content of the sediment expressed either as VS or COD.

After primary and secondary treatment was installed in most of the U.S. point wastewater discharges, the dominance of diffuse urban and rural sources became evident. Recent investigations have revealed that SOD includes oxygen demand from several separate processes (Bowman and Delfino, 1980; DiToro et al., 1990): (1) biological respiration and oxygen consumption of all living organisms residing in the upper aerobic benthic zone; (2) chemical oxidation of reduced substances in the sediments, such as reduced iron, manganese, sulfides after they come to oxygen-containing water; (3) biochemical oxidation of methane and ammonia (simultaneous nitrification–denitrification) that evolve from the lower anaerobic sediments.

Rudd and Taylor (1980) found that most of the SOD is attributed to the biochemical oxidation of evolving methane. Since methane has low solubility in water, the SOD portion due to methane oxidation by aerobic bacteria may be limited since only dissolved methane is bioavailable. The simultaneous nitrification–denitrification may be controlled by the growth rate of nitrifiers. The oxidation of methane and the simultaneous nitrification–denitrification in the interstitial layer were explained in the preceding section.

The *diagenesis model* proposed for SOD estimation by DiToro and coworkers (DiToro et al., 1990; DiToro, 1986) relates SOD to the input of particulate organic matter into the bottom sediment layer and its decomposition. In this concept an oxygen demand equivalent for reduced species (electron donors), such as organic carbon ($CH_4(aq.)$), HS^-, Fe^{2+}, and NH_4^+, is estimated. A mass-balance equation of the oxygen demand equivalents is then used to calculate their flux in the sediment–water interface, a consequence of which is the SOD. The production of soluble reduced end products in the sediments occurs via the bacterial breakdown of particulate organic matter. The most important products of the breakdown of organic matter are methane and ammonia. Inorganic reduced compounds may be significant in cases where insufficient organic matter is deposited into the sediment layer. Both methane and ammonia will exert an oxygen demand in the upper aerobic layer of the sediment in the sediment–water interface. The inorganic portion of SOD may amount to $0.3\,g/m^2$-day (Walker and Snodgrass, 1986). Figure 12.9 shows the diagenesis concept of the SOD. The diagenesis model and extensive field data measured in the Milwaukee River (DiToro et al., 1990) led to several important conclusions on the mechanism of SOD:

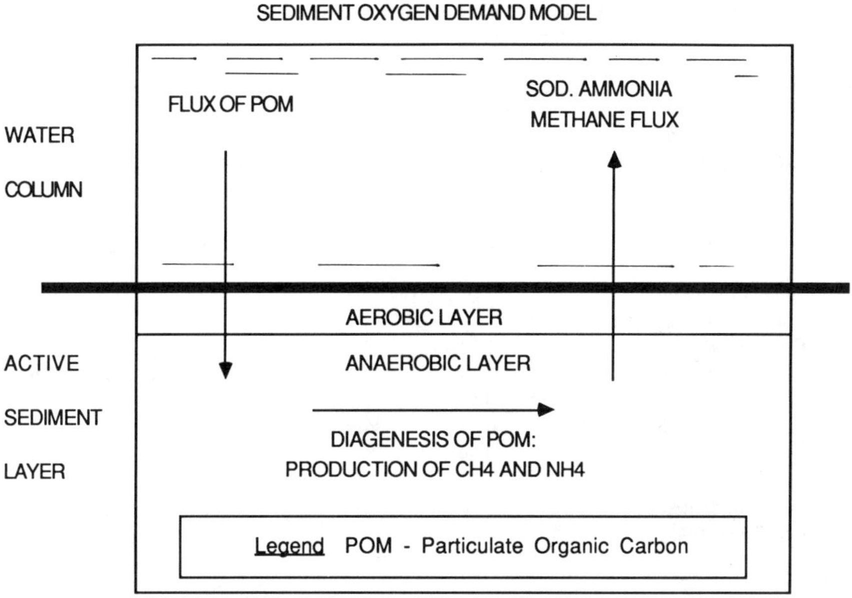

SEDIMENT OXYGEN DEMAND MODEL

```
WATER        | FLUX OF POM            SOD. AMMONIA
COLUMN       |                        METHANE FLUX
             |     ↓                       ↑
─────────────────────────────────────────────────
                    AEROBIC LAYER
ACTIVE              ANAEROBIC LAYER
SEDIMENT            DIAGENESIS OF POM:
LAYER               PRODUCTION OF CH4 AND NH4

             Legend  POM - Particulate Organic Carbon
```

FIGURE 12.9. Diagenesis concept of sediment oxygen demand. (After DiToro et al., 1990, by permission of ASCE.)

1. Since SOD is a result of the oxidation of end products of an anaerobic diagenesis process in the sediment, it is related to the availability of oxygen in the sediment–water interface, which is related to the concentration of dissolved oxygen in water. The dissolved oxygen dependency of SOD has been observed for dissolved oxygen concentrations that are much larger than concentrations known to limit oxidation (10% of saturation).
2. SOD is related to the solubility of methane in water, which is relatively low. For this reason the dependence of SOD on the reduced carbon (methane) flux from the sediments is not linear. DiToro et al. (1990) developed a square root dependency function between SOD and carbon fluxes.

The SOD model by DiToro et al. is complex and requires calibration with field data that are not commonly measured (e.g., gas fluxes from the sediment). Typical in situ measured SOD data are given in Table 12.7.

Temperature Effects

Almost all of the reaction rates in the oxygen balance process are temperature dependent. The relationship is expressed in the following form:

TABLE 12.7 In-Situ Measured Sediment Oxygen Demand

Benthic Deposit	Sediment Oxygen Demand (g/m²-day)	Temperature (°C)	Investigator
Rivers and estuaries			
River sludges	0.9		Oldaker et al.[a]
River muds—2 cm	3.4		McDonnel and Hall (1969)
25 cm	6.4		
Sandy bottom	0.2–1.0		Thomann (1972)
Mineral soils	0.05–0.1		
Sphaerotilus covered bottom 10 g dry wt/m²	7.0		
Estuarine mud	1–2		
River Benaviken	1.44	2	Edberg and Hofsten (1973)
	0.68	10	
River Sjomosjon	0.31	0	
Jaders Bruk	0.84	9	
	1.4	2	
Lower Milwaukee River	2.6–6.7	20	Kreutsberger et al. (1980)
Potomac estuary	2.5	15	O'Connel and Wales[a]
Lakes			
Lake Erken	2.6	18	Edberg and Hofsten (1973)
	0.43	4	
Lake Ramsen	2.3	17	
Lake Michigan–Green Bay	1.6–1.9	12	Paterson, Epstain, and McEnvoy (1975)

[a] Quoted in Filos and Molof (1972).

$$K_T = K_{20}\theta^{(T-20)} \tag{12.11}$$

where
K_T = reaction rate at temperature T (T in °C)
K_{20} = reaction rate at the reference temperature of 20°C
θ = thermal factor, which has the following accepted values

Deoxygenation rates K_r, K_d, K_1 $\theta = 1.047$
Reaeration rate K_a $\theta = 1.025$
Sediment oxygen demand, S_B $\theta = 1.05-1.06$

Temperature also affects the oxygen saturation value, which can be approximated as follows (Krenkel and Novotny, 1980):

$$c_s = 14.652 - 0.41022T + 0.007991T^2 - 0.000077774T^3 \tag{12.12}$$

Temperature conditions selected for the waste assimilative capacity evaluation should correspond to the average temperature of the warmest month of the year (e.g., August). Statistically rare temperatures in combination with statistically rare low flow would lead to excessively rare and unrealistic conditions.

Example 12.2: Dissolved Oxygen Computation*

A medium-sized city with a population of 50,000 and a city area of 20 km^2 is discharging its treated sewage and storm runoff by separate sewer systems into an impounded river. The length of the impounded reach is $X = 20$ km. Lateral (diffuse) flow increase is about $q_r = 0.1$ m^3/sec-km, with an average BOD$_5$ concentration of 3 mg/l.

Calculate the dissolved oxygen concentration in the river during a storm that would follow a critical dry period. The storm resulted in a runoff event with a volume of $R = 1.5$ cm that lasted 2 hours. The measured sediment oxygen demand was 2 g/m^2-day. The per capita BOD$_5$ loads in sewage is assumed as 54 g/cap.-day, and sewage flow is 400 l/cap.-day (Novotny et al. 1989).

Other given parameters:

Low flow (7 days duration—10 years expectancy) $Q = 15$ m^3/sec
Average depth in the impoundment $H = 3$ m
Average width of the impoundment $B = 64.33$ m
Water temperature $T = 25°C$
Deoxygenation rate (at 20°C) $K_r = K_d = 0.25$ day^{-1}
Upstream BOD$_5$ concentration = 1.2 mg/l
Upstream dissolved oxygen concentration $c_0 = 7.6$

Solution   From Equation (12.12) the oxygen saturation concentration at 25°C becomes $c_s = 8.4$ mg/l.

Sewage characteristics

Flow

$$q = 0.4 \text{ m}^3/\text{cap.-day}$$
$$0.4 \times 50,000 \text{ (pop)} \times 1 \text{ (day)}/86,400 \text{ (sec)} = 0.23 \text{ m}^3/\text{sec}$$

---

*This example is presented to illustrate the concepts under steady-state assumption. Such an assumption may not be appropriate for diffuse-pollution discharges driven by storm-water events. Dynamic models (see the section on models in this chapter) are more proper for this type of analysis.

BOD$_5$ discharge (sewage) = 54 g/cap.-day
= 54 (g/cap.-day) × 50,000 (pop) × 1 (day)/86,400 (sec)
= 31.25 g/sec

Assuming 90% removal in the sewage treatment plant, the BOD$_5$ loading from sewage is

$$0.1 \times 31.25 = 3.12 \text{ g/sec}$$

Storm characteristics

Flow from storm sewers

$$Q_s = 1.5 \text{ (cm)} \times 0.01 \text{ (m/cm)} \times 20 \text{ (km}^2\text{)} \times 10^6 \text{ (m}^2\text{/km}^2\text{)}$$
$$\times \tfrac{1}{2} \text{(hr)} \times 1\text{(hr)}/3600\text{(sec)}$$
$$= 41.67 \text{ m}^3\text{/sec}$$

(Note that the flow of sewage—dry-weather flow—is insignificant when compared to the storm water—wet-weather—flow). From Table 1.4 the average BOD$_5$ concentration of urban storm water is 30 mg/l(g/m^3). Hence the BOD load from storm water is

BOD$_5$ discharge (storm) = 41.67 m^3/sec × 30 (g/m^3) = 1250 g/sec

Note again that the BOD load by treated sewage during the wet-weather period is small when compared to the BOD load by urban runoff.

Total initial river flow

$$Q_0 = \text{river flow} + \text{storm water flow} + \text{sewage flow}$$
$$= 15 + 41.67 + 0.23 = 56.9 \text{ m}^3\text{/sec}$$

Terminal (at the end of the reach) flow

$$Q_e = Q_0 + X \times q_r = 56.9 + 20 \times 0.1 = 58.9 \text{ m}^3\text{/sec}$$

Average flow

$$Q_{av} = (56.9 + 58.9)/2 = 57.9 \text{ m}^3\text{/sec}$$

Average velocity

$$u_{av} = Q_{av}/(B \times H) = 57.9/(63.33 \times 3) = 0.3 \text{ m/sec} = 25.92 \text{ km/day}$$

According to Meadows, Weeter, and Green (1978), the steady-state dissolved oxygen model (Eqs. (12.7) and (12.8)) can be used when

$$X = 20 \text{ km} < 0.05 \, Q_0/q_r = 0.05 \times 56.9/0.1 = 28.45 \text{ km}$$

BOD input from lateral diffuse flow

$$L_r = \frac{1.4 (\text{lateral flow} \times \text{BOD}_5 \text{ conc})}{\text{cross-sectional area}}$$

$$= \frac{1.4 \times 0.1 (\text{m}^3/\text{km-sec}) \times 0.001 (\text{km/m}) \times 86{,}400 (\text{sec/day}) \times 3 (\text{g/m}^3)}{[3 \times 64.33] \text{m}^2}$$

$$= 0.19 \text{ g/m}^3\text{-day}$$

Initial Conditions Initial ultimate BOD concentration downstream from the city

$$L_0 = 1.4 \times \left[\frac{\text{sewage BOD}_5 \text{ load} + \text{storm-water BOD}_5 \text{ load}}{Q_0} \right.$$

$$\left. + \text{ upstream BOD}_5 \right]$$

$$= 1.4 \times \left[\frac{3.12 + 1250}{56.9} + 1.2 \right] = 23.22 \text{ mg/l}$$

Initial oxygen deficit

$$D_0 = c_s - c_0 = 8.4 - 7.6 = 0.8 \text{ mg/l}$$

The coefficient of reaeration can be calculated by Equation (12.10) with the O'Connor/Dobbins parameters (Table 12.6)

$$K_a = 3.92 U^{0.5} H^{-1.5} S_e^0 = 3.92 \times 0.3^{0.5} \times 3^{-1.5} \times 1.0 = 0.41 \text{ day}^{-1}$$

Correct K_a for 25°C

$$K_{a25°C} = 0.41 \times 1.25^{(25-20)} = 0.46 \text{ day}^{-1}$$

Correct K_r for 25°C

$$K_{r25°C} = 0.25 \times 1.047^{(25-20)} = 0.31 \text{ day}^{-1}$$

The dissolved oxygen deficit at the end of the 20-km-long reach is then computed by Equation (12.8). In the calculation substitute the travel time t for X/u or

$$t = 20\,\text{km}/25.92\,\text{km/day} = 0.77\,\text{days}$$

$$D = 0.8e^{-0.46\times 0.77} + \frac{0.31}{0.46 - 0.31}(e^{-0.31\times 0.77} - e^{-0.46\times 0.77})23.22$$
$$+ \frac{0.31}{0.46 \times 0.31}\bigg[(1 - e^{-0.46\times 0.77})$$
$$- \frac{0.31}{(0.46 - 0.31)0.31}(e^{-0.31\times 0.77} - e^{-0.46\times 0.77})\bigg]0.19$$
$$+ (1 - e^{-0.46\times 0.77})\frac{2}{0.46 \times 3} = 6.20\,\text{mg/l}$$

At the end of the reach the dissolved oxygen concentration will be

$$c = c_s - D = 8.4 - 6.2 = 2.2\,\text{mg/l}$$

For the conditions presented in this example the dissolved oxygen standard would be violated.

Several environmental texts contain more detailed discussions of the dissolved oxygen balance model, its fundamentals, and variations (Krenkel and Novotny, 1980; Orlob, 1983; Thomann and Mueller, 1987; Velz, 1984, and others).

Oxygen Balance Model for Tidal Rivers and Estuaries

In tidal rivers and estuaries the tides induce longitudinal mixing and dispersion of the constituents. The measure of the tidal mixing is the tidal dispersion coefficient, E, which has a unit of length^2/time. Development of the dissolved oxygen balance equations for estuaries is covered in a publication by O'Connor (1960). The form of the solution is similar to the classic dissolved oxygen equation (Eqs. (12.7) and (12.8)). Neglecting the lateral inflow, the dissolved oxygen balance equation becomes

$$L = L_0 \exp\left[\frac{U}{2E}(1 + \alpha_r)\right] \qquad (12.13)$$

$$D = D_0 \exp(j_a X) + \frac{K_r L_0}{K_a - K_r}\big[\exp(j_r X) - \exp(j_a X)\big]$$
$$+ \frac{S_B}{K_a H}[1 - \exp(j_a X)] \qquad (12.14)$$

where
$$j_a = 0.5\bar{U}(1 - \alpha_a)/E$$
$$j_r = 0.5\bar{U}(1 - \alpha_r)/E$$
$$\alpha_r = \sqrt{1 + \frac{4K_r E}{\bar{U}^2}}$$
$$\alpha_a = \sqrt{1 + \frac{4K_a E}{\bar{U}^2}}$$

For a single-point input

$$L_0 = W/(Q\alpha_r)$$

Also
W = point source total BOD load (g/day)
E = tidal dispersion coefficient (m$_2$/day)
Q = freshwater input in the estuary (m$_3$/day)
$\bar{U} = Q/A$ = freshwater advance velocity averaged over the tidal cycle
A = cross-sectional area of the estuary

In the preceding equation X is measured in the direction of flow toward the sea. It should be noted that if Equation (12.10) is used for estimating the reaeration coefficient K_a, it should be based on the average tidal velocity, u_t, and not on the averaged advance velocity $\bar{U}$.

The longitudinal tidal dispersion coefficient may be estimated from the salinity distribution in the estuary by plotting the salinity versus distance on a semilog paper (log salinity versus arithmetic distance). The slope of the line is then $\bar{U}/E$. Since $\bar{U}$ is known, then $E = \bar{U}(X_2 - X_1)/\ln(s_2/s_1)$, where 1 and 2 are arbitrarily chosen points on the plot and s and X are respective corresponding salinities and distances (see Thomann and Mueller, 1987 and Krenkel and Novotny, 1980).

Effect of Nutrients on Dissolved Oxygen Balance of Streams and Estuaries

The nutrients (nitrogen and phosphorus) increase the productivity of surface-water bodies, including that of streams and estuaries. The productivity impacts the dissolved oxygen balance of streams by (a) production of the photosynthetic oxygen, (oxygen demand by respiration of algal and macrophyte biomass, and (b) oxygen demand by decomposing dead algal biomass (exhibited most commonly as SOD since the dead biomass may settle into the bottom sediment layer). In addition, the unoxidized organic and ammoniacal nitrogen present in the water column impose the dissolved oxygen demand by nitrification.

Nitrification in the water column. As pointed out previously in this chapter, the process of nitrification in the bottom sediment may affect the sediment oxygen demand. The same process that is oxidation of ammonia to nitrite and nitrate by *Nitrosomonas* and *Nitrobacter*, may also be of significance to the oxygen balance of streams. It was also pointed out that since the nitrifying organisms associate themselves with solids and solid surface, their concentrations in free-flowing water may not be significant. Tuffey, Hunter, and Matulevich (1974) surveyed the available literature and arrived at the conclusion that a significant impact of nitrification on the oxygen balance of the streams can be only found for shallow streams with a solid bottom (rocky, boulders) where attached growths are dominant and for estuaries where sufficient detention (more than five to seven days) enables the development of sufficient densities of nitrifying organisms. Requiring some treatment plants to nitrify in the treatment process in order to reduce the ammonia toxicity of their effluents may result in enough nitrifying organisms being introduced with the effluent into the receiving water body to stimulate nitrification.

The occurrence of nitrification in a stream section is characterized by a simultaneous disappearance of ammonia and an increase of nitrate concentration. Ammonia decrease alone may not be a sufficient indication, since ammonia may be adsorbed by clay and organic particulates and settle into sediments. In addition, both ammonia and nitrates are nutrients to photosynthetic organisms (see the next section).

The nitrification process in free-flowing water and its impact on the dissolved oxygen regime is described in great detail in Krenkel and Novotny (1980). Mathematically, the nitrogenous oxygen demand (NOD) is treated in the same way as BOD. In a laboratory BOD bottle or stream, the dissolved oxygen balance model NOD equals approximately $4.33 \times$ TKN (total Kjeldahl nitrogen) = organic nitrogen + ammonia. The oxygen deficit increase due to nitrification is then

$$\Delta D \text{ (nitrification)} = \frac{K_N L_{0N}}{K_a - K_N} [\exp(-K_N X/u) - \exp(-K_a X/u)] \quad (12.15)$$

where
L_{0N} = initial nitrogenous oxygen demand (NOD = $4.33 \times$ TKN)
K_N = first-order nitrification rate, day^{-1}

Consequently, the NOD reduction in the stream is

$$L_N = L_{0N} \exp(-K_N X/u)$$

For estuaries and tidal streams the term $-K_N X/u$ would be replaced by $j_N X$, where

$$j_N = 0.5\bar{U}(1 - \alpha_N)/E$$

and

$$\alpha_N = \sqrt{1 + \frac{4K_N E}{\bar{U}^2}}$$

The magnitude of the nitrification reaction rate coefficient has been reported to range from 0.1 to 15.8 day^{-1} (Ruane and Krenkel, 1975). Several authors have proposed a zero-order model as better describing the nitrification process when TKN concentrations are higher. Both models (first-order for lower TKN concentrations and zero-order for higher nitrogen concentration) may be unified by the Michaelis–Menten equation describing substrate removal by microorganisms. Temperature affects nitrification differently than other processes. The thermal factor for nitrification is around $\theta = 1.1$.

Example 12.3: Effect of Nitrification

The oxygen balance of the stream described in Example 12.2 is also affected by nitrification of 0.6 mg of TKN present in the stream. The nitrification rate was estimated as $K_N = 0.5\,\text{day}^{-1}$. Estimate the impact on the dissolved oxygen at the end of the 20-km section of the river.

Solution Recall that the average stream velocity in the section was $u = 25.92$ km/day and the reaeration coefficient at 25°C was $K_a = 0.46\,\text{day}^{-1}$. Calculate the initial NOD

$$L_{0N} = 4.33 \times \text{TKN} = 4.33 \times 2.6\,\text{mg/l}$$

Adjust K_N

$$K_{N 25°C} = 0.5 \times 1.1^{(25-20)} = 0.8\,\text{day}^{-1}$$

Then by introducing the remaining parameters in Equation (12.14) the increase of the dissolved oxygen deficit, D_N, due to nitrification becomes

$$\Delta D_N = \frac{0.8 \times 2.6}{0.46 - 0.8}[\exp(-0.8 \times 20/25.92) - \exp(-0.46 \times 20/25.92)]$$
$$= 0.99 \, \text{mg/l}$$

The total dissolved oxygen deficit at the end of the reach will then become

$$D = D_c + \Delta D = 6.2 + 0.99 = 7.19 \, \text{mg/l}$$

and the dissolved oxygen concentration at the end of the reach is

$$c = C_s - D = 8.4 - 7.19 = 1.21 \, \text{mg/l}$$

Productivity. Odum (1956) has shown that nutrients affect the productivity of streams and have an impact on the dissolved oxygen balance. Odum pointed out that the ratio of the photosynthetic organic matter production and the respiration (decay) in relation to the concentration of the organic matter itself is the key in determining whether there is a net increase in the organic matter or a net decrease. The fundamental relation of the organic matter balance was postulated by Odum (1956) as

$$I + P = E + R \qquad (12.16)$$

where
I = import rate of organics from allochthonous sources and/or from upstream
P = production rate of organic matter
R = decomposition and/or respiration
E = export of organic matter downstream

Decomposition is a function of the organic matter concentration, while production depends on nutrient level and light, provided that the source of inorganic carbon (alkalinity) is sufficient and not limiting.

It was pointed out previously that based on the Stumm and Morgan relationship, 1 atom of phosphorus or 16 atoms of nitrogen will generate 154 molecules of oxygen during photosynthesis, and the same amount of oxygen will be used during decomposition in the form of BOD.

Originally, the trophic status concept was applied only to lakes; today all bodies of water can be evaluated according to their trophic status. Žáková (1989) classified stream trophic status according to the algal biomass concentration, as shown below:

Trophic Degree	Algal Biomass (mg/l)
ultraoligotrophic	<5
oligotrophic	5–50
oligomezotrophic	50–100
mezotrophic	100–200
mezoeutrophic	200–350
eutrophic	350–500
polytrophic	500–1000
hypertrophic	>1000

The trophic status terminology for lakes is explained in the next section. For streams the term *oligotrophic* would imply a stream with a low concentration of nutrients (*oligo* means *few*), while *eutrophic* would mean nutrient-enriched streams. *Ultraoligotrophic* would typically mean a stream with little biological activity, such as a stream draining a glacier or a stream affected by toxics or acid rainfall or snowmelt (an unhealthy situation).

Using the previous concepts of stream enrichment by productivity stimulated by nutrient inputs, Novotny and Bendoricchio (1989) have described the linkage of nutrients and primary productivity to organic carbon production, its deposition, and subsequent impact on oxygen resources as follows:

- Oxygen produced by photosynthesis above saturation in streams is released to the atmosphere and is lost from the nutrient-oxygen cycle.
- In flowing waters, photosynthesis and respiration do not occur at the same time and in the same place. Photosynthesis takes place on sunny days in shallow reaches and euphotic zones, while respiration takes place when light is limited (cloudy days and night hours) and in profundal zones. The fact that lack of light commonly accompanies storm runoff events may explain some of the dissolved oxygen anomalies mentioned in papers by Field and Turkeltaub (1981), and Keefer, Simons, and McQuivey (1979).

In streams where shallow productive reaches are followed by deeper sections, oxygen deficiency becomes the most troublesome (Novotny and Zheng, 1988; Novotny and Bendoricchio, 1989). Such situations include rivers entering reservoirs or bays, and estuaries or larger slower moving rivers. The algal biomass produced in shallow well-lighted reaches in which most oxygen above saturation was lost, may enter river sections that are deeper and in which only the upper layer is productive and

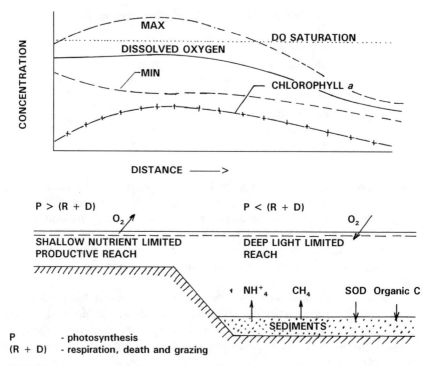

FIGURE 12.10. Concept of the impact of primary productivity on stream dissolved oxygen.

capable of producing photosynthetic oxygen while the algae in the deeper layer are respiring. Figure 12.10 illustrates this concept.

Novotny and Bendoricchio (1989) concluded that:

- Primary productivity resulting from nutrient inputs during medium and small flows represent the link between nonpoint pollution inputs and water quality deterioration. Large masses of organic matter can be produced in urban and suburban nutrient-enriched streams, which may subsequently impose dissolved oxygen demand on the receiving aquatic systems. This secondary oxygen demand caused by nutrients may be higher than the primary oxygen demand from discharged BOD in storm runoff and combined sewer overflows.
- The primary productivity process in streams and to a lesser degree in stagnant bodies of water is generally oxygen deficient because a significant portion of the oxygen produced above saturation levels may escape from the system. In most adverse cases, very little of the

photosynthetically produced oxygen may be included in the dissolved oxygen balance of the body of water.
- Control of nutrient inputs from erosion during high-flow runoff events generated by large but hydrologically rare storms, may not be effective since primary productivity in streams is most active during low flows and algae and macrophytes may actually be flushed out during the high flows. Also sediment inputs and the resulting turbidity cut down on light penetration and inactivates part of the phosphorus and ammoniacal nitrogen.

NUTRIENT PROBLEM OF LAKES: EUTROPHICATION

The problem of nutrient inputs is especially important for lakes, impoundments, and shallow estuaries (e.g., Chesapeake Bay). In these bodies of water the classic dissolved oxygen balance model may not work in evaluating the waste-assimilative capacity of these bodies of water, and the production of organic matter (primary productivity) by phytoplankton and large plants (macrophytes) may greatly exceed the allochthonous point and nonpoint BOD and NOD contributions. Oxygen levels are also affected by photosynthesis and respiration, and BOD concentrations are

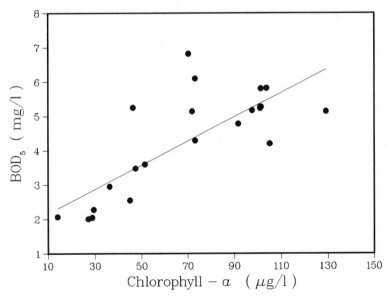

FIGURE 12.11. Relation of chlorophyll-*a* to BOD in the Milwaukee River. (Data from the Metropolitan Milwaukee Sewerage District.)

affected by the density of planktonic organisms and their residues (Fig. 12.11).

The production of organic matter in lakes, shallow estuaries, and reservoirs depends on the trophic status of the impoundment. Such bodies of water have experienced accelerated eutrophication in the last few decades, caused by among other factors, increased use of agricultural and lawn "care" chemical fertilizers. Streams and lakes that 50 years ago were reasonably clean are today infested by algae to a point that their color is rich green with little light penetration, instead of transparent water. This situation is especially troublesome for Eastern and Central Europe. An example of the Lake Balaton situation in Hungary was featured in Chapter 1, and Table 12.3 documented extremely high levels of nitrates in the rivers of the Czech Republic.

In this context the term *eutrophication* represents a process by which a body of water progresses from its origin (for example, a glacial lake) to its extinction, which is dry land. Eutrophication is not synonymous with pollution; however, pollution can accelerate the rate of eutrophication. It is a dynamic morphological process that has variable rates, which differ from year to year, season to season, and even hour to hour. It progresses in several stages. The net increase of organic matter (net productivity) that drives eutrophication takes place in a surface-water body in which organic matter production (primary productivity) nourished by nutrients and minerals exceeds its loss by respiration, decay, grazing by higher organisms, and outflow. The rate of eutrophication (productivity) depends on a number of factors, many of them uncontrollable. In today's context "Eutrophication refers to the natural and artificial addition of nutrients to bodies of water and to the effect of these added nutrients on water quality" (Rohlich, 1969). The stage of eutrophication is determined by the trophic state of the body of water. The trophic status may be related to the algal biomass or chlorophyll-*a* concentrations and other parameters, as shown in Table 12.8.

TABLE 12.8   Trophic Status of Lakes

Water Quality	Oligotrophic	Mezotrophic	Eutrophic	Source
Total P (μg/l)	<10	10–20	>20	U.S. EPA (1974)
Chlorophyll-*a* (μg/l)	<4	4–10	>10	U.S. EPA (1974)
Secchi disc depth (m)	>4	2–4	<2	U.S. EPA (1974)
Hypolimnetic oxygen (percent of saturation)	>80	10–80	<10	U.S. EPA (1974)
Phytoplankton production (g of organic C/m^2-day)	7–25	75–250	350–700	Mason (1991)

The term *oligotrophic* is used to describe a clean body of water, typically the youngest stage of lake formation. This body of water has a very low mineral content, so productivity and the algal content are low, and the water is highly transparent. Ultraoligotrophic bodies of water commonly refer to bodies of water whose productivity was eliminated, for example, by a severe drop of pH due to acid rainfall or by toxicity. Unlike the oligotrophic bodies of water whose biota may be healthy although few, ultraoligotrophic bodies of water may be unhealthy.

As the nutrient and mineral content of water is increased by runoff and/or by wastewater disposal, the photosynthetic (autotrophic) organisms (the producers) increase in number. As the productivity increases the organic matter content of the body of water becomes *mezotrophic*, and when the nutrient and organic contents are high the body of water becomes *eutrophic*. The *polytrophic* and *hypertrophic (hypereutrophic)* classification may be again added to describe bodies of water that are severely impacted by nutrient inputs, mostly from the excessive use of fertilizing chemicals, which in many parts of the world have caused more than an order of magnitude increases of nutrient levels of receiving bodies of water and very severe algal infestations. Chlorophyll-*a* concentrations of hypertrophic lakes may exceed 200 µg/l. For lakes and reservoirs, the final stages of their existence are silted pond, swamp, marsh, or other types of wetland. Figure 12.12 shows symptoms of eutrophication, while Figure 12.13 clearly demonstrates aesthetic impairment of water quality caused by the excessive growth of phytoplankton and aquatic weeds in a hypertrophic reservoir.

The photosynthetic process of algae and macrophytes in eutrophic water bodies can be recognized by cyclic fluctuations of the dissolved oxygen concentrations in the euphotic zone (a zone where light can penetrate). Oxygen is produced during the day by photosynthesis and consumed during the night and cloudy daytime hours by endogenous respiration. On bright sunny days during the production season, this often results in supersaturation of the euphotic zone with oxygen during afternoon hours and a significant drop of dissolved oxygen concentrations during the late night and early morning hours. Similar dissolved oxygen fluctuations are also common in eutrophic streams.

The related physical and chemical changes caused by advanced eutrophication (pH variations, oxygen fluctuations or lack of oxygen in the lower hypolimnetic layer of stratified bodies of water, waste excretions by some algae, etc.) may interfere with recreational and aesthetic uses of the body of water and may also cause a shift in fish and shellfish populations from better quality fish to rough fish. In addition, possible taste and odor problems caused by algae can make water less suitable for potable use

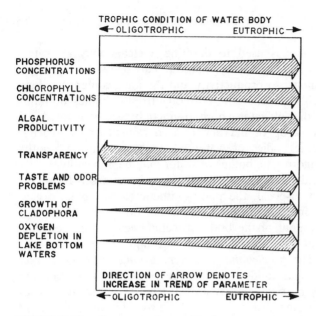

FIGURE 12.12. Symptoms of eutrophication.

FIGURE 12.13. Impact of eutrophication on a shallow reservoir. (Lake Sinissippi in Wisconsin in the 1970s; in the late 1980s the water quality of the lake was restored by the Wisconsin Department of Natural Resources.)

and contact recreation. When the concentration of phytoplankton algae exceeds a certain threshold nuisance value during the late summer period, the situation is termed *algal bloom*.

Hutchinson (1969) pointed out that the terms eutrophic, oligotrophic, mezotrophic, etc., should not only be used to describe water quality but should also be related to the drainage area and sediment and nutrient transport. Also Žáková (1989) pointed out that the trophic status of a lake or a reservoir is related to the trophic status in the tributaries; however, this relation is also affected by other factors, such as depth of the reservoir. In this context the term *available nutrients* should be used to describe eutrophication potential.

The necessity of including tributaries (watershed) in eutrophication systems caused the recategorization of lakes and reservoirs (and to a lesser degree large estuaries) into *autotrophic*, that is bodies of water that receive a major portion of the nutrients from internal sources (sediment, atmosphere), and *allotrophic*, which receive a major portion of the nutrients from external sources. *Allochthonous* nutrients are those originating from the watershed, that is, from point and nonpoint sources, while *autochthonous* nutrient sources include nutrients stored in the lake water and sediments.

Trophic indices and trophic status. Determination of the trophic status of lakes and other bodies of water based on limnological observations, taxonomy or the distribution of organisms and their productivity, and chemical water quality, is relatively difficult and requires experience and often subjective judgment. Table 12.8, relating trophic status to the concentration of organic (photosynthetic) matter, is only one of the possible judgmental classifications. A lack of a precise definition of "trophic status" makes it difficult to develop an accurate engineering tool that would enable estimation of the stage of the eutrophication process of a given body of water. However, modeling of the eutrophication process advanced in the 1980s to the point where a reasonably reliable judgment could be made on the relation of nutrients to productivity and, hence, the trophic status of a water body.

Some methods of estimating the trophic status based on few indicators have evolved and have been published and/or used for classifying the lakes. Many of the techniques were relative systems in which lakes were classified and ranked only in respect to each other or to an average water quality of a given group of lakes and not to some objective independent scale. These systems often give different weights to various parameters similar to those shown on Figure 12.12, and some of these are then included trophic status indices. The most frequently used ones are dissolved oxygen (if the water body is stratified, then dissolved oxygen in the lower hypolimnetic layer is considered), total phosphorus, transparency by Secchi disc (the depth at which a white disc can no longer be seen by the observer above the water surface), inorganic nitrogen, and chlorophyll-*a* concentration.

Trophic Index by Carlson (1977). This index was developed for lakes that are phosphorus limited. Carlson based his index on the fact that there are intercorrelations between the transparency expressed by the Secchi disc depth, algal concentrations expressed by chlorophyll-*a*, and vernal (spring) or average annual phosphorus concentrations. The trophic status index (TSI) was then defined as

$$\text{TSI(SD)} = 10\left(6 - \frac{\ln SD}{\ln 2}\right) \tag{12.17}$$

where SD = Secchi disc depth in meters. From the correlations between the chlorophyll-*a* concentrations, total phosphorus, and Secchi disc depth, the other two expressions for the TSI became

$$\text{TSI}(Chl) = 10\left(6 - \frac{2.04 - 0.68 \ln Chl}{\ln 2}\right) \tag{12.18}$$

and

$$\text{TSI(TP)} = 10\left(6 - \frac{48/TP}{\ln 2}\right) \tag{12.19}$$

where
Chl = concentration of chlorophyll-*a* (µg/l)
TP = concentration of total phosphorus (µg/l)

The method offers three indices instead of one single value. The best indicator of the trophic status varies from lake to lake and from season to season. Secchi disc values may be erroneous in lakes where turbidity is caused by factors other than algae. The Carlson index works well for north temperate lakes, but performs poorly in lakes with excessive weed problems (North American Lake Management Society, 1990).

Based on the observations of several northern lakes, most of the oligotrophic lakes had TSIs below 40, mezotrophic lakes had TSIs between 35 and 45, while most eutrophic lakes had TSIs greater than 45. Hypertrophic lakes, on the other hand, can have TSI values above 60 (Sloey and Spangler, 1978; Krenkel and Novotny, 1980).

Trophic status derived from nutrient loading of lakes. During investigations of eutrophication problems of lakes in Madison, Wisconsin, Sawyer (1947) noted that algal blooms occurred when concentrations of inorganic nitrogen (NH_4^+, N_2^-, NO^{2-}) and inorganic phosphorus exceeded respective values of 0.3-mg N/liter and 0.001-mg P/liter. It should be noted that due to the uptake of nutrients by algae, very low concentrations would be measured during the productive summer period; therefore, the critical nutrient concentrations should be evaluated during the spring (vernal) season while the spring overturn is in progress. Similarly, most of the lakes were eutrophic if the phosphorus concentration was above 20 μg/l.

Vollenweider (1975, 1976) then developed a simple input–output model phosphorus model for a completely mixed lake as follows (Thomann and Mueller, 1987):

$$V\frac{dp}{dt} = W - v_s A_s p - Qp \qquad (12.20)$$

where
V = volume of the lake (l^3)
p = phosphorus (total) in the lake (M/l^3), for example, μg/l
Q = outflow (l^3/time)
A_s = lake surface area (l^2)
W = allochthonous source of phosphorus (M/time), for example, g/sec
v_s = settling rate of phosphorus, l/time

At a steady-state $dp/dt = 0$, and dividing both sides by the surface area, Equation (12.19) becomes

$$p = \frac{W'}{q_s + v_s} \qquad (12.21)$$

where
q_s = hydraulic overflow rate $Q/A_s = H\rho$
$\rho = Q/V = 1/\tau_w$ = flushing rate
τ_w = detention time in the lake

The problem with using Equation (12.21) is lack of adequate understanding on the settling (removal) rates of phosphorus. Vollenweider (1976) substituted the following approximation

$$v_s = H\sqrt{\rho}$$

which led to

$$p = \frac{W'}{H\rho(1 + \sqrt{1/\rho})} \tag{12.22}$$

The relationship of annual loading of phosphorus per unit lake area and the hydraulic parameters for respective phosphorus concentrations of 10 µg/l and 20 µg/l by Vollenweider is plotted on Figure 12.14. The lake depth, H, is expressed in meters, and the flushing rate $\rho = Q/V$ in years^{-1}.

Assuming that phosphorus is the limiting nutrient, Lee, Rust, and Jones (1978) have expanded the Vollenweider relationship to include chlorophyll-a, Secchi disc depth, and hypolimnetic oxygen depletion, thus closing the circle between the nutrient loading, water quality, and trophic status. Figures 12.15 through 12.17 show the results of Lee, Rust, and Jones. On the abscissa, $L(P) = W' =$ annual surface phosphorus loading in mg/P/m²-year, hydraulic overflow loading, q_s, is in meters/year, mean depth of the lake, H, is in meters, and the hydraulic residence time, $\tau_w =$ water body volume/annual flow, is in years.

Example 12.4: Lake Eutrophication

The annual loadings for a lake with 28 km² surface area and an average depth of 13 meters are given in Table 12.9. Estimate the trophic level of the lake using Vollenweider and Lee et al. concepts.

Solution   Annual flushing rate

$$\rho = \frac{\text{Annual flow}}{\text{Lake volume}} = \frac{78{,}000{,}000 \ (\text{m}^3/\text{yr})}{28 \ (\text{km}^2) \times 13 \ (\text{m}) \times 10^6 \ (\text{m}^2/\text{km}^2)} = 0.21 \ \text{yr}^{-1}$$

Annual phosphorus loading

$$L_p = \frac{23{,}300 \ (\text{kg/yr}) \times 10^3 \ (\text{g/kg})}{28 \ (\text{km}^2) \times 10^6 \ (\text{m}^2/\text{km}^2)} = 0.83 \ \text{g/m}^2\text{-yr}$$

Annual nitrogen loading

$$L_N = \frac{236{,}000 \ (\text{kg/yr}) \times 10^3 \ (\text{g/kg})}{28 \ (\text{km}^2) \times 10^6 \ (\text{m}^2/\text{km}^2)} = 8.43 \ \text{g/m}^2\text{-yr}$$

Nutrient Problem of Lakes: Eutrophication   791

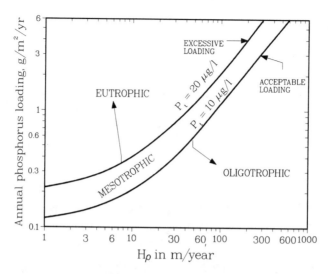

FIGURE 12.14. Relationship between nutrient loading and lake trophic conditions. (Adapted from Vollenweider [1975]. Courtesy *Schweizerische Zeitschrift für Hydrologie*.)

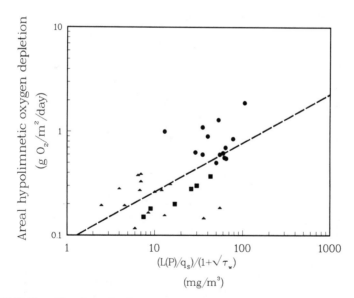

FIGURE 12.15. Phosphorus loading characteristics and hypolimnetic oxygen depletion in lakes. (Redrawn from Lee, Rust, and Jones [1978] with permission. Copyright 1978 by the American Chemical Society.)

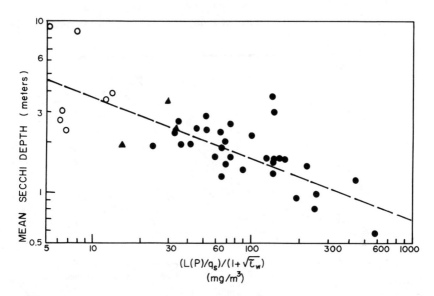

FIGURE 12.16. Phosphorus loading characteristics and mean Secchi disc depths for lakes. (Redrawn from Lee, Rust, and Jones [1978] with permission. Copyright 1978 by the American Chemical Society.)

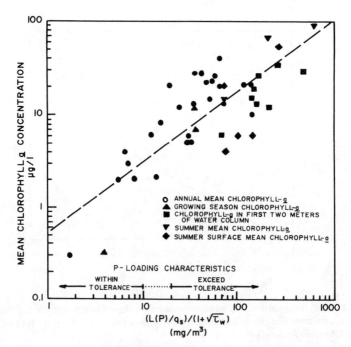

FIGURE 12.17. Phosphorus loading characteristics and mean chlorophyll-*a* relationship in lakes. (Redrawn from Lee, Rust, and Jones [1978] with permission. Copyright 1978 by the American Chemical Society.)

Even though the nitrogen-to-phosphorus load ratio is about 10:1 and nitrogen may be limiting, controlling phosphorus is the only practical solution, considering that more than 50% of nitrogen originates from nitrogen in ground water (suspected sources include septic tank discharges and overuse of fertilizers in agriculture).

Vollenweider concept From Figure 12.14 (Vollenweider's chart) for $H\rho = 13 \times 0.21 = 2.73$ m/yr the acceptable loading (to maintain oligotrophic conditions) is about 0.12 g P/m²-yr and the excessive loading (eutrophic conditions) is about 0.23 g P/m²-yr. Since these phosphorus loadings are exceeded, the lake is considered to be eutrophic.

Lee et al. concept Use, for example, Figure 12.16 (Secchi disc) and apply the following values.

$$q_s = H\rho = 2.73 \text{ m/yr}$$

$$\tau_w = 1/\rho = 1/0.21 = 4.78 \text{ yr}$$

$$(L_p/q_s)/(1 + \sqrt{\tau_w}) = \frac{0.83 \text{ (g/m}^2\text{-yr)} \times 10^3 \text{ (mg/g)}}{2.73 \text{ (m/yr)}} \Big/ (1 + \sqrt{4.78}) = 95.4$$

The expected Secchi disc depth would be around 2 meters. Then from Equation (12.17)

$$\text{TSI(SD)} = 10\left(6 - \frac{\ln 2}{\ln 2}\right) = 50$$

indicating again borderline eutrophic conditions.

TABLE 12.9 Annual Loadings

Source	N (kg/yr)	P (kg/yr)	Flow (mil m³/yr)
Urban sewage	23,000	8,000	1.0
Urban runoff	15,000	4,000	7.0
Rural runoff	26,000	10,000	10.0
Precipitation	47,000	1,000	21.0
Ground water	125,000	300	47.0
Evaporation	—	—	−8.0
Total	236,000	23,300	78.0

Example 12.5: Maximum Nutrient Loadings to Maintain Dissolved Oxygen Standard

A lake is receiving nutrient inputs from runoff and treated municipal wastewater. Estimate the maximum loading of phosphorus to the lake that would keep the lake below eutrophication status and maintain summer dissolved oxygen concentrations in the lower hypolimnion layer above 5 mg/l.

Lake characterization

$$\text{Surface area} = 10\,\text{km}^2$$
$$\text{Average depth} = 8\,\text{m}$$
$$\text{Annual inflow including precipitation} = 30 \times 10^6\,\text{m}^3/\text{yr}$$

Annual flushing rate

$$\rho = \frac{\text{Annual flow}}{\text{Lake volume}} = \frac{30{,}000{,}000\,(\text{m}^3/\text{year})}{10\,(\text{km}^2) \times 8\,(\text{m}) \times 10^6\,(\text{m}^2/\text{km}^2)} = 0.375\,\text{year}^{-1}$$

From Figure 12.14 (Vollenweider chart) the admissible and dangerous loadings of phosphorus for $H\rho = 8 \times 0.375 = 3.00$ are

$$\text{Dangerous P loading} = 0.28\,\text{g}/(\text{m}^2\text{-year})$$
$$\text{Acceptable P loading} = 0.13\,\text{g}/(\text{m}^2\text{-year})$$

If $0.20\,\text{g}/(\text{m}^2\text{-year})$ is selected, the loading will be

$$\text{max P load} = 0.2\,[\text{g}/(\text{m}^2\text{-year})] \times 10\,(\text{km}^2) \times 10^6\,(\text{m}^2/\text{km}^2) \times 0.001\,(\text{kg/g})$$
$$= 2000\,\text{kg/year} = 5.48\,\text{kg of P/day}$$

Solution To estimate the maximal loading to maintain the dissolved oxygen standard using the Lee et al. concept, several assumptions must be made:

1. Oxygen is used in the hypolimnion (approximately 5 meters depth from the bottom, $H_b = 5$ meters) and there is no oxygen interchange

between the upper (epilimnion) and lower (hypolimnion) layers. This is a reasonable estimate since the transfer across the epilimnion–hypolimnion boundary (so-called thermocline) is greatly reduced and is possible only by molecular diffusion (see Krenkel and Novotny (1980) for a discussion).
2. The average temperature of the hypolimnion is 15°C and the initial oxygen concentration at the beginning of the summer productive (growing) season is near saturation. The oxygen saturation, c_s, at 15°C is 10.2 mg/l.

Then the allowable drop of the dissolved oxygen during the growing season between the spring and summer overturn (assume 4-month growing period) is

$$\Delta C = c_s - \text{DO standard} = 10.2 - 5.0 = 5.2 \text{ mg/l } [= \text{g/m}^3]$$

The allowable oxygen depletion rate (4-month period) is

$$\frac{\Delta C}{\Delta t} H_b = \frac{5.2 \text{ (g/m}^3)}{122 \text{ days}} \times 5 = 0.21 \text{ g/(m}^2\text{-day)}$$

From the Lee et al. chart (Fig. 12.15) the loading parameter for the maximal rate of reduction of the hypolimnetic oxygen of 0.21 g/(m²-day) is

$$[L_p/q_s]/(1 + \sqrt{\tau_w}) = 7 \text{ mg/m}^3$$

Herein

$$q_s = H\rho = 3.00 \quad \text{and} \quad \tau_w = 1/\rho = 1/0.375 = 2.67 \text{ years}$$

Then

$$L_p = 7q_s(1 + \sqrt{\tau_w}) = 7 \times 3.00 \times (1 + \sqrt{2.67}) = 55 \text{ mg/m}^2\text{-yr}$$

The corresponding P loading is then

$$\text{max P load} = \frac{55 \text{ [mg/(m}^2\text{-yr)]} \times 10^{-6} \text{ [kg/mg]} \times 10 \text{ [km}^2\text{]} \times 10^6 \text{ [m}^2\text{/km}^2\text{]}}{365 \text{ [days/year]}}$$

$$= 1.51 \text{ kg of P/day}$$

MODELING FATE OF CONVENTIONAL POLLUTANTS AND NUTRIENTS

One has to realize that water quality management and pollution abatement are typically designed for hydrologically rare events. Fifty or so years ago when point source abatement was the only objective and biodegradable pollutant discharges the only type of pollution considered, steady-state simple models of the dissolved oxygen balance were the only type of models used by planners and pollution abatement authorities. Including eutrophication and toxic components into the water quality picture and also considering diffuse sources in addition to point source pollution in an integrated manner made the simple steady-state dissolved oxygen models obsolete. Also with the advent of fast and efficient yet small desktop or lap (personal or work station) computers, far more complex and detailed modeling is now required and has become common.

To describe mathematics and the software design of present dynamic water quality models is beyond the scope of this book. The reader is referred to publications by Thomann and Mueller (1987), O'Connor (1988), Orlob (1983), or by the senior author (Krenkel and Novotny, 1980), and to the various manuals describing the published models.

A dynamic (time variable) compartmentalized water quality model must simultaneously solve three basic mathematically described physical–chemical–biological relationships:

1. *The equation of continuity or water volume conservation*, which describes the physical law, stating that a change of water mass within a compartment must be balanced by inflow (inputs) and outflow, or

$$\frac{dV_i}{dt} = Q\mathrm{in}_i - Q\mathrm{out}_i \qquad (12.23)$$

where
V_i = volume of the compartment i
$Q\mathrm{in}_i$ = sum of flows into the compartment from adjoining compartments and from allochthonous sources (waste discharges, runoff, precipitation, etc.)
$Q\mathrm{out}_i$ = sum of all flows leaving the compartment (to downstream compartment, water withdrawn, water lost by evaporation, etc.)

For one-dimensional rectangular channel segments, the equation continuity may be modified to

Modeling Fate of Conventional Pollutants and Nutrients 797

$$\frac{\partial H}{\partial t} = -\frac{1}{B}\frac{\partial Q}{\partial X} \tag{12.24}$$

where
H = depth
B = width of the channel
X = distance

The water volume (mass) continuity equation predicts water heights (depths) within the system.

2. *The equation of hydraulic motion*, which is based on Newton's second law of mechanics, stating that forces acting on the body of water are balanced by acceleration of the mass within the body. The forces include internal forces (gravity, pressure, and turbulent friction) and external forces, such as wind drag. Many variations of this equation have been published in the literature, including the St. Venant equation, characteristic equations, and kinematic wave approximation.

   Typically the hydrodynamic equations of motion solve the velocity and flow distribution within the system. The kinematic wave approximation, the most common algorithms in dynamic water quality models describes the propagation of a long wave through a shallow-water system while conserving both momentum (energy) and volume (mass).

3. The *pollutant mass balance equations* for each pollutant and compartment, which can be generally written as (O'Connor, 1988)

$$V_i \frac{dc_{ij}}{dt} = J_{ij} + \Sigma R_{ij} + \Sigma T_{ij} + \Sigma S_{ij} \tag{12.25}$$

where
c_{ij} = concentration of the pollutant j in compartment i
J_{ij} = transport of the pollutant j through the compartment
R = reactions within the system (biological degradation, growth, chemical modification, etc.)
T = transfer from one phase to another (for example, volatilization, biological uptake)
S = allochthonous (boundary) sinks and sources of the substance (for example, settling, atmospheric aeration, wastewater, and nonpoint source inputs).

A dynamic model solves these equations simultaneously assuming a small time interval, Δt, and a small element with a volume ΔV and dimension ΔX. In these numerical solutions, finite-difference terms $\Delta E/\Delta t$

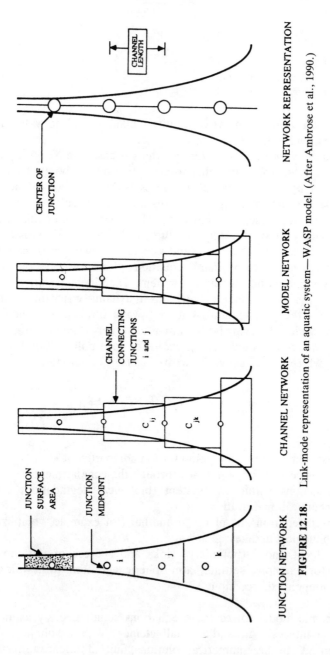

FIGURE 12.18. Link-mode representation of an aquatic system—WASP model. (After Ambrose et al., 1990.)

and $\Delta F/\Delta X$ replace equivalently the derivatives dE/dt and dF/dX, respectively. Herein, E and F are computational variables (depth, concentrations, flow, etc.).

A link-node representation has been developed for such computational schemes (Ambrose et al., 1990). In this representation (Fig. 12.18) the system is divided into a series of interconnected uniform channels (links) connected by junctions (nodes). Each junction is a volumetric unit that acts as a receptacle for the water transported through its connecting channels. In water quality simulations junctions are equivalent to segments in the water quality model, whereas channels correspond to segment interfaces. Channel flows are used to calculate mass transport between the segments in the water quality model. Junction volumes are used to calculate pollutant concentrations within each quality segment. Each segment is assumed to be completely mixed. Figure 12.19 shows that for water quality simulations, the network may be subdivided three-dimensionally, that is, laterally, vertically, as well as longitudinally and

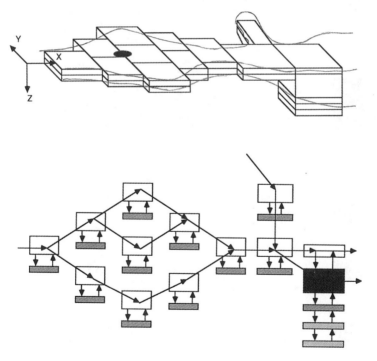

FIGURE 12.19. Layered division of the body of water system—WASP model.

benthic segments can be included along with water column segments. Segment volumes ΔV (= $B \times H \times \Delta X$) and simulation time Δt are directly related. As one decreases or increases, the other must do the same in order to preserve the stability of the solution.

The WASP4 Model

The preceding schematics and modeling concepts are a representation of the U.S. EPA model WASP4 (Ambrose et al., 1990). The schematics of the WASP4 model is shown on Figure 12.20. The WASP4 model consists of two separate units, the hydrodynamics submodel DYNHYD5, which simulates the hydraulics of the water body, and WASP, which simulates the fate of pollutants. The water quality model WASP then

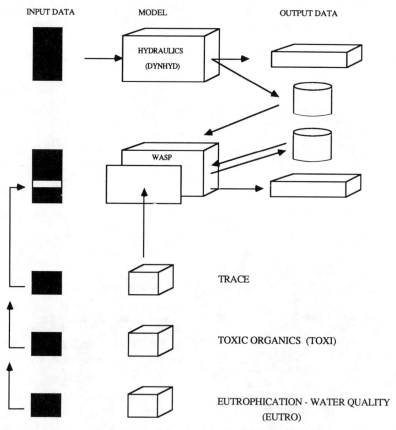

FIGURE 12.20. Block schematics of the WASP model. (From Ambrose et al., 1990.)

includes two kinetic packages: EUTRO4 simulates the dissolved oxygen, nutrients, and eutrophication, and TOXI4 simulates the fate of toxic pollutants. EUTRO4 and TOXI4 may involve both water column and sediment segment simulations. The EUTRO4 and TOXI4 submodels use the output of flows, velocities, and depths from the DYNHYD5 model. Alternatively, the hydrodynamic conditions may be specified by the user. Only concepts of the EUTRO4 submodel will be discussed in this chapter. The fate of the toxic compounds described in the TOXI4 submodel is briefly introduced in the next chapter.

Modeling eutrophication and dissolved oxygen. The separate steady-state models for dissolved oxygen, nutrients (nitrogen and phosphorus)

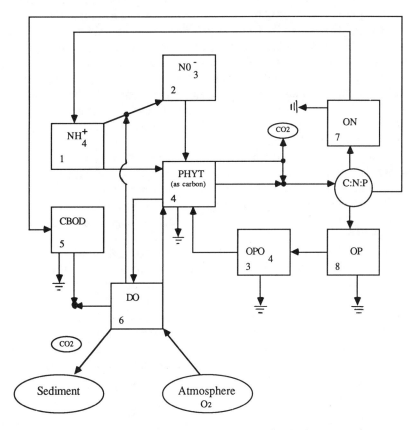

FIGURE 12.21. EUTRO4 coupled modeling concept of dissolved oxygen and eutrophication simulation. *Key*: 1. Ammonia—N; 2. nitrate—N; 3. orthophosphate—P; 4. phytoplankton biomass; 5. biodegradable organics (BOD); 6. dissolved oxygen; 7. organic nitrogen; 8. organic phosphorus. (From Ambrose et al., 1990.)

...phication (phytoplankton or chlorophyll-*a* densities) discussed ... preceding sections are replaced in the WASP4 model by a coupled dynamic system shown on Figure 12.21. The eutrophication model of WASP is called EUTRO4 and it is a simplified version of the Potomac Eutrophication Model (PEM) (Thomann and Fitzpatrick, 1982; Thomann and Mueller, 1987).

As shown on Figure 12.21, the pollutant mass balance equation (Eq. (12.25)) in the EUTRO4 submodel is replaced by a set of coupled differential equations that includes BOD balance, ammonia, nitrite, nitrate, dissolved oxygen, phosphate, sediment, and chlorophyll-*a*, phytoplankton biomass. Six levels of modeling complexity are then identified by the creators of the model: (1) Simple BOD–DO steady-state balance (Streeter–Phelps equation), (2) modified Streeter–Phelps equation (similar to Eqs. (12.7) and (12.8)), (3) full linear dissolved oxygen balance, (4) simple eutrophication kinetics, (5) intermediate eutrophication kinetics, and (6) intermediate eutrophication kinetics with benthos.

Benthos and suspended sediment interactions may play an important role in the eutrophication process. Consider, for example, both ammonia and phosphorus, which can exist in the water environment as adsorbed on sediments (particulate form) (in this form they are not available for algal growth) and in available dissolved form.

Sediment–water interactions can be considered at two levels of complexity: in the first level the user specifies the fluxes of ammonia, phosphate, and SOD in the sediment–water interface. These inputs may be specified as either constant or time variable (seasonal). In the second level these fluxes are calculated from a modeling diagenesis process in the sediment layer.

The particulate nutrients as well as produced organic matter may settle in reaches with low velocity. Subsequently the nutrients stored in benthos can be released back into the water column. The decomposition of organic matter in benthic sediment can have a profound effect on the concentration of the dissolved oxygen and nutrients in the overlying water. The diagenesis processes thus determine the sediment oxygen demand.

The growth of phytoplankton is stimulated by the levels of available nutrients (ammonia, nitrate, and dissolved phosphates) and by light. The reaction rates are affected by temperature.

pH AND ACIDITY

The pH that expresses the molar concentration of the hydrogen ion as its negative logarithm ($pH = -\log[H^+]$) is one of the primary indicators used for evaluation of surface-water quality and the suitability for various beneficial uses. Most aquatic biota are sensitive to pH variations. Fish-kills and reduction and change of other species result when the pH is altered outside their tolerance limits. Biochemical degradation and transformation (for example, nitrification) are also pH sensitive and diminish when the pH reaches acidic levels. Most of the aquatic species prefer a pH near neutral but can withstand a pH in the range of about 6 (7 for nitrifiers) to 8.5.

The toxicity of many compounds can also be altered if the pH is changed. The solubility of many metals as well as other compounds is affected by pH, resulting in increased toxicity in the lower pH range. As pointed out in Chapter 6, leaching of some metals (aluminum) from soils caused by elevated acidity of precipitation represents a "time bomb" and a threat to sustainability of soil resources.

A change of pH and acidity of surface waters resulting from diffuse inputs can occur mainly from two sources: (1) acid precipitation, and (2) mine acid drainage. Both sources have similar origins and have been explained in previous chapters. Acid drainage from mines is a result of mine water being in contact with sulfur-bearing minerals, while acidity of precipitation is caused by atmospheric sulfur and nitrogen oxide emissions. Oxidation and hydrolysis of these compounds in surface or atmospheric water produces sulfuric and nitric acids, which then dissociate to H^+, SO_4^{--}, and NO_3^- ions.

Acid rain, which is defined as rain with a pH of less than 5.6, is a result of sulfuric (SO_2) and nitrate (NO_x) emissions from urban, industrial, transportation (automobile), and electric utility fuel-burning operations that use sulfur- and nitrogen-containing fuels. The process of acid rain formation was explained in Chapter 4. The lethal and sublethal effects of acid rain or acid drainage from mines have been noticed both in North America and Europe (Scandinavia) (Almer et al., 1974; Beamish and Harvey, 1972; Likens and Borman, 1974). Undesirable "oligotrophication" (a severe loss of productivity by low pH conditions) and fish-kills are the most visible and dangerous consequences of acidification. Loss of the natural fish population due to acidic rain and snowmelt inputs in the lakes of New York's Adirondack Mountains and many other pristine lakes of North America (primarily Canada) and Scandinavia have been documented and widely publicized. The damage to the fish population of these lakes was brought about by both long-term exposure to low pH and short-term pH shocks by runoff and snowmelt events. Table 12.10 shows the biological effects of decreased pH on lakes.

Many watersheds and surface-water systems have a natural ability to neutralize the excess acidity (although it was pointed out in Chapter 6 that this buffering may be limited and may be exhausted in some places if

TABLE 12.10   Biological Impact of Acidification

pH Range	General Biological Effects
6.5 to 6.0	Small decrease in species diversity of plankton and benthic invertebrate communities but no measurable change in total community abundance or production. Some adverse effects may be noticed for highly acid-sensitive fish species (e.g., fathead minnow, striped bass).
6.0 to 5.5	Loss of sensitive species of minnows and dace; in some waters decreased reproduction of lake trout and walleye. Visual accumulation of filamentous green algae in the littoral zone. Distinct decrease in species diversity of plankton and benthic invertebrate composition, but no appreciable decrease in total community biomass or productivity.
5.5 to 5.0	Loss of several important sport fish species, including lake trout, walleye, rainbow trout, and smallmouth bass, as well as additional nongame species such as creek chub. Further increase in the extent and abundance of filamentous green algae in lake littoral areas and streams. Continued shift in the species composition and decrease in species diversity of plankton, periphyton, and benthic invertebrate communities, decrease in total abundance and biomass of benthic invertebrates and zooplankton in some waters. Inhibition of nitrification.
5.0 to 4.5	Loss of most of fish species, including most important sport fish species, such as brook trout and Atlantic salmon, few fish species able to survive and reproduce below pH 4.5. Measurable decline in the whole system rates of decomposition of some forms of organic matter, potentially resulting in decreased rates of nutrient cycling. Substantial decrease in the number of species of zooplankton and benthic invertebrates and further decline in the species richness of the phytoplankton and periphyton communities, measurable decrease in the total community biomass of zooplankton and benthic invertebrates in most waters. Reproductive failure of some acid-sensitive species of amphibians such as salamanders and some frogs.

Source: After Baker et al. (1990).

the acidic inputs remain unabated or even increase in the future). During the overland flow, rain and snowmelt dissolve calcium- and magnesium-containing rocks, leach aluminum from soils, and are enriched by mineral and organic salts, such as phosphates and humates. In urban areas, buffering is provided by the infrastructure, as is pointed out in Chapter 8. The elutriated constituents from the soils, minerals, and infrastructure often provide enough buffering capacity to maintain the pH of surface runoff within the acceptable ranges. At this time, rain acidity does not seem to have a noticeable adverse effect on larger bodies of water, such as the Great Lakes, that have elevated alkalinity, hardness, and salt content. However, it has also been pointed out that leached dissolved aluminum (not a problem in nonacidic bodies of water) can reach toxic levels in acidic bodies of water (Baker and Schofield, 1982).

The ability of surface waters to neutralize acidic inputs depends primarily on carbonate (CO_3^{2-}) and bicarbonate (HCO_3^-) content that is expressed as *alkalinity*. Some North American and Scandinavian lakes are particularly sensitive to acidic inputs. These lakes have watersheds underlain by siliceous hard rocks, such as granite, some gneisses, quartzites, and quartz sandstones. These materials are resistant to weathering and produce waters that contain very low concentrations of neutralizing compounds (alkalinity less than 30 mg $CaCO_3$/l). When acid rain falls on such watersheds the acids are not neutralized during the overland flow, and streams and lakes become acidified. A relationship between suphur loads in acidic rainfall and the acidity of Swedish lakes is shown in Figure 12.22.

Areas of highly siliceous bedrock are widespread on the Precambrian Fennoscandian Shield in Scandinavia, the Rockies, the Canadian Shields, New England, the Adirondack Mountains of New York, the Appala-

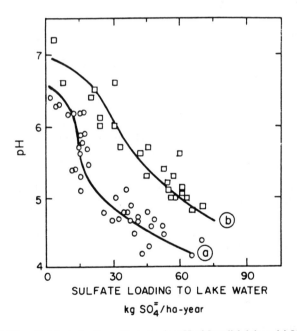

FIGURE 12.22. Sulphate loading effects on the pH of Swedish lakes. (a) Very sensitive lake systems. (b) Less sensitive lake systems. (Reprinted from Glass, Glass, and Renie [1979] with permission; copyright 1979 by the American Chemical Society. Data from National Environmental Protection Board, Jolna, Sweden.)

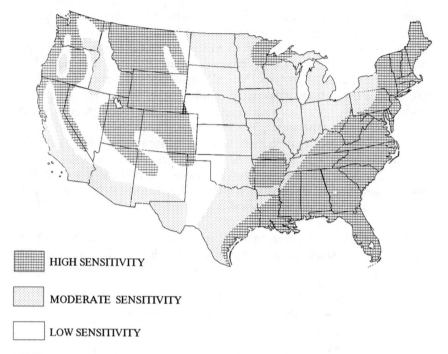

FIGURE 12.23. Areas of the conterminous United States sensitive to acid deposition based on soils, climatic conditions, geology, and types of vegetation. (*Source*: U.S. Environmental Protection Agency.)

chians, and smaller areas elsewhere. These areas contain lakes that are very sensitive to acid rainfall inputs (Fig. 12.23).

As reported by Likens et al. (1979) the acidification of thousands of freshwater lakes and streams in southern Norway has affected fish populations in a 33,000-km² area. As shown in Table 12.10, other adverse effects attributed to acidification must also be considered. Effects of acidification and acid deposition on surface-water bodies were also discussed by Haines (1981), Huckabee et al. (1989), Stokes (1986), and others.

BACKGROUND (NATURAL) WATER QUALITY

As defined in Chapters 1 and 7, background water quality represents the chemical and biological composition of surface waters that would result from natural causes and factors. However, some natural impacts may be transient and devastating to water quality, such as the eruption of Mount

St. Helen, which had afterward deposited an enormous amount of ash in the streams and lakes of the U.S. Northwest. Hence, for planning purposes and water quality evaluations one should consider more or less steady processes and/or common expectable events.

Approaching the natural or background water quality is also the goal of integrated water pollution abatement planning. It is impossible and legally unenforceable to eliminate all sediment from rivers, as well as it is impossible to bring BOD down to zero or oxygen to saturation values, notwithstanding the fact that such measures could have severe water quality and other consequences. Estimating the background (natural) water quality is also a key in use attainability studies since, legally, use-based water quality standards cannot be enforced if their violation is caused by natural causes (see Chapter 16).

The U.S. Geological Survey has established the National Hydrologic Benchmark Network of monitoring stations. These freshwater stations are located in watersheds that are among the least affected by the activities of man that can be found across the United States. Most of these stations are located in national parks, wilderness areas, state parks and forests, and similar areas protected from development. Water quality measured at these stations may provide the best approximation within the ecoregional context of the background (natural) levels of contaminants in U.S. surface waters.

A study by the Midwest Research Institute compiled the data of the average concentrations of various pollutants from undisturbed streams. The values from 57 stations composing the National Hydrologic Benchmark Network showed distinct regional distribution as summarized in Table 12.11. However, it should be noted that the ranges for BOD_5, nitrate, suspended solids, phosphate, and bacteria reported in the table are based on approximately only one station per state.

TABLE 12.11 Approximate Regional Natural Water Quality–Average Annual Concentrations

Parameter	U.S. Regions				
	Eastern	Midwest	Great Plains	Mountains	Pacific
Suspended sediment (mg/l)	5–10	10–50	20–100	5–20	2–5
BOD_5 (mg/l)	1.0	1–3	2–3	1–2	1
Nitrate N (mg/l)	0.05–0.2	0.2–0.5	0.2–0.5	0.1	0.05–0.1
Total P (mg/l)	0.01–0.02	0.02–0.1	0.1–0.2	0.05	0.05–0.1
Total coliforms (MPN/100 ml)	100–1000	1000–2000	500–2000	100	100–500

Source: From McElroy et al. (1976).

Background levels of some toxic metals and background mineral contributions were discussed in Chapter 7. It must be remembered that there are no background levels for organic chemicals (pesticides, PAHs, halogenated hydrocarbons, volatile organic chemicals, etc.) since these substances are man-made and are alien to nature. The apparent background levels of these substances mostly originate from atmospheric fallout.

The natural levels of biodegradable organic pollution as well as sediment loads depend on the type of stream, the character of the basin, and geographical regions. There are four general types of undisturbed types of native lands (excluding mountains): *arid land (including deserts), prairie, wetland,* and *woodland.* These four basic types have various forms, depending on the geographical location and elevation. It is obvious that the loads of constituents and background water quality in these four natural lands will vary but there are some common characteristics:

1. Streams from *arid lands* are often ephemeral and typically have very high sediment content during intense but infrequent storm events. The salinity of these streams may also be elevated. Nutrient (N and P) content is very low.
2. *Prairie streams* have elevated solids content during wet-weather flow events. Nutrient loading is low. Actually prairies represent a sink of atmospheric nutrient loads.
3. *Forested land*
   a. *Mountain forests* generally exhibit the best water quality; also they have low mineral and almost no organic contents.
   b. *Lowland forest* streams have higher organic content. Often these streams originate in wooded wetlands and water may contain residues of organic decomposition occurring in soils and wetlands. These streams would have measurable BOD and COD contents.
4. *Headwater wetlands,* that is, streams draining wetlands and wetland bodies of water themselves, have higher organic content and may also have low dissolved oxygen content caused by the decomposition of organic matter. Humic substances increase color and turbidity of these water bodies.

Northern wetlands have two distinct seasons. During the productive season, wetlands are sinks of nutrients and other water contaminants. During the dormant season, nutrients and contaminants may be released. Typically stream flow from northern wetlands is more acidic, contains organic acids and appreciable levels of nutrients and other contaminants, and has relatively low dissolved oxygen levels (Table 12.12). Natural

TABLE 12.12 Average Concentration of Contaminants in Runoff from Natural Minnesota Peatlands

Characteristic	Units	Peatland Type		
		Bog	Transition	Fen
Dissolved oxygen	mg/l	6.4	5.4	5.6
O_2 saturation	%	56	50	53
pH		5.6	6.5	7.0
Alkalinity as $CaCO_3$	mg/l	10	29	75
Color	mg/l	311	260	242
Aluminum	mg/l	0.55	0.25	0.25
Mercury	µg/l	6	3	5
Arsenic	µg/l	2	3	3
Selenium	µg/l	1	1	1
Total phosphorus	mg/l	0.06	0.05	0.08
Total Kjeldahl nitrogen[a]	mg/l	1.5	1.6	1.8
Nitrate—N	mg/l	0.06	0.05	0.08
Humic acid	mg/l	11	9	9
Fulvic acid	mg/l	100	104	97
Chemical Oxygen Demand	mg/l	118	104	97

Source: From Clausen and Brooks (1983).

[a] Ammonia plus organic nitrogen.

wetlands can be classified as bogs or fens (see Chapter 14). Bogs are acidic and nutrient poor because they are isolated from regional ground water, while fens are nourished by regional ground water. Many wetlands are transitional (Clausen and Brooks, 1983).

It is important to note that the background (natural) contaminant loads are not related to the designated water use of the body, while the existing receiving water quality standards are. The natural water quality and constituent loads can be related to morphological, geological, and geographical characteristics, land cover, soil type, and other ecological factors. *Ecoregions* represent regions with the same natural water quality and ecological characteristics (Omernik, 1987; Gallant et al., 1989). Figure 12.5 showed the distribution of the major ecoregions of the conterminous United States. Apparently, natural (background) water quality within an ecoregion will vary less than between the ecoregions.

Generally, natural background water quality should be measured rather then estimated. In many ecoregions there are still some headwater stream sections that may be considered as unaffected or undisturbed by man's activities. As pointed out in the section of this chapter titled "The Acquatic Ecosystem," these stream sections can then be considered as *reference stream sections.* Background water quality represents the limit of the waste assimilation process that polluted waters should eventually approach during recovery.

References

Almer, B., C. Ekstrom, E. Hornstorm, and V. Miller (1974). Effect of acidification on Swedish lakes, *Ambio* **3**:30–36.

Ambrose, R. B., Jr., T. A. Wool, J. P. Connolly, and R. W. Schranz (1990). *WASP4, A Hydrodynamic and Water Quality Model—Model Theory, User's Guide, and Programmer's Guide*, revision 1, Environmental Research Laboratory, U.S. Environmental Protection Agency, Athens, GA.

Anonymous (1987). *A Water Resources Management Plan for the Milwaukee Harbor Estuary*, Southeastern Wisconsin Regional Planning Comm., Waukesha, WI.

Baker, J. P., and C. L. Schofield (1982). Aluminum toxicity to fish in acidic waters, *Water Air Soil Pollut.* **18**:289–309.

Baker, J. P., D. P. Bernard, S. W. Christensen, and M. J. Sale (1990). Biological effects of changes in surface water acid-base chemistry, in *National Acid Precipitation Assessment Program, Acidic Deposition: State of Science and Technology*, Vol. II, September 1990, State of Science and Technology Report 13.

Beamish, R. J., and H. H. Harvey (1972). Acidification of the La Cloche Mountain Lakes: Resulting fish mortalities, *J. Fish Res. Board Can.* **29**(8):1131–1143.

Bowman, B. G. T., and J. J. Delfino (1980). Sediment oxygen demand techniques: A review and comparison of laboratory and in-situ systems, *Water Res.* **14**(5):491–500.

Carlson, R. E. (1977). A trophic state index for lakes, *Limnol. Oceanogr.* **22**: 361–369.

Churchill, M. A., H. L. Elmore, and R. A. Buckingham (1962). The prediction of stream reaeration rate, in *Advances in Water Pollution Research*, Vol. 1, Proceedings, IAWPR International Conference, London, Pergamon Presss, Oxford.

Clausen, J. C., and K. N. Brooks (1983). Quality of runoff from Minnesota peatlands. 1. A characterization, *Wat. Resour. Bull.* **19**(5):763–767.

Committee on Restoration of Aquatic Ecosystems (1992). *Restoration of Aquatic Ecosystems*, National Academy Press, Washington, DC.

DiToro, D. M. (1986). A diagenetic oxygen equivalent model of sediment oxygen demand, in *Sediment Oxygen Demand, Processes, Modeling, and Measurement*, K. J. Hatcher, ed., University of Georgia Press, Athens, GA, pp. 171–208.

DiToro, D. M., P. R. Paquin, K. Subburamu, and D. A. Gruber (1990). Sediment oxygen demand model: Methane and ammonia oxidation, *J. Environ. Eng. ASCE* **116**(5):945–986.

Edberg, N., and B. V. Hofsten (1973). Oxygen uptake of bottom sediments studied in situ and in the laboratory, *Water Res.* **7**(9):1285–1294.

Fair, G. M., E. W. Moore, and H. A. Thomas, Jr. (1941). The natural purification of river muds and pollutional sediments, *Sew. Works J.* **13**:270–307, 756–778.

Fausch, K. D., J. R. Karr, and P. R. Yant (1984). Regional application of an index of biotic integrity based on stream-fish communities, *Transact. Am. Fish. Soc.* **113**:39–55.

Field, R., and R. Turkeltaub (1981). Urban runoff receiving water impact: Program overview, *J. Environ. Eng. Div. ASCE* **107**(EE1):83–100.

Filos, J., and A. Molof (1972). Effect of benthal deposits on oxygen and nutrient economy of flowing waters, *J. WPCF* **44**(4):644–662.

Foree, E. G., W. J. Jewell, and P. L. McCarty (1971). The extent of nitrogen and phosphorus regeneration from decomposing algae, in *Proceedings, Fifth International Conference on Water Pollution Research*, IAWPR, San Francisco, CA, Pergamon Press, Oxford.

Gallant, A. L., et al. (1989). *Regionalization as a Tool for Managing Environmental Resources*, EPA 600/3-89/060, U.S. Environmental Protection Agency, Corvallis, OR.

Gardinier, R. D., M. T. Auer, and R. P. Canale (1984). Sediment oxygen demand in Green Bay (Lake Michigan), in *Proceedings, 1984 Specialty Conference of ASCE*, American Society of Civil Engineers, New York, pp. 514–519.

Glass, N. R., G. E. Glass, and P. J. Renie (1979). Effect of acid precipitation, *Environ. Sci. Technol.* **13**:1350–1355.

Haines, T. A. (1981). Acidic precipitation and its consequences for aquatic ecosystems: A review, *Trans. Am. Fish. Soc.* **110**:669–706.

Heaney, J. P., and W. C. Huber (1984). Nationwide assessment of urban runoff impact on receiving water quality, *Water Resour. Bull.* **20**(1):35–42.

Huckabee, J. W., et al. (1989). An assessment of the ecological effects of acidic deposition, *Arcg. Environ. Contam. Toxicol.* **18**:3–27.

Hutchinson, G. E. (1969). Eutrophication, past and present, in *Eutrophication: Causes, Consequences, Correctives*, G. A. Rohlich, ed., National Academy of Sciences, Washington, DC.

Kadlec, R. H. (1988). *Denitrification in Wetland Treatment Systems*, paper presented at 61st Conference of the Water Pollution Control Federation, Session 26, Dallas, TX.

Karr, J. R., K. D. Faush, P. L. Angermeier, P. R. Yant, and I. J. Schlosser (1986). *Assessing Biological Integrity of Running Waters. A Method and Its Rationale*, Illinois Natural History Survey, Spec. Publ. No. 5, Champaign, Ill.

Karr, J. R., and D. R. Dudley (1981). Ecological perspectives on water quality goals, *Environ. Manage.* **5**:55–68.

Keefer, T. N., R. K. Simons, and R. S. McQuivey (1979). Dissolved oxygen impact from urban storm runoff, EPA 600/2-79-156, U.S. Environmental Protection Agency, Cincinnati, OH.

Keeney, D. R. (1973). The nitrogen cycle in sediment-water systems, *J. Environ. Qual.* **2**(1):15–19.

Keeney, D. R., S. Schmidt, and C. Wilkinson (1975). *Concurrent Nitrification-Denitrification at the Sediment-Water Interface as a Mechanism for Nitrogen Losses from Lakes*, Tech. Rep. No. 75-07, Water Resources Center, University of Wisconsin, Madison, WI.

Kolkwitz, R., and M. Marson (1908). Ökologie der pflanzlichen Saprobien, *Berl. Deutscher Botan. Ges.* **261**:505–519.

Krebs, C. J. (1989). *Ecological Methodology*, Harper and Row, New York.
Krenkel, P. A., and V. Novotny (1980). *Water Quality Management*, Academic Press, New York.
Krenkel, P. A., and G. T. Orlob (1962). Turbulent diffusion and the reaeration coefficient, *J. Sanit. Eng. Div. ASCE* **88**(3):53–83.
Kreutsberger, W. A., T. L. Meinholz, M. Harper, and J. Ibach (1980). Predicting the impact of sediments on dissolved oxygen concentrations following combined sewers overflow events in the Lower Milwaukee River, *J. WPCF* **52**: 192–201.
Lee, G. F., W. Rust, and R. A. Jones (1978). Eutrophication of water bodies: Insight of an age-old problem, *Environ. Sci. Technol.* **12**:900–908.
Likens, G. E., and F. H. Borman (1974). Acid rain: A serious regional environmental problem, *Science* **184**:1176–1179.
Likens, G. E., R. F. Wright, J. N. Galloway, and T. J. Butler (1979). Acid rain, *Sci. Am.* **241**:43–51.
McDonnell, A. J., and S. D. Hall (1969). Effect of environmental factors on benthal oxygen demand, *J. WPCF* (res. supplement) **41**:353–363.
McElroy, A. D., S. Y. Chiu, J. W. Nebgen, A. Aleti, and F. W. Bennett (1976). *Loading Functions for Assessment of Water Pollution from Nonpoint Sources*, EPA 600/2-76/151, U.S. Environmental Protection Agency, Washington, DC.
Mason, C. F. (1991). *Biology of Freshwater Pollution*, 2d ed., Longman Sci. & Technical, Wiley, New York.
Meadows, M. E., D. W. Weeter, and J. M. Green (1978). Assessing nonpoint water quality for small streams, *J. Environ. Eng. Div. ASCE* **104**(EE6):1119–1133.
Ministry of the Environment (1991). *Environment of the Czech Republic*, Publishing House EkoCentrum, Brno, Czechoslovakia.
Myers, C. F., J. Meek, S. Tuller, and A. Weinberg (1985). Nonpoint sources of water pollution, *J. Soil Water Conserv.* **40**(1):14–18.
Niedzialkowski, D., and D. Athayde (1985). Water quality data and urban nonpoint source pollution: The Nationwide Urban Runoff Program, in *Perspectives on Nonpoint Source Pollution*, Proceedings, Nationwide Conference, U.S. Environmental Protection Agency, Washington, DC, pp. 437–441.
Nonpoint Source Control Task Force (1983). Nonpoint source pollution abatement in the Great Lakes basin—An overview of post-PLUARG development, International Joint Commission, Windsor, Ontario.
North American Lake Management Society—NALMS (1990). *The Lake and Reservoir Restoration Guidance Manual*, EPA 440/4-90-006, U.S. Environmental Protection Agency, Washington, DC.
Novotny, V. (1985). Urbanization and nonpoint pollution, in *Proceedings, Fifth World Congress on Water Resources*, International Water Resources Association, Paper No. 17, Asp. 6, Brussels, Belgium.
Novotny, V., and G. Bendoricchio (1989). Water quality—Linking nonpoint pollution and deterioration, *Water Environ. Technol.* **1**(Nov.):400–407.
Novotny, V., and P. A. Krenkel (1975). A waste assimilative capacity model for a shallow, turbulent stream, *Water Res.* **9**:233–241.

Novotny, V., and S. Zheng (1988). Impact of nonpoint pollution on a Great Lakes freshwater harbor estuary, in *Proceedings, Spec. Group Seminar, Water Quality Impact of Storm Sewage Overflows on Receiving Waters*, J. B. Ellis, ed., 14th Biennial Conf., IAWPRC, Brighton, England, pp. 27–36.

Novotny, V., K. Imhoff, M. Olthof, and P. Krenkel (1989). *Karl Imhoff's Handbook of Urban Drainage and Wastewater Disposal*, Wiley, New York.

O'Connor, D. J. (1960). Oxygen balance of an estuary, *J. Sanit. Eng. Div. ASCE* **86**(SA3):35–55.

O'Connor, D. J. (1988). Models of sorptive toxic substances in freshwater systems, *J. Environ. Eng. Div. ASCE* **114**(3).

O'Connor, D. J., and W. E. Dobbins (1958). Mechanism of reaeration in natural streams, *Trans. ASCE* **123**:641–666.

Odum, H. T. (1956). Primary productivity of flowing streams, *Limnol. Oceanogr.* **1**(2):102–117.

Ohio Environmental Protection Agency (1987). *Biological Criteria for the Protection of Aquatic Life*, Ohio EPA, Columbus, OH.

Omernik, J. M. (1987). Ecoregions of the conterminous United States, *Ann. Assoc. Am. Geol.* **77**(1):118–125.

Orlob, G. T., ed. (1983). *Mathematical Modeling of Water Quality*, Wiley, New York.

Owens, M., R. W. Edwards, and J. W. Gibbs (1964). Some reaeration studies in streams, *Int. J. Air Water Pollut.* **8**:469.

Painter, H. A. (1970). A review of literature on inorganic nitrogen metabolism in microorganisms, *Water Res.* **4**:393–450.

Paterson, D. J., E. Epstain, and J. McEnvoy (1975). *Water Pollution Investigation: Lower Green Bay and Lower Fox River*, EPA 600/3-77/105, U.S. Environmental Protection Agency, Washington, DC.

Phelps, E. B. (1944). The oxygen balance, in *Stream Sanitation*, Wiley, New York.

Plafkin, J. L., M. T. Barbour, K. D. Porter, S. K. Gross, and R. M. Hughes (1989). *Rapid Bioassessment Protocols for Use in Streams and Rivers: Benthic Macroinvertebrates and Fish*, EPA/444/4-89-001, U.S. Environmental Protection Agency, Washington, DC.

Rankin, E. T., C. O. Yoder, and D. Mishne (1990). *Ohio Water Resources Inventory*, executive summary and Vol. 1, Ohio Environmental Protection Agency, Columbus, OH.

Robertson, C. K. (1979). Quantitative comparison of the significance of methanogenesis in the carbon cycle of two lakes, *Arch. Hydrobiol. Ergebnis Limnol.* **12**:123–135.

Rohlich, G. A. (1969). *Eutrophication: Causes, Consequences, Correctives*, National Academy of Science, Washington, DC, pp. 3–7.

Ruane, R. J., and P. A. Krenkel (1975). Nitrification and other factors affecting nitrogen in the Holston River, in *Proceedings, IAWPR Conference on Nitrogen as a Water Pollutant*, Copenhagen, Denmark.

Rudd, J. W. M., and C. D. Taylor (1980). Methane cycling in aquatic environments, in *Advances in Aquatic Microbiology*, M. R. Droop and H. W. Jannach,

eds., Vol. 2, Academic Press, New York, pp. 77–150.
Sawyer, C. H. (1947). Fertilization of lakes by agricultural and urban drainage, *New Engl. Water Works Assoc.* **61**:109–127.
Schindler, D. W. (1985). The coupling of elemental cycles by organisms: Evidence from whole lake chemical perturbations, in *Chemical Processes in Lakes*, W. Stumm, ed., Wiley, New York, pp. 225–250.
Sládeček, V. (1979). Continental systems for the assessment of river water quality, in *Biological Indicators of Water Quality*, A. James and L. Evison, eds., Wiley, Chichester, England, New York, 3.1–3.32.
Sloey, W. E., and F. L. Spangler (1978). Trophic status of the Winnibego pool lakes, in *Proceedings, Second Annual Conference, Wisconsin Section, AWRA*, WRC-Univ. of Wisconsin, Madison.
Sonnen, M. B. (1980). Urban runoff quality: Information needs, *J. Tech. Counc. ASCE* **106**(TC1):29–41.
Stokes, P. (1986). Ecological effects of acidification on primary producers in aquatic ecosystems, *Water Air Soil Pollut* **30**:421–424.
Streeter, H. W., and E. B. Phelps (1925). *A Study of the Pollution and Natural Purification of the Ohio River*, Public Health Bull. 146, U.S. Public Health Service, Washington, DC.
Stumm, W., ed. (1985). *Chemical Processes in Lakes*, Wiley, New York.
Stumm, W., and J. J. Morgan (1962). Stream pollution by algal nutrients, in *Transactions, Twelfth Annual Conference of Sanitary Engineers*, University of Kansas.
Thomann, R. V. (1972). *Systems Analysis and Water Quality Management*, McGraw-Hill, New York.
Thomann, R. V., and J. J. Fitzpatrick (1982). *Calibration and Verification of a Mathematical Model of the Eutrophication of the Potomac Estuary*, prepared for Department of Environmental Services, Government of District of Columbia, Washington, DC.
Thomann, R. V., and J. A. Mueller (1987). *Principles of Surface Water Quality Modeling and Control*, Harper & Row, New York.
Thompson, P., R. Adler, and J. Landman (1989). *Poison Runoff*, National Resource Defense Council, Washington, DC.
Torno, H. C. (1984). The nationwide urban runoff program, in *Proceedings, Third International Conference on Urban Storm Drainage*, Chalmers University of Technology, Göteborg, Sweden.
Tuffey, T. J., J. V. Hunter, and V. A. Matulewich (1974). Zones of nitrification, *Water Resour. Bull.* **10**(3):555–564.
U.S. Environmental Protection Agency (1974). *The Relationships of Phosphorus and Nitrogen to the Trophic State of Northeast and North-central Lakes and Reservoirs*, National Eutrophication Survey Working Paper No. 23, U.S. EPA, Washington, DC.
U.S. Environmental Protection Agency (1983). *Results of the Nationwide Urban Runoff Program*, Vol. 1. *Final Report*, WH-554, Water Planning Division, U.S. EPA, Washington, DC.

U.S. Environmental Protection Agency (1987). *Quality Criteria for Water 1986*, U.S. EPA, Washington, DC.

U.S. Environmental Protection Agency (1990). *National Water Quality Inventory—1988 Report to Congress*, EPA 440-4-90-003, Office of Water, U.S. EPA, Washington, DC.

Velz, C. J. (1984). *Applied Stream Sanitation*, 2d ed., Wiley, New York.

Vollenweider, R. A. (1975). Input-output models with special reference to the phosphorus loading concept in limnology, *Schweiz. Z. Hydrol.* **37**:53–83.

Vollenweider, R. A. (1976). Advances in defining critical loading levels for phosphorus in lake eutrophication, *Mem. Ist. Ital. Hydrobiol.* **33**:53–83.

Walker, R. R., and W. J. Snodgrass (1986). Model for sediment oxygen demand in lakes, *J. Environ. Eng. ASCE* **112**(1):25–43.

Wezernak, C. T., and J. J. Gannon (1968). Evaluation of nitrification in streams, *J. Sanit. Eng. Div. ASCE* **94**(SA5):883–895.

Žáková, Z. (1989). Phytoplankton of reservoirs in relation to the trophic potential of inflow water, *Arch. Hydrobiol. Beih. Ergebn. Limnol.* **33**:373–376.

13

Toxic Pollution and Its Impact on Receiving Waters

The presence of toxic substances, such as organic and inorganic chemicals, heavy metals and radionuclides, has become a major environmental problem in recent years. The substances are present in varying degrees in all phases of the environment, air, water and land. They are transferred between and among these media, undergo transformation within each and accumulate in viable and nonviable constituents.

Donald J. O'Connor, Manhattan College, Bronx, NY

TOXICITY CONCEPTS

Surface waters are the primary recipients of waste materials. As long as the consequences of waste discharges are not injurious to the uses of the body of water or otherwise do not impair its integrity, then the body of water is not polluted. The uses that can be adversely affected by toxic pollution include, for example, water supply, fishing, irrigation, navigation, or recreation. The injury of the integrity of the body of water implies a harm done to the aquatic ecosystem (biota within the system including fish, insects, planktonic microorganisms, and aquatic plants). In order to protect the uses or the aquatic ecosystem from injury the contaminant levels must be kept below a certain "safe" threshold. However, almost any compound, including even kitchen salt (NaCl) has a toxic threshold which, if exceeded, will cause harm to the aquatic ecosystem and/or man (Krenkel and Novotny, 1980). But the same example of toxicity of sodium chloride clearly shows that the toxicity threshold varies among the species. Freshwater organisms would die in salt or even brackish environment and, vice versa, saltwater organisms would die if salinity is reduced. Diversion of Mississippi River flood waters into brackish Louisiana coastal bodies of water during high floods results in die-off of shell fish accustomed to the higher salt concentrations of these waters. Hence, in some cases organisms can adapt to higher concentrations of some potentially toxic compounds.

The presence and exposure of organisms to the toxic levels of contaminants can have effects such as fish-kills and disease and contamination of fish flesh to levels that make fish unfit for human consumption. In addition toxic materials can damage or stop the biological processes occurring in the aquatic ecosystems, including long-term inhibition of growth, reproduction, and migration of organisms, and adverse effects on the rate of degradation of biodegradable contaminants.

The definition of toxicity in the Clean Water Act (Section 502(13)) is very specific:

> The term "toxic pollutants" means those pollutants, or combinations of pollutants, including disease-causing agents, which after discharge and upon exposure, ingestion, inhalation or assimilation into any organism, either directly from the environment or indirectly by ingestion through food chains, will, on the basis of information available to the administrator, cause death, disease, behavioral abnormalities, cancer, genetic mutation, physiological malfunctions (including malfunctions in reproduction) or physical deformations, in such organisms or their offspring.

This definition of toxicity encompasses essentially all the possible adverse effects of pollutants on all types of organisms (Weber, 1981).

The concept of "biological integrity" was defined in Chapter 12. The biological integrity is related to the habitat of an unimpaired body of water. It also includes such impacts as habitat alteration by channel modification that do not result from toxic discharges or contamination. Typically "toxicity" is determined by bioassays in which test organisms are exposed in laboratory or in-situ (field) conditions to various doses or concentrations of the pollutant, while "biological integrity" is determined by in-situ examination of the biological community itself and comparing it to an unimpacted (reference) biotic composition within the same ecoregion.

Even though almost all water quality constituents can become toxic at high enough levels, there is a definite number of compounds that are either toxic at relatively low levels or at levels that may result from waste discharges. These potentially toxic compounds have been designated by the Environmental Protection Agency as *priority pollutants*. Some of the priority pollutants are *carcinogenic*, that is, they can increase the risk of cancer to the human population and/or to fish. The list of priority pollutants is continuously under review and periodic updating and is growing. At the time of writing this book it contained 129 pollutants, of which 13 were toxic elements (metals) and the remaining were mostly organic chemicals. The priority pollutants are listed in the Appendix B.

In Europe, the European Community has listed the most dangerous toxic substances on the so-called "Black List" and less dangerous pollutants on the "Gray List" (Mason, 1991).

Toxicity is generally considered a man-made (cultural) phenomenon resulting from discharges of contaminants into surface waters. However, sometimes toxicity may be caused by natural phenomena. Hydrogen sulfide or hydroxylamine, both known byproducts of natural processes taking place in natural waters, can be toxic to aquatic organisms. The periodic table includes over 90 elements from hydrogen to transuraniums, and all but 20 can be characterized as metals. As many as 59 of these elemental metals can be classified as "heavy metals" and potentially toxic. However, only 17 of these metals are considered both very toxic and available in places at concentrations that exceed toxicity levels. Of these 17 toxic metals (Table 13.1), nine are being mobilized into the environment by man at rates greatly exceeding those of natural geological processes (Chapman, 1978; Weiss et al., 1975).

Figure 13.1 shows the effect of toxic discharges from point and diffuse sources on human health and the integrity of aquatic life. Almost all toxic organic compounds are man-made and have been introduced since the

TABLE 13.1 Toxic Metals of Particular Environmental Concern

Very Toxic and Readily Accessible	Man-Induced Mobilization Higher than Natural Rate
Co, Bi	Ag^a
Ni, Cu	Cd
Zn, Sn^a	Cu
$Se,^a Te^a$	Hg^a
$Pd,^a Ag^a$	Ni
Cd, Pt^a	Pb^a
$Ag,^a Hg^a$	Sb
$Tl,^a Pb^a$	Sn^a
Sb	Zn

From Chapman (1978).

[a] Metal alkyls stable in aqueous environment, reported to be biomethylated in sediments.

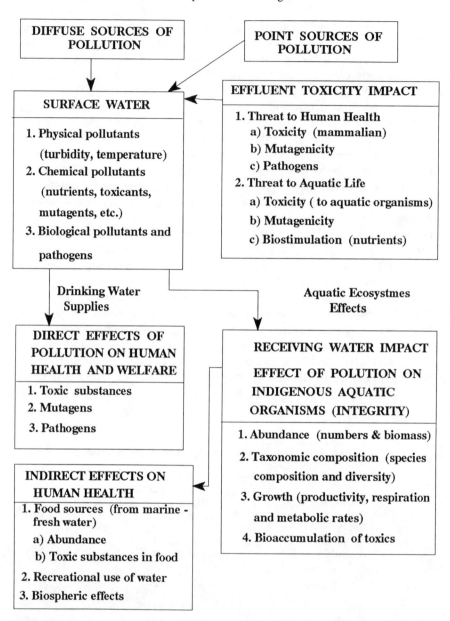

FIGURE 13.1. Impact of toxic emissions on human health and integrity of aquatic biota. (After Weber, 1981.)

1900s. With one or two exceptions, there are no background (natural) levels of these contaminants.

The sources of toxic pollutants are so numerous that an attempt to list them may omit some of them. Sources of diffuse nature have been listed in the preceding chapters. As pointed out in Chapter 7, toxic contamination from mining metals has been occurring for 2000 years. But it was the introduction of toxic organic chemicals during and after World War II that made the distribution of toxics globally widespread and threatening to global and regional ecological systems. Without control the widespread use of toxic compounds by man, including the use of household chemicals, pesticide and fertilizer use in agriculture, industrial emissions, and disposal of waste, may ultimately be threatening to the existence of life on this planet. On the other hand, life in today's society is impossible without the use of chemicals. Many economists claim that high yields of agricultural crops needed to feed an increasing population throughout the world can only be sustained with the use of chemicals. A sustainable use of chemicals that would not pose a harm to bodies of water and to terrestrial ecosystems must emerge in a relatively short time.

The toxic pollutants affecting receiving bodies of water include (Mason, 1991):

1. Metals, such as cadmium, copper, chromium, lead, mercury, nickel, and zinc, that arise from industrial operations, ores, mining, deteriorating infrastructures, traffic, agricultural use, and others sources.
2. Organic compounds, such as pesticides, PCBs, solvents, petroleum hydrocarbons, surfactants, organometallic compounds, phenols, formaldehyde, biochemical methylation of metals in aquatic sediments, etc.
3. Dissolved gases, such as chlorine and ammonium.
4. Anions, such as cyanides, fluorides, sulfides, and sulphates.
5. Acids and alkalis.

TOXICITY AND ITS MEASUREMENT

Toxicity may be defined as an alteration or impairment of the normal functions of organisms caused by an exposure to or ingestion of a compound or mixture of compounds. The effects of contaminants on living organisms are generally evaluated using the basic concepts that were originally formulated by Sprague (1969).

1. *Acute toxicity*. The exposure of organisms to a compound or a mixture of compounds will result in a crisis, usually in a short time during or following the exposure.

2. *Chronic toxicity*. The exposure will have a sublethal damaging impact on the organisms occurring over a longer period of time up to the entire life cycle.
3. *Lethal toxicity*. Exposure will result in death of the organism.
4. *Sublethal toxicity*. Exposure is damaging to the organism, but it will not result in death.
5. *Cumulative toxicity*. The effects on the organisms are brought about, or increase in strength, by successive exposure.

There are two considerations of the toxic levels of contaminants in aquatic water bodies. The first consideration is to protect human health; the second is to protect the well-being of aquatic life. The acute toxicity concept has been applied primarily to aquatic organisms and protection of aquatic biota. Toxicity levels (criteria) for the protection of human health are related to "acceptable" risk to the population of contracting an additional case of cancer (for carcinogenic toxicants) and/or a debilitating disease. These accepted risks are extremely low and below risk levels related to others of man's activities, such as driving a car, flying on a commercial jet plane, participating in sports, or walking on a street.

Toxicity Tests

The toxicity levels of compounds for aquatic organisms have been established by scientists in toxicity bioassay tests. Ideally the test organisms should include representatives from four groups, that is, microorganisms, plants, invertebrates, and fish. At least eight different families of species specified by states are required for a complete test of toxic effects. The test organisms should be amenable to captivity, accurately identified, relatively uniform in size and healthy, and acclimated to laboratory conditions.

The test organisms are placed in containers with various dilutions of the toxicants plus one container with test water only. The number of organisms surviving and/or unaffected (for example, by observing the mobility or respiration rate of the organisms) after the specified time periods (24, 48, 72, 96 hours) is recorded. The data are then used for establishing a functional dose (concentration)–response relationship. It is obvious that in the test the organisms will not respond uniformly to the dose of the toxic compounds because of different sensitivities to the compound. Therefore the most important parameter of interest in the toxicity bioassay test is the dose or concentration at which, after the specified test period, 50% of the test organisms survive or their life functions are not affected by the dose. A dose is defined as the amount of the toxic compound ingested by the test organism divided by the body mass of the organism (for example, mg/kg). In the test of aquatic organisms the toxicity is related to concentration (mg of contaminant/ volume of water, i.e., mg/l or μg/l) rather than to dose because of the inherent difficulty in establishing a dose.

The *lethal dose* or *concentration* (*LD* or *LC*) implies that an exposure of the test organism has resulted in death. The 50% survival dose or concentration value then represents so-called LD_{50} (dose) or LC_{50} (concentration), and it is a representative of the acute toxicity of the compound or of the waste containing a mixture of potentially toxic substances. The time of exposure is important in toxicity studies as well as in toxicity risk assessment studies of toxic discharges into receiving bodies of water. The $LC_{50}(48)$ is the concentration of a toxic material at which 50% of the test organisms died after 48 hours of exposure.

The *effective dose* or *concentration* (*ED* or *EC*) is a term used when other than lethal effects are considered such as impact on reproduction or respiratory stresses. The terms ED_{50} or EC_{50} are then equivalently used to describe such adverse effects in 50% of the test organisms within the prescribed test period.

The *chronic criteria* are not based on a dose–response relationship obtained in an bioassay, but rather on an observed long-term impact of the contaminants on the life functions of organisms. The endpoint of the chronic toxicity test is the *no observed effect concentration* (*NOEC*) and the *low observed effect concentration* (*LOEC*). The NOEC is the highest concentration of toxicant to which the test organisms are exposed that causes no observable effects. The effects measured may include a decrease in reproduction and growth. The LOEC is the lowest concentration of toxicant to which the test organisms are exposed that causes an observed effect.

Review of Toxicity Tests

Currently, the water quality based pollution controls specify either chemically specific water quality limitations (such as those included in Appendix B) or whole body toxicity testing. The whole body toxicity test (conducted both on the effluent and the receiving water) appropriately indicates the effect of the wastes on the aquatic life. Toxicity tests are classified based on the method of adding test solutions and the duration of the test (APHA, 1989):

1. *Static test*. In this test, the test organism remains in the same water for the entire duration of the test. The tested (effluent or receiving body) and dilution waters are mixed in a chamber to the desired concen-

tration. The test organism is then placed in the chamber containing the diluted tested water. The static nature of the test may cause erroneous results under certain conditions. For example, a high BOD content may cause a depletion of dissolved oxygen in the mixture, which will result in the death of the organisms. Also, some toxic compounds may degrade or volatilize during the test, thus distorting the toxicity effect on the test organisms.

2. *Recirculation test.* In this test the mixture in which organisms are residing is pumped through an apparatus, such as a filter, to maintain water quality but not to reduce the concentration of the test material. This type is not routinely used because it is expensive and the results may be distorted.

3. *Renewal test.* The renewal test is similar to the static test except the test solutions and control water are periodically renewed and the test organisms are transferred to chambers with freshly prepared mixtures of test and control water or by replacing test mixtures in the original chambers.

4. *Flowthrough test.* The test solutions and control water flow into and out of the chambers in which the test organisms are maintained. The flow may be intermittent or continuous. Stock solutions of the test material can be continuously mixed with the dilution water in different proportions. Flowthrough tests are desirable for high BOD samples and for those containing volatile or unstable substances. Organisms with high metabolic rates are difficult to maintain in standing water, whereas flowthrough tests provide well-oxygenated test solutions, stable concentrations, and continuous removal of metabolic wastes.

Several factors should be considered in making the choice of toxicity test system. On-line continuous flowthrough testing can sample and measure "peaks" of toxicity as they may occur during the test period. This may be of concern when testing runoff from industrial sites that exhibits flushes of toxics. If the discharge toxicity is highly variable and continuously discharges (industrial effluents) either a flowthrough or a renewal test would be appropriate (U.S. EPA, 1991). However, the sample should be a composite collected over the period of discharge. If the effluent is not considered variable, such as a discharge from a pond with a long detention time, then a static or renewal test using a grab or 24-hour composite sample would be an appropriate test system. Flowthrough tests are more costly and require a complex delivery system that is not typically available for measuring the toxicity of diffuse sources.

Duration of the test determines what type of toxicity is been investigated. *Acute toxicity tests* are short-duration tests, typically representing a small fraction of the lifetime of the organisms. The concentrations are higher and the impact on the organisms is severe, usually death. The test is usually completed in less than 4 days (96 hours). In *chronic toxicity tests*, organisms are exposed to lower concentrations, preferably for the entire reproductive life of the organism. In *subchronic tests*, the exposure of the organisms lasts for a time period that is less than a complete reproductive life cycle, but longer than that for acute toxicity testing. Testing involves the exposure of organisms during sensitive life cycle periods, such as early stages of development, critical life stages, embryo–larval, or frog egg tests (Rand and Petrocelli, 1985). Typically, a 7-day period is assumed unless specified otherwise.

Test Organisms

The species used in characterizing the toxicity of an effluent or runoff will depend on the requirements of the regulatory agencies. The U.S. Environmental Protection Agency (1991) recommends as a minimum three species (for example, a vertebrate-fish, an invertebrate, and a plant) should be tested; however, the EPA recommends against selecting "*a most sensitive species*" for toxicity testing. Species that have been widely used in toxicity tests and are acceptable test organisms in freshwater are listed below (after U.S. EPA, 1985):

Vertebrates

1. Cold Water
    a. Brook trout (*Salvelinus fontinalis*)
    b. Coho salmon (*Oncorhyncus kisutch*)
    c. Rainbow trout (*Salmo gairdneri*)
2. Warm Water
    a. Bluegill (*Lepomis macrochirus*)
    b. Channel catfish (*Ictalurus punctatus*)
    c. Fathead minnow (*Pimephales promelas*)

Invertebrates

1. Cold Water
    a. Stoneflies (*Pteronarcys* spp.)
    b. Crayfish (*Pacifastacus Ieniusculus*)
    c. Mayflies (*Baetis* spp. or *Ephemerella* spp.)
2. Warm Water
    a. Amphipods (*Hyalella* spp., *Gammarus lacustris* or *G. fasciatus* or *G. pseudolimnaeus*)
    b. Cladocera (*Daphnia magna* or *D. pulex* or *Ceriodaphnia* spp.)

c. Crayfish (*Orconectes* spp. or *Cambarus* spp. or *Procambarus* spp.)
d. Mayflies (*Hexagenia limbata* or *H. bilineata*)
e. Midges (*Chironomus* spp.)

Table 13.2 lists the duration and observable endpoints for several test species in short-duration chronic toxicity tests.

Conducting toxicity tests using three species quarterly for 1 year is recommended by the EPA to adequately assess the variability of toxicity in waste discharges. Analysis of species sensitivity ranges found in the national water quality criteria indicates that if tests are conducted on three particular species (*Daphnia magna*, *Pimephales promelas*, and *Lepomis macrochirus*) the most sensitive of the three will have an LC_{50} within one order of magnitude of the most sensitive of all species.

TABLE 13.2   Short-Term Chronic Toxicity Methods

Species/Common Name	Test Duration	Test Endpoints
Freshwater species		
Cladoceran *Ceriodaphnia dubia*	Approximately 7 days (until 60% of control have 3 broods)	Survival, reproduction
Fathead minnow *Pimephales promelas*	7 days	Larval growth, survival
	9 days	Embryo–larval survival, percent hatch, percent abnormality
Freshwater algae *Selenastrum capricomutum*	4 days	Growth
Marine/estuarine species		
Sea urchin *Arbacia punctulata*	1.5 hours	Fertilization
Red macroalgae *Champia parvula*	7–9 days	Cystocarp production (fertilization)
Mysid *Mysidopsis bahia*	7 days	Growth, survival, fecundity
Sheepshead minnow *Caprinodon variegatus*	7 days	Larval growth, survival
	7–9 days	Embryo–larval survival, percent hatch, percent abnormality
Inland silverside *Menidia beryllina*	7 days	Larval growth, survival

Source: After U.S. EPA (1991).

Functional relationships. The simplest way to find a functional toxicity relationship is by plotting the percent survival versus logarithms of the dose or concentration of the contaminant in the container). Other methods involve fitting the data to a mathematical function.

Figure 13.2 shows a graphical representation of a toxicity bioassay test. Besides the graphical plotting the dose (concentration) relationship has been analyzed using several mathematical models. The most popular method, which also appears to be the most theoretically sound, is the *probabilistic* or *probit method*. A second widely applied method is the *log-logistic model*.

In the *probit method* the variation in the organisms' sensitivity to the toxic compound is represented by a probability density function. In the *log-logistic* model the fraction of the population that responds to a certain dose of the compound is represented by an empirical function that can be linearized and plotted on semilog paper or the parameters of the equation can be determined by a linear regression. Computer

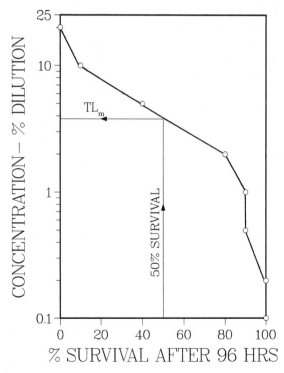

FIGURE 13.2. A functional relationship of toxicity of a compound vs. mortality in a bioassay test.

programs are also available for determining LC_{50} (LD_{50}), which were described by Sprague (1973), Rosiello, Essigmann, and Wogar (1977), and by Altshuler (1981).

Effect of bioavailability. It was shown in Chapter 6 that some components in the soil environment are strongly tied to the soil particles or precipitate as solids. Such compounds include potentially toxic compounds, such as most of organic chemicals, toxic metals, ammonia, and nontoxic phosphates. The same is true in the aquatic environment when solids are present, such as benthic layer or in flows with higher content of suspended solids.

Potentially toxic compounds may affect organisms via two pathways. The first pathway involves adsorption or the bonding of the compound by the organic matter of the organism. This essentially affects the cell tissues inside the body of the organism. For higher organisms the adsorption and bonding differ between the various organs and are higher for lipids (fats) and the liver of fish and other aquatic organisms. Only dissolved and dissociated (ionized) toxic compounds are available for such bonding to cell tissues. The second route is by ingestion of contaminated food or sediments. Contaminated sediments may pass through the digestive system of the organism and become available due to changed chemistry inside the digestive tract. Figure 13.3 outlines the processes involved in the toxic interactions of various forms of metals. It has been observed that in the presence of solids aquatic organisms and plants respond differently to toxic exposure expressed as total concentration. The ingestion route is applicable only for higher trophic level organisms.

Essentially for lower aquatic organisms and for fish the adsorbed fraction is biologically unavailable. Hence the toxicity data for sediments and solids flows may not be correlated to the total concentration of the toxic compound, but the concentration–response curve could be correlated to the dissolved fraction concentration in the water or interstitial (pore) water of sediments. For this reason a new unit has been introduced for high solids media (DiToro et al., 1991), that is

$$\text{Sediment toxic unit (STU)} = \frac{\text{pore water concentration}}{\text{water only } LC_{50}}$$

It is presumed that benthic organisms residing in sediments with different total concentrations of the toxicants, but with the same STU would exhibit the same toxic response. An STU of one would occur when the pore water concentration of the toxic compound equals LC_{50}, even though the total concentration of the toxicant in the sediment may be much higher than LC_{50}.

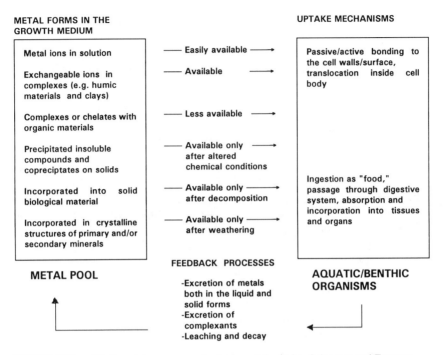

FIGURE 13.3. Biotic uptake processes for toxic metals. (After Salomons and Förstner, 1984.)

As elucidated previously (Chapter 6) the relation between the dissolved (pore) water concentration of an adsorbable compound and the adsorbed (precipitated) fraction of the pollutant can be expressed by an isothermic equation. For most organic toxic chemicals a linear partition isotherm is used (see Chapter 6 for a more detailed discussion of partitioning of toxic chemicals on sediments), or (Eq. (6.6))

$$r = \Pi c_d$$

where
c_d = dissolved (pore water) concentration of the chemical, µg/l
r = solid phase (adsorbed) chemical concentration, µg/g
Π = partition coefficient, l/g

The total concentration of the pollutant is then the sum of the dissolved (c_d) and particulate concentrations (c_p). Hence, if $c_p = m_{ss} \times r$, where m_{ss} is the concentration of solids in g/l and θ is the water content (θ = porosity for sediments and $\theta = 1$ for water), then (Eq. (6.7))

830 Toxic Pollution and Its Impact on Receiving Waters

$$c_T = \theta c_d + c_p = c_d + m_{ss} \times r = c_d(\theta + \Pi m_{ss})$$

The relation between the dissolved (pore water) concentration of the pollutant and the total concentration in the presence of appreciable solids load is then simply (Eq. (6.8))

$$c_d = \frac{1}{\theta + \Pi m_{ss}} c_T$$

The coefficient for nonionic organic chemicals is related to the organic carbon related partition coefficient K_{oc} such as $\Pi \times m_{ss} = K_{oc} \times m_{oc}$, where m_{oc} is the organic carbon concentration of the sediment or suspended solids. On the other hand the magnitudes of the K_{oc} can be related to the octanol partition coefficient K_{ow}, as shown in Chapter 6. The magnitude of the octanol partition coefficient for some chemicals is given in Appendix C.

Example 13.1: Bioavailability of a Toxic Compound

Calculate the bioavailable fraction of a potentially toxic compound whose partition coefficient $\Pi = 10 \text{ l/g}$ (related to the total solids concentration) in water ($m_{ss} = 50 \text{ mg/l} = 0.05 \text{ g/l}$ and $\theta = 1$), and in the sediment ($m_{ss} = 150{,}000 \text{ mg/l} = 150 \text{ g/l}$ and $\theta = 0.9$).

In water

$$\frac{c_d}{c_T} = \frac{1}{1 + 10 \times 0.05} = 0.666$$

or two-thirds of the total concentration is dissolved and may be considered bioavailable and toxic.

In sediment

$$\frac{c_d}{c_T} = \frac{1}{0.9 + 10 \times 150} = 0.00066$$

Less than 0.1% of the compound in the sediment is bioavailable, therefore its toxicity is far less than that in water. On the other hand, sediments are long-term sinks of toxic compounds and the total concentration of the toxicant after prolonged contact with the contaminated overlying water can become much higher than in the water column.

The bioavailability effect explains why some potentially heavily contaminated sites with organic carbon sediments (such as some wetlands) can still support relatively viable biota.

Bioaccumulation and Biomagnification

In the 1960s when the full impact of DDT use was realized, biologists noted a large discrepancy between the DDT levels in organisms of different trophic levels. The higher the trophic level of the organism, the higher the bodily concentration of DDT and some other compounds (Table 13.3). It is interesting to note that the concentration of DDT was $4.14/0.003 = 1380$ times magnified throughout the trophic levels. If water concentrations had been reported in this work, the differences in the bodily concentrations and those of DDT in water would have most likely been several orders of magnitude, that is, the DDT concentrations in water could have been below the detection limit.

At the onset of the discussion it is necessary to define the terms *biomagnification*, *bioaccumulation*, and *bioconcentration*. Both bioaccumulation and bioconcentration are caused by an imbalance in the organism body between the intake of the toxic compound in water and/or food and epuration (excretion), resulting in a progressive increase of the bodily content of the toxic compound. The intake of the compound is due to water transfer across the gills, surface sorption, and ingestion of contaminated food. Depuration of the compound from the body is due to desorption, metabolism, excretion, and growth.

Biomagnification means that the concentration of the contaminant is increasing with the trophic level of the organism, as pointed out in Table 13.3, while *bioaccumulation* can be independent of the trophic level. Not all bioaccumulating toxic pollutants will also exhibit biomagnification, although, on the other hand, biomagnification cannot take place without bioaccumulation. The food web (trophical levels) of organisms was

TABLE 13.3 Concentration of DDT in a Marine Environment

Species	Trophic Level	DDT (mg/kg)
Oar weed	1	0.003
Sea urchin	2	0.05
Lobster	3	0.024
Shag liver	4	2.8
Cormorant liver	5	4.14

Source: After Robinson et al. (1967). Reprinted with permission; copyright 1967 by Macmillan Magazines Limited.

presented in the preceding chapter and shown on Figure 12.2. A *bioconcentration factor* is then defined as the ration of the concentration of the toxic pollutant in the bodily tissue of the organism to that in water. A *bioaccumulation factor* is defined as involving contaminant uptake from both water and food. A *biomagnification factor* is a ratio of the bodily concentration of a compound to that in lower trophic level organisms.

Simple mass balance will explain the mechanism in a single aquatic first trophic level organism (such as phytoplanktonic alga) that is being exposed to a compound in water (Thomann, 1991a, 1991b; Thomann and Mueller, 1987)

$$\frac{dwv}{dt} = k_u wc - Kvw \tag{13.1}$$

where
v = body concentration of the chemical, µg/g weight of the organism
k_u = uptake sorption and/or transfer rate, 1/day-g(weight)
w = body weight of the organism, g(weight)
c = bioavailable water concentration of the compound, µg/l
K = desorption and excretion rate, day^{-1}
t = time in days

The equation, which at this moment ignores intake of the contaminant in food, basically states that if the bodily intake given by $k_u c$ is greater than the bodily loss by depuration expressed by Kwv', then the bodily concentration of the compound will increase. The product vxw is the total bodily burden of the compound. To obtain the rate of the bodily concentration, the deferential on the left side of the equation can be expanded as

$$\frac{d(wv)}{dt} = v\frac{dw}{dt} + w\frac{dv}{dt} \tag{13.2}$$

Letting $(dw/dt)/w = G$, where G (day^{-1}) is the net specific growth of the organism, and $K' = K + G$, then Equation (13.1) becomes

$$\frac{dv}{dt} = k_u c - K'v \tag{13.3}$$

At a steady state (no net increase of the whole body mass burden of the contaminant)

$$v = \frac{k_u c}{K'} \tag{13.4}$$

A bioconcentration factor (BCF) or N_w is defined as the ratio of the bodily concentration of the compound to its water (ambient) concentration at a steady state, when the intake of the toxicant is by biosorption only, or

$$N_w = \frac{v}{c} = \frac{k_u}{K'} \tag{13.5}$$

The units for the bioconcentration factor N_w are $(\mu g/g)/(\mu g/l) = 1/g$.

Thomann (1991a, 1991b) has pointed out that for lower trophic level aquatic organisms the sorption rates are generally much faster than the uptake and excretion rates of higher levels of the food chain. Then an instantaneous equilibrium may be assumed, and from Equation (13.5)

$$v = N_w c \tag{13.6}$$

For species that are above the first trophic level, uptake of toxicant due to ingestion of contaminated food must be considered. This uptake is a function of the contaminant concentration in the food (prey), rate of consumption of the food, and the degree to which the ingested toxicant in the food is actually accumulated by the organism (Connolly and Thomann, 1991b). Adding the food intake into Equation (13.1) yields

$$\frac{dv_2}{dt} = k_u c + \alpha G v_f - K' v_2 \tag{13.7}$$

where
α = the fraction of ingested toxicant that is assimilated
v_f = toxicant concentration in the food (prey), $\mu g/g$ (weight)
v_2 = bodily concentration of the toxicant

If Equation (13.7) represents the bioaccumulation of a second trophic level organism, then the concentration of the toxicant in the food will be $v_f = N_w c$. Identical equations can be written for higher than second trophic levels. The steady-state solution for a simple food web is then (Thomann, 1991b)

$$v_1 = N_{w1} c \tag{13.8a}$$

and

$$v_j = N_{wj}c + \frac{a_{j,j-1}C_{j,j-1}}{K'_j}v_{j-1} \quad \text{for} \quad j > 1 \qquad (13.8b)$$

where
$C_{j,j-1}$ = specific food intake rate of the organism, g(weight) prey per g(weight) predator per day
N_{wj} = bioconcentration factor due to biosorption only, 1/g(weight)

In terms of the bioaccumulation factor $N_j = v_j/c$, Equation (13.8) becomes by successive substitutions for v_{j-1}

$$N_1 = N_{w1}$$
$$N_2 = N_{w2} + f_{21}N_{w1}$$
$$N_3 = N_{w3} + f_{32}N_{w2} + f_{32}f_{21}N_{w1}$$
$$N_4 = N_{w4} + f_{43}N_{w3} + f_{43}f_{32}f_{21}N_{w1} \qquad (13.9)$$

(and so forth if higher trophic levels are considered), where

$$f_{j,j-1} = \frac{a_{j,j-1}C_{j,j-1}}{K'_j}$$

The classic growth equation for organisms is

$$G = aC - r \qquad (13.10)$$

where
a = food assimilation efficiency
r = respiration rate, day^{-1}

Then

$$C_{j,j-1} = (G_j + r_j)/a_{j,j-1} \qquad (13.11)$$

The bioconcentration factor, BCF = N_{wj}, is due to biosorption of the toxic compound only, primarily through the gills or through the lipoprotein membranes of organisms. The bioaccumulation factor, BAF = N_j = v_j/c, includes both biosorption and intake from the food.

The bioconcentration factor N_w (and the bioaccumulation factor N) has a dimension of liter per gram of bodily weight. Since many toxic chemicals tend to accumulate primarily in the lipid tissue of the organisms, the bioaccumulation factor may be normalized by the weight fraction of the lipid, such as $N'_w[\text{l/g(lipid)}] = N_w[\text{l/g}(w)] \times [\text{g}(w)/\text{g(lipid)}]$. For phytoplankton the bioaccumulation coefficient may be normalized by the organic carbon (approximately 50% of dry weight) content. It was then found that the lipid-normalized bioaccumulation factor BAF = N'_j (and also the bioconcentration factor BAC = N'_{wj}) may be correlated to the octanol partition coefficient for the chemical, K_{ow} (see Chapter 6 for an explanation of partitioning of organic chemicals between water and organic solids and the Appendix for the magnitudes of K_{ow}). The correlation is different between the trophic-level organisms due to the different effects of growth rate and food ingestion on the accumulation of components with widely varying biosorption capabilities. The relation of N_j to K_{ow} is shown on Figure 13.4. For the first trophic level (phytoplankton), the normalized (organic carbon) bioconcentration factor $N'_1 = N'_{w1} = K_{ow}$ for all nonpolar organic toxic chemicals; for higher trophic levels, the

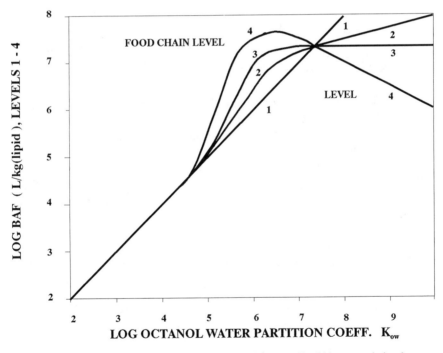

FIGURE 13.4. Relation of the lipid (organic content) normalized bioaccumulation factor, N_j, to the octanol partitioning coefficient, K_{ow}. (After Thomann, 1991a.)

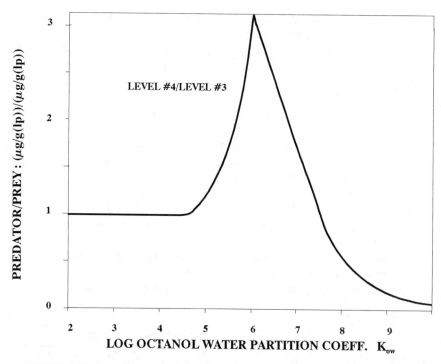

FIGURE 13.5. Predator–prey bodily concentration ration. (After Thomann, 1991a.)

equality $N'_{wj} = K_{ow}$ holds for chemicals with $K_{ow} < 5$. Figure 13.5 shows the predator–prey bodily concentration ratio (biomagnification factor), which can also be related to the biosorption characteristic of the chemical, the octanol partitioning coefficient.

Table 13.4 shows the ratio of BAF to the BCF as a function of the trophic level of an aquatic organism, and the $\log K_{ow}$. The BAF/BCF ratio ranges from 1 to 100, with the highest ratios applying to organisms in higher trophic levels, and to chemicals with $\log K_{ow}$ close to 6.5. For chemicals with $\log K_{ow}$ greater than 7 (l/kg), there is some uncertainty regarding the degree of bioaccumulation.

Trophic level-four organisms are typically the most desirable species for sport fishing, and therefore the multipliers corresponding to this trophic level should be used in setting water quality standards. In rare situations where only lower trophic-level organisms are found (for example, oyster beds) the BAF/BCF factors for lower trophic levels can be used. BAF/BCF factors are very difficult to measure accurately, the process is very expensive, and the results may be problematic and subject to uncertainty (U.S. EPA, 1991).

The steady-state assumption in this concept is crude and neglects the

TABLE 13.4 Estimated Food Chain Multipliers FM = BAF/BCF

log K_{ow}	Trophic Levels		
	2	3	4
3.5	1.0	1.0	1.0
4.0	1.1	1.0	1.0
4.2	1.1	1.1	1.1
4.4	1.2	1.1	1.1
4.6	1.2	1.3	1.3
4.8	1.4	1.5	1.5
5.0	1.6	2.1	2.6
5.1	1.7	2.5	3.2
5.2	1.9	3.0	4.3
5.3	2.2	3.7	5.8
5.4	2.4	4.6	8.0
5.5	2.8	5.9	11
5.6	3.3	7.5	16
5.7	3.9	9.8	23
5.8	4.6	13	33
5.9	5.6	17	47
6.0	6.8	21	67
6.1	8.2	25	75
6.2	10	29	84
6.3	13	34	92
6.4	15	39	98
6.5	19	45	100
>6.5	19.2[a]	45[a]	100[a]

Source: After U.S. EPA (1991).

[a] These values are conservative best estimates.

growth and life span of the organisms. The steady state during the duration of an experiment and/or the life span of organisms can only be reached if the duration of the experiment and life span of the organisms are long enough to reach equilibrium. For some chemicals with very strong biosorption (high K_{ow}) and/or shorter life span, the steady state may not be reached at higher trophic levels. Consequently, a time-variable model should be used.

The state of the art of modeling bioaccumulation in the food chain of aquatic organisms has evolved to the point where reasonably accurate (order of magnitude) simulations are possible. The EPA model WASP includes the food chain bioaccumulation model (at the time of writing this book the food chain model was not yet incorporated into the public domain version of the WASP model). The reader is referred to pertinent EPA publications on WASP and to the publications of the Manhattan

College, Bronx, New York, team that developed the food chain bioaccumulation models (Connolly and Winfield, 1984).

Example 13.2: Biomagnification Calculation

Estimate the PCB concentration in Lake Michigan invertebrates (*Mysis*—trophic level 2), which have a life span of about 1 year. A constant dissolved PCB concentration (note that adsorbed PCBs on sediments is not bioavailable) is $1\,\text{ng/l}$ (= $10^{-3}\,\mu\text{g/l}$). The following additional information is also given (adapted from Connolly and Thomann, 1991b):

The octanol partitioning coefficient for PCB's: $K_{ow} = 10^{6.5}\,\text{l/kg} = 3000\,\text{l/g}$
Fraction of dry weight of phytoplankton: 0.1
Organic carbon/dry weight ratio: 0.5
Fraction of dry weight of invertebrates: 0.2

PCB Concentration in the Phytoplankton $(N_1 = N_w = K_{ow})$

$$v_1 = N_1 \times c = 3000\,[\text{l/g(g of org. C)}] * 0.001\,[\mu\text{g/l}]$$
$$= 3.0\,\mu\text{g/g(org. C)} \times 0.5\,[\text{g(org. C)/g(dry weight)}]$$
$$\times 0.1\,[\text{g(dry)/g(total weight)}]$$
$$= 0.15\,(\mu\text{g/g}) = 0.15\,\text{mg/kg}$$

Since 1 liter of phytoplankton (total) biomass weighs approximately 1 kg (weight), the concentration of PCBs in the phytoplankton is 15,000 times greater than that in water.

PCB Concentration in the Invertebrates   From Figure 13.4 for $\log K_{ow} = 6.5$ and trophic level 2, $\log N_2' = 7.0$ or $N_2 = 1.0 \times 10^7\,\text{l/kg} = 10^4\,\text{l/g}$ (lipid or carbon). By a calculation similar to that for the phytoplankton

$$v_2 = 10{,}000 \times 0.001 \times 0.5 \times 0.2 = 1.0\,(\mu\text{g/g}) = 1.0\,\text{mg/kg}$$

Hence the steady-state PCB concentration in *Mysis* is about 10^6 times greater than that of PCBs in water.

WATER QUALITY AND SEDIMENT CRITERIA AND STANDARDS FOR TOXICITY

Water Quality Criteria and Standards

The legal difference between criteria and standards has already been explained in Chapter 1. The criteria from which standards were adopted by the states have developed primarily from the toxicity bioassay conducted on different organisms and from studies of the effects of toxic compounds on man. The federal water quality criteria adopted in the United States specify maximum exposure concentrations that will provide protection of aquatic life and human health. Generally, however, the water quality criteria describe the quality of water that will support a particular use of the water body. These criteria may be then used by the states as a basis of enforceable water quality standards. The water quality standards apply to all waters of the United States, including most of the natural wetlands. The latest edition of the federal water quality criteria was published in 1987 (U.S. EPA, 1987a), followed by publication of specific criteria in several issues of the *Federal Register*.

The purpose of the following discussion is (1) to provide the rationale behind the federal water quality criteria and extrapolate the logic of how they apply to sediments, (2) provide guidance to the development of site-specific criteria. The water (sediment) quality standards are paramount to integrated pollution abatement, since it is expected that most of the effluent and waste load limitations based on the National Pollution Discharge Elimination System will be related more to the quality of the receiving water and will be more stringent than those based on the required waste elimination technology. Furthermore, the same logic and methodology may be applied to the assessment of the risks of contaminated sediments and water to aquatic biota and man.

The most important uses of surface-water bodies is to provide water for human consumption and contact recreation, and for aquatic life protection and propagation. For the protection of aquatic life a two-value criterion has been established to account for acute and chronic toxicity of waste compounds. The human health criterion specifies the risk incurred with exposure to the toxic compounds at the criterion concentration. The latter (human health) criterion is associated with the increased risk of contracting a debilitating disease, such as cancer.

Aquatic Life Protection Criteria

The water quality standards regulation in the United States allows the individual states to develop numerical criteria of their own or modify the EPA's recommended criteria to account for site specificity or other scientifically defensible factors (U.S. EPA, 1991). The criteria may be based on chemically specific numeric values for the priority pollutants or on the whole effluent toxicity (the term effluent applies to point discharges regardless of whether these are of diffuse or traditional point origin).

Chemical-Based Numerical Criteria

The LC_{50} or EC_{50} acute toxicity values and chronic toxicity observations determined for various key aquatic organisms have been used as a basis for the development of the single-chemical numerical water quality criteria. The development of site-specific toxicity criteria based on a complete toxicity bioassay performed on at least three species of test organisms follows the following procedures (U.S. EPA, 1991; Connolly and Thomann, 1991a).

Acute toxicity criterion or criterion of maximum concentration (CMC). Using acute toxicity data (LC_{50} or EC_{50}) with at least three different families of species (specified by state guidelines) calculate:

1. Geometric mean of all LC_{50} (EC_{50}) values for each species, yielding "species mean value (SMAV)"
2. Geometric mean of all SMAV values within a genus, "genus mean acute value (GMAV)"
3. Assuming that GMAV values are log-normally distributed, determine the 5% GMAV (that is, the concentration exceeded by 95% of the GMAV values) yielding the "final acute value (FAV)"
4. The acute toxicity criterion (CMC) is then CMC = α × FAV, where the α multiplier corrects the FAV values derived from the 50% lethality value LC_{50} rather than from a threshold-lethal (zero mortality) effective concentration (U.S. EPA, 1991). The recommended value for this procedure is $\alpha = 0.5$ (Connolly and Thomann, 1991a).

Chronic toxicity criterion or criterion continuous concentration (CCC). The chronic toxicity criteria also evolved from toxicity tests. Two values are determined in the test as follows: (1) a lower chronic toxicity limit is the highest tested concentration that did not cause an "unacceptable" amount of adverse effects, and (2) an upper chronic toxicity limit is the lowest tested concentration that did not cause an "unacceptable" amount of an adverse effect. The effects are related to the long-term viability of the species and include loss of reproductive capability, reduction in mobility, change in feeding, and reduction of metabolic rates.

To establish CCC the following procedure is outlined:

1. Compute chronic values from chronic toxicity data. A chronic value is the geometric mean of the lower and upper chronic limits. The geometric mean of the two values is thus called the *chronic value.*
2. If sufficient chronic values are available (at least three species are required), then they are analyzed in the same way as outlined in the procedure given earlier for CMC.

3. In the absence of a large enough chronic toxicity data base, ratios of LC_{50} or LD_{50} to available chronic toxicity data (ACR) are then used to establish the chronic toxicity criteria. The ACR is the ratio of acute LC_{50} or EC_{50} to the chronic value. ACR values are needed from at least three families including one fish, one invertebrate, and one acutely sensitive species.
4. A final ACR (FACR) is calculated from species geometric mean ACR values. This procedure depends on whether the ACR is a function of the SMAV defined previously. The minimum allowable ACR is 2.

Bioaccumulation, additivity, antagonism, synergism, and persistence of the chemical may be important in determining the magnitude of the standard.

Traditional "point source" application of receiving water standards was related to extremely rare low flows, such as the 10 years expectancy–7 days exceedence low flow (Q_{7-10}). Such low-flow concepts are of little use in storm-water management or in water-quality-based controls of discharges of priority pollutants. Statements such as "the standard should not be exceeded at all times" are also inappropriate, since concentrations represent statistical time series for which only infinitesimally large values would have a 100% statistical probability of not being exceeded. Thus a standard (or a criterion) for a harmful substance must have three components (U.S. EPA, 1991):

- Magnitude: The amount of a pollutant (or pollutant parameter, such as toxicity), expressed as concentration, is allowable.
- Duration: The period of time (averaging period) over which the instream concentration is averaged for comparison with criteria concentrations. The specification limits the duration of concentration above the criteria.
- Frequency: How often criteria can be exceeded.

The permissible frequency of excedence of toxicity criteria based on the July 29, 1985, issue of the *Federal Register* is:

- *Acute toxicity criteria*: One hour average concentration (essentially a grab sample), not to be exceeded more than once in three years on an average
- *Chronic toxicity criteria*: Four day average concentration, not to be exceeded more than once in three years on the average.

Since most of the water quality constituent concentrations from a sufficiently long record follow a log-normal distribution, the acute toxicity criterion (standard) would be violated if the 99.9 percentile of *maximum*

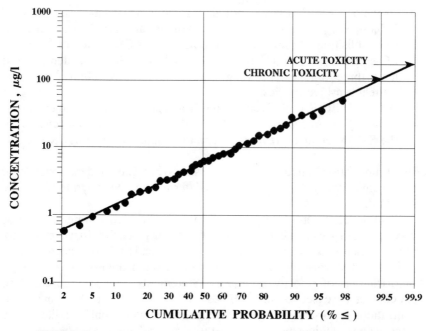

FIGURE 13.6. Probability concept of criteria for acute and chronic toxicity.

daily concentrations, arranged in ascending order of magnitude on the log-cumulative probability plot, exceeded the criterion (Fig. 13.6). Similarly, the chronic toxicity criterion would correspond to the 99.6 percentile log-cumulative probability characteristic of *average daily concentrations*. A one hour average value for acute toxicity would imply grab samples taken randomly. For chronic toxicity, composite samples (over a 24 hour period) are more appropriate. For planning studies and modeling testing for chronic toxicity violations, both in water and sediments, resulting from known and anticipated point and nonpoint discharges will suffice. The three year recurrence was derived from observations of the length of recovery of ecosystems after a toxic spill.

Appendix B presents the water quality criteria for recognized pollutants and other water quality parameters adapted as standards in the United States. It should be pointed out that the values in Appendix B were derived from toxicity bioassays performed in laboratory conditions, with organisms residing mostly in water with no or minimum sediment. Thus the criteria would logically correspond to bioavailable concentrations of the toxicant. Legal interpretation of the toxicity standards are not clear as to whether total (adsorbed + dissolved-bioavailable) or bioavailable fractions should be considered.

Whole Effluent Toxicity Criteria

Whole effluent toxicity is the total toxic effect of an effluent measured directly by a toxicity test. In such measurements the toxicity is expressed in toxicity units, both chronic and acute.

The *acute toxic unit* (TU_a) is the reciprocal of the effluent concentration in the dilution water (in percent) that causes 50% of the organisms to die by the end of the acute exposure period, or

$$TU_a = 100/LC_{50}$$

The *chronic toxic unit* (TU_c) is the reciprocal of the effluent concentration (in percent in dilution water) that causes no observable effects on the test organisms by the end of the chronic exposure period, or

$$TU_c = 100/NOEC$$

For example, a waste discharge with an acute toxicity of an LC_{50} in 5% effluent (5% wastewater in 95% dilution) is discharge containing 20 TU_a. The toxic units enable specifying water quality criteria based upon the whole toxicity. For comparative purposes, a waste discharge that contains 20 TU_c is twice as toxic as a discharge containing 10 TU_c.

The whole effluent (body of water) test is conducted as delineated before. At least three different species are required for the test. To protect aquatic biota the EPA recommends that the criterion maximum concentration (CMC) be set to 0.3 TU_a for the most sensitive of at least three test species. The selection of the test species is not critical (see the list on page 825 and Table 13.2 for selection) provided that species from ecologically diverse taxa are used (i.e., a fish, an invertebrate, and a plant). The factor of 0.3 is again used to adjust the typical LC_{50} (50% mortality), LC_0 (virtually no mortality). For chronic protection, the CCC should be set at 1.0 toxic units (TU_c).

Human Health Protection Criteria

Human health criteria are based on the average daily dosage of a potential toxicant. Man can receive toxic dosages originating in surface (ground) water via two pathways: (1) by drinking contaminated water, and (2) by eating contaminated fish and shellfish. All criteria are chemical specific and contain only a single expression of allowable magnitude: a criterion concentration that protects humans against long-term chronic health effects during an average lifetime period, which was set as 70 years.

The daily intake of a toxicant is

$$I = \phi_f v + \omega c \tag{13.12}$$

where
I = daily intake of the toxic compound by man, mg/day
ϕ_f = average daily consumption of fish, kg
v = fish chemical concentration, mg/kg
ω = average daily water intake, l
c = the water concentration, mg/l

The EPA-recommended values of daily average fish consumption are given below (compiled by U.S. EPA, 1991):

Intake	Explanation
6.5 g/day	Represents an estimate of average consumption of fish and shellfish from estuarine and freshwater by the entire U.S. population. This is an average of both consumers and nonconsumers of fish (most states).
20 g/day	Represents an estimate of the average consumption of fish and shellfish from marine, estuarine, and freshwater by the U.S. population. This level also includes consumers and nonconsumers of fish (Arizona, Illinois, Louisiana, and Wisconsin).
165 g/day	Represents consumption of fish and shellfish by the 99.9 percentile of the U.S. population consuming the most fish and seafood.
180 g/day	This is a reasonable "worst case" based on the assumption that some individuals would consume fish at a rate equal to the combined consumption of red meat.

Using a bioaccumulation factor N as

$$N_{\text{lipid}} = \frac{v}{c}$$

The bioaccumulation factor is normalized by the lipid content of the fish, hence the unit of N_{lipid} is liters/kilogram of lipid. Then the daily intake is

$$I = [\phi_f N + \omega]c \tag{13.13}$$

where $N = N_{\text{lipid}} \times$ (% lipid in fish)/100.

At a low dose level the incremental risk (for example, of contracting cancer) is proportional to the dose of the toxic compound, or

$$R = qI \qquad (13.14)$$

where
R = incremental risk over the background (no contamination) risk of contracting a disease
q = risk–dosage coefficient obtained from toxicity studies

Then for an average person weighing W kg, the maximum daily intake of the toxicant is

$$I_{\max} = \frac{W*R}{q} \qquad (13.15)$$

Substituting Equation (13.15) into Equation (13.13) the maximum water concentration (criterion) that will provide human health protection at risk R becomes

$$c_{\max} = \frac{WR}{q(\phi_f N + \omega)} \qquad (13.16)$$

There has been considerable controversy among environmentalists and scientists over the acceptable risk level for human health protection. By accepting a very small risk factor $R = 10^{-6}$, one still de facto accepts the fact that probably one person in a million will get a cancer or a debilitating disease because of water contamination. Again those not familiar with the statistical characteristics of natural time series may advocate no risk, which would imply an absolute never exceeding zero concentration (an impossible task). For a comparison the following activities pose about the same risk of death or debilitating injury to man (1 death increase in a population of one million): riding 16 km on a bike or driving 500 km in a car or flying 1600 km by a commercial jet plane or travelling 6 minutes by a canoe or drinking one-half liter of wine (J. P. Connolly, 1991, pers. comm.).

Example 13.3: Toxicity Criterion Estimate for PCBs

Estimate a water quality criterion for PCBs that would provide human health protection at an acceptable risk level of $R = 10^{-6}$ (an increase of one cancer per one million population over the background level). The following information has been obtained:

Bioaccumulation factor for PCBs: $N_{lipid} = 3 \times 10^6$ l/kg (lipid)
Average body weight of a person: $W = 70$ kg
Average daily fish consumption: $\phi_f = 20$ g $= 0.02$ kg
Average lipid content of fish: 5%
Average daily water intake: $\omega = 2$ l
Risk-dose coefficient of $q = 3 \times 10^{-4}$

Solution Calculate the biomagnification factor

$$N = N_{lipid} \times (\% \text{ lipid})/100 = 3 \times 10^6 \times 3/100 = 9 \times 10^4 \text{ l/kg}$$

Then from Equation (13.6)

$$c_{max} = \frac{70 \times 0.000001}{0.0003 \times (0.02 \times 9 \times 10^4 + 2)} = 1.3 \times 10^{-4} \text{ mg/l}$$

Example 13.4: Water Quality Standard Based on Maximal Body Intake

From epidemiologic studies it was found that a daily intake of a carcinogenic compound has the following impact on the test human population:

Daily Dosage (mg)	Number of Additional Cancers/Population
10	5/100,000 (5×10^{-5})
50	2/10,000 (2×10^{-4})
100	5/10,000 (5×10^{-4})

The compound has $\log K_{ow} = 6$.

Estimate the water quality criterion for a compound that would protect the population at the risk level of 1 additional cancer per 1,000,000.

Solution Assume

Bioaccumulation factor for carnivorous fish (trophic level 4) from Figure 13.3 for

$$K_{ow} = 10^6 \quad N_{lipid} = 3 \times 10^6 \text{ l/kg,}$$

Average body weight of a person: $W = 70$ kg
Average daily fish consumption: $\phi_f = 20$ g $= 0.02$ kg

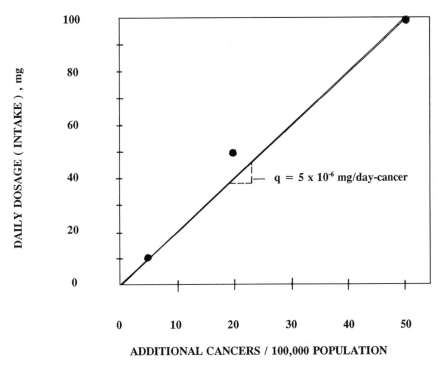

FIGURE 13.7. A plot of cancer deaths increases vs. bodily toxic compound daily dosage above the ambient dose for Example 13.4.

Average lipid concentration of fish: 4%
Average daily water intake: $\omega = 2l$

By plotting the dosage–cancer increase numbers in Figure 13.7 the risk–dose coefficient is approximately $q = 5 \times 10^{-6}$ mg/day. The corrected bioaccumulation factor is $N = N_{\text{lipid}} \times$ (lipid content of fish) $= 3 \times 10^6 \times 4/100 = 1.2 \times 10^5$.

Repeating the calculation similarly to the previous example for $R = 10^{-6}$, the human health protection criterion is

$$c_{\max} = \frac{70 \; 0.000001}{5 \times 10^{-6} \; (0.02 \times 1.2 \times 10^5 + 2)} = 0.0113 \, \text{mg/l}$$

It may be interesting to note that for these cases most of the compound intake and risk of cancer are due to eating contaminated fish.

The U.S. Environmental Protection Agency recommends using the most current risk information when updating or generating criteria. For this purpose the agency has developed the Integrated Risk Information System (IRIS), which is an electronic on-line data base that provides chemical exposure and estimated human health effects (U.S. EPA, 1987b). The right side of Table B.3 in Appendix B contains the human health protection water quality criteria.

Sediment Water Quality Criteria

The role of sediments and their impact on toxicity of many priority pollutants must be considered. It has already been pointed out that suspended solids and their organic, and to a lesser degree clay fractions, have an affinity to adsorb constituents. Adsorbed and/or precipitated pollutants may not be bioavailable, especially to lower trophic-level organisms of the food web. Example 13.1 has shown that the concentrations of a toxic compounds in the pore water of sediments may be much smaller than those measured in the overlying water. This is due to partitioning between dissolved (bioavailable) and adsorbed–precipitated (nonbioavailable) fractions of the compound.

The importance of the bioavailability and partitioning phenomena in sediments can be overlooked. This will impact directly on the LC_{50} of organisms residing in the sediment. The observations have shown that the concentration–response curve for the biological effect of a constituent could not be correlated to the total concentration of the chemical in the sediment, but to a certain parameter that reflects on the bioavailability of the compound in the sediment. It was shown that this parameter was the dissolved pore water concentration of the compound (DiToro et al., 1991). This parameter is then expressed by the sediment toxic unit (STU). The response curves of the toxicity test and the LC_{50} for various compounds and organisms correlated surprisingly well to the STUs of the sediments, as shown in Figure 13.8. Furthermore, it was shown that the benthic organisms have about the same sensitivity to toxic compounds as the water column organisms. This apparent equality between the sensitivities of benthic and aquatic water column organisms enables one to use water toxicity criteria to also protect the organisms residing in the benthos (DiToro et al., 1991).

Nonionic organic toxic chemicals. Since the adsorptivity of the nonionic organic chemicals is related to the organic carbon particulate fraction of the sediment, the sediment organic carbon is the primary route of exposure of the organisms. Recall that the relation between the sediment chemical concentration (microgram of chemical/gram of sediment) and the pore water concentration (in micrograms/liter) is $r = \Pi c_d$, also $\Pi \approx$

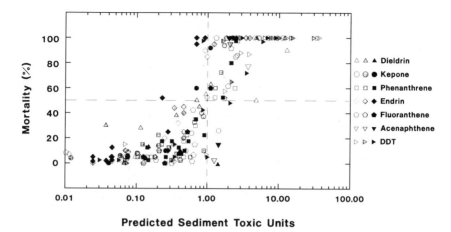

FIGURE 13.8. Correlation of the mortality of aquatic benthic organisms vs. calculated sediment toxic units. (After DiToro et al., 1991a, reprinted with permission, copyright 1991, SETAC.)

$(m_{oc}/m_{ss}) \times K_{ow}$. Then substituting the sediment quality criterion (SQC) for r and the water quality criterion (WQC) for c_d, the relationship between the sediment toxicity and water quality criteria is the same simple partition equation (DiToro et al., 1991)

$$\text{SQC} = \Pi \text{WQC} \approx f_{oc} K_{oc} \text{WQC} \qquad (13.17)$$

where $f_{oc} = m_{oc}/m_{ss}$ is the fraction of organic carbon in the sediment.

DiToro et al. (1991) proposed the use of chronic water toxicity criteria for WQC in order to protect benthic biota, thus implying that compliance with the chronic toxicity criterion will also satisfy the acute toxicity criterion. This concept was accepted by the U.S. EPA. The estimated sediment toxicity unit is then approximately

$$\text{STU} = \frac{c_s(f_{oc} \times K_{oc})}{\text{water only WQC}} \qquad (13.18)$$

where c_s is the concentration of the chemical in the sediment expressed in micrograms per gram of dry weight of the sediment. The limit of applicability of the preceding relationships is the organic carbon fraction, $f_{oc} > 0.1\%$. Below $f_{oc} = 0.1\%$ other factors that influence partitioning, such as particle size, pH, and the electrical charge of the minerals in the sediment, influence the equilibria between the sorbed and dissolved fractions of the chemicals.

Observations also indicated that in addition to the particulate carbon in the sediments, chemicals may be made nonbioavailable by dissolved organic carbon, such as that contained in humic compounds (McCarty and Jimenez, 1985; Landrum et al., 1985, 1987; DiToro et al., 1991). The presence of dissolved organic carbon in the pore water may distort the partitioning between the adsorbed and dissolved toxic compounds, and the apparent partitioning coefficient $K'_p = c_s/c_d$ may not agree with the expected partitioning coefficient $f_{oc} \times K_{oc}$ (DiToro et al., 1991). The differences may amount to an order of magnitude. However, such differences between observed and estimated data are common and should be expected in modeling toxicity relationships. The reliability of the estimates then coincides with the risk in the formulation of the toxicity limits.

Example 13.5: Relation between Sediment and Water Column Toxicities

Estimate maximal acceptable concentration of lindane in an organic sediment. The following information is given:

Octanol partition coefficient of lindane (Appendix C): $K_{ow} = 10^{3.7} = 5000\,l/kg$
Organic content of the sediment: $f_{oc} = 0.1$ (10%)
Chronic toxicity criterion (Appendix B): WQC = $0.08\,\mu g/l$

Solution From Equation (6.20a) the carbon partition coefficient $K_{oc} = 0.63 \times 5000 = 3150\,l/kg = 3.15\,l/g$. The sediment toxicity criterion is then

$$SQC \approx f_{oc} K_{oc} WQC = 0.10 \times 3.15 \times 0.08 = 0.025\,\mu g/g \text{ of dry weight}$$

Example 13.6: Sediment Toxicity Estimation

Chemical analysis by gas chromatograph–mass spectrometer apparatus yielded the sediment concentration of methoxychlor as $0.2\,\mu g/g$ (= mg/kg) of sediment. The sediment organic content is 10% on a dry weight basis. Make a judgment as to whether the sediment contamination is toxic to the benthic biota. Given are:

Methoxychlor sediment concentration: $c_s = 0.2\,\mu g/g$ of methoxychlor
Octanol partitioning coefficient for methoxychlor (Appendix C): $K_{ow} = 10^{4.78} = 60{,}000\,l/kg$
Chronic water quality criterion (Appendix B): WQC = $0.03\,\mu g/l$
Sediment organic carbon fraction: $f_{oc} = 0.1$

$$K_{oc} = 0.45 \times 60{,}000^{0.989} = 23{,}922 \, \text{l/kg} = 23.92 \, \text{l/g}$$

The organic carbon partitioning coefficient is

$$K_{oc} = 0.63 \times K_{ow} = 60{,}000 \, \text{l/kg} = 60 \, \text{l/g}$$

The toxicity in sediment toxicity units (STU) is (Eq. (13.18))

$$\text{STU} = \frac{c_s/f_{oc} \times K_{oc}}{\text{WQC}} = \frac{0.2/(0.1 \times 60)}{0.03} = 1.1$$

Since STU ≈ 1.0 one could judge the sediment as mildly toxic (based on Fig. 13.6) to the benthic biota.

Example 13.8: Volumetric Concentration Toxicity Limit for Sediments

Using the information from the previous Example, estimate the toxicity limit for lindane in the sediment in micrograms/liter of wet sediment. The following information is given:

Porosity: $p = 50\%$
Volatile (organic) fraction of the sediment: $VC = 10\%$
Specific gravity of organic fraction: 1.1
Specific gravity of mineral fraction: 2.4
Estimated chronic SQC for lindane (Example 13.5): 0.025 µg/g of dry weight

Volume (ml/g) = (dry weight in grams)
$\times [(1 - VC)/2.4 + VC/1.1]/(1 - \text{porosity})$
$= 1 \, \text{g} \times [(1 - 0.1)/2.4 + 0.1/1.1]/(1 - 0.5) = 0.93 \, \text{ml/g}$

Hence

$\text{SQC}(\mu\text{g/ml} = \text{mg/l}) = 0.025(\mu\text{g/g})/0.93(\text{ml/g})$
$= 0.027 \, \mu\text{g/ml}$
$[= \text{mg/l of sediment}]$

Note that the value of SQC just given is about 1000 times greater than the corresponding water column criterion from which it was derived.

MODELING THE FATE OF TOXIC COMPOUNDS IN RECEIVING WATER BODIES

Basic Equations and Fundamentals

The fundamentals of modeling the fate of toxic compounds and sediment–water interactions were explained in Chapter 6. The water quality modeling concepts were presented in the preceding chapter. The mass balance continuity equation for toxic substances was introduced by O'Connor (1988) in the preceding chapter as

$$V\frac{dc_i}{dt} = J_i + \Sigma R_i + \Sigma T_i + \Sigma W \qquad (13.19)$$

in which
c_i = concentration of the chemical in compartment, i
J = transport through the system
R = reactions within the system
T = transfer from one phase to another
W = inputs

A chemical in water can exist either dissolved or as a particulate (adsorbed). As pointed out previously, only a dissolved fraction is considered bioavailable, and hence, toxic. Expanding Equation (13.19) for dissolved and adsorbed fractions, substituting appropriate reaction formulas for transformation and decomposition, and dividing both sides of the equation by V, yields (O'Connor, 1988)

$$\frac{dc_d}{dt} = \frac{W_d}{V} - \frac{c_d}{t_0} - (K_1 + K_c + K_a)c_d + K_2 C_p \qquad (13.20a)$$

$$\frac{dc_p}{dt} = \frac{W_p}{V} - \frac{c_p}{t_0} + K_1 c_d - (K_2 + K_p + K_s)c_p \qquad (13.20b)$$

where
C_d = dissolved fraction concentration of the compound
c_p = adsorbed fraction concentration
V = volume of the computational segment (water column or sediment)
t_0 = detention time within the segment = V/Q
Q = flow through the segment
K_1 = adsorption rate coefficient

Modeling the Fate of Toxic Compounds in Receiving Water Bodies   853

K_2 = desorption rate coefficient
K_c = decay rate coefficient of the dissolved fraction of the compound
K_a = transfer (volatilization) rate coefficient of the dissolved fraction
K_p = decay coefficient of the particulate (adsorbed) fraction
$K_s = v_s/H$ = sediment transfer (settling or resuspention) rate coefficient
W_c, W_p = mass input of the dissolved and adsorbed compound fractions
v_s = settling (resuspension) velocity of the particulates

In addition to the water continuity equation (Eq. (13.24)), the sediment mass-balance continuity equation must also be considered, or

$$\frac{dm}{dt} = \frac{m_i}{t_0} - m\left[\frac{1}{t_0} + K_s\right] \quad (13.21)$$

where
m = concentration of suspended solids in the segment
$m_i = W_s/Q$ = average concentration of suspended solids in the input
W_s = mass input of suspended solids into the segment

The process of sedimentation and resuspension of cohesive sediments was explained in Chapter 5.

To simplify Equations (13.20) the following assumptions were made:

1. An instantaneous adsorption–desorption is reached. From Equation (6.6) $c_d = \Pi c_p/m$ (because c_p and m are typically expressed in milligrams or micrograms per liter, r in micrograms/gram, respectively). This eliminates the adsorption–desorption rate coefficients K_1 and K_2.
2. From Equations (6.7) and (6.8) the dissolved and particulate fractions of the compound related to the total concentration, c_T, are

$$f_d = \frac{c_d}{c_T} = \frac{1}{\theta + \Pi m}$$

and

$$f_p = \frac{c_p}{c_T} = \frac{\Pi m}{\theta + \Pi m}$$

where θ = dimensionless water content (θ = porosity for sediments, for water θ = 1). Note that for bottom sediments, $\Pi m \gg \theta$, hence, $f_{pb} \approx 1$ and $f_{db} \approx 1/(\Pi_b m_b)$, where the subscript b denotes bottom conditions.

Adding Equations (13.20a) and (13.20b) and considering the preceding simplifications results in

$$\frac{dc_T}{dt} = \frac{W_T}{V} - \frac{c_T}{t_0} - f_d(t)[K_c + K_a]c_T - f_p(t)[K_p + K_s]c_T \quad (13.22)$$

The magnitudes of the constants and rate coefficients for modeling the fate of toxic compounds were compiled by Schnoor et al. (1987). The processes involved in modeling the fate of toxic compounds when solids are present were explained in Chapter 6.

The toxic chemical model. The toxic chemical submodel TOXI4 included in the WASP model (Ambrose et al., 1991) is a dynamic compartmental model of the transport and fate of toxic chemicals. Its concepts and components were introduced in Chapter 12. This model considers several physical–chemical processes, and can handle up to three chemicals and up to three types of particulate matter per each simulation. The chemical can be independent or linked by reaction yields (for example, the DDT to DDE transformation). Each chemical can be considered in up to five forms, including the natural form and ionic species, and each of them can exist in five phases (dissolved, sorbed to organic carbon, and sorbed to three species of solids). The equilibrium between the solids adsorbed and dissolved phases is described by the partition equilibrium equation (Eq. (6.6)).

The transfer and transformation reactions and processes that the toxic chemicals can undergo in aquatic environments include sorption, volatilization, ionization, photolysis, hydrolysis, biodegradation, and chemical oxidation. WASP (TOXI4) describes sorption and ionization as equilibrium reactions, while the remaining reactions are described as concentration change rate equations.

TOXI4 can be used to simulate toxic metals and organic chemicals. However, it should be pointed out that metal modeling requires site-specific judicious selection of certain key parameters (Ambrose et al., 1991). As with EUTRO4, TOXI4 can be implemented at different levels of complexity that take into consideration increasing sophistication and detail in the description of solids behavior, and kinetic and equilibrium reactions.

Modeling Strategies for the Fate of Toxic Chemicals in Aquatic Systems

Typically, the waste-assimilative capacity of a receiving water body for organic pollutants discharged from point sources is determined using calibrated and verified steady-state models (or dynamic models executed as steady state) applied to an extreme low-flow condition, such as the 7-day duration–10-year recurrence interval flow characteristic (so-called Q_{7-10}).

However, the formulation of toxic standards does not allow a simple application of steady-state models. As pointed out in the section titled "Water Quality and Sediment Criteria and Standards for Toxicity" of this chapter, toxic criteria are related to the probability (frequency) of the exceedance of the concentrations, which for acute toxicity allows a one-hour average concentration of a priority pollutant to exceed the criterion once in 3 years, while for chronic toxicity the 4-day average concentration can exceed the criterion once in 3 years (on average). Hence the modeling requires some statistical considerations.

The EPA (1991) guidance manual states that the hydrologically based low-flow, 7-day-exceedance–10-year-recurrence-interval-flow Q_{7-10} is similar to the biologically based 4-day exceedance–3-year-recurrence flow. Therefore in steady-state studies of a traditional point source discharge, Q_{7-10} may be used for chronic toxicity impact calculations using the CCC criterion. Using the Q_{7-10} for acute, CMC estimations would lead to an excessive number of water quality criteria exceedances. Using hydrologically based low-flow characteristics for storm-water-based discharges is inappropriate and makes no sense.

The toxic concentrations in runoff leaving a source area (the event mean concentrations (EMCs)) or in a point discharge are statistically distributed. The statistical characteristics of EMCs in urban runoff were presented in Chapter 8. It was shown that EMCs are log-normally distributed with a coefficient of variation (CV) (= log standard deviation/log mean) of around 1. There is no significant correlation between the magnitude of runoff and other parameters, and EMC magnitude for most toxic parameters.

In lieu of a single worst case steady-state modeling three dynamic techniques were recommended by the U.S. Environmental Protection Agency (U.S. EPA, 1991; Ambrose et al., 1988): (a) Monte Carlo simulation, (b) log-normal probability modeling, and (c) dynamic continuous modeling.

The in-stream (lake, estuary) response of the concentration of a toxic compound can be schematically represented by an input–output transformation concept, or

$$c = \mathbf{F(p)L}_i \qquad (13.23)$$

where
c = in-stream concentration response to a toxic input
$\mathbf{L}$ = toxic load vector from point and nonpoint sources (including sediment)
$\mathbf{F(p)}$ = transfer function or a model for the receiving water system
$\mathbf{p}$ = parameters–coefficients of the model
i = source subscript

The form of transfer function may range from a simple dilution formula for a soluble conservative compound to a complex dynamic model such as WASP. In this sense, Equation (13.23) is a schematic representation and does not imply a simple multiplication of $\mathbf{L}$ by $\mathbf{F(p)}$. Both $\mathbf{L}$ (load) and $\mathbf{F}$ (transfer function) can be considered to be deterministic or stochastic (probabilistic).

In the *Monte Carlo methodology* (Marr and Canale, 1988) a simplified steady-state toxic compound–sediment fate model may be used. The inputs (loads) and system parameters of the model are statistically analyzed and their characteristics (mean, coefficient, and variation and probability distribution) are obtained.

The *log-normal probability modeling* (DiToro, 1979; Mancini, 1983) calculates the mean and standard deviation of in-stream concentration

TABLE 13.5 Toxicants Fate and Transport Models

Model	Environment	Time Domain	Spatial Domain	Chemical
DYNTOX	River	Dynamic	Far-field 1-dimensional	Organic metal
EXAMS-II	Lake, river, estuary	Steady-state quasi-dynamic	Far-field 3-dimensional	Organic
WASP4	Lake, river, estuary	Steady-state dynamic	Far-field 3-dimensional	Organic metal
HSPF	River	Dynamic	Far-field 1-dimensional	Organic metal
SARAH-2	River	Steady-state	Treatment plant, near field	Organic
MINTEQA2	Lake, river, estuary	Steady-state	—	Metal

Source: After U.S. EPA (1991).

response of a toxic compound from the mean and standard deviation of the inputs (runoff volume and EMC) and stream flows assuming lognormal distribution.

Table 13.5 presents a summary of available models for modeling the fate and transport of toxic chemicals. These models are mostly public domain models, available from the U.S. EPA laboratory in Athens, Georgia. User's training seminars are periodically organized by the EPA. Most water quality models were developed with an emphasis on the dynamics in the water column and water column concentrations. Two models listed in Table 13.5 (EXAMS-II and WASP4) are capable of simulating water-column–sediment interactions, including resuspension, settling, and diffusion. However, additional work needs to be completed on the mechanisms of sediment–water-column exchange before the models can be validated for predictive applications involving sediments (U.S. EPA, 1991).

References

Altshuler, B. (1981). Modeling of dose-response relationships, *Environ. Health Perspect.* **42**:23–27.

Ambrose, R. B., et al. (1988). Toxic wastes: Waste allocation simulation models, *J. WPCF* **60**(9):1646–1655.

Ambrose, R. B., T. A. Wood, S. L. Martin, J. P. Connolly, and R. W. Schanz (1991). *WASP5.x, a Hydrodynamic and Water Quality Model—Model Theory, User's Manual, and Programmer's Guide*, U.S. Environmental Protection Agency, Athens, GA.

APHA (1989). *Standard Methods for Examination of Water and Wastewater*, 17th ed., American Public Health Association, Washington, DC.

Chapman, G. (1978). Toxicological considerations of heavy metals in the aquatic environment, in *Toxic Materials in the Aquatic Environment*, Oregon State University, Water Resources Institute, Corvallis, OR.

Connolly, J. P., and R. V. Thomann (1991a). Toxicity, in *Contaminated Sediments: Models for Criteria, Exposure and Remediation*, Continuing Education Seminar notes, Manhattan College, Bronx, NY.

Connolly, J. P., and R. V. Thomann (1991b). Food chain, in *Contaminated Sediments: Models for Criteria, Exposure and Remediation*, Continuing Education Seminar notes, Manhattan College, Bronx, NY.

Connolly, J. P., and R. P. Winfield (1984). *WASTOX, a Framework for Modeling the Fate of Toxic Chemicals in Aquatic Environments*. Part 2: *Food Chain*, Environmental Engineering Program, Manhattan College, Bronx, NY.

DiToro, D. M. (1979). *Statistics of Receiving Water Response to Runoff*, paper presented at National Conference on Urban Stormwater and Combined Sewer Overflow Impact on Receiving Water Bodies, University of Central Florida, Orlando, FL.

DiToro, D. M., et al. (1991). Technical basis for establishing sediment quality criteria for nonionic organic chemicals using equilibrium partitioning, *Environ. Toxicol. Chem.* **10**(12):1541–1583.

Krenkel, P. A., and V. Novotny (1980). *Water Quality Management*, Academic Press, New York.

Landrum, P. F., S. R. Nihart, B. J. Eadie, and L. R. Herche (1987). Reduction in bioavailability of organic contaminants to the amphipod Pontoporeia hoyi by dissolved organic matter of sediment interstitial waters, *Environ. Toxicol. Chem.* **6**:11–20.

Landrum, P. F., M. D. Reinhold, S. R. Nihart, and B. J. Eadie (1985). Predicting the bioavailability of organic xenobiotics to Pontoporeia hoyi in the presence of humic and fulvic materials and natural dissolved organic matter, *Environ. Toxicol. Chem.* **4**:459–467.

McCarty, J. F., and B. D. Jimenez (1985). Reduction in bioavailability to bluegills of polycyclic aromatic hydrocarbons bound to dissolved humic material, *Environ. Toxicol. Chem.* **4**:511–521.

Mancini, J. L. (1983). Development of methods to define water quality effects of urban runoff, EPA 600/2-83-125, U.S. Environmental Protection Agency, Cincinnati, OH (NTIS Publ. 2584-122928, Springfield, VA).

Marr, J. K., and R. P. Canale (1988). Load allocation for toxics using Monte Carlo techniques, *J. WPCF* **60**(5):659–666.

Mason, C. F. (1991). *Biology of Freshwater Pollution*, 2d ed., Longman Sci. & Technical, Wiley, New York.

O'Connor, D. J. (1988). Models of sorptive toxic substances in freshwater systems, Part I to III, *J. Environ. Eng. Div. (ASCE)* **114**(3).

Robinson, J., et al. (1967). Organo-chlorine residues in marine organisms, *Nature* **214**:1307–1311.

Rosiallo, A. P., J. M. Essigmann, and G. N. Wogar (1977). Rapid and accurate determination of the median lethal dose (LD_{50}) and its error on a small computer, *J. Toxicol. Environ. Health* **3**:797–809.

Salomons, W., and U. Förstner (1984). *Metals in the Hydrosphere*, Springer Verlag, Berlin, New York.

Schnoor, J. L., C. Sato, D. McKechnie, and D. Sahoo (1987). *Processes, Coefficients, and Models for Simulating Toxic Organics and Heavy Metals in Surface Waters*, EPA 600/3-67/015, U.S. Environmental Protection Agency, Athens, GA.

Sprague, J. B. (1969). Review paper. Measurement of pollutant toxicity to fish. I. Bioassay methods for acute toxicity, *Water Res.* **3**:793–821.

Sprague, J. B. (1973). The ABC's of pollutant bioassay using fish, in *Biological Methods for the Assessment of Water Quality*, J. Cairns and K. L. Dickson, eds., American Society for Testing and Materials, Philadelphia, PA.

Thomann, R. V. (1991a). Bioconcentration and depuration by aquatic organisms, in *Contaminated Sediments: Models for Criteria, Exposure and Remediation*, Continuing Education Seminar notes, Manhattan College, Bronx, NY.

Thomann, R. V. (1991b). Bioaccumulation model of organic chemical distribution in aquatic food chain, in *Contaminated Sediments: Models for Criteria*,

Exposure and Remediation, Continuing Education Seminar notes, Manhattan College, Bronx, NY.

Thomann, R. V., and J. A. Mueller (1987). *Principles of Surface Water Quality Modeling and Control*, Harper and Row, New York.

U.S. Environmental Protection Agency (1985). *Methods for Measuring the Acute Toxicity of Effluents to Freshwater and Marine Organisms*, EPA 600/4-85/013, U.S. EPA, Washington, DC.

U.S. Environmental Protection Agency (1987a). *Quality Criteria for Water 1986*, U.S. EPA, Washington, DC.

U.S. Environmental Protection Agency (1987b). *Integrated Risk Information System*. Vol. 1. *Supportive Documentation*, EPA 600/8-86/032a, Office of Health and Environmental Assessment, U.S. EPA, Washington, DC.

U.S. Environmental Protection Agency (1991). *Technical Support Documents for Water Quality-Based Toxics Control*, EPA 505/2-90-001, Office of Water, U.S. EPA, Washington, DC.

Weber, C. I. (1981). Evaluation of the effects of effluents on aquatic life in receiving waters—An overview, in *Ecological Assessment of Effluent Impacts on Communities on Indigenous Aquatic Organisms*, ASTM STP 730, J. M. Bates and C. I. Weber, eds., American Society for Testing and Materials, Philadelphia, PA, pp. 3–13.

Weiss, H., K. Bertine, M. Kolse, and F. D. Goldberg (1975). The chemical compositions of Greenland glacier, *Geochim. Cosmochim. Acta.* **29**:1–10.

14

Wetlands

Wetlands are described as "the kidneys of the landscape."

W. J. Mitsch and J. G. Gosselink

Until the last century the drainage basin of the Lagoon of Venice in Italy was composed primarily of lowland marshes and wetlands transected by canals and tributary streams. Marshes and wetlands served as sinks for nutrients that promoted the growth of lush vegetation and at the same time protected the lagoons and other waterways from the symptoms of eutrophication and hypereutrophication, which are now serious problems.

The drainage work that began in the basin of the Lagoon of Venice in approximately 1880 and has continued until today, has transformed the wetlands of the basin into agricultural and urban drylands. A complex network of drainage canals with pumping stations has significantly lowered the ground-water levels throughout the basin, and large quantities of drainage and irrigation return flows rich with nutrients and residues of other agricultural and industrial chemicals are now directed toward the Lagoon. By draining the wetlands, the Lagoon was deprived of a natural buffering system for nutrients that before the beginning of the drainage works were retained therein. Increased use of fertilizers by the agricultural sector, and the switch from organic (manure) fertilizers to chemicals, are another factor contributing to the greatly increased transport of nutrients from the basin to the Lagoon.

Originally, the Lagoon itself was surrounded by tidal brackish marshes, which still remain in some parts of the Lagoon. These marshes constituted a natural buffer that shielded the Lagoon from the influx of nutrients by runoff and shallow ground-water flows from surrounding lands. Impounding the marshes and forming the fish ponds (*valli da pesca*) limited this buffering capacity.

Loss of wetlands is not limited only to the watershed of the Lagoon of Venice. Until recently in the United States, as well as throughout the world, wetlands were considered a source of disease (malaria) and an obstacle to man's use of land resources for growth, agriculture, and economic development. Early "wetland management programs" in the United States—both governmental and private efforts—concentrated on drainage, filling, and conversions to agriculture and urban uses. The "success" of these early programs has been documented by the U.S Fish and Wild Life Service (a government agency under the U.S. Department of Interior), which estimated that more than one-half of the approximately one million square kilometers of wetlands in the lower 48 states were lost between the arrival of the first settlers and the present time. The rate of loss of wetlands in the years before 1980 was about 1000 to 2500 km^2 per year. In Illinois alone, over 40,000 km^2 of wetlands, representing about 27% of the total state area, were drained and converted to agriculture and urban lands since the mid-1800s.

Today, wetlands are viewed in a very different light. Numerous studies have found that wetlands are essential for healthy hydrology and the ecology of watersheds. They prevent floods, cleanse waters, protect shoreline, and recharge ground-water aquifers. They retain nutrients and other pollutants, which are then incorporated in the wetland biomass. For example, studies of urban and rural watersheds near Minneapolis, Minnesota, by Oberts (1982) documented that nutrient export from watersheds with a significant wetland area (10% to 20% of the watershed area) were much lower than the export from watersheds where the wetlands were drained. Wetlands are also ecological assets because they provide habitat for waterfowl, animals, and vegetation. Control of mosquitoes can be achieved by planting mosquito-eating fish (*Gambusia affinis*) or by fish management.

The Congress, the federal government, and many states have realized the very high ecological and hydrological values of wetlands and enacted wetland protection and rehabilitation acts. For example, in Florida developers must now reestablish 2 hectares of artificial wetland for each hectare of natural wetland lost due to development. Draining of wetlands is now prohibited in many states and efforts are on the way to reestablish formerly drained wetlands. Through the efforts of the Wetland Foundation in Chicago programs aimed at wetland restoration are promoted. An example of such programs is the Des Plaines experimental wetland located north of Chicago where scientists study wetland restoration, their incorporation into the landscape, habitat creation and enhancement, and diffuse-pollution control (Hey et al., 1989). Restoration of riparian wetlands and the construction of new wetlands play an important role in diffuse-pollution abatement.

DEFINITIONS AND TYPES OF WETLANDS

There are a variety of lands and bodies of water that are called wetlands. Wetland definitions include the following features (Mitsch and Gosselink, 1986):

1. Wetlands are distinguished by the presence of water.
2. Wetlands often have unique soils that differ from adjacent uplands.
3. Wetlands support vegetation adapted to the wet conditions (hydrophytes), and consequently are characterized by the absence of flooding intolerant vegetation.

Other less distinct features have also been pointed out, such as:

1. The depth of standing water in wetlands may vary throughout the year or be absent for some time.
2. Wetlands are often at the margin (boundary) between deep water and terrestrial uplands, and are affected by both systems.
3. Wetlands vary in size from small—one to a few hectares of prairie wetlands—to large expansive wetlands of several thousands of square kilometers (example, the Everglades in Florida).

In general terms, wetlands are defined (Cowardin et al., 1979) as lands where saturation with water is the dominant factor determining the nature of soil development and the types of plant and animal communities living on the surface.

Due to the existence types of various of wetlands, a precise definition for wetlands that would satisfy all is not possible (Mitsch and Gosselink, 1986). The most comprehensive definition for wetlands was advanced by the U.S. Fish and Wildlife Service:

> Wetlands are lands transitional between terrestrial and aquatic systems where the water table is usually at or near the surface or the land is covered by shallow water. Wetlands must have one or more of the following attributes: (1) at least periodically, the land supports predominantly hydrophytes, (2) the substrate is predominantly undrained hydric soils, or (3) the substrate is nonsoil

(organic matter) with water or covered by shallow water at some time during the growing season each year.

Since 1975 the U.S. Fish and Wildlife Service has been conducting an inventory of wetlands throughout the United States. U.S. wetland terminology recognizes a variety of types and terms that are applied to wetlands (Mitsch and Gosselink, 1986). The most important types of wetlands are as follows.

Coastal or Marine Wetlands

The marine system extends from the outer edge of the continental shelf to the high water or spring tides. *Marine subtidal* includes areas that are continuously submerged, while *marine intertidal* includes areas in which the substrate (soil and organic bottom matter) is alternately exposed and flooded by tides.

The estuarine system consists of deepwater tidal habitats that are semienclosed by land but have open, partially obstructed, or sporadic access to the open sea or ocean and in which sea water is at least occasionally diluted by freshwater runoff from the land. *Estuarine subtidal*

FIGURE 14.1. Tidal marsh inside the Lagoon of Venice in Italy. (Photo: V. Novotny.)

Definitions and Types of Wetlands   865

is that portion that is continuously submerged (considered deepwater habitat), while *estuarine intertidal* is the portion that is by turns exposed and flooded by tides. The marshes within the tidal portion of the Lagoon of Venice could be categorized as estuarine intertidal or subtidal wetlands (Fig. 14.1).

Depending on salinity and type of vegetation, marine wetlands can be categorized as:

Tidal salt marsh. Contains primarily salt or brackish water.
Tidal freshwater marsh. More distant from the coast; experiences some tidal effects, but mostly contains freshwater.
Mangrove wetlands. Saltwater marshes in tropical and subtropical areas (Florida). Salt-tolerant trees and brush usually dominate these wetlands.

Inland Wetlands

The *palustrine wetlands* include all nontidal wetlands excluding those adjacent to or that are a part of river channels and littoral zones of larger

FIGURE 14.2.   Forested wetland: Huricon marsh in Wisconsin. (Photo: A. Capodaglio.)

and distinct lakes and other impoundments. Most definitions divide the palustrine wetlands into the following groups:

Forested and scrub land. Wetlands dominated by the presence of woody vegetation (Fig. 14.2).
Emergent. Wetlands with primary erect, rooted herbaceous plants typically found in wet environments and other palustrine areas.
Nonvegetated wetlands. Small inland bodies of water and wetlands dominated by aquatic beds.

An arbitrary depth of 2 meters has been used by the U.S. Fish and Wildlife Service to distinguish between wetland and deepwater systems.

Riverine wetlands (Fig. 14.3) adjoin deepwater stream channels, while *lacustrine wetlands* (Fig. 14.4) are littoral wetlands situated in topographic depressions or surrounding lakes or impounded rivers (reservoirs). The surface area of the lake or impoundment should be more than 8 hectares (20 acres), or be deeper than 2 meters, or have an active wave-formed or bedrock shoreline feature. The term *riparian wetlands* is commonly used for lacustrine and riverine wetlands.

The following terms are commonly associated with nontidal wetlands:

Swamp. Wetland dominated by trees and shrubs.
Marsh. Frequently or continually inundated wetland characterized by emergent vegetation adapted to saturated soil conditions. In European terminology, a marsh does not accumulate peat.
Bog. A peat-accumulating wetland that has no significant inflows or outflows and supports acidophilic mosses.
Fen. A peat-accumulating wetland that receives some drainage. Fens and bogs are generically called peatlands.
Slough. A swamp or shallow lake system, or a slowly flowing swamp or marsh.

Other Wetlands

Rice fields, some *disposal of wastewater on land*, and *artificial wetlands* represent most of the remaining land surfaces that can be categorized as wetlands. An artificial wetland is a man-made transformation of previously dry upland into a wetland, and is accomplished by one or more of the following measures:

1. Disconnecting and abandoning previously installed drainage.
2. Permanent or periodic flooding of the land with water from nearby bodies of water.

Definitions and Types of Wetlands   867

FIGURE 14.3. Meandering stream and wetland in northern Wisconsin. (Photo: University of Wisconsin.)

FIGURE 14.4. Lacustrine riparian wetland in northern Wisconsin. (Photo: University of Wisconsin.)

3. Diverting urban storm water and/or treated or partially treated effluents onto the land, which creates saturated soil conditions (similar to land-disposal practices for wastewater).
4. Grading and excavating top soil layers to the ground-water table.

Las Vegas Wash, a wetland located between the City of Las Vegas and Lake Mead in Nevada, was unintentionally created by increased effluent discharges from a rapidly developing metropolis into a previously ephemeral stream and surrounding land in desert climatic conditions. Generally, however, urbanization results in loss of wetland.

Creating an artificial wetland is an engineering and ecological endeavor that must be planned and executed properly to avoid development of unwanted species (for example, obnoxious algal mats in shallow standing waters or nuisance growths of water hyacinths in subtropical and tropical areas). *Ecological engineering* is a new branch of environmental engineering that deals with protection and restoration of ecological systems, including wetlands (Mitsch and Jørgensen, 1989). To ensure that proper vegetation will develop, the bottom surface of the constructed wetland should be covered by a mulch brought from an existing nearby wetland, and the wetland should be seeded with proper wetland vegetation. Typical wetland vegetation is shown in Table 14.1. All plant species reported in the table and shown in Figure 14.5 are common throughout the world. However, Mitsch (1990) pointed out that the list of possible wetland plants is very large and that an almost infinite number of combinations of plants exists. It should also be pointed out that water hyacinth and duckweed are suitable in aquaculture ponds. However, their use in wetlands is questionable and even undesirable.

Creation of wetlands and their use for water-pollution control is now

TABLE 14.1 Typical Emergent Wetland Plants

Common Name	Scientific Name	Temperature Range (°C)		Maximum Salinity Tolerance (g/l)	Effective pH Range
		Desirable	Seed Germination		
Cattail	*Typha* spp.	10–30	12–24	30	4–10
Common reed	*Phragmites communis*	12–23	10–30	45	2–6
Rush	*Juncus* spp.	16–26		20	5–7.5
Bulrush	*Scirpus* spp.	16–27		20	4–9
Sedge	*Carex* spp.	14–32			5–7.5

Source: After U.S. EPA (1988).

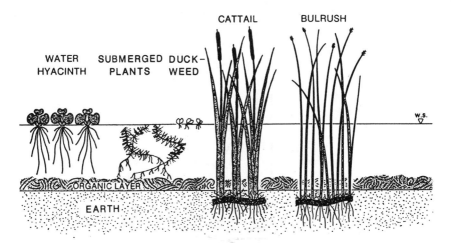

FIGURE 14.5. Common aquatic plants.

enhanced by recent efforts to take polluting agricultural lands out of production. In the United States such lands are commonly converted to so-called "buffer strips" or high slope woodland. Creation of riparian wetlands in combination with overland-flow buffer strips could provide an improved water-pollution control benefit. Large projects to restore riverine wetland systems are being carried out on the Des Plaines River in Lake County, Illinois (Hey, 1988), the Milwaukee River in Wisconsin (Gayan and D'Antuono, 1988), and several other rivers in the United States. A large number of wetland restoration projects have been carried out in Florida, which enacted the most far-reaching wetland protection and restoration laws (Fisk, 1989). In southeastern Wisconsin, a 36-ha (90-A) man-made wetland has been created to protect and enhance the water quality of the Delavan Lake (810 ha). The wetland was constructed as a critical component of a major restoration project and is very important to the long-term management of the lake.

WETLAND FUNCTION

In contrast to early views that regarded wetlands as a nuisance, beneficial uses of wetlands are numerous and should be considered in any wetland protection, enhancement, use, and creation endeavor. The most important uses and ecological functions of wetlands are as follows.

- *Flood storage and conveyance.* Riverine wetlands and adjacent floodplain form natural corridors in which floodwater is stored and held

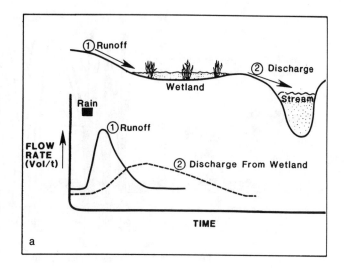

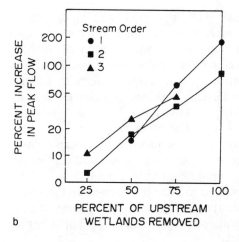

FIGURE 14.6. (a) Effect of wetlands on stream flow. (After Mitsch and Gosselink, 1986.) (b) Impact of wetland drainage on peak flow and flooding. (After Ogawa and Male, 1983.)

back. Palustrine wetlands that receive surface runoff from a surrounding watershed will attenuate the runoff peaks (Fig. 14.6).

- *Stream flow modification.* Wetlands have an impact on other flows as well. Their tendency to retain water is most evident during the summer, fall, and winter periods, while during spring when wetlands are commonly flooded they tend to release more flow than the surrounding upland.

- *Erosion reduction and sediment control.* Riverine and lacustrine wetland vegetation slows runoff and flood flow, thus mitigating shoreline erosion. The vegetation in all types of wetlands slows down the flow, thus enhancing sedimentation. Vegetation stems also filter sediments from water, and the roots of the vegetation then bind and stabilize the deposited silt.
- *Ground-water recharge/discharge.* Wetlands and ground water are interconnected; however, the connection is poor since the very existence of the wetland usually implies highly impervious substrate subsoils. Possible connections are shown on Figure 14.7, taken from Novitzki (1979). If shallow ground water is discharging into a wetland, it can bring nutrients (primarily nitrate nitrogen) and minerals, which are then used by the wetland vegetation. The ground-water zone can then be recharged with water from the wetland with less nitrate contamination. If a wetland is recharging (the flow is from the wetland toward ground water), it is a natural barrier preventing some mobile pollutants from entering the ground-water zones.
- *Pollution control.* The foregoing discussion on pollution prevention of ground-water resources and sediment control can be extended to other bodies of water with which wetlands interact. Wetland vegetation and microorganisms residing on the stems and roots of vegetation and in sediments can filter, absorb, and decompose suspended and dissolved organic matter, and convert organic and ammoniacal nitrogen to nitrate (nitrification) with subsequent denitrification of nitrate to nitrogen gas. Some phosphorus can be used by vegetation and/or adsorbed onto sediments. The accumulated residual organic matter can also immobilize toxic pollutants and essentially bury them without harm to the biota. Extensive research is being conducted on the use of natural and man-made wetlands as tertiary treatment facilities for municipal and industrial wastewater and for storm-water runoff pollution control.
- *Wildlife habitats.* The land–water interface, including upland buffer zones, is among the richest wildlife habitats in the world. This is because of the abundance of water, nutrients, and shelter provided by the wetlands and their vegetation. Many endangered species of animals rely on wetlands for their habitats. For example, in Illinois 40% of the state's threatened and endangered species depend on the remaining wetlands, which now represent less than 3% of the state's area. Many subtropical species in Florida and Louisiana and many other areas would vanish with the loss of wetlands.
- *Recreation and enjoyment.* Over twenty million sport and commercial fishers nationwide use wetlands, along with millions of hunters. Millions more just enjoy watching wetland birds, wildlife, and plants.

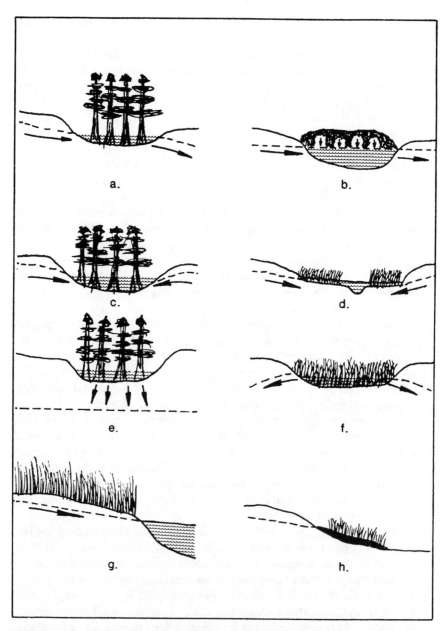

FIGURE 14.7. Wetland–ground-water interactions. (a) Both inflows and outflows of ground water through the marsh. (b) Underflow of raised bog. (c) Swamp as ground-water depression wetland. (d) Marsh as ground-water depression wetland. (e) Perched swamp or surface-water depression wetland. (f) Marsh as ground-water source (recharging). (g) Ground-water flow through saltwater marsh or riparian wetland. (h) Ground-water seep wetland or ground-water slope wetland. (After Novitzki, 1979.)

Use of Wetlands for Water Quality Control

A large amount of the recent scientific and technical literature focused on the application of wetlands—natural or artificial—for water-pollution control (for example, Reed, Middlebrooks, and Crites, 1988; Hammer, 1989a). An article in *Civil Engineering* (Dawson, 1989) pointed out the benefits of using wetlands for treatment or posttreatment of municipal wastewater in several U.S. locations. A treatise on ecological considerations and mitigation of wetlands that are used for water quality control and disposal of wastewater has been published by Godfrey et al. (1985).

Following Reed, Middlebrooks, and Crites (1988) and Hammer (1989a), the major uses of wetlands in combination with the disposal of wastewater effluents and storm-water runoff are:

- Disposal of treatment effluents into natural wetlands.
- Use of natural wetlands for further wastewater renovation.
- Use of effluents or partially treated wastewater for the enhancement, restoration, or creation of wetlands.
- Use of constructed (artificial) wetlands as a wastewater treatment process.
- Use of natural or man-made wetland for storm-water management and treatment.
- Potential use of constructed wetlands for immobilization of toxics and their permanent safe burial.

In the United States, natural wetlands are considered as any other receiving surface-water body; hence, they are protected from excessive discharges of pollution and any discharge requires a permit. This generally precludes discharges of untreated or partially treated (with a primary treatment only) wastewater or polluted storm water (runoff) into natural wetlands. Most states (except those that enacted special wetland standards) make no distinction between a wetland with standing water and adjacent surface water. One may expect that similar legislation is in place or will be enacted in other developed countries. In most cases it would mean that the discharges of wastewater into wetland systems must receive treatment.

The use of natural wetlands is beneficial and applicable for nutrient control and removal, for effluent polishing, for removal of nutrients and other pollutants from urban storm-water runoff, for in-stream waste-assimilative capacity enhancement (for example, diverting a part of the flow of a polluted stream onto an adjacent riparian wetland), and as buffers between agricultural areas and surface-water bodies. However, as is true for any surface-water body, the ability of wetlands to receive and assimilate pollution is not limitless, and those who plan to discharge pollution into natural wetlands as well as the regulatory agencies must make sure that the waste-assimilative capacity of the wetland into which the discharge is planned is not exceeded. The waste-assimilative capacity of wetlands can be estimated by modeling (Mitsch, Straskraba, and Jørgensen, 1989). Since the hydraulics of wetland systems is the same as for surface-water bodies with a significant sediment component, the U.S. EPA's water quality model WASP may be used, with or without modifications, for wetland modeling.

On the other hand, artificial (constructed) wetlands (Figs. 14.8 to 14.10) are generally not subjected to surface-water quality standards and can be conveniently and inexpensively used for effluent treatment, polishing, nutrient removal, and storm-water management and treatment. Constructing a wetland in a place where there is none now would bypass in most cases the regulations restricting discharges into surface-water bodies and only the outflow from the wetland would have to meet the effluent standards (which as was stated previously can be effluent or stream limited). Unlike for natural wetlands, where both influent into the wetlands and the outflow are regulated and must comply with environmental standards, only outflow from artificial wetlands is commonly

FIGURE 14.8. Wetland restoration project near Tampa, Florida. Wetlands are incorporated into a new residential development. (Photo: V. Novotny.)

Wetland Function 875

FIGURE 14.9. Restored wetland used for storm-water treatment in the Minneapolis-St. Paul area. (Photo: G. Oberts.)

FIGURE 14.10. Riparian wetland restoration, Des Plains River, Lake County, Illinois. (Photo: V. Novotny.)

regulated. Also water quality control performance can be significantly improved since the vegetation and wetland management of the artificial wetland system can be optimized for the most efficient removal of pollutants.

Organic Matter and Nutrient Mass Balance in Wetlands

Natural and well-constructed man-made wetlands contain an abundance of diverse vegetative and bacterial populations that are very effective in removing a variety of pollutants, including decomposition and immobilization of toxics. Both emerged and submerged plants and algae are usually present. In addition to the biological and biochemical absorption and degradation of pollutants, physical processes also attenuate pollution. The processes that contribute to the removal of pollutants by wetlands can be grouped as follows.

1. Physical processes
   a. Sedimentation
   b. Filtration
2. Physical–chemical processes
   a. Adsorption of pollutants on plants and soil and organic substrates
3. Biochemical processes
   a. Aerobic biochemical degradation of organic matter by bacteria in water, attached to plants, stems, in the top layer of sediments, and in aerobic pockets near roots and rhizomes of plants
   b. Nitrification by nitrifying organisms primarily residing on plant stems, in the top layer of humic sediments and on the roots and rhizomes of plants
   c. Denitrification by bacteria residing in anaerobic water and sediments
   d. Anaerobic decomposition of organic matter in sediments and anaerobic water
   e. Uptake of nutrients and some pollutants by the plants and their incorporation in the plant biomass

The richness and variety of the processes is the primary asset of wetland treatment. Due to very low velocities of water and dense vegetation, sedimentation and filtration are very effective in removing suspended solids and the pollutants attached on them. Bacteria attached to plant stems and in the upper part of sediments are the major factor for BOD removal. In a sense, wetlands in their function of BOD removal resemble a trickling filter rather than a free-standing body of water.

Kadlec and Alvord (1989) pointed out that there are three major processes that participate in the removal of waterborne pollutants entering the wetland: *biomass increases*, *burial* and *gasification*. The nutrients that enter the wetland are incorporated into the biomass, primarily vegetation and algae (autotrophic biota); however, some nutrients as well as biodegradable organics are incorporated into the biomass of heterotrophic organisms. The production of new biomass due to the nutrient and organic input is a temporary but often long-term sink for assimilated pollutants.

Formation of new organic soils and peat that contain organics and nutrients represent a more permanent sink. Note that coal deposits were formed millions of years ago by the deposition of organic matter in wetlands. Assimilated nitrogen, carbon, and sulphur after biological transformation can be released into the atmosphere and permanently lost from the system. Organic solids and other chemicals can adsorb and/or immobilize and even decompose toxic materials. If toxic loads are below the assimilative capacity of the wetland the constructed wetlands could safely dispose of the toxics and bury the end products. Unfortunately, the state-of-the-art knowledge on assimilation of toxics by wetlands is still not advanced enough to make any definite conclusions and recommendations.

Production, Retention, and Removal of Biodegradable Organic Matter

The concepts of organic productivity and the roles of nutrients and oxygen in aquatic bodies of water and underlying sediments were explained in Chapter 12. Processes occurring in soils were the subject of Chapter 6. The same concepts are applicable to wetlands since wetlands are considered a transition between land and bodies of water.

Both tidal marshes and inland wetlands are the most productive ecosystems in the world. Up to 2 tonnes per hectare of organic matter can be produced annually by these systems; however, typical values of organic matter production by primary productivity in tidal marshes are between 50 and 300 kg/ha-year, while productivity of inland wetlands is somewhat higher, ranging from 90 to 600 kg/ha-year. This is higher than the productivity of intensively cultivated agricultural fields (Mitsch and Gosselink, 1986). In this context, an allochthonous (= from sources outside the wetland, such as runoff and wastewater discharges) influx of organics may not be significant in comparison with the productivity of the wetland. Table 14.2 presents the compiled data on the organic productivity

TABLE 14.2 Production of Organic Matter (Net Primary Productivity) in Natural Wetlands

Wetland Type	Net Primary Productivity (g/m²-year) (Range)	Total Biomass (kg/m³)	Nitrogen Loading (g N/m²-yr)	Plant Nitrogen Uptake (g/m²-yr)
Northern bog	500 (153–1943)	53	0.8	9
Inland fresh marsh	1980 (1070–2860)	46	22	48
Tidal fresh marsh	1370 (780–2100)	46	75	54
Salt marsh	1950 (330–3700)	46	30–100	25
Swamp forest	870 (390–1780)	52	900	14
Riparian forest	1040 (750–1370)	37	10,000	17
Mangrove	1500 (0–4700)	60	30	24

Source: After Mitsch and Gosselink (1986).

of various types of wetlands. These data are important when considering the assimilative capacity of wetlands for various pollutants.

The removal of organic matter is by sedimentation, filtration, and biological absorption by microorganisms residing in water, attached to vegetation, and living in sediments. Figure 14.11 shows the processes in water and sediment columns. The biodegradation can be both aerobic and anaerobic, with aerobic degradation taking place primarily in the water column and in the upper part of the sediments, and anaerobic degradation occurring in sediments and in water if the oxygen supply is exhausted. However, it has been pointed out by Reed, Middlebrooks, and Crites (1988) that some rooted plants (for example, the water hyacinth) can transfer oxygen to their root zone and rhizomes and in this way create small pockets for aerobic degradation and nitrification (Fig. 14.12).

About half of the organic matter is retained and decomposed in tidal marshes and the second half is transported to the downstream sea or estuary system (Mitsch and Gosselink, 1986). The export of organic matter from inland wetlands varies greatly and depends on the type of the wetland, its hydrology, and many other factors.

Immobilization of Toxic Compounds by the Organic Substrate of Wetlands

Organic matter production in wetlands may also be a key to their use in disposing (sink) of low quantities of toxic compounds. Most toxic com-

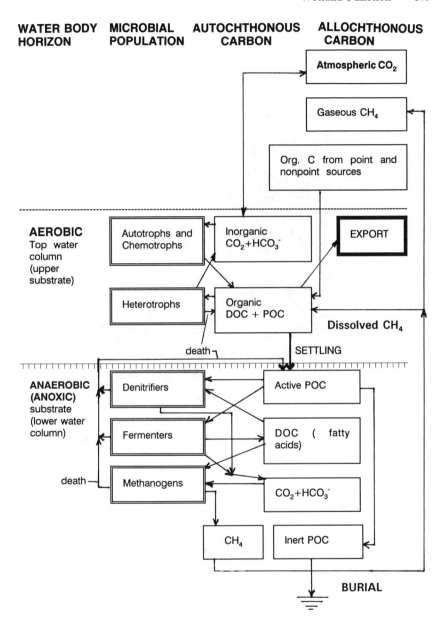

FIGURE 14.11. Water–sediment column processes. *Key*: POC-particulate organic carbon; DOC-dissolved organic carbon.

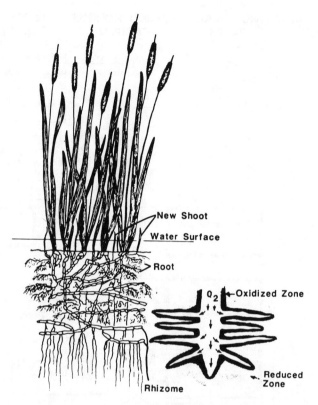

FIGURE 14.12. Transport of oxygen by roots and rhizomes in anaerobic wetland substrate. (After Hammer and Bastian, 1989, with permission from Lewis Publishers, a subsidiary of CRC Press, Boca Raton, Florida.)

pounds, such as toxic metals and organic chemicals, have a strong affinity for immobilization by particulate and colloidal organic matter (toxic metals and organics) and by sulfide ligants (toxic metals) generated by anaerobic decomposition of organic matter, both in sediments and wetland substrates. Complexed toxic compounds are unavailable to aquatic biota, including plankton, plants, and animal (invertebrate and higher) forms. Complexed and adsorbed toxic compounds can also be filtered and/or settled out in wetlands. Quantitatively, this function of wetlands is generally unknown. Some concepts and calculations will be presented in the next section of this chapter.

The assimilative capacity of wetlands for toxic compounds is related to the primary productivity of the wetland that is stimulated by the nutrient inputs and recycling.

Nutrient Transformation and Removal

The best use of wetlands for water quality control, in addition to sediment and particulates filtering, is for nutrient removal. The high production of organic matter by wetlands requires proportionate sources of nutrients that may come from autochthonous (inside the ecosystem) or from allochthonous (outside the boundary of the ecosystem) sources. Autochthonous sources imply nutrient recycling within the wetland whereby nutrients (N and P) are taken up by plants and released back after death and decomposition. Allochthonous sources include the influx of nutrients from the tributaries and from the atmosphere. Whether a wetland becomes an effective sink of nutrients depends on the buildup of inert residual organic matter and on the permanent burial rate in the sediment of the produced organic matter (organic forms of nutrients),

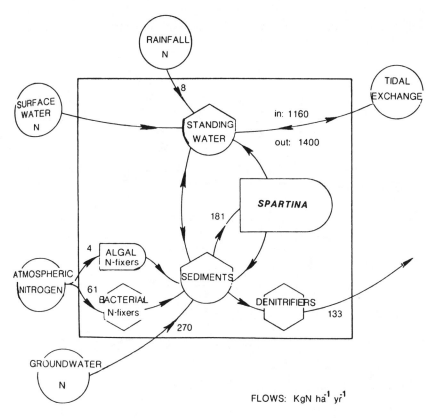

FIGURE 14.13. Nutrient balance of the Great Sippewissett salt marsh in Massachusetts. (After Mitsch and Gosselink, 1986.)

their permanent adsorption and immobilization by the clay and organic components of the sediment (ammonia and phosphates), and on the gasification rate of nitrate–nitrogen by denitrification.

Kadlec and Alvord (1989) have shown that a northern wetland (located in Michigan) responds dramatically to elevated nutrient inputs from allochthonous sources; in this case, from biologically treated wastewater. Increases in nitrogen and phosphorus caused an expansion of the biomass and a buildup of several millimeters of new organic soil from the organic residues of plants grown in the wetland, which then buried significant quantities of nutrients. Nitrification was minimal during growing season because of the lack of oxygen in the sediments and bottom layers. Denitrification of nitrate was almost complete and the nitrate was normally exhausted in the wetland during the growing season.

A nutrient budget for a saltwater marsh located in Massachusetts is shown on Figure 14.13. Denitrification in saltwater marshes can be high and the marshes can become nutrient sinks, especially when high nutrient waters pass through them (Valiela, Teal, and Sass, 1973). Many coastal wetlands are nitrogen limited.

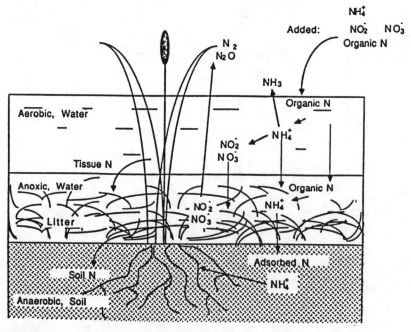

FIGURE 14.14. Routes and transformation of nitrogen in a pallustrine wetland receiving wastewater input. (After Kadlec, 1988.)

Figure 14.14 shows routes and transformations of nitrogen in a palustrine wetland used for wastewater treatment. Due to the heavier organic load and high productivity of the wetland, the bottom water layer is mostly anaerobic. Thus, the entire substrate is anaerobic with the exception of possible small aerobic pockets around the roots of plants. In another situation, especially on bright sunny days during photosynthetic oxygen production, the water column becomes aerobic and the oxygen supply creates a thin aerobic top sediment layer that allows the nitrification process.

Mitsch and Gosselink (1986) compiled data on the performances of natural wetlands for the removal of nutrients. The performance or treatment efficiency of wetlands is greatly affected by the geographical locations. Northern wetlands located in colder climatic zones have two distinct seasons: the growing season and the dormant season. During the growing season the nutrients are effectively removed by the plants and by simultaneous nitrification–denitrification. However, during the dormant season, nutrients are released and the wetland may become a source of

TABLE 14.3 Examples of Natural Wetlands Receiving Wastewater Inputs and Nutrients Removal Efficiencies

Type of Wetland	Location	Loading (Pop./ha)	Substrate	Nutrient removal (percent)	
				Total N	Total P
Northern peatland					
Bog	Wisconsin	30	O	98	78
Nontidal freshwater marsh					
Cattail marsh	Wisconsin	17	O	80	88
Lacustrine marsh	Ontario	—	—	38	24
Deepwater marsh	Florida	99	O	—	97
Lacustrine marsh	Hungary	—	—	95	—
Riverine swamp	South Carolina	—	O	—	50
Tidal freshwater marsh					
Deepwater marsh	Louisiana	—	O	51	53
Complex marsh	New Jersey	198	I	40	0
Tidal salt marsh					
Brackish marsh	Chesapeake Bay	—	O/I	0	1.5[a]
Salt marsh	Georgia	Sludge	O/I	50	—
Salt marsh	Massachusetts	Sludge	O/I	85	—

Source: Compiled by Mitsch and Gosselink (1986).
Note: O: organic substrate; I: inorganic substrate; (—): information not given.
[a] Load given in g/m^2-year.

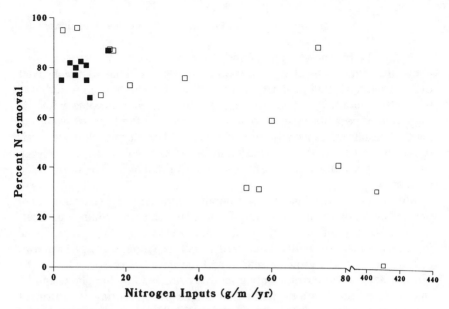

FIGURE 14.15. Nitrogen removal efficiency by wetlands as a function of nitrogen loading rate. (Based on data from Mitsch and Gosselink, 1986; Richardson and Nichols, 1985; and Knight, Windchester, and Highman, 1984.)

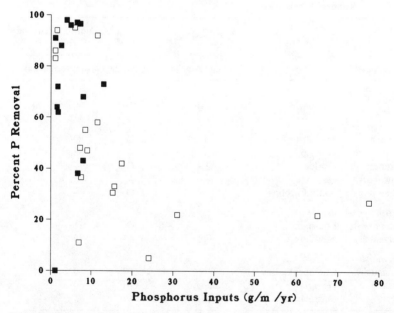

FIGURE 14.16. Phosphorus removal efficiency by wetlands as a function of phosphorus loading rate. (Based on data from Mitsch and Gosselink, 1986; Richardson and Nichols, 1985; and Knight, Windchester, and Highman, 1984.)

the nutrients. The ranges of removal efficiencies compiled by Mitsch and Gosselink (1986), shown in Table 14.3 and in Figures 14.15 and 14.16 (related to the nutrient loading), were very wide, both for phosphorus and nitrogen.

From the foregoing discussion and from literature findings the following rules have become apparent:

1. The wetland removal efficiency for nitrogen is high if the nitrogen is in a nitrate (oxidized) form. In this case, both nitrogen uptake by plants with subsequent burial of dead plants and algae, and denitrification are responsible for nitrogen removal. In some cases, 98% to 99% nitrogen removal efficiencies have been measured in such systems. Nitrate nitrogen is not adsorbed by soil organic particulates. Generally, wetlands require a start-up period for full development of denitrification (Kadlec, 1988). Maintaining aerobic pools in the wetland enhances nitrification but suppresses denitrification.

Stengel and Schulz-Hock (1989) have shown that denitrification in constructed wetlands was possible throughout the entire year in the climatic conditions of central Europe (similar to the northeastern United States).

2. If nitrogen arrives in the wetland in an ammoniacal or organic form, the primary removal mechanisms are by ammonification of the organic nitrogen and by the uptake of the ammonia by plants and heterotrophic microorganisms. Burial of the organic nitrogen and phosphorus with the produced organic matter and adsorption of ammonia and phosphate onto the soil (clay) and organic particulates then represent the primary removal mechansms for the nutrients (both nitrogen and phosphorus). Unless specifically managed to enhance nitrification (for example, by alternate draining and flooding of the wetland), it may be difficult to nitrify the ammoniacal nitrogen; hence, gasification of the nitrogen by denitrification may be minimal or greatly reduced. Expected total nitrogen removal could be around 40% to 60% on an annual basis, depending on the geographical location (see 3 below).

As stated before and shown on Figure 14.12, limited nitrification can occur in wetlands even when sediment and the sediment–water interfacial area are anaerobic. Under these circumstances, photosynthetic oxygen is supplied by plant roots and rhizomes that penetrate into the anaerobic zones and provide a limited aerobic environment for nitrification and nitrifying microorganisms. In such cases the depth of the anaerobic zone and the type of plant can make a significant difference (Reed, Middlebrooks, and Crites, 1988; Hammer and Bastian, 1989).

3. If the wetland is located in northern climatic conditions, part of the nutrients accumulated in the substrate is released and the wetland may become a source of the nutrients (both nitrogen and phosphorus) during the dormant season.
4. Removal of phosphate is by assimilation into new organic matter and by adsorption on clay, iron, aluminum, and organic particles with subsequent burial. Phosphorus removal in many wetlands is not effective due to the limited opportunity of phosphates to interact with soils and other adsorbing media. If the pH is low, as it is in many peat wetlands (bogs and fens), the phosphate may precipitate. In this sense the mechanism of phosphate removal and retention by wetlands is similar to that of saturated soils.

From lake sediment studies it is known (Mortimer, 1971) that phosphate is released from the settled organic matter and sediments under anaerobic conditions in the sediments and sediment–water interface. If the sediment–water interface is aerobic, the release is to a great degree blocked. The efficiency of wetlands to remove phosphorus is generally lower than for nitrogen.

5. Tidal and marine marshes are less effective in their nutrient removal efficiencies than the freshwater wetlands. This has been reported to be due to salt stresses on the microorganisms (Mitsch and Gosselink, 1986). Tidel brackish marshes with a high variability of salt concentrations may exhibit the lowest removal efficiencies. For example, brackish marshes around Chesapeake Bay showed no nitrogen removal capability (Bender and Correl, 1974).
6. Management of wetlands for water quality control and pollution removal is very important, as are the type of vegetation and water depths. Precise guidelines for the control of these factors have not been developed and are now only emerging. With the exception of one wetland located in Michigan (Kadlec and Alvord, 1989) no long-term experience (ten years or more) has been gathered for wetland management for water-pollution control.

The following list contains typical design loadings and expected performance of natural wetlands receiving secondary effluents (after Reed, Middlebrooks, and Crites, 1988).

Pretreatment needs	*Secondary treatment*
Climatic conditions	Warm
Detention time, days	10
Depth, m	0.2–1
Hydraulic loading, m^3/ha-day	100
Expected effluent quality	
BOD$_5$, mg/l	5–10
TKN, mg/l	5–10

Due to variations in storm-water characteristics and poor understanding of the wetland processes that remove pollutants under highly variable conditions, treatment efficiency predictions of a wetland storm-water system are not possible (Livingston, 1989).

Retention of Sulfur

Few studies are available that describe the retention and transformation of sulfur in wetlands, yet, from the water quality management viewpoint, sulfur may play a very significant role in the process of immobilization and detoxification of toxic metals. Sulfur transformation in wetlands is biologically mediated (Faulkner and Richardson, 1989) and is related to and governed by redox and pH interactions. Major transformations (shown on Fig. 14.17) are reduction of sulfate (SO_4^-) to sulfide (S^{--}) compounds in an anoxic environment, aerobic sulfate and elemental sulfur oxidation, and mineralization of organic sulfur to inorganic sulfate.

Sulfate reduction is accomplished by heterotrophic bacteria that use sulfate as an electron acceptor instead of oxygen. It should be pointed out that in the absence of oxygen bacteria preferentially use nitrate over sulfate; nevertheless, in wetlands an anoxic substrate environment, both processes can take place almost simultaneously. One mole of sulfate is reduced in the anoxic environment to sulfide for every two moles of organic carbon oxidized, as shown in the following reaction (Hedin, Hammock, and Hyman, 1989; DiToro et al., 1990)

$$SO_4^{2-} + 2CH_2O^* \rightarrow 2CO_2 + S^{2-} + 2H_2O$$

where CH_2O^* represents a simple organic molecule such as acetate. Depending on the chemical environment, hydrogen sulfide is released as gas, ionizes to HS^- and S^{--}, or precipitates as a polysulfide, elemental sulfur, or iron sulfides. It is the metal sulfide form that immobilizes and detoxifies the toxic metals. The precipitation of iron sulfide is

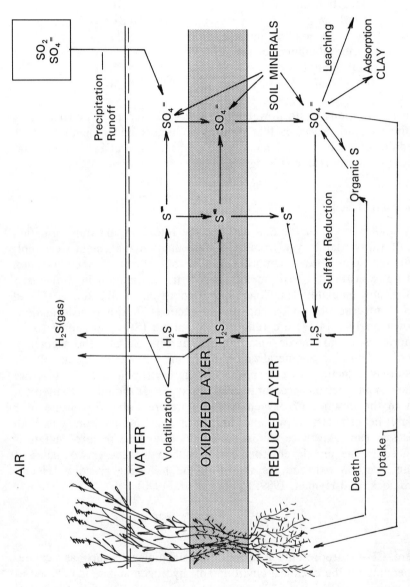

FIGURE 14.17. Sulphur transformation in wetlands. (After Mitsch and Gosselink, 1986.)

$$Fe^{2+} + S^{2-} \rightarrow FeS(s)$$

The measure of this immobilization capability is the acid volatile sulfide (AVS) test (DiToro et al., 1990, 1991b). The iron sulfide precipitate is the major component of the AVS (DiToro et al., 1990). As was reported in the preceding chapter for each mole of AVS present in the substrate one mole of toxic metal is immobilized (complexed) and, therefore, detoxified.

Wetlands function as effective sulfur sinks. Sulfur inputs originate from the atmosphere (acid rainfall) or from water inputs and wastewater effluents, such as acid mine drainage, and from natural sources. Up to 87% sulfur removals have been reported; however, if the wetland substrate and overlying waters become aerobic subsequent oxidation of reduced S to SO_x^{2-} can release sulfur back to the water, from where the sulfur can be flushed from the system.

Measured sulfur removals in natural wetlands ranged from 20% to 90% at sulfur loading rates between 20 and 50 g of S/m^2-year (Faulkner and Richardson, 1989). Long-term sulfur removals and retention by wetlands requires strongly reduced conditions to enhance SO_4^{--} reduction. Wetlands constructed to remove and store S (for example, those used for acid mine drainage disposal and/or from disposal and treatment of toxic metals containing wastewater and runoff) should be designed to promote such conditions.

Metal sulfide complexes (including potentially toxic metals) can be destroyed by acidic and aerobic conditions. Making the substrate aerobic will promote carbonic acid formation in pore water, which will reduce the pH, and concurrent oxidation of sulfides will reduce AVS to such levels that sulfides become ineffective for metal complexation. Under anaerobic conditions only very large acid inputs, such as those typical for overloaded constructed wetlands treating acid mine drainage may volatilize the sulfide–metal complexes. The most common cause of sulfide destruction is oxidation. When sulfides are oxidized and acid volatile sulfide concentrations become low, as they would be in fully aerobic substrate, then other properties of the substrate, such as organic matter and clays, will control the toxic metal activity (toxicity).

CONSTRUCTED WETLANDS

Constructed wetlands are man-made systems that imitate the functions of the natural systems; however, they provide a better chance for management and control. They can include swamps (wet regions dominated by trees, shrubs, and other woody vegetation) or bogs (low-nutrient, acidic waters dominated by mosses). Most common constructed wetlands are marshes, which are shallow water bodies dominated by emergent herbaceous vegetation, such as cattails (*Typha*), bulrushes (*Scirpus*), rushes (*Juncus*), and reeds (*Phragmites*). As pointed out by Hammer (1989b), marshes can be adapted to a tremendous variety of soils, climatic conditions, and to wide fluctuations of water quality and hydrological conditions.

Unlike the overland-flow wastewater treatment systems where vegetation is not essential (for example, in rapid infiltration systems briefly introduced in Chapter 6), vegetation in wetland systems is the predominant purification medium. The principal function of vegetation in created (and natural) wetland systems is to provide an additional environment for microbial populations, in addition to slowing down the flow and filtering the solids. The plants provide a substantial amount of reactive surfaces for microbes. As pointed out plants also increase the amount of aerobic microbial environments in the substrate due to a unique ability of wetland plants to transfer air to their root and rhizomes.

Due to the previously mentioned limitation imposed on natural wetlands by regulations and by their function, constructed wetlands are more amenable to be used as a treatment facility and can be operated and magaged more efficiently. Essentially, constructed wetlands retain the positive characteristics of natural wetlands and, with proper management and some maintenance, can minimize the drawbacks. Constructed wetlands can be and have been used for the following water-management and pollution-control tasks:

- Disposal of treated or partially treated municipal effluents (Kadlec, 1988; Kadlec and Alvord, 1989; U.S. EPA, 1988; Mitsch, 1990).
- Wastewater polishing and renovation (Reed, Middlebrooks, and Crites, 1988; U.S. EPA, 1988).
- Urban and suburban storm-water management and treatment (Palmer and Hunt, 1989; Esry and Cairns, 1989; Livingston, 1989; McArthur, 1989; Mitsch, 1990).
- River water quality restoration and nonpoint pollution control (Hey, 1988; Hey et al., 1989).
- Mine acid drainage disposal (Reed, Middlebrooks, and Crites, 1988; Brodie et al., 1989a).
- Disposal of some industrial wastes, such as ash pond seepage (Brodie et al., 1989b), refinery wastewater (Lichfield and Schutz, 1989), and pulp mill effluents (Thut, 1989).
- Flood control (Mitsch, 1990).
- Detoxification of wastewaters and runoff containing toxic metals and adsorbable organic chemicals.

Unlike natural wetlands that are limited by location, soils, and other factors, constructed wetlands can be built almost anywhere, including lands that would have limited use. Conversion of marginal agricultural lands, or lands that were taken out of production under the farm conservation programs, to riparian wetlands is feasible and an attractive alternative for control of pollution from diffuse (nonpoint) sources. They also provide superior performance and flexibility.

Types of Constructed Wetlands

The EPA (U.S. EPA, 1988) and Water Pollution Control Federation (WPCF) (1990) manuals and other treatises on the design of constructed wetlands (Reed, Middlebrooks, and Crites, 1988; Hammer, 1989a; Kusler and Kentula, 1989; Mitsch, 1990) recognize two fundamental types of constructed wetlands.

Free water surface (FWS) system with emergent plants: A FWS system (Fig. 14.18) typically consists of basins or channels with a natural or constructed subsurface barrier of clay or impervious geotechnical material (lining) to prevent seepage. The basins are then filled with soils to support emergent vegetation. Water flows slowly over the soil surface. Figure 14.19 shows an outline of a typical storm-water pond-wetland system recommended by the Maryland Department of Natural Resources for urban storm-water quality control (Livingston, 1989; Mitsch, 1990).

If the soil is brought from an existing wetland, wetland vegetation will emerge without seeding; however, seeding and planting of vegetation are a part of the construction process. Mitsch (1990) referred to the former type of constructed wetland as a *self-designed wetland*, and to the latter as a *designer wetland*. To develop a wetland that is ultimately a low maintenance one, the natural successional process needs to be allowed to proceed. Often this may mean some inital period of invasion by undesirable species, but if proper hydrologic and nutrient loads are maintained, this invasion in usually temporary.

Subsurface flows systems (SFS) with emergent vegetation: An SFS wetland, also known as vegetated submerged bed (VSB), consists of a trench or bed underlaid with an impermeable layer of clay or synthetic material. As shown in Figure 14.20, the system is built with a slight inclination of 1% to 3% between the inlet and outlet. The media used include crushed stone or rock fill, gravel, and different soils. The water flows through the medium and is purified by filtration, absorption by microorganisms and adsorption onto soils and organic matter, and by the roots of the plants. The subsurface medium is generally anaerobic; however, as specified previously, plants can convey by their roots and rhizomes some oxygen into the subsurface layer, thus creating therein small aerobic pockets.

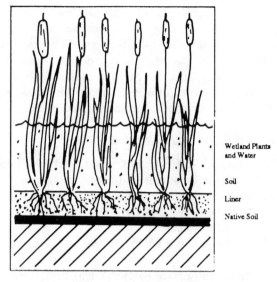

FIGURE 14.18. Free water surface wetland schematics. Water level is above the surface; vegetation is rooted and emergent above the water surface; water flow is primarily above ground; vegetation may be planted or allowed to colonize voluntarily.

As pointed out by Reed, Middlebrooks, and Crites (1988), both wetland-type treatment systems are "attached-growth" biological reactors, comparable in function to trickling filters and rotating biological contactors. Hence, their performance depends on the detention time of the waste in the system, the loading rates and the condition of the biota that develops within the system, and on the oxygen availability. If denitrification is an objective, alternate aerobic–anaerobic conditions should be promoted within the system and the anaerobic–denitrification zones should have an ample supply of organic carbon, which can be provided by primary productivity and from organic residues of plants grown in the wetland.

Subsurface-flow (SFS) constructed wetlands have been used extensively in Europe under the name *root-zone method*, *hydrobotanical system*, *soil-filter trench*, *biological macrophytic* and *marsh beds* (Reed, Middlebrooks, and Crites, 1988) or *vegetated submerged bed* (VSB) systems (Water Pollution Control Federation, 1990). The root-zone method was developed in Germany. The reed-bed system (Cooper and Hobson, 1989) used in the United Kindom evolved from the German system. The performances of some constructed wetland systems in the United States and Canada are shown in Table 14.4.

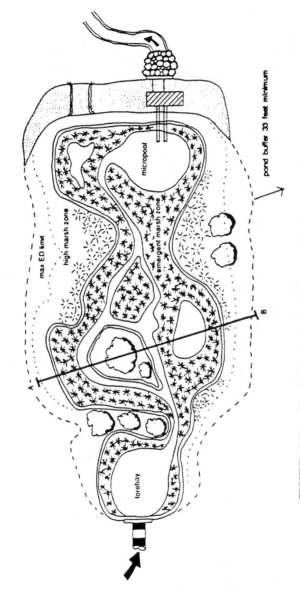

FIGURE 14.19. Maryland pond–wetland system. (After Schueler, 1992.)

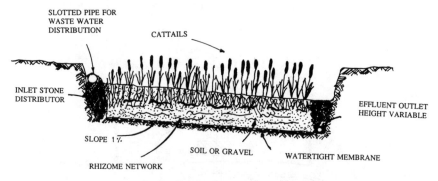

FIGURE 14.20. Submerged flow system (SFS). (After U.S. EPA, 1988.)

Nitrogen and phosphorus removal efficiencies related to loading rates were published recently in a comprehensive treatise by Hammer (1989a). The removal efficencies related to nutrient loading rates are shown on Figures 14.15 and 14.16.

Design Considerations and Parameters for Constructed Wetlands

Mitsch and Jørgensen (1989) pointed out that ecological engineering insists on taking advantage of our ever-increasing knowledge of ecology and its principles (for example, succession, energy flow, self-design) in order to design a system that will be as close to the natural features of the landscape as possible and that will require a minimum amount of maintenance. This means resisting the ever-present temptation to overengineer.

The basic principles of ecological engineering of wetland restoration

TABLE 14.4 Performance of Pilot-Scale Constructed Wetlands

		Effluent Concentration (mg/l)					
Location	Wetland Type	BOD	SS	NH_4^+	NO_3^-	Total N	Total P
Listowel, Ontario, Canada	Open water, channel	10	8	6	0.2	8.9	0.6
Arcata, California	Open water, channel	<20	>8	<10	0.7	11.6	6.1
Santee, California	Gravel-filled channels	<30	<8	<5	<0.2	—	—
Vermontville, Michigan	Seepage basin wetland	—	—	2	1.2	6.2	2.1

Source: From Read, Middlebrooks, and Crites (1988).

and creation were outlined by Mitsch (1990), Mitsch and Jørgensen (1989), and Odum (1989) as follows.

1. Design the system for mimimum maintenance. The system of plants, animals, microbes, substrate, and water flows should be developed for self-maintenance and self-design (Mitsch and Jørgensen, 1989; Odum, 1989).
2. Design a system that utilizes natural energies, such as the potential energy of streams.
3. Consider the landscape when developing a system. The best sites are locations where wetlands previously existed or where nearby wetlands still exist. Do not overengineer wetland designs with structures, unnatural shapes of basins, uniform depths, and regular morphology. Ecological engineering promotes mimicking nature.
4. Design a system as an ecotone. This means including a buffer strip around the site, but also means that the wetlands site itself needs to be viewed as a buffer between the upland and the aquatic system to be protected.
5. Consider the surrounding lands and future land-use changes. For example, in agricultural zones planned idling of land for soil-erosion control may obviate the need for a wetland to control pollution by nonpoint runoff.
6. Hydrologic conditions are paramount. Without water for at least a part of the growing season, a wetland is impossible. A detailed surface and ground-water study is therefore necessary.
7. Give the system time. Wetlands are not created overnight.
8. The soils should be surveyed. Highly permeable soils do not support viable wetland systems.
9. The chemical composition of feed waters, including ground-water discharge, can be significant to wetland productivity and/or bioaccumulation of toxic materials.
10. Riparian wetlands present a particular problem because their flooding also causes scouring, sediment shifts, erosion, and deposition. Convex sides of river channels may be preferable to concave sides because of the higher erosive forces on the latter.

Other factors to be considered during design include site accessibility (site access should be controlled to avoid vandalism), land ownership, land prices and water-use rights (if water is to be diverted from the stream onto the wetland), and wildlife and fishery considerations (water fowl migratory routes, fish spawning areas). Few design parameters and limited experience are available for constructed wetlands used for diffuse-

TABLE 14.5 Wetland Design Parameters

Design Considerations	Constructed		Natural
	Free Water Surface (FWS)	Submerged Bed Systems (SFS)	
Minimum size requirement ha/1000 m^3/d	2–4	1.2–1.7	5–10
Hydraulic loading,[a] cm/day	2.5–5	5.8–8.3	1–2
Maximum water depth, cm	50	Water level below ground surface	50; depends on native vegetation
Bed depth, cm	NA	30–90	NA
Minimum aspect ratio	2:1	NA	1:4
Minimum hydraulic residence time, days	5–10	5–10	14
Minimum pretreatment	Primary; secondary is optional	Primary	Primary; secondary; nitrification; TP reduction
Configuration	Multiple cells in parallel and series	Multiple beds in parallel	Multiple discharge sites
Distribution	Swale, perforated pipe	Inlet zone (>0.5 m wide) of large gravel	Swale, perforated pipe
Maximum loading, kg/ha-day			
BOD$_5$[b]	100–110	80–120	4
Suspended solids	up to 150		
TKN	10–60	10–60 up to 15	3
Phosphorus	?		0.3–0.4
Additional considerations	Mosquito control with mosquitofish; remove vegetation	Allow flooding capability for weed control	Natural hydroperiod should be >50%; no vegetation harvest

Sources: After Water Pollution Control Federation (1990; copyright Water Environment Federation) and Mitsch (1990).

[a] Hydraulic loading is a reciprocal of the minimum-size requirement after conversion.
[b] Applicable only to wastewater treatment systems.

pollution control. The data in Table 14.5 contain typical design parameters for constructed wetlands treating wastewater effluents.

Hydrology

Important hydrologic factors and parameters in the design of constructed (restored) wetlands include hydroperiod, hydraulic loading rate (hydrologic size requirement), hydraulic residence time, infiltration capacity, and the overall balance (Water Pollution Control Federation, 1990; Mitsch, 1990). The hydrologic factors that affect these parameters are dependent on climate and the seasonal patterns of stream flow and runoff (both surface and subsurface) feeding the wetland. D'Avanzo (1989) and Mitsch (1990) have pointed out that is improper hydrology that leads to most failures of created wetland systems. On the other hand, if hydrology is favorable, chemical and biological conditions will also have a favorable and healthy response. Hydrology is less self-correcting and forgiving than the ecology of the wetland (plants and animals), which may self-correct initial mistakes (Mitsch, 1990).

Hydroperiod is defined as the depth of water over time. This parameter is important mostly for natural wetlands. Typical hydroperiods are shown on Figure 14.21. In addition to depth the key component of the hydroperiod factor is the duration of flooding. One has to realize that soil chemistry during flooding turns rapidly anaerobic and that each plant has different specific capabilities to adapt and survive in low-oxygen soils. Wetlands with variable depths have the most potential for developing a diversity of plant and animal species. Table 14.6 summarizes hydroperiod tolerance ranges for a variety of wetland plants and species. This information can be used in checking appropriate native wetland habitats for general suitability and for the design of constructed systems (Water Pollution Control Federation, 1990).

Alternate flooding and aeration of solids promotes nitrification–denitrification. Deepwater areas, devoid of emerging vegetation, offer habitats for fish (for example, *Gambusia affinis*, the mosquito eating fish). Water levels can be controlled by inflow and outflow structures, including weirs and feed pumps.

Several states have developed guidelines for bottom profiles of wetlands to provide a variety of depths (remember the third rule of ecological engineering specified previously). For example, Florida regulations on wetlands require a littoral shelf with gently sloping sides of 6:1 or flatter extending out from a point 60–77 cm below the water surface. They further recommend less than 70% of open water surface. A mean depth of 1–3 meters is recommended for permanent pools. Maryland's regulations for wetlands used for storm-water treatment specify that 75% of the wetland should have a depth under 30 cm (50% less than or equal to 15 cm and 25% 15–20 cm; 25% should have a depth ranging from 60 cm to 100 cm).

Infiltration capacity and soil texture. As pointed out previously, highly permeable soils are not suitable for wetland creation. As a matter of fact, most natural wetlands resulted because of poor soil permeability conditions. Consequently, permeability (infiltration rates) must be kept

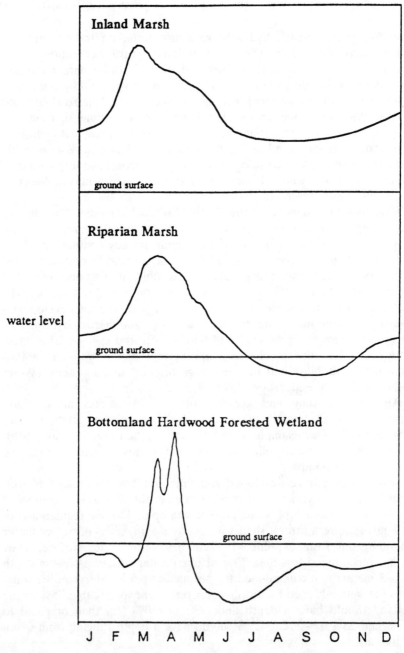

FIGURE 14.21. Typical hydroperiods for inland marsh, riparian marsh, and bottomland hardwood forested wetlands. (After Mitsch, 1990.)

TABLE 14.6 General Hydroperiod Tolerance Ranges for Selected Wetland Plant Communities

Wetland Type	Typical Species	Average Water Depth (m)	Average Hydroperiod (%)
Floating deep	Hyacinths, pennywort		
Floating rooted aquatic	Water lily, water dock, water shield	0.5–2	70–100
Submerged aquatic	Hydrills, egeria, water millfoil, sagittaria, naiad	0.5–3.0	80–100
Emergent marsh	Cattails, pickerelweed, sawgrass, bulrush, sedge, maidencane	0.1–1	40–100
Floodplain	Red maple, black gum, cabbage palm, pond cypress, oaks, pines, bald cypress, ash	0.2–0.3	10–50
Swamp forest	Bald cypress, ash, black gum, tupelo, gum, red maple	0.3–1	50–80
Cypress dome	Pond cypress, red maple, black gum, Dahoon holly	0.1–0.3	50–75
Wet prairie	St. Johns wort iris, sagittaria	0.1–0.2	20–50
Oak, palm, hammock	Oak, cabbage palm	<0.1	5–10

Source: From Water Pollution Control Federation (1990), copyright Water Environment Federation.

below a certain threshold value, which obviously may vary depending on local site-specific and geographic conditions. Measured infiltration rates for natural wetlands range from zero for a fen in Michigan (Haag, 1979) to about 0.5 cm/day (Water Pollution Control Federation, 1990).

The maximum substrate (soil) infiltration rate for wetlands should be around 1 mm/hr (Water Pollution Control Federation, 1990). This rate essentially implies a substrate or subsoil rich in clay. High clay content of the active substrate may limit root and rhizome penetration and may be impermeable for plant roots. Therefore the use of local coarser texture (loam) soils underlain by an impermeable clay layer is often the best design (Mitsch, 1990).

Hydraulic loading rates. Table 14.5 lists the suggested hydraulic loading rates (HLR) that are commonly given as the depth of water per area per time, or

$$\text{HLR} = Q/A \tag{14.1}$$

An inverse of HRL is the area requirement per unit flow. Most of what is known about the hydraulic loading rates has been gathered from observations of wetlands receiving wastewater. Mitsch (1990) pointed out that the hydraulic loading rates used for wetlands treating wastewater would be too low for riparian wetlands used for runoff and stream quality control. The maximum hydroperiod tolerance range of the wetland plant community must also be considered when a suitable HRT is being determined.

Retention time. Table 14.5 also provides the optimum retention time of water in the wetland. The retention time for free water surface (FWS) systems can be calculated from a simple formula that considers the water volume of the wetland and average flow, or

$$\text{HRT} = pV/Q \tag{14.2}$$

where
HRT = hydraulic residence time, days
V = $W*H*L$ = active volume of the wetland, m^3
p = porosity or (water volume)/(total volume) ratio
  $p = 0.9$ to 1 for free water surface wetlands depending on the growth density of vegetation, and
  p = void fraction of the substrate for subsurface flow systems (SFS)
W = width of the system, m
H = average depth of the system, m
L = length of the wetland (m) = $W*\text{AR}$
AR = aspect ratio (length/width)
Q = average flow, m^3/day
A = surface area, m^2

Alternatively, the hydraulic residence time for SFS systems may be calculated from the Darçy's law as

$$\text{HRT} = p\frac{V}{Q} = \frac{L}{K_p S} \tag{14.3}$$

where
L = length of the bed in meters
K_p = hydraulic (saturated) permeability of the wetland, m/day
S = bed slope, m/m

Normal design bed depths for SFS systems are between 0.6 and 0.9 meters (Water Pollution Control Federation, 1990). For time-variant intermittent diffuse sources, the HRT concept makes sense only if the hydraulic load is average over a longer time period (season or year). A simple relationship between the hydraulic loading rate (HLR), porosity, p, depth, H, and hydraulic residence time is HLR = $p \times H$/HRT.

System Configuration
The length–width ratio (also called the *aspect ratio* (*AR*)) for wetlands treating wastewater should be at least 4:1 to 6:1 (Water Pollution Control Federation, 1990). However, Mitsch (1990) pointed out that more elongated wetland cells (with an aspect ratio of at least 10:1) are needed in riparian settings and/or when receiving larger flows such as storm-water runoff.

In constructing a wetland for treatment purposes, the wetlands are arranged in cells with or without ponds. Cells can be parallel so that alternate drawdown management can be practiced to enhance nitrification, oxidation, mosquito control, and to maintain optimum hydroperiods for a variety of plants. The ponds remove pollutants and assist in the operation of the system. In an SFS system the ponds can interrupt short circuiting and reestablish uniform flow through the cells. They can also help in mosquito control by providing a living environment for *Gambusia*, which can migrate from the pond into the shallow wetlands.

Generally, the components of a wetland system include a treatment or pretreatment unit (note that it is neither advisable nor feasible to discharge raw untreated wastes into the wetland system), the wetland cells, and ponds. Pretreatment units may not be needed if wetlands are used for storm-water management. In such cases, natural (swale) drainage can provide enough filtering and pretreatment of storm water.

Possible design configurations of constructed wetlands were included in a paper by Steiner and Freeman (1989) and are shown on Figure 14.22. Mitsch (1990) presented nine actual design of wetlands located throughout the United States. Figures 14.23 to 14.27 show possible design alternatives of constructed or restored riparian wetlands.

Constituent Loadings
For wetlands treating wastewater the organic load expressed as kilograms of BOD_5/ha-day (or a similar unit such as lb/acre/day or g/m^2-day) is of major concern. Wetlands are extremely efficient at assimilating the BOD and also suspended solids loads. However, the organic load from allochthonous sources may be small when compared to the primary productivity (autochthonous organic matter production) of the wetland itself. For wetlands treating nonpoint pollution the organic loading is of less importance, and the parameters of concern are suspended solids, nutrients, and toxic components.

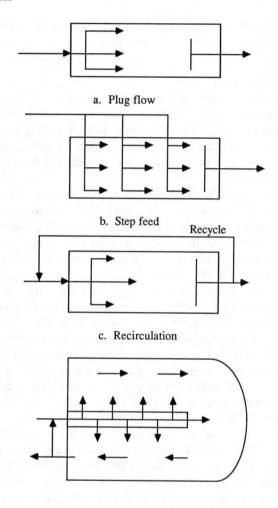

FIGURE 14.22. Basic flow patterns and configurations for constructed wetlands. (After Steiner and Freeman, 1989, with permission from Lewis Publishers, a subsidiary of CRC Press, Boca Raton, Florida.)

Suspended solids. Very high SS removals have been noticed in constructed and natural wetlands. The removal efficiency for suspended solids is inversely proportional to influent SS concentrations and the surface area of the wetland and the detention time, which is consistent. As pointed out previously solids are removed by settling, filtration, and adsorption. Of

Constructed Wetlands   903

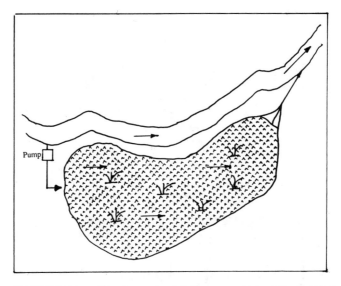

FIGURE 14.23.   Riparian wetland fed by pump. (After Mitsch, 1990.)

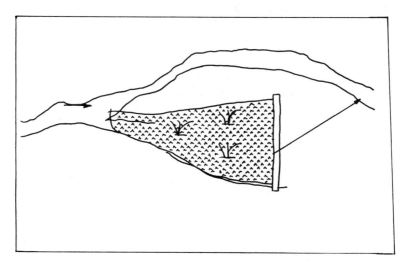

FIGURE 14.24.   Wetland built by incorporating diversion. (After Mitsch, 1990.)

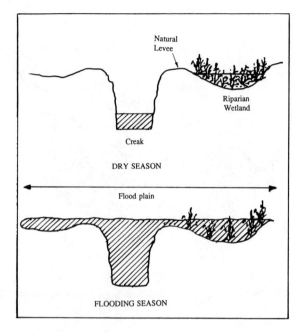

FIGURE 14.25. Riparian wetland design fed by natural flooding over the stream banks. (After Mitsch, 1990.)

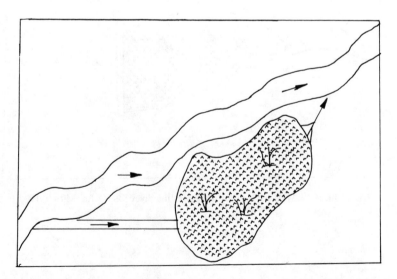

FIGURE 14.26. Riparian wetland design fed by gravity flow. (After Mitsch, 1990.)

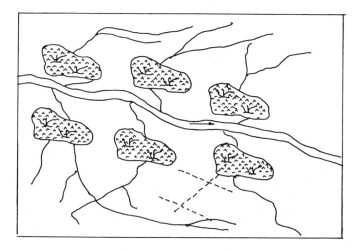

FIGURE 14.27. Schematics of small wetlands intercepting small polluted tributaries of a stream. (After Mitsch, 1990.)

concern are extremely high loads that could clog up the ecosystem by clay and silt particles. The critical HRT for achieving TSS removal efficiencies above 70% is about 5 days.

The operational experience gathered at the Des Plaines, Illinois, experimental wetland shows that while the influent suspended solids were primarily composed of soil particles in concentrations of tens to hundreds at milligrams/liter, effluent solids primarily included organic solids produced by the wetland, with effluent concentrations of only a few milligrams/liter. The influent is typically turbid, while the effluent color is from the humic substances of the wetland. The effluent character is comparable to native prairie streams.

Biodegradable organics. If the constructed wetland is used for the treatment of wastewater, or if the diffuse input contains appreciable concentrations of biodegradable organics (such as flows from concentrated animal operations), BOD removals may be considered. For FWS and SFS wetlands BOD removal may be described by a plug flow equation of the type

$$\frac{C_e}{C_o} = \exp(-k_T \text{HRT}) \qquad (14.4)$$

where
C_e = effluent BOD_5 (mg/l or mass load in kg/day)
C_o = influent BOD_5 (mg/l or kg/day)
k_T = temperature-dependent first-order BOD removal coefficient, day^{-1}

The temperature-dependent BOD removal coefficient (equivalent in concept to the deoxygenation coefficient in streams; see the preceding chapter) is commonly reported at a reference temperature of 20°C. Equation (12.11) with the thermal factor $\theta = 1.06$, may be used to adjust the coefficient for other design temperature (Water Pollution Control Federation, 1990). Most of the available data on BOD removals are for municipal wastewater and may not be applicable to diffuse loads from agricultural or urban sources.

Nutrients. Both nitrogen and phosphorus are necessary nutrients for wetland biota growth and are retained by it. The process of nitrogen and phosphorus assimilation was discussed previously. It was also pointed out that nitrogen assimilation depends on the form of the nitrogen in the input with the nitrate nitrogen resulting in the highest removal efficiencies.

TKN (total Kjeldahl nitrogen = organic N plus $NH_4^+ - N$) removal is correlated to loading. Loads as high as 10 kg/ha-day still provide removal efficiencies between 75% and 95%. With loading rates within the 10 to 80 kg/ha-day range, TKN removal rates are highly dependent on HRT and decrease significantly at design HRTs of less than 5 days. At HRTs of 5 days or more nitrate nitrogen in the inflow is almost completely assimilated by denitrification anaerobic processes occurring in the wetlands.

Based on a regression analysis of available data from the operation of natural and constructed wetlands, the Water Pollution Control Federation (1990) manual reported the following equation relating effluent and influent total nitrogen (TKN = organic + ammoniacal) concentrations:

$$C_e = 0.193 \times C_o + 1.55 \, \text{HLR} - 1.75 \tag{14.5}$$

The correlation coefficient for this relationship is $r^2 = 0.79$.

Phosphorus removals are not as high as those for nitrogen. A compilation of wetland systems contained in the Water Pollution Control Federation (1990) manual revealed phosphorus removal efficiencies between 30% and 50% for most of the created systems studied. Total phosphorus removal efficiency increases with higher input concentrations, form of the phosphorus (particulate adsorbed onto the sediment vs. dissolved), phosphorus partitioning between the dissolved and suspended forms (which is related to the concentration of suspended solids) and the subsequent burial of the adsorbed phosphates with the sediment and organic matter in the substrate, oxic–anoxic conditions in the water–substrate interface, and on the hydraulic retention time. In colder climatic conditions, nutrients are released during the dormant season. Nitrogen and phosphorus removals for natural wetlands were shown on Figures 14.15 and 14.16.

Example 14.1: Free Water Surface Wetland Design

Design a free water surface (FWS) wetland for treament of agricultural drainage that has average total nitrogen concentrations of 15 mg/l. Most of the nitrogen is either organic or ammoniacal. The desired effluent concentration is about 5 mg/l (66% reduction). The design drainage flow is about $0.1 \, m^3/sec = 8640 \, m^3/day$. Select a shallow active depth of water in the wetland (0.4 meters) to provide for nitrification–denitrification.

Solution From Equation (14.5) the hydraulic loading rate is

$$\text{HLR} = \frac{C_e - 0.193 C_o + 1.75}{1.55} = \frac{5 - 0.193 \times 15 + 1.75}{1.55}$$
$$= 2.48 \, \text{cm/day} = 0.025 \, \text{m/day}$$

The corresponding surface is then

$$A = Q/\text{HLR} = (8640 \, [m^3/\text{day}])/(0.025 \, [m/\text{day}]) = 345{,}600 \, m^2 = 35 \, \text{ha}$$

As a rule of thumb, the Water Pollution Control Federation manual recommends that to attain total nitrogen removals of 50% or greater, the designer should use a minimum constructed area of wetland of about $4 \, \text{ha}/1000 \, m^3/\text{day}$ of flow.

Assuming the average depth of 0.4 meter, the hydraulic residence time in the wetland is $\text{HRT} = p \times H/\text{HLR} = 0.9 \times 0.4/0.025 = 14.4$ days. This detention time will provide for nitrification–denitrification nitrogen removal processes in the wetland (HRT should be greater than 7 days).

With a selected aspect ratio (based on the terrain configuration) of 4:1, the width of the wetland is $W = \sqrt{A^2/(1+4)} = \sqrt{350{,}000/5} = 264 \, m$ and the length of the wetland is $350{,}000/264 = 1326 \, m$. At least four cells are recommended in two parallel series. The first cell may be shallower FWS (nitrification), the second deeper FWS or SBS (denitrification).

Note: This design is based on the assumption that most of the nitrogen is in ammoniacal and organic forms. If nitrate nitrogen is the dominating compound, the wetland area and volume may be smaller, as shown in the following example.

Example 14.2: Wetland Design for Nitrate-N Removal

Experimental data indicate that 95% removal may be achieved if the loading rate of nitrate is about 10 kg/day or less, while if the loading exceeds 80 kg/ha-day the wetland nitrate removal efficiency is marginal. By interpolation, at between 10 and 80 kg of nitrate N/ha-day, 66% removal may be achieved (very approximately) using a loading rate of about 33 kg/ha-day. Select 25 kg/ha-day for safety reasons.

Solution The daily nitrate load is then

$$\text{Load} = Q \times C = 8640\,\text{m}^3/\text{day} \times 15\,\text{mg/l}\,(\text{g/m}^3) \times 0.001\,\text{kg/g}$$
$$= 129.6\,\text{kg/day}$$

and the surface area becomes

$$A = 129.6\,[\text{kg/day}]/25\,[\text{kg/ha-day}] = 5.2\,\text{ha} - \text{select 6 ha} = 60{,}000\,\text{m}^2$$

The hydraulic loading is then

$$\text{HLR} = Q/A = 8640\,[\text{m}^3/\text{day}]/60{,}000\,[\text{m}^2] = 0.14\,\text{m/day}$$

which is high (Table 14.5). Select an HLR of 8 cm/day = 0.08 m/day. Then the adjusted surface area is

$$A_{\text{adjusted}} = Q/\text{HLR} = 8640/0.08 = 108{,}000\,\text{m}^2 = 10.8\,\text{ha}$$

Since the nitrogen is in nitrate form, the dominant component removal process should be anoxic (anaerobic) to assure denitrification. In this case, either deeper FWS ($H \approx 0.9\,\text{m}$) or saturated SFS systems should be selected, and the hydraulic retention time should be above 5 days.

The hydraulic residence time requirement gives the estimate of the depth of the bed. Select a minimal HRT of 5 days. Then the required volume of the bed is

$$V = Q \times \text{HRT}/p = 8640 \times 5/0.5 = 86{,}400\,\text{m}^3$$

and the required depth of the bed becomes

$$H = V/A = 86{,}400/108{,}000 = 0.8\,\text{m}$$

If the hydraulic conductivity of the bed is known, one can also estimate the velocity in the bed. A judgment may then be made as to whether the flow will be fully submerged (in the bed) or whether ponding will occur. Say that the hydraulic conductivity of the bed is 10 m/day and the slope of the bed is 5%. Then the velocity in the bed is (from Darçy's equation)

$$v = K_p \times S = 10\,[\text{m/day}] \times 0.05 = 0.5\,\text{m/day}$$

This velocity is too low to accommodate the flow solely in the bed. The designer would therefore have to design the wetland as a combination of the FWS and SFS systems.

Metals and Toxics

Due to very high concentrations of organic matter in the substrate, wetlands could conceivably have a higher assimilative capacity for toxics and metals. As with aquatic sediment conditions, toxicity, adsorption (immobilization), and the burial of toxic compounds depend on the substrate characteristics and adsorption–desorption and the precipitation processes occurring in the substrate. DiToro (1989) and DiToro et al. (1990) pointed out that the toxicity of metals and organics correlates to their activity, which is more-or-less related to the dissolved and ionic concentration of the constituent in the pore water of the substrate.

In the reduced environment of the wetland substrate, sulfide precipitation of metals is the predominant process of inactivation of toxic metals, while the organic particulate matter content of the substrate controls immobilization of nonionic organic compounds (PCBs, DDT, PAHs, many pesticides) and organic mercury (DiToro et al., 1990). The assimilative capacity for organic chemicals is also related to the chemical's biodegradability and volatility. However, models for estimating the assimilative capacity of wetlands for toxic compounds are not available and the estimates are crude, at best. One may speculate, however, that if constructed wetlands are to be used for the disposal of small quantities of toxic compounds, the preferable design would be SFS, which provides higher concentrations of organic matter in the bed as compared to concentrations of particulate organic matter in the water in FWS systems.

Mass Balance of Toxic Compounds in a Submerged Bed Wetland

The capability of wetlands to safely assimilate toxic materials is related to the fate of the toxic materials in the wetland system. The fundamental concepts were explained in the preceding two chapters. The primary productivity of the wetland autotrophic biomass provides organic solids that may effectively immobilize the toxic compounds and make them biologically unavailable. It was also pointed out that in sediment the toxicity of a compound should be related to the pore water concentration of the compound and not to the total concentration of the compound in the sediment volume. The same postulate applies even more to wetlands.

Several investigations have shown that wetland substrates may appear heavily contaminated by toxics, and yet the wetland biota appears either unaffected or only marginally impacted. This apparent paradox will be explained by the partitioning of the pollutants between the dissolved (toxic) and immobilized components in the wetland substrate. In contrast to lake and river sediments, wetland substrate is richer in particulate carbon and generally is anaerobic, both of which situations contribute to the reduction of the compound's toxicity to the wetland biota.

These concepts can be illustrated in an example of a simple, single-basin submerged bed system. For multiple basins in a series the same considerations will apply for the first basin receiving the full load. For modeling the fate of toxics in a more complex free surface system, one can use the EPA model WASP described in the preceding chapter.

The following assumptions are made in the subsequent calculation:

1. There is a steady buildup of organic matter in the wetland due to the inputs of the allochthonous (influent) and autochthonous (primary productivity) organic matter.
2. Mass-balance considerations are made over a relatively long time period. Diurnal and seasonal variations are not considered.
3. A significant portion of the organic input is nonrefractory (inert) and constitutes the inert organic carbon buildup in the wetland. According to the information provided by Foree, Jewell, and McCarty (1971) this inert fraction ranges between 40% to 60% in a mostly anaerobic environment.
4. The primary immobilizing (adsorbing) component is particulate inert organic carbon.
5. The wetland substrate is compacted and relatively uniform.

Following O'Connor's (1991) concepts the mass balance of the inert organic solids and toxics in the SFS wetland becomes:

A. Balance of Particulate Organics

$$\frac{dM_b}{dt} = \frac{d(V_b m_b)}{dt} = Q_{in}POM_{in} + NPP \times A_b - K_b V_b m_b - Q_{out}POM_{out}$$

$$= \phi_r Q_{in}POM_{in} + NPP \times A_b - K_b V_b m_b \tag{14.6}$$

where
M_b = total mass balance of organic particulate carbon in the wetland, kg

V_b = volume of the wetland bed (substrate), m³
m_b = concentration of the particulate organic matter in the substrate, kg/m³
POM_{in} = particulate organic matter concentration of the influent, g/l = kg/m³
NPP = net primary productivity of the particulate organic carbon by the wetland, kg/m²-day
A_b = surface area of the wetland, m²
Q_{in} = influent flow, m³/day
Q_{out} = effluent flow, m³/day
POM_{out} = effluent concentration of the particulate organic carbon, g/l
K_b = decay rate of the organic carbon in the wetland, day⁻¹
ϕ_r = removal rate of organic solids in the lagoon, dimensionless

In order to protect biota in the wetland the pore water concentration of the toxic should be related to inert organic matter, especially in cases when the toxic compound is conservative. This is because the decomposition process of the deposited particulate organic matter in the wetland is only qualitatively known. Hence let m_{oc} be the inert (nonbiodegradable) carbon fraction of the organic particulate matter. Also the effluent POC concentration is negligible when compared to the sum of the influent organic solids load and the net primary productivity in the wetland. Then Equation (14.6) becomes

$$\frac{dV_b m_{oc}}{dt} = f_{NRC}(\phi_r Q_{in} POM_{in} + NPP \times A_b) \qquad (14.7)$$

Here $f_{NRC} = m_{oc} m_b$ is a fraction of the particulate organics, which is inert carbon, and m_{oc} is the concentration of inert organic carbon in the substrate in kg/m³. Expand the left side deferential to

$$\frac{dV_b m_{oc}}{dt} = V_b \frac{dm_{oc}}{dt} + m_{oc} \frac{dV_b}{dt} \qquad (14.8)$$

Since an assumption was made that the concentration of the inert organic matter remained constant over time, that is, $dm_{oc}/dt = 0$, Equation (14.7) can be modified to express the increase in the wetland volume due to annual deposition of inert particulate organic matter

$$\frac{dV_b}{dt} = \frac{f_{NRC}}{m_{oc}}(\phi_r Q_{in} POC_{in} + NPP \times A_b) \qquad (14.9)$$

The preceding equation represents a uniform linear increase of the volume of the wetland. This relationship will be recalled in the further development of the allowable load of a toxic compound to the wetland.

B. Toxic Chemical Balance in the Wetland

$$\frac{dM_{Ch}}{dt} = \frac{dV_b C_T}{dt}$$
$$= W - K_a A_b C_d - K_d C_d V_b p - Q_{out}(C_d - \text{POM}_{out} r) \quad (14.10)$$

where
M_{Ch} = total mass of the chemical in the wetland, mg
C_T = concentration of the total chemical in the wetland, µg/l = mg/m^3
W = total load of the chemical, mg/day
K_a = rate of volatilization of the chemical, m/day
C_d = concentration of the dissolved chemical in pore water, µg/l
K_d = decay rate of the chemical in the wetland by photolysis, chemical hydrolysis, and biochemical degradation, day^{-1}
p = porosity (pore volume) of the wetland, dimensionless

The total mass of the chemical is a composite of the chemical dissolved in the pore water and that adsorbed on or complexed by the sediment, or

$$M_{Ch} = V_b C_T = p V_b C_d + (1 - p) V_b \rho_s r \quad (14.11)$$

where
ρ_s = is the dry weight density of the sediment, kg/m^3
r = adsorbed or complexed particulate chemical, mg/kg = µg/g

Introducing the partition relationship between the dissolved and particulate phase concentrations ($r = \Pi C_d$) presented in the preceding chapter and dividing both sides by V_b, one obtains

$$C_T = p C_d + (1 - p) \rho_s \Pi C_d \quad (14.12)$$

It was also shown in Chapters 6 and 13 that the partition coefficient for a chemical may be related to the octanol partition coefficient, K_{ow}, for the chemical and the particulate organic carbon fraction of the sediment. Then $\Pi \approx f_{NRC} \times K_{ow}$, where in this particular case f_{NRC} is the inert particulate organic carbon fraction of the wetland substrate (sediment). Recall also that the unit for Π and K_{ow} is l/g = m^3/kg. If K_{ow} is reported

Constructed Wetlands   913

in liters/kilogram, divide the reported value by 1000. The relationship between the total concentration of the toxicant commonly measured in micrograms/liter of the total volume of the wetland substrate and the concentration of the dissolved toxicant in pore water is then

$$\frac{C_d}{C_T} = \frac{1}{p + \rho_s\Pi(1-p)} = \frac{1}{p + \rho_s f_{\text{NRC}} K_{\text{ow}}(1-p)} = \frac{1}{p + m_{\text{oc}} K_{\text{ow}}} \quad (14.13)$$

This equation, similar to Equation (6.8) in Chapter 6, is very important. In a fashion similar to the consideration of sediment toxicity discussed in Chapter 6 one can substitute the aquatic toxicity standard for C_d and obtain the maximum total concentration limit (C_T) of the chemical in the substrate of the wetland.

Equation (14.13) can also be introduced into Equation (14.10) to relate the load, W, to the pore water concentration. Then

$$(p + m_{\text{oc}} K_{\text{ow}}) \frac{dV_b C_d}{dt} = W - K_a A_b C_d - K_d C_d V_b p$$

$$\quad - Q_{\text{out}} C_d (1 + \text{POM}_{\text{out}} f_{\text{NRC}} K_{\text{ow}}) \quad (14.14)$$

Expand the differential on the left side to

$$\frac{dV_b C_d}{dt} = V_b \frac{dC_d}{dt} + C_d \frac{dV_b}{dt}$$

and drop the first term on the right side because C_d should be kept constant (meaning it should not rise over a period of one year or longer). Also Equation (14.9) is substituted for an increase in the wetland volume, dV_b/dt. Then the dissolved, steady-state concentration of the toxicant in the wetland substrate is

$$C_d = \frac{W}{\{(p + m_{\text{oc}} K_{\text{ow}})(f_{\text{NRC}}/m_{\text{oc}})(\phi_r Q_{\text{in}} \text{POM}_{\text{in}} + \text{NPP} \times A_b)\} + \{K_a A_b + K_d V_b p + Q_{\text{out}}(1 + \text{POM}_{\text{out}} f_{\text{NRC}} K_{\text{ow}})\}} \quad (14.15)$$

For a wetland, $m_{\text{oc}} \times K_{\text{ow}} \gg p$. Also assume that in humid regions $Q_{\text{in}} \approx Q_{\text{out}}$, hence, $\text{POM}_{\text{out}} = (1 - \phi_r)\text{POM}_{\text{in}}$. Then by dropping p and by dividing the numerater and denominator of the right side of Equation (14.15) by A_b, the relation between the dissolved (bioavailable) concentration of the toxicant, and the specific load of the toxic compound, W' (mg/m$_2$-day) becomes

$$C_d \approx \frac{W'}{K_{ow} \times f_{NRC}(\text{HLR} \times \text{POM}_{in} + \text{NPP}) + (K_a + K_d Hp + \text{HLR})} \quad (14.16)$$

where
HLR = hydraulic loading rate, m/day
H = average depth of the substrate, m

Note that the first part of the right side of the denominator reflects the immobilization of the toxic compound by the input and buildup of the inert organic carbon, and the second part represents the loss of the toxicant by volatilization, decay, and dissolved loss in the effluent.

For a conservative highly adsorbable (complexed) compound

$$C_d = \frac{W'}{K_{ow} f_{NRC}(\text{HLR} \times \text{POM}_{in} + \text{NPP})} \quad (14.17)$$

Example 14.3: Maximal Permissible Concentration of Lindane in SFS Wetland

Extimate the maximum volumetric concentration of lindane that would protect biota. The aquatic water quality criterion for lindane is WQC = 0.08 µg/l, the octanol partitioning coefficient is K_{ow} = 5 l/g, and the organic particulate concentration (measured, for example, as volatile suspended solids) in the wetland substrate is 1000 kg/m³ (= g/l), of which 50% is inert (f_{NRC} = 0.5). The porosity of the substrate is 40% (p = 0.4).

Solution Estimate inert organic carbon concentration

$$m_{oc} = 0.5 \times 1000 \, [\text{kg/m}^3] = 500 \, \text{kg/m}^3$$

From Equation (14.9) the maximal total concentration limit is

$$C_T = \text{WQC}(p + m_{oc} K_{ow}) = 0.08(0.4 + 500 \times 5) = 200 \, \mu\text{g/l}$$

Example 14.4: Maximal Load and Influent Concentrations of Toxic Compounds

The runoff entering the wetland contains the pesticide lindane and PCBs. O'Connor (1991) and Schnoor and O'Connor (1991) provided the following absorption and degradation characteristics for the chemicals in aquatic environments:

Lindane
Decay rate of the chemical: $K_d \approx 0.06$ day$_{-1}$
Volatilization rate: $K_a \sim 0.00001$ m/day
Octanol partition coefficient: $K_{ow} \sim 5000$ l/kg = 5 l/g

PCBs
Decay rate: $K_d \approx 0$
Volatilization: $K_a \approx 0.13$ m/day
Octanol partition coefficient: $K_{ow} \approx 10^6$ l/kg = 1000 l/g

The aquatic toxicity water quality criterion for lindane is WQC = 0.08 μg/liter and that for PCBs is WQC = 0.14 μg/l.

Estimate the maximal load of lindane and PCBs per square meter of the wetland that has the following parameters:

Depth of the bed: $H = 0.5$ m
Porosity: 35% $p = 0.35$
Net primary productivity (Table 14.2): NPP = 2 kg/m^{2}year = 0.0055 kg/m^{2}day
Hydraulic loading: HLR = 0.02 m^3/m^2-day
Influent volatile suspended solids: POM$_{inf}$ = 50 mg/l = 0.05 g/l
Fraction of inert carbon versus organic solids: $f_{NRC} = 0.3$

Lindane The maximal load is calculated by Equation (14.16) as

$$\max W' = \text{WQC}[K_{ow} \times f_{NRC}(\text{HLR} \times \text{POM}_{in} + \text{NPP}) \\ + (K_2 + K_d Hp + \text{HLR})]$$
$$= 0.08[5 \times 0.3 \times (0.02 \times 0.05 + 0.0055) \\ + (0.0001 + 0.006 \times 0.5 \times 0.4 + 0.02)]$$
$$= 0.0025 \text{ mg/m}^2\text{-day}$$

Since $W' = C_{in} \times Q_{in}/A_b = C_{in} \times \text{HLR}$, the maximal influent concentration limit becomes

$$C_{in} = \max W'/\text{HLR} = 0.0025\,[\text{mg/m}^2\text{-day}]/0.02\,[\text{m}^3/\text{m}^2\text{-day}] \\ = 0.125 \text{ mg/m}^3 \, (= \mu\text{g/l})$$

The wetland would provide about $100(0.125 - 0.08)/0.125 = 36\%$ removal of lindane. Note that because of relatively low K_{ow}, K_d, and K_a parameters lindane will not be effectively removed and stored in the wetland.

PCBs

$$\max W' = 0.014[1000 \times 0.3(0.02 \times 0.05 + 0.0055) + (0.13 + 0 + 0.02)]$$
$$= 0.029 \, \text{mg/m}^2\text{-day}$$

and the allowable influent concentration limit of PCBs is then

$$C_{in} = 0.029/0.02 = 1.47 \, \mu\text{g/l}$$

The wetland would provide $100(1.47 - 0.014)/1.47 = 99\%$ environmentally safe removal of the PCB load.

For the safe burial of toxic metals in the wetland or in the riverine and lacustrine sediments, the decay and volatilization rates are zero. Alternatively, for SFS wetlands or anaerobic sediments, the complexation of metals can be related to the availability of reduced sulfides, hence, sulfur compounds balance must be performed. If the sulfur input in the influent (mostly as sulphates), expressed in moles per time, is substantially greater than the input moles of toxic metals per time and the wetland or sediment system is mostly anaerobic, most of the metal input will be complexed and made nonbioavailable by the sulfides formed in the system.

Other Considerations
The advantages of constructed wetlands include low construction cost— essentially for grading, mulching, and planting wetland vegetation, with few structural components—and very low operation–maintenance costs for monitoring water levels, flows, the quality and vitality of vegetation, and grounds maintenance. Based on the experience of the Tennessee Valley Authority, Hammer (1989b) does not recommend plant harvesting except in special applications involving the water hyacinth or duckweed ponds incorporated within the wetland system (see the subsequent section on aquaculture). Plant harvesting will substantially increase operation and maintenance costs. Furthermore harvesting would eliminate the valuable organic matter needed for immobilization of toxics.

As said throughout this chapter wetland operation is related to the treatment objective. If nitrification and organic load removal is the objective, a substantial portion of the wetland system should be aerobic. Shallow free water systems are most suited for this purpose. Denitrification requires anoxic conditions. Also immobilization and detoxification of metals are best accomplished in an anaerobic environment. If an SFS wetland is heavily loaded by metals, clogging from substantial deposits of precipitated metals may occur (Watson et al., 1989). Only very limited and short-term (less than three years) data on the removal of metals and other toxics by wetlands were reported in the literature at the time of the writing of this book.

Disadvantages of constructed wetland systems include the land requirement and a long start-up period. Current design recommendations specify 1.3 to 5.5 hectares per $1000\,m^3$/day of treated wastewater (Table 14.5), depending on pretreatment and the design discharge limits (Hammer, 1989b). The start-up period before full removal efficiencies are achieved may extend over several growing seasons. However, considering that a typical structural installation of a conventional treatment plant may also take several years, this drawback may not be important.

As pointed out before, improperly designed and operated wetland systems can also cause a pest problem (mosquitoes and rodents) that may be a nuisance to the surrounding population. Both can be simply and easily avoided with proper design and operation.

In conclusion, constructed and, to a lesser degree, natural wetlands provide an effective control of pollution from various sources, including control of nutrients and sediments from diffuse sources. These are the most natural pollution control facilities, requiring in most cases no chemicals or energy other than light. Yet, their efficiency is comparable to expensive structural treatment plants. Constructed wetlands can be installed anywhere, including the lands that were formerly used for agricultural production, especially those that represent former drained wetlands. Such systems are now considered as a part of ecological engineering and remediation approaches to pollution problems (Mitsch and Jørgensen, 1989).

Mosquitoes should not be a problem for subsurface (SFS) systems. For free water systems *Gambusia* can be used, provided that some water pools (ponds) remain aerobic and fish can migrate from the pools into the other parts of the wetland. Nichols (1988) described a method developed by Hruby, an ecologist, who proposed a wetland–ditch combination. The ditches create narrow reservoirs in which water levels are manipulated to permit a fish population to enter marshes to spawn. The newly hatched fish larvae then feed on the mosquito larvae. Up to 97% mosquito control can be achieved.

AQUATIC PLANT SYSTEMS

Aquatic plant systems are shallow ponds with floating or submerged aquatic plants. As pointed out in the preceding section, ponds are commonly a part of natural wetlands system and are recommended for inclusion in constructed systems. Generally, after some time most ponds will develop some kind of aquatic vegetation. Typical types of aquatic vegetation were shown in Figure 14.5. In the context of aquatic plant systems, consider ponds into which aquatic vegetation was introduced or planted specifically for water-pollution abatement. These systems include two types based on the dominant plant types. The first type includes floating plants, and the second employs submerged plants.

Floating Plant Systems

The *water hyacinth* (*Eichhornia crassipes* (Fig. 14.5)) has been the most common floating plant used in aquatic treatment systems. It is a nuisance plant that can develop in large quantities and densities in subtropical regions and may completely cover a body of water with a dense layer (Fig. 14.28).

The water hyacinth has been used extensively for the treatment and postreatment of wastewater. The major characteristics of water hyacinths that makes them attractive for the removal of pollutants are their extensive root system, which provides a medium for bacteria, and a rapid growth rate. However, these plants are temperature sensitive and will die rapidly during winter freezing conditions. Such systems may not be suitable for

FIGURE 14.28. Water hyacinth pond.

year-round applications in most of Europe, with the exception of the most southern parts (such as Sicily), and a majority of the United States, with the exception of Florida and the southern portions of California, Texas, and Arizona. However, Žáková and Veber (1990) documented that in the colder climatic conditions of central Europe, water hyacinth plants can be put into greenhouses during the winter and replanted in the spring. The best winter survival was achieved when plants were winterized in peat. For year-round operation, water hyacinth treatment lagoons were installed in Austin, Texas in a covered greenhouse.

Duckweed (Fig. 14.5) are small green freshwater plants (*Lemna* sp., *Spirodela* sp., *Wolffia* spp., and several others) that range in size from a millimeter to a few millimeters. Duckweed has a very fast growth rate, believed to be faster than that of water hyacinths. Effluents from duckweed lagoons (Fig. 14.29) should exceed the performance of conventional facultative treatment lagoons for BOD, suspended solids, and nitrogen removal.

Plant uptake (about 25% of the removal), ammonia volatilization, and nitrification–denitrification are the primary mechanisms for nitrogen removal in duckweed lagoons (Reed, Middlebrooks, and Crites, 1988). Overall nitrogen removal follows a first-order kinetics described by the following equation

FIGURE 14.29. Lemna pond. (Courtesy Lemna, Inc.)

$$\frac{N}{N_0} = e^{-kt} \qquad (14.18)$$

where
N_t = total nitrogen in system effluent, mg/l
N_0 = total nitrogen in the influent, mg/l
k = a rate constant dependent on temperature and plant density
t = time

Reed, Middlebrooks, and Crites (1988) and Zirschky and Reed (1988) emphasized that frequent harvesting of plants was needed to sustain high levels of nitrogen removal. Due to the fact that the water environment of the lagoon is mostly anaerobic and the duckweed cultures do not have the capability of transferring oxygen because their roots are very small and do not extend into the water, nitrification is limited or does not occur. Therefore if the input of nitrogen is in an ammoniacal form, the denitrification process may not have an adequate supply of nitrate–nitrogen and can be suppressed. Due to the fact that water in the duckweed ponds is anaerobic, the BOD removal is about the same as that in anaerobic facultative lagoons. Anaerobic effluent should be aerated (for example, on cascades).

Phosphorus removal in duckweed lagoons, just as in wetland systems, is not very high, and if removal of P is the object, the use of precipitating chemicals, such as alum or ferric chloride, is recommended (Reed, Middlebrooks, and Crites, 1988).

The use of duckweed for pollution abatement is not as well developed as is the use of water hyacinths. Mosquitoes should not be a problem as long as a thick mat of duckweed plants is maintained on the surface. Mosquito larvae cannot survive in the anaerobic environment created in the lagoon. The mat of duckweed also prevents algae from developing. Unharvested dead and dying plants can create a significant BOD and suspended solids load on the lagoon and release the nutrients accumulated in their tissues. For this reason, as stated before, the duckweed crop should be periodically harvested to ensure rapid regrowth and effective nutrient uptake. Special harvesting equipment is commercially available.

The harvested plant biomass has a higher protein content than other crops used for feeding farm animals. It also contains about 1.5% of phosphorus and about 6% of nitrogen (based on dry weight). The harvested plants can be used directly as feed if transportation is not an issue, or the biomass can be dried and/or composted. In the context of sustainable environmental management, research directed toward converting the harvested plants from the aquaculture lagoons and wetlands should be initiated and the organic fertilizer derived from the biomass should be substituted for chemical fertilizers. The benefit of this substitution would be less ground-water contamination by nitrates and subsequently less nutrient input to surface-water bodies. This may involve the cooperation of the fertilizer manufacturing industries.

Submerged Plants

The potential for using submerged plants is limited due to their tendency to be shaded by algae and their sensitivity to anaerobic conditions (U.S. EPA, 1988). Also clogging of roots and stems by filamentous organisms can severely limit their functioning. The best results for BOD, suspended solids, and nitrogen removal were obtained in an experimental study (Eighmy and Bishop, 1985) with a common water weed (*Elodea* sp.).

In summary, the water hyacinth aquaculture plants may not be suitable for most of North American and European climatic zones. Duckweed lagoons can be used in colder zones and will provide effective removal of organics and nitrogen. A wetland–aquaculture lagoon combination can be considered, but should be investigated in a pilot study. In contrast to water hyacinth lagoons, duckweed ponds and lagoons may not provide enough nitrification because the environment in the lagoon may be mostly anaerobic. This may also cause an odor problem. These problems can be mitigated and alleviated.

References

Bender, M. E., and D. L. Correl (1974). *The Use of Wetlands as Tertiary Treatment Systems*, National Science Foundation report NSF-RAE-74-033, Washington, DC.

Brodie, G. A., D. A. Hammer, and D. A. Tomijanivich (1989a). Treatment of acid drainage with constructed wetland at the Tennessee Valley Authority 950 Coal Mine, in *Constructed Wetlands for Wastewater Treatment*, D. A. Hammer, ed., Lewis Publ., Chelsea, MI, pp. 201–210.

Brodie, G. A., D. A. Hammer, and D. A. Tomijanivich (1989b). Constructed wetlands for treatment of ash pond seepage, in *Constructed Wetlands for Wastewater Treatment*, D. A. Hammer, ed., Lewis Publ., Chelsea, MI, pp. 211–220.

Cooper, P. F., and J. A. Hobson (1989). Sewage treatment by Reed Bed System: The present situation in the United Kingdom, in *Constructed Wetlands for Wastewater Treatment*, D. A. Hammer, ed., Lewis Publ., Chelsea, MI, pp. 153–172.

Cowardin, L. M., V. Carter, F. C. Golet, and E. T. LaRoe (1979). *Classification of Wetlands and Deepwater Habitats in the United States*, U.S. Fish and Wildlife Service pub. FWS/OBS-79/31, Washington, DC.

D'Avanzo, C. (1989). Long-term evaluation of wetland creation project, in *Wetland Creation and Restoration: The Status of the Science*, Vol. I, J. A. Kunsler and M. E. Kentula, eds., EPA 600/3-89-038a, U.S. Environmental Protection Agency, Corvallis, OR, pp. 75–84.

Dawson, B. (1989). High hopes for cattails, *Civil Eng.*, no. 4, pp. 48–50.

DiToro, D. M., et al. (1989). *Briefing Report to the EPA Science Advisory Board on the Equilibrium Partitioning Approach to the Sediment Quality Criteria*, EPA 440/5-89.002, Office of Water and Regulations and Standards Division, U.S. Environmental Protection Agency, Washington, DC.

DiToro, D. M., et al. (1990). Toxicity of cadmium in sediments: The role of acid volatile sulfide, *Environ. Toxicol. Chem.* **9**:1487–1502.

DiToro, D. M., et al. (1991). Technical basis for establishing sediment quality criteria for non-ionic organic chemicals using equilibrium partitioning, *Environ. Toxicol. Chem.* **10**(12):1541–1583.

Eighmy, T. T., and P. L. Bishop (1985). *Preliminary Evaluation of Submerged Aquatic Macrophytesin: A Pilot-Scale Aquatic Treatment System*, Department of Civil Engineering, University of New Hampshire, Durham, NH.

Esry, D. H., and D. J. Cairns (1989). Overview of the Lake Jackson restoration project with artificially created wetlands for treatment of urban runoff, in *Wetlands: Concerns and Successes*, D. W. Fisk, ed., American Water Resources Association, Bethesda, MD, pp. 247–257.

Faulkner, S. P., and C. J. Richardson (1989). Physical and chemical characteristics of freshwater wetland soils, in *Constructed Wetlands for Wastewater Treatment*, D. A. Hammer, ed., Lewis Publ., Chelsea, MI, pp. 41–72.

Fisk, D. W. (1989). *Wetlands: Concerns and Successes*, American Water Resources Association, Bethesda, MD.

Foree, E. G., W. J. Jewell, and P. L. McCarty (1971). The extent of nitrogen and phosphorus regeneration from decomposing algae, in *Proceedings, International Conference on Water Pollution Research*, IAWPR, San Francisco, CA, Pergamon Press, Oxford.

Gayan, S. L., and J. D'Antuono (1988). An integrated experience—The Milwaukee River East-West Branch Watershed Resource Management Plan, in *Proceedings, Nonpoint Pollution: 1988—Policy, Economy, Management, and Appropriate Technology*, V. Novotny, ed., American Water Resources Association, Bethesda, MD, pp. 165–172.

Godfrey, P. J., E. R. Kaynor, S. Pelczarrski, and J. Benforado (1985). *Ecological Considerations in Wetlands Treatment of Municipal Wastewaters*, Van Nostrand Reinhold, New York.

Haag, R. D. (1979). *The Hydrology of the Houghton Wetland*, University of Michigan Wetlands Ecosystem Research Group, Ann Arbor, MI.

Hammer, D. A. (1989a). *Constructed Wetlands for Wastewater Treatment*, Lewis Publ., Chelsea, MI.

Hammer, D. A. (1989b). *Constructed Wetlands for Wastewater Treatment—An Overview of Emerging Technologies*, paper presented at Urban Stream Corridor and Stormwater Management Symposium, March 14–16, Colorado Springs, CO.

Hammer, D. A., and R. K. Bastian (1989). Wetland ecosystems: Natural water purifiers? in *Constructed Wetlands for Wastewater Treatment*, D. A. Hammer, ed., Lewis Publ., Chelsea, MI, pp. 5–20.

Hedin, R. S., R. Hammack, and D. Hyman (1989). Potential importance of sulfate reduction processes in wetlands constructed to treat mine drainage, in *Constructed Wetlands for Wastewater Treatment*, D. A. Hammer, ed., Lewis Publ., Chelsea, MI, pp. 508–514.

Hey, D. L. (1988). Wetlands: A future nonpoint pollution control technology, in *Proceedings, Nonpoint Pollution: 1988—Policy, Economy, Management, and Appropriate Technology*, V. Novotny, ed., American Water Resources Association, Bethesda, MD, pp. 225–236.

Hey, D. L., M. G. Cardamore, J. H. Sather, and W. J. Mitsch (1989). Restoration of riverine wetlands: The Des Plaines River wetland demonstration project, in *Ecological Engineering. An Introduction to Ecotechnology*, W. J. Mitsch and S. E. Jørgensen, eds., Wiley, New York, pp. 159–184.

Kadlec, R. H. (1988). *Denitrification in Wetland Treatment Systems*, paper presented at the 61st Conference of the Water Pollution Control Federation, Session 26, Dallas, TX.

Kadlec, R. H., and H. Alvord, Jr. (1989). Mechanisms of water quality improvement in wetland treatment system, in *Wetlands: Concerns and Successes*, D. W. Fisk, ed., American Water Resources Association, Bethesda, MD.

Keeney, D. R., S. Schmidt, and C. Wilkinson (1975). *Concurrent Nitrification–Denitrification at the Sediment—Water Interface as a Mechanism for Nitrogen Losses from Lakes*, Tech. Rep. no. 75-07, Water Resources Center, University of Wisconsin, Madison, WI.

Knight, R. L., B. R. Windchester, and J. C. Highman (1984). Carolina Bays—feasibility for effluent advanced treatment and disposal, *Wetlands* 4:177–204.

Kusler, J. A., and M. E. Kentula (1989). *Wetland Creation and Restoration: The Status of the Science*, Vol. I, EPA 600/3-89-038a, Environmental Research Laboratory, U.S. Environmental Protection Agency, Corvallis, OR.

Lichfield, D. K., and D. D. Schutz (1989). Constructed wetlands for wastewater treatment at Amoco Oil Company's Mandan, North Dakota refinery, in *Constructed Wetlands for Wastewater Treatment*, D. A. Hammer, ed., Lewis Publ., Chelsea, MI, pp. 233–238.

Livingston, E. H. (1989). Use of wetlands for urban stormwater management, in *Constructed Wetlands for Wastewater Treatment*, D. A. Hammer, ed., Lewis Publ., Chelsea, Mi, pp. 253–263.

McArthur, B. H. (1989). The use of isolated wetlands in Florida for stormwater management, in *Wetlands: Concerns and Successes*, D. W. Fisk, ed., American Water Resources Association, Bethesda, MD, pp. 185–194.

Mitsch, W. J. (1990). *Wetlands for the Control of Nonpoint Source Pollution: Preliminary Feasibility Study for Swan Creek Watershed of Northwestern Ohio*, Ohio Environmental Protection Agency, Columbus, OH.

Mitsch, W. J., and J. G. Gosselink (1986). *Wetlands*, Van Nostrand Reinhold, New York.

Mitsch, W. J., and S. E. Jørgensen, eds. (1989). *Ecological Engineering. An Introduction to Ecotechnology*, Wiley, New York.

Mitsch, W. J., M. Straskraba, and S. E. Jørgensen (1989). *Wetland Modeling*, Elsevier, Amsterdam.

Mortimer, C. H. (1971). Chemical interchanges between sediments and water in the Great Lakes, *Limnol. Oceanogr.* **16**(2):387–404.

Nichols, A. B. (1988). A vital role for wetlands, *J. WPCF* **60**(7):1215–1221.

Novitzki, R. P. (1979). Hydrologic characteristics of Wisconsin's wetlands and their influence on floods, stream flow, and sediment, in *Wetland Function and Values: The State of Our Understanding*, P. E. Greeson, J. R. Clark, and J. F. Clark, eds., American Water Resources Association, Minneapolis, MN.

Oberts, G. L. (1982). Impact of wetlands on nonpoint source pollution, in *Proceedings, International Symposium on Urban Hydrology, Hydraulics and Sediment Control*, University of Kentucky, Lexington, KY, pp. 225–231.

O'Connor, D. J. (1991). Basic concepts and equations, in *Contaminated Sediments, Models for Criteria, Exposure, and Remediation*, 36th Institute in Water Pollution Control, Manhattan College, Riverdale, NY.

Odum, H. T. (1989). Ecological engineering and self-organization, in *Ecological Engineering: An Introduction to Ecotechnology*, W. J. Mitsch and S. E. Jørgensen, eds., Wiley, New York, pp. 79–101.

Ogawa, H., and J. W. Male (1989). *The Flood Mitigation Potential of Inland Wetlands*, Water Resources Research Center Publ. No. 138, University of Massachussetts, Amherst, MA.

Palmer, C. N., and J. D. Hunt (1989). Greenwood urban wetland: A manmade stormwater treatment facility, in *Wetlands: Concerns and Successes*, D. W. Fisk, ed., American Water Resources Association, Bethesda, MD, pp. 205–214.

Reed, S. C., E. J. Middlebrooks, and R. W. Crites (1988). *Natural Systems for Waste Management and Treatment*, McGraw-Hill, New York.

Richardson, G. J., and D. S. Nichols (1985). Ecological analysis of wastewater management criteria in wetland ecosystems, in *Ecological Considerations in Wetlands Treatment of Municipal Wastewaters*, P. J. Godfrey, et al., eds., Van Nostrand Reinhold, New York.

Schnoor, J. L., and D. J. O'Connor (1991). Kinetics and transfer processes, in *Contaminated Sediments, Models for Criteria, Exposure, and Remediation*, 36th Institute in Water Pollution Control, Manhattan College, Riverdale, NY.

Schueler, T. R., P. A. Kumble, and M. A. Heraty (1992). *A Current Assessment of Urban Best Management Practices. Techniques for Reducing Non-point Source Pollution in the Coastal Zone*, Technical Guidance Manual, Metropolitan Washington Council of Governments, Office of Wetlands, Oceans, and Watersheds, U.S. Environmental Protection Agency, Washington, DC.

Steiner, G. R., and R. J. Freeman, Jr. (1989). Configuration and substrate design considerations for constructed wetlands for wastewater treatment, in *Constructed Wetlands for Wastewater Treatment*, D. A. Hammer, ed., Lewis Publ., Chelsea, MI, pp. 363–378.

Stengel, E., and R. Schulz-Hock (1989). Denitrification in artificial wetlands, in *Constructed Wetlands for Wastewater Treatment*, D. A. Hammer, ed., Lewis Publ., Chelsea, MI, pp. 484–492.

Stumm, W., and J. J. Morgan (1962). Stream pollution by algal nutrients, in *Transactions, Twelfth Annual Conference on Sanitary Engineering*, University of Kansas, Lawrence.

Thut, R. N. (1989). Utilization of artificial marshes for treatment of pulp mill effluents, in *Constructed Wetlands for Wastewater Treatment*, D. A. Hammer, ed., Lewis Publ., Chelsea, MI, pp. 239–244.

U.S. Environmental Protection Agency (1988). *Constructed Wetlands and Aquatic Plant Systems for Municipal Wastewater Treatment. Design Manual*, EPA 625/1-88/022, U.S. EPA, Washington, DC.

Valiela, I., J. M. Teal, and W. Sass (1973). Nutrient retention in salt marsh plots experimentally fertilized with sewage sludge, *Estuar. Coast. Marine Sci.* **1**:261–269.

Water Pollution Control Federation (1990). *Natural Systems for Wastewater Treatment*, Manual of Practice FD-16, WPCF, Alexandria, VA.

Watson, J. T., S. C. Reed, R. H. Kadlec, R. L. Knight, and A. E. Whitehouse (1989). Performance expectations and loading rates for constructed wetlands, in *Constructed Wetlands for Wastewater Treatment*, D. A. Hammer, ed., Lewis Publ., Chelsea, MI, pp. 319–352.

Žáková, Z., and K. Veber (1990). *Biologické základy pěstováni a využívání vodního hyacintu v ČSFR* (Biological fundamentals of use of water hyacinth in CSFR), Academia, Prague, Czech Republic.

Zirschky, J., and S. C. Reed (1988). The use of duckweed for wastewater treatment, *J. WPCF* **60**(7):1253–1258.

15

Management and Restoration of Streams, Lakes, and Watersheds

Accomplishing restoration means ensuring that ecosystem structure and function are recreated or repaired and that natural dynamic ecosystem processes are operating effectively again.

Committee on Restoration of Aquatic Ecosystems,
National Research Council

A large proportion of the surface waters of the United States, especially streams and lakes, have been degraded from chemical, biological, and physical habitat destruction as a result of urbanization, deforestation, overgrazing, industrialization, agricultural practices, mining, flood-control projects, channelization, reservoir and dam construction, diversions, dredging, and other land uses. On the other hand, because of improvements in water quality from control of point source discharges, restoration of the biological and physical habitat in these waters could produce large improvements in the structure and function of the biological communities beyond those gained by improving water quality alone through control of point sources. When used in conjunction with elimination or reduction of nonpoint sources in the watershed, ecological restoration can lead to significant water quality benefits. Improvements in the physical and biological habitats of surface waters also can lead to improvements in water quality by increasing the capacity of aquatic ecosystems to process contaminants.

The National Research Council (Committee on Restoration of Aquatic Ecosystems, 1992) concluded that habitat degradation is a primary factor in limiting the attainment of beneficial uses of the nation's surface waters.

The researchers also concluded that an accelerated effort toward restoration of aquatic ecosystems is needed and that failure to restore aquatic ecosystems promptly will, in many cases, result in sharply increased environmental costs later through the extinction of species or ecosystem types or permanent ecological damage.

Stream restoration projects often are labor intensive; therefore, the use of funds for such projects has the additional benefit of creating numerous jobs, potentially in urban and rural areas with high unemployment rates. This chapter describes the necessary steps for planning a restoration project and the numerous approaches to restoration and management for streams, rivers, lakes, ponds, and their watersheds.

Although there are separate sections discussing river/stream and lake/reservoir systems, an integrated approach to restoration is recommended. Proposed restoration projects should consider the major ecological interactions in a watershed. Rivers, streams, lakes, and reservoirs are interconnected parts of a watershed that also often include ground water, estuaries, and wetlands. A practical and effective approach to restoration of aquatic ecosystems would include the consideration of all significant ecological elements on a watershed scale. In this way, the cumulative impacts of an ecosystem can best be evaluated.

RESTORATION VERSUS RECLAMATION, REHABILITATION, AND MANAGEMENT

The fundamental goal of restoration is to return the ecosystem to a condition that approximates its condition prior to disturbance (Cairns, 1988; Committee on Restoration of Aquatic Ecosystems, 1992). The terms restoration, reclamation, rehabilitation are often used interchangeably, but their meanings are different. *Restoration* is the return of an ecosystem to a close approximation of its former condition. *Reclamation* is a process designed to adapt a resource to serve a new or altered use. This could mean a process to convert a native ecosystem to agricultural uses or to convert a disturbed resource, such as a mined surface area, to productive use, such as pasture land. *Rehabilitation* is a term that describes putting a severely disturbed and/or damaged resource back into good working order. It is often used to indicate improvements that are primarily of a visual nature. Hence, restoration is a holistic process achieved through manipulation of all relevant elements, including the reintroduction of plants and animals. It is the most ambitious of the processes and is not possible in all situations; and where possible, it may not be economically feasible.

Whether an aquatic ecosystem is restored, reclaimed, rehabilitated, or preserved, it usually needs to be managed. *Management* is used here to mean the manipulation of an ecosystem to ensure the maintenance of one or more conditions or functions (Committee on Restoration of Aquatic Ecosystems, 1992). Watershed management is a term used to describe the management practices within an entire watershed that are designed to control or reduce inputs of pollution. From a technical standpoint, the watershed is the most logical scale on which to undertake restoration or management of aquatic resources. Unfortunately, political boundaries do not usually coincide with watershed boundaries and other institutional constraints can make restoration and management at the watershed scale difficult.

Restoration Project Planning

Planning a restoration project requires consideration of four basic items

1. *Project mission.* The overall general purpose, such as the restoration of a lake and perhaps an adjoining wetland.
2. *Goals.* The outcome desired, such as improving lake fisheries.
3. *Objectives.* Derived from the goals, such as the improvement in populations of specific fish species in the lake.
4. *Performance indicators.* Specific measurable quantities that are taken from each objective, such as an increase in the standing crop of fish.

Developing appropriate indicators requires selecting characteristics from a large number of possible ecological assessment criteria. These indicators are taken from certain structural, functional, and holistic criteria. The criteria listed below have been adapted from Berger (1990) and the National Research Council (Committee on Restoration of Aquatic Ecosystems, 1992).

Structural Criteria
- *Water quality.* Parameters include dissolved oxygen, temperature, pH, dissolved cations and anions, toxic contaminants, and suspended matter.
- *Soil condition.* Parameters include soil chemistry, erodibility, permeability, organic content, and soil stability.
- *Geological condition.* Includes surface and subsurface rock.
- *Hydrology.* Includes discharge, surface-flow properties, sediment load, retention times (for lakes), and ground-water flow and exchange with surface water.

- *Topography and morphology.* Surface contours, relief, configuration of site surface features, and the subsurface features.
- *Flora and fauna.* Presence and characteristics of the species, including evidence of biotic stress.
- *Waste assimilative capacity (loading capacity).* Includes nutrient availability, nutrient flux, and food web support.

Functional Criteria

- Surface and ground-water storage, recharge, and supply
- Floodwater and sediment retention
- Transport of organisms, nutrients, and sediments
- Humidification of atmosphere
- Oxygen production
- Nutrient cycling
- Biomass production, food web support, and species maintenance
- Provision for shelter for biota
- Detoxification of waste and purification of water
- Reduction of erosion and mass wastage
- Energy flow

Holistic Criteria

- *Resilience.* The ability of the ecosystem to recover from perturbations.
- *Persistence.* The ability of the ecosystem to survive or undergo natural successional processes without human management intervention.
- *Verisimilitude.* A characteristic of the restored ecosystem that reflects its similarity to a reference standard or its conditions prior to restoration.

Performance indicators can be constructed from a review of the preceding criteria to determine which are important in evaluating the particular project. For example, in the case of a project involving the improvement in water quality, criteria might include indicators such as pH, toxic contaminants, and suspended sediment.

The National Research Council (Committee on Restoration of Aquatic Ecosystems, 1992) presented a checklist to be considered before, during, and after any restoration activity (see below). Each of the questions in the checklist should be answered before proceeding to the next step in any restoration project.

Project Planning and Design

1. Has the problem requiring treatment been clearly understood and defined?

Restoration Versus Reclamation, Rehabilitation, and Management

2. Is there a consensus on the restoration program's mission?
3. Have the goals and objectives been identified?
4. Has the restoration been planned with adequate scope and expertise?
5. Does the restoration management design have an annual or midcourse correction point in line with adaptive management procedures?
6. Are the performance indicators—the measurable biological, physical, and chemical attributes—directly and appropriately linked to the objectives?
7. Have adequate monitoring, surveillance, management, and maintenance programs been developed along with the project, so that monitoring costs and operational details are anticipated and monitoring results will be available to serve as input in improving restoration techniques used as the project matures?
8. Has an appropriate reference system (or systems) been selected from which to extract target values of performance indicators for comparison in conducting the project evaluation?
9. Have sufficient baseline data been collected over a suitable period of time on the project ecosystem to facilitate before-and-after treatment comparisons?
10. Have critical project procedures been tested on a small experimental scale in part of the project area to minimize the risks of failure?
11. Has the project been designed to make the restored ecosystem as self-sustaining as possible to minimize maintenance requirements?
12. Has thought been given to how long monitoring will have to be continued before the project can be declared effective?
13. Have risk and uncertainty been adequately considered in project planning?

During Restoration

1. Based on the monitoring results, are the anticipated intermediate objectives being achieved? If not, are appropriate steps being taken to correct the problem(s)?
2. Do the objectives or performance indicators need to be modified? If so, what changes may be required in the monitoring program?
3. Is the monitoring program adequate?

Post-Restoration

1. To what extent were project goals and objectives achieved?
2. How similar in structure and function is the restored ecosystem to the target ecosystem?
3. To what extent is the restored ecosystem self-sustaining, and what are the maintenance requirements?
4. If all natural ecosystem functions were not restored, have critical ecosystem functions been restored?
5. If all natural components of the ecosystem were not restored, have critical components been restored?
6. How long did the project take?

7. What lessons have been learned from this effort?
8. Have those lessons been shared with interested parties to maximize the potential for technology transfer?
9. What was the final cost, in net present value terms, of the restoration project?
10. What were the ecological, economic, and social benefits realized by the project?
11. How cost-effective was the project?
12. Would another approach to restoration have produced desirable results at lower cost?

In order to address some of the questions presented in the checklist, the following items must be considered.

- *Project duration.* Each project should have a realistic schedule to allow enough time for evaluation. The duration of the project and its accompanying monitoring should be sufficient to address the particular environmental conditions of the site. For example, regions where environmental conditions include frequent floods or drought should be monitored long enough so that these variables do not interfere with the accurate assessment of project success.
- *Project scale.* The areal scale of the project must also be sufficient to account for spatial factors and for interaction between the target system to be managed or restored and the surrounding landscape.
- *Budgetary factors.* Not all projects have adequate funding to accomplish complete restoration. When funding conditions are inadequate or uncertain, efforts should be made to prioritize activities to best meet the overall project goals and objectives.

Generally, the goals of water body restoration–rehabilitation efforts are (a) restoration of damaged aquatic ecosystems, and (b) increasing the waste-assimilative capacity for increased safe disposal of wastes.

RESTORATION TECHNIQUES FOR RIVERS AND STREAMS

Much progress has been made in the past few decades in restoring rivers and streams affected by point sources of pollution through a variety of federal, state, and local programs. Unfortunately, there is no comparable suite of programs to deal with the restoration of rivers and streams affected by diffuse sources. Yet development pressures, agriculture, and changing land use have resulted in diffuse sources becoming the major fraction of wasteload to surface waters. And unlike slow-moving lakes, rivers and streams have historically served as a convenient and inexpensive means of waste disposal because the flow carries away or assimilates the pollutants.

The idea of stream waste-assimilative capacity enhancement was first proposed and implemented by Karl Imhoff, a pioneer of European wastewater management, in the 1920s in the industrialized Ruhr district of Germany (see Chap. 1 for a discussion of river basin management agencies (verbände) in the Ruhr area). He recognized the fact that the small rivers of the district would not be capable of safely assimilating municipal and industrial wastewater discharges, even after a full (primary plus secondary biological) treatment. The waste-assimilative capacity and water quality enhancement measures implemented therein included (Imhoff and Imhoff, 1990; Novotny et al., 1989):

1. Building stream impoundments. The objectives are (a) to enlarge the water surface, (b) to prolong the detention time, and (c) to induce deposition and retention of suspended solids and sludges. Loading of solids should be allowed in a limited amount. Deep reservoirs that stratify in summer are unsuitable for assimilation of large quantities of wastewater. When the impoundment fills with solids it must be dredged or abandoned and restored. As will be pointed later abandonment and restoration of old river impoundments to a more natural streamlike state is now being practiced in several places.
2. In-stream aeration by turbines and aerators to increase the reaeration capacity of the streams.
3. Lining of a small stream (Emscher River) that was heavily loaded by partially treated discharges of municipal and industrial wastewater. The lining increases velocity and decreases depth, resulting in a greatly increased aeration.
4. Back-pumping of the Rhine River into the Ruhr River (by the power plants on the Ruhr reservoirs) to increase low flows during times of extreme droughts.
5. Low-flow augmentation or increase of low-water flows.

The present Ruhr River basin management agency also implemented

6. Weed cutting and nutrient inactivation by chemicals in the impoundments.
7. Dredging of accumulated bottom sediments from the impoundments.

The agencies are also involved in stream restoration activities (see Chap. 1).

This section provides an overview of the various techniques available for the restoration of rivers and streams. Several types of river and stream problems are frequently encountered, and these can be categorized in seven main areas: (1) sedimentation; (2) low dissolved oxygen levels resulting in fish kills; (3) overgrazing of riparian areas; (4) stream channelization; (5) drinking water taste, odor, color, and organics; (6) poor fishing; and (7) acidic conditions. For each of these major problem areas, several in-stream techniques have been found to be effective, long-lasting, and generally without significant negative impact when used properly. These procedures will be described with regard to their underlying ecological principles and modes of action, effectiveness (including brief case histories), potential negative impacts, and additional benefits and costs. The reader will be referred to further reports in the basic scientific literature (for example, Committee on Restoration of Aquatic Ecosystems, 1992; Krenkel and Novotny, 1980). The less-studied or less-effective procedures will also be briefly described.

Basic Assumptions

The following discussions of in-stream technique effectiveness, except where explicitly stated, always assume that loadings of nutrients, silt, and organic matter to the watercourse have already been controlled. Most instream procedures will be quickly overwhelmed by continued accumulation of these substances. The watercourse and its watershed are coupled. In-stream programs can complement watershed efforts; however, such problems as bank erosion and sedimentation may persist despite load reductions or diversion projects unless an in-stream procedure is also used.

As for restoration and management techniques that are not mentioned in this section, in nearly every case these procedures have not been described in the open scientific literature, and therefore have not had the benefit of testing, discussion, explanation, and criticism that is so vital to the development of techniques of proven effectiveness and minimal negative impact. Caution should be exercised in the use of a procedure not listed here.

Selecting an Appropriate Stream-Restoration Technique

Selecting the wrong type of action may make the aquatic ecosystem worse off than no intervention at all. In addition, to be effective the action must address key limiting factors or major problems affecting the ecosystem. Otherwise, money and effort will be expended with little to no positive returns. For these reasons, none of the methods described here should be attempted without first consulting a professional familiar with ecosystem restoration and management. Furthermore, some of these activities require permits or are prohibited by some state regulatory authorities.

This section provides a comprehensive review of most possible stream-restoration techniques; its scope is not limited by any specific objectives or philosophy. Nevertheless, in selecting the techniques to apply in any given situation, careful thought and analysis are necessary to ensure that the program (a) addresses the cause of the problems and major use limitations to ensure an effective expenditure of funds and effort, (b) will not result in unexpected adverse effects on other water uses or ecosystem components, and (c) is consistent with public values and the long-term goals for the watercourse and its watershed.

Costs are another important consideration in selecting an approach. The cost of implementing a given technique can vary widely, depending on the area of the country, watershed size, accessibility, and other site-specific conditions and factors. Because of this variability, cost estimates are not provided here, although relative costs are discussed for some techniques. As part of the planning process, cost estimates should be obtained for the subset of options being considered. Both implementation costs and long-term operation and maintenance costs over the life of the project must be evaluated. Certain questions must be answered (see the section on project planning). How long will the treatment or beneficial effects last? How often will management actions need to be repeated? How much effort and expense will be required for routine upkeep and monitoring?

The size of the river or stream and associated watershed can have a major influence on the feasibility and effectiveness, as will the costs of various techniques. Restoration in larger rivers is more problematic than in smaller systems because of the size and complexity of the systems and often the different types and sources of problems. For rivers interconnected to other rivers, streams, and lakes, special consideration must be given to the potential for effects downstream or in adjacent waters. For example, exotic plants and animals may rapidly disperse throughout a drainage basin, becoming a regionwide nuisance.

Also, the seasonal and annual fluctuations in water quantity and the quality of flowing waters often require a longer time series of data than is available or funded in connection with a restoration project. Without an adequate time series of data, restoration projects are often hampered by their inability to determine success or failure.

Important concepts related to the management and restoration of rivers and streams were reviewed by the National Research Council

(Committee on Restoration of Aquatic Ecosystems, 1992) and are presented below:

- Flow and retention
- Openness
- Dynamism
- Patchiness
- Resistance and resilience

These attributes are integral to understanding the connection between rivers and streams and their ecosystems. Large rivers and floodplains are intimately linked and, therefore, should be restored and managed as a single ecosystem. For small streams, the stream and its riparian zone are linked. River and stream problems requiring restoration or management can be in the river or stream channels and pools, the riparian zone or floodplain, or the watershed. The types of problems that can occur in river–floodplain or stream–riparian systems are listed below (adapted from the Committee on Restoration of Aquatic Ecosystems, 1992):

Acidity
Bank erosion
Blockage of stream or river channel
Braided channel
Dissolved oxygen deficiency
Excessive flooding
Food scarcity for biota
Loss of fish refugia
Nutrient loss or excess
Pool deficit
Poor spawning success
Sediment loss
Siltation
Species (extinct, threatened, or endangered)
Stream cover (deficit or overgrown)
Water quality (turbidity or chemical pollution)
Water velocity (too high or too low)

Stream Restoration Techniques

Assuming that watershed inputs of pollution are controlled, the various approaches that can be used in river and stream restoration can be divided into structural and nonstructural techniques. Nonstructural techniques are broadly defined as any method that does not require either physical alteration of the watercourse or construction of a dam or other structure. Structural methods range from simple biotechnical approaches, such as the use of tree trunks and branches along banks to slow water velocity (sometimes referred to as "soft" engineering techniques), to "hard" engineering approaches, such as the use of concrete or riprap for bank stabilization.

Nonstructural Techniques

Nonstructural techniques do not require physical alteration of the watercourse and include those administrative or legislative policies and procedures that limit or regulate some activity. Nonstructural methods are listed and discussed below.

Techniques	*Possible Approaches to Technique*
Flow regulation	Legislative
	Administrative
Plantings	Trees
	Brush
	Herbaceous vegetation
	Grass
Pollution abatement	Instream controls
	Riparian zone controls
	Watershed controls
Propagation facilities	Incubation
	Spawning
Land acquisition	Greenways
	Buffer strips
	Parks and other lands
Land use regulation	Instream
	Riparian zone
	Watershed
Biomanipulation	Game fish stocking
	Control of undesirable fish species and stunted fish populations
	Prey enhancements to supplement food supplies

Flow Regulation

Regulating the flow through legislative or administrative approaches is a viable nonstructural technique for restoration. Although reserving or reclaiming flow for instream uses (fish, wildlife, and recreation) is usually thought of as an important restoration technique in the arid western parts of the United States, the approach is becoming more widely used elsewhere as well. For example, droughts in the Tennessee Valley region in the late 1980s resulted in requests to modify water withdrawal practices and potential conflicts with the health of aquatic life in the Tennessee River system.

Plantings

A common problem in developed streams and rivers is the clearing of riparian zones. Plantings in the riparian zone that do not alter the watercourse can be considered nonstructural restoration activities. The buffer zones can be gradually reforested over time through plantings of trees, brush, herbaceous vegetation, and grass. Often these plantings can be performed by volunteers, and have been very effective when local governments arrange the logistics, assemble the sites, and obtain the planting stock. Bernstein and Kumble (1992) described a successful volunteer effort in the Washington, D.C., metropolitan area. Herson (1992) presented practical guidance on riparian reforestation in an urban watershed. Important considerations in reforestation of stream riparian zones include:

- Site assessment
- Soil preparation
- Species selection
- Planting techniques
- Long-term maintenance

Pollution Abatement

Nonstructural approaches to pollution abatement include modifying practices in the stream, riparian zone, and watershed to prevent pollution, and include changing practices, such as lawn maintenance, construction phase activities, and recreation. For example, simply phasing construction sequencing to limit the amount of disturbed area at any given time will greatly reduce the downstream suspended-sediment levels. Streams that are affected by nutrient loads from lawn fertilization can be improved by changing the type of fertilizer used (such as a low-phosphorus mixture) or the frequency and timing of fertilizations. Nonstructural pollution abatement activities also include the use of fencing around streams and riparian zones. Fencing Sheep Creek in Colorado was effective in excluding livestock and humans (Stuber, 1985). Compared to unfenced areas, the fencing resulted in a narrower and steeper stream, increased streambank vegetation, and twice the estimated trout population.

Propagation Facilities
The use of facilities that propagate aquatic species through incubation and spawning is a common approach where it is desirable to maintain fishing in an area that otherwise would not have a sustainable sport fish population. Hatchery raised fish, however, may not always successfully survive in the wild.

Land Acquisition
Acquiring land near rivers and streams can protect the watercourse through the maintenance of buffer zones and prevention of potentially destructive land uses in the watershed. Examples of land-acquisition approaches include establishment of greenways, buffer strips, and parks. These can be purchased by the government or special foundations and trusts specially designed to provide such protection. The state of Wisconsin provides funds for the acquisition of lands (buffer zones and riparian wetland) along the Wisconsin and other protected rivers.

Land Use Regulation
Regulating land uses in the riparian zone and watershed through legislative or administrative approaches is a viable non-structural technique for restoration.

Biomanipulation
Many fisheries management techniques involve the direct manipulation of the fish community and other organisms that may serve as prey for or be predators or competitors with the fish species of interest. Three types of activities include (1) game fish stocking, (2) control of undesirable fish species and stunted fish populations, and (3) prey enhancements to supplement food supplies.

Structural Techniques
Structural techniques are those that require some type of physical alteration of the watercourse, and may include amendments to existing man-made structures, such as dams and levees. Administrative or legislative policies and procedures that limit or regulate some activity are often inadequate because natural restorative processes can take decades. In these instances, structural techniques are needed to speed up natural restoration processes. These methods generally help in two main areas: (1) stabilization of streambanks to reduce erosion, and (2) improvement of aquatic habitat through the creation or improvement of certain morphological procedures. Karouna (1992) categorized stream restoration techniques into three areas: (1) bioengineering techniques, (2) bank

armoring techniques, and (3) aquatic habitat improvement methods. The specific techniques under each category are listed and described below (list adapted from Karouna, 1992):

Bioengineering Techniques	Bank Armoring Techniques	Aquatic Habitat Techniques
Live stakes	Gabions	Log drop structures
Live fascines	Riprap	V-notch gabions
Brush mattresses	Joint plantings	Log wedge
Branch packings	Root wads	K-dams
Brush layering		Jack dams
Vegetated geogrids		Boulder "S" dams
Live cribwalls		Boulder clusters
		Deflectors
		Constrictors
		Channel block
		Overhead or inclined logs
		Brush bundles
		Bank covers

A manual by the Metropolitan Washington Council of Governments (Kumble and Schueler, 1992) presents a large variety of alternative means of stream restoration and aquatic habitat protection:

1. *Bioengineering techniques.* Bioengineering methods use plants to mimic natural streambank stabilization in situations where the streambank has been eroded or lacks vegetation through some destructive process (Figs. 15.1–15.4). Plantings can provide more ecological benefits than erosion control through the addition of stream cover, shade, and improvement of bank soil conditions. Native plant species are recommended because they are generally better adapted to local environmental conditions.
2. *Bank armoring techniques.* Bank armoring methods use rock, wood, steel, and other conventional construction materials to stabilize streambanks (Fig. 15.5). These methods rarely provide significant ecological benefits other than erosion control unless they are combined with bioengineering techniques.
3. *Aquatic habitat improvement methods.* These techniques involve improving aquatic habitat through the installation of certain instream structures (Figs. 15.6–15.9). Disturbed streams often lack diverse morphological features. Habitat improvement structures can add these features:

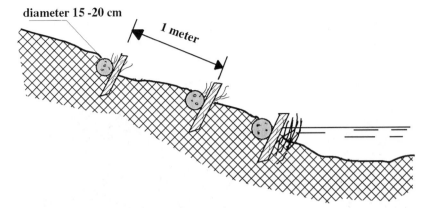

FIGURE 15.1. Live fascines for stream-bank stabilization. Live fascines are bundles of live cuttings wired together and secured into the streambank with live or dead stakes. Live facines are used to protect banks from washout and seepage, particularly at the edge of a stream, and where water levels fluctuate moderately. (From Kumble and Schueler, 1992; drawings courtesy Metropolitan Washington Council of Governments.)

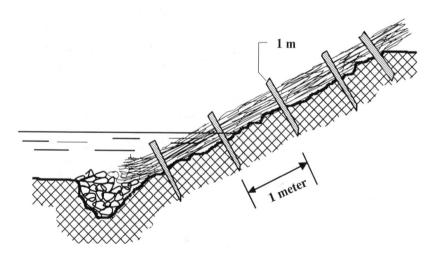

FIGURE 15.2. Brush mattresses for stream-bank stabilization. Six-foot willow switches are wired together to form a mat, which is then secured to the bank by stakes, fascines, poles, or rock fill. The toe of the slope is reinforced with a brushlayer anchored by a live fascine or riprap. The matress should lie perpendicular to the water. (From Kumble and Schueler, 1992; drawings courtesy Metropolitan Washington Council of Governments.)

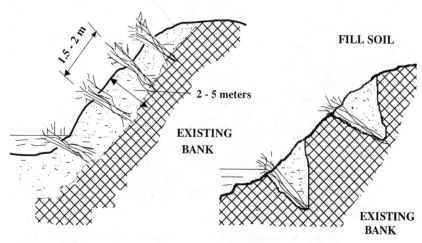

FIGURE 15.3. Brush layering for stream-bank stabilization. Live branches are placed in regular arrays on the face of a slope. The branches are oriented perpendicular to the face of the slope. On fill slopes, the plants are placed on prepared earth lifts. On cut slopes, narrow trenches are dug in the slope and plant material is placed there. Branches are 1.3–7.2 cm in diameter. A fill layer is used on heavily eroded slopes. (From Kumble and Schueler, 1992; drawings courtesy Metropolitan Washington Council of Governments.)

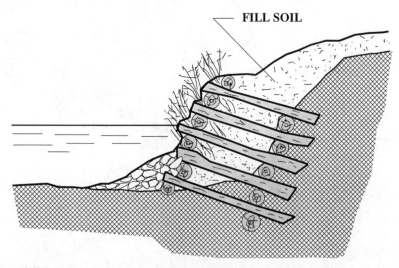

FIGURE 15.4. Live cribwall for stream-bank stabilization. The live cribwall is a rectangular framework of logs, rock, and woody cuttings. It is used to protect an eroding streambank especially at the outside bends of main channels where strong currents are present and locations where an eroding bank eventually forms a split channel. (From Kumble and Schueler, 1992; drawings courtesy Metropolitan Washington Council of Governments.)

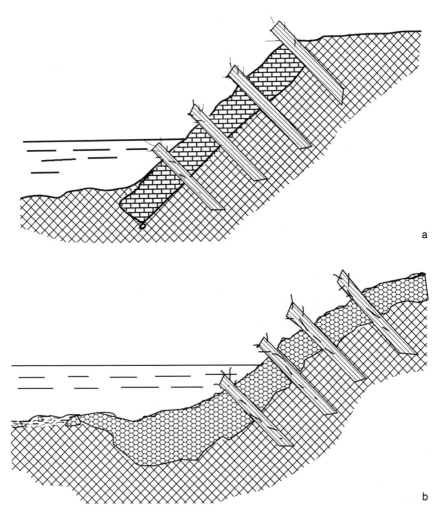

FIGURE 15.5. Joint planting–bank armoring. Willow cuttings are placed in the joints of (a) gabions or (b) riprap. (From Kumble and Schueler, 1992; drawings courtesy Metropolitan Washington Council of Governments.)

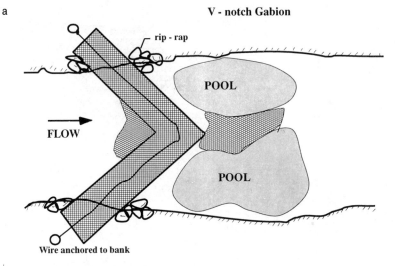
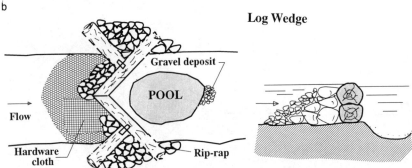

FIGURE 15.6. Aquatic habitat improvement. (a) V-Notch gabion. (b) Log wedge. These structures are designed to create and maintain large in-stream pools. They will also cause small gravel deposits to form that are used for spawning purposes. (From Kumble and Schueler, 1992; drawings courtesy Metropolitan Washington Council of Governments.)

a. Gravel beds
b. Structural complexity
c. Restricted flow
d. Riffles and pools

These features are important to the creation of spawning and rearing areas for aquatic life. The selection of a particular technique or combination of techniques depends on current habitat deficiencies, watershed conditions, and the current morphology and hydrology of the watercourse. Karouna (1992) described several general guides for the construction of habitat improvement devices:

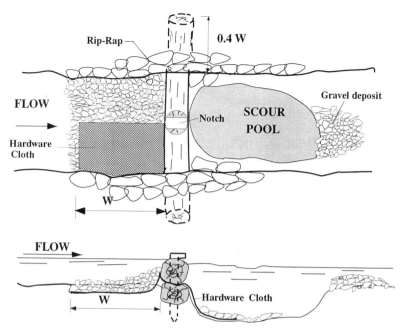

FIGURE 15.7. Aquatic habitat improvement: log-drop structures (check dams). Log-drop structures create scour pools downstream of the structure by directing water down into the stream bottom. The scour pools provide cover for fish. Gravel will deposit upstream of the log drop and also downstream of the scour pool. These deposits are often used as spawning areas. (From Kumble and Schueler, 1992; drawings courtesy Metropolitan Washington Council of Governments.)

FIGURE 15.8. Log dam for stream restoration. (Photo: H. Olem.)

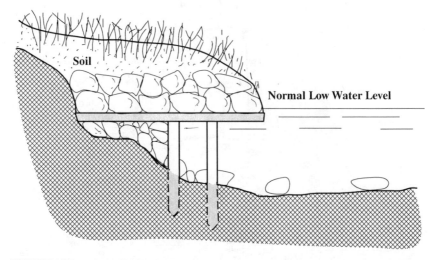

FIGURE 15.9. Aquatic habitat restoration—bank cover. These structures are installed to create an undercut bank effect, which provides cover for fish. Bank covers also serve to stabilize eroding banks. (From Kumble and Schueler, 1992; drawings courtesy Metropolitan Washington Council of Governments.)

    a. The device should be most effective during low flows, while having negligible effects during high-flow periods.
    b. The device should be located according to habitat requirements. For example, pools form naturally at bends, while riffles form along straight sections.
    c. The structures should not form a barrier to fish migration.
    d. The habitat created should not adversely affect another.
    e. Structures should not alter the stream flow of the watercourse and adversely affect unprotected streambanks.
    f. Structures should not be built during certain inappropriate time periods. For example, periods of fish spawning and incubation should usually be avoided.

4. *Low-flow augmentation.* Low-flow augmentation is an accepted measure that provides cleaner diluting flow during times of water quality emergencies. The source of the diluting flow may include upstream reservoirs, pumping from a nearby larger body of water or from another watershed, or by recycling (by back pumping) water from cleaner and more diluted downstream flows. Interbasin transfer of water can be difficult in states that follow a riparian water rights doctrine (see Chap. 2 for the legal ramifications resulting from water rights doctrines.) The most widely known case of low-flow augmentation as a water-pollution control measure is the diversion of the

Chicago River, which originally flowed into Lake Michigan, into the Illinois River, which is a tributary of the Mississippi River. This measure, implemented in the early 1900s diverted sewage from the lake (which is a source of drinking water for the metropolis) into the Illinois River. In Milwaukee, Wisconsin, cleaner Lake Michigan water is pumped into harbor sections of the Milwaukee River (see also Chap. 1) during times of low oxygen levels in the harbor. The dissolved oxygen depression (often to anoxia) is mainly caused by algal respiration. The cause of algal biomass is upstream productive reaches stimulated by nutrients from diffuse pollution (Novotny and Bendoricchio, 1988). The amount of dilution water needed can be estimated by water quality models (see Chap. 12).

5. *In-stream aeration*. In-stream aeration is a feasible measure for maintaining adequate dissolved oxygen levels for streams as well as for lakes and reservoirs. This alternative may be feasible for streams that are dystrophic and exhibit low dissolved oxygen levels, because they originate from wetlands or naturally eutrophic water bodies. In-stream aeration is accomplished by turbine aeration in power plants or by the installation of floating or submerged aerators. Also more natural cascades, spillways, and water falls provide additional oxygen (Krenkel and Novotny, 1980). In-stream aeration is simpler than aeration of stratified impoundments, which are discussed in the subsequent section, because streams are typically vertically mixed without stratification. Aeration is a temporal measure that can be used (a) when summer low oxygen levels drop even for a short time (e.g., during the night and early morning hours) below the dissolved oxygen limit for fish protection (4 to 6 mg/l, depending on the type of fish; or (b) during the winter when stream aeration is reduced by ice cover (also used for lakes and reservoirs). Aeration can be useful for streams draining wetlands that exhibit dystrophic conditions (low dissolved oxygen levels caused by high rates of decomposition of organic matter in the wetlands). The efficiency of aeration increases as the oxygen deficit increases. The Lippe River Association in the Ruhr area of Germany used aeration during the extremely dry summer of 1959 to keep an overloaded stretch of the river aerobic and odorless (Imhoff and Imhoff, 1990). Using the turbines of hydropower plants decreases the overall efficiency of the power plant. On the average 0.8–2.5 kg of O_2 can be supplied per kilowatt of power loss. Investigations into the aeration of lake Baldeney and the lower Ruhr reaches examined the efficiency and economy of floating centrifugal aerators, diffuser pipes, turbine aeration, and aeration from weirs and spillways. It was concluded that the turbine and weir aeration schemes provided the cheapest alternatives (Imhoff, 1969). The experience of

the Tennessee Valley Authority with turbine aeration was described by Davis et al. (1983).

The oxygen concentration increase at the point of aeration (the *sag point* on the dissolved oxygen longitudinal profile) is computed according to the following equation (Novotny et al., 1989)

$$\Delta C_0 = \frac{B_L D}{180 \times Q} K' \tag{15.1}$$

where
ΔC_0 = oxygen concentration increase at the point of aeration, mg/l
B_L = aeration power input in kilowatts [for weir and spillway aeration, substitute 9.81 × flow (m³/s) × fall height (m)]
D = average oxygen deficit, %
Q = flow, m³/s
K' = coefficient

The coefficient, K', represents the oxygen yield per kilowatt-hr at 50% average oxygen deficit (mean of the upstream and downstream deficit) at the point of aeration at 20°C. When accurate data are not available, K', can be estimated from the following:

	K' (kg O_2/kW-hr)
Cascades and rapids	1.5
Sharp-crested weir	0.6
Weirs and spillways	0.4
Turbine aeration	1.0
Surface aeration by floating aerators	0.5
Diffuse aeration	0.4

6. *Removal of river impoundments and stream channel restoration.* Removal of stream structural devices can be considered restoration techniques, the same as construction of new devices or modification of existing structures. For example, the Maine legislature passed a resolution in 1990 calling for the removal of the Edwards Dam on the Kennebec River by the year 2000 (Committee on Restoration of Aquatic Ecosystems, 1992). Despite some dam modifications to allow fish passage, the dam still blocks the migration of Atlantic salmon and impairs eleven other species of fish.

Many dams were built on streams for various purposes more than a hundred years ago. These were the structures that provided hydraulic head for hydropower plants, mills, and navigation. Sediment accumulated behind these dams in the last 80 or so years originated from urban and rural diffuse sources, wastewater discharges, and combined sewer overflows. In many cases this sediment is contaminated by toxic pollutants and exhibits high sediment oxygen demand (SOD). Often the impoundment is filled with sediments and thus has essentially ceased to function.

The time may have arrived for many such installations to be removed and the river rehabilitated to a new, more natural use. The problem is what to do with the sediments that are contaminated. The North Avenue Dam in Milwaukee, Wisconsin, is an example. This small dam (about 10 meters high) was built in the second half of the nineteenth century. Its original use was to provide a navigation head for a canal between Lake Michigan and a tributary of the Mississippi River that was never built. Additional uses of the dam in the first half of the twentieth century included power production and limited recreation. After more than a hundred years of existence the dam became a water quality problem. Turbidity (Secchi disc depth) in the impoundment decreased to 0.25 meters. The dissolved oxygen in the flow typically drops by 2 mg/l while the flow passes through the impoundment.

In 1990 the dam was open for repair and left open. In a few months a new channel was eroded in 10 meters of clayey-organic sediments (Figs. 15.10 and 15.11) and the exposed mudflats were experimentally revegetated. Although the accumulated sediments have a very high content of toxic metals (lead, zinc, and others), PAHs, and PCBs, the new river bed sediments are more sandy and far less toxic than the accumulated sediments. A study has been conducted to determine the remediation of the sediments (the study was not completed by the time of the writing of this book). The options are (1) protect and stabilize the new channel and cap the exposed mudflats with a cleaner fill, followed by revegetation; (2) protect and stabilize the new channel and convert the mudflats to a riparian wetland (wetlands have higher waste assimilative capacity for toxic compounds; see Chap. 14); (3) dredge the contaminated mudflats and dispose of the dredge spoils in a safe manner (most likely in an engineered disposal site located in the harbor or the lake) and establish a new channel; (4) fence off the contaminated mudflats. The Wisconsin Department of Natural Resources has identified a number of other candidate river impoundments that have ceased their original function and now pose a water quality problem.

RESTORATION TECHNIQUES FOR LAKES AND RESERVOIRS

This section provides an overview of the various techniques available for the restoration of lakes and reservoirs. Much of the information in this

950   Management and Restoration of Streams, Lakes, and Watersheds

FIGURE 15.10. North Avenue impoundment on the Milwaukee River immediately after drawdown with exposed mudflats. (Photo: W. Wawrzyn, Wisconsin Department of Natural Resources.)

FIGURE 15.11. North Avenue impoundment after drawdown. The mudflats revegetated. (Photo: W. Wawrzyn, Wisconsin Department of Natural Resources.)

section is taken from Cooke and Olem (1990) and Baker et al. (1993). Six types of lake or reservoir problems are frequently encountered by lake users. These are (1) nuisance algae; (2) excessive shallowness; (3) excessive rooted plants ("weeds" or macrophytes) and their attached algae mats; (4) drinking water taste, odor, color, and organics; (5) poor fishing; and (6) acidic conditions. For each of these major problem areas, several in-lake techniques have been found to be effective, long-lasting, and generally without significant negative impact when used properly. These procedures are described in detail in Olem and Flock (1990), with regard to their underlying ecological principles and modes of action, effectiveness (including brief case histories), potential negative impacts, and additional benefits and costs. The most promising techniques are summarized here. The less well-studied or less-effective procedures are also briefly described.

Basic Assumptions

The effectiveness of each in-lake technique described in Cooke and Olem (1990), except where explicitly stated, always assumes that loadings of nutrients, silt, and organic matter to the lake have already been controlled. Most inlake procedures will be quickly overwhelmed by continued accumulation of these substances. The lake and watershed are coupled. In-lake programs can complement watershed efforts; however, such problems as algae, turbidity, and sedimentation often remain despite load reductions unless an in-lake procedure is also used.

As for restoration and management techniques that are not mentioned in this section, in nearly every case these procedures have not been described in the open scientific literature and therefore have not had the benefit of testing, discussion, explanation, and criticism needed to develop techniques of proven effectiveness and minimal negative impact. Caution should be exercised in the use of a procedure not listed here.

Selecting an Appropriate Lake-Restoration Technique

Selecting the wrong type of action may make the lake ecosystem worse off than no intervention at all. In addition, to be effective, the action must address key limiting factors or major problems affecting the lake ecosystem. Otherwise, money and effort will be expended with little to no positive returns. For these reasons, none of the methods described here should be attempted without first consulting a professional familiar with ecosystem restoration and management. Furthermore, some of these activities require permits or are prohibited by some state regulatory authorities.

A number of the techniques described are suitable only in limited or certain circumstances. For example, in many regions of the country, fertilizing a lake to increase fish productivity would be considered highly undesirable and not a viable option because other lake uses and water quality would be adversely affected. This section provides a comprehensive review of most possible approaches; its scope is not limited by any specific objectives or philosophy. Nevertheless, in selecting the techniques to apply in any given lake, careful thought and analysis are necessary to ensure that the program (a) addresses the cause of the lake's problems and major use limitations to ensure an effective expenditure of funds and effort, (b) will not result in unexpected adverse effects on other lake uses or ecosystem components, and (c) is consistent with public values and the long-term goals for the lake and its watershed.

Costs are another important consideration in selecting an approach. The cost of implementing a given technique can vary widely, depending on the area of the country, lake size, accessibility, and other lake-specific conditions and factors.

The size of the lake, watershed, and associated drainage system can have a major influence on the feasibility and effectiveness, as well as the cost of various techniques. In-lake treatments (such as alum additions to reduce internal nutrient loads) may have relatively little effect in lakes with short hydraulic residence times and can be very costly in large lakes. The larger and more diverse a lake's watershed, the more difficult it may be to control nonpoint source inputs of nutrients and sediment. For lakes interconnected to other lakes, rivers, and streams, special consideration must be given to the potential for effects downstream or in adjacent waters. For example, exotic plants and animals may rapidly disperse throughout a drainage basin, becoming a regionwide nuisance. If undesirable fish species, such as carp, also occur in adjacent, connected waters, efforts to reduce or eliminate the species in just one lake are likely to fail.

In many cases, restoration, management, and protection begin on the land, in the surrounding watershed. Olem and Flock (1990) include a detailed discussion on watershed management, including the use of best management practices to reduce erosion and the export of sediments, nutrients, and toxic contaminants to receiving waters. Best management practices have been developed for agricultural, silvicultural, urban, and construction activities (U.S. EPA, 1987). The effectiveness, cost, and chance of negative side effects associated with selected watershed best management practices are summarized in Table 15.1.

TABLE 15.1 Summary of the Effectiveness, Costs, and Chance of Negative Side Effects Associated with Selected Watershed Best Management Practices

	Effectiveness					Chance of Negative Effects
	Sediment	Nitrogen	Phosphorus	Runoff	Cost	
Agriculture						
Conservation tillage	G-E	P	F-E	G-E	F-G	F-G
Contour farming	F-G	U	F	F-G	G	P
Contour stripcropping	G	U	F-G	G-E	G	P
Range and pasture management	G	U	U	G	G	P
Crop rotation	G	F-G	F-G	G	F-G	P
Terraces	G-E	U	U	F	F-G	F
Animal waste management	N/A	G-E	G-E	N/A	P	F
Urban						
Porous pavement	F-G	F-G	F-G	G-E	P-G	F
Street cleaning	P	P	P	P	P	U
Silviculture						
Ground cover maintenance	G	G	G	G	G	P
Road and skid trail management	G	U	U	U	P	F
Construction						
Nonvegetative soil stabilization	E	P	P	P-G	F-G	F
Surface roughening	G	U	U	G	F	P
Multicategory						
Streamside management zones	G-E	G-E	G-E	G-E	G	F
Grassed waterways	G-E	U	P-G	F-G	F-G	P
Interception or diversion practices	F-G	F-G	F-G	P	P-F	P
Streambank stabilization	G-E	P	P	G-E	F-G	F
Detention/sedimentation basins	G	U	U	P	P-G	F

Note: E = Excellent; F = Fair; G = Good; P = Poor; U = Unknown.

Control of Toxic Contaminants

Toxic contaminants in lakes can include metals, pesticides, and oils, in agricultural, industrial, and urban runoff. Elevated levels of these substances can degrade water quality and impact aquatic organisms. These pollutants may bioaccumulate in fish tissues, limiting the suitability of fish for human consumption. Possible corrective actions include:

- Eliminating the source of the contaminants, by applying best management practices
- Dredging and removing contaminated lake sediments
- Isolating contaminants concentrated in the bottom sediments from the overlying water column, by covering the sediments with a relatively impermeable layer, such as bentonite (a form of clay) or a plastic liner
- On-site water treatments, for example, diluting the contaminated water with "clean" water pumped in from other sources or withdrawing and treating the lake water (e.g., by chemical precipitation/coagulation and filtration), and then returning the treated water to the lake
- The addition of chemicals, such as alum (aluminum sulfate), to the lake, which may accelerate the precipitation of toxic substance(s) out of the water column into the bottom sediments
- Deep-water aeration for contaminants (e.g., some metals and ammonia) that precipitate or become nontoxic in the presence of dissolved oxygen
- Controlling changes in water level, if the exposure and suspension of contaminated sediments tend to increase the solubility and mobilization of the toxic substance
- Biomanipulation, if the potential for human exposure to bioaccumulated toxics can be reduced by altering the food chain or targeting fish species for fisheries management (e.g., avoiding game fish, such as lake trout, that are top predators and have high levels of body fat).

Relatively few field tests have been conducted to evaluate the long-term effectiveness of most of these techniques. In addition, a number of them can have potentially serious negative side effects (e.g., dredging operations may resuspend toxic contaminants and actually increase bioavailability). Not infrequently, no action is the most environmentally sound and cost-effective approach, allowing natural processes, such as sedimentation, to gradually reduce the concentration and availability of toxic substances after the source of contaminants has been eliminated.

Reducing Nutrient Loads and Nutrient Availability

Eutrophication can decrease dissolved oxygen levels, increase water temperatures, decrease water clarity, and alter the types and abundance of phytoplankton, zooplankton, and other organisms available as fish prey. Increased nutrient loadings generally increase lake productivity, in many cases including the total production of fish. However, the accompanying changes in physical and chemical habitat often result in an undesirable shift in the types of fish species able to survive and flourish in these waters. As an indirect effect of high nutrient loads, lakes may be unsuitable for game species of particular interest to local anglers.

Olem and Flock (1990) describe methods for evaluating and controlling problems with lake eutrophication. Two basic approaches for reducing nutrient inputs and availability are described: (1) managing the watershed, and (2) various in-lake treatments. "Managing the watershed" may involve changes in land use or land-use practices in the watershed, in particular the application of best management practices (see Table 15.1). The relative merits and effectiveness of each depend on the relative importance of nutrient sources to the lake.

Watershed management reduces external nutrient loadings, while in-lake treatment procedures eliminate internal nutrient sources or reduce nutrient availability. In general, in-lake treatments will only be effective over the long-term if accompanied or preceded by efforts to reduce external nutrient loads. In conjunction with a plan for improved watershed management, in-lake treatments may serve to accelerate the process of lake recovery.

Many of the in-lake treatments just noted for reducing toxics are also effective for reducing nutrients. The following types of in-lake treatments summarized in Baker et al. (1993) have been demonstrated to effectively reduce nutrient availability in at least some circumstances:

Chemical treatments for phosphorus. Aluminium salts, such as aluminum sulfate (alum) and sodium aluminate, have a strong affinity to adsorb and absorb inorganic phosphorus and remove phosphorus-containing particulate matter from the water column as part of the "floc," or loose precipitate, that forms. The result, after the floc settles, is not only a reduction in phosphorus availability, but also, generally, a substantial increase in water clarity. Adverse effects may occur, however, if the dosage is too high. Especially in softer, more acidic waters, excessive inputs of aluminum salts can decrease lake pH and result in concentrations of dissolved aluminum in the water column that are toxic to fish and other biota. Example applications of alum treatments to reduce nutrient levels in lakes include Horseshoe and Snake Lakes, Wisconsin (Peterson et al., 1973; Garrison and Knauer, 1984), Medical Lake, Washington (Gasperino et al., 1980; Soltero et al., 1981), Annabessacok Lake, Maine (Dominie, 1980), and Dollar and West Twin Lakes, Ohio (Cooke et al., 1982).

Sediment removal. Sediments with high concentrations of phosphorus or nitrogen can serve as an internal nutrient source. If nutrients are concentrated within the upper sediment layers, dredging and the removal of these nutrient-rich sediments may substantially reduce nutrient recycling and availability (Olem and Flock, 1990). At the same time, however, dredging operations can cause extensive damage to the benthic community, which may be an important food source for fish, and may disturb fish spawning habitats, if not carefully designed and implemented. Example applications include Lake Trummen, Sweden (Andersson, Berggren, and Hambrin, 1975; Bengtsson et al., 1975; Cronberg, Gelin, and Larsson, 1975) and Lilly Lake, Wisconsin (Dunst et al., 1984).

Dilution and flushing. For small lakes, it may be possible to reduce nutrient concentrations in the water column by adding sufficient amounts of low-nutrient waters from other sources, thereby diluting and flushing the high-nutrient water out of the lake. Applications of this technique are limited, however, by the general absence of suitable alternative water supplies. Examples include Moses Lake, Washington (Welch and Patmont, 1980; Perkins, 1983).

Aeration. Phosphorus remobilization from lake sediments is generally higher in anaerobic waters (with no dissolved oxygen) than in well-aerated waters (because the oxygenated forms of iron and manganese in natural waters form an insoluble precipitate with phosphorus). Thus, aeration techniques that increase oxygen levels in deeper waters may decrease nutrient recycling and availability. Increases in dissolved oxygen also have a direct positive effect on fish. Specific aeration methods are discussed later in this chapter.

Sediment oxidation. Oxidation of the lake's sediments may also reduce the remobilization of phosphorus into the water column. Rather than add air, however, a solid, such as calcium nitrate, is added to the sediments as an oxidizing agent. The procedure is termed RIPLOX, after its originator (Ripl, 1976), and is still experimental. Example applications include Lake Lillesjon, Sweden (Ripl and Lindmark, 1978) and Long Lake, Minnesota (Willenbring, Miller, and Weidenbacher, 1984).

Water withdrawals from the hypolimnion. The colder, deeper layers of a thermally stratified lake or reservoir generally have higher nutrient concentrations than waters in the epilimnion or metalimnion. Selectively withdrawing these nutrient-rich waters, using a siphon or deepwater outlet in a dam, can decrease the quantity of nutrients stored and recycled in the waterbody. However, hypolimnetic withdrawals may also trigger thermal instability and lake turnover. In addition, the discharge of these nutrient-rich, often anaerobic waters from the hypolimnion of eutrophic lakes can cause adverse effects in downstream receiving waters. Example applications include Lake Wononscopomuc, Connecticut (Kortmann et al., 1983) and Lake Mauensee in Switzerland (Gachter, 1976; Cooke et al., 1993).

All of these in-lake treatments require significant expenditures on equipment, chemicals, and labor. Case studies, as well as additional information on the methods, costs, and potential negative side effects of each approach, can be found in Olem and Flock (1990) and Cooke et al. (1993).

Increasing Dissolved Oxygen Levels

Low levels of dissolved oxygen may occur in lakes as a result of natural conditions, as well as cultural eutrophication. The lowest concentrations tend to occur in the deeper waters of the hypolimnion during thermal stratification in late summer, during long periods of snow and ice cover in winter, or in dense macrophyte beds at night or following long periods of cloud cover.

One important option to consider for lakes with problems with low dissolved oxygen is to manage the fisheries for species able to tolerate relatively low levels of oxygen, or that do not inhabit areas of the lakes (such as the hypolimnion) that experience oxygen depletion. For example, many salmonid species require both relatively high levels of dissolved oxygen and, because of their intolerance of warmer water temperatures, must reside in the cooler waters of the hypolimnion during summer. Therefore, sustaining a fisheries for salmonids may require a costly and continuing effort to aerate the lake's hypolimnion during some or all of the summer. Alternatively, if the problems with oxygen depletion are moderate, the lake may be able to support a cool-water or warm-water fisheries without extensive restoration efforts.

Problems with low dissolved oxygen can also be alleviated by one or more of the following:

- Decreasing the quantity of organic matter decomposed in the lake (the major process consuming oxygen) by (a) limiting the export of organic materials from the watershed to the lake, in particular excessive exports associated with human activities, such as runoff from feedlots or direct discharges of sewage wastewaters, (b) dredging to remove organic-rich sediments, and (c) decreasing in-lake productivity, by reducing nutrient loads and nutrient availability

- Increasing photosynthesis (an oxygen-generating process), especially during times and in locales (deeper waters of the hypolimnion) subject to oxygen depletion, primarily by increasing light penetration
- Destratifying the lake (artificial circulation), bringing low oxygen waters in the hypolimnion in contact with the lake surface and the well-oxygenated waters of the epilimnion
- Direct aeration

The approaches and optimal design criteria for lake aeration projects vary somewhat for systems installed to alleviate problems with winterkill (oxygen depletion during winter) as opposed to low levels of dissolved oxygen in the hypolimnion during summer. During winter, the goal is not to aerate the entire body of water, but instead to create an oxygen-rich refuge area for fish, generally near the lake surface. Major design concerns include problems with equipment ice-up and the need, for safety reasons, to minimize the loss or weakening of the lake's ice cover. Hypolimnetic aeration systems must deal with the more difficult problem of aerating waters at greater depths. Where the objective is to establish or maintain a cold-water fishery, hypolimnetic aeration must be achieved without disturbing the lake's thermal stratification. Otherwise, low levels of dissolved oxygen can be avoided by preventing thermal stratification, that is, through artificial circulation of the water column. In both cases, during both winter and summer, maximum reliability at minimal cost is an important design feature.

Pump and Baffle Aeration System
One type of winter aeration system is the pump and baffle. Oxygen-poor water is extracted from a nearshore area of the lake, pumped to the top of a chute located on shore, and then allowed to cascade over a set of baffles (constructed of wooden boards). The turbulence created as the water passes over the baffles helps to reaerate the water. The reoxygenated water is then returned to a different part of the lake, away from the intake area, creating a zone of oxygen-rich water.

Generally, approximately 10% of the lake's volume should be aerated (S. McComas, 1993, pers. comm.). Thus, for a lake with a 40-ha area with an average depth of 1.8 m (volume 72,000 m^3), the objective would be to aerate 7000 m^3 of water. A typical rig uses a 7.5-kW (10-hp) motor with a 150-mm or 200-mm (6- or 8-in.) pump, that delivers between 10.0 and 190 m^3/sec (1600 and 3000 gallons per minute or 2.3 to 4.3 million gallons per day). Thus, a suitable refuge area for the example lake could be created within five to ten days. For larger lakes, multiple rigs may be needed and, as a result, the pump and baffle technique may be too costly.

The velocity of water discharged into the lake must be neither too low nor too high. At low velocities, the amount of water released would be too small to produce an oxygen-rich zone of sufficient size. At high velocities, the reoxygenated water discharged will mix too thoroughly with the oxygen-poor lake water, elevating the entire lake's oxygen level only slightly (and too little to significantly improve fish survival), rather than creating a smaller refuge area with higher oxygen levels. The level of oxygen that must be maintained depends on the fish species in the lake.

Freeze-up can be a major problem with pump and baffle systems, especially during cold winter days. Freeze-up can occur at the water intake, on the chute, or at the water discharge. For example, the chute may become top heavy, due to ice buildup, and fall over. Thus, the system must be checked daily to ensure proper operation.

Pump and baffle systems have several major advantages relative to other aeration techniques. In particular, when properly operated, only a small area of the lake's ice cover is opened. Open areas and thin ice are safety hazards, for which the operator of an aeration system is liable. All of the major pieces of equipment are on shore. In addition, the chute can be mounted on a trailer and moved from one lake to another or to different areas of the lake as needed. Generally, to prevent winterkill, aeration will be required for about two months, depending on winter conditions. By monitoring dissolved oxygen levels in the lake, the system can be operated only during those times and winters when needed.

Pump and baffle systems have been built by lake associations, or may be purchased as a unit from a number of manufacturers.

Artificial Circulation

Artificial circulation eliminates thermal stratification, or prevents its formation, either by mechanical pumping or through the injection of compressed air from a pipe or ceramic diffuser at the lake's bottom (Fig. 15.12). The rising column of bubbles, if sufficiently powered, will produce lakewide mixing. As a result, the conditions that create hypolimnetic oxygen depletion (isolation of the deeper waters from the atmosphere, with little to no primary production in these deeper, darker waters) are eliminated.

The most common cause of failure in artificial circulation projects is the lack of sufficient air flow to produce satisfactory mixing. On average, about $1.1\,m^3/min$ of air flow is required per hectare of lake surface to adequately mix the lake and elevate levels of dissolved oxygen (see Lorenzen and Fast, 1977). In general, it is easier and more effective to apply the mixing energy early enough to prevent stratification, rather than attempting to turnover an already stratified lake (Burns, 1988).

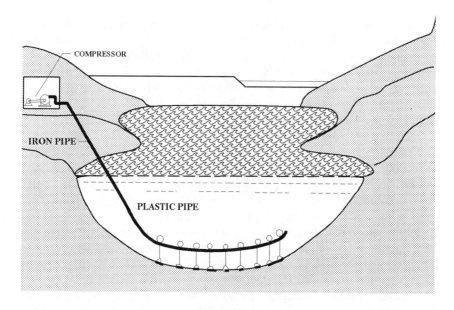

FIGURE 15.12. Schematic of lake (reservoir) destratification by aeration.

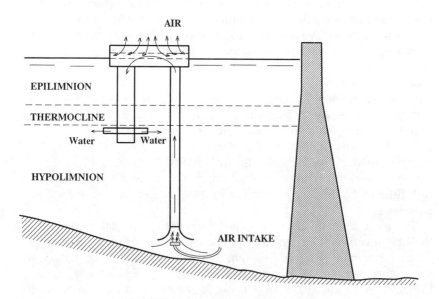

FIGURE 15.13. Hypolimnetic aeration.

Artificial circulation is one of the most commonly used lake restoration techniques (Cooke et al., 1993). Examples of its utility for improving fisheries yields include Parvin Reservoir, Colorado (Lackey, 1972) and Corbett Lake, British Columbia (Halsey, 1968). The technique is best used in lakes that are not nutrient limited; nutrient concentrations are often higher in the hypolimnion and, as a result, mixing can stimulate increased algal growth. In addition, artificial circulation is not a viable option for cold-water fish species, which use the hypolimnion as a thermal refuge during summer.

Hypolimnetic Aeration
Hypolimnetic aerators can be used to increase oxygen levels in the hypolimnion without disturbing the lake's thermal stratification. An airlift device is used to bring cold hypolimnetic water to the surface. The water is aerated by contact with the atmosphere; gases such as methane, hydrogen sulfide, and carbon dioxide, which accumulate under anaerobic conditions, are lost; and then the water is returned to the hypolimnion (Fig. 15.13).

Hypolimnetic aerators require a large hypolimnion to work properly and are generally ineffective in shallow lakes and reservoirs. Costs depend on the amount of compressed air needed, which is a function, in turn, of the area of the hypolimnion, the rate of oxygen consumption in the lake, and the degree of thermal stratification (Kortmann, 1989). Example applications include Waccabuc Lake, New York (Fast, Dorr, and Rosen, 1975), Larson and Mirror Lakes, Wisconsin (Smith, Knauer, and Wirth, 1975), and Tory Lake, Ontario (Taggart and McQueen, 1981).

Oxygen Injection
Recent studies have shown that it is often more cost-effective and practical to inject pure oxygen into the hypolimnion, as opposed to air injections or aerating the hypolimnion via air-lift systems (Aquatic Systems Engineering, 1990). At Richard B. Russell Reservoir, Georgia, dissolved oxygen levels in the hypolimnion have been increased from <3 mg/l to >9 mg/l, with an oxygen transfer efficiency of about 75% (Gallagher and Mauldin, 1987; Mauldin et al., 1988). Liquid oxygen is stored in tanks on-site and connected to several supply heads submerged and anchored in the reservoir. Flexible membrane diffusers mounted on the supply heads are used to maximize absorption efficiency and minimize maintenance requirements. Flexible membrane systems should last 2–6

years (or 10 years or more if operated ≤6 months per year); the compressor and distribution system should last substantially longer (estimated 30-year life).

Snow Removal to Increase Light Penetration

Snow removal from the lake surface, to increase light penetration and photosynthesis (oxygen generation) under the ice, is a low-tech, low-cost alternative to aerators that may be sufficient to prevent winterkill in lakes with marginal levels of dissolved oxygen (S. McComas, 1993, pers. comm.). Snow is a much more effective absorber of light than is ice. While 85% of the available light will penetrate 12.5 cm of clear ice, 12.5 cm of snow over 7.5 cm of ice will block out almost all light. Even thin layers of snow can greatly decrease light penetration, decreasing primary productivity, and thus leading to oxygen depletion and winterkill.

Using volunteers with snowplows, snow removal can be completed at relatively low cost. Alternating strips of snow removal are recommended, rather than clearing the entire area and stockpiling the accumulated snow in a single location. Thirty percent or more of the lake surface should be cleared. In general, snow removal is more effective at preventing winterkill in shallow lakes with abundant rooted macrophytes than in deep lakes with phytoplankton as the dominant primary producer (S. McComas, 1993, pers. comm.).

Additional information on lake aeration systems can be found in Olem and Flock (1990), Cooke et al. (1993), and Lorenzen and Fast (1977). The most cost-effective approach for increasing dissolved oxygen levels depends on the size (area and depth) of the lake, nature and causes of the problem, and fisheries management objectives.

Liming Acidic Lakes

Waters may be naturally acidic, for example, in regions with naturally acidic soils and large inputs of organic acids, or acidified as a result of acid mine drainage, acidic deposition, or other anthropogenic sources of acids. Extensive research, using a variety of neutralizing agents as well as application techniques, has been conducted in recent years to refine and test methods for neutralizing acidic lake waters. There are five basic approaches to treating acidic lakes:

- *Limestone addition to the lake surface.* Small limestone particles, limestone powder, or a limestone slurry are dispersed via boat, plane, or helicopter over the lake. (Or, during winter, the limestone can be spread on the ice by truck, entering the lake in the spring as the ice melts.) Direct addition of limestone to the lake surface is the most commonly employed method to date for decreasing lake acidity. Because limestone is used for agricultural liming, it is usually available at low cost. However, the cost of the limestone dispersal can be significant, particularly for remote lakes without road access. In addition, repeated applications are needed; lakes with short water-retention times may need to be treated annually.

- *Injection of base materials into the lake sediment.* Limestone, hydrated lime, or sodium carbonate can be injected into the lake sediments, resulting in a gradual decrease in lake acidity. This technique is largely experimental, however, and limited to small, shallow lakes with soft organic sediments and road access for transport of the application equipment. The treatment may last substantially longer than surface applications, but at the same time the lake's benthic community is disturbed, turbidity may increase, and the costs are higher.
- *Mechanical stream doser.* Lake acidity may be decreased by neutralizing the acidic waters in upstream tributaries. Mechanical dosers are automated devices that release dry powder or slurred limestone directly into the stream, with the quantity of material added controlled by monitors of stream flow or stream chemistry. The treatment is continuous, expensive, and generally not recommended for lakes unless all other alternatives have been ruled out.
- *Limestone addition to the watershed.* Limestone is spread on all or parts of the lake's watershed, decreasing the acidity of runoff and shallow ground-water flow into the lake. Although the costs of one application are higher, the overall costs may be lower than for limestone applications to surface waters, because the effects are much more long-lasting. Watershed liming may be especially appropriate for lakes with short retention times (less than six months).
- *Pumping of alkaline ground water.* Where abundant supplies of alkaline ground water are available, these waters may be discharged directly into lakes or lake tributaries, decreasing acidity. Applications of this method have been limited, however, and the costs and effectiveness of this approach are not well known.

Several books have been published recently on methods for liming lakes and streams, including Olem (1991), Olem et al. (1991), and Brocksen, Marcus, and Olem (1992).

Spawning Habitat Management

Fish production in some waters may be limited by the availability of suitable sites for natural reproduction, or the poor quality of available sites, resulting in relatively low reproductive success. The types of habitats required for spawning and the factors that influence spawning and reproductive success vary greatly among species. Thus, the methods employed for spawning-site management are also, to some degree, species-specific. The types of activities fall within three broad categories: (1) efforts to protect existing spawning habitat, (2) spawning habitat improvements, and (3) the construction of new spawning sites.

The first task required to either protect or improve spawning habitat is to locate existing sites in the lake or associated streams used by target fish species. Some species spawn in dispersed areas throughout the lake, while others concentrate activities within fairly localized sites used consistently year after year. For fish species that spawn in shallow waters, finding spawning sites can be relatively easy. Potential spawning areas should be visited regularly during the spawning season by trained observers looking for spawning adults, spawning nests, or egg masses. For species that spawn in deeper waters or that provide little noticeable evidence of spawning activity (e.g., that do not build nests or lay their eggs in visible masses), finding spawning areas will be much more difficult and time-consuming, requiring SCUBA, for underwater observations, or sampling gear to detect the presence of eggs in the bottom sediments or water column.

Protection of Spawning Habitat

Many fish species spawn in relatively shallow nearshore areas, subject to disturbance from swimmers, boat traffic, and runoff from the adjacent watershed. It may be desirable to limit construction, land uses, or fertilizer applications in watersheds adjacent to critical spawning areas (including tributary streams used for spawning), or to divert and treat runoff that is high in suspended solids (to prevent excessive siltation that would decrease egg and larval survival).

Fish species that guard their spawning nests, such as largemouth and smallmouth bass and most sunfish species, are particularly susceptible to disturbances from swimming and boating. If repeatedly disturbed, males may eventually desert their nests, resulting in poor survival of eggs and fry. During spawning season, important spawning sites can be identified with buoys, and boat traffic and swimming restricted from these areas. Waves rebounding off retaining walls may also drive off bass and sunfish males guarding nests. Shoreline structures that better absorb wave energy, such as riprap or vegetation, are preferable near important spawning areas.

Spawning Habitat Improvements

Siltation is a major cause of degraded spawning habitat. Some species, such as bass, crappies, and bluegill, sweep the nest area off with their tails before spawning, and thus moderate siltation is not a major concern. Others, however, such as walleye, do not; the build up of silt in walleye spawning areas can significantly decrease egg survival and may limit or prevent spawning activity.

Silt can be removed from spawning areas by using a water pump to blow the silt and algal growth off the rocks. For example, a 7.5-cm (3-in.) pump can generate enough force to remove silt from the rock face or turn cobble-sized rocks over to expose a new face. If mounted on a pontoon or raft, several spawning sites can be cleaned in a half-day (S. McComas, 1993, pers. comm.).

Brook trout have fairly restrictive spawning requirements and, as a result, the availability of suitable spawning areas is often an important factor limiting brook trout natural reproduction and productivity. Brook trout prefer to spawn in areas with upwelling, well-oxygenated ground water, in the littoral zone, tributary streams, or the lake outlet. Plants and sediments can accumulate on these underwater springs, obstructing ground-water flow and preventing brook trout spawning. The locations of underwater springs can be confirmed by inserting a small-diameter pipe (e.g., a 5-cm (2-in.) PVC-type pipe) into the lake bottom. If upwelling ground water is present, the water level in the pipe will rise above the lake level. Small-scale dredging techniques can then be used to remove the blanket of material obstructing the flow. Carline (1980) provides examples of the success of this technique for spawning-site improvements in Wisconsin lakes.

Finally, some fish species, including muskellunge and northern pike, spawn in flooded marshes and other heavily vegetated areas in bays or river floodplains. These heavily vegetated flooded areas also serve as important nursery areas for the young fish of many species. Therefore, in lakes and reservoirs with controlled water levels, reproductive success can be increased by raising water levels during the spring to coincide with fish spawning and the occurrence of early life stages. The role of water-level management for managing fisheries in lakes and reservoirs is discussed further later in this chapter.

Aquatic Plant Management

In lakes and reservoirs where thick beds of macrophytes cover a high proportion of the lake bottom, an aquatic plant control program may be needed to improve yields of large, predatory game fish and increase the growth rates of panfish species, such as bluegill and white and black crappie. For uses such as swimming and boating, minimizing macrophyte beds are desired. For fisheries management, on the other hand, moderate growths of aquatic plants enhance the fisheries. The complete elimination of macrophyte beds may be as harmful to fisheries as are excessive plant growths.

The objective of aquatic plant management is therefore to provide the appropriate amount of aquatic plants, taking into account the effects of macrophytes on fish communities, other lake uses (e.g., swimming and boating), nutrient cycles, and aesthetics. Macrophytes and terrestrial vegetation also help to stabilize the lake bed and shoreline, reducing problems with lakeshore erosion and high turbidity.

Excessive plant growths, as a result of eutrophication or the inadvertent introduction of an exotic macrophyte species, are a common lake problem. Olem and Flock (1990) provide a thorough discussion of various approaches to controlling nuisance plant growths:

Sediment removal and sediment tilling. Lakes can be dredged to remove sediments and deepen the lake, so that less of the lake bottom receives adequate light for macrophyte growth. The maximum depth at which macrophytes are able to grow depends on water transparency and the plant species. Hydrilla, a nuisance exotic plant in southern waters, can grow at lower light intensities than native plants (Canfield et al. 1985), making control through lake deepening a difficult task. Reductions in nutrient loads, to control eutrophication, can increase lake transparency, increasing the depth at which macrophytes can grow and countering the effectiveness of dredging to reduce macrophyte growth. Sediment removal and tilling [e.g., rototilling using cultivation equipment (Newroth and Soar, 1986)] can also be used to disturb the lake bottom, tearing out plant roots for short-term macrophyte control. Both dredging and tilling can have negative side effects, including destruction of the benthic community and an increase in the turbidity and siltation.

Water-level drawdown. In lakes where water levels can be controlled, lake levels can be lowered to expose macrophytes in the littoral zone to prolonged drying and/or freezing. Some species of plants are permanently damaged by these conditions, killing the entire plant, including roots and seeds, with exposures of 2–4 weeks. Other plant species are unaffected or even increase. Water-level management to control macrophyte growths and for other purposes is discussed further in the next section.

Shading and sediment covers. Covers can be placed on the water or sediment surface, as a physical barrier to plant growth or to block light (Engel, 1984). Sediment covers, made of polypropylene, fiberglass, or other similar material, can effectively prevent growths in small areas, such as near docks and swimming areas, but are generally too expensive to install over large areas. Applications of silt, sand, clay, or gravel have also been used, but plants eventually root in them. Shading, to reduce growth rates, can be provided by floating sheets of polyethylene (Mayhew and Runkel, 1962) or by planting evergreen trees along the lakeshore (Engel, 1989).

Introduction of grass carp. Grass carp is an exotic fish species that feeds on macrophytes. Grass carp do not consume aquatic plant species equally readily. Generally, they avoid alligatorweed, water hyacinth, cattails, spatterdock, and water lily. The fish prefer plant species that include elodea, pondweeds (*Potamogeton* spp.), and hydrilla. Low stocking densities can produce selective grazing on the preferred plant species, while other less preferred species, including milfoil, may even increase. Overstocking, on the other hand, will eliminate the weeds. Biological control has the objective of achieving long-term control of plants without introducing expensive machinery or toxic chemicals. However, the use of grass carp for aquatic plant management is only allowed in certain states.

Introduction of insects that infest macrophytes. Several exotic insect species have been imported to the United States and approved by the U.S. Department of Agriculture for use in macrophyte control. Each insect species grows and feeds on only select target plant species. In particular, insects have been used in southern waters to aid in the control of alligatorweed and water hyacinth (Sanders and Theriot, 1986; Haag, 1986). Because insect populations tend to grow more slowly than the plants, insects work best when used in conjunction with another plant control technique (e.g., harvesting or herbicides). No significant negative side effects from insect infestations have been documented (Olem and Flock, 1990).

Mechanical harvesting. Mechanical harvesters, constructed on low-draft barges, can be used to cut and remove rooted plants (Fig. 15.14) and floating water hyacinths (Fig. 15.15). Cutting rates range from 0.1 to 0.3 ha/hr, depending on machine size; Cooke et al. (1993) provide a listing of commercially available plant cutters. Harvesters can effectively clear an area of vegetation, although the benefits are only temporary. Rates of plant regrowth can be rapid (within weeks), but can be slowed if the cutter blade is lowered into the upper sediment layer (Conyers and Cooke, 1983). Cut plants are removed from the lake, eliminating an internal source of nutrients and organics with potential long-term benefits. However, some plant species, such as milfoil (Nicholson, 1981), may be fragmented and dispersed, and actually increase in abundance after harvesting operations. Also, small fish

FIGURE 15.14. Small aquatic weed harvester.

FIGURE 15.15. Water hyacinth harvesting.

can be caught and killed by mechanical harvesters. Thus, harvesting operations should precede spawning periods and/or avoid important spawning and nursery areas.

Herbicides. Herbicides used to kill aquatic plants include Diquat, endothall, 2,4-D, glyphosate, and fluridone (Olem and Flock, 1990). Although herbicide treatments can rapidly reduce macrophyte growths, the benefits are short-term and the potential for negative side effects is high. Plants are left in the lake to die and decompose, releasing plant nutrients and, in some cases, causing oxygen depletion and algal blooms. Plants generally regrow after several weeks or months, or may be replaced by other, more tolerant macrophyte species. Most chemicals currently approved are toxic to aquatic organisms and humans only at relatively high doses. Little information is available, however, on the long-term ecological consequences of herbicide use. Herbicide applicators must be licensed, have adequate insurance, wear protective gear, use only EPA-approved chemicals, and follow label directions exactly. Generally, because herbicides do not remove nutrients or organics from the lake or address the cause of the aquatic plant problem, herbicides should be used only where other techniques are unacceptable or ineffective. Westerdahl and Getsinger (1988) provide guidelines for herbicide use and application.

The relative effectiveness, costs, and potential for negative side effects for each of the preceding techniques is summarized in Table 15.2. In some lakes, especially newly impounded reservoirs and acidic bodies of water, the problem may be too few rather than too many macrophytes. In

TABLE 15.2 Comparison of Lake Restoration and Management Techniques for Control of Nuisance Aquatic Weeds

Treatment One Application	Short-Term Effectiveness	Long-Term Effectiveness	Cost	Chance of Negative Effects
Sediment removal	E	E	P	F
Drawdown	G	F	E	F
Sediment covers	E	F	P	L
Grass carp	P	E	E	F
Insects	P	G	E	L
Harvesting	E	F	F	F
Herbicides	E	P	F	H

Source: From Olem and Flock (1990).

Note: E = Excellent; F = Fair; G = Good; P = Poor; H = High; L = Low, based on the consensus judgment of 12 lake restoration experts (Olem and Flock 1990).

new impoundments, it may be desirable to introduce (transplant) suitable plants, such as sago pondweed, water celery, and lily pads, or to raise water levels to flood vegetated areas. In lakes with fluctuating water levels, terrestrial plants, such as winter wheat or ryegrass, can be planted in the exposed lake bed during periods of water drawdown. These plants provide both erosion control during drawdown and fish cover when the lake is reflooded, with no risk of developing nuisance levels or excessive plant growths. Howells (1986) and Evans (1989) provide guidelines for establishing plants in water-fluctuation zones and for managing plant communities in southern reservoirs.

Water-Level Management

Controlled changes in water level, at the right time, place, and magnitude, can be beneficial and provide an important management tool for maintaining and improving a lake ecosystem and fisheries. Baker et al. (1993) describe the use of water-level management for a number of purposes:

- Water-level drawdowns for vegetation control
- Water-level drawdowns for overcrowded, slow-growing fish populations
- Water-level drawdowns to enhance fish cover
- High-water levels in the spring to provide increased spawning and nursery areas
- Water-level drawdowns during spawning to reduce the reproductive success of undesirable fish species

In lakes with a well-maintained outlet structure and drawdown capability, the dollar costs of a water level-management program are minimal and any or all of the preceding actions can be useful for fisheries management. However, potential adverse side effects include (a) impacts on other lake uses, such as swimming and boating, (b) damage to shore banks, shorelines, and shoreline retaining walls, (c) reduced numbers and diversity of benthic invertebrates in the littoral zone, which are important prey items for some fish species, and (d) an increased likelihood of winterkill during winter drawdowns. Drawdowns should be conducted only in lakes with a steady water inflow, sufficient to refill the lake when needed. Ploskey (1982) presents additional information on the benefits and design of water-level management programs.

Reservoir Construction

Physical features that can be altered very little once a lake is constructed can drastically affect the fish community and the cost of maintaining quality fishing. Factors such as watershed area, watershed usage, and the erodibility of soils in the watershed will impact lakewater quality. Lake volume, mean and maximum depth, watershed topography, shoreline development, and basin slope all influence the degree of thermal stratification and other lake characteristics that indirectly affect the suitability of the lake habitat for fish. For example, Hill (1986) identified three primary physical characteristics associated with the quality of fishing in small reservoirs in Iowa: (1) the mean basin slope, (2) the watershed-to-lake-area ratio, and (3) an adjusted siltation index based on the watershed-to-lake-area ratio, soil erosion rates for soil types in the watershed, and the proportion of the watershed farmed using approved soil-conservation practices. All of these physical features can be determined prior to lake construction and should influence site selection, lake design, and construction methods.

Baker et al. (1993) describe a number of important factors to consider in the design and construction of small lakes and reservoirs, relevant specifically to fisheries management:

- The watershed should be well vegetated to prevent excessive siltation, and free of pesticides and other pollutants. Runoff from row crops, livestock operations, and industrial sites (or from areas where erosion cannot be controlled) should be treated or diverted around the lake.
- The optimal watershed size (and watershed-to-lake-area ratio) depends on the lake volume, rate of rainfall, topography of the watershed, and land uses in the watershed. In Georgia, the recommended watershed size for a small, one-acre lake is about ten acres of pasture or 25 acres of forested watershed (Georgia Department of Natural Resources, 1988). The state of Illinois recommends 9 hectares of watershed per surface acre of water as a rule of thumb (Illinois Department of Conservation, n.d.).
- The amount of timber and brush that should be cleared from the lake basin before filling depends on the lake's size and planned uses. In small lakes (<1.4 ha), all trees, stumps, and brush should be removed to produce the maximum harvestable fish biomass. In larger reservoirs, on the other hand, it is often desirable to leave small blocks of well-placed timber and brush near the shoreline in coves and embayments to provide fish cover. Timber should be cleared, however, in the area of the dam, in recreational and boating areas, and in the deeper parts of the reservoir, where it is unlikely to improve fishing and may aggravate problems with oxygen depletion as the organic matter left behind gradually decomposes. Too much standing timber and brush can provide excessive fish cover, and should be avoided.
- If it is necessary to supplement natural spawning areas and fish cover,

spawning sites and artificial reefs can be constructed more easily before the lake fills than afterward.
- Extensive shallow areas will encourage the growth of aquatic macrophytes, an undesirable feature in most small lakes designed for fishing and fish production. Recommended slope ratios along the shoreline are generally 2:1 or 3:1. Also, sand blankets, gravel beds, or fiberglass mats can be used to inhibit rooted plant growth in selected areas.
- Water depths should be sufficient to minimize problems with low levels of dissolved oxygen in the winter and late summer. The state of Illinois, for example, recommends that lakes be at least 0.45 ha in area and 3 to 3.6 m deep in 25% of the basin for successful fish management (Illinois Department of Conservation, 1989).
- The dam and water overflow should be constructed to allow easy control of water levels, using a drain pipe or gate value. In most cases, a standpipe overflow is recommended that draws water from deeper areas of the lake.
- Upstream migration of undesirable fish and, to a lesser degree, the downstream migration of game fish can be hindered by constructing an emergency spillway with a substantial vertical fall (at least 1.2 m) and sufficiently wide so that water depths over the spillway never exceed 5 cm.

Additional guidelines on the design and construction of small lakes and reservoirs are available from most state fisheries agencies (e.g., Oklahoma Department of Wildlife Conservation, 1984; Illinois Department of Conservation, n.d.) and the U.S. Department of Agriculture, Soil Conservation Service (1982).

Biomanipulation

Many fisheries management techniques involve the direct manipulation of the fish community and other organisms that may serve as prey for or be predators or competitors of the fish species of interest. Three types of activities include: (1) game fish stocking, (2) control of undesirable fish species and stunted fish populations, and (3) prey enhancements to supplement food supplies.

INTEGRATED AQUATIC RESTORATION: CASE STUDIES

Aquatic ecosystem restoration projects should be undertaken using an integrated approach. This type of approach attempts to consider the major ecological actions in a watershed and respond holistically to cumulative ecological impacts. Fragmentation of the ecological restoration and management efforts has been commonplace in government and industry. There are many barriers to integrated aquatic ecosystem restoration. Watershed boundaries often overlap with political boundaries. Even within watersheds, there is often administration of ecological concerns by different agencies.

Fortunately, over the last 20 years there has been an increase in the formation of agencies with jurisdictions at the watershed or ecoregion level. For example, the International Joint Commission established joint U.S. and Canadian goals for the Great Lakes. Throughout the United States, river basin and interstate commissions have been created to deal with environmental issues at the watershed level. These regional planning agencies include such organizations as the Interstate Commission on the Potomac River Basin, the Ohio River Sanitation Commission, and the Tahoe Regional Planning Agency.

This section presents two case studies of aquatic restoration efforts where there was a concerted effort to integrate restoration and promote the philosophy of watershed protection. These include the Merrimack River Watershed Protection Project and the Canaan Valley, West Virginia, Project (U.S. EPA, 1993). The integrated plans to restore the Milwaukee River in Wisconsin, the Emscher River in Germany, Lake Balaton in Hungary, and the Lagoon of Venice in Italy were discussed in Chapter 1.

Merrimack River, Massachusetts and New Hampshire

The Merrimack River watershed covers 13,000 km^3 in parts of Massachusetts and New Hampshire. More than 300,000 people rely on the river for drinking water. The river also provides water for industrial and agricultural uses, and serves to assimilate waste and generate electricity. Many people use the river and its shores for relaxation and recreation.

Wastewater discharges, toxic contaminants, urban runoff, increased water withdrawal, and wetlands loss are the primary threats to long-term water quality and ecological integrity in the river. Project participants include several federal agencies (EPA, USDA, Department of the Interior, Army Corps of Engineers); state governments (Massachusetts and New Hampshire); regional planning agencies; local governments, industries, and utilities; agricultural, environmental, and recreational organizations; and universities. The participants are resolving both water quality and quantity issues by developing data-management systems and striving to balance competing needs within the watershed. A few projects underway aim to:

- Provide decisionmakers with information on the extent and condition of wetlands to protect the most valuable areas
- Help light industries (such as auto repair shops, dry cleaners, or photofinishers) understand what steps they can take to prevent pollution
- Provide decisionmakers with information about potential contamination of water supplies, helping them to focus regulatory activities (such as inspections and permitting) on preventing pollution and planning for emergency response if spills occur

Canaan Valley, West Virginia

The 15,000-ha Canaan Valley in West Virginia, designated as a National Natural Landmark in 1975, comprises fragile wetlands areas containing a unique and irreplaceable boreal ecosystem. The Blackwater River, originating in the valley's southern end, is an important source of drinking water and the largest stream network in the state with a self-sustaining brown trout population.

The valley and its resources attract a wide spectrum of interests. For example, a power company proposes flooding 3150 ha of the valley; real estate developers plan to increase the number of vacation homes, golf courses, ski slopes, and condominiums; a major off-road vehicle race, called the Blackwater 100, is held in the valley annually; and natural resource conservationists strive to protect rare plants and wildlife habitat, including wetlands.

In 1990, federal, state, and local participants formed the Canaan Valley Task Force to resolve a variety of issues, ensuring long-term environmental protection while allowing reasonable, sustainable economic growth. Early accomplishments include:

- A study of the impact of off-road vehicles
- A study of the economic impact of the proposed Canaan Valley National Wildlife Refuge
- Suspension of certain nationwide general permits for discharges of dredged or fill material in wetlands in the valley
- Advanced identification of wetlands
- Establishment of a wetlands surveillance program
- Implementation of a public outreach program

CONCLUSIONS

Many methods are available for management and restoration. The best approach for any given situation will vary, depending on the target species, nature of the problem, management goals and objectives, and the characteristics of the body of water. There are no single solutions, although experience has demonstrated that some methods tend to work better in certains situations and for some problems than others.

Unnecessary actions or overmanagement are as undesirable as no activity at all. Monitoring should be conducted routinely to continually assess the effectiveness of restoration and management actions, and the plan altered appropriately. The goal should be to select and apply the approach(es) that will achieve the desired goals for the lowest cost and effort and also ensure the long-term sustainability of the ecosystem.

An integrated approach to restoration is recommended. Proposed restoration projects should consider the major ecological interactions in a watershed. Rivers, streams, lakes, and reservoirs are interconnected parts of a watershed that also often include ground water, estuaries, and wetlands. A practical and effective approach to restoration of aquatic ecosystems would include the consideration of all significant ecological elements on a watershed scale. In this way, the cumulative impacts of an ecosystem can best be evaluated.

References

Andersson, G., H. Berggren, and S. Hambrin (1975). Lake Trummen restoration project. III. Zooplankton macrobenthos and fish, *Verh. Int. Ver. Limnol.* **19**:1097.

Aquatic Systems Engineering (1990). *Assessment and Guide for Meeting Dissolved Oxygen Water Quality Standards in Hydroelectric Plant Discharges*, EPRI GS-7001, Electric Power Research Institute, Palo Alto, CA.

Baker, J. P., H. Olem, C. S. Creager, and B. P. Parkhurst (1993). *Fish and Fisheries Management in Lakes and Reservoirs*, EPA-841-R-93-002, U.S. Environmental Protection Agency, Washington, DC.

Bengtsson, L., S. Fleischer, G. Lindmark, and W. Ripl (1975). Lake Trummen restoration project. I. Water and sediment chemistry, *Verh. Int. Ver. Limnol.* **19**:1080–1087.

Berger, J. J. (1990). Evaluating Ecological Protection and Restoration Projects. A Holistic Approach to the Assessment of Complex Multi-attribute Resource Management Problems, Ph.D. dissertation, University of California, Davis.

Bernstein, G., and P. Kumble (1992). Anacostia watershed restoration small habitat improvement program: Checklist and design matrix, in *Watershed Restoration Sourcebook: Collected Papers Presented at the Conference; Restoring Our Home River: Water Quality and Habitat in the Anacostia*, November 6 and 7, 1991, College Park, MD, P. Kumble and T. Schueler, eds., Publication No. 92701, Metropolitan Washington Council of Governments, Washington, DC.

Brocksen, R., M. Marcus, and H. Olem (1992). *Practical Guide to Managing Acidic Surface Waters and Their Fisheries*, Lewis Publ., Chelsea, MI.

Burns, F. L. (1988). *Aeration of Lakes and Reservoirs in Australia*, presented at North American Lake Management Society annual meeting, St. Louis, MO.

Cairns, R. F. (1988). Increasing diversity by restoring damaged ecosystems, in *Biodiversity*, E. O. Wilson, ed., National Academy Press, Washington, DC, pp. 333–343.

Canfield, D. E., Jr., K. A. Langeland, S. B. Linda, and W. T. Haller (1985). Relations between water transparency and maximum depth of macrophyte colonization in lakes, *J. Aquat. Plant Manage.* **23**:25–28.

Carline, R. F. (1980). Features of successful spawning site development for brook trout in Wisconsin ponds, *Trans. Am. Fish. Soc.* **109**:453–457.

Committee on Restoration of Aquatic Ecosystems (1992). *Restoration of Aquatic Ecosystems*, National Research Council, National Academy Press, Washington, DC.

Conyers, D. L., and G. D. Cooke (1983). A comparison of the costs of harvesting and herbicides and their effectiveness in nutrient removal and control of macrophyte biomass, in *Lake Restoration, Protection, and Management*, EPA 440/5-83-001, U.S. Environmental Protection Agency, Washington, DC, pp. 317–321.

Cooke, G. D., and H. Olem (1990). Lake and reservoir restoration and management techniques, in *The Lake and Reservoir Guidance Manual*, 2d ed. EPA 440/4-90-006, North American Lake Management Society for U.S. Environmental Protection Agency, Washington, DC, pp. 117–159.

Cooke, G. D., R. T. Heath, R. H. Kennedy, and M. R. McComas (1982). Change in lake trophic state and internal phosphorus release after aluminum sulfate application. *Water Resour. Bull.* **18**:699–705.

Cooke, G. D., E. B. Welch, S. A. Peterson, and P. R. Newroth (1993). *Restoration and Management of Lakes and Reservoirs*, 2d ed., Lewis Publ., Boca Raton, FL.

Cronberg, G., C. Gelin, and K. Larsson (1975). Lake Trummen restoration project. II. Bacteria, phytoplankton, and phytoplankton productivity, *Verh. Int. Ver. Limnol.* **19**:1088–1096.

Davis, J. L., et al., eds. (1983). *Proceedings, International Conference on Water Power 1983*, University of Tennessee, Knoxville, TN, pp. 1326–1335.

Dominie, J. R. (1980). Hypolimnetic aluminum treatment of softwater Annabessacook Lake, in *Restoration of Lakes and Inland Waters*, EPA 440/5-81-010, U.S. Environmental Protection Agency, Washington, DC, pp. 417–423.

Dunst, R. C., J. G. Vennie, R. B. Corey, and A. E. Peterson (1984). *Effect of Dredging Lilly Lake, Wisconsin*, EPA 600/3-84-097, EPA Cooperative Agreement No. R804875, Environmental Research Laboratory, U.S. Environmental Protection Agency, Corvallis, OR.

Engel, S. (1984). Evaluating stationary blankets and removable screens for macrophyte control in lakes, *J. Aquat. Plant Manage.* **22**:43–48.

Engel, S. (1989). Lake use planning in local efforts to manage lakes, in *Enhancing States' Lake Management Programs*, North American Lake Management Society, Washington, DC, pp. 101–105.

Evans, J. W. (1989). *Objectives and Guidelines for Aquatic Plant Management in Georgia*, Game and Fish Division, Georgia Department of Natural Resources, Fort Valley, GA.

Everhart, W. H., and W. D. Youngs (1981). *Principles of Fishery Science*, Cornell University Press, Ithaca, NY.
Fast, A. W., V. A. Dorr, and R. J. Rosen (1975). A submerged hypolimnion aerator, *Water Resour. Res.* **11**:287-293.
Gächter, R. (1976). Die Tiefenwasserableitung, ein Weg zur Sanierung von Seen, *Schweiz. Z. Hydrol.* **38**:1-28.
Gallagher, J. W., and G. V. Mauldin (1987). Oxygenation of releases from Richard B. Russell Dam, in *Proceedings, CE Workshop on Reservoir Releases*, U.S. Army Corps of Engineers, Vicksburg, MS, pp. 121-124.
Garrison, P. J., and D. R. Knauer (1984). Long-term evaluation of three alum treated lakes, in *Lake and Reservoir Management*, EPA 440/5-84-001, U.S. Environmental Protection Agency, Washington, DC, pp. 513-517.
Gasperino, A. F., M. A. Beckwith, G. R. Keizur, R. A. Soltero, D. G. Nichols, and J. M. Mires (1980). Medical Lake improvement project: Success story, in *Restoration of Lakes and Inland Waters*, EPA 440/5-81-010, U.S. Environmental Protection Agency, Washington, DC, pp. 424-428.
Georgia Department of Natural Resources (1988). *Management of Georgia Fish Ponds*, Game and Fish Division, GDNA, Atlanta, GA.
Haag, K. H. (1986). Effective control of waterhyacinth using Neochetina and limited herbicide application, *J. Aquat. Plant Manage.* **24**:70-75.
Halsey, T. G. (1968). Autumnal and overwinter limnology of three small eutrophic lakes with particular reference to experimental circulation and trout mortality, *J. Fish. Res. Board Can.* **25**:81-99.
Herson, L. (1992). Riparian restoration: A guide to reforestation planning, in *Watershed Restoration Sourcebook: Collected Papers Presented at the Conference; "Restoring Our Home River: Water Quality and Habitat in the Anacostia."* November 6 and 7, 1991, College Park, MD, P. Kumble and T. Schueler, eds., Publication No. 92701, Metropolitan Washington Council of Governments, Washington, DC, pp. 217-231.
Hill, K. R. (1986). Classification of Iowa lakes and their fish standing stock, *Lake Reserv. Manage.* **2**:105-109.
Howells, R. G. (1986). *Guide to Techniques for Establishing Woody and Herbaceous Vegetation in the Fluctuation Zones of Texas Reservoirs*, Federal Aid in Sport Fish Restoration Project F-31-R-12, Texas Parks and Wildlife Department, Austin, TX.
Illinois Department of Conservation (n.d.). *Small Lakes and Ponds: Their Construction and Care*, Fishery Bulletin No. 3, Division of Fisheries, IDC, Springfield, IL.
Illinois Department of Conservation (1989). *Management of Small Lakes and Ponds in Illinois*, 2d ed., Division of Fisheries, IDC, Springfield, IL.
Imhoff, K. R. (1969). *Advances in Water Pollution Research*, Pa. II-7, Pergamon Press, Oxford.
Imhoff, K., and K. R. Imhoff (1990). *Taschenbuch der Stadtenwässerung*, Oldebourg Verlag, Munich.
Karouna, N. (1992). Techniques for restoring urban streams, in *Watershed Restoration Sourcebook: Collected Papers Presented at the Conference; "Restoring*

Our Home River: Water Quality and Habitat in the Anacostia." November 6 and 7, 1991, College Park, MD, P. Kumble and T. Schueler, eds., Publication No. 92701, Metropolitan Washington Council of Governments, Washington, DC, pp. 103–122.

Kortmann, R. W. (1989). Aeration: Technologies and sizing methods, *Lake Line* **9**:6–7, 18–19.

Kortmann, R. W., E. Davis, C. R. Frink, and D. D. Henry (1983). Hypolimnetic withdrawal: Restoration of Lake Wonoscopomuc, Connecticut, in *Lake Restoration, Protection and Management*, EPA 440/5-83-001, U.S. Environmental Protection Agency, Washington, DC, pp. 46–55.

Krenkel, P. A., and V. Novotny (1980). *Water Quality Management*, Academic Press, New York.

Kumble, P., and T. Schueler, eds. (1992). *Watershed Restoration Sourcebook: Collected Papers Presented at the Conference; "Restoring Our Home River: Water Quality and Habitat in the Anacostia."* November 6 and 7, 1991, College Park, MD. Publication No. 92701, Metropolitan Washington Council of Governments, Washington, DC.

Lackey, R. T. (1972). Response of physical and chemical parameters to eliminating thermal stratification in a reservoir, *Water Resour. Bull.* **8**:589–599.

Lorenzen, M. W., and A. W. Fast (1977). *A Guide to Aeration/Circulation Techniques for Lake Management*, EPA 600/3-77-004, U.S. Environmental Protection Agency, Washington, DC.

Mauldin, G., R. Miller, J. Gallagher, and R. E. Speece (1988). Injecting an oxygen fix, *Civ. Eng.* **58**(Mar.):54–56.

Mayhew, J. K., and S. T. Runkel (1962). The control of nuisance aquatic vegetation with black polyethylene plastic, *Proc. Iowa Acad. Sci.* **69**:302–307.

Newroth, P. R., and R. J. Soar (1986). Eurasian watermilfoil management using newly developed technologies, *Lake Reserv. Manage.* **2**:252–257.

Nicholson, S. A. (1981). Changes in submersed macrophytes in Chautauqua Lake, 1937–1975, *Freshw. Biol.* **11**:523–530.

Novotny, V., and G. Bendoricchio (1989). Water quality—Linking nonpoint pollution and deterioration, *Water Environ. Technol.* **1**(Nov.):400–408.

Novotny, V., K. R. Imhoff, M. Olthof, and P. A. Krenkel (1989). *Karl Imhoff's Handbook of Urban Drainage and Wastewater Disposal*, Wiley Interscience, New York.

Oklahoma Department of Wildlife Conservation (1984). *Pond Management in Oklahoma*, Fish Division, ODNC, Oklahoma City, OK.

Olem, H. (1991). *Liming Acidic Surface Waters*, Lewis Publ. Chelsea, MI.

Olem, H., and G. Flock, eds. (1990). *The Lake and Reservoir Restoration Guidance Manual*, 2d ed. EPA 440/4-90-006, North American Lake Management Society for U.S. Environmental Protection Agency, Washington, DC.

Olem, H., R. K. Schreiber, R. W. Brocksten, and D. B. Porcella (1991). *International Lake and Watershed Liming Practices*, Terrene Institute, Washington, DC.

Perkins, M. A. (1983). *Limnological Characteristics of Green Lake: Phase I*

Restoration Analysis, Department of Civil Engineering, University of Washington, Seattle, WA.

Peterson, J. O., J. T. Wall, T. L. Wirth, and S. M. Born (1973). *Eutrophication Control: Nutrient Inactivation by Chemical Precipitation at Horseshoe Lake, Wisconsin*, Tech. Bull. 62, Wisconsin Department of Natural Resources, Madison, WI.

Ploskey, G. R. (1982). *Effects of Water-level Changes on Reservoir Ecosystems, Fish, and Fishing, with Guidance on Reservoir Operations to Enhance Fisheries*, Tech. Rep. 81, U.S. Fish and Wildlife Service, Washington, DC.

Ripl, W. (1976). Biochemical oxidation of polluted lake sediment with nitrate—A new restoration method, *Ambio* **5**:132–135.

Ripl, W., and G. Lindmark (1978). Ecosystem control by nitrogen metabolism in sediment, *Vatten* **34**:135–144.

Sanders, D. R., Sr., and E. A. Theriot (1986). *Large-scale Operations Management Test (LSOMT) of Insects and Pathogens for Control of Waterhyacinth in Louisiana. II. Results for 1982–83*, Tech. Rep. A-85-1, U.S. Army Corps of Engineers, Vicksburg, MS.

Smith, S. A., D. R. Knauer, and T. L. Wirth (1975). *Aeration as a Lake Management Technique*, Tech. Bull. No. 87, Wisconsin Department of Natural Resources, Madison, WI.

Soltero, R. A., D. G. Nichols, A. F. Gasperino, and M. A. Beckwith (1981). Lake restoration: Medical Lake, Washington, *J. Freshw. Ecol.* **1**:155–165.

Stuber, R. J. (1985). Trout habitat, abundance, and fishing opportunities in fenced vs. unfenced riparian habitat along Sheep Creek, Colorado, in *Riparian Ecosystems and Their Management: Reconciling Conflicting Uses*, Gen. Tech. Rep. RM120, U.S. Forest Service, Ft. Collins, CO, pp. 310–314.

Taggart, C. T., and D. J. McQueen (1981). Hypolimnetic aeration of a small eutrophic kettle lake: Physical and chemical changes, *Arch. Hydrobiol.* **91**:150–180.

USDA Soil Conservation Service (1982). *Ponds—Planning, Design, and Construction*, Agric. Handb. No. 590, U.S. Government Printing Office, Washington, DC.

U.S. Environmental Protection Agency (1987). *Guide to Nonpoint Source Pollution Control*, Office of Water, U.S. EPA, Washington, DC.

U.S. Environmental Protection Agency (1993). *The Watershed Approach. Annual Report 1992*, EPA-840-S-93-001, Office of Water, U.S. EPA, Washington, DC.

Welch, E. B., and C. R. Patmont (1980). Lake restoration by dilution: Moses Lake, Washington, *Water Res.* **14**:1317–1325.

Westerdahl, H. E., and K. D. Getsinger, eds. (1988). *Aquatic Plant Identification and Herbicide Use Guide*, Vol. I and II, Aquat. Plant Control Res. Progr. Tech. Rep. A-89-9, U.S. Army Corps of Engineers, Vicksburg, MS.

Willenbring, P. R., M. S. Miller, and W. D. Weidenbacher (1984). Reducing sediment phosphorus release rates in Long Lake through the use of calcium nitrate, in *Lake and Reservoir Management*, EPA 440/5-84-001, U.S. Environmental Protection Agency, Washington, DC, pp. 118–121.

16
Integrated Planning and Control of Diffuse Pollution— Watershed Management

The environment functions as an integrated whole and each part is to some degree dependent on the other. Recognition of this inter-relatedness would improve our ability to constrain and reduce pollution.

United Kingdom, Department of the Environment,
Integrated Pollution Control: A Consultation Paper, July 1988

Until recently, emphasis by environmental engineers and decisionmakers has been directed toward the treatment of traditional point source pollution (in this sense, classification of point source pollution included mostly discharges from municipal and industrial wastewater sources, without considering wet-weather inputs). Billions of dollars were spent in the United States on the point source cleanup mandated by the earlier versions of the Clean Water Act (1972 and 1977). Similar policies were in place in several industrialized countries of Europe. As a result of these policies, marked improvements of water quality of some bodies of water were noticed. The River Thames in London, which for decades exhibited anoxic faulty conditions during warm summer periods, is alive again and can support a viable recreational fishery. The dissolved oxygen levels of the Potomac River and estuary near Washington, D.C., have risen from zero to about half-saturation during the summer. The successes and failures of the environmental efforts implemented in the United States under the Clean Water Act have been documented by Wolman (1988), who noted that focusing on the point source abatement has only maintained more-or-less a status quo in the majority of water quality monitoring stations throughout the United States. It has to be pointed out, however, that most of the monitoring stations operated by the U.S. Geological Survey may not be located in places where the most profound water quality changes have occurred. Hence the recent experience of emphasis on point source abatement indicates that focusing on one type of pollution (municipal and industrial wastewater discharges) may not be efficient and that an integrated approach that would address both point and nonpoint (diffuse) sources is needed.

The elimination or reduction of pollution sources may require excessive expenses. Furthermore, many diffuse sources cannot be regulated (see Chapter 1) and enforcement of control is not feasible. Therefore, the integrated solutions should also address the magnitude of the waste-assimilative capacity and its enhancement wherever possible (primarily in limited water quality situations). For example, in Milwaukee, Wisconsin, after spending over $2 billion on point source cleanup and sewer rehabilitation, the receiving waters of the lower Milwaukee River and estuary still remain unacceptably polluted due to diffuse pollution from upstream rural sources and urban runoff from the Milwaukee metropolitan area. Focusing on point source abatement in the Boston harbor, Puget Sound, and other coastal watersheds is not sufficient. A water quality plan for the Milwaukee harbor (lower Milwaukee River) suggested in-stream measures (dilution, in-stream aeration) that would cost only about 0.2% of the price of the point source cleanup and would bring the water quality within the standards for conventional pollutants (dissolved oxygen, bacteria, etc.) prescribed for this body of water. The dilution water is conveyed into the Milwaukee harbor from Lake Michigan by so-called flushing tunnels that were built in 1888 and have become an engineering landmark. The dilution remedy is not a universal solution. As a matter of fact, it is quite unusual and not available in many places. However, other solutions for enhancing the waste-assimilative capacity of receiving bodies of water, such as reestablishing riparian wetlands, may be feasible.

As was pointed out in the preceding chapter, in the 1920s Karl Imhoff was the first prominent environmental engineer and planner who recognized that a water-pollution problem cannot be resolved by point source cleanup only. In the Ruhr area of Germany several very small rivers carry the entire burden of water supply and wastewater disposal from one of the most industrialized regions of the world, with a resident population of about 8 million. The integrated approach used both innovative institutional arrangement (see, for example, Kneese and Bower, 1968, for a discussion) by creating watershedwide agencies (authorities) with great powers to control pollution, and in-stream measures for enhancing the waste-assimilative capacity, such as (Imhoff and Imhoff, 1990):

- Dilution (low-flow augmentation) by increasing flows from upstream reservoirs during critical water quality periods and by back-pumping Rhine River water into the small receiving streams (tributaries) of the Rhine
- Building river reservoirs to act as polishers during posttreatment
- In-stream aeration by power plant turbines and floating aerators
- Dredging accumulated sediment to reduce sediment oxygen demand and improve water quality
- Cutting weeds in reservoirs affected by higher nutrient inputs

Today, promising in-stream water quality enhancement methods include restoration of riparian (water bordering) wetlands, installation of buffer strips, polishing lagoons and ponds, and even fish management (Mitsch and Jørgensen, 1989; Novotny and Chesters, 1981). The concepts of integrated pollution control are covered in a publication by Haigh and Irwin (1990).

POLLUTION LOAD TRADE-OFFs AND THE PERMIT PROCESS

To the receiving body of water whose integrity is impaired by pollution discharges it does not matter whether the pollution originates from point or nonpoint sources. In either or both cases damage is done to the uses of the body of water, including its use for fish and wildlife protection and propagation and/or for recreation, water supply, or other beneficial and designed uses. For example, if water quality deterioration is caused by discharges of phosphorus, all sources of phosphorus must be considered. In addition, the forms of phosphorus and their availability to stimulate the photosynthetic algal and plant growth are secondary factors to be considered.

In the planning–permit preparation phase the waste-assimilative capacity of the receiving bodies of water is generally determined by theoretically and mathematically sound models that have been calibrated and verified by extensive field data. The use of models is necessary because the waste-assimilative capacity is either determined for hydrological conditions that are statistically rare, or are defined in terms of probabilities. Furthermore the pollution inputs are projected into the future. Modeling is also used for assessing the possibilites of the in-stream and off-stream waste-assimilative capacity enhancement, such as the impact on water quality of riparian wetlands and buffer strips.

In the United States, legislatures generally attempt to treat polluters uniformly. Unfortunately most of the legislative control regulations contain added exceptions for special interests, such as the exclusions from the NPDES permits of some agricultural and industrial sources, and diffuse-pollution sources from smaller urban areas. Under the uniform control scenario of the NPDES permit system, if the waste-assimilative capacity of the receiving body of water is exceeded, the allowable pollution load is distributed by the regulating agency uniformly among the polluters (those covered by the permit requirement). The sources of diffuse pollution from urban and nonurban areas are well known and have been described in great detail throughout this book.

The planning studies that will lead to the allocation of pollutant loads among the sources must consider whether the pollutant mass load or pollutant concentration is the cause of the problem. Pollutant mass loading from all sources can be a problem in bodies of water undergoing accelerated eutrophication. In this case, nutrient loads (in kilograms per year or season) from all sources, *regardless of whether they are point or nonpoint*, must be restricted. In a *trade-off approach* the regulatory agency will issue the permits based on the estimated maximal pollutant load that will not impair the integrity of the receiving body of water. That implies nonviolation of water quality standards. Even the present water quality criteria (standards in some states) for nitrogen and phosphorus are based on seasonal limits rather than short-term exposures.

The maximal long-term load of some toxic components should be considered if these toxic components are incorporated into the sediment or tissues of biota in the affected bodies of water (bioaccumulation). At the present time, fish tissue standards are in place, but sediment standards are being implemented. For example, PCB concentrations in water of affected water bodies are commonly undetectable even though sediment and fish tissue concentrations are unacceptable. In these and similar cases, the long-term loads from all sources (including point and nonpoint and atmospheric, dry- and wet-weather sources alike) and not short-term effluent concentrations must be restricted. These standards should be based on long-term chronic toxicity and appropriate safety factors. The allowable long-term (for example, annual or seasonal) loads can only be determined by relatively complex dynamic long-term simulation effects or other modeling methodologies described in Chapters 12 and 13. At present, the state of the art of available models is still evolving, and application of the models is very expensive, requiring extensive field data and calibration and verification efforts.

The short-term chemical-based concentration standards are typically based on lethal toxicity obtained from toxicity tests lasting from two to four days. These standards are applied to water, and the loads are then determined from mass-balance models. Due to the fact that only exposure to the maximal concentration during a relatively short period is considered, the models are more simple, more reliable. Typically, steady-state receiving water quality models suffice for dry-weather point sources, supplemented with event-oriented loading models or estimates for diffuse wet-weather sources.

WATER USES AND USE ATTAINABILITY

In the United States, water quality standards define water quality goals that are related to the designated use of the body of water. The law requires states to adopt water quality standards to protect public health or welfare, enhance the quality of water, and serve the purposes of the Clean Water Act. "Serve the purposes of the act" means that water quality standards should

- Include provisions for restoring and maintaining the chemical, physical, and biological integrity of state waters
- Wherever attainable, provide water quality for the protection and propagation of fish, shellfish, wildlife, and recreation in and on the water
- Consider the use and value of state waters for public water supplies, propagation of fish and wildlife, recreation, agricultural and industrial purposes, and navigation.

When states designate uses that will support aquatic life propagation, contact recreation and human health (drinking water supply) considerations must be given to whether such uses can be attained. If the state does not designate uses that would comply with the goals of the Clean Water Act, a use attainability analysis must be performed.

A *designated use* is a term that is specified in water quality standards for a body of water or a segment of a body of water. Typical uses include public water supply, propagation of fish and wildlife, recreational purposes, agricultural, industrial, and navigational. The EPA does not recognize waste transport and assimilation as an acceptable water use (U.S. EPA, 1988).

According to the Clean Water Act, the desirable water uses that must be protected include public water supply, recreation, and propagation of fish and wildlife. Each state develops its own use classification system based on the generic uses cited in the Clean Water Act. The states may differentiate and subcategorize the types of uses that are to be protected, such as cold-water or warm-water fisheries, or specific species that are to be protected, such as trout, salmon, or bass. States may also designate special uses to protect sensitive or valuable aquatic life or habitat.

An existing use is a use that was achieved on a body of water on or after November 28, 1975. It can be modified or changed only if new important uses of the body of water have been added that require more stringent criteria. *Attainable uses* are those uses (based on the state's system of water-use classifications) that can be achieved when effluent limits under the Clean Water Act, Sections 301(b)(1)(A) and (B) and Section 306, are implemented for point source discharges and when cost-effective and reasonable best management practices are implemented for nonpoint sources.

The Clean Water Act (Section 305(b)) requires states to submit a water quality assessment report (U.S. EPA, 1991a), which should contain an evaluation of the extent of support of designated uses in their surface-water bodies. The 1990 summary of use attainment is presented in Table 16.1. The so-called 305(b) process is "the principal means by which EPA, Congress, and the public evaluate water quality, the progress made in maintaining and restoring water quality and the extent to which the problem remains" (U.S. EPA, 1991a). The law requires that the States submit their water quality assessment reports to the EPA administrator every two years.

Statutory water uses. The specific water uses of a body of water are

TABLE 16.1 Support of Designated Uses in U.S. Waters in 1990

	Rivers (×1000 km)	Lakes (×1,000 ha)	Estuaries (km²)	Great Lakes Shoreline (km)	Coastal Shoreline (km)
U.S. total	2573	15,944	91,162	8200	30,874
Assessed	1041	7483	69,135	7723	6758
	(64%)	(47%)	(75%)	(94%)	(22%)
Fully support	655	3308	38,860	137	6074
designated use	(23%)	(21%)	(42%)	(2%)	(89%)
Threatened	69	1175	7905	111	78
	(2%)	(7%)	(9%)	(2%)	(1%)
Partially support	216	1405	17,024	2276	466
designated use	(7%)	(9%)	(18%)	(29%)	(7%)
Do not support	100	1603	5345	5290	187
designated use	(3%)	(10%)	(6%)	(68%)	(3%)

determined by the states. The statutory body of water uses specified in the Section 305(b) reports include:

- Fish consumption (commercial fishing)
- Shell fishing (commercial)
- Aquatic life support
- Swimming and other contact recreation
- Secondary water contact (boating)
- Drinking water supply
- Agriculture (irrigation)
- State defined uses

Types of bodies of water. The guidelines for the Section 305(b) reports (U.S. EPA, 1991a) recognize the following types of bodies of water: streams, lakes, estuaries, coastal waters, wetlands. As shown in Table 16.1 the classification of the attainability of water uses of U.S. surface-water bodies categorizes the surface-water bodies as

Use attained

- fully supporting—the designated use is attained
- fully supporting but threatened

Use not attained

- partially supporting
- not supporting

The *outstanding national resource waters (ONRW)* are high-quality or ecologically sensitive unique waters, such as those within the jurisdiction of national and state parks and wildlife refuges. The primary intent of designating the bodies of water ONRW is to protect or attain the highest quality and/or unique waters. The ONRW category also protects waters of ecological significance for which water quality expressed by some traditional parameters may not be high (for example, dissolved oxygen in wetlands and some productive estuaries). The states establish the ONRW criteria in order to protect the characteristics that prompted a body of water to be designated ONRW. In many cases, the ONRW represent the ecoregional reference reaches for establishing goals of water quality pollution abatement. This represents another reason why a national policy must be developed and implemented that would protect the ONRW (Committee on Restoration of Aquatic Ecosystems, 1992).

Use Attainability

If the body of water has been classified as partially supporting or not supporting the designated use, the states or designated agencies perform use attainability analysis (UAA) to determine the proper use of the body of water. Based on the UAA it is possible to modify or change nonexisting designated water use, if attaining the use is not possible due to the existence of one or more of the following factors (U.S. EPA, 1988):

1. Naturally occurring contaminant concentrations that prevent the attainment of the use
2. Natural, intermittent, or low flow or water levels that prevent the attainment of the use
3. Human-caused conditions or sources of pollution that prevent the attainment of the use
4. Dams, diversions, or other hydrologic modifications that preclude attainment of the use
5. Physical conditions associated with the natural features of the body of water, unrelated to quality, which impede attainment of aquatic life protection uses
6. More stringent controls than those required by Sections 301(b) and 306 of the Clean Water Act would be needed to attain the use and implementation if such controls would result in substantial and widespread adverse social and economic impact

There are basically two reasons why UAA is needed for water bodies where use is not attained. One reason is to determine what levels of quality are possible to attain by implementation of various feasible point and nonpoint source abatement measures. For water quality abatement of streams and estuaries, EPA regulations and guidelines recognize only the source-control measures. However, in addition to the pollution source controls, water-body restoration and waste-assimilative capacity enhancement measures should also be considered (Committee on Restoration of Aquatic Ecosystems, 1992). These could include a variety of measures, such as in-stream aeration, dredging of contaminated sediments, low-flow augmentation, and restoration of riparian wetlands and buffer strips with tall shoreline vegetation. These measures could be made a part of an overall ecosystem restoration effort for the body of water. Ecosystem restoration and waste-assimilative capacity enhancement concepts for lakes have been considered and recommended by the U.S. EPA (NALMS, 1990).

The second purpose is to determine the most desirable use. This concept involves a study of the socioeconomic impact of the attainment of a specified use. The states are required to periodically review their water quality standards and revise them if appropriate. Concurrently, the UAA analysis is to be performed for bodies of water for which the designated use is not attained. The UAA is a multifaceted assessment of the physical, chemical, biological, and economic factors that affect the attainment of a use. The methodology of the UAA is described in pertinent manuals (U.S. EPA, 1983; Novotny et al., 1994). The UAA generally answers the following questions about the conditions of the water body:

1. What is the existing use to be protected?
2. What is the extent to which pollution (as opposed to physical factors) contribute to the impairment of the use?
3. What is the level of point source control required to restore or enhance the use?
4. What is the level of nonpoint source control required to restore or enhance the use?

Two other questions can be added:

5. What are the needed stream-restoration (waste-assimilative capacity enhancement) measures that would alter the adverse physical conditions of the receiving body of water that is impacting the aquatic habitat as well as water quality?
6. What is the optimal use of the body of water that would not impose a widespread adverse socioeconomic impact on the population involved and society as a whole?

A UAA consists of a survey and assessment of the body of water, a wasteload allocation (total maximum daily load process), and socioeconomic analyses (Fig. 16.1). It is a comprehensive process of evaluation, in its ecoregional context, of a body of water.

Once a designated use is attained it must be maintained. The Clean Water Act and ensuing antidegradation regulations do not allow a change to a lower quality use. However, states may modify nonexisting designated uses when it can be demonstrated through a UAA that attaining the higher designated use is not feasible.

In the integrated approach to water quality based pollution control there is another avenue open for the control of pollution discharges. Chemical-specific criteria are scientifically based numerical limits developed from laboratory bioassays with a safety factor. The EPA (1991b)

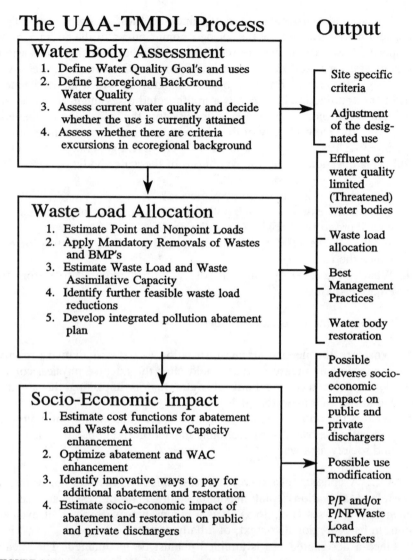

FIGURE 16.1. Three components of the water quality planning process. (Adapted from Novotny et al., 1994.)

has documented by experimental stream bioassay studies that such criteria provide nearly full protection of most aquatic species. Yet these criteria and standards are designed to protect most but not all of the species most of the time. Furthermore, natural conditions and irreversible modifications of the habitat may have caused a violation of a nationwide chemical standard, yet no impairment of integrity is noticed. In this case, site-specific criteria that could be related to the biological integrity of the receiving body of water could be adopted.

Waste-Assimilative Capacity: Loading Capacity

As previously defined the objective of water quality planning and evaluation efforts is to achieve and/or preserve the physical, chemical, and biological integrity of the receiving bodies of water. This is typically interpreted as maintaining water quality that would comply with the ambient water quality standards specified for the designated use of the body of water. Consequently, wastewater discharge from any source that does not result in a violation of a standard or will not impair the biological integrity of the body of water may be considered as noninjurious by the regulatory agencies (although in a court of law compliance with standards is not a defense against a liability if damage has occurred). The quantity of waste and contaminant loads that can be discharged into the environment without damage or use impairment of a receiving body of water, atmosphere, or land, minus the natural load is then called *a waste-assimilative capacity (WAC)*.

Cairns and Orvas (1989) defined the WAC as that range of concentration of a substance or a mixture of substances that will cause no deleterious effects upon the receiving ecosystems. The WAC is another important concept in our understanding of the process. If a waste discharge load into a receiving body of water is below the WAC, it does not violate the water quality standards and could be considered an acceptable waste load. The WAC is an economical asset that should always be considered, although it is not recognized by the EPA as a beneficial use of a body of water.

The EPA's Water Quality Planning and Management Regulation expressed in a guidance manual (U.S. EPA, 1991c) defines *loading capacity (LC)* as the greatest amount of loading a water can receive without violating water quality standards. Apparently, the loading capacity, as defined by the EPA, is almost synonymous with the waste-assimilative capacity defined herein; however, LC only refers to violations of the established standards, while the broader WAC definition involves the integrity of the receiving body of water. Typically, the WAC (or LC) of surface-water bodies may be higher for decomposable organic matter, but it is very low to nil for some toxic chemicals that bioaccumulate in tissues of aquatic organisms and become injurious to animals and man using them as food. These differences are reflected in the magnitude of the standards.

According to these definitions, a waste constituent becomes a pollutant if it is discharged into the environment in quantities that are injurious to or impair beneficial uses of the environmental resources. Many waste constituents in small quantities are not injurious, and some of them may even be beneficial in low quantities and become injurious (toxic) only in quantities that exceed the WAC. For example, some metals that in high concentrations are known to be toxic, such as zinc, are necessary nutrients to aquatic life in smaller trace quantities.

Water quality management has become increasingly more complicated. The introduction of *priority pollutants* into the framework of water quality abatement planning and biological, physical, and chemical integrity concepts for the receiving bodies of water is causing planners and pollution control authorities to determine the required pollution controls using water quality based waste load allocations rather than technology-based effluent limitations. For this reason the EPA has been promoting and enforcing the *total maximum daily load (TMDL)* process, which establishes the allowable loadings for a body of water based on its WAC, and thereby provides the basis for the states to establish water quality based controls (U.S. EPA, 1991b, 1991c). According to the EPA's guidelines, the TMDL is the sum of *allowable loads* that include loads from both point (WLA) and nonpoint (LA_{NP}) sources as well as the natural background loads (BL). Hence

$$TMDL \leq LC \tag{16.1}$$

and

$$TMDL = WLA + LA_{NP} + BL$$

Note: Previously in many water quality control plans and studies for waste load allocation, the terms TMDL and WLA (allowable point source loads) were incorrectly used interchangeably instead of considering both LA and WLA as components of TMDL.

Margin of Safety

The margin of safety (MOS) is normally incorporated into the conservative assumptions used to develop the TMDL that is subsequently approved by regulatory agencies. If the safety margin is required to be larger than that incorporated in the assumptions of TMDL, additional MOS is added as a separate component, or

$$\text{TMDL} = \text{WLA} + \text{LA}_{NP} + \text{BL} + \text{MOS} \qquad (16.2)$$

The WAC (LC) of most receiving bodies of water is not fixed, but can be enhanced (increased) by various management measures, including stream and lake aeration, low-flow augmentation, sediment dredging, nutrient inactivation, restoration and implementation of riparian wetlands, and planting shade trees. Returning a previously lined and straightened or impounded stream to its more natural state typically increases its capacity to receive and safely assimilate pollutants.

Waste-assimilative capacity enhancement should be a part of an overall ecosystem restoration effort, whereby the entire ecosystem and its boundary stresses are considered (Fig. 12.4). A simple aeration or waste dilution may cause a particular water quality standard to be met, but may not restore the ecosystem's integrity. Both stream restoration and pollution input reduction must be considered and their efficiency weighed against the overall ecosystem restoration goal, of which water quality is perhaps the most important component. For example, restoring a stream to a more natural state by removing unnecessary impoundments and stream linings and by providing shade trees along the stream may have a better water quality impact than an excessive reduction of nutrient levels in the effluents.

FINANCING DIFFUSE-POLLUTION ABATEMENT PROGRAMS: WHO SHOULD PAY?

Although Congress has made CSOs and a part of urban and agricultural storm-water management mandatory, no significant financial resources have been allocated for future projects; however, some CSO abatement is still receiving federal grants authorized by the previous amendments of the Clean Water Act (1972 and 1977). The General Accounting Office (U.S. General Accounting Office, 1990) noted that for nonpoint source abatement *"a key contributing factor to these resource constraints is that available funds are overwhelmingly oriented toward point sources (traditional) rather than nonpoint sources...." "EPA has not allocated the amounts required to meet the most basic elements of its nonpoint source pollution agenda...."* In addition, the EPA has not requested and appropriated funding for Section 319 programs as authorized by

Congress. As pointed out in the U.S. General Accounting Office (1990) study, out of $400 million authorized by Congress between 1988 and 1991 only $22 millions were requested by the EPA. It is quite apparent that under present budgetary constraints and EPA policies the bulk of funding for urban diffuse-source abatement must come from state and local sources.

Some states (for example, West Virginia and New Hampshire) are tying the implementation of nonpoint pollution programs to the availability of federal funding. Other states (Wisconsin, New Jersey) have established or plan to establish their own state funds for implementation of NP programs. Typically, states in their Section 319 reports expected that the bulk of the programs would be funded by public financing and cost sharing.

Generally, there are three fundamental means of recovering the costs of pollution abatement.

1. Placing the cost burden on polluters alone ("polluters pay principle").
2. Putting the economic burden of pollution abatement on the beneficiaries of improved water quality. Those individuals benefiting will then provide grants and subsidies to the polluters to implement the technology necessary to reduce or eliminate pollutant loads. This is called the "benefits received" approach.
3. A combination of the two systems.

Polluters Pay

Polluters bear the chief responsibility for abatement and its direct cost. This is the most equitable payment plan. It has been primarily used to control pollution from industrial sources and has been implemented through the NPDES permit system. As pointed out in the previous section, this system requires the reclassification of diffuse sources as point sources. The payments are clear in the case of industrial polluters, who pay for pollution from their revenues. The form of payment of municipal polluters is not as clear since several types of payments are common. These include property taxes, user fees, income and sales taxes, and general revenues from other sources.

The difference between taxes and user fees is that taxes are compulsory, while charges are optional. The use of user fees to finance storm-water and CSO management implies that the service is private goods; however, storm-water management in urban areas historically has been regarded as public goods and has generally been financed by property taxes. Without regulation it usually has been considered a low-priority service, so funding has been inadequate (Lindsay, 1988).

Regulations
Under the "polluter pays" scenario, polluters are usually required to meet specific abatement standards, subject to penalties if falling short. This is again the fundamental enforcement principle incorporated into the NPDES permit system by the states. Many have questioned why regulation and enforcement are needed in pollution abatement, specifically in nonpoint pollution programs where past (unsuccessful) practice was to rely on voluntary participation.

In the absence of regulation, pollution abatement must rely on other means to persuade polluters to install abatement devices. These means were listed by Novotny (1988), and include moral persuasion, court litigation, and financial grants. Economists have demonstrated that market forces do not work in pollution abatement (market failure) because of the external character of pollution discharges (Solow, 1978; Gaffney, 1988). An economic externality is an economic activity or action of an individual or group that economically impairs others without their consent. A technological external diseconomy (Kneese and Bower, 1968) is a situation where a particular action produces economical results on an independent entity. Diffuse pollution, both from urban and nonurban sources, is a classic example of external diseconomics (Novotny, 1988): pollutants from diffuse sources are transferred downstream where they impair downstream uses of bodies of water, yet the sufferers of downstream pollution have no economic mechanism to recover the damages from the polluters. Those responsible for the pollution of the Boston Harbor and adjacent coastal areas, or the Chesapeake Bay and other coastal bodies of water, damage the fishing and recreation industries (users), but, without regulation and its enforcement, there is no market mechanism to recover the cost of these damages. An externality problem can cross state and even international boundaries. The water quality problems of the Chesapeake Bay are caused by pollution discharges from several states, yet those who suffer the consequences of the pollution (excluding, of course, transient vacationers) live in Virginia and Maryland. A well-publicized case of coastal pollution in the San Diego, California, area[1] documents international externality problems.

---

[1] San Diego Bay is being polluted by waste discharges from both San Diego and Tijuana, Mexico.

Fiscal Federalism

In general terms, the U.S. federal system allocates the cost of government programs in proportion to benefits received. Where local benefits predominate, mainly local funds are used; where state or national benefits are significant, costs should be borne by state and federal governments (Braden, 1988a). Diffuse-pollution control programs should reflect these attributes of federalism. Litter control and street sweeping are a local problem, and its control is paid for locally by either taxes or user fees or both. Federalism and the recent reluctance of the federal government to commit adequate funds for diffuse-pollution abatement puts the financial burden on states and local funding. However, the federal government must play a role in financing and/or regulating and enforcing diffuse-pollution programs where interstate externality problems would impair efficient abatement. Such is the case of coastal pollution.

Uniformity

The U.S. Constitution precludes the arbitrary application of laws. In the Clean Water Act this led to (effluent) standards that are uniformly applied to all polluters, regardless of the capability of the receiving water bodies to assimilate pollution and/or the extent of damage or absence of damage imposed on downstream users. This scheme, if applied to diffuse-pollution abatement, most likely would not lead to a least cost solution that would maximize water quality improvement. Furthermore, facing the huge cost of the abatement, the uniform approach could result in a situation where pollution of some waters would remain unacceptable, while unnecessary funds would be spent in places where they would not be needed. Using uniform standards mostly implies the "polluter pays" principle, where the cost of pollution abatement is borne by the polluter and the abatement is enforced by a discharge permit.

"Targeting" is used for nonuniform abatement (Braden, 1988a) in water quality limited watersheds. Targeting diffuse-pollution abatement for inland and coastal watersheds that have a diffuse-pollution problem will result in a more efficient use of funds and will bring more water quality improvement benefits for less money. Wisconsin's or North Dakota's "Priority Watershed Program" is an example of targeting specific watersheds that have a diffuse-pollution problem. State funds in the form of grants and cost sharing (up to 70%) are then allocated for abatement in those targeted watersheds. The polluter pays principle and its enforcement are difficult to implement when targeting is used, although the mechanism for enforcement exists in some states if the polluters fail to implement diffuse-pollution abatement. For example, the "bad actors"[2] legal doctrine was proposed in Wisconsin to provide further incentives to rural polluters to voluntarily implement nonpoint pollution-control practices. After sign-up and a grace period, it was proposed that the Wisconsin Department of Natural Resources (DNR) would be able to mandate participation of critical nonparticipants in the program whose nonpoint pollution was causing the DNR to fail to meet its water quality goals. Those bad actors would have to participate at a reduced subsidy (from 75% to 25%) from the state. Similar provisions for agricultural nonpoint sources are incorporated in the Food Security Act.

Targeting is also used when, for example, specific water quality improvement and waste-assimilative capacity enhancement projects are proposed, such as wetland restoration and in-stream water quality control measures. These projects require funding that should be recovered from both the beneficiaries and polluters. The funds may be recovered from the polluters, for example, in the form of fines for excessive pollution, from user charges for storm-water disposal, from fees paid by developers for their erosion-control plans, from a part of the gasoline tax, or from motor vehicle license fees. The use of stream standards and a definition of water quality limited "impaired" bodies of water, similar to point source abatement, may also be used to enforce the polluter pays principle for the abatement of diffuse pollution in targeted watersheds. The distribution of allowable waste loads based on the waste-assimilative capacity of the body of water in question can be made initially in a uniform fashion by a regulatory agency (state or the EPA). Then the polluters can be allowed to negotiate the discharge loads among themselves as long as the waste-assimilative capacity of the body of water is not exceeded. This is the principle behind the *transferable discharge permits* or *bubble approaches* (Braden, 1988b; Tietenberg, 1985).

Benefits Received Approach

In this approach the beneficiaries of improved water quality either tax themselves or pay user fees. The revenues are then allocated for pollution abatement. This can be accomplished in several ways. The beneficiaries can provide grants and subsidies to polluters to install abatements or the funds are used for mitigation of adverse water quality effects (e.g., weed control or pollution diversion).

This concept presumes that the sufferers of adverse water quality are

---

[2] In 1992 the "bad actor" provision was included in the Wisconsin Nonpoint Pollution Control Legislation, but was vetoed by the governor.

willing to pay for the water quality improvement benefits. The willingness to pay is not the same as the damage (cost) of pollution, but it is related to it. Nor should the willingness to pay be equated with the ability to pay because of the very important premise that those asked to pay for the benefits of improved water quality either suffer or can potentially suffer from adverse water quality. Thus, users of beaches or recreational fishers are willing to pay for improved water quality and, consequently, for diffuse-pollution abatement.

The problem with *willingness to pay* arises when the beneficiaries and polluters are not a part of the same group and there is a distance between the pollution source and the location where the benefits will be received. Suburban owners of septic systems (including those that are failing) located far from a body of water are generally unwilling to connect to a sewer system if they do not receive water quality benefits, while owners of water shoreline properties are willing to tax themselves in return for improved water quality. Therefore, the polluter pays principle is most equitable when polluters and beneficiaries of improved water quality come from the same population group, for example, a large urban area on a body of water that is in or reasonably close to it. The Boston harbor, located within metropolitan Boston, is such an example. The construction industry and its clients (future homeowners) living in the area affected by construction pollution represent another example where the polluters pay principle should be strictly adhered to.

Considering the magnitude of the diffuse-pollution problem, both payment schemes may be considered. The *polluter pays* funding approach should be applied to most local problems, using regulation and standards. Such standards should apply uniformly to all sources of urban pollution, category by category. These standards can be in the form of *design (performance) standards* that establish that certain pollution-abatement technology is to be applied uniformly to all sources within the source category (e.g., construction erosion and sediment control ordinances, mandatory street sweeping, mandatory collection and treatment of a specified portion of the average surface runoff, or limiting the number of CSOs) or *effluent standards* that limit or prohibit certain polluting substances in effluents, CSOs, runoff, and subsurface flows.

In "targeted watersheds," that is, in areas with a serious and extended water quality problem and with the important premise that beneficiaries are separated from the polluters, a two-level approach may be used (Novotny, 1988). In the first level, the polluters pay principle, using mandatory pollution control, should be enforced by a NPDES-type permit system.

The NPDES permit program currently being implemented for the control of urban stormwater is not uniform nor is it equitable (see Chapter 2 for a discussion of equity and other imperatives of pollution control). It also contains many exclusions such as small urban areas, interurban freeways and highways, military facilities and training grounds, suburban zones with septic tanks, small construction sites, and other urban and nonurban sources. Such exclusions could preclude an effective diffuse-population program. If these controls do not suffice, a combination of the benefits received and polluter pays approaches could be used to achieve the water quality goals. The benefits received approach is then used to secure additional necessary funding for abatement. This can be accomplished by a combination of grants from state and federal funds, from taxes on received benefits related to the improved water quality, and user fees, such as money collected for fishing licenses, boat launching fees, and the fees for the use of public beaches. Taxing riparian land owners who receive disproportionally large benefits from the use of bodies of water is a viable alternative.

These funds can then be allocated to the targeted areas. The polluter pays approach can be incorporated into the discharge permit in the targeted areas. This permit should be related to the waste-assimilative capacity of the body of water or coastal area in question. The polluters (urban areas, industries, storm-water management utilities) should then be allowed to renegotiate the discharge levels among themselves according to the transferable discharge permits scheme (Braden, 1988b; Tietenberg, 1985). The equity of proportioning the additional cost of abatement in the targeted areas between the polluters and users (beneficiaries) must be incorporated in the plans of abatement.

Many authors (Gaffney, 1988; Braden, 1988b; Novotny, 1988) warn against the excessive reliance on subsidies in diffuse-pollution abatement programs. Although subsidies are politically attractive to lawmakers, it is a well-established fact that in the absence of regulation and enforcement, polluters will do nothing unless and until they receive a full subsidy for the cost of abatement. Furthermore, allocating the subsidies to only certain types of abatement may lead to inefficient solutions. Braden (1988b) also pointed out that subsidies actually encourage pollution activities by shielding them from their true cost to society. He concluded that subsidies are too costly and too prone to perversion. On the other hand, since taxing riparian and other beneficiaries of improved water quality is a viable alternative for raising revenues, partial cost sharing and incentives may be considered. The bulk of the funding raised from the beneficiaries and from fines for pollution, however, should be used for actions such as ecological restoration and the protection of coastal wetlands, and not as grants to polluters.

The first level of implementation of diffuse pollution abatement does not require a great deal of reliance on water quality standards; however, it is paramount that wet-weather water quality standards be considered when the second-level approach is used. Adequate wet-weather water quality standards are needed in situations when mandatory uniform abatement does not result in the satisfactory water quality improvement of local bodies of water, because the level of abatement must then be related to the waste-assimilative capacity of the body of water in question, which is determined (primarily by water quality modeling) by using the accepted and enforced water quality standards.

The "Bubble Approach" and the Bubble Dimension

In a "classic" watershed planning approach the benefits of the project are weighted against its cost, while other objectives of the development are also considered, such as the resource conservation and preservation and well-being of people affected by the development (see Maas et al., 1970 for a most authoritative discussion on the objectives and benefit–cost analysis of water resource development projects). In the water-pollution control area, the objectives were given by Congress in the Clean Water Act (1972 Amendments) and are well known and followed. The major objectives of the Clean Water Act are fish and aquatic wildlife protection and preservation and the suitability of receiving bodies of water for contact recreation. These objectives were translated into water quality criteria, which were then implemented by the states as standards. Furthermore, the quality of point sources must comply with the equity, irreversible impact technology-based standards.

The use of dual standards resulted in two planning scenarios with interchangeable objectives and constraints. In the first scenario, the waste-assimilative capacity of the receiving body of water is greater than the combined pollution loads from all sources after the effluent standards for point sources are implemented. This leads to the so-called "effluent controlled" or "nontargeted" planning scenario. The objective in this scenario is to minimize the cost at each individual source, with the avoidance of violation of the effluent standard as a constraint. Point source effluent discharge permits are then based on the technology-based effluent standards. Under the present regulations, incentives for nonpoint source control are minimal and the abatement approaches rely mostly on voluntary approaches. This violates the "principle of uniformity," which states that all polluters should be treated equally.

In the second scenario, the waste-assimilative capacity is less than the combined pollution load *from all sources*, resulting in a violation of the water quality standards. This is the so-called "stream (receiving body of water) -limited" or "targeted" situation, whereby effluent discharge permits are based on the waste-assimilative capacity of the receiving body of water. The objective of water quality abatement is thus to meet the stream water quality standards at minimum cost or under the economic and fiscal constraints.

Hence, the waste-assimilative capacity of the receiving body of water is the most important constraint that determines the maximum allowable pollution load that cannot be exceeded. This is the maximum size of the "bubble," as it is referred to in the jargon of pollution-abatement literature. If the pollution load exceeds the waste-assimilative capacity, the receiving body of water is damaged and the downstream water uses are impaired—the bubble has burst. However, the WAC may be increased by instream WAC enhancement measures. Hence, the "bubble" is flexible and can expand. Under the "effluent-limited" situation, however, the primary objective of the abatement efforts is minimization of cost. Hence the size of the "bubble" is then related to the budgetary constraints of the agencies responsible for financing the abatement.

The policy options that can be used to achieve the water quality goal (i.e., to attain the designated use) include pollution taxes, water quality standards, effluent standards, tradeable discharge permits, process design standards, liability for damages, and abatement subsidies. Incentives can be both positive and negative, that is, rewarding abatement or penalizing pollution. No single option is available to achieve the goal, hence the use attainability analysis must consider the best combination of policy options. It should be pointed out that the United States has made comparatively little use of fiscal incentives for pollution control (Braden, 1993).

An important source of inefficiency in pollution control occurs in the waste loads allocation process where high-cost abaters are forced to meet the same standards as low-cost abaters, which can lead to wasteful abatement. It is true that using differential treatment for polluters, based on the cost of abatement, is causing legal problems and may be difficult or impossible to implement. However, using the so-called *transferrable discharge permits* to transfer the cost from high-cost to low-cost polluters is feasible. Here the polluters negotiate among themselves the waste allocation subject to the constraint of the limited loading capacity of the receiving body of water. It appears that "pollution trading" may be on the verge of becoming a major tool for the EPA and states to attack the nation's biggest source of water pollution—the diffuse-source pollution loads (*Clean Water Report*, June 9, 1992). Hence, both point/point and point/nonpoint source abatement trading is feasible.

Under a proposal considered by the U.S. EPA in the early 1990s, a point/point source (P/P) trading program would allow a discharger to avoid a part of treatment or treatment upgrade by paying or arranging for at least an equal reduction in discharges from other facilities releasing in the same body of water. All facilities would still have to comply with the technology-based (equity) effluent limitations. Hence, these trading concepts are applicable to water quality limited bodies of water.

Targeting water quality limited streams implies the selection and application of pollution-control policies that promote economic efficiency. There are three aspects of targeting. First, the cost of administering policies should be balanced against the benefits of policy refinement. Second, targeting enables differentiation between the sources and may promote the use of transferrable discharge permits. Third, it allows differentiation in space and time, that is, the bodies of water that require abatement most and/or are most desirable will be abated ahead of some other less desirable bodies of water. The third aspect is dictated by budgetary and other economic constraints imposed on pollution abatement and by the fact that there is a tremendous variability in physical and ecological characteristics among the nation's surface and coastal waters. Consequently, different systems respond in different ways to pollution stresses.

The Committee on Restoration of Aquatic Ecosystems of the National Academy of Engineering (1992) recommends a national body of water restoration effort that would include four elements:

1. National restoration goals and assessment strategies for each ecoregion.
2. Principles for priority setting and decision making.
3. Policy and program redesign for federal and state agencies to emphasize restoration.
4. Innovation in financing and use of land and water markets.

In light of existing and anticipated budgetary constraints, innovative ways to finance restoration efforts are necessary.

INSTITUTIONAL ISSUES

As pointed out previously, water quality is a response to various sources of potential contaminants, including natural and cultural sources. The cultural sources are then divided into point and nonpoint. Equity and irreversible impact criteria (technology-based effluent and performance standards) must be formulated and enforced in an integrated fashion both on point and nonpoint sources. Excluding certain sources would violate the equity criterion. Third, the receiving water standards and criteria determine the maximum permissible load TMDL = WQS × Q, where WQS is the standard and Q is the flow. The standards are related to the designated use of the body of water. In some cases, the natural load, BL, may be greater than TMDL. If this is the case, then either the designated use is changed or no waste discharges should be permitted. This may happen if the body of water in question is designated as an outstanding natural resource or the water is used for water supply.

In most cases, however, there may be an excess load between the TMDL and the background load (BL). This excess permissible load then constitutes the waste-assimilative capacity WAC or WAC = TMDL − BL = F^{-1} (WQS). The F^{-1} is a functional relationship (model) that relates the water quality to allochthonous waste discharges. If the total load from cultural sources (TDL) is less than WAC, then the body of water is not water quality limited and cannot be considered a targeted system. The waste-discharge permits can be issued based on equity and irreversible impact standards. Such permits can be issued by state or even local (county) land-use and water quality agencies and could be based on design and/or performance criteria. Tying these permits to water quality and requiring excessive monitoring from individual dischargers would be wasteful and could present an administrative nightmare.

If, however, the total load (TDL) is greater than WAC, then the receiving body of water is limited and should be considered as a potentially *targeted system*. A targeted watershed with a major diffuse pollution problem can be declared "impaired." Waste-discharge permits must then be related to the waste-assimilative capacity of the body of water in question (for example, an estuarine system). The management options in targeted (impaired) watersheds are:

1. Enforce more stringent effluent and performance standards
2. Improve WAC (e.g., by dredging in-place sediments, low-flow augmentation)
3. Change receiving water standards by proving by the use attainability analysis that the standards cannot be attained

Based on these options, it is evident three types of institutions are needed.

The first category of institution is *regulators*. These institutions are already in place in the form of the U.S. Environmental Protection Agency, to execute federal policies, and state pollution-control agencies (Departments of Environmental Protection, Natural Resources, etc.), for intrastate pollution control. These institutions carry out the legislative policy mandates, specify standards and criteria, provide oversight, arbitration, research, provide and distribute grants when federal and state interests in water quality remediation are involved, among other activities. In the integrated pollution-abatement scenario these agencies decide whether a body of water is a targeted (impaired) system or a national outstanding resource. They also designate the uses of the bodies of water, specify the standards and execute use attainability studies. They provide oversight for water management and pollution abatement agencies. In nontargeted watersheds where permits are based on effluent (point sources) design–performance (diffuse sources standards), the agencies may delegate their permit-issuing authority to local (county) agencies or establish their own permit-issuance branch. They can also provide arbitration in disputes between pollution-abatement and water quality management institutions. This agency *should not manage the water resources or pollution abatement*. Financing the agencies should be derived from state or federal sources, and should not use funds obtained from polluters or users.

The second category of institution includes the *waste dischargers*. Present municipal sewerage agencies and industrial polluters are covered by this category. These agencies are responsible for urban, CSO, and point source abatement. As an alternative to municipal sewerage districts, which in most cases are responsible for urban point and diffuse-pollution abatement and which utilize taxes for financing, many communities have recently created storm-water utilities financed by user fees (Lindsay, 1988; American Society of Civil Engineers, 1985). Users are people and organizations that generate storm-water runoff that is discharged into the utility drainage system. In rural areas, drainage and irrigation districts and utilities, in addition to individual farm operators, are examples of potential organization of waste discharges.

Financing for these agencies and their pollution-abatement efforts should be mostly based on the *polluter pays principle* and regulated by a permit. For example, fees can be based on estimates of the amount of runoff or drainage (irrigation return) flow that leaves a user's property. Rate factors can be correlated with land use. With the utility approach, users are differentiated from beneficiaries, who are people protected from the damage of runoff or who receive other benefits, such as improved water quality. For wastewater treatment the charges are typically based on the measured or estimated flow and pollutant load. In some states, existing water management organizations (Florida) are now assuming the role of a storm-water management utility (Viessman, 1988).

In nontargeted watersheds, as an alternative approach to issuing individual permits for each pollution source, the pollution-abatement agency can ask the regulating institution for an umbrella technology-based permit for each category of pollution source within the its jurisdiction. The pollution-abatement agency will then issue permits and/or monitor compliance. In nontargeted watersheds the two agencies (regulators and pollution abatement) are the only institutions needed.

In targeted watersheds, a new concept of water quality management agencies should be established. This regional watershedwide agency would be responsible for planning the actions, managing discharges and their control, managing water quality, and raising revenues for carrying out the operation and capital investments for waste-assimilative capacity enhancement measures, such as restoration and creation of riparian wetlands or the management of buffer strips.

The decision whether a watershed is a targeted (impaired) or nontargeted is based on a scientific–engineering study that would estimate the total maximum daily load (TMDL) and waste-assimilative capacity, use attainability analysis, and on the public surveys of use impairment and use attainability. Such studies should be performed by the regulatory agency or by planning agencies (consultants) designated by the regulatory agency (e.g., the EPA for interstate and coastal water bodies and state regulatory agencies for interstate watersheds). Funding authorization for these studies can be derived from Section 319 or 208 of the Clean Water Act. Unlike previous Section 208 studies, if the body of water is declared a targeted or impaired body of water, the planning activity must be followed by an implementation that includes the establishment of watershedwide agencies responsible for the implementation.

Diffuse-pollution problems are ubiquitous, their dimension and origin not well established, often fitting watershed boundaries rather than political ones (Viessman, 1988). By analyzing the problems of diffuse pollution in a regional rather than a local context, it is possible to identify more efficient options that might not be recognized or available otherwise. For example, successes with the regional approach to diffuse pollution management in the Washington, D.C., area demonstrated that gains can be substantial (Viessman and Welty, 1985). Other examples of successful regional water and water quality management organizations include the Florida Water Management Districts,[3] the Water Pollution

---

[3] The Florida Water Resources Act of 1972 provided for the creation of water management districts. These regional management agencies have broad powers relative to the allocation and use of Florida's waters. In the early days the districts were largely concerned with water quantity issues, water supply, flood control, and drainage. With the passage of time, however, they have moved solidly into the water quality management area. Furthermore, they have taken on an environmentally slanted orientation.

For example, the core mission of the South Florida Water Management District is:

Management Associations in the Ruhr industrial district of Germany (Kneese and Bower, 1968), the Nebraska Natural Resources Districts and British Water Authorities.

These regional organization, which have the broad authority to raise revenues and implement management, would be more efficient than fragmented local and state organizations, which in many cases have limited authority to regulate, but do not have the authority to conduct in-stream water quality management (e.g., wetland restoration or low-flow augmentation) or collect user fees from the beneficiaries of improved water quality. Also discharge permit trading (the bubble approach to pollution discharge permitting) can only be accomplished if a regional watershedwide authority is a "referee" in the trading process and the same authority will make sure that the final discharge loads agreed upon by the polluters do not exceed the waste-assimilative capacity. Subsequently, the agency will continually reissue the permits.

The management of diffuse pollution also requires interagency cooperation and integrated approaches. Programs currently or potentially related to diffuse pollution in coastal areas are carried out by the U.S. Environmental Protection Agency, NOAA, U.S. Department of Agriculture, and many state and local agencies. The Florida Water Management Organizations again provide a good example of how a regional umbrella watershed management organization can cooperate and coordinate these multiagency efforts. Organizational arrangements and responsibilities of the regional water quality management organizations should be incorporated in the planning process (for example, Section 319 planning).

The core mission of the South Florida Water Management District is to manage water and related resources for the benefit of the public and in keeping with the needs of the region for the purpose of providing:

— Environmental protection and Enhancement
— Water Supply
— Flood Protection
— Water quality protection

This is being implemented through the coordination of operations, planning, public involvement, regulation, and construction. Inherent in the mission is the responsibility to assist the public and government officials in growth management by identifying water resource impact of land-use decisions and by advising on options for reducing adverse impacts and protecting water resources.

The water management districts have dealt with a variety of nonpoint pollution problems and in doing so have employed a number of nonconventional approaches, including Save Our River programs, Lake Okeechobee restoration, and the Lake Apopka land swamp proposal (Viessmann, 1988).

Water quality management today is heavily influenced by regulatory measures, all of which come from several directions (the EPA, Department of Interior programs and authorizations, Department of Agriculture, states, etc.). Viessman (1988) correctly pointed out that water management guided by narrowly devised regulatory measures, that is, by regulatory actions that only call for "planning," but do not provide for implementation (such as the planning activities of Section 208 of the Clean Water Act), are destined to be mediocre at best and destructive at worst.

Strategy of Abatement

Generally, mitigation of diffuse pollution from urban and nonurban areas relies on structural and nonstructural measures, or the so-called best management practices. These are well known and have been published in numerous EPA and state manuals, as well as in several chapters of this book. The strategy, however, means selecting the most optimal alternatives under the given legislative and fiscal constraints.

As pointed out in many articles (for example, see Novotny [1988] for a summary), as well as earlier in this chapter, abatement requires regulation and enforcement. As we have also seen, under the current legislative means of enforcement, only those diffuse sources that have been legally classified as "point sources" have a reasonable chance for clean-up. Furthermore, additional mitigation and land protection can be achieved using present or near-future wetland protection laws. This still leaves a large number of sources for which chances of abatement are minimal.

It is evident that the land-disturbing activities (construction, wetland drainage, stream channelization, etc.) during land-use transition are responsible for the highest loads of pollutants. Hence the strategy of mitigation should first focus on their control. However, enforcement is not available or is insufficient.

In established urban lands without significant construction, the NURP study established that there is not much difference between the strength (EMCs) of urban runoff from the major land uses (see Chapter 8). However, since the degree of imperviousness affects the runoff volume, the highest loads can be expected from highly impervious urban lands. Therein, structural facilities, such as detention–retention, followed by treatment, seem the most appropriate. For more pervious areas, typically for low- to medium-intensity residential areas with storm sewers, less structurally intensive measures, such as buffer strips along the streams; floodplain management; retention, storage, and sedimentation in suburban ponds and lakes; and public education to reduce or eliminate (by switching to xeriscape) the use of fertilizers and pesticides on suburban lawns, are appropriate.

In suburban zones that are served by septic tanks and do not have sewers, controls aimed at the reduction of ground-water protection from nitrate contaminations and surface pollution by surfacing septic tank effluents, should be strengthened. Since construction activities yield the highest number of sediments, effective construction erosion-control regulations must be implemented on a national scale.

The best and most efficient results in controlling pollution by urban diffuse sources is achieved if the management is incorporated into the development plans. However, the recent report to the Congress by the General Accounting Office (U.S. GAO, 1990) pointed out that the EPA regulations aimed at the control of pollution by urban runoff are not sufficient. Many areas of the country experiencing the most rapid growth are not included, and by delaying the implementation of storm-water control programs, the U.S. Environmental Protection Agency "is losing a valuable opportunity to prevent stormwater problems rather than rely on expensive structural controls after development has occurred."

In agricultural areas, the programs mandated by the Food Security Act (the Farm Bill) should concentrate on lands and practices that have the most adverse water quality impact, with soil and water conservation as a constraint. Reliance on purely voluntary approaches is ineffective and typically results in very low participation of polluters in the programs.

References

American Society of Civil Engineers (1985). Street users fees, *Civ. Eng.*, p. 43.

Braden, J. B. (1988a). Financing nonpoint pollution abatement—the peculiar case of agriculture, in *Political, Institutional and Fiscal Alternatives for Nonpoint Pollution Abatement Programs*, V. Novotny, ed., Marquette University Press, Milwaukee, WI.

Braden, J. B. (1988b). Nonpoint pollution policies and politics: The role of economic incentives, in *Proceedings, Nonpoint Pollution 1988—Policy, Economy, Management and Appropriate Technology*, V. Novotny, ed., American Water Resources Association, Bethesda, MD, pp. 57–65.

Braden, J. B. (1993). Categorization of water bodies—targeting, in *Use Attainability Analysis Manual*, prepared for Water Environment Research Foundation, Alexandria, VA, AquaNova Internat., Mequon, WI.

Cairns, J., Jr., and D. Orvas (1989). Ecological consequence assessment: Predicting effects of hazardous substances upon aquatic ecosystems, in *Ecological Engineering: An Introduction to Ecotechnology*, W. Mitsch and S. E. Jørgensen, eds., Wiley, New York, pp. 409–442.

Committee on Restoration of Aquatic Ecosystems (1992). *Restoration of Aquatic*

Ecosystems—Science, Technology, and Public Policy, National Academy Press, Washington, DC.

Gaffney, M. (1988). Nonpoint pollution: Tractable solutions to intractable problems, in *Political, Institutional and Fiscal Alternatives for Nonpoint Pollution Abatement Programs*, V. Novotny, ed., Marquette University Press, Milwaukee, WI.

Haigh, N., and F. Irwin (1990). *Integrated Pollution Control in Europe and North America*, The Conservation Foundation, Washington, DC.

Imhoff, K., and K. R. Imhoff (1990). *Taschenbuch der Stadtenwässerung*, Oldenbourg Verlag, Munich.

Kneese, A. V., and B. T. Bower (1968). *Managing water quality: Economics, Technology, Institutions*, Johns Hopkins University Press, Baltimore, MD.

Lindsay, G. (1988). Equity and efficiency aspects of alternate stormwater management financing systems, in *Proceedings, Nonpoint Pollution 1988—Policy, Economy, Management and Appropriate Technology*, American Water Resources Association, Bethesda, MD, pp. 87–96.

Maas, A., et al. (1970). *Design of Water Resources Systems*, Harvard University Press, Cambridge, MA.

Mitsch, W., and S. E. Jørgensen, eds. (1989). *Ecological Engineering: An Introduction to Ecotechnology*, Wiley, New York.

North American Lake Management Society (1990). *Lake and Reservoir Restoration Guidance Manual*, EPA 440/4-90-006, U.S. Environmental Protection Agency, Washington, DC.

Novotny, V. (1988). Diffuse (nonpoint) pollution—a political, institutional, and fiscal problem, *J. WPCF* **60**(8):1404–1413.

Novotny, V., and G. Chesters (1981). *Handbook of Nonpoint Pollution: Sources and Management*, Van Nostrand Reinhold, New York.

Novotny, V., J. Braden, D. White, H. Jones, and R. Schonter (1993). *Integrated Approach to Water Quality Abatement—Role of the Use Attainability Analysis*, paper presented at Surface Water Quality Symposium, Water Environment Federation Annual Conference, Anaheim, CA, October, 1993.

Novotny, V., et al. (1994). *Use Attainability Analysis—A Key Component of Integrated Water Quality Planning Efforts*, Water Environment Research Foundation, Alexandria, VA.

Solow, R. M. (1978). The economist's approach to pollution and its control, in *Pollution Taxes, Effluent Charges and Other Alternatives for Pollution Control*, U.S. Congress Rep. No. 95-5, Washington, DC, pp. 137–142.

Tietenberg, T. H. (1985). *Emissions Trading*, Resources for the Future, Washington, DC.

U.S. Environmental Protection Agency (1983). *Technical Support Document for Waterbody Surveys and Assessments for Conducting Use Attainability Analysis*, Vol. I–III, Office of Water Regulations and Standards, U.S. EPA, Washington, DC.

U.S. Environmental Protection Agency (1988). *Introduction to Water Quality Standards*, Office of Water, U.S. EPA, Washington, DC.

U.S. Environmental Protection Agency (1991a). *Guidelines for the Preparation of the 1992 State Water Quality Assessments*, 305(b) reports, Office of Wetlands, Oceans and Watersheds, U.S. EPA, Washington, DC.

U.S. Environmental Protection Agency (1991b). *Technical Support Document for Water Quality-Based Toxics Control*, EPA 605/2-90-001, Office of Water, U.S. EPA, Washington, DC.

U.S. Environmental Protection Agency (1991c). *Guidance for Water Quality-Based Decisions: The TMDL Process*, EPA 440/4-91-001, Assessment and Watershed Protection Division, U.S. EPA, Washington, DC.

U.S. General Accounting Office (1990). *Water Pollution—Greater EPA Leadership Needed to Reduce Nonpoint Source Pollution*, report to Congress, GAO/RCED-91-10, Washington, DC.

Viessman, W. (1988). Regional options for managing and regulating nonpoint pollution, in *Political, Institutional and Fiscal Alternatives for Nonpoint Pollution Abatement Programs*, V. Novotny, ed., Marquette University Press, Milwaukee, WI.

Viessman, W., and C. Welty (1985). *Water Management: Technology and Institutions*, Harper and Row, New York.

Wolman, M. G. (1988). Changing national water quality policies, *J. WPCF* **60**(10):1774–1781.

Epilogue

People can be attracted by new ways of ordering their lives, as well as driven by the recognition of what will happen if they do not change.

Herman Daly and John Cobb (quoted in
State of the World 1990, L. R. Brown)

We began this book with the gloom of threatening environmental problems caused by some forms of diffuse pollution. We would like to end with a more optimistic closure.

In 1992/1993 a two-year experiment was conducted in Arizona. In the experiment eight scientists lived in a hermetically sealed environmental structure called *Biosphere II*. Supposedly the world outside of this structure is Biosphere I. During the experiment living processes generated waste; however, the waste and energy in it were recycled back in the ecological cycle. Essentially, Biosphere II did not generate a substantial net increase of waste.

It was not a precisely scientific experiment and some criticized it as a commercialized, publicity project. Nevertheless, this experiment has brought us to the third millennium. If man is to survive on this planet, a sustainable ecology is a must. Figures 1 and 2 show the contrast between the ecological cycles before the year 2000 and those expected to occur after the year 2000. They show change from the traditional energy-demanding and waste-producing living process to a sustainable low-waste, less energy-demanding ecology. Ecological engineering introduced in the Prologue and Chapter 1 will be a key in achieving this goal.

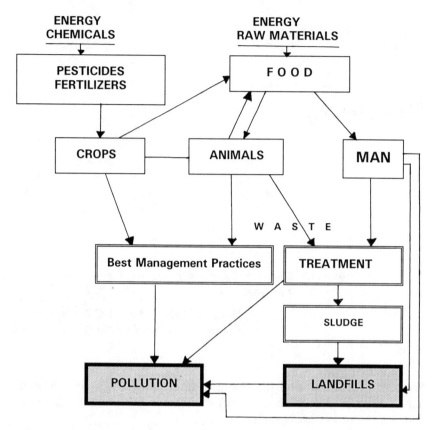

FIGURE 1. Ecological cycles of intensive mass flows of energy, raw materials, and waste, producing high levels of pollution.

Hence, the environmental policies of the third millennium will be different from those of the second millennium, and especially of the second half of the twentieth century. The fact that diffuse pollution is now both global as well as local requires broad international cooperation. With the adoption of the *sustainable development imperative* by the OECD nations, the overall approach to environmental policy is changing. Attention is now being focused on ensuring that all dimensions of environmental problems are considered, including pollution control, resource management, and broad quality of life; on developing more effective institutional arrangements for the formulation and implementation of environmental policies; on promoting technological changes

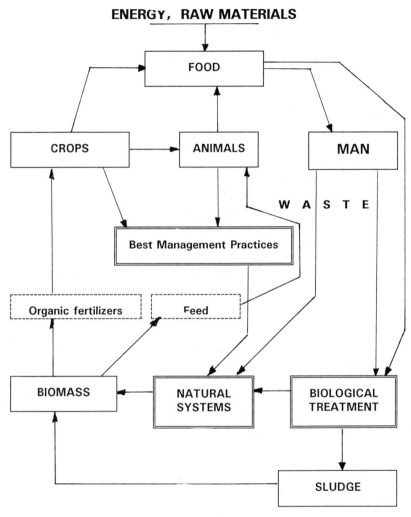

FIGURE 2. Ecological cycle of mass flows of energy, raw materials, and byproducts in a closed-loop sustainable mode with minimum pollution.

toward "clean, green" growth; on using economic instruments to provide market signals for environmental protection that better reflect relative scarcities; on streamlining the regulatory instruments for greater efficiency and cost effectiveness; on modifying production/consumption patterns to maintain a stock of scarce resources and reduce pollution; and on analyzing linkages between the environment and the economy and developing environmental indicators for measuring environmental performance. Today's and future approaches also emphasize (OECD, 1991):

- Broader objectives for environmental policy based upon the concept of sustainable development, and focus upon resource conservation as well as pollution control
- More effective institutional arrangements for the formulation and implementation of environmental policy
- More use of economic instruments to provide appropriate market signals for environmental protection; and improvements to the efficiency and cost effectiveness of regulatory instruments
- Wider use of anticipatory approaches to the formulation and implementation of environmental policy
- Development of more integrated approaches to environmental controls, both within the environmental sector and between the environmental sector and other sectors of the economy.

As pointed out in the *State of the World 1990* (Brown, 1990), building a more environmentally stable (sustainable) world society requires a vision of it. If the acid rain producing fossil-fuel power plants are to be replaced, then with what? If forests are not to be cleared for intensive agriculture, then where will the food be coming from? If a throwaway culture leads to pollution, then what should replace it? All these activities produce diffuse pollution.

A sustainable society is the one that satisfies its needs without jeopardizing the prospects of future generations. It is becoming increasingly apparent that the future economy cannot be driven by fossil-fueled power plants, heating, and vehicular traffic that can have catastrophic climatic consequences. It has been estimated that to stabilize the climate the per capita fossil-fuel consumption must be cut to about one-eighth of what it was in the late 1980s. If this is achieved, it obviously will also reduce in about the same proportion associated acid rainfall and toxic (mostly diffuse) emissions. Most likely solar, geothermal, and to a lesser degree (safe) nuclear energy sources will replace fossil fuel. Electric automobiles and people movers have already been introduced and successfully tested. Electric trains have long been the dominant means of transportation in many advanced countries (however, note that the production of electricty for electric trains and cars causes acid rain and other forms of diffuse pollution).

Reuse and recycling will replace throwaway waste. Biological and natural treatment systems provide the link between the waste production and reusable biomass. Biomass harvested from buffer strips and lands put into the conservation reserve (wood and hay) can be reused. Similarly, organic fertilizers and soil conditioners can be produced from the biomass that is harvested from the wetlands and aquaculture systems used for assimilating the waste. Organic fertilizers are less likely to pollute groundwater aquifers and, with proper management, their use will not threaten surface-water resources. In some other ecological waste-disposal systems the produced organic biomass may immobilize and help to decompose residual toxic chemicals that will still be used.

In the third millennium, land could be used in accordance with the rules and laws of sustainability. There should be no loss of wetlands. As a matter of fact, lowlands surrounding bodies of water that have been drained will be restored to their wetland status. High slope eroding lands will be reforested or terraced. Agriculture will rely more on nitrogen-fixing plants and organic fertilizers and less on chemicals.

Farmers could be planting less monoculture crops and will be relying more on crop rotation and soil conservation to maintain soil moisture and to reduce soil loss and pesticide and fertilizer application rates. It is already apparent that farming in arid and semiarid areas, which relies heavily on mining ground-water aquifers, will cease in a few decades. New, less water-demanding crops will replace the thirsty and polluting crops in arid areas. This will result in less discharge of heavily polluted irrigation return flows.

In urban areas, diffuse-pollution-generating urban sprawl should be controlled, and suburbanites in developed countries will consider xeriscape rather than planting environmentally damaging lawns. Diffuse-pollution abatement will be incorporated into the landscape of urban environments, along with the retardance and water-conservation drainage.

As pointed out in this book, the solution of the problem of diffuse pollution is closely tied to political and economic factors. The new political order now emerging will spend less on military and more on environment, less on heavy polluting industries, oil refineries, and vehicular traffic and more on solar energy, recycling, and environmental protection. The *State of the World 1990* concluded that "as the transition to a more environmentally benign economy progresses, sustainability will gradually reclipse growth oriented economy as the focus of economic policy making." As far as the environmental engineering aspects of this change is concerned, the present book has tried to open the door.

References

Brown, L. R. (1990). *State of the World 1990*, W.W. Norton, New York.
Organization for Economic Co-operation and Development (1991). *The State of the Environment*, OECD, Paris.

Appendixes

APPENDIX A

CONVERSION FACTORS

	U.S. Customary to S.I. (Metric)		
			S.I.
U.S. Customary Units	Multiply by to obtain	Symbol	Name
acre	0.405	ha	hectare[a]
acre-ft	1233.5	m^3	cubic meter[b]
acre-in	102.79	m^3	
foot (ft)	0.3048	m	meter[c]
ft/s (fps)	0.3048	m/s	meter per second
ft^2 (sqft)	0.0920	m^2	square meter[d]
ft^3 (cuft)	0.0283	m^3	cubic meter
ft^3/s (cfs)	0.0283	m^3/s	cubic meters per second
°F	0.555 (°F-32)	°C	degrees Celsius
gallon-U.S. (gal)	3.785	l	liter
gal/acre	9.353	l/ha	liter per hectare
gal/ft^2	40.743	l/m^2	liters per square meter
gal/ft^2	0.0407	m^3/m^2 = m	meter
gal/ft^2-day	4.72×10^{-7}	m/s	meter per second
gal/ft-day	1.438×10^{-7}	m^2/s	square meter per second
hp	0.746	kW	kilowatts
inch (in.)	2.54	cm	centimeter
in.^{2}	6.452	cm^2	square centimeter
in.^{3}	16.39	cm^3	cubic centimeters
in.^{3}	0.0164	l	liter
pound (lb)	0.454	kg	kilograms[e]
lb	454	g	grams

Continued

U.S. Customary to S.I. (Metric)

U.S. Customary Units	Multiply by to obtain	S.I. Symbol	S.I. Name
lb/acre	1.121	kg/ha	kilograms per hectare
lb/ft^3	16,042	g/m^3	grams per cubic meter[f]
lb/in.^{2} (psi)	0.0703	kg/cm^2	kilograms per square cm[g]
lb/Mgal	0.120	mg/l	miligrams per liter
lb/mi	0.282	g/m	grams per meter
mile (mi)	1.609	km	kilometer
mile/hour (mph)	0.447	m/s	meter per second
million gallons (Mgal)	3785	m^3	cubic meters
million gallons per day (Mgd)	0.0438	m^3/s	cubic meters per second
mi^2 (sqmi)	2.59	km^2	square kilometer
parts per billion (ppb) in water	1.0	μg/l	micrograms per liter
parts per million (ppm) in water	1.0	mg/l	miligrams per liter
ton (short)	0.907	t	tonne
tons/acre	2240	kg/ha	kilograms per hectare
tons/sqmi	3.503	kg/ha	kilograms per hectare
yard (yd)	0.914	m	meter
yd^3	0.765	m^3	cubic meter

[a] 1 ha = 10,000 m^2 = 0.01 km^2
[b] 1 m^3 = 1000 l
[c] 1 m = 100 cm = 0.001 km = 1000 mm
[d] 1 m^2 = 10,000 cm^2 = 10^{-6} km^2
[e] 1 kg = 1000 g = 0.001 tonne
  1 g = 1000 mg = 10^6 μg = 10^9 ng
[f] 1 g/m^3 = 1 mg/l
[g] 1 kg/cm^2 = 0.968 atm = 0.981 bars

APPENDIX B

U.S. EPA Water Quality Criteria for Dissolved Oxygen Concentrations

	Cold-Water Biota		Warm-Water Biota	
	Early Life Stages[a,b]	Other Life Stages	Early Life Stages[b]	Other Life Stages
30-day mean	NA[c]	6.5	NA	5.5
7-day mean	9.5 (6.5)	NA	6.0	NA
7-day mean minimum	NA	5.0	NA	4.0
1-day minimum[d]	8.0 (5.0)	4.0	5.0	3.0

[a] Recommended water column concentrations to achieve the required intergravel dissolved oxygen concentrations shown in parentheses. The 3-mg/l difference is discussed in the criteria document. The figures in parentheses apply to species that have early life stages exposed directly to the water column.
[b] Includes all embryonic and larval stages and all juvenile forms to 30 days following hatching.
[c] NA—not applicable.
[d] All minima should be considered as instantaneous concentrations to be achieved at all times. Further restrictions apply for highly manipulative discharges.

Criteria for Ammonia

Acute Toxicity Criterion (1 hr average, once in 3 yr exceedence)

$$\text{Criterion (in mg/l)} = \frac{0.52}{2 * FT * FPH}$$

where $FT = 10^{0.03(20-TCAP)}$ for $TCAP \le T \le 30$
and
$\qquad FT = 10^{0.03(20-T)}$ for $0 \le T \le TCAP$
$TCAP$ = 20°C where salmonids or other sensitive coldwater species are present and 25°C where salmonids and other sensitive coldwater species are absent
$\qquad T$ = water temperature, °C

Chronic Toxicity Criterion (4 day average, once in 3 yr exceedence)

$$\text{Criterion (in mg/l)} = \frac{0.8}{\text{Ratio} * FT * FPH}$$

where FT and FPH are calculated using the same formula as given above. $TCAP$ is determined as follows:

$TCAP$ = 15°C where salmonids or other sensitive coldwater species are present and 20°C where salmonids and other coldwater species are absent
$\quad$ Ratio = 16   for $7.7 \le pH \le 9$
and
$\quad$ Ratio = $\{(24)[10^{(7.7-pH)}]\}/[1 + 10^{(7.4-pH)}]$ for $6.5 \le pH \le 7.7$

Source: From *Ambient Water Quality Criteria for Ammonia—1984*, U.S. EPA 440/5-85-001, Washington, DC.

U.S. EPA Selected Water Quality Criteria in Micrograms per Liter for Priority Pollutants

	Aquatic Life				Human Health 10^{-6} Risk		Priority Pollutant/ Carcinogenic
	Freshwater		Marine				
Chemical	Acute[a]	Chronic[a]	Acute[a]	Chronic[a]	Water and Fish Ingestion	Fish Ingestion Only	
Aldrin	3		1.3		7.4×10^{-5}	7.9×10^{-5}	Y/Y
Alkalinity		>20,000					N/N
Antimony	9000	1600			146	45,000	Y/N
Arsenic					0.0022	0.0175	Y/Y
Arsenic (penta)	850	48	2319	13			Y/Y
Arsenic (tri)	360	190	69	36			Y/Y
Asbestos					30,000[b]		Y/Y
Barium					1000		N/N
Benzene	5300		5100	700	0.66	40	Y/Y
Cadmium	c	c	43	9.3	10		Y/N
Carbon tetrachloride	35,200		50,000		0.4	6.94	Y/Y
Chlordane	2.4	0.0043	0.09	0.004	0.00046	0.00048	Y/Y
Chlorinated benzenes	250	50	160	129			Y/Y
Chloroform	28,900	1240			0.19	15.7	Y/Y
Chlorphenoxy herbicides 2,4,5-T					10		N/N
Chromium (hexavalent)	16	11	1100	50	50		Y/N
Chromium (trivalent)	c	c	10,300		170,000	3,433,000	N/N
Copper	c	c	2.9	2.9			Y/N
Cyanide	22	5.2	1	1	200		Y/N
DDT	1.1	0.001	0.13	0.001	0.000024	0.000024	Y/Y
DDE (DDT metabolite)	1050		14				Y/Y
Demeton		0.1		0.1			Y/N
Di-2-ethylhexylphthalate					15,000	15,000	Y/N
Dibutylphthalate					35,000	154,000	Y/N
Dichlorobenzenes	1120	763	1970		400	2600	Y/N

Dichlorobenzidine	118,000				0.02	Y/Y
Dichloroethane 1,2	11,600	20,000	113,000		0.01	Y/Y
Dichloroethylene	2020		224,000		0.94	Y/Y
Dichlorophenol 2,4	23,000	365			0.33	N/N
Dichloropropane	6080	5700	10,300	3040	3090	Y/N
Dichloropropene	2.5	244	790			Y/N
Dieldrin		0.0019	0.71	0.0019	87	Y/Y
Diethylphthalate	2120				0.000071	Y/N
Dimethylphenol 2,4					350,000	Y/N
Dimethylphthalate	330	230	590			Y/N
Dinitrotoluene	0.01	0.00001		370	313,000	N/Y
Dioxin (2,3,7,8-TCDD)	270				70	Y/Y
Diphenylhydrazine 1,2					1.3×10^{-8}	Y/N
Dinitro-o-cresol 2,4					0.0423	Y/N
Dinitrotoluene 2,4					13.4	Y/N
					0.11	
Endosulfan	0.22	0.056	0.034	0.0087	74	Y/N
Endrin	0.18	0.0023	0.037	0.0023	1	Y/N
Ethylbenzene	32,000		430		1400	Y/N
Fluoranthene	3980		40	16	42	Y/N
Guthion		0.01		0.01		N/N
Haloethers	360					Y/N
Halomethanes	11,000	122	12,000	6400	15.7	Y/Y
Heptachlor	0.52	0.0038	0.053	0.0036	0.00029	Y/Y
Hexachlorobenzene					0.19	Y/N
Hexachlorobutadiene	90	9.3	32		0.00028	Y/Y
Hexachlorocyclohexane (Lindane)	2	0.08	0.16		0.00072	Y/Y
					0.45	
					50	

U.S. EPA Selected Water Quality Criteria in Micrograms per Liter for Priority Pollutants (continued)

Chemical	Aquatic Life				Human Health 10^{-6} Risk		Priority Pollutant/ Carcinogenic
	Freshwater		Marine		Water and Fish Ingestion	Fish Ingestion Only	
	Acute[a]	Chronic[a]	Acute[a]	Chronic[a]			
Hexachlorocyclohexane (technical)					0.012	0.041	Y/Y
Hexachlorocyclohexane (α)					0.0092	0.031	Y/Y
Hexachlorocyclohexane (β)					0.016	0.055	Y/Y
Hexachlorocyclohexane (γ)							
Hexachlorocyclopentadiene	7	5.2	7		206		Y/N
Hexahchloroethanes	980	540	940		1.9	8.74	N/Y
Iron		1000			300		N/N
Isophorone	117,000		12,900		5200	5,200,000	Y/N
Lead	c	c	140	5.6	50		Y/N
Malathion		0.1		0.1			N/N
Manganese					50	100	N/N
Mercury	2.4	0.0012	2.1	0.025	0.144	0.146	Y/N
Methoxychlor		0.03		0.03	100		N/N
Mirex		0.001		0.001			N/N
Monochlorobenzene					488		Y/N
Napthalene	2300	620	2350		13.4	100	Y/N
Nickel	c	c	75	8.3	13.4		Y/N
Nitrates					10,000		N/N
Nitrobenzene	27,000		6680		19,800		Y/N
Nitrophenols	230	150	4850		13.4	7.65	Y/N
Nitrosamines	5850		3,300,000				Y/Y
Nitrosobutylamine N					0.0064	0.587	Y/Y
Nitrosodiethylamine N					0.008	1.24	Y/Y

Nitrosodipenthylamine N						Y/Y
Nitrosopyrrolidine N						Y/Y
Parathion	0.065	0.013				N/N
PCB's	2	0.014	10	0.03	4.9	Y/Y
Pentachlorinated Ethanes	7240	1100	390	281	0.16	N/N
Pentachlorobenzene					0.000079	0.000079	N/N
Pentachlorophenol	20	13	13	7.9	74	85	N/N
Phenol	10,200	2560	5800		1010		Y/N
Phthalate esters	940	3	2944	3.4	3500		Y/N
Polynuclear aromatic hydrocarbons			300		0.0028	0.031	Y/Y
Selenium	260	35	410	54	10		Y/N
Silver	c	0.12	2.3		50		Y/N
Sulfide-hydrogen sulfide		2		2			N/N
Tetrachlorinated ethanes	9320						Y/N
Tetrachlorobenzene 1,2,4,5					38	48	Y/N
Tetrachloroethane 1,1,2,2		2400	9020		0.17	10.7	Y/Y
Tetrachloroethylene	5280	840	10,200	450	0.8	8.85	Y/Y
Tetrachlorophenol 2,3,5,6				440			Y/N
Thallium	1400	40	2130		13	48	Y/N
Toluene	17,500		6300	5000	14,300	424,000	Y/N
Toxaphene	0.73	0.0002	0.21	0.0002	0.00071	0.00073	Y/Y
Trichlorinated ethanes	18,000				0.6	41.8	Y/Y
Trichloroethane 1,1,1			31,200		18,400	1,030,000	Y/N

U.S. EPA Selected Water Quality Criteria in Micrograms per Liter for Priority Pollutants (*continued*)

Chemical	Aquatic Life Freshwater Acute[a]	Aquatic Life Freshwater Chronic[a]	Aquatic Life Marine Acute[a]	Aquatic Life Marine Chronic[a]	Human Health 10^{-6} Risk Water and Fish Ingestion	Human Health 10^{-6} Risk Fish Ingestion Only	Priority Pollutant/ Carcinogenic
Trichloroethane 1,1,2		9400			0.6	41.8	Y/Y
Trichloroethylene	45,000	21,900	2000		2.7	80.7	Y/Y
Trichlorophenol 2,4,5		970			2600		N/N
Trichlorophenol 2,4,6					1.2	3.6	Y/Y
Vinyl Chloride	c	c			2	525	Y/Y
Zinc			95	86			Y/N

Sources: U.S. Environmental Protection Agency (1987); *Quality Criteria for Water 1986*; and *Federal Register*.

[a] Permissible exceedence: acute toxicity criteria—1-hr average concentration, not to be exceeded more than once in three years on the average; chronic toxicity criteria—4-day average concentration, not to be exceeded more than once in three years on the average.
[b] Expressed in fibres/liter.
[c] Metals toxicity related to hardness by the relationship AT or CT (μg/l) = exp{α ln[hardness(mg $CaCO_3$/l) + β}

Metal	Acute Toxicity AT[b] α	Acute Toxicity AT[b] β	Chronic Toxicity CT[b] α	Chronic Toxicity CT[b] β
Cadmium	1.128	−3.828	0.7852	−3.490
Chromium (trivalent)	0.819	+3.688	0.8190	+1.561
Copper	0.9422	−1.464	0.8545	−1.465
Lead	1.266	−1.416	1.266	−4.661
Nickel	0.8460	+3.3612	0.8460	+1.1645
Zinc	0.8473	+0.8604	0.8473	+0.7614

Toxicity Characteristics for Contaminated Sediments in the TCLP Test

Priority Constituent	Maximum Concentration (mg/l)
Arsenic	5.0
Barium	100.0
Benzene	0.5
Cadmium	1.0
Carbon tetrachloride	0.5
Chlordane	0.03
Chlorobenzene	100.0
Chloroform	6.0
Chromium	5.0
Cresols	200.0
2,4-D	10.0
1,4-Dichlorobenzene	7.5
1,2-Dichloromethane	0.5
1,1-Dichloroethylene	0.7
2,4-Dinitrotoluene	0.1
Endrin	0.02
Heptachlor (and its epoxide)	0.008
Hexachlobenzene	0.1
Hexachloro-1,3-butadiene	0.5
Hexachloroethane	3.0
Lead	5.0
Lindane	0.4
Mercury	0.2
Methoxychlor	10.0
Methyl ethyl ketone	200.0
Nitrobenzene	2.0
Pentachlorophenol	100.0
Pyridine	5.0
Selenium	1.0
Silver	5.0
Tetrachloroethylene	0.7
Toxaphene	0.5
Trichloroethylene	0.5
2,3,5-Trichlorophenol	400.0
2,4,6-Trichlorophenol	2.0
2,4,5-TP (Silvex)	1.0
Vinyl chloride	0.2

Source: U.S. EPA, *Federal Register* (March 29, 1990).

APPENDIX C

Environmental Characteristics of Priority Pollutants

Priority Pollutant	Solubility (mg/l @°C)	Vapor Pressure (mm Hg @°C)	Henry's Constant (atm-m^3/M [@25°C])	log K_{ow} (l/kg)	Biochemical Decay Coefficient	Parameters Affecting K_p
Halogenated Aliphatic Hydrocarbons						
Bromoform	3,200 @30	5.6 @25	5.32 E−4	2.30	1	V LB
Carbon tetrachloride	800 @20	113 @25	2.93 E−2	2.73	2	V LB
Chlorodibromomethane	NA	15 @10.5	7.83 E−4	2.09	0	V LB
Chloroethane	5,740 @20	2,660 @25	1.11 E−2	1.43	NA	V LB
Chloroform	9,300 @25	160 @20	3.39 E−3	1.97	2	V
Dichlorobromomethane	4,000 @25[a]	50 @20	2.12 E−3	1.88	0	V LB
Hexachlorobutadiene	2 @20	0.15 @20	0.0256	4.78	NA	V S
Hexachlorocyclopentadien	1.8 @25	0.08 @25	0.016	3.99	2	V LS
Hexachloroethane	50 @22	0.40 @25	9.85 E−3	4.14	2	S[b]
Methyl bromide (bromomethane)	1.75 E4 @20	1,420 @20	NA	1.1	NA	V LB
Methyl chloride (chloromethane)	6,360 @20	3,800 @20	8.82 E−3	0.91	NA	V
Methylene chloride	1.67 E4 @25	429 @25	3.19 E−3	1.25	2	V LB
Tetrachloroethylene	150 @25	19 @25	2.87 E−2	2.53	1	V B
Trichloroethylene	1,100 @25	77 @25	1.17 E−2	2.53	1	V B
Vinyl chloride (chloroethene)	1.1 @25	2,580 @25	2.78 E−2	0.60	NA	V
1,1,1-Trichloroethane	4,400 @25	100 @20	4.08 E−3	2.47	1	V LB
1,1,2,2-Tetrachloroethane	2,900 @20	4 @25	3.8 E−4	2.39	0	V LB
1,1,2-Trichloroethane	4,500 @20	25 @25	7.42 E−4[a]	2.17	1	V LB
1,1-Dichloroethane	5,500 @20	234 @25	5.45 E−3	1.79	1	V LB
1,1-Dichloroethylene	210 @25	591 @25	1.49 E−2	1.48	2	V LB
1,2-Dichloroethane	8,690 @20	79 @25	1.10 E−3	1.45	1	V LB

1,2-Dichloropropane	2,700 @20	42 @20	2.82 E−3	1	V LB
1,2-*trans*-Dichloroethylene	600 @20	331 @25	5.32 E−3	1	V LB
1,3-Dichloropropylene	2,700 @25	43 @25	3.5 E−3	1	V LB

Phenolic Compounds

P-Chloro-*m*-cresol	2.5 E4 @25[a]	0.11 @20	NA	NA	[c]
Pentachlorophenol	14 @20	1.1 E−4 @20	2.8 E−6	1	S B
Phenol	8 E4 @25	0.35 @25	1.3 E−6	2	LV HB
2,4,6-Trichlorophenol	800 @25	1 @76.5	4 E−6	2	B
2,4-Dichlorophenol	4,500 @25	1 @25	2.8 E−6	2	HB
2,4-Dimethylphenol	590 @25[a]	98 @104	1.7 E−6	2	[d]
2,4-Dinitrophenol	5,600 @18	1.49 E−5 @25[a]	6.45 E−10	2	B
2-Chlorophenol	2.85 E−4 @20	40 @82	1.03 E−5	NA	B
2-Nitrophenol	2,100 @20	1 @49.3	7.56 E−6[a]	2	B
4,6-Dinitro-o-Cresol	130 @20	1 E−4 @25	1.4 E−6	0	B
4-Nitrophenol	16,000 @25	2.2 @146	2.5 E−6[a]	2	B

Polycyclic Aromatic Hydrocarbons

Acenaphthene	3.42 @25	10 @131	2.41 E−4	2	LS HB
Acenaphthylene	3.93 @25	0.029[a]	1.14 E−4	2	S HB
Anthracene	1.29 @25	1.7 E−5[a]	8.6 E−5	1	V S HB
Benzo (a) anthracene	0.010 @24	2.2 E−8 @20[a]	1 E−6[a]	0.0528	LV S HB
Benzo (a) pyrene	0.0038 @25	5 E−9 @20	4.9 E−7	0	LV S B
Benzo (b) fluoranthene	0.014 @25[a]	5 E−7 @20[a]	1.22 E−5[a]	NA	LV S HB
Benzo (ghi) perylene	2.6 E−4 @25[a]	1 E−10 @20[a]	1.44 E−7[a]	0.1310[a]	LV HS HB
Benzo (k) fluoranthene	4.3 E−3 @25[a]	9.59 E−11 @20	1.22 E−5[a]	0.1111[a]	LV HS HB
Chrysene	0.006 @25	6.3 E−9 @25	1.05 E−6	1	LV S HB
Dibenzo (a,h) anthracene	5 E−4 @25	1 E−10 @20[a]	7.3 E−8	0.1310[a]	LV S HB
Fluoranthene	0.265 @25	5.0 E−6 @25	6.5 E−6	2	LV S HB
Fluorene	1.9 @25	10 @146	1.17 E−4	1	S HB
Indeno (1,2,3-cd) pyrene	5.3 E−4 @25[a]	1.1 E−10 @20[a]	6.95 E−8	NA	LV S HB
Naphthalene	30 @25	0.082 @25	4.83 E−4	2	HB[e]
Phenanthrene	0.816 @21	1 @118.2	3.93 E−5	2	LV S HB
Pyrene	0.16 @26	2.5 @200	5.1 E−6	2	LV S B

Environmental Characteristics of Priority Pollutants (continued)

Priority Pollutant	Solubility (mg/l @°C)	Vapor Pressure (mm Hg @°C)	Henry's Constant (atm-m^3/M) [(@25°C)]	log K_{ow} (l/kg)	Biochemical Decay Coefficient	Parameters Affecting K_p
Halogenated Ethers						
Bis-(chloromethyl) ether	2.2 E4 @25	30 @22	2.1 E−4	−0.38	NA	—
Bis-(2-chloroethyl) ether	1.02 E4	0.71 @20	1.3 E−5	1.58	2	LB^e
Bis-(2-chloroethoxy) methane	8.1 E−4 @25	<0.1 @20	2.8 E−7^a	1.26	0	^{f}
Bis-(2-chloroisopropyl) ether	1.7 E3^a	0.85 @20^a	1.1 E−4^a	2.10	2	LB^e
2-Chloroethyl vinyl ether	6,000 @25	26.75 @20	2.5 E−7^a	1.28	2	V^f
4-Bromophenyl phenyl ether	4.8 E−4 @25^a	1.5 E−3 @20^a	1.0 E−4^a	4.28	0	LS LB^e
4-Chlorophenyl phenyl ether	3.3 @25	0.0027 @25	2.19 E−4	4.08	0	LS LB^e
Phthalate Esters						
Bis (2-ethylhexyl) phthalate	0.4 @25	1.2 @200	3 E−7	5.3	1	LV S B
Butyl benzyl phthalate	2.9	8.6 E−6 @20	8.3 E−6^a	4.78	2	LV B
Diethyl phthalate	896 @25^a	0.05 @70	1.2 E−6	2.96	2	LV B
Dimethyl phthalate	5,000 @20	<0.01 @20	2.15 E−6	1.87	2	LV B
Di-*n*-butyl phthalate	400 @25	0.1 @115	2.8 E−7	5.2	2	LV S B
Di-*n*-octyl phthalate	3 @25	<0.2 @150	1.7 E−5	9.2	1	LV HS B
Monocyclic Aromatics						
Benzene	1,780 @20	95 @25	5.55 E−3	2.13	2	V LB
Chlorobenzene	488 @25	11.8 @25	3.93 E−3	2.84	2	V LB
Ethylbenzene	152 @20	10 @25.9	6.44 E−3	3.15	2	V^g
Hexachlorobenzene	0.11 @24	1 @114.4	1.70 E−3	5.47	0	LV S LB

Nitrobenzene	1,900 @25	0.15 @20	2.38 E−5	1.85	2	LB
Toluene	515 @25	28 @25	5.92 E−3	2.69	2	V^s
1,2,4-Trichlorobenzene	19 @22	1 @38.4	1.42 E−3	3.98	1	V LS LB
1,2-Dichlorobenzene	145 @25	1.5 @25	1.94 E−3	3.38	1	V LB
1,3-Dichlorobenzene	123 @25	1.0 @12.1	2.63 E−3	3.38	1	V LB
1,4-Dichlorobenzene	79 @25	1.8 @30	2.72 E−3	3.39	1	V LB
2,4-Dinitrotoluene	270 @22	0.0013 @59	4.5 E−6	2.01	1	LB
2,6-Dinitrotoluene	270 @22	6 @150	7.9 E−6^a	2.05	1	LB
Pesticides						
Acrolein	2.08 E−5 @20	210 @20	5.66 E−5	−0.09	2	LV B
Aldrin	0.017 @25	2.3 E−5 @20	4.96 E−4	5.11	0	V S HB
BHC-alpha	1.63 @25	0.06 @40	NA	3.81	0	B^h
BHC-beta	0.70 @25	0.17 @40	NA	3.8	0	B^h
BHC-delta	21.3 @25	0.02 @20	NA	4.14	0	LS B^h
BHC-gamma	7 @20	9.4 E−6 @20	4.93 E−7	3.24	2	B^h
Endosulfan-alpha	0.32 @22	1 E−5 @25	1.0 E−5	3.55	0	V B
Endosulfan-beta	0.33 @22	1 E−5 @25	1.9 E−5	3.62	0	V B
Chlordane	0.0156 @25^a	1 E−5 @25	4.79 E−5	2.78	0	V LB
Dieldrin	0.186 @20	1.8 E−7 @25	5.84 E−5	4.09	0	LS
Endosulfan sulfate	0.22	1 E−5 @25^a	2.6 E−5	3.66	0	B^i

Environmental Characteristics of Priority Pollutants (*continued*)

Priority Pollutant	Solubility (mg/l @°C)	Vapor Pressure (mm Hg @°C)	Henry's Constant (atm-m³/M [@25°C])	log K_{ow} (l/kg)	Biochemical Decay Coefficient	Parameters Affecting K_p
Endrin	0.26 @25	2 E−7 @25	4 E−7	5.6	0	S B[i]
Endrin aldehyde	50 @25[a]	2.0 E−7 @25[a]	2 E−9	3.15	NA	B[i]
Heptachlor	0.056 @25	3 E−4 @25	1.48 E−3	4.41	0	V LS LB
Heptachlor epoxide	0.35 @25	3 E−4 @25[a]	3.16 E−5	2.65	0	LB[e]
Isophorone	1.2 E−4	0.38 @20	5.75 E−6[a]	1.7	2	[f]
p,p′-DDD	0.16 @24	1 E−7 @30	2.2 E−8	5.99	0	V S LB
p,p′-DDE	0.040 @20	6.5 E−6 @20	2.2 E−5	5.69	0	V S LB
p,p′-DDT	0.0031 @25	1.5 E−7 @20	3.89 E−5	6.19	0	V HS LB
Toxaphene	0.5 → 3 @25	0.2 → 0.4 @25	4.89 E−3	3.3	0.0161[a]	B[j]
2,3,7,8-Tetrachlorodibenzo-p-dioxin (TCDD)	1.93 E−5	7.4 E−10 @25	2.1 E−3	6.64	NA	HS LB
PCBs and Related Compounds						
PCB-1016	0.049 @24	4 E−4 @25	1.35 E−2	4.38	0	V S B[k]

PCB-1221	0.59 @24	6.7 E−3 @25	2.28 E−4	4.09	2	V S B[k]
PCB-1232	1.45	4.06 E−4 @25	NA	>4.54	2	V S B[k]
PCB-1242	0.10 @24	4.06 E−4 @25	3.43 E−4	4.11	0	V S B[k]
PCB-1248	0.054	4.94 E−4 @25	4.4 E−4	5.75	0	V S B[k]
PCB-1254	0.057 @24	7.71 E−5 @25	8.37 E−3	6.03	0	V HS B[k]
PCB-1260	0.080 @24	4.05 E−5 @25	3.36 E−4	>6.11	0	V HS B[k]
2-Chloronaphthalene	6.74 @25	0.017 @20	5.4 E−4[a]	4.12	2	V S B[k]

Metals and Cyanide

Antimony	IN	1 @886	NA	NA	V S B[l]
Arsenic	IN	1 @372	NA	NA	V S B[m]
Beryllium	IN	NA	NA	NA	LV
Cadmium	IN	1 @393	NA	NA	S
Chromium	IN	1 @1616	NA	NA	LS
Copper	IN	1 @1628	NA	NA	S LB[n]
Lead	IN	1 @970	NA	NA	S LB[o]
Mercury	0.056 @25	1.2 E−3 @20	1.14 E−2	NA	V S B[p]
Nickel	IN	1 @1800	NA	NA	HS[q]
Selenium	NA	NA	NA	NA	V S B[r]

Environmental Characteristics of Priority Pollutants (*continued*)

Priority Pollutant	Solubility (mg/l @°C)	Vapor Pressure (mmHg @°C)	Henry's Constant (atm-m³/M [@25°C])	log K_{ow} (l/kg)	Biochemical Decay Coefficient	Parameters Affecting K_p
Silver	IN	1 @1310	NA	NA		S^s
Thallium	IN	1 @825	NA	NA		S^t
Zinc	IN	1 @487	NA	NA		S^u
Miscellaneous						
Acrylonitrile	7.35 E4 @20	137 @30	8.8 E−5	−0.92	2	V
Asbestos	NA	NA	NA	NA		S
Benzidene	400 @12	5 E−4[a]	3 E−7[a]	1.81	NA	6^f
Cyanide	NA	NA	NA	NA		HV LS HB[v]
N-Nitrosodi-n-propylamine	9900 @25	8 @25	NA	1.31	0	LB
N-Nitrosodimethylamine	MISCIBLE	0.1	NA	−0.47	0.0451[a]	LB
N-Nitrosodidipherylamine	35	2.6 E−5 @25[a]	3.13	2.79	2	B
1,2-Diphenylhydrazine	221	1 E−5 @22[a]	3.4 E−9	3.03	1	f
3,3'-Dichlorobenzidene	3.1 @25		8 E−7[a]	3.51	NA	u

Key to Rate Summary: 0 = No significant degradation rate. 1 = 0.05 day⁻¹ < k_B < 0.5 day⁻¹; use 0.05 day⁻¹. 2 = k_B > 0.5 day⁻¹; use 0.5 day⁻¹

Notes: Unless noted, data (except K_p data) obtained from U.S. EPA Risk Reduction Engineering Laboratory (RREL), Cincinnati, Ohio, Treatability Database. K_p data obtained from Michael A. Callahan et al., *Water-related Environmental Fate of 129 Priority Pollutants.* EPA-440/4-79-029, 1980. NA = Data not available; IN = Insoluble; LV = little/somewhat volatile; V = moderately volatile; HV = highly volatile; LS = little/somewhat sorbed; S = moderately sorbed; HS = highly sorbed; LB = little/slowly biodegradable; B = moderately biodegradable; HB = highly biodegradable.

[a] Data obtained from U.S. EPA, *Processes, Coefficients, and Models for Simulating Toxic Organics and Heavy Metals in Surface Waters.* EPA/600/3-87/015, June 1987.
[b] No specific data on V nor B found in the literature.
[c] Biodegradation demonstrated in soil samples and acclimated sewage sludge, but uncertain in ambient surface waters.
[d] B = information inconclusive.
[e] V = unknown.
[f] B = unknown.
[g] S and B = importance not determined.
[h] V = unknown—information contradictory; B = varies with environment.
[i] V and S = no information.
[j] B = only in anaerobic sediments.
[k] V = depressed by presence of organic solids; S = strongly adsorbed by presence of organic solids, especially with high organic content; B = only important for those with fewer than 4 chlorines per molecule.
[l] Biomethylation may occur.
[m] Metabolized by a number of organisms to organic arsenicals having increased mobility.
[n] Some copper complexes may be metabolized. Organic ligands are important in sorption and complexation processes.
[o] Biomethylation in sediments can remobilize lead.
[p] Sorption strongest onto organic materials. Can be metabolized by bacteria to methyl and dimethyl forms that are quite mobile.
[q] Most mobile of heavy metals.
[r] Volatilization via biomethylation or formation of H_2Se. Hydrous metals strongly sorb Se. Metabolism may result in methylation with subsequent volatilization.
[s] Strongly sorbed by hydrous Mn and Fe oxides, clay minerals, and organics.
[t] Adsorbed to clay minerals and hydrous metal oxides.
[u] Strong affinity for hydrous metal oxides, clays, and organic matter. Sorption increases with pH.
[v] At pH ≤ 10, most free cyanide is HCN, which is quite volatile.

Index

accumulation function, 464–466
acid and methane fermentation in sediments, 746
acid-base reactions, 408
acidic surface waters, 204–208
acidification of soil, 313
acidity, 41, 802–806
 rain, natural, 202
 of urban precipitation, 480–481
acid mine drainage, 40, 208, 803
acid rain, 102, 185, 202, 578, 803–804
 areas sensitive to acidic precipitation, 806
 biological impact on receiving water bodies, 804
 buffering in urban areas, 481
 effect on drinking water, 211–215
 natural, 202
 sulfuric acid formation, 210
 sulphate loading of lakes, 805
acid volatile sulfides (AVS), 357–358, 360, 889
adsorbing compounds, for metals in soils and sediments, 355
adsorption-desorption reactions, 410
adsorption
 of pollutants
  incorporation in models, 530
  in soils, 342–345, 352–357
  in wetlands, 882
 of viruses by soils, 378
advection, in groundwater zones, 398–399
aeration for water quality improvement, 947–948
 in-lake aeration, 957–962

aerobic decomposition, 745–746, 878
aerobic processes, 745
aggradation, 243
 floodplain, 293
agricultural and silvicultural pollution, 574–690
 sources, 681–689
agricultural chemicals, 10
agricultural conservation programs, 678
agricultural diffuse pollution models, 540–546
agricultural land use, 679
Agricultural Nonpoint Source Pollution Model (AGNPS), 540–541, 546
agricultural pollution, 6–7, 9, 29–34
agricultural production, 676
Agricultural Research Service (ARS), 541
Agricultural Runoff Model (ARM), 536, 541
agriculture
 average farm size, 676
 commodity programs, 677–679
 extent, 674–680
 intensive, 29
 management, 674–690, 712–715
 private land use, 77, 679–680, 725
 sustainable, 3, 680
air pollution sources, 188
air turbulence, 463–465
aldrin, 6
 in soils, 340
 in urban runoff, 480
algal biomass, 781, 784
algal bloom, 53, 787
algal respiration, 747, 766

1035

1036   Index

algicides, 340
alkalinity, 805
aliphatic hydrocarbons, 344, 390
alum, addition to lakes, 955–956
aluminum toxicity, due to acidic inputs, 805
American Public Works Association, 447
ammonium, 760, 778
   adsorption, 369–371, 762
   ammonification, 362, 368
   fixation, 364, 762
   release from sediments, 295
   unionized, 41
   volatilization, 209, 364, 371, 919
animal waste management, 9, 48, 703–705
anoxic (anaerobic) decomposition, 746, 878
antecedent soil moisture conditions. *See* SCS Runoff Curve model
anthracene, 642
antidegradation regulation and principle, 989
apatites, 334
appropriation water rights, 83
aquaculture, 683
aquatic ecosystem, 740–764
   biotic components, 741–742
   integrity of, 749–759
aquatic habitat, 4–5, 12
   destruction, 5, 927–928
   erosion effects on, 250
   management, 963–965
   protection, 90
   restoration, 927–929, 940–946
aquatic life protection criteria, 839–843
aquatic plants
   management, 965–970
   systems, 917–921
aquatic weed control, 969
aquifer, 172, 177, 413
   artesian (confined), 174
   composition, 392–394
   mining, 178
   recharge
     by infiltration from basins and trenches, 600
     by land application of wastewater, 423
     by using pervious pavement, 590
     by wetlands, 175, 871
     hydrological, 174–177
     representation, 173
   safe yield, 178

Areal, Nonpoint Source Watershed Environmental Response Simulation (ANSWERS) model, 540, 546
Areawide Stormwater Model (ABMAC), 532, 546
arid lands, 26, 32
   water quality, 808
ARMA models, 166–170, 546–550, 655
AROCHLOR. *See* polychlorinated biphenyls
aromatic hydrocarbons, 345
arsenic
   arsenides-mineral sources, 412
   in natural base flow, 407
   in soils, 359
   in surficial materials, 319
   in urban runoff, 42, 477–479
   source minerals, 412
artificial wetlands, 866, 889–917. *See also* constructed wetlands
asbestos, 41
   in urban runoff, 477–479
   traffic emission, 456
Aswan high dam, 14
atmosphere–surface water body integrations, 198–202
atmospheric contamination, by chemicals from soils, 341
atmospheric deposition, 7, 10, 444, 451–454, 469, 808
   control, 578
   dry, 28–29, 453, 578
   mercury, 220
   monitoring, 559
   pesticides, 227–231
   sources, 187
   wet, 28, 453, 578
atmospheric emissions, 28, 189, 195
atmospheric nitrogen, impact on surface waters, 216
atmospheric particulates, deposition rate of, 452
atmospheric pollution, by toxic compounds, 217–231
atmospheric reaeration, 766, 768
   coefficient, 768–769
atrazine, 230, 340, 684
attainable water use, 986
automatic samplers, 561
availability factor for washoff of pollutants, 472

bacground (natural) water quality, 13–14, 32, 806–810

bank armoring techniques, 940
base flow
  hydrologic definition, 103–105, 172
  of urban streams 441
  quality (natural), 406–416
bedload, 245, 246, 294
benthic macroinvertebrates, 758
benthos, 744, 748
benzene, in urban runoff, 478–479
benzoperylene, in urban runoff, 480
Best Available Technology Economically Achievable (BATEA), 81, 90
Best Management Practices
  agricultural, 689–711
  definition, 18
  efficiency of agricultural BMPs, 697–711
  hydrological, 109
  reliability and longevity, 575
  selection for agriculture, 692–693
  urban, 573–575
bioaccumulation, 4, 16, 325, 745, 831–838
bioaccumulation factor, 832–835
  lipid normalized, 835–836
bioavailability of toxic compounds, 325, 828–829
biochemical oxygen demand (BOD), 21
  in CSOs, 499
  in dissolved balance of surface waters, 765–768
  in feedlot runoff, 688
  in landfill leachate, 430
  in septic tank effluents, 418
  in solids from urban areas, 448
  NURP studies, 492–493
  removal in ponds, 627
  removal in wetlands, 905–906
  traffic emission, 456
bioengineering, 939–943
biological criteria, 759
biological degradation, 330–332
biological treatment, 641
biomagnification, 745, 831–838
biomagnification factor, 832–835
  food chain multiplier, 837
  mass blance model, 832–834
  predator–prey concentration ratio, 836
biomanipulation, 972
biotic indices, 753–759
  index of biotic integrity, 754–758
  invertebrate community index, 758
  rapid bioassessment protocol, 759
  saprobien index, 753–754

biotic integrity of water bodies, measures, 753–759, 818
biotransformation, 330
black list of toxic pollutants, 819
buffering, of acid rain in urban areas, 481
buffer strips, 52, 94, 602
buffer zones, 601–602
buildup and washoff, 463–469, 470–473, 519–520, 522, 536–537
Bureau of Land Management, 679

cadmium
  in natural base flow, 407
  in precipitation, 453–454
  in soils, 359
  in solids from urban areas, 448
  in surficial materials, 319
  in urban runoff, 42, 477–479
  source minerals, 412
Canaan Valley, 974
capillary soil water zone, 172
carbonate rocks, 394
carbon/nitrogen ratio, effect on nitrogen in soils, 369
catch basins, 602, 612–613
cesium-137 use for soil loss estimates, 284–286
channel erosion, 460
  control, 603–604
channel stabilization, 603–611
Chebotarev, Ignaovich, Souline sequence of groundwater quality, 411
check dams, 607
chemical-based numerical criteria, 840–843
chemical oxygen demand (COD), 21
  in CSOs, 499
  in feedlot runoff, 688
  in landfill leachate, 430
  in septic tank effluents, 418
  in solids from urban areas, 448
  NURP studies, 492–493
  removal in ponds, 627
  surface waters impairment, 736
  traffic emission, 456
chemical time bomb, 313, 803
chemicals application control, 588
Chemicals, Runoff and Erosion from Agricultural Management Systems (CREAMS) model, 268–269, 540–541, 546
Chesapeake Bay, 46–48, 783, 886
  pesticide input from atmosphere, 230

Chicago River, diversion of, 947
chlordane, 6
    in urban runoff, 478–480
chlorinated hydrocarbons, from land wastewater disposal systems, 422
chloroform, in urban runoff, 478–479
chlorophyll-a
    in lakes, 784–790
    relation to BOD, 783
chromium
    in natural base flow, 407
    in precipitation, 453–454
    in soils, 360
    in surficial materials, 319
    in urban runoff, 42, 477–479
    source minerals, 412
    surface waters impairment, 736
    traffic emission, 456
chrysene, in urban runoff, 478–479
clay, 395, 396
    dispersion index, 290
clear-cutting, 686–687
coastal wetlands, 864–865. *See also* tidal marshes
Coastal Zone Management Act, 91
coefficients
    atmospheric reaeration, 768–769
    deoxygenation, 768
    dispersion, 403
    Manning's, 154–155
    partitioning, 339, 356–357, 359, 361, 404, 829–830
        octanol, 326, 342–345, 404–405, 642–647, 830, 835, 850, 912
    permeability (hydraulic conductivity), 396
    removal, for street pollutants, 463–464
    runoff, 440–441
        for Rational Formula, 146–148
    tidal dispersion, 776–777
    urban washoff, 472
    variation, 487
collection system control, 602–616
coliforms, 8
    in CSOs, 36, 476, 499
    in soils, 376
    in treatment plant effluent, 36
    in urban runoff, 36, 476
    surface water impairment, 736
    combined sewer overflows (CSO), 3, 37–40, 50, 440, 495–500
        bottlenecks in CSO control, 650
        characterization of quality, 36, 496–500
        chemical oxygen demand in, 499
        coliforms in, 36, 476, 499
        control of pollution, 89–90, 577–602, 612–613, 627–635
        control strategy, 651–654
        critical rainfall intensity for initiation of, 39, 498–499
        dry period effects on loads, 482
        first flush in, 499, 629
        frequency and duration, 496–498
        microorganisms in, 476, 497–499
        nitrogen and phosphorus in, 36, 499
        quality control basins, sizing, 618–619
        sewer flushing, 613
        storage facilities, 618
        suspended solids, 36
combined sewers, 7, 496
Commerce Clause of U.S. Constitution, 77, 84, 88
Commodity Credit Corporation, 677
complexation reactions, 410
computer aided design (CAD), 512, 526
confined animal operations, 29, 34, 683
    pollutant loads, 688
    management, 703–705
Conservation Reserve programs, 93
conservation tillage, 9, 94, 691–697
constructed wetlands, 866, 889–917
    basic flow patterns, 902
    components, 901
    configuration, 901
    design parameters, 894–909
    loading, 901–902
    pollutant removals, 902–907
    problems, 917
    retention, time, 899–900
    subsurface flow, 891–893
    types, 891–894
construction site erosion, 29, 248, 483
    USLE estimates, 262, 264
consumers (organisms), 745
continuous models, 513, 517, 532–536, 562
copper
    from NURP studies, 492–493
    in landfill leachate, 427
    in natural base flow, 407
    in precipitation, 453
    in soils, 360
    in surficial materials, 319
    in urban runoff, 42, 477–489
    source minerals, 412
    traffic emission, 456
coprecipitation, 328
corrosive waters, 212

cost sharing, 728
cribwall, for stream bank stabilization, 942
criteria and standards, 14–16, 839–848
   acceptable risk level for human health protection, 845
   acute toxicity, 823
   based on equity imperative, 80–81
   biological, 759
   chronic toxicity, 823
   effluent, 96, 81
   groundwater quality, 391–392
   human health protection, 843–848
   performance, 96
   permissible exceedence, 841
   sediment, 848–852
   site specific, 759
   whole effluent toxicity, 843
cumulative probability distribution, 485–486, 489
curb-side pollutant loads, 461–470
curb
   height, effect on accumulation of street solids, 467–468
   length density, relation to imperviousness, 446
   loading of pollutants 448
cyanides, 41
   in street salt, 460
   in urban runoff, 42, 477

Darcy's law, 395
data sources and retrieval, 557–558
DDT
   in atmospheric deposition, 229
   in soil environment, 340
   in urban runoff, 480
   levels in organisms, 831
   residence in atmosphere, 232
deaeration, 743
deamination, 760
decomposers, 745
deep well injection of wastewater, 388
deforestation, 10–11, 23, 26–28, 246–247
degradation, 243–244, 293
deicing
   chemicals, 459, 469
   operations, 126, 459, 469
   salts, 588
      impact on groundwater resources, 388–389
delivery ratio, 239, 270–271, 521, 544, 577, 603
   for lead, 361

in streams, 293–294
overland flow effects, 274–275
denitrification, 367–368, 421, 542, 760–764, 871, 882
denudation, 238
deoxygenation coefficient, 768
deposits (geological), 392–394
designated use, 16, 985–987
design storm, 129–131
detention–retention facilities, 616–637
   ponds, 621
deterministic models, 509–547
diazinon, in urban runoff, 480
di-benzo-thiopene, 642
dieldrin, 340, 478–480
diffuse pollution, 18
   abatement
      strategy of, 1007–1008
      targeting, 996–998, 1002
   programs
      financing, 993–1002
      voluntary, 7, 48, 726
   urban
      magnitude of, 443–445
      winters, 459–460
diffuse sources, definition, 21
discharge permits, marketable or transferable, 75, 999, 1001
disinfection, 641
dispersion
   coefficient of, 403
   of groundwater contaminants, 398–403
dissolution–precipitation equilibria, in groundwater, 409
dissolved oxygen
   balance of estuaries, 776–777
   balance of streams, 747, 765–776
   dystrophic conditions, 947
   effect of nutrients, 777
   in Milwaukee River, 51
   fluctuations in eutrophic water bodies, 785
   problem, 764–783
   standard, 764–765
   Streeter–Phelps models, 766–769, 802
   WASP model, 802
drainage, 54
   and irrigation districts, 1004
   natural (swale), 35
   systems, 576, 577
dry weather flow, 3, 166
dry (french) wells, 593, 601
duckweed, 868–869, 919–920

dustfall, 454, 469
dystrophic conditions, 947

Earth Day, 9
ecological balance, 23
ecological engineering, 868
ecological system, 22–23
ecologic cycle, 1013
economic externality, 995
economic incentives and instruments, 75, 78, 724, 993–1002, 1014
  benefits-received approach, 997–1000
  polluter pays principle, 994–996, 1004
ecoregions, 751–753, 809, 987
ecosystem, 740–741
  management, 928–929
effluent-limited water bodies, 441, 1000
Emscher River, 60–62, 933
enrichment ratio, 246, 286–292
  for urban snowmelt, 474
  pesticides, 349
  soil organic matter, 322
  soil phosphorus in runoff, 338
enrichment to delivery relation, 289, 290, 291, 292
environmental abatement objectives, 74
environmental corridors, 601–602. *See also* buffer zones
environmental degradation, 11
environmental laws, 72, 85–96, 725
  judicial, 96
  state, 95
  statutory, 86–96
environmental policies, 74–80, 94
  compromises, 76
  imperatives, 77–80
environmental political parties, 76
ephemeral gullies, 545
ephemeral streams, 104, 175, 387, 441
equation of continuity, 796
equation of motion, 797
equilibrium curb loads, 466
equity imperative, 78, 80–81
erodibility, 239
erosion, 237–244
  agricultural, 248
  construction site erosion, 262
  control, 8, 94, 686
  critical shear stress, 299–300
  definition, 238–244
  deforestation, 26
  effect of land use on, 248
  effect of vegetation, 243
  gully, 241
  highway, 248
  in urban areas, 442, 444, 460
  land use impact, 247
  of cohesive sediments, 299–304
  of topsoil, 684
  rill and interrill, 240
  silvicultural, 248
  streambank, 241, 249
  strip mines, 249
  topographical factors, 253
  upland, 153
  urban, 248
  wind, 243
erosion blankets, 701
estuaries, longitudinal mixing and dispersion in, 776
ethers, 643
euphoitic layer, 743–744
eutrophication, 30, 53, 361, 741
  lake and reservoir, 783–796
  modeling, 798–802
  symptoms of, 785–786
evaporation and evapotranspiration, 120–126
  pan, 121
  potential, 124
Evaporation Index method, 124–125
event mean concentration (EMC), 488, 855
  mean values for pollutant load estimates, 493
  of urban runoff, 466, 477, 484–485
event-oriented models, 513, 517, 531–536, 562
expected value, definition, 487
expert systems, 535
externality, definition, 73

fascines, for stream bank stabilization, 941
fecal coliform die-off in soils, 376–377
fecal coliforms/fecal streptococci ratio, 477
feedlots. *See* confined animal operations
fertilizers, 32
  application
    impact on groundwater, 388
    in urban areas, 442
  organic, 322
  rates, 312, 361–362
  management, 9
Fick's law, 330
field capacity soil moisture, 111
filters, 639, 710
filter strips, 589, 593–598, 603, 706–707
  critical distance, 593–595
  designs, 594–598
  model of trap efficiency, 595–598

use for pretreatment, 600
vegetation, 589, 593–598
first flush
  control, 629, 642
  effect, 482, 484
    in urban snowmelt, 474
    in CSOs, 499, 629
  retention, 632–635
  sediment flow, 283
fish community assessment method, 755.
  See also biotic indices
fish consumption, 844
fish kills, 58, 764, 803, 818, 958
fish ponds, 55
flexible membrane diffusers, for oxygen
    injection, 961
flood control, 144, 869
flooding potential, 441
Florida Water Management Districts,
    1005–1006
flow
  components, 102
  dividers, 496–497, 629
  for waste assimilative capacity
    determination, 855
  measurement and devices, 559–560
  sediment-carrying capacity of, 276
flow-weighted composite sample, 484,
    560–562
fluidsep (vortex) separator, 614
fluoranthene, in urban runoff, 480
fluorocarbons, 10
food chain multiplier, 837
Food, Agriculture, Conservation and
    Trade Act, 678, 1008
Food Security Act, 93
food web, 744
forest land, 683
forests, water quality, 808
frequency and probability distribution
    functions, 485–486
Freundlich adsorption isotherm, 327,
    370
fugitive dust emission, 453
fugitive particulate losses from streets,
    464
functional dose (concentration),
    response in toxicity tests, 822,
    827–828
fungicides, 340

gabions, 611, 701, 789
galena, 412. *See also* lead minerals
Gaussian normal distribution, 486
Gelhar–Wilson groundwater model,
    433–435

geographical information systems (GIS),
    507, 512, 526–528
grab samples, 560
grass carp, 967
grassed waterways, 603–611
  design velocities, 607
  Manning's roughness factor, 609
  vegetative cover, 608
grass strips. *See* filter strips
gravity soil water storage, 111
gray list of toxic pollutants, 819
Great Lakes, 4, 8
  pesticide inputs, 230
  sources of PCBs, 352
  studies, 6
  Water Quality Agreement, 7
Green–Ampt infiltration equation, 542
Green function, in stochastic models,
    549
greenhouse effect, 10, 26
groundwater
  cleanup, 390–391
  contamination, 32, 388–390, 392
  control of landfill leachate, 431–432
  discharge, 387, 394, 406
  dispersion, 398–403
  dissolution–precipitation equilibria
    in, 409
  flow, retardation of pollutants in,
    404–405
  hardness, 409
  hydrological balance, 177–178
  mining of, 103
  models, 179, 432–435
  movement, 171, 392–406, 433
  natural, 406–416
    water quality, processes, 408–411
  pollution, 104, 312
    by soil nitrogen, 362
    by land wastewater disposal
      systems, 422
    by leaching of chemicals from soils,
      341
    by solid-waste disposal sites, 429–
      432
    sources, 389, 415–432
  quality
    models, 432–435
    natural (background), sources of,
      411–415
  recharge, 387, 394
  retardation of pollutants, 404–405
  safe yield, 387, 394
  soil water zones, 172–173
  systems, 171–179
  zones, fate of chemicals in, 392

1042   Index

Groundwater Loading Effects of Agricultural Management Systems (GLEAMS) model, 542, 546

halogenated aliphatics
  in groundwater, 419
  in urban runoff, 478–479
hazardous lands, 522
hazardous waste, disposal sites, 430
helical bend regulator/concentrator, 615
Henry's constant, 329
heptachlor, 340
herbicides, for weed control of impoundments, 969
highway drainage model, 535, 539
human cancer criteria, urban runoff violations of, 480
human health protection criteria, 843–848
  acceptable risk, 845
humic acids, 407, 410
hydraulic conductivity, 109–110, 396–397
  coefficient, 396
  of frozen soils, 120
  of pervious pavements, 590
hydrogen sulfide, 887
hydrograph, 163–165, 284
hydrologic activity, 35
hydrologically active areas, 142, 525
hydrological models, 165, 527–546
  agricultural (rural) diffuse pollution, 540–546
  calibration and verification, 142, 535, 554–556
  classification, 510
  components, 529
  continuous, 513
  deterministic, 509–546
  distributed parameter, 509–512, 540
  erosion and soil loss, 252–270
  event oriented, 513, 517
  lumped parameter, 508–512, 540–541
  reliability and usefulness, 513–516
  SCS runoff curve model, 521
  selection, 516–519
  single reservoir, 513
  use of models, 515–516
  watershed models, 526
hydrological procedures, 507
hydrologic cycle, 101–103
Hydrologic Simulation Program—FORTRAN (HSP-F) model, 536–539, 546

hydromulching, 586, 701
hydroseeding, 586–587
hypolimnetic aeration, 960–961
hypolimnetic oxygen depletion, 791, 958

Illinois Urban Drainage Area Simulator (ILLUDAS), 533
impaired water bodies, 1002–1003
impervious area, directly connected, 141–142, 464, 592–593
imperviousness, 440
impervious urban surfaces, pollutant loads, 519–521
impoundments
  harvesting of weeds in, 967–968
  water-level drawdown, of, 966
Indian water rights, 85
infiltration, 109–121, 589, 598–602
  control, 119
  estimating, vegetation factor for, 114
  Green-Ampt equation, 118
  Holtan formula, 113–115
  Horton formula, 113
  into frozen soils, 119–120
  of rooftop runoff, 601
  on site, 63
  Philip's model, 116–118
  problems with appropriation water rights, 85
  rates, 110–120
  wetlands, 897
infiltration basins, 593, 600
infiltration devices, 598–602
  design parameters, 601
  into sewers, 632
  maintenance, 601
  rapid infiltration systems, 421–422
  vegetation factor, 114
infiltration trench, 599–600
inland wetlands, 865–869
institutions, 9, 982–983, 1002–1008
in-stream aeration, 947–948
in-stream water quality enhancement, 983
integrated aquatic restoration, 972–974
integrated pest management, in agriculture, 697
Integrated Risk Information System (IRIS), 848
integrated water quality management, 16, 79, 750, 807, 982, 989, 1004
intensity–duration models for rainfall, 130–131
interbasin water transfer, 83

interflow, 171, 317
invertebrates, 748
ion-exchange reactions, 410
irreversible impact doctrine, 78
irrigated cropland, 9, 34, 424–425, 682
irrigation return flow, 34, 53, 85, 682, 684, 709, 1015
  impact on groundwater resources, 388–389, 424–429
irrigation water management, 709
isopluvial map, 130

karst, 394
kinematic wave model, 152, 156, 536

lacustrine wetlands, 866
Lagoon of Venice, 52–57, 861
lagoons, 641, 702–705
Lake Balaton, 57–59, 784
Lake Erie, 4
lake (impoundment) restoration, 949–972
  addition of chemicals, 954–955
  aeration, 956–962
  artificial circulation, 959–960
  control of toxic contamination, 954–955
  deep water aeration, 954
  destratification, 959–961
  dilution and flushing, 956
  exotic plant and animal control, 952
  in-lake treatment, 952
  liming acidic lakes, 962–963
  sediment removal and control, 956
  spawning habitat management, 963–965
  techniques, 951–954
  water level management, 870
  watershed management, 952–953
  water withdrawal from hypolimnion, 956
lakes
  chemical treatment of, 955–956
  injection of liquid oxygen into, 961
  with low dissolved oxygen, 957
land application
  of sludge, 423
  of water and wastewater
    impact on groundwater, 419–424
    problems and restrictions, 422–423
land cover factor (C), 586
land drainage, 6
landfills, 429–432
  impact on groundwater resources, 388–389, 429–432
  leachate management, 430, 431–432

land use, 25–41
  effects
    on erosion and sediment loads, 248–249
    on event mean concentrations, 491
    on urban diffuse loads, 439–451
Langmuir adsorption isotherm, 327, 335, 339
leaching
  requirement and ratio, 427–428
  septic tanks, 418, 444
lead
  emissions, global, 218
  from traffic, 218, 456
  health effects, 220
  in atmospheric deposition, 217–220
  in CSOs, 36, 499
  in groundater, 213, 407
  in landfill leachate, 427
  in natural base flow, 407
  in NURP studies, 491–493
  in potable water, 213
  in precipitation, 453–454
  in soils, 360
  in solids from urban areas, 448
  in surficial materials, 319
  in urban runoff, 36–37, 42, 477–479
  in treatment plant effluents, 36
  surface waters impairment, 736
lead minerals, 412
legal doctrines, 77–80, 96
ligands, 326, 353–357, 645, 880
limestone, 394–396
  addition to lakes, 962
  addition to watersheds, 963
liming acidic lakes, 962–963
lindane, 41
  in soils, 340
  in urban runoff, 478–480
linear isotherm, 328
linear particle accumulation process, 463, 469
linear particle removal concept, 471–472
linear watershed models, 150
link-node representation, for water quality models, 798–799
litter control programs, 578–579
loading function and models, 317–319, 373, 508, 522–524
log-drop structures, for stream restoration, 945
log-normal probability distribution, 484–487, 491, 563
Love Canal incident, 389
low-flow augmentation, 946

malaria control, 5
malathion, in urban runoff, 480
mangrove wetland, 865
manhole, modification for increased infiltration, 599
Manning's formula, 278, 608–609
Manning's coefficient, 154–155
  for grassed waterways, 609
  for ripraps, 612
median, definition, 487
melting point, 129
  of saline snowmelt, 473
mercury
  agriculture, 221
  atmospheric deposition, 220
  environmental effects, 226
  exposure limits, 223–225
  human exposure, 221–226
  in drinking water, 224
  in soils, 361
  intake limits, 224
  lowest observed effects on humans, 226
  methyl mercury, 223
  occupational exposure, 224
  pathways in human uptake, 222
  poisoning, Minamata, 76, 358
  residence time, 221
  soil contamination, 227
  sources, 220
  toxic effects, 226
Merrimack River, 973
metals. *See also* toxic metals
  activity in soils and sediments, 352–353
  chemical equilibria in soils and sediments, 356
  complexation, 353
  coprecipitation, 353
  interaction between solid and dissolved species, 354
  methylation, 358–359, 361
methane evolution, from sediments, 736, 748
  effect on BOD, 738
methanogenesis, in sediments, 736
methylene chloride, in urban runoff, 478–479
Michaelis–Menton equation, 331, 779
microorganisms. *See also* pathogenic microorganisms
  autotrophic, 330
  chemotropic, 364
  from agriculture, 684
  in CSOs, 476, 497–499
  in nitrogen cycle, 362–364, 366–367
  in septic tank effluent, 418
  in soils, 374–378
  in street refuse, 448
  in urban runoff 476
Milwaukee River, 50–52, 974, 982
Minamata mercury poisoning, 76, 358
mineral loads, in base flow, 413
minerals, dissolution of, 410
mining nonpoint pollution, 40
mixing layer, 323
Model Enhanced Unit Loads (MEUL), 522–523
modeling, black box concepts, 50
Model of Urban Runoff and Sewer Flow (MOURSEF), 532–533, 539
models, 507–546. *See also* hydrological models
  accuracy and reliability of, 513–516
  agricultural, 540–546
  application of, 515–516
  calibration and verification, 535, 554–556
  continuous, 513, 517, 532–536, 562
  conventional pollutants and nutrients, 796–802
  deterministic, 509–547
  EPA computerized loading procedures, 321
  event-oriented, 513, 517, 531–536, 562
  groundwater, 432–435
  highway drainage, 535, 539
  infiltration, 113–118
  kinematic wave, 152, 156, 536
  linear watershed, 150
  neural network, 550, 656
  rainfall, intensity–duration, 130–131
  regional regression, 524–525
  screening, 522–528
  selection of, 516–519
  statistical routines and screening, 509, 517–528
  steady state, 796
  stochastic, 166, 540–550
  surface and groundwater, 507–546
  toxic chemicals, 852–857, 910–914
  toxicity, 827
  types, 508–513
  urban runoff, 471, 531–540
  urban washoff, 472
  water quality, 766
    link-node representation, 798–799
models, other
  Areal Nonpoint Source Watershed Environmental Response Simulation, 540, 546

Areawide Stormwater, 532, 546
ARMA, 166–170, 546–550, 655
Chemicals, Runoff and Erosion from Agricultural Management Systems, 268–269, 540–541, 546
Gelhar–Wilson groundwater, 433–435
Model of Urban Runoff and Sewer Flow, 532–533, 539
Monte Carlo, 547, 855–856
Negev erosion, 269
Nonpoint Simulator, 536
Pesticide Root Zone, 543, 546
Pesticide Runoff Simulator, 542
Real Time Control, of sewerage, 655–656
SCS Runoff Curve, 134–141, 158–163, 521, 532, 541–543
Simulator for Water Resources in Rural Basins (SWRRB), 542, 546
Stanford Watershed, 107, 127, 536
Storage-Treatment-Overflow Runoff, 532, 539
STORM, 463
Stormwater Management (SWMM), 463, 533–534, 539
Streeter–Phelps, 766–769, 802
TVA base flow quality, 413–417
TVA-HYSIM, 532
Unit Hydrograph, 148–156
USGS Distribute Routing Rainfall Runoff, 535
Vollenweider's for lakes, 789–793
WASP, 798–802, 854–855
WASSP-QUAL sewer flow, 533, 538–539
monitoring and data acquisition, 556–564
flow, 556–560
quality, 560–562
monitoring station, 558–561
Monte Carlo models, 547, 855–856
mosquito control, 5, 862, 897, 917
Mount St. Helen, eruption of, 807
mulching and protective covers, 586–587, 701

naphthalene, 642
National Estuary Program, 88
National Hydrologic Benchmark Water Quality Monitoring Network, 807
National Nonpoint Pollution Control Program, 10
National Oceanic and Atmospheric Administration (NOAA), 91, 121, 556

National Pollution Discharge Elimination System (NPDES), 37, 87–90, 92, 839, 994, 1005
Nationwide Urban Runoff Project (NURP), 7, 41, 444, 446, 451, 476–480, 484, 491–495, 557, 574, 585, 738–739
natural flow, 103
Negev erosion model, 269
nekton, 748
neural network models, 550, 656
nickel
  in natural base flow, 407
  in precipitation, 453–454
  in roof runoff, 480
  in surficial materials, 319
  in urban runoff, 42, 477–479
  source minerals, 412
  traffic emission, 456
Nile River, 14
nitrate, 418, 427
  natural content of groundwater, 412–413
  pollution, 31–32, 362
    contamination of groundwater, 388, 412–413
    in Long Island, NY, 418
    of Central Europe, 388
nitrification, 363, 542, 741, 760–764, 766, 778–779
  clean-up of nitrate pollution of groundwater, 391
  in aquaculture ponds, 919
  in land application systems, 421
  in soils, 365–367
  reaction rate, 779
  simultaneous with denitrification in sediment-water interface, 763–764, 770
  wetlands, 871, 882
nitrogen, 21
  enrichment in overland flow, 291
  fixation, 362–363
  forms in waters, 760–762
  from agriculture, 682
  immobilization, 369
  in atmospheric deposition, 215–217
  in CSOs, 36, 499
  in feedlot runoff, 34, 688
  in landfill leachate, 427
  in precipitation, 453
  in soils, 361–374
  in solids from urban areas, 448
  in urban runoff, 36
  limiting nutrient, 742
  loads to Lagoon of Venice, 56

nitrogen (*cont.*)
  NURP studies, 491–493
  removal in land disposal systems, 421
  removal in ponds, 627
  removal in wetlands, 882–885
  selection of BMPs for control, 692–693
  sources, 361–362
  surface waters impairment, 736
  traffic emission, 456
nitrogen cycle, 362–363, 761
nitrogen oxide emissions, 189
nitrogen to phosphorus ratio, 742, 792
Nitrosomonas and *Nitrobacter*, 741, 762, 778
nonpoint pollution loads, suburban, 483
Nonpoint Simulator Model (NPS), 536
nonpoint sources, definition, 20–21
North Avenue Dam, Milwaukee, 949–950
NPDES Stormwater Permit Program, 92, 483, 563
nutrient control, 59
  agricultural management, 698
  in lakes, 954–956
  transformation and removal in wetlands, 881–887, 906–908
nutrient loading of lakes, 789–793
nutrient loads, 56, 61
nutrient losses, 30
nutrients
  allochthonous and autochthonous, 787
  Chesapeake Bay, 46
  effect on dissolved oxygen balance, 777
  impact on surface waters, 738–739
  limiting, 741

octanol partitioning coefficient, 326, 342–345, 404–405, 642–647, 830, 835, 850, 912
Office of Underground Storage tanks (U.S. EPA), 419
oil, surface water impairment, 736
on-site analyses and monitoring, 560–561
organic chemicals
  control by wetlands, 909–917
  in soils, 339–352
  mobility in soils and sediments, 341–347
  volatile, emissions, 189
organic matter, 316
  production
    in surface water bodies, 742
    in wetlands, 876–878
  transformation of, 745
Organization for Economic Cooperation and Development (OECD), 79
Outstanding National Resource Waters (ONRW), 987, 1003–1004
overflow frequency to unit storage volume relationship, 619–620
overflow rate, 624, 789
overland flow
  routing, 144–171, 536
  sediment transport capacity of, 275–280
  treatment and disposal systems, 421
oxidation ponds, 641
oxygen. *See also* dissolved oxygen
  depletion in lakes, 957
  injection to lakes, 961
  sinks and sources in surface waters, 766
  supersaturation by photosynthetic action, 743
ozone depletion, 10

Pareto optimality, 73
Parshall flume, 559
particulate organic carbon, 643–644
partitioning, of chemicals and metals, 325–332, 829–831, 910–914
partitioning coefficient, 326, 339, 342–345, 356–357, 359, 361, 404, 829–830, 912. *See also* octanol partitioning coefficient
pasture, 29–32, 663, 686
  management, 710
pathogenic microorganisms, 8
  control in agriculture, 693
  in CSOs, 497–499
  in soils, 375–378
  in urban runoff and CSOs, 476–477
pavement conditions, impact on pollutant loads, 456
pentachlorophenol, in urban runoff, 478–479
perched water table, 397
perennial streams, 104, 175, 387
permeability, 109–110, 554
  coefficient of (hydraulic conductivity), 396
  of animal feedlots, 688
  of geological materials, 395–396
  relation to soil texture, 110
pervious (porous) pavements, 589–592
  benefits, 590
  connected, decreasing, 592
  construction cost, 592

drainage requirements, 590
groundwater contamination potential, 592
hydraulic conductivity, 590
installation, 591
load bearing strength, 590
longevity, 592
urban, control of, 586–589
peryphyton, 748
Pesticide Root Zone Model (PRZM), 543, 546
Pesticide Runoff Simulator (PRS) model, 542
pesticides, 41, 345–347
adsorption, 326–330
application rate, 312
atmospheric deposition, 227–231
chemical breakdown in soils, 348
concentration in precipitation, 229
contamination of groundwater, 388–390
control in agriculture, 694
cycle, 226
enrichment in overland flow, 291
from agriculture, 32–33
inputs to Chesapeake Bay, 230
inputs to Great Lakes, 230
inputs to wetlands, 230
losses from soils, 348
microbial metabolism, 348
persistence in soils, 347
photodecomposition, 346
residence in atmosphere, 231
source of pollution by urban runoff, 478
wet deposition, 227
washoff, 227
pH
effect on ammonia volatilization, 371
effect on nitrification, 365
effect on precipitation of metals, 357
of landfill leachate, 427
of rainfall
in North America, 203
worldwide, 205–206
of urban snowmelt, 474
phenanthrene, in urban runoff, 478–479
phosphorus, 21
control, 50
lake restoration, 955–956
removal in ponds, 627
removal in wetlands, 886
enrichment in overland flow, 291
from agriculture, 682
in CSOs, 36, 499
in feedlot runoff, 688

in soils, 334–339
in street salt, 460
in urban runoff, 36
limiting nutrient, 742
mobility in soil, 336
natural base flow, 407
NURP studies, 491–493
release from sediments, 295
removal in aquaculture systems, 920
selection of agricultural BMPs, 692–693
solubility in soil water, 336
surface water impairment, 736
traffic emission, 456
photosynthesis, 741–745
phthalate esters, in urban runoff, 478–479
phytoplankton, 741–745
piezometric pressure, 173
plankton, 748
plant available soil water capacity, 111
plant systems, floating, 918–921
plant uptake of chemicals and nutrients, 332
point sources, 34
definition, 18–22
control, 87
impact on stream ecology, 927
pollutants. *See also* adsorption, of pollutants; nitrate pollution; priority pollutants
accumulation, near curb, 447, 461–470
analyses, 562
atmospheric transport of, global, 196–200
black list of, 819
carcinogenic, 4, 41–42, 345, 818–819
conservative, 317
curb loading of, 448
delivery of, 145, 371, 602–616
gray list of, 819
half-life of, 334
highway loads of, 457, 479
in precipitation, 452
in soils
mobility of, 324
pathways of, 317–318
loads from urban areas, 442–443
mass balance equation, 797, 802, 852, 919–911
overland transport of, 317
removal of, 627, 638
by precipitation (washoff), 470–473
by snowmelt in urban areas, 473–476

1048   Index

pollutants (cont.)
  sorption of, in solids and sediments, 332–333
  unit loads of, 507, 519–522
  washoff, availability factor, 472
  wet deposition of, 200–217
  winter accumulation of, 469–470
polluter pays principle, 79, 994–996, 1004. See also economic incentives and instruments
pollution. See also groundwater pollution; toxic pollution
  agricultural, 6–7, 9, 29–34
  abatement, uniformity in, 996
  air, sources, 188
  atmospheric, by toxic compounds, 217–231. See also atmospheric pollution
  atmospheric transport of, global, 196–200
  definition, 11
  diffuse, 18. See also diffuse pollution
  economic and social causes of, 70–74
  in urban areas, annual variations, 470
  loads, 39, 64, 80, 482–483
  manure, hazard from disposal of, 689
  market impacts on, 75–76
  on impervious surfaces, 461–470
  prevention, 80
  removal efficiencies in models, 525
  sediment, 23, 250, 294, 479
Pollution from Land Use Activities Reference Group (PLUARG), 740
polychlorinated biphenyls (PCB), 4, 37, 43–46
  biodegradability in soils and sediments, 352
  distribution, 44–45, 642
  properties, 44–45
  fish contamination, 46
polycyclic aromatic hydrocarbons (PAHs), 37, 41, 586, 642
  in atmospheric deposition, 229
  in soils, 340–341
  in urban runoff, 478–479
ponds, 620–629
  agricultural, 702
  dual use for quality and flood control, 629
  enhanced high efficiency wet pond system, 623
  hydraulic detention time, 628
  pond-wetland system for urban runoff pollution control, 893

potency factors, 318
  removal efficiency, 622, 627
  sizing and dimensions, 628
potency factors, 449, 521–523
Potomac Eutrophication Model. See WASP water quality model
prairies, 29
  streams, water quality, 808
precipitation
  composition of, 200–217
  pollutant content, 452
  reactions in soils and sediments, 325–332
  runoff relationship, 102–106
priority pollutants, 8, 41–46
  in urban runoff, 477–480
  removal by Best Management Practices, 642–647
Priority Watershed Program, 996
probability of exceedence, 485–486
probability paper, 486–487
producers, 741
productivity, 741, 748, 780–785, 878–879
profundal zone, 744
pump and baffle aeration of lakes, 958–959
pyrene, 642–643
  in urban runoff, 478–480

radar, use in Real Time Control of urban sewerage, 553
rain gauges, 559
rainfall, excess, 103, 129–142
  estimation, 521
  from impervious areas, 141–142
  from pervious areas, 131–141
  hydrologically active areas, 142, 525
rainfall intensity
  critical for initiation of CSO, 498–499
  design storm, 129–131
  duration curves, 130–131
rainfall runoff transformation process, 106–107
  excess rain, 129–142
  initial subtraction, 108
  overland routing, 144–171
  peak runoff by SCS method, 160
  schematics, 105
range land, 29, 32, 683, 686
  management, 710
rapid infiltration wastewater disposal system, 421–422
rational formula, 145–148, 532

Real Time Control (RTC) of urban
    sewerage, 552–554, 649–659
  bottleneck control, 650
  components of, 654–655
  existing systems, 656–658
  models, 655–656
  strategy, 651–654
receiving water criteria, 80
redox reactions, 410
reference water bodies, 753–756, 810,
    987
regional regression models, 524–525
regulated waters, definition, 89
regulation, 72–79, 729, 995
  wetland, 897
regulators and concentrators, 613–616
removal coefficient, for street
    pollutants, 463–464
reservoir construction parameters, 970–
    972
riparian buffer zones, 84, 94, 708, 869
riparian owners, 84
riparian water rights doctrine, 82–84,
    947
riparian wetlands, 84, 869, 904
ripraps, 611, 709
river impoundments, removal of, 948–
    949
riverine wetlands, 866
river restoration, 61
roadside swales, 603–607
rock-reed microbial filter, 710
roof runoff contamination, 480
runoff, 476. *See also* urban runoff
  and pollutant routing, 530
  enrichment of
    by clays, 286–292
    by nitrogen, 371
    by soil organic matter, 322
    by soil pollutants, 317
  variations, 291
  surface, 129–142
runoff coefficient, 440–441
  for Rational Formula, 146–148
runoff pollution, generation process in
    urban areas, 461–476
Rural Clean Water Program (RCWP), 9

Salmonella, in urban runoff, 476
salt, 427, 459
samples, 560–563
sand, sandstone, 394–396
sand filter, 639
sanitary landfills, leaching, 483
saprobien biotic index, 753

Sartor et al. equation for washoff, 472–
    473
screening, 616, 639
  models, 522–528
SCS hydrologic soil groups, 109
SCS Runoff Curve model, 134–141,
    158–163, 521, 532, 541–543
SCS soil maps, 109, 315
SCS soil texture classification, 289
SCS unit hydrograph, effluent of
    imperviousness, 160–161
secchi disc, 788
section 208 studies, 87, 522, 557
section 319 plans, 88, 92
sediment barriers and silt fences, 611–
    612
sediment control basins, 620
  design, 624–626
  dredging and removal of
    contaminated sediments, 91, 954,
    966
  ideal settling basin, 625
sediment delivery, 239, 249, 270–286,
    521, 544, 577, 603
  effect of drainage, 282–283
  estimated by cesium, 137, 284–286
  factors affecting delivery, 272–282
  overland flow effects, 274–275
sediment flow, discharge relationship,
    rating curve for, 252, 282
sediment loads, from urban areas, 449
sediment oxygen demand (SOD), 747,
    764, 766, 769–772
  in situ measured, 772
  remediation, 948–949
sediment particle sizes, 145
sediments
  acid and methane formation in, 746
  bedload and washload, 769
  cohesive, 245–246, 275
    affinity to adsorb pollutants, 295
    aggregation, 297
    physical rates of transport, 295–304
  contaminated, dredging and removal
    of, 53, 954
  control in agriculture, 692
  criteria, 848–852
  deposition of, 293–294
  diagenesis and methanogenesis, 746,
    770–771
  effects on toxicity and bioavailability,
    251
  inert or refractory fraction, 746
  loss from feedlots, 34
  methane evolution from, 736, 748
  noncohesive, 245, 275

1050 Index

sediments (cont.)
  organic-rich, 295
  pickup and transport by runoff, 471
  pollution, 23, 250, 294, 479
  rating curve, 251
  sorption of pollutant, 245
sediment toxic unit, 828, 849
sediment transport
  capacity of overland flow, 275–280
  cohesive sediments, 295–304
  critical tractive force for, 278
  hysteric loops in, 284
  in streams, 292–304
  overland, 270–292
  Shields diagram for, 278
  storage of transported sediment in streams, 293
sediment traps, 606
sediment-water interface
  carbon and nutrient cycles in, 747
  processes, 747, 764, 802, 879
sediment yield, 239, 247, 251–253
sedimentary rocks, 394
sedimentation, 251, 301–304
selenium, 94, 319
septic systems, leaking into storm sewers, 483
septic tanks, 415–418
  failures, 35
  impact on groundwater resources, 388–389, 415–418
  sources of nitrogen pollution, 365
settling, 297, 302
settling velocity, 297, 299
  of individual particles, 595, 624
  particles associated with urban runoff, 627
  phosphorus in lakes, 789
sewerage agencies, 1004
sewers. See also storm sewers
  combined, 37. See also combined sewer overflows
  flushing of, 613
  separated, 37, 62
  solids, 444, 481–483
    control, 579
    effect on CSO loads, 499
Shakiji River, 62–64
sheet and rill erosion, 460
Shields diagram for sediment transport, 278
Shirako River, 62–64
silviculture, 686–688
simazine, 230
Simulator for Water Resources in Rural Basins (SWRRB) model, 542, 546

slow rate land application systems of waste, 419–420
snow enrichment by pollution, 126
snowmelt, 126–129
  degree-day formula, 126
  enrichment of, 291, 474–476
  runoff 126, 460
  winter rates, 473
snow removal, 469
  to increase light penetration into lakes, 962
Soil Conservation Service (SCS), 93, 521, 558
soil loss, 6
  measured by cesium-137, 284–286
  urban, 35
soil map, 109, 315
soil particles, specific surface of, 287
soil profile, 314–316
soils
  acidification, 313
  chemical stabilization of, 586
  classification, 109–110
  conservation, 8–9
  contamination, 4, 313, 317–319, 323–334
  detatchment of organic matter, 322
  moisture characteristics, 111
  moisture distribution during infiltration, 112
  permeability, 396
  porosity, 395
  sorptivity, 117–118
  water storage, 111
soil texture, 109, 253
soil water zone, 172–173
spawning habitat management in lakes, 963–965
surface runoff, 129–142
standard deviation, definition, 487
standards, 14–16. See also criteria and standards
  technology-based, 81
Stanford Watershed Model, 107, 127, 536
statistical routines and screening models, 509, 517–528
statutory water use, 986
steady-state models, 796
stochastic (ARMA) models, 540–550
Stokes law for settling, 297, 595, 624
storage, 575. See also surface storage; underground storage tanks
  depression, 106, 442
  hydrological, 106, 589
  interception, 107

Index   1051

storage basins. *See also* ponds; tunnels
   agricultural, 702–703
   dimensioning, 632–635
   flow balancing, 634–636
   for CSO control, 39, 61, 629–636
   in combination with treatment, 637–638
Storage-Treatment-Overflow-Runoff Model (STORM), 532, 539
STORET data management system, 555
STORM model, 463
storm sewers, 442
   conttrol of pollution by runoff, 577–602
   historic outlook, 2, 3
   nonstormwater discharges into, 483
stormwater
   management
     for quality control, 575
     utilities, 1004
   permits, 90
   reuse, 62
Stormwater Management Model (SWMM), 463, 533–534, 539
   calibration, 555
   data needs, 535
straw mulch application, 586
stream bank stabilization, 709, 939–943
stream restoration, 932–949
   biomanipulation, 939
   flow regulation, 937–938
   land acquisition and regulation, 939
   nonstructural, 937–939
   plantings, 938
   pollution abatement, 938
   propagation facilities, 939
   techniques, 936
streams
   effluent dominated, 104
   enrichment by productivity of organic matter, 781
   ephemeral, 387
   order, 754, 757
   perennial, 104, 175, 387
Streeter–Phelps equation, 766–769, 802
street flushing, 579
street refuse
   deposition, 454
   particle size distribution, 457–458
street salt additives and composition, 460
street solids
   accumulation, 462, 466
   urban, vegetative input, 454–455
street sweepers, sweeping, 463, 578–586
   efficiency, 581–582, 585
   particle removal by, 582–583
   types, 579–582
   water quality benefits, 584–585
submerged plant systems, 921
subsidies, 73, 999
sulfides
   immobilization of metals, 889. *See also* acid volatile sulfides
   minerals containing, 410
sulfur, in wetlands, 887–889
sulfur dioxide emissions, 189
sulphate, 805, 886
Superfund sites. *See* hazardous waste, disposal sites
surface and groundwater models, 507–546
surface runoff, definition, 103
surface storage, 554, 592
surface waters, contamination, 312
suspended solids, 36
   removal, 627, 902
sustainable agriculture, 3, 680
sustainable development, 24, 1012–1014
swirl-flow regulator, solids concentrator, 613–616

taxes, 994
TCLP toxicity analysis, 432
telemetered monitoring and monitors, 559
temperature effects
   on dissolved oxygen balance, 771–773
   on nitrification, 365
terraces, 698–699
Thames River, 3
thermal factor, 772, 779, 906
thermal stratification, 961
tidal dispersion coefficient, 776–777
tidal marshes, 864–865, 882
tidal water exchange, 56
time of concentration, 145, 153–156
toluene, in urban runoff, 478–479
Total Maximal Daily Load, 989–993, 1005
toxic compounds, 6
toxic chemicals
   sources, 478
   model, 852–857
toxic dissolved gases, 821
toxicity
   acute, 821–823
   acute toxicity unit, 843
   chemically specific tests, 823–825
   chronic, 822–823
   chronic toxicity unit, 843

toxicity (*cont.*)
  concepts, 817–821
  criteria, 840–843
    frequency (permissible) of exceedence of, 841
  cumulative, 822
  definition, 818
  functional dose (concentration) response relationship, 822, 827
  lethal, 822
  measurement, 821–838
  natural, 819
  of urban runoff, 477
  reduction by Best Management Practices, 642–647
  sediment toxic unit, 828
  TCLP analysis, 432
  test organisms, 825
  tests, 822–831
  whole body toxicity testing, 823–825, 843
toxic loads, Chesapeake Bay, 47
toxic metals, 41, 645, 819, 821
  atmospheric, 186–195
  biotic uptake of, 829
  in septic tank effluents, 418
  in soils and sediments, 319–320, 352–362
  interaction between solid and dissolved species, 354
  in urban runoff, 477–479
  in wetlands, 909
toxic organic chemicals, 821
toxic pollution, 6, 21, 41–46
  in the atmosphere, 217–231
  biotic uptake, 829
  control, 52, 642–649
  global, 821
  impact on human health, 820
  impact on integrity of water bodies, 820
  partitioning of toxic chemicals, 829–831
  subsurface disposal, 388
traffic
  contribution to diffuse pollution, 456
  emissions, 188
    lead, 218
  exhaust emissions, 455–456
  impact on urban pollutant loads, 455–459
  source of urban runoff pollution, 4, 478
transfer function, 549–551
treatment plant bypasses, 89
trophic degree, 781
trophic indices for lakes, 787–788
trophic levels, 744–745
trophic status, 780–781, 784–789
tunnels (underground), 39, 50, 629–631, 637
TVA base flow quality model, 413–417
TVA-HYSIM model, 532

unconfined animal operation, 29
underground formations, 395, 397, 398
underground storage tanks, 418–419, 483
unit hydrograph, 148–156, 163–171, 541, 549–551
  convolution integral, 149
  deconvolution, 163–165
  direct estimation, 165
  effect of urbanization, 155
  instantaneous, 150
  in stochastic models, 166
  SCS method, 158–161
  statistically estimated, 163–171
  synthetic, 151–158
unit loads
  definition, 446
  from impervious surfaces, 520
  from pervious urban areas, 521–522
  from urban areas, 444–451
  in screening procedures, 522
  Model Enhanced Unit Loads (MEUL), 522–523
  of pollutants, 507, 519–522
Universal Soil Loss Equation (USLE), 254–270, 521, 528, 530, 542–543, 555, 586, 696
  cropping management factor, 261–263
  erosion control practice factor, 263
  modifications, 268–270
  rainfall energy factor, 254–257
  reliability, 263–265
  slope length factor, 259–260
  soil erodibility factor, 257–259
urban drainage, 576
urban erosion 460
urbanization, 34–39
  hydrological impacts of, 104
urban land uses, 443–444
urban runoff. *See also* runoff
  control and management, 575–602
  control of pervious areas, 586–589
  effects of runoff volume on event mean concentration, 491
  filtration of, 639–640
  historic, 1
  loads to Lagoon of Venice, 56
  models, 471, 531–540

Index   1053

pesticides, 478
pollutant strength, 36
pollution, 7–8, 491–495, 738–740
pollution carrying, frequency of, 442
quality, 491–495
quality from street flushing, 579
source control measures, 577–602
sources of pollution, 445, 451–461
statistical quality characteristics of, 484–495
storage strategies for control, 619
surface waters pollution from, 738
toxicity, 477
treatment of, 637–642
use of wetland control, 891–893
urban sediment sources, 250
urban sewerage system, 2, 650
urban stormwater runoff, 439
urban washoff coefficient, 472
urea, hydrolysis of, 368
use attainability, 16, 807, 988–994
Use Attainability Analysis (UAA) 988–994, 1001, 1005
user fees, 994
U.S. Army Corps of Engineers diffuse pollution control programs, 93, 719
U.S. Department of Agriculture, 675
  agricultural pollution management programs, 714–717
U.S. Department of Interior, diffuse pollution programs, 718–719
U.S. Environmental Protection Agency (U.S. EPA), 558, 818
  agricultural pollution management programs, 712–715
U.S. Fish and Wildlife Service, 5, 862–863
U.S. Geological Survey (USGS), 556
USGS Distribute Routing Rainfall Runoff Model (DR3 QUAL), 535

vadose-aerated zone, 172–174, 392
variation, coefficient of, 487
vehicle pollutant emissions, 456, 463
velocity, design for grassed waterways and swales, 607
velocity sensors, 559
Venice canals, 54, 55
vinyl chloride, 586
viruses, in soils, 377
volatilization, 328–330
Vollenweider's model, for lake eutrophication, 789–793
vortex flow/solids separator, 614–615

washload, 243, 249, 294–295
washoff
  model (urban), 472
  of pollutants from impervious surfaces, 439, 472, 520
washout functions (atmospheric), 200
WASP water quality model, 798–802, 854–855
WASSP-QUAL sewer flow model, 533, 538–539
waste assimilative capacity, 16, 85, 982–984, 990–993, 1000–1003, 1005
  definition and determination, 16
  enhancement, 933, 982–983
  for toxic chemicals, 855
  of soils, 313
  of wetlands, 874, 880, 909–917
waste distribution, 3
wastewater treatment, 4
  by land application, 419–424
  by wetlands, 871–873
  reuse, 419
water
  bicarbonate content, 408–409
  conservation, 589
water bodies
  biotic integrity, measures, 753–759, 818
  fate of toxic chemicals in, 852–857
  impaired, 1002–1003
  management of, 929–1003
  reference, 753–756, 810, 897
  restoration, 16, 927–972, 993
    goals, 928–929
    lakes and reservoirs, 949–972
    planning, 929
    rivers and streams, 932–949
    structural criteria, 929–932
  types of, 987
  uses, 5, 985–993
    support of designated uses, 736–738
waterborne epidemics, 2
Water Erosion Prediction Project (WEPP) Hilslope Profile Model, 269, 543–546
water hyacinth, 868, 918, 968
water pollution
  causes, 737
  definition, 11–18
  of coastal waters, 738
Water Pollution Control Act Amendments. *See* Clean Water Act
water quality
  abatement, 5

water quality (*cont.*)
  approaches, 81
  background (natural), 13–14, 32, 806–810
  control, detention volume for, 617
  criteria and standards, 838–848, 984
    acute or chronic toxicity, 840
    chemical-based numerical, 840–843
    sediment, 848–852
  ecoregional approach to, 413
  goals, policy options, 69, 74, 1001
  impact of diffuse sources, 737
  improvement
    aeration for, 947–948
    socioeconomic impact of, 989
  limited water bodies, 1001
  management
    institutions, 3, 1002–1007
    integrated, 16, 79, 750, 807, 982, 989, 1004
    programs, 47–48, 51–52, 57, 59, 88
  models, 83, 766, 796–802
  of urban runoff, 492–493
  planning process, 81, 990
    margin of safety, 992
  problems, 735–740
watersheds
  hydrologic, modification of, 144, 589–601
  hydrological response, 440
  hydrology, 102–106
  management, for lake restoration, 952–955
  response to rainfall, 153
  transition, 25–41
weathering, chemical, 239
weirs, flow measuring, 559
wet deposition of pollutants, 200–217
Wetland Foundation, 862
wetland–groundwater interaction, 872
wetlands, 483
  coastal, 864–865
  constructed. *See* constructed wetlands
  definition, 863
  design parameters, 894–909
  drainage, 5–6, 28, 55, 93, 317
  filling, 93
  free water surface, 896–892, 907
  function, 869–873
  hydraulic loading of, 899–901
  hydrology, 896–901
  hydroperiod of, 897–899

  inland, 865–869
  lacustrine, 866
  loss, historical, 862, 868
  management programs, 5–6
  mangrove, 865
  metals and toxics control, 862
  nutrient transformation, 881–887
  organic matter production, 876–887
  pollutant removal, 92, 876, 905–909
  protection, 52, 92–94
  regulation, 897
  restoration, 869, 874–875, 894–909
  retention time, 899–900
  riparian, 84, 869, 904
  streams, water quality of, 808
  submerged bed, vegetated, 892
  sulfur retention and transformation, 887–889
  types, 863–889
  use for agricultural runoff control, 707
  use for stormwater control, 636–637, 871
  use for water quality control, 873–876, 881
  vegetation, 868–869
wet weather flow, 37
wildlife habitat, 871
Wisconsin Priority Watershed Programs, 48–52

xeriscape, 460, 588–589, 1015

Yalin sediment movement equation, 276–280, 470, 542
Yellow River, 14

zinc
  in landfill leachate, 427
  in natural base flow, 407
  in NURP studies, 491–493
  in precipitation, 453–454
  in roof runoff, 480
  in soils, 361
  in urban runoff, 42, 477–479
  source minerals, 412
  surface waters impairment, 736
  traffic emission, 456
zones
  cohesive sediment transport, 301–302
  groundwater flow, 171–179
  groundwater quality, 411